FISHES
of
Idaho

A Natural History Survey

John W. Sigler
Donald W. Zaroban

Illustrated by
Joseph R. Tomelleri
&
Emily S. Damstra
Cover Design by
Renai C. Brogdon
Additional Art by
Link Jackson

Caxton Press

Fishes of Idaho: A Natural History Survey

Caxton Press, 312 Main Street, Caldwell, ID 83605.

First Edition

Library of Congress Control Number: 2018936692

CIP Information available at http://lccn.loc.gov

Printed in China through Four Colour Print Group - Louisville, Kentucky

CAXTON PRESS - Caldwell, Idaho

200735

DEDICATIONS

With my deepest gratitude, I dedicate this book to Bonnie Jean Zaroban. My lovely bride of nearly 40 years and mother of our three wonderful children, Stephanie, Thomas, and Stacie.

—Donald W. Zaroban

In the Acknowledgements section of this text, we note our appreciation for the many explorers, scholars, fisheries biologists and ichthyologists who collected and presented fisheries data over the preceding two centuries, all of which had an impact on this text by conveying and passing on knowledge and ideas.

My father, William F. Sigler, mentored many graduate students at Utah State University. One of those graduate students assembled the information noted below and I provide it here in celebration of the sciences associated with ichthyology and fisheries management and to honor the memory of my father.

It is my hope and my belief that the progression noted below is ongoing and will continue to benefit the professional fisheries field.

Louis Agassiz 1807-1873

David Starr Jordan 1851-1931

Carl L. Hubbs 1894-1979

Reeve M. Bailey 1911-2011

William F. Sigler 1909-1995

I dedicate this book to my wife of some forty years, Betty Jean Sigler and to my three great children. Thank you.

—John W. Sigler

CONTENTS

FOREWORD
Cecil Andrus

It is my pleasure to introduce you to *Fishes of Idaho: A Natural History Survey*. Fishing has been in my blood for as long as I can remember. My father took me, as a little boy, to watch Indian fishermen at Celilo Falls on the Columbia River. Still-plentiful salmon were a vital part of the family diet of an Oregon boy growing up during the Great Depression. Later, when I was a young husband and sawmill manager in Orofino, Idaho, fishing became my time for respite and thinking.

This book celebrates Idaho: her waters, her fish and her fishing. I feel privileged to introduce Idaho's fishing bounty to you. More than 70 percent of the Earth's surface is water-covered, and the oceans hold about 96.5 percent of all Earth's water. In Idaho we are blessed with 95,000 miles of flowing water; roughly 26,000 miles of fishable streams and over 500,000 acres of lowland lakes and reservoirs. Across Idaho, there are more than 8,900 and rivers and streams and over 200 named lakes and reservoirs. From the smallest tributary in the desert southwest drainages of Owyhee County to the wondrous Salmon River, or from the mighty Snake River to the crystal waters of Bear Lake and Lake Pend Oreille, fishable water is everywhere in Idaho.

Over the 60 years I have lived in Idaho, I have had the great pleasure of hiking, riding, hunting and fishing over most of the state. In the mid-1950s, about four hundred thousand salmon returned up the Snake River system each year, most heading for such unspoiled tributaries of the Salmon, Clearwater, and Imnaha Rivers. I would picnic with my young family at Bruce's Eddy on the North Fork of the Clearwater River and wade into the river with hopes of hooking a Chinook Salmon or Steelhead, the Northwest's fighting and famously tasty species of seagoing trout.

Now, at the risk of being provincial for just a minute, I have to tell you that in my opinion we have in the state of Idaho probably the very best fishing in the entire world. For a person who lives to use a fly rod it is just a little bit of Heaven on earth.

Many of you know, or have fished, the great Henry's Fork. And during my stint as Governor we were able to significantly advance the goal of protecting that great fishery for future generations. The work continues with the advocacy and science of the Henry's Fork Foundation and community involvement through the Henry's Fork Watershed Council. Through the work of the Nature Conservancy and others Silver Creek has been protected and will continue to challenge and delight fishermen, I hope for a thousand years.

I had the pure pleasure of fishing Kelly Creek in north central Idaho. That too must be considered one of the crown jewels of all of our trout fisheries. Those cutthroat were so aggressive I had to hide behind a rock to change flies. Kelly Creek is protected as a roadless area, and someday may be designated a Wilderness area—the gold standard of protection.

Speaking of protection, I remember during my first term as Governor when we worked to get better protection of the Cutthroat Trout by getting catch and release rules in place for the Middle Fork of the Salmon River.

So if you are like me, you try as often as you can to visit your favorite trout stream, or cherish that one week a year when you can pack in to a high mountain lake and tangle with a fish who, just maybe, has never seen anyone in a plaid shirt and has never tasted an artificial fly.

I have enjoyed battles with the great sea run salmon and Steelhead in the Clearwater and Salmon Rivers, I've fought monster sturgeon in Hells Canyon, and I've caught bass until my arm is sore on CJ Strike Reservoir. It seems like fishing has just added years to my life. Some of my most satisfying fishing has been while I relaxed near the shore of a pristine mountain lake. Nothing is better than a couple fat trout sizzling in a frying pan while they are so fresh that they still have wiggle in them! Another way I've enjoyed them is wrapped in aluminum foil with onions, carrots, Idaho potatoes, and seasoning, and then buried in the coals of the campfire to slowly cook to perfection. Oh, but I digress as I reflect back on my personal fishing experiences; the message I want to impart here is that as I explored our great state, I gained a particular perspective on Idaho and her waters. It is that perspective which I want to share with you.

This book presents life history information, distribution, and biology on nearly 100 fish. I won't try to tell you I've caught them all, for many of those fish are the small and secretive ones that regular anglers never see. Yet they are part of our wildlife heritage. That is why this book intrigues me as an avid outdoorsman, anybody can learn more about Idaho by opening it and discovering some nuance of a fish or its surroundings. If one knows a lot about Idaho and her fishing, this book provides a bonus in the detailed information on some of the best fishing and water recreation spots in Idaho. The "Idaho Aquatic Gems" section provides information for 17 of Idaho's most notable waters on when, where, how and what to fish for while exploring Idaho's great diversity of fishing opportunity. If you know very little about Idaho's fish, this book can teach you much.

When I think of these Idaho Aquatic Gems I am reminded of a famous quote from U.S. Supreme Court Justice Oliver Wendell Homes: "A river is more than an amenity. It is a treasure." If you want to know what he meant, go to the sixty-nine mile stretch of the South Fork of the Snake River between Palisades Dam and the confluence with the Henry's Fork. It's no secret that I see the South Fork as a treasure. It is one of the great Cutthroat Trout and Brown Trout fishing streams of the world. And in recent years the work of Idaho Fish and Game and Trout Unlimited and others have continued to secure that top rank status. Float on the river, which you can do for miles. Stand beneath the gently sawing cottonwoods and poplars. Or scramble the bluffs above the river, take note of its braided channels, and scan the canyon rim for eagles.

I think this is the place to offer some historical perspective about this latest Idaho fish book. It is fourth in a line of publications about Idaho fish and fishing and summarizes the thousands of hours of research and the hundreds of scientific papers written about the topic since statehood. Nearly seventy five years ago, the Idaho Fish and Game Department, in conjunction with the Idaho Department of Education, published what may have been the first official publication about Idaho wildlife; "Wildlife of Idaho" by William Rush in 1942. In the book's dedication, Governor Chase A. Clark, wrote:

> *"To The Boys and Girls of Idaho: Our Fish and Game Commission is giving you this book so that you may better know the mammals, birds and fish of Idaho. Idaho is blessed with a great many forms of wildlife. It is as rich in wild things as it is in minerals, forest, range lands, cultivated fields, lakes and rivers.*

The wildlife is yours. It belongs to the boys and girls of Idaho. The older folks are only taking care of it for you until you reach the age of manhood and womanhood when you will take over the reins yourselves. You must watch to see that these older people do not neglect your wildlife. You must guard against their killing too many of the game mammals, and birds or catching too many fish. You must see to it that the wild creatures are well taken care of, that they have good homes with plenty to eat, whether it be in the forest, fields or waters. You should have more animals, rather than less so that there will be wildlife to pass on to your children and grandchildren. I sincerely hope you will enjoy studying this book and will profit greatly from it."

Governor Clark's words were written when I was a boy, ten years old. I was raised during the Depression on a small farm in logging country. Those days, everyone was so poor that there was no money to use on non-essentials as going into town to a movie. Yet, I was very fortunate. I had a father who understood the land and its resources. He taught me how and when to hunt and fish, when to take and when not to. But, those activities not only provided some food for the table, but were about our only means of recreation.

Today I consider it a great gift. I want other children to have the same gift and learn the same lessons. But if we don't provide the habitat and the water quality necessary to protect our fish and game, they – the wildlife as well as our children – are not going to have the chance.

The next book on Idaho fish and fishing was the 1982 "Fishes of Idaho" book by my friend Jim Simpson and his co-author Richard Wallace. Simpson was the Chief of Fisheries for the Idaho Department of Fish and Game during my first two terms as Governor, and Wallace was the ichthyology professor at the University of Idaho. For more than 30 years this was the only reference book scientists and avid anglers had readily available for biology, life history and distribution information on Idaho's fish.

Professor Wallace and Dr. Donald Zaroban authored the "Native Fishes of Idaho" in 2013 which laid the ground work for the book you have in your hands. It describes about half of the fish swimming in our state today. It also introduces the reader to excellent color illustrations of fish that look like what you catch when you are enjoying the outdoors in Idaho.

In this new volume, fisheries professionals and an interested angling public can extract a myriad of valuable information as either reference material or as direction in pursuit of their favorite fish or water body. Don Zaroban and John Sigler, and a cast of workers in the background, have gathered, recorded, and refined information on more fish than any avid angler could know about or catch. They have laid it out in a readable and provocative format that invites further scrutiny for the curious, and enhances the fisheries sciences.

To top this presentation off with style and flair, the color illustrations of each fish provide an accurate and superbly detailed rendition of each fish. I have not met Joseph Tomelleri, but he is undoubtedly the best fish illustrator alive today. He brings exquisite detail to these fish and enables us, to a degree that is astounding, to feel and see Idaho's aquatic beauties. See for yourself if these authors and this illustrator have not captured the essence of Idaho's fish. This comprehensive publication is a long overdue outgrowth of the works that preceded it. Please enjoy this book with me.

ACKNOWLEDGEMENTS

Idaho Power Company

Idaho Chapter American Fisheries Society

Idaho Department of Fish and Game

Idaho Department of Environmental Quality

Orma J. Smith Museum of Natural History, The College of Idaho

Idaho Trout Unlimited Council

US Bureau of Reclamation

US Fish and Wildlife Service

Caxton Press

To paraphrase the dedication from Jordan and Evermann's 1896 publication, *The Fishes of North and Middle America: a Descriptive Catalogue of the Species of Fish-Like Vertebrates Found in the Waters of North America, North of the Isthmus of Panama*, we gratefully acknowledge the efforts of all the researchers, explorers and other individuals who participated in the collection, analysis, and cataloging of fishes in North America and the United States since the 1700's.

In particular, with a specific perspective on the Pacific Northwest and Idaho, we acknowledge:

Reeve M. Bailey, Robert J. Behnke, Theodore C. Bjornn, Carl E. Bond, Carl H. Eigenmann, Rosa S. Eigenmann, Barton Warren Evermann, Charles Henry Gilbert, Charles Frederic Girard, John (Jack) S. Griffith, Carl Leavitt Hubbs, David Starr Jordan, W. M. Keil, Allan D. Linder, Douglas F. Markle, John D. McPhail, Robert Rush Miller, Virgil S. Pratt, Leonard P. Schultz, William F. Sigler, James C. Simpson, Gerald Ray Smith, Lyle M. Stanford, Richard L. Wallace, Richard R. Whitney, and Richard S. Wydoski.

Completion of this text could not have been achieved without the amazing amount of digitized material available from a range of online providers. We thank the Smithsonian Institution, University of Chicago, Stanford University, Biodiversity Heritage Library, Harvard University, and Google, as well as those organizations that provided funding for the digitization of historic writings. Additional access to historical materials was graciously provided by the Library and the Department of Fish and Wildlife Sciences at the University of Idaho.

Sponsors and Major Contributors

Idaho Power Company, Boise, Idaho: sponsorship funds, in particular Jim Chandler.

Idaho Chapter, American Fisheries Society: sponsorship funds and philosophical insights, in particular, Jim Chandler.

Idaho Department of Environmental Quality: sponsorship funds and in-kind personnel costs, in particular Barry Burnell.

Idaho Department of Fish and Game: Boise and across the State of Idaho: in-kind donation of time and talent in formatting the electronic version of the manuscript and professional reviews of various chapters and approaches to information presented.

In particular:

Kelly Kennedy Yokoyama and Emalee Lathouwers: format, graphic design, layout, e-publishing; Jeff Dillon: State Fisheries Manager, Gems Contributor and internal coordination extraordinaire; Lance Hebdon: chapter reviews, internal coordination; Scott Grunder: chapter reviews, internal coordination; Ed Schriever: Chief of Fisheries, Idaho Department of Fish and Game, internal IDFG coordination; Jim Fredericks: Chief of Fisheries, Idaho Department of Fish and Game, Gems contributor and internal coordination; Evan Brown: distribution map production and layout coordination; Paul Bunn: intranet support.

Authors of the "Idaho Aquatic Gems:"

Bear Lake
David Teuscher, Regional Fisheries Manager, Southeast Region
Idaho Department of Fish and Game

Big Wood River
Russell F. Thurow, Fisheries Research Biologist, US Forest Service: Rocky Mountain Research Station, Boise, ID
and
Fred E. Partridge, Regional Supervisor, Idaho Department of Fish and Game (retired), Jerome, ID

CJ Strike Reservoir
Joe Kozfkay, Regional Fishery Manager, Southwest Region
Idaho Department of Fish and Game

Clearwater River
Brett Bowersox, Fisheries Staff Biologist
Idaho Department of Fish and Game

Coeur d'Alene Lake
Jim Fredericks, Chief of Fisheries
Idaho Department of Fish and Game

Hells Canyon
Joe DuPont, Regional Fishery Manager, Clearwater Region
Idaho Department of Fish and Game

Henrys Fork
Rob Van Kirk, Senior Scientist
Henry's Fork Foundation

Henrys Lake
Jessica Buelow, Regional Fisheries Biologist, Henrys Lake
Idaho Department of Fish and Game

Lake Cascade
Dale Allen, Regional Fisheries Manager, Southwest Region
Idaho Department of Fish and Game

Lake Pend Oreille
Andy Dux, Regional Fishery Manager, Panhandle Region
Idaho Department of Fish and Game

Lucky Peak Reservoir
Joe Kozfkay, Regional Fishery Manager, Southwest Region
Idaho Department of Fish and Game

Middle Fork Salmon River - Frank Church River of No Return Wilderness
Russell F. Thurow, Fisheries Research Biologist, US Forest Service: Rocky Mountain Research Station, Boise, ID;
and
Bart L. Gamett, Fish Biologist, US Forest Service: Salmon-Challis National Forest, Mackay, ID;
and
Richard N. Williams, Conservation Biologist, Department of Biology, The College of Idaho, Caldwell, ID

Priest Lake
Jim Fredericks, Chief of Fisheries
Idaho Department of Fish and Game

Saint Joe River
Chip Corsi, Regional Supervisor, Panhandle Region
and
Matt Corsi, Principal Fishery Research Biologist
Idaho Department of Fish and Game

Silver Creek
Chad Chorney, Big Wood Project Manager
Trout Unlimited

South Fork Boise River
Jeff Dillon, State Fishery Manager
and
Art Butts, Regional Fisheries Biologist, Southwest Region
Idaho Department of Fish and Game

South Fork Snake River / Palisades Reservoir
Dan Garren, Regional Fisheries Manager, Upper Snake Region
Idaho Department of Fish and Game

Other Contributors and Supporters

Idaho Council of Trout Unlimited: sponsorship funds, in particular Andy Brunelle.

Henry's Fork Foundation: sponsorship funds

Bureau of Reclamation, US Department of the Interior: sponsorship funds

The College of Idaho, Caldwell, Idaho: administrative support.

Department of Fish and Wildlife Sciences, College of Natural Resources, University of Idaho, Moscow, Idaho: library access and privileges.

Quinney College of Natural Resources, Utah State University, Logan, Utah: library access and privileges.

Orma J. Smith Museum of Natural History, The College of Idaho: support of the fish collection.

Smith-Root, Incorporated, Vancouver, Washington: contributions to the Ichthyology chapter; in particular, Jeff Smith, CEO/President; Carl Burger, Senior Scientist; Phong Nguyen, Scientist.

Caxton Press: pre-publication and printing support, in particular Scott Gipson.

Major Segment Contributors

Cecil Andrus-Foreward

Gerald R. Smith, Curator Emeritus and Professor Emeritus, University of Michigan, Ann Arbor, Museum of Paleontology: Keys to Idaho Fishes, Biogeography Chapter, and much encouragement and insight.

Ernest Keeley, Professor and Fisheries Researcher, Department of Biology, Idaho State University: Biogeography Chapter, immeasurable assistance with chapter reviews, philosophical approaches, and methodologies for literature research.

Paul Link, Geologist and Professor, Department of Geology, Idaho State University: Biogeography Chapter.

Glenn Thackray, Geologist and Professor, Department of Geology, Idaho State University: Biogeography Chapter.

Joseph R. Tomelleri, artist extraordinaire, Joseph R. Tomelleri - Cimarron Trading Company, Leawood, Kansas: full color illustrations of fishes, [www.americanfishes.com].

Emily S. Damstra, Natural Science Illustration, Guelph, Ontario Canada: all fish key illustrations. [www.emilydamstra.com].

Renai C. Brogdon, graphic designer, desktop publisher: cover design and cover art work.

Link Jackson, artist: art work.

Support and Assistance

Bill Clark, Director of the Orma J. Smith Museum of Natural History, The College of Idaho: manuscript review.

Bill Horton, retired, Idaho Department of Fish and Game, Boise, Idaho: colleague and organizer/fixer extraordinaire, graduate school compadre in Moscow, Idaho.

Chris Luecke, Dean, Quinney College of Natural Resources, Utah State University, Logan, Utah: chapter reviews, long-time philosophical guidance, and encouragement.

Diana Sweet Miller (and crew), Wyoming Department of Game and Fish, fisheries biologist: assistance with Snake River Sucker collection at Jackson Lake, Grand Teton National Park, Wyoming, June 2014.

Brett High, Idaho Department of Fish and Game, Regional Fisheries Biologist, Upper Snake Region, Idaho Falls, Idaho: collection of *Chasmistes*-like suckers in Palisades Reservoir.

Douglas F. Markle, Professor Emeritus. Department of Fisheries & Wildlife. Oregon State University, Corvallis Oregon: photographs of Cascadian Sucker and Umatilla Dace.

Patsy Palacios, Director of Library Services, Quinney College of Natural Resources, Utah State University, Logan: library searches, literature reviews, data management, complete bibliography development, and formatting.

Krissy Wilson, Fisheries Biologist, Utah Division of Wildlife Resources, Cypriniformes species literature.

David Bennett, Fisheries Biologist and Professor of Fisheries Science, University of Idaho (retired): review of Cypriniformes chapter.

Jim Chandler and Ken Lepla, Idaho Power Company: significant contributions to the White Sturgeon life history.

Jay Hesse, Fisheries Program Manager, Nez Pierce Tribe, Lapwai, Idaho.

Annette Wenda, editor and grammarian, independent contractor.

Zotero. Zotero was an amazing asset throughout the development of this book. Zotero is a project of the Roy Rosenzweig Center for History and New Media, and was initially funded by the Andrew W. Mellon Foundation, the Institute of Museum and Library Services, and the Alfred P. Sloan Foundation.

*"If there is magic on the planet,
it is contained in the water."*

Loren Eiseley (1907-1977)

Link Jackson ©

INTRODUCTION

This book celebrates Idaho and the diverse and amazing natural resources contained within her borders. Because this is a "fishes of" book, the authors have placed a great deal of emphasis on both the fish found in Idaho and the water resources within the state that fish depend upon.

In addition to detailed life history sections for all of the taxa (species and subspecies) of fish known to exist in Idaho, we have provided information on some of the cultural and historical aspects of water management and fisheries management, including some descriptive materials that precede statehood. Our intent with respect to these materials is to provide readers with "pointers" toward more detailed historical, cultural and philosophical information should they desire a more in depth understanding of the evolution of fish, fishing, and fisheries management in Idaho. Detailed reference materials are provided for each segment of the book so that the interested reader can further pursue areas of interest.

We have purposely provided a significant amount of graphic material. From the individual taxon illustrations of Joe Tomelleri, to Emily Damstra's illustrations of meristic and morphometric characteristics associated with the taxonomic keys, to Evan Brown's distribution maps, the reader is presented with a pictorial of Idaho's fishes and some of the many amazing aspects of Idaho waters.

For readers interested in the practices associated with fisheries management in the State of Idaho, we have included a chapter on the mechanisms and protocols used to study fish and methods and practices for reporting fisheries information. Provided references point the interested reader to extensive literature on current fisheries management practices as well as associated topics.

Biogeography is discussed in a separate chapter and provides a discussion that addresses the relationship between Idaho's impressive geologic history and the effect those events had on the distribution of fishes throughout Idaho. This chapter provides the reader with an introductory overview of geologic events that influenced fish distribution and a discussion of several species' current distribution as related to past geologic events.

A separate chapter, "Idaho's Aquatic Gems," provides the reader with detailed descriptions, including "how to catch the fish," for 17 of Idaho's best fishing spots. These "Gems,' written by individuals who know them best, provide both a text and a pictorial presentation of where to fish, how to fish and what to expect at premier fishing waters from Bear Lake in the south to Lake Pend Oreille and Priest Lake in the north.

Following the "Gems" section is a summary presentation of many of the state's Urban Fishing Opportunities. This section provides the reader with location, species, and other information on where to fish near Idaho's urban centers across all regions of the state. Additional information on this and other Idaho Department of Fish and Game (IDFG) recreational fishing programs can be found on the Department web site.

Included in the text are dichotomous keys which will allow the reader to "key out"or identify any fish known to occur in Idaho waters. Because the key is highly illustrated, readers will be able to quickly learn what to look for in the species identification process.

Included at the end of the book is a Glossary of many of the terms used throughout the text. In many cases, these terms are used in the dichotomous keys.

Finally, the book includes over 700 cited and supplemental references on many of the topics presented, allowing the reader to pursue more in depth information on a wide variety of topics. Many of the references provided are available electronically or can be found at local libraries.

The purpose of this book is to help students, citizens, anglers, researchers, and managers understand and appreciate Idaho's fish fauna. We have a number of species, each of which has different characteristics and uses; and to organize our knowledge and communicate that knowledge, each species is identified with a formal species name and a common species name. The formal species name is Latinized and consists of a genus name and a species name, which is unique within that genus. For example *Oncorhynchus clarkii* is the formal name of the Cutthroat Trout in the genus of Pacific salmon and trout, *Oncorhynchus*. Rainbow Trout are a member of that same genus.

Common names, like "Cutthroat Trout," are chosen by a committee of fish biologists to try to apply a stable name that everybody can understand. The Cutthroat Trout species is made up of individual fishes, but it, itself, is an individual in the sense that it lives in a particular place where it evolved or migrated to, and its members are all related in some sense. It is also an individual in the sense that if it dies out, it cannot be replaced. A species' formal scientific name can change if its relationships are discovered by careful study to be different than previously thought. For example, the Cutthroat Trout was once classified in *Salmo*, the genus name of Atlantic Salmon and trout, but study of bones and DNA showed that to be a wrong interpretation of relationships. Fortunately, the common name is meant to apply unambiguously to all of the fishes that have the diagnosable characteristics and ecology of that population, and is meant to be stable for a long time to facilitate communication. It does not bear the burden of keeping up with interpretations of relationships.

The American Fisheries Society and the American Society of Ichthyologists and Herpetologists *Committee on Names of Fishes* (Names Committee) issues an updated list of accepted names and formalizes changes in a new list every 10 years. Therefore, some fish names temporarily exist in two versions, the old and the new. In this book, we present and explain some cases of names in flux. Fortunately, the Names Committee has chosen and stabilized long-standing common names so that they change less and facilitate communication with less disruption.

Idaho has roughly 26,000 miles of fishable streams and more than 200 major lowland lakes and reservoirs, providing nearly limitless opportunities for anglers and other recreationalists. Habitat conditions in these waters have a great influence on fish populations and strongly dictate the species and numbers that can be supported. Lowland lakes and reservoirs and large rivers generally support many kinds of fish, including bass, trout, and nongame species. Small colder streams and high-elevation lakes typically support only a few species of fish. Different management strategies involving stocking and special fishing rules are used to best provide the diversity of fishing that anglers want.

Across Idaho, there are more than 8,900 and rivers and streams. From the smallest tributary in the desert southwest drainages of Owyhee County to the mighty Snake River to the wondrous Salmon and the River of No Return Wilderness area to the crystal waters of Bear Lake, Lake Pend Oreille, and Lake Coeur d'Alene, surface water is plentiful in Idaho. Due to its abundance and distribution, water in Idaho has been a driving force in both the settlement and development of industry and the economy in Idaho for more than 200 years. The landform-scale tectonic events that contributed to both the distribution and the abundance of surface water in Idaho coupled with human's more recent activities are varied and complex, and most of them contributed in some fashion to the current distribution of fishes throughout Idaho.

We hope the material presented in this book helps you in your enjoyment of the Gem State.

"You need to know where you have been before you can move forward."

Mick Hoover

1

AN OVERVIEW OF PRECEPTS FOR IDAHO FISHERIES MANAGEMENT

This chapter highlights a variety of aspects associated with management of Idaho's waters, fish, and fishing. It provides a brief summary of fisheries management principles in Idaho. Fisheries management within the IDFG is guided by a variety of technical plans and processes. These plans are supported by field and laboratory research. The *Compass* is the strategic plan developed by the Department in 2005 to guide all aspects of fisheries management. Currently, the Department is operating under guidance outlined in the Strategic Plan 2013-2018. This material is excerpted from the IDFG website. The current Strategic Plan, as well as other technical reports are available to the public.

Mission Statement (Idaho Code Section 36-103)

All wildlife, including all wild animals, wild birds, and fish, within the state of Idaho, is hereby declared to be the property of the state of Idaho. It shall be preserved, protected, perpetuated, and managed. It shall only be captured or taken at such times or places, under such conditions, or by such means, or in such manner, as will preserve, protect, and perpetuate such wildlife, and provide for the citizens of this state and, as by law permitted to others, continued supplies of such wildlife for hunting, fishing and trapping.

Working under the guidance of the Commission, the Department manages the fish and wildlife of the state. Idaho fisheries programs are managed in the context of the Department Mission Statement and with the impetus of the overall economics of recreational fisheries adding to the mix.

The Idaho Department of Commerce estimates that the recreation and tourism industry is the third largest in the state. Sport fishing comprises a substantial part of this business. IDFG conducted a fishery economics survey in 2011 by mailing out 59,200 surveys to Idaho fishing license holders. In 2011, Idaho's population was 1,584,986 with one in five eligible people residing in Idaho purchasing a fishing license. Based on the results of almost 26,000 completed responses from this survey, the Department estimated that 425,415 anglers spent more than 3.6 million days (nearly 2.8 million trips) on Idaho waters. Fishing in Idaho generated $548,351,483 in statewide retail sales in 2011 with an additional $14,962,572 for fishing licenses and permits. Comparatively, recreational fishing in Idaho generated $437,631,735 in statewide retail sales in 2003 with an additional $12,289,806 for fishing licenses and permits. IDFG conducted an economic survey following the 2010 Chinook Salmon season that generated an estimated direct angler expenditure of $39.6 million (IDFG, unpublished data). This survey did not measure any indirect economic activity or jobs created. Approximately 103,407 days of effort were expended during nearly 75,000 angler trips. The average cost per trip was over $500.

Vision

The Idaho Department of Fish and Game shall work with the citizens of Idaho in providing abundant, diverse fish and wildlife and ensuring a rich outdoor heritage for all generations.

Statewide Fisheries Programs

Within the Bureau of Fisheries, professional staff are organized into operational sections including: 1) resident fisheries management, 2) anadromous fisheries management, 3) hatcheries (resident and anadromous), 4) research, and 5) habitat.

Resident and Anadromous Fisheries Management

The primary responsibility of the resident and anadromous fisheries management sections of the Bureau of Fisheries is to monitor and manipulate fish populations to maintain/create public fisheries, protect and enhance fish habitat, develop angler access, provide information to anglers, coordinate with the general fishing public, and develop fishing rules. Most of the management effort involves Department field staff (biologists, technicians, and others) working in coordination with headquarters staff, and with personnel of state and federal agencies, Indian tribes, and non-governmental organizations. The programmatic function of habitat protection and technical assistance is currently implemented through the Director's Office of the Department. Regional and headquarters fisheries staff supply data and provide technical support to regional environmental staff biologists and the Director's Office.

Resident Fish Management

Resident fish management practices include a number of broad, over-arching principles as well as species-specific practices. In general, IDFG is increasing protection of native trout populations by discontinuing stocking of reproductively-viable nonnative species (Brook Trout) and strains (coastal Rainbow Trout), increasing catch limits on nonnative fishes, and decreasing catch limits on native "wild trout."

Management practices are being applied to reduce genetic introgression of native Cutthroat Trout by Rainbow Trout. Trend monitoring of Redband Trout populations will be continued to ensure that connectivity between metapopulations is reestablished or maintained.

For more sensitive species, IDFG coordinates with the U.S. Fish and Wildlife Service (USFWS) and U.S. Forest Service (USFS) in conservation and recovery planning for Bull Trout, continues to implement the management plan for Big Lost River basin Mountain Whitefish, continues to participate on the Kootenai Sturgeon Recovery Team for White Sturgeon, and implement the Snake River White Sturgeon management plan.

Hatchery supplementation of Largemouth Bass, Smallmouth Bass, Black Crappie, White Crappie, Bluegill, Yellow Perch, Channel Catfish, Walleye, Northern Pike, and Tiger Muskie will be continued. Species specific management programs will continue such that in ecosystems where trout and bass are co-managed for a quality fishery, increases in trout size will occur in populations where less than 20 percent of the fish exceed 16 inches; where trout and bass are co-managed for a trophy fishery, increases in catch rates and sizes are such that 20 percent or more of the trout exceed 16 inches. In the same context, native nongame fish species will be managed for long-term persistence and sufficient numbers to perform ecological functions.

Alpine lake management addresses a series of goals such that some lakes will be managed as fishless to allow for maintenance of natural conditions for native fauna while lakes in wilderness and national recreational areas will be managed in cooperation with the land management agency. Across the state, some alpine lakes will be managed to maintain self-sustaining populations of native trout and in all cases management will strive to maintain biotic integrity of amphibian and invertebrate populations as well as reduce or eliminate impacts of nonnative fish species as part of alpine lake management plans.

Anadromous Fisheries Management

The Snake River upstream from Lewiston historically produced an estimated 55 percent of the summer Steelhead, 40 percent of the spring Chinook Salmon, and 45 percent of the summer Chinook Salmon in the Columbia River. Historically, Idaho was also a key production area for fall Chinook Salmon. Lesser numbers of Sockeye Salmon and Coho Salmon inhabited the Snake River drainage although Snake River Coho Salmon were declared extinct in the mid-1980s. Runs of naturally reproducing salmon and Steelhead in Idaho have generally improved since historic low abundances experienced in the mid-1990s, but they are still lower than the 1960s and early 1970s. The declines in run sizes led to federal Endangered Species Act listings of Snake River salmon and Steelhead starting in 1992. The three Snake River federally-listed Evolutionarily Significant Units (ESUs) and the listing status of each are: Snake River Sockeye Salmon (endangered), Snake River spring/summer Chinook Salmon (threatened), and Snake River fall Chinook Salmon (threatened). Also, the Snake River Steelhead distinct population segment (DPS) is listed as threatened. Forty-three of the 62 independent populations within the listed ESUs and DPS are entirely contained in waters managed by IDFG, and two populations occur in waters jointly managed with Oregon and Washington. The federal Endangered Species Act listings add complexity to state management. IDFG's long-range goal of the anadromous fish program is to recover and preserve Idaho's salmon and Steelhead runs to provide benefits for all users. This management goal complements the *Compass* objectives to: 1) maintain or improve game populations to meet the demand for hunting, fishing, and trapping; and 2) ensure the long-term survival of native fish, wildlife, and plants. The anadromous management program will employ many of the key *Compass* strategies consistent with these objectives, generally described below. Key management objectives to achieve the management goal are: 1) maintain genetic and life history diversity and integrity of both naturally- and hatchery-produced fish; 2) rebuild naturally producing populations of anadromous fish to utilize existing and potential habitat at an optimal level; 3) achieve equitable mitigation benefits for losses of anadromous fish caused by development of the hydroelectric system on the Snake and Columbia rivers; 4) improve overall life cycle survival sufficient for delisting and recovery by addressing key limiting factors identified in all "H's" of hydropower, habitat, harvest, and hatchery effects; 5) allow consumptive harvest by sport and treaty fishers; and 6) coordinate Pacific Northwest regional management with Idaho anadromous management to ensure achievement of Idaho management objectives and the long-range program goal. To help meet the anadromous program long-range goal and federal recovery goals, Idaho's anadromous fish management utilizes both natural and hatchery production. In natural production waters, populations are sustained by spawning and rearing in the natural habitat without human intervention, regardless of the parentage of the spawners (i.e., naturally produced progeny of hatchery or wild/natural origin fish). Hatchery production occurs in a controlled artificial spawning and rearing environment with hatchery fish released to provide fishing opportunity or to sustain populations. For management purposes, IDFG classifies three groups of salmon and Steelhead based on definition of production and broodstock history: wild, natural, and hatchery fish.

Fish Hatchery Programs

The hatcheries section raises fish to meet specific fishery management objectives. New technologies are developed to raise healthy fish in the most cost-effective manner. The hatcheries section includes a fish health unit designed to identify and treat various fish diseases and to improve the health of particular stocks. Fisheries projects that benefit directly from the hatchery program include put-and-take and put-grow-and-take fish stocking programs, supplementation of salmon and Steelhead natural production, supplementation of reduced populations of resident fish, and production of other game fishes to produce sport fisheries. Both anadromous and resident fish hatchery programs are discussed in more detail below.

Fisheries Research Program

The fisheries research section has five organizational components: anadromous fish mitigation, resident fish species mitigation (both 100 percent Federal funding), discretionary research, program management/technical support (both funded 75 percent with federal sport fish restoration or Dingell Johnson funds), and lastly a genetics program (funded via a variety of Federal, state and private (e.g., Idaho Power Company) funds. IDFG operates a fish genetics laboratory to provide an efficient, cost-effective means of generating detailed genetic information necessary for the improved management of Idaho's native fish species. Importantly, information gained from genetic research studies completed in the laboratory are being used by managers to assess current and future genetic risks, preserve existing genetic variability, delineate and prioritize populations for management purposes, identify suitable populations for translocations and reintroductions, identify suitable populations for broodstock development, and address genetic concerns in Endangered Species Act petitions. The Endangered Species Act status of anadromous Sockeye Salmon and Chinook Salmon mandates research/mitigation efforts on those two species are to be intricately tied to Federal recovery processes. Currently, two research projects (captive Chinook Salmon and Sockeye Salmon research) are focusing on long- and short-term implementation of captive technologies for maintenance of genetic diversity and restoration of wild stocks to harvestable levels. Resident fish species mitigation research is designed to mitigate for fish populations impacted by development of the federal Columbia River hydropower system. Population monitoring, evaluation, and other findings are used to recover populations of Kootenai River White Sturgeon, Kokanee, Rainbow Trout, Bull Trout, Burbot, and other species that have been adversely impacted by hydropower systems. The general direction of these research activities is coordinated with other resource agencies, provincial governments, Indian tribes, and federal or utility funding entities and set through funding contracts. To provide direction for the remainder of IDFG's fisheries research program, a combination of management, hatchery and research personnel identified needed information and tools that would enhance fisheries management in Idaho. These research questions are being addressed by three separate Dingell-Johnson funded projects entitled *Hatchery Trout*, *Wild Trout*, and *Lake and Reservoir Research*.

Fisheries Habitat Program

The fish habitat section designs and implements fish habitat restoration projects focused on addressing the primary factor(s) limiting native fish production and productivity. Projects are designed to be ecologically sustainable over the long-term. Staff interact with a variety of state and federal agencies, non-governmental organizations, and private landowners to develop and implement projects. The majority of funding comes from federal or private sources. Because of this, staff also develop proposals and compete for available funding.

Relevant Federal Statutes

Fisheries management in this century involves not only state guidance and regulation but Federal mandates as well. Federal statutes that have significant impacts to fish and fishing in Idaho include the Endangered Species Act and the Clean Water Act.

Endangered Species Act Of 1973 (Public Law 93-205, Approved Dec. 28, 1973, 87 Stat. 884)

> *AN ACT To provide for the conservation of endangered and threatened species of fish, wildlife, and plants, and for other purposes.*
>
> *SEC. 2 (a) FINDINGS.—The Congress finds and declares that (1) various species of fish, wildlife, and plants in the United States have been rendered extinct as a consequence of economic growth and development untempered by adequate concern and conservation; (2) other species of fish, wildlife, and plants have been so depleted in numbers that they are in danger of or threatened with extinction; (3) these species of fish, wildlife, and plants are of esthetic, ecological, educational, historical, recreational, and scientific value to the Nation and its people; (4) the United States has pledged itself as a sovereign state in the international community to conserve to the extent practicable the various species of fish or wildlife and plants facing extinction.*

The Endangered Species Act as amended was passed by Congress in 1973. Its purpose is to provide a means of ensuring the preservation of animal and plant species that are in danger of extinction. An endangered species is any species that is in danger of extinction throughout all or a significant portion of its range, whereas a threatened species is any species which is likely to become an endangered species within the foreseeable future throughout all or a significant portion of its range. Species may be broadly defined to include isolated breeding populations that are ecologically significant such as the Kootenai River White Sturgeon, which is comprised of a single spawning population. Recovery of a species to a level safe from extinction is the key goal of the act. Typically, restoration actions are guided by a recovery plan, and the tools of recovery may range from captive breeding to land acquisition. Critical habitat is identified for listed species in order to provide special protection for key spawning and rearing areas.

Clean Water Act *Federal Water Pollution Control Act (33 U.S. C. 1251 et seq.)*

> *AN ACT To provide for water pollution control activities in the Public Health Service of the Federal Security Agency and in the Federal Works Agency, and for other purposes.*
>
> *Be it enacted by the Senate and House of Representatives of the United States of America in Congress assembled,*
>
> *Title 1 – Research and Related Programs: Declaration of Goals and Policy*
>
> *SEC. 101. (a) The objective of this Act is to restore and maintain the chemical, physical, and biological integrity of the Nation's waters. In order to achieve this objective it is hereby declared that, consistent with the provisions of this Act—*

> *(1) It is the national goal that the discharge of pollutants into the navigable waters be eliminated by 1985; It is the national goal that wherever attainable, an interim goal of water quality which provides for the protection and propagation of fish, shellfish, and wildlife and provides for recreation in and on the water be achieved by July 1, 1983;*
>
> *(2) It is the national policy that the discharge of toxic pollutants in toxic amounts be prohibited;*
>
> *(3) It is the national policy that Federal financial assistance be provided to construct publicly owned waste treatment works;*
>
> *(4) It is the national policy that area wide treatment management planning processes be developed and implemented to assure adequate control of sources of pollutants in each State;*
>
> *(5) It is the national policy that a major research and demonstration effort be made to develop technology necessary to eliminate the discharge of pollutants into the navigable waters, waters of the contiguous zone and the oceans; and*
>
> *(6) It is the national policy that programs for the control of nonpoint sources of pollution be developed and implemented in an expeditious manner so as to enable the goals of this Act to be met through the control of both point and nonpoint sources of pollution.*
>
> *(b) It is the policy of the Congress to recognize, preserve, and protect the primary responsibilities and rights of States to prevent, reduce, and eliminate pollution, to plan the development and use including restoration, preservation, and enhancement) of land and water resources, and to consult with the Administrator in the exercise of his authority under this Act. It is the policy of Congress that the States manage the construction grant program under this Act and implement the permit programs under sections 402 and 404 of this Act. It is further the policy of the Congress to support and aid research relating to the prevention, reduction, and elimination of pollution, and to provide Federal technical services and financial aid to State and interstate agencies and municipalities in connection with the prevention, reduction, and elimination of pollution.*

The Clean Water Act establishes the basic structure for regulating discharges of pollutants into the waters of the United States and has empowered states to establish water quality standards to ensure those waters are maintained or restored to fishable and swimmable status wherever attainable. In 1948 the Federal Water Pollution Control Act became law and began paving the way for future federal actions to clean up the nation's waters. The act was significantly reorganized and expanded in 1972, beginning the modern era of the Clean Water Act.

Over the years, the focus of the Clean Water Act and the regulatory agencies managing it has shifted several times. Initially, focus was on setting wastewater-discharge standards for industry. This era was highlighted by the famous "River on Fire," when a picture of the flaming Cuyahoga River in Ohio graced the pages of a 1969 edition of Time (a truly arresting image showing flames leaping from the water, completely engulfing a ship). As it turned out, that photo was actually from a serious fire in November 1952 (the river had been on fire numerous times since the turn of the century). No picture of the 1969 river fire is known to exist, but it helped concentrate efforts to develop a national program to clean up rivers across the country.

In the 1960s and '70s, much emphasis was placed on point-source discharges from municipal wastewater facilities (both large and small), and millions of dollars were expended to upgrade existing facilities or build treatment works for previously untreated or minimally treated discharges. Under the CWA, the US Environmental Protection Agency (EPA) has implemented pollution-control programs such as establishing wastewater standards for industry and has also set water-quality standards for the majority of contaminants entering surface waters.

The CWA made it unlawful to discharge any pollutant from a point source (discrete conveyances such as pipes or man-made ditches) into navigable waters, unless a permit was obtained through the EPA's National Pollutant Discharge Elimination System (NPDES). Individual homes that are connected to a municipal system, use a septic system, or do not have a surface discharge do not need an NPDES permit; however, industrial, municipal, and other facilities must obtain permits if their discharges go directly to surface waters.

In the last decade of the 1900s and continuing into the 21st century, CWA focus and increased efforts by the states have shifted to addressing nonpoint sources of pollution that have been shown to cause a large proportion of stream and river impairment around the country. Particular emphasis in Idaho has been on addressing storm-water runoff (municipal, industrial, and construction related) and runoff from agricultural operations, including irrigation-return flows, animal-feeding operations, and livestock grazing.

The larger metropolitan areas in Idaho, generally cities or groups of adjacent smaller towns with populations exceeding 50,000 people, are required to have approved NPDES permits in place. These permits describe and dictate how storm-water facilities are managed prior to discharge into a jurisdictional water of the United States. Idaho's rivers, lakes, streams, reservoirs, and wetlands not only provide great natural beauty but also, in concert with groundwater supplies, provide the water necessary for drinking, industry, and agriculture, as well as supporting aquatic life and fishing opportunities across the state.

As the state agency tasked with implementing the federal Clean Water Act in Idaho, the Department of Environmental Quality (DEQ) is required by Section 303(e) of the federal act to develop a Continuing Planning Process that describes the ongoing processes and planning requirements of the state's water-quality program. One tool the DEQ uses to implement its water quality program is its Water Quality Management Plan. In essence, the plan is a compilation of the guidance and programs the DEQ uses to implement the Clean Water Act. The state of Idaho's commitment to water-quality protection is articulated in Title 39 of the Environmental Protection and Health Act of 1972 and codified in Idaho's Administrative Rules.

In essence, the DEQ uses water quality standards to evaluate if Idaho's water bodies are adequately protected. Idaho's water quality standards define the goals that have been set for a water body by designating beneficial uses for the water, setting criteria necessary to protect those uses, and preventing degradation of water quality. Beneficial uses are the desired uses that water bodies generally support and include water supply (domestic, agricultural, and industrial), recreation (such as swimming, boating, and fishing), and aquatic life, including fishes. Each beneficial use has a unique set of water quality requirements or criteria that must be met for the use to be supported. Most water bodies have multiple beneficial uses. A water body is considered impaired when it does not meet the water quality criteria needed to support one or more of its beneficial uses. The DEQ develops restoration or watershed-improvement plans (often called total maximum daily load plans, or TMDLs) that allocate responsibility for pollutants to various entities or land uses that are causing water-quality impairment. Through the process of working with local stakeholders and other interests to implement best management practices, improved wastewater treatment, or both, the DEQ ensures these water bodies will meet water quality standards and the fishable and swimmable goals of the Clean Water Act. It is the DEQ's responsibility to ensure that beneficial uses are maintained or moving toward timely restoration in the vast array of watersheds across Idaho. From the forested northern and central mountains to the broad Snake River plain, the DEQ works to engage stakeholders that have a vested interest in clean water. It is with this in mind that we find Idaho a great place to live, work, and perhaps bump into one or more of the many fishes found in the pages herein.

Other Fisheries Management Concerns

Invasive Species Introductions and the Threat of Hybridization

In North America, freshwater fishes are probably the most threatened group of vertebrates other than amphibians. A recent review of the literature found that exotic species introductions (illegal and intentional) were a factor in 68 percent of fish extinctions in North America, and the rates of introductions have increased dramatically in the past 50 years, affecting nearly every major watershed in the United States. Fisheries management programs have transplanted popular sport fishes, such as Northern Pike, Brown Trout, Brook Trout, Rainbow Trout, and Lake Trout, to provide recreational opportunities, while other aquatic organisms have been transplanted to enhance fish growth. Introductions of exotic species threaten the biodiversity of native faunas throughout the world[1].

During the last 100 years, and well after the establishment of the Endangered Species Act, extinctions of three genera, 27 species, and 13 subspecies of fishes from North America have been documented. Extinctions are recorded from all areas except northern Canada and Alaska. Regions suffering the greatest loss are the Great Lakes, Great Basin, Rio Grande Valley of Mexico, and Parras Valley in Mexico. More than one factor contributed to the decline and extinction of 82 percent of the fishes. Physical habitat alteration was the most frequently cited causal factor (73 percent). Detrimental effects of introduced species also were cited in 68 percent of the extinctions. Chemical habitat alteration (including pollution) and hybridization each were cited in 38 percent of the extinctions, and overharvesting adversely affected 15 percent of the fishes. This unfortunate and unprecedented rate of loss of the fishery resource is expected to increase as more of the native fauna of North America become endangered or threatened[2].

Invasive species threaten the biodiversity of aquatic ecosystems worldwide[3] and are considered the second greatest threat to biodiversity loss in North America[4]. Hybridization (interbreeding of individuals from genetically distinct populations) between introduced and native taxa can be a major consequence of species introductions, especially when nonnative species hybridize with rare or endangered species, threatening their persistence[5].

Hybridization is more common in fish than in any other vertebrate taxa. Many fishes have external fertilization and similar mating behaviors, facilitating interbreeding. In undisturbed ecosystems, reproductive isolation is maintained primarily by spatial and temporal isolation during reproduction. However, such barriers to interbreeding are lost in the face of habitat disturbance and introduction of nonnative species. This is particularly true for salmonids in which widespread introgression (gene flow between populations with individuals that hybridize) among native and nonnative taxa has replaced native species over large areas throughout their native ranges, which is suspected to lead to a loss of behavioral and ecological adaptations in native populations[2,5].

Hybridization with introduced species has been a major factor in the decline and extinction of many native fishes throughout North America[2]. Although natural hybridization can lead to evolutionary novelty and speciation in many plants and some vertebrate species, human-mediated hybridization and introgression can lead to the extinction of native genotypes and the loss of locally adapted gene complexes and ecological adaptations. Therefore, understanding the invasive patterns and mechanisms that permit and constrain hybridization is required for conservation of aquatic biodiversity and developing recovery and management programs for native species at risk of hybridization with introduced taxa[5].

In a study of Westslope Cutthroat Trout in the upper Flathead drainage of Montana, for example, distribution of spawning locations varied geographically among four groups of fishes. Westslope Cutthroat Trout spawned in headwater tributaries, Rainbow Trout and their hybrids spawned in the lower drainage, and Westslope Cutthroat Trout hybrids generally spawned in the middle portion of the drainage, overlapping with Westslope Cutthroat Trout spawning locations. Evidence that hybridization is spreading upstream via long upstream movements of a few individuals with a high proportion of Rainbow Trout admixture was documented. Two Rainbow Trout males made long-distance migrations and spawned in tributaries used by Westslope Cutthroat Trout during the study. One Rainbow Trout migrated 73 km (45 mi) upriver and spawned in the lower section of Hay Creek, the same general location that a Westslope Cutthroat Trout used for spawning. Similarly, a Rainbow Trout made an upriver movement of 67 km (42 mi) and spawned in Lincoln Creek, the same stream used by a Westslope Cutthroat Trout and a Westslope Cutthroat Trout hybrid for spawning in 2007[5].

Finally, individuals with a high proportion of Rainbow Trout admixture made long-distance movements to spawning areas that Westslope Cutthroat Trout hybrids used for spawning; four fishes migrated a mean distance of 55 km (34 mi) and spawned in Dutch and Anaconda Creeks. However, the majority of Rainbow Trout and their hybrids spawned in tributaries in the lower portions of the drainage[5].

Hatchery Uses and Management Applications

In varying management situations throughout the cold-water, cool-water, and warm-water fisheries of North America and Europe, fish hatcheries are an important tool used to help achieve fisheries management objectives and were probably the earliest tool used in North American fisheries management. As early as the 1870s, federal policy makers at the US Commission of Fish and Fisheries turned to stocking programs to replenish depleted stocks, which were mostly marine at that time[6].

Historical Overview—United States

Fish culture had its beginnings in ancient China and has been practiced by the Chinese from time immemorial. Development of techniques associated with artificial fertilization of trout eggs is usually credited to a French monk, Dom Pinchon, who lived in the 14th century and was the first to employ hatching boxes involving modern fish-culture principles. The first successful attempt at artificial propagation of fish in the United States has been generally credited to Theodatus Garlick, who in 1857 published a treatise on his efforts, which are mostly concerned with Brook Trout. At about the same time, techniques using dry fertilization were developed in Russia and the United States, and this procedure (which is still used today) greatly increased the number of viable eggs produced. By 1865 fish culture was providing a substantial income to several individuals in the United States. In many instances, the local demand for eggs and small fish exceeded production capacities, and few if any fish made it to the market as food[6].

Most states had established state fishery commissions by the mid-1800s, and declines in fish abundance were a concern of these agencies. The historical record holds relatively little documentation of lost inland fisheries or of well-defined reasoning for such losses. Acknowledged and likely contributors to declines in fisheries included the depletion of timber due to logging, industrial, and agricultural operations; increased runoff and erosion associated with logging and agriculture; alteration of streams for irrigation and installation of dams for power; and overfishing. Regardless of the cause of declining fish populations, the use of fish culture to increase fish stocks seemed a logical and direct action to address the issue[6].

In the western United States, particularly the Great Basin areas, early fish introductions were made primarily for the purpose of increasing the region's food supply, and stocking whatever species were available was encouraged. While Common Carp and other exotic species were being distributed to easily accessible waters of the western United States, "hatching stations" and salmonid aquaculture practices were being developed and refined[7].

Rainbow Trout proved amenable not only to captive rearing, but also to selective breeding programs, exhibiting adaptability over just a few generations. Over several decades, Rainbow Trout can be bred to grow faster and mature earlier, and efforts have also been made to select for disease resistance, fecundity, and even such traits as color, shape, and fighting ability. These characteristics made the Rainbow Trout a good candidate for fish production and offered the promise that a single species might be adapted to deliver a variety of outcomes desired by fisheries resource managers. This potential of Rainbow Trout and other salmonids led to significant investment in the construction and operation of state, tribal, and federal fish hatcheries[8].

Today, fish hatcheries and rearing operations meet a much broader array of recreational and conservation objectives. Directly providing recreational sport fisheries with traditional put-and-take and put-grow-and-take programs remains important. Other hatchery programs have been established to conserve or supplement native fish stocks or to mitigate the effects of habitat alteration on fisheries resources. To meet these evolving demands, modern fish culture incorporates a science-based approach to evaluate potential improvements to feeding regimens, broodstock selection, disease control, and new rearing techniques and partners with fisheries researchers and managers to assess post-release performance.

To effectively use products of fish hatcheries, fisheries managers must first determine what product is required. Product definition can be viewed as the identification of the fish best suited to the intended use and may be more refined than simple designation of a species or sex. Product definition for fish hatcheries may include survivability, reproductive capability, catchability, competitive ability, and angler acceptability, which can be defined as catch challenge, status, and taste or palatability[6]. A detailed summary of the role of hatcheries in modern fisheries management may be found in Trushenski et al.[9], as well as several of the references presented in that publication[8–14].

Fish Hatchery Use In Idaho Fisheries Resource Management

Fish hatcheries have been present in Idaho for more than 100 years; the first, Hayspur Fish Hatchery, was constructed in 1907. In 1934 the state of Idaho and the Civilian Conservation Corps constructed the first Rainbow Trout rearing facility in American Falls. The purpose of the facility was to improve the availability of fish for sportsmen.

Over the century since hatcheries were initially introduced as a tool to increase fish populations, there has been great advancement in scientific knowledge of our fisheries, as well as in associated areas of study such as aquatic ecology and population genetics. This knowledge has informed not only an improved public understanding of fisheries resources, but also an evolution of the public's conservation ethic regarding those resources. As a result, hatcheries and their products have been retooled to achieve harvest and conservation objectives while minimizing potential negative impacts of hatchery fish upon natural populations. Idaho fisheries programs have been at the forefront in balancing these resource management objectives.

Conservation Hatcheries

Conservation hatcheries are an extreme form of supplementation hatcheries and follow protocols to intensively manage the genetic integrity of the broodstock as well as the overall fitness of the progeny. Culture methods are typically modified to mimic natural conditions to the extent feasible. Fish originating from conservation hatcheries have been raised to be as genetically and behaviorally similar to wild fish as possible and are best suited to management plans focused on the restoration of imperiled populations. Conservation hatcheries also serve increasingly important roles as refugia for rare species or genetic profiles[15].

Many hatcheries are functional "hybrids," operating as harvest-augmentation, supplementation, or conservation hatcheries by turns or simultaneously to produce various fish in a manner consistent with their intended uses. Clear and well-documented objectives are essential for all hatchery programs, especially for facilities rearing fish for different uses[9,15].

Site-Selective Stocking Practices

A relatively straightforward practice that limits hatchery fish interaction and influence upon desirable naturally reproducing fish populations is providing spatial separation between the two. Idaho's fisheries managers provide a diversity of fishing opportunities and conserve natural fish stocks by limiting the release of hatchery fish into habitat already occupied by other valued species. For example, hatchery Rainbow Trout are not typically stocked into waters that support wild populations of Westslope or Yellowstone Cutthroat Trout. This fiscally and environmentally conservative approach allows abundant harvest of hatchery fish in waters with limited natural productivity while reserving productive fish habitat for wild or native stocks that have adapted to specific locations and environmental conditions. The shift toward more wild trout management in Idaho has led fisheries managers to reduce the stocking of flowing waters. Currently, hatchery-produced resident trout are stocked into only about three percent of Idaho's fishable stream miles, a level less than half that of historical highs.

A total of 21 endemic spring and summer Chinook Salmon populations remain in Idaho: six (29 percent) of these populations have associated hatchery programs; the remaining 15 (71 percent) populations are managed as wild fish sanctuaries. Likewise, 18 endemic Steelhead populations exist in Idaho, with 11 (61 percent) being managed as wild fish production areas.

Stocking Sterile Trout

Historical stocking of nonnative resident trout, primarily coastal-origin Rainbow Trout, has led to introgression with native trout in many locations across the West. Fortunately, Idaho still has robust populations of genetically intact native fish across much of the state, and current-day stocking practices put relatively few hatchery trout in waters containing wild native fish. As an additional safeguard against genetic impacts, the IDFG developed techniques to produce sterile triploid hatchery Rainbow Trout and implemented this in statewide stocking programs in 2000. Currently, more than half of the trout stocked annually by IDFG are sterile triploid fish. Similarly, triploid Rainbow Trout × Cutthroat Trout hybrids are produced for the Henrys Lake fishery, and triploid programs for several other cultured species are currently in the development or evaluation phase. While Idaho pioneered the development of sterile trout for stocking programs, many other states have now adopted this approach to enhance sport-fishing opportunities while minimizing risks to native species.

Segregated/Integrated Anadromous Programs

Contemporary salmon and Steelhead supplementation programs aim to augment depressed fish populations by allowing hatchery-produced adult fish to migrate into spawning habitat to interact with naturally produced fish. The increase in the number of spawning adults results in an increase in the number of juveniles produced. In order to maintain the genetic fitness of supplemented natural populations, supplementation programs manage the proportion of hatchery-origin spawners introduced to the environment and therefore manage the genetic influence of hatchery fish upon the natural population. In this model, three subgroups are derived from a single stock of origin: "segregated" hatchery fish, which are the progeny of two hatchery-spawned parents; "natural" fish, which are the progeny of two fish that spawned in the environment; and "integrated" hatchery fish, which are the progeny of a hatchery × wild spawning cross or a cross of two integrated parents. Discrete marks are assigned to hatchery-produced smolts so that returning adult fish can be identified as segregated, integrated, or naturally produced. Thus, the hatchery can maintain desired levels of segregated and integrated production, and, more important, escapement of segregated, integrated, and naturally produced adults into the spawning habitat can be managed to ensure that hatchery influence upon the genetic composition of the natural run falls within acceptable limits.

Parentage-Based Genetic Management

Even more complex spawning models have been developed to maximize the genetic diversity retained within severely depleted fish populations. In such at-risk populations, each fish is important to the future viability of the stock, and individual fish are tagged so that they can be identified by their parentage and unique genetic profile. In the case of the Snake River Sockeye Salmon recovery program, which has pulled the stock back from the brink of extinction, the individual fish parentage data are factored into a unique annual multi-family spawning matrix so that the program not only increases fish abundance, but also preserves the genetic integrity and diversity of the stock. A similar genetics-based approach is applied by an integrated team of fisheries managers, researchers, and hatchery operators to the Bear River Bonneville Cutthroat Trout conservation program in southeastern Idaho.

Idaho's Fish Production Programs

Public-sector hatchery programs focused on fisheries resource management in Idaho can be broadly characterized as either resident or anadromous fisheries programs. Both are discussed briefly below.

Resident Fish Hatcheries in Idaho

Almost all resident fish hatcheries in Idaho are operated by IDFG. They are financially supported primarily through the sale of fishing licenses, but also receive funds from the federal Dingell-Johnson Sport Fish Restoration program. Additionally, several IDFG fish stocking projects are funded by Pacificorp, Avista Utilities, the Bureau of Reclamation, and municipalities across the state. IDFG resident (nonanadromous) hatcheries rear and stock multiple species to meet requests of fisheries managers. The overwhelming majority of the fish stocked are Rainbow Trout and Kokanee, but other species and subspecies of trout, as well as Arctic Grayling, White Sturgeon, and Chinook Salmon, are also produced by IDFG hatcheries to provide angling opportunity and conserve fish populations in resident waters. The IDFG facilities that focus primarily on trout production are the Nampa Fish Hatchery, American Falls Fish Hatchery, Hagerman State Fish Hatchery, Grace Fish Hatchery, and Mackay Fish Hatchery. The Hayspur Fish Hatchery serves as the program's primary Rainbow Trout broodstock station. The Cabinet Gorge Fish Hatchery is the hub of the statewide Kokanee stocking program; the Mackay Hatchery also plays an important role in Idaho Kokanee production. Two other facilities, the Sandpoint Hatchery and Mullan Hatchery, serve as seasonal redistribution stations, receiving transferred Rainbow Trout that were reared at other facilities and stocking them into nearby waters. Several anadromous hatcheries also play a redistribution role for IDFG's resident stocking program.

Several tribal fisheries management programs in the state of Idaho also utilize hatchery fish to support program objectives. Highly regarded trout fisheries on the Shoshone-Paiute Tribes' Duck Valley Indian Reservation are supported by stocking of hatchery fish. As of this date, the Shoshone-Bannock Tribes have completed a master plan to design and evaluate production options for the Crystal Springs Fish Hatchery, which will support anadromous fish production as well as provide resident fish to support Fort Hall Indian Reservation fisheries. In northern Idaho, the Kootenai Tribe of Idaho is working with state and federal agencies to recover populations of Kootenai River White Sturgeon and Burbot with production from the Tribe's Twin Rivers Hatchery and Tribal Fish Hatchery. The Nez Perce Tribe stocks hatchery-reared Rainbow Trout to maintain put-and-take fisheries (resident fish substitution) at Tunnel, Talmaks, and Mud Springs ponds. The Idaho Power Company (IPC) and the USFWS also stock resident fish to partially mitigate fisheries impacts of hydroelectric dam construction and operation[16].

Although stocking figures vary from year to year, current annual stocking objectives for statewide resident fish are approximately 1.7 million "catchable"-size put-and-take trout in the 254 to 305 mm (10 to 12 in) range, 3.1 million fingerling put-grow-and-take trout in the 51 to 152 mm (2 to 6 in) range, and up to 12 million early- and late-run Kokanee fingerlings in the 51 to 76 mm (2 to 3 in) range[16].

Anadromous Fish Hatcheries in Idaho

Anadromous fish production in Idaho is a complex cooperative venture with many contributing partners. Fisheries managers, hatchery owners and operators, funding entities, and fishery monitors include the IDFG, the Nez Perce Tribe, the Shoshone-Bannock Tribes, the USFWS (including the Lower Snake River Compensation Plan [LSRCP] office), the IPC, the Bonneville Power Administration (BPA), and the US Army Corps of Engineers (USACE). The latter four partners provide the majority of the funding for anadromous fish production in the state. The objectives of the multifaceted program are to partially mitigate for anadromous fisheries and harvest losses associated with the Columbia River and Snake River hydroelectric system and to recover depressed anadromous fish stocks.

The IPC owns four fish hatcheries dedicated to anadromous production. These facilities are operated by the IDFG and are funded by the IPC as a condition of the Federal Energy Regulatory Commission permit for operation of the Hells Canyon Dam Complex (Oxbow, Brownlee, and Hells Canyon Dams) on the Snake River. The Rapid River Fish Hatchery traps and spawns adult spring Chinook Salmon and rears resulting progeny to release as smolts. The Pahsimeroi Fish Hatchery traps and spawns adult summer Chinook Salmon and rears resulting progeny to release as smolts and also traps and spawns adult Steelhead, transferring resulting eyed eggs to Niagara Springs Fish Hatchery. The Oxbow Fish Hatchery traps and spawns adult Steelhead to produce eyed eggs for transfer to Niagara Springs Fish Hatchery and also traps adult spring Chinook Salmon that are transferred to Rapid River Fish Hatchery to supplement broodstock. The Niagara Springs Fish Hatchery rears and releases Steelhead smolts resultant from Pahsimeroi and Oxbow eyed-egg transfers. The IPC also funds production of fall Chinook Salmon at the Irrigon Fish Hatchery in Oregon; these fish are released in the Snake River as subyearling smolts.

The USFWS owns six anadromous fish hatcheries in the state of Idaho, along with associated satellite facilities dedicated to the trapping of returning adult fish, the prerelease acclimation of juvenile fish, or both. Operation of these facilities is supported primarily by BPA funds administered through the USFWS Lower Snake River Compensation Plan, a program created by Congress in 1976 to partially mitigate for anadromous fisheries and harvest losses associated with four lower Snake River dams (Ice Harbor, Little Goose, Lower Granite, and Lower Monumental Dams). Four of the USFWS hatcheries and their satellites are operated by IDFG: the Clearwater Fish Hatchery, which traps and spawns spring and summer Chinook Salmon eggs, releasing progeny as smolts, and rears and releases Steelhead smolts cultured from transferred eggs; the Magic Valley Steelhead Hatchery, which rears and releases Steelhead smolts cultured from transferred eggs; the McCall Fish Hatchery, which traps and spawns summer Chinook Salmon eggs, releasing progeny as smolts; and the Sawtooth Fish Hatchery, which traps and spawns adult summer Chinook Salmon and Steelhead, rearing resulting Chinook Salmon to smolt and transferring Steelhead eggs to be reared at other LSRCP facilities. The USFWS owns Kooskia Hatchery, which is managed by the Nez Perce Tribe; this facility rears and releases spring Chinook Salmon and Coho Salmon smolts and also traps adult spring Chinook Salmon, Coho Salmon, and Steelhead, transferring broodstock to Dworshak National Fish Hatchery. The USFWS staffs and operates the Hagerman National Fish Hatchery, which rears and releases Steelhead smolts cultured from eggs collected at other facilities. The LSRCP also funds Shoshone-Bannock Tribes' in-stream weir-and-egg-box operations in the Yankee Fork River and Panther Creek[16].

The BPA owns the Nez Perce Tribal Hatchery, which is managed by the Nez Perce Tribe. The facility rears and releases both fall Chinook Salmon subyearling smolts and spring Chinook Salmon parr, presmolts, and smolts. Most fish reared at the Nez Perce Tribal Hatchery are acclimated and released off-station. The BPA funds the Nez Perce Tribe to operate the Fall Chinook Acclimation Ponds project, which receives subyearling and yearling fall Chinook Salmon from Lyons Ferry Hatchery for acclimation and release into the Clearwater and Snake Rivers. The BPA funds the Nez Perce Tribe to operate the Johnson Creek Artificial Propagation Enhancement project, which is a summer Chinook Salmon conservation effort. The Tribe collects adult broodstock and releases smolts in Johnson Creek; fish are spawned and reared at the IDFG-operated McCall Fish Hatchery facilities[16].

The USACE owns the Dworshak National Fish Hatchery, which is co-managed by the USFWS and the Nez Perce Tribe. The facility traps and spawns adult Steelhead and spring Chinook Salmon, distributing some eggs to be reared at other hatcheries and rearing remaining progeny to release as smolts. Steelhead production is funded through the USACE and Chinook Salmon production through LSRCP. Releases occur both on-station and in Clearwater subbasin tributaries. Dworshak also traps and rears adult Coho Salmon, rearing progeny to release as smolts; this program is supported by Pacific Coastal Salmon Recovery funding.

The IDFG owns and operates two hatcheries dedicated to the recovery of the Snake River Sockeye Salmon, which is on the Endangered Species List. These facilities, as well as a weir-and-smolt trap on Redfish Lake Creek, are funded by the BPA. The Eagle Fish Hatchery serves as a captive-brood facility for the critically depressed Sockeye Salmon stock and also traps and spawns returning adult fish. The majority of the eggs collected by Eagle staff from returning adults and captive brood are transferred to the Springfield Fish Hatchery. Springfield, constructed in 2013, has the capacity to rear and release up to one million smolts.

Although stocking figures vary from year to year, current annual anadromous juvenile release objectives in Idaho's Snake River and Salmon River basins are approximately 14 million spring and summer Chinook Salmon smolts, five million fall Chinook Salmon smolts, eight million Steelhead smolts, one million Coho Salmon smolts, and one million Sockeye Salmon smolts.

Based on data from a 2011 IDFG economic survey, it is estimated that direct spending by Idaho anglers targeting hatchery-produced anadromous Steelhead and Chinook Salmon contributed more than $90 million to the state economy and accounted for approximately 16 percent of all fishing-related spending in 2011. Compared to the estimated Steelhead and Chinook Salmon production budget across all funding entities of approximately $14 million, angler spending in Idaho is estimated to provide a return-on-investment ratio of greater than 6:1. It must be noted that these economic-benefit figures accrue to angler spending in Idaho only; significant economic benefits associated with treaty and downriver fishing effort in Washington and Oregon are not captured.

Chapter Citations

1. Muhlfeld, et al., 2008. 2. Miller, Williams and Williams, 1989. 3. Mack, et al., 2000. 4. Mooney and Cleland, 2001. 5. Muhlfeld et al., 2009. 6. Sigler and Sigler, 1990. 7. Sigler and Sigler, 1986. 8. Halverson, A. 2010. 9. Trushenski, Flagg, and Kohler, 2010. 10. Stroud, 1986. 11. Nickum, et al., 2005. 12. Paquet, et al., 2011. 13. Weber and Brown, 2011. 14. Halverson, 2008. 15. Trushenski, et al., 2015. 16. Hesse, 2015.

2

ICHTHYOLOGY 101—THE STUDY OF FISHES IN IDAHO

Introduction—Unwritten Laws of Fisheries

Ahead of our discussions on protocols, mechanisms and rationales for studying fish in Idaho, we present the section below which is a much-abbreviated version of information presented in "The Unwritten Laws of Fisheries[1]." This excerpt was prepared by the "Laws" author and is a fitting introduction to the myriad issues associated with fisheries management in Idaho.

As a fisheries professional in Idaho we each have very enviable jobs. We chose our profession for reasons that can be incredibly difficult to articulate—unless you have it—it is difficult to understand the passion each of us have for the resource we manage. It is part of our core—it is part of who we are. When entering the fisheries field, as either a temporary or permanent employee, we shoulder responsibilities that are hard to appreciate until you have worked in the field for some time. As fishery professionals working in Idaho we have all been entrusted with the proper management of the State's fishery resources. Part of that "trust" afforded to us by the public includes maintaining the highest standard of integrity and using the most scientifically valid and defensible methods available to us when handling fish. Our work should be necessary, purposeful, and useful to our organization and to others. Anything less is an affront to our employer, the resource, and the public's we serve. Our reported results should always be repeatable, trusted, peer reviewed if feasible, and based on verified scientific observation and not biased by preconceived conclusions or personal opinion.

The brotherhood and sisterhood of fisheries professionals is anchored by the confidence in one another's ability and desire to collect the most reliable and accurate data. The credibility of fishery managers is essential for us to maintain if we are to effectively carry out our responsibility to the public trust. Our profession relies on each of us always producing the best scientific information possible. This high standard is crucial for our profession to continue to be held in high regard by the public that we serve. Maintenance of that relationship depends on each of us "patrolling our ranks" and holding each other accountable to the highest standards of professional integrity and developing and using the best scientific techniques available. Trust is very difficult to earn and can be easily lost.

The ever increasing demands on our fisheries and aquatic resources will continue to lead to an ever increasing scrutiny of our actions, our data, and our fisheries profession. This scrutiny is healthy and should be embraced as it challenges us to constantly rethink our assumptions and it demands that we constantly improve our science. The challenges for modern day fisheries managers are likely more demanding and complex than in any other time in fish management history. During an era of information overload and ready electronic access to information, constituents can come to any meeting partially armed with any type of information they so desire. An informed public is the desire of any fish management agency, however, helping the public decipher between scientifically valid information and personal opinion in a blog is a delicate matter. Managers are now required to wear multiple hats every day and can unwittingly cross lines between policy making, decision making, economist, educator, and information disseminator. This must be done with great care, as there are few examples where a resource agency has all of the information necessary to make the best decision and is able to accurately predict the outcome of that decision. The time is rare when there are enough financial and physical resources available to make definitive conclusions regarding a proposed project. Our challenge is to help guide the public, decision makers, and policy makers by providing timely, accurate, and valid data so that informed choices can be made.

The public's perception of resource agencies is continually evolving. One day we are held in high regard and the next we are kicked to the curb. To maintain a healthy perception of our profession requires that each of us continually engage with the public so that they are familiar with our programs and how they relate to them. Assuming that the public always understands what we do (and why we do it) is the greatest fallacy—without constant contact with the public our credibility will slowly erode.

In summary, our work should always be above reproach. We should always be loyal to the resource our charge is to protect, be truthful regarding our knowledge (and lack thereof) when making decisions, and strive to continually improve ourselves relative to continuing education and assisting our fellow biologists in becoming the best professional they can be.

Introduction to Fisheries Science

Our modern use of the word "ichthyology" is derived from the Greek *ikhthus*, meaning "fish," and *logos*, meaning "study" or "study of"; hence, ichthyology is the modern study of fishes, and ichthyologists are those scientists who study fish in particular. Students and fish enthusiasts alike may wish to review Moyle and Cech's *Fishes: An Introduction to Ichthyology*[2], Moyle's *Fish: An Enthusiast's Guide*[3] or Helfman et al.[4] for both a presentation of the science of ichthyology and the methods for the serious study of fishes and an informative guide for the curious. In addition to these texts, the American Fisheries Society (AFS) has, over the years, published a number of excellent books on methods, guides, and study mechanisms for fishes. These books more particularly address fisheries management issues, a topic generally outside of the realm of the current publication, but references are included at the end of the chapter[5–9].

Understanding Fish Names and Taxonomy

If you flip through any modern "Fishes of" book, such as *Native Fishes of Idaho*[10], *Fishes of Utah*[11], *Fishes of the Great Basin,*[12] or *Trout and Salmon of North America*[13], you will find, among whatever other information the authors deemed important, surnames and dates associated with each fish species mentioned. If, for instance, you look up information on Northern Pike, you will find, in most fish books, the associated informational tidbit *Esox lucius*—Linnaeus 1758. This information conveys that this species, *lucius*, was first described by Linnaeus in 1758 and was assigned by Linnaeus to the genus *Esox*. Both genus and species names are presented in italics by convention. Some fisheries books, such as the classic *Freshwater Fishes of Canada*, by W. B. Scott and E. J. Crossman, published in 1973[14], provide additional information on the etymology or the

Our current binomial (two-term) system of fish names is based on work completed by Carl Linnaeus, who lived from 1707 to 1778 and is often referred to as the "Father of Taxonomy." Linnaeus developed this system over many years and is responsible for the binomial naming of hundreds of species. Much of his system, with some modifications, is still in use today.

origin, meaning, and history of the genus and species names and changes that have been made to the nomenclature since the original description, if any.

Each scientific name of a fish is composed of two or three parts in Latin or in Latinized form and is printed in italics. The first of these names is the genus or generic name, and the second is the species name. When the scientific name of the fish is introduced into a species account, it is followed by the name of the person who first properly ascribed the species name to that fish. When other workers have moved the species to a genus other than that in which it was originally described, the name of the first author still accompanies the species designation, but is placed in parentheses to call attention to the fact that a shift in generic assignment has occurred. Alterations in scientific names and their assignment and usage follow the rules of the International Code of Zoological Nomenclature. The American Society of Ichthyologists and Herpetologists and the American Fisheries Society have standardized common names[15] making common names actually more stable than scientific names.

Most people, when discussing any issue related to fish or fishing, often refer to fish by their vernacular names. These names can vary according to region. For example, Coho Salmon are sometimes referred to as "silver salmon," whereas Chum Salmon are sometimes called "dog salmon." This common name derives from the observation of early European explorers that Native Americans mainly caught Chum Salmon to feed to their dogs, saving the other species of salmon for themselves. In the past, the tradition of spelling common names for fishes was to use lowercase letters for names unless the name included a well-established proper noun. Hence, "rainbow trout" was presented in lowercase, but "Atlantic" was capitalized in the first part of the name for Atlantic Salmon because "Atlantic" is a proper noun. In 2013 The Committee on the names of fishes (Committee) decided to change this tradition and use capitalization of all official common names of fishes[15]. Hence, "rainbow trout" are now referred to as "Rainbow Trout" in AFS publications.In this text, we have expanded the use of capitalization to apply to all proper nouns; hence we use Kokanee rather than kokanee, Redband Trout rather than redband trout and Steelhead rather than steelhead. This practice is a departure from Commitee conventions but the authors felt it promotes consistency in this book.

Both the content and the format for the presentation of information on fish species such as Northern Pike are presented based on an established protocol. In North America, the best authority for fish names and issues related to phylogenetic issues of recent (Holocene) fishes is the *Common and Scientific Names of Fishes from the United States, Canada, and Mexico*[15], the current edition of which was published in 2013. This 384-page guide to both fish nomenclature and phylogenetic structure explains the rules for fish names and for changes to fish names in the geographic areas noted. It provides a beginning basis for understanding when a fish species was first described, who completed the description, and what changes, if any, have occurred since that original description. An additional excellent source of information on all fishes of the world, including those in Idaho, is Nelson's 2006 *Fishes of the World*[16]. Not all individuals interested in studying fish agree with the Committee's phylogenetic analyses.

As an example of a much less straightforward nomenclature issue, we turn to the well-known Rainbow Trout. In the 1987 publication of *Fishes of the Great Basin*, the species name in the life history of the Rainbow Trout was presented as follows:

Rainbow trout *Salmo gairdneri*—Richardson 1836

This information conveys the understanding that the original description of the Rainbow Trout was completed by Richardson, and the Committee reference book for nomenclature attributed this description to Richardson in 1836. The AFS publication cited in the 1987 *Fishes of the Great Basin* was the fourth edition of the list, published in 1970, and was 174 pages long.

In the 1996 *Fishes of Utah*, the life history heading for Rainbow Trout was as follows:

Rainbow Trout *Oncorhynchus mykiss* (Walbaum, 1792)

It relied on the 1991 fifth edition of the Committee list[17]. Walbaum and the 1792 date, separated by a comma, appear in parentheses, which is a Committee convention indicating that while Walbaum was the author of the original description associated with *mykiss*, subsequent phylogenetic conventions have assigned this species to the genus *Oncorhynchus*, not the *Salmo* classification presented by Richardson, now reserved for Atlantic Salmon, Brown Trout, and related species. This apparent confusion results from the fact that Richardson, describing a fish from the mouth of the Columbia River, was unaware of the earlier description completed by Walbaum, which was based on fish from Kamchatka, Russia. Several iterations later, and with the assistance of genetic markers (DNA), the official list published in 2013[15] categorizes the Rainbow Trout as *Oncorhynchus mykiss* (Walbaum, 1792). However, biologists, being human, sometimes disagree on what constitutes a different species or sometimes a finer level of differentiation, a subspecies. Hence, in some cases there are still needed refinements of taxonomic names that are yet to be completed. An example involving an iconic Idaho species follows. Unofficially, a subspecies, which the Committee list does not yet address, is also accepted by many ichthyologists—*Oncorhynchus mykiss gairdneri*, the Redband Trout. And the scientific names do not even begin to address the confusion between Rainbow Trout, Redband Trout, and Steelhead, which remains unresolved. Behnke[13,18,19] provides some insight into the vagaries of the common names associated with this taxa and a great deal of anecdotal information based on his nearly 50 years of research on North American trout. Additional information on these scientific vagaries is provided in the fish life history sections for *O. m. gairdneri*.

Identifying Fish

Identifying an individual fish, regardless of how it was captured, requires an understanding of basic fish anatomy and, depending on which species is being identified, some level of skill in both specimen examination and understanding of physical characteristics. Everyone has some basic concept of the shape of a fish, but the range of details on exactly how those shapes are described provides the secret to being able to identify an individual fish to the species level.

So we begin with a few basics. In this text, we include a *Glossary* as well as a dichotomous key (Chapter 5) with numerous illustrations, including figures and examples of fish body shapes and associated fin locations and descriptions for various characteristics.

Fish body shapes range from cylindrical or round, in instances where the length of the fish's body is several times its depth (height), such as trouts, catfishes, pikes, suckers, and some minnows, to a body form such that the fish is nearly as deep as it is long, such as sunfish. Perhaps the next most obvious characteristic about a particular fish is the fins. Illustrations for the placement and size of various descriptive characteristics are included in the keys for a particular fish family or species. Nominally, many freshwater fish would appear to have eight fins: dorsal, adipose, caudal, anal, and two sets of paired fins, the pelvic and pectoral.

Fins themselves have a variety of useful characteristics for assisting in species identification. For instance, do the fins have spines or only soft rays? If spines are present, are they serrated or smooth? And if only soft rays, how many are present? An additional useful characteristic associated with fins is their placement on the body. Placement of the pectorals is in close proximity to the breast of the fish and just behind the head. For pelvic fins, placement on the body can be abdominal (on the midbody) or thoracic (near the throat or head). In addition to placement of the fins, their absence is also a valuable piece of information. For instance, a fish without an adipose fin (the fleshy protuberance located on the back between the dorsal fin and the caudal peduncle) is not, at least in Idaho, a trout (Salmonidae), catfish (Ictaluridae), or a Sand Roller (Percopsidae). Illustrations for each of these characteristics are provided in the keys.

Taxonomic classification is accomplished by counting and measuring body parts, in addition to the morphological traits noted above. Dichotomous keys are an artificial categorization of traits used to separate and identify organisms. Fish, and other organisms, can be "keyed" or identified using a system of paired statements that are mutually exclusive, forcing the user to select one or the other trait to move to the next segment. Both meristic (features that can be counted) and morphological (attributes of form and shape that can be measured) characteristics of a species are used because they are external and easily observable, such as fin placement, number of rays in the dorsal fin, lateral line curve, or the presence and number of spines in the fins. Other characteristics are internal and require sacrificing individual fish, such as numbers of pharyngeal teeth or basibranchial teeth. Where available, genetic comparisons of fish species (and particularly species that are closely related and may have suffered genetic introgression, or subspecies) can provide useful information on the identity of individual fish.

Fish Sampling Techniques
Nets, Poisons, and Electrofishing

Anglers may collect individual fish specimens or groups of fishes from a particular body of water as part of a recreational effort, a technique not unheard of among professional fisheries biologists. But for large-scale, long-term, or comprehensive studies of aquatic ecosystems, more effective and rigorous collection methods must be employed. Here we present a short overview of methods for collecting fish species from a variety of aquatic ecosystems. It is imperative that professional and student ichthyologists remember that in almost all instances, state, federal, or other permits must be obtained prior to collection of fishes from natural aquatic ecosystems, regardless of the type of gear used. Readers are referred to the reference list for discussions of various fish-capture techniques, limitations associated with the techniques, and other useful information. Excellent standardized and comprehensive techniques for collecting in North America are provided in the AFS publication *Standard Methods for Sampling North American Fishes*[7]. The majority of the text below is excerpted from *Recreational Fisheries: Management, Theory, and Application*[20].

Nets

Nets of various types and configurations have been used to collect fish from marine and freshwater for thousands of years. Biblical references include Jesus walking by the Sea of Galilee and calling to the fishermen. Properly set and cared for, nets of various kinds are capable of capturing large numbers of fish with minimal effort.

Fishery research makes use of various types of nets to capture fish for a variety of reasons. For instance, nets can be used in situations where other collection gear, such as electrofishing or toxicants, is inappropriate or not allowed. Nets and associated traps offer certain advantages over other types of collection gear. Among these advantages are that personnel are not generally required to stay with the net or trap while it is fishing, large numbers of nets or traps can be set to run on a rotating basis by the same crew, net size and mesh size can be varied to allow capture of a particular size or species of fish, and nets or traps can be fished in areas of known fish concentrations. Principal disadvantages of net gear used in fisheries research, such as gill nets, are that few fish are recovered alive. This drawback can be lessened by more frequent retrieval of the nets or through use of trap nets that do not entangle fish.

Types of nets can be separated by a variety of characteristics. Nets are made from either natural or synthetic fibers, though most are now made of synthetic fiber because of increased resistance to rotting. Two types of nets are generally used for both commercial and fishery research: passive nets (entrapment or entanglement nets) and active-haul nets. There are six basic types of nets: gill, trap, trammel, seine, trawl, and cast.

One other consideration of fish-collection gear deserves discussion. All of the common types of fish-collecting gear, such as gill nets, seines, hook and line, spears and jibs, traps, diversions, chemicals, explosives, electrofishing, demonstrate gear selectivity and introduce bias in fish collections. We point out some of these selectivities in the following sections.

Passive Net Gear

Passive net gear includes nets that are placed in one location and rely on fish moving past that location for capture. There are two basic types of capture mechanism, either entrapment or entanglement. Net types included in this group are the gill net, the trammel net, and the trap net. The selectivity of passive sampling nets has been reviewed by Hubert et al.[21]

Retrieving gill nets on Palisades Reservoir, July 2016. Photo by JWS.

Gill nets are designed to catch fish by the opercles or other body projections as they swim into the net. Fish are trapped by the mesh when they make an effort to escape. Gill nets may be hung in various ratios. Hanging ratio determines the shape of the net mesh when it is set. Hanging ratio is the relationship between the length of the float line and the length of the stretched mesh hanging from the float line. Effectiveness of the mesh shape varies with the morphology of fish being sought and, obviously, with fish size. Generally, gill nets are set off the bow of a boat in either a floating or a sinking (bottom-set) configuration. The most obvious bias of the gill net is related to mesh size. For a given a mesh size, fish of the appropriate size for that mesh become entangled, fish that are smaller than the mesh usually swim through without becoming entangled, and fish that are much larger than the mesh will not be able to penetrate the net far enough to become entangled.

Trammel nets are constructed by hanging three sets of net webbing onto a single top and bottom line. The two outside webs are of larger mesh and thread size than the inner web. Fish are captured when they go

Fisheries field work often provides some perfect scenery. Palisades Reservoir, summer 2016. Photo by JWS.

Gill nets are typically set from the bow with the operator backing the boat away from the net. Photo by JWS.

Retrieving gill nets is typically accomplished from the bow of the boat while the operator maintains boat position relative to the net. Photo by JWS.

Retrieving gill nets can be a challenge in harsh weather conditions. Photo courtesy IDFG.

through the outside or large-size web on one side and then push the inner smaller webbing through the outside of the opposite side. The pocket that is formed is an effective trap and holds the fish. Trammel nets can be used in large river habitats. Trammel nets are used infrequently in fisheries stock assessment: they are selective for fish species which have rough surfaces and body protrusions such as catfish and temperate basses. Trammel nets, however, do not generally damage captured fish as much as gill nets but do have similar sampling bias.

Trap nets are made of netting and consist mainly of leads (fence) and a heart. The lead or fence directs the fish into the V-shaped throat and works because of the natural tendency of fishes to follow the webbing that leads into the trap. Fish traps may vary considerably in design and size, and there are several generally recognized types such as the Alaskan, the Great Lakes, and hoop and fyke nets.

Weirs are fence-like guidance devices that divert fish across the stream section and into a trap. Weir sections or some other form of diversion may be built across an entire stream width to be most effective.

Active Net Gear

Major differences between active and passive fish-collection gear are that active gear is moved by the operators often with the assistance of mechanical devices, such as boats. Active net gear includes seines, trawl nets, cast nets, plankton nets, and lift nets, some of which are discussed below. Seines include any net of any size with a top float line and a weighted bottom line (lead line) that is pulled through the water, either by human power or by some mechanical device. Larger seines typically have bags or purses in the center. Beach or

Hoop nets may be useful for non-lethal sampling or in locations where gill nets are ineffective. Photo courtesy IDFG.

Setting the first in a series of tandem hoop nets. Photo courtesy IDFG.

Baited hoop nets are useful for non-lethal sampling of benthic fishes, like these Channel Catfish. Photo courtesy IDFG.

Fish weirs are used to trap migrating fish, like these Kokanee, within a river system. Photo courtesy IDFG..

Beach seines are an effective fish collection tool where shallow littoral zones allow access. Photo by JWS.

Captured fish are funneled into the bag end of the seine during the haul. Photo by JWS.

haul seines are fished primarily in shallow water and on beaches that have a smooth, relatively unobstructed bottom. Purse seines are large seines used in open water. A purse line, which runs through rings attached by rope bridles, closes the bottom of the seine after it has been cast in a circle around a school of fish. The lead lines are hoisted on the deck of the boat, and the netting can then be pulled aboard the boat and the fish removed. Active nets are also selective in terms of species, size, and swimming ability of individuals in the population being sampled[22].

Cast nets are an ancient art of catching fish and is practiced by both commercial and sport anglers in many parts of the world. A cast net is a circular piece of netting with lead weights secured around the perimeter that help form the pocket during retrieval, and the net has a rope rigged to the center for retrieving. A cast net can be flung over fish by anglers wading in shallow water or from a dock or boat.

Poisons

Rotenone, arguably the best-known fish toxicant, is a chemical compound extracted from the roots of several species of plants in the bean family (Leguminosae) and has been used for centuries by native South America peoples to capture fish for food. Rotenone has been used in fisheries management in the United States since at least 1934. It is extremely toxic to most fishes, but is not particularly toxic to any animal without gills. Rotenone can be used in a variety of aquatic ecosystems and on a variety of fish species with significant effectiveness. Rotenone is easy to apply with a variety of applicators, from the shore or from boats. Dry rotenone is a respiratory irritant for humans, and a mask should be worn when there is chance for contact. Several rotenone formulations are registered for fisheries use. Fisheries investigators should confirm that the use of rotenone is permitted in the target ecosystem and that appropriate public relations announcements have been made to ensure no public outcry when dead fish begin appearing in a river or lake.

Antimycin is an alternative to rotenone and sodium cyanide can be used in some situations. A variety of other broad-spectrum and specific fish toxicants have been used over the past several decades in various aquatic ecosystems.

Electricity

Electrofishing equipment can be separated into four general categories. The smallest electrofishing equipment is typically a backpack electrofisher, which can be further divided into battery-operated units and motor-driven units. The next size of electrofisher is a unit considered to be tow-along and is dragged behind the crew in a small nongrounded skid or small boat of some sort. Shore-based units are generally quite large (i.e., typically driven by 3.5- to 7.5-kilowatt generators) power sources, since weight and size are not restrictive issues. Finally, boat-mounted boom electrofishers are typically used on larger bodies of water and are capable of carrying a crew of several members[20].

Electricity has been used to gather or control aquatic organisms for at least 100 years. The effects of electrical current on the orientation and movement of fish were first described by Mach in 1875[23], Hermann in 1885[24], and Hermann and Matthias in 1894[25].

British patents for electric fishing equipment were granted in 1863 and 1895. A variety of manufacturers now provide sophisticated backpack, shore-based, and boat-mounted electrofishing apparatuses. Researchers are cautioned that using electrofishing equipment can be hazardous, and all crew members should be fully trained and cognizant of the dangers and the safeguards afforded by the equipment[20]. As with any modern technology, proper operation and professional training are important ingredients for safe applications and desirable outcomes, for both the resource to be sampled and the user of the equipment. Because current electrofishing technology relies on the use of pulsed current (designated in pulses per second, or Hz), users should always commence their field applications by starting low (using surplus or surrogate fish if possible), and gradually increasing power settings to achieve safe sampling outcomes (for example, start at pulse frequencies of 10, 15 or 20 Hz before using settings of 30 Hz or higher). Many variables affect the power transferred to fish, but pulse frequency is among the main factors that can either promote a successful field sampling effort or result in fish injury, if improper settings are used or the operator is not properly trained.[26]

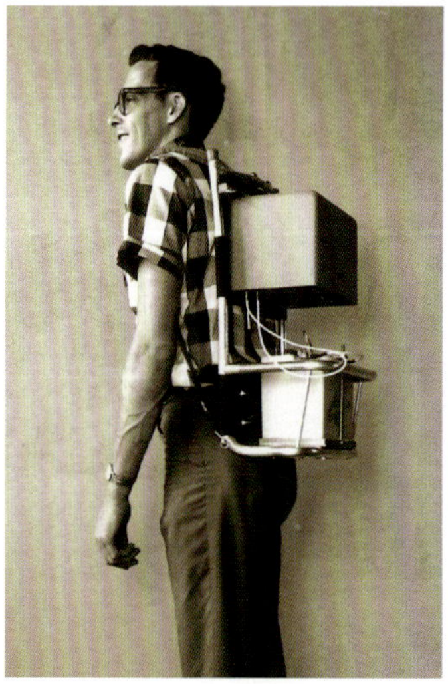

Early back pack electrofishing equipment was functional. Photo courtesy Smith-Root Corporation.

Back pack electrofishing effectively captures fish in small streams. Photo courtesy Smith-Root Corporation.

The purpose of using an electric field is sometimes not to catch fish directly but to enhance the effectiveness of other sampling methods, particularly in large rivers and reservoirs. Seines, trap-nets, and trawls can all incorporate electrodes designed to modify fish behavior and improve catch rates. Even fish counters can use electric fields (to assess a change in electrical impedance as fish move through the counter tunnel and distort the field, because of differential conductivities between fish and their surrounding medium).

In this section we address many of the important efficiency and safety issues associated with the uses of electricity in fisheries research and management. It is not a complete or in-depth presentation. Readers are directed to the references throughout the text and in particular Reynolds and Kolz most recent revised material[27]. A useful overview with important safety considerations is provided by Reynolds and Harlan[28].

Commercially, there are two basic types of electricity, alternating current (AC) and direct current (DC). Standard household current in the United States is 120-volt, 60-hertz (cycle) AC. Every electric circuit has a source of electrical energy (power) usually referred to as a generator. Generators have two terminals: a positive or anode and negative or cathode. Conventional practice assumes electric current flows from the positive terminal into the load, which is water in electrofishing operations. All substances have resistance to the flow of electricity through them which is the concept affecting the efficiency of electrofishing. The unit of energy associated with a generator is a volt (V) and one volt of potential energy will drive one ampere (A)

Raft mounted electrofishing equipment allows access to a variety of stream locations. Photo courtesy IDFG.

Multiple backpack electrofishing units working together can sample medium streams. Photo courtesy IDFG.

Boat mounted electrofishing units with sophisticated electrode arrays are effective in large streams and lakes. Photo courtesy Smith-Root Corporation.

Back pack electrofishing equipment allows collection from individual habitat types. Photo courtesy Idaho Department of Environmental Quality.

of electric current through one ohm (*w*) of resistance. Duty cycle or pulse width refers to the time current flows during each second, essentially a measure of on-off[11].

AC is appropriately named because of the "alternating" polarity of the flow of electrons. In AC electrofishing operations, electrodes are not designated as positive (anode) or negative (cathode) because the polarity switches automatically and continuously. In pulsed DC or straight DC current configurations, electron flow is from the cathode to the anode. General electric waveform information is provided under "Electrofishing Waveforms[11]." Several specific terms associated with electrofishing technology, such as *electrical current, electrical conductivity, narcosis, electrotetany, galvanotaxis,* and others, are included in the *Glossary* of this volume. Researchers should be familiar with these terms and their functional meanings in the field, as well as with all safety aspects of electrofishing equipment operation. Training courses from knowledgeable experts should be considered a valuable investment[20].

Electrical conductivity of water (i.e., water conductivity) is based on the amount of dissolved salts and is an indicator of ionic concentration in the water. Water conductivity is the most important environmental variable influencing electrofishing efficiency and should always be noted when conducting electrofishing sampling[27]. Conductivity can be calculated mathematically if the kind and amount of dissolved salts are known, though for fisheries work, measurement with a conductivity meter is generally sufficient[20].

Canoe-mounted electrofishing gear with portable anodes is a good option for medium size rivers too shallow for drift or power boats. Photo courtesy IDFG.

Various authors have divided natural waters into groups based on their content of dissolved salts. One designation is provided in the table below.

Water Designation	Salt Content – g/liter
Freshwater	Up to 1
Brackish water	1 – 25
Water with marine salinity	25-50
Salt waters (higher than marine)	Above 50

In a medium such as water, with three-dimensional volume, resistance in a given direction from the source is not necessarily uniform. The distribution of current in water is described by current density, J (A/cm^2), or current per unit area where J = Joules. The nature of J and E, where E is the electric field vector are critical to effective electrofishing and are consistently related by Ohm's law $J = \sigma E$, where σ is conductivity.

Electrofishing can be useful in many types of fisheries investigations. In streams and lakes or ponds (for example, in the littoral zone), electrofishing can be utilized to capture or direct movement of fish for a variety of reasons. While electrofishing is an excellent qualitative tool, its use in quantitative studies must be preceded by a determination of the effect of the electrofishing itself, that is, electrofishing bias (or sampling efficiency), including physiological changes to fish collected. The primary bias of electrofishing is related to its size selectivity (large fish of a given species are more affected than small ones) and species selectivity (the ability of some species to more readily sense and thereby escape the electrical field)[20].

In qualitative samples, it may not be important if fish are affected physiologically or when an unknown number escape capture. But in quantitative sampling, knowing both the numbers and the species caught, as well as those missed, is important, therefore making physiological changes and responses also important[20,29].

The theoretical basis for electrofishing is that application of an electric field into water affects nearby fish in such a fashion that they can be collected. In the presence of an electrical field, electrical stimulus immobilizes fishes (electronarcosis or electrotetany) by inhibiting the nervous system[30]. Fish may, however, be immobilized at depths beyond which they can be recovered or even seen. In a DC field, fish near the anode may be drawn to it by galvanotaxis. Pulsed DC is more effective than continuous-flow DC because of the short recovery time between pulses that allows for sequentially drawing the fish to the positive electrode by forced swimming[20]. Ideally, fish are captured between the stages of galvanotaxis and narcosis. Tetany should be avoided as there is a greater chance for injury (i.e., spinal injury and hemorrhaging) as a result of intense muscular contraction[31,32].

In general, fish responses are characterized in the following order:

1) escape behavior (fish can sense the field before it is too intense, and are able to make an escape)

2) galvanotaxis (fish start to swim towards the anode, whether it is weak or intensely coordinated swim movement)

3) galvanonarcosis (fish have become limp, the hypothesis is that the brain is inhibited)

4) forced swimming (forced swimming actually comes after narcosis, in which fish have lost equilibrium, and tetany-like swim movements are inconsistent, and directed towards the anode)

5) and finally tetany (fish are completely stiff as a result of short muscular fibers being directly excited)[31,32]. Some studies list death after tetany.

AC power is most injurious to fish because of the nature of change in polarity of the current. Continuous DC is the least injurious to fish, but drains a battery much too quickly to be practical in the field. Continuous DC is best for attracting fish to the anode. Pulsed DC essentially represents a middle ground between AC and continuous DC. It is more injurious to fish, but does not consume as much battery power.

Conceptually, using an electrical field to collect fish incorporates the theory of power density at three levels: the electrical field perceived by fish (perception level), a higher level of power density at which galvanotaxis or electrotetanus occurs (effective level), and an even higher level of current density at which complete galvanonarcosis (DC fields) occurs immobilizing the fish or drawing it to the anode. Electrotetany can occur in AC and pulsed DC fields. Lethal effects can occur in intense electric fields or prolonged exposure, included pulsed DC and straight DC fields depending on the duration of exposure. AC tends to be more lethal, particularly for salmonids. The most effective electrofishing (the maximum number of fish are collected with minimum numbers of deaths or physiological damage) occurs when the electrical field has a very small perception-level zone, a very large effective-level zone, and a small or absent lethal-level zone[20].

Four major uses of electrofishing equipment can be described: 1) capture, which includes harvesting and censusing or fish rescue; 2) movement or control, which includes controlling fish movement and performing activities using electrical screens; 3) electrosedation, which includes narcosis, for fish handling or transportation or tagging; and 4) electroslaughter of catches to improve quality of fish flesh, as with sardines[20].

Besides the power source, the design of electrodes for particular operations is highly important. There have been a large number of experimental electrodes (e.g., spheres, rings, plates, etc.) reported in the literature over the past 50 years. The intensity of the voltage gradient near an electrode is dependent on the size or diameter of the electrode[27]. Small-diameter electrodes (e.g., cables) create intense voltage gradients around the electrodes, but the electric field dissipates quickly within a short radial distance[27]. Conversely, electric energy stemming from electrodes with large cross sections dissipates less quickly; therefore the electric field extends farther from the electrodes and the voltage gradient is less intense compared to small-diameter electrodes. Accordingly, electrodes with large cross sections have a larger effective radius, and the reduced voltage gradient reduces the incidence of fish injury. Researchers should consult their specific equipment supplier to ascertain the best electrode array for their use[20].

Electrofishing devices that yield short but powerful voltages repeatedly are often used in electrofishing because they can attain large current and voltages from comparatively low (small) power supplies. Many pulse forms have been developed over the past several decades, and researchers are advised to discuss specific requirements or desires with their particular device manufacturer for best results. Pulses include rectangular, trapezoidal, triangular, bell shaped, exponential, or capacitive. Some issues of significant damage to fish physiologically have been reported. Researchers are advised to review available literature or contact subject-matter experts to determine the potential negative effects of various pulse forms[20].

A theoretical and historical assumption in the electrofishing community that the field of an electrofishing device is homogenous has been discredited. In the non-homogenous fields used in electrofishing operations, fish reactions differ from those in a theoretical field. Important considerations are the vector and initial swimming speed of the fish when entering the electric field and the direction of the fish's motion relative to the electric current field lines, taking into account the vectors of the electrodes (e.g., movement of the boat or operators)[20].

Additionally, many early studies hypothesized that a fish in water was a homogeneous ellipsoid with a given (if unknown) conductivity. It is, however, particularly when dealing with multiple fish species, more correct to regard individual fish bodies as having a given conductivity, enclosed in a thin envelope (skin) with another given but different electrical conductivity. These variables may have practical application in some electrofishing uses such as removal of injured fish[20].

Recovery times are addressed by Adams et al.[33], Holliman et al.[34] and Robb et al.[35]. Physiological changes associated with electroshocking are addressed by Barton and Dwyer[36], Bouck and Ball[37] and Burns and Lantz[38]. Noncaptured fish recovery times have been addressed by Dalbey et al.[39].

Electrofishing-related technology includes the graduated fields of pulsed DC that have been used to successfully guide or to deter fish in various resource management applications. For example, a series of generator-powered electrodes mounted on a stream bottom have been used to block range extensions by invasive species and to prevent undesirable fish from reaching spawning habitat[40,41]. Similar types of graduated-field fish barriers have also been deployed to redirect the migrations of Pacific salmonids into fish ladders and bypass channels, and to protect fish by preventing their access into hydropower tailraces and turbines[42]. Larger fish sense the field first; smaller cohorts and life stages are guided or deterred by the graduated nature of the field (which becomes more intense as fish move further into the electrode array). The electrodes are placed perpendicular to stream flow. As fish attempt to move upstream, they experience the strongest gradient from snout to tail when oriented upstream (parallel with flow of electricity between each electrode pair). Fish turn sideways to experience a weaker electric gradient and in so doing, become less hydrodynamic and are flushed downstream to safety (usually towards a fish-pass channel). Thus, the graduated electric barrier leverages the use of stream velocity to achieve the desired fish-guidance outcome. These electric guidance arrays (designed to be non-lethal) can also be vertically suspended in a water column if needed, and where floating debris or boat traffic is not an issue. Some 65 graduated-field fish barriers have been used around the world in various fishery management applications[42].

Electricity can also have non-capture related purposes in a fishery. In some situations it is essential to have a barrier to block, limit or even enhance fish movements. In these cases the use of an electric barrier or an electric deterrence or guidance technology may be advantageous for fisheries management, especially in flowing water applications where river or creek discharge must not be obstructed.

Creel Surveys

At times, it is easier for fisheries investigators to collect data from anglers in the field rather than collect fish individually. It has been noted that both large and small businesses keep, or attempt to keep, accurate records of all of their transactions. Such records are used to improve operations in terms of profit and cost. In a similar fashion, fisheries biologists and resource managers require records that can be used to measure fisheries' resource values and improve management. Freshwater fisheries resources in the western United States are the basis for one of the largest businesses in existence. Managers need accurate information on recreational values, numbers and types of fish being caught, and the amount of time expended in that effort. Creel surveys are mechanisms whereby fisheries investigators can contact recreational anglers to determine information not only on their catch, but also on their general satisfaction for the fishing at that particular site. A complete census of anglers on any particular ecosystem is generally economically or physically infeasible because of the number of field personnel required. Creel surveys, therefore, are designed as samples rather than a true census. Statisticians, public relations and social science investigators can be of immense help in designing not only the level of effort required to collect accurate and adequate data, but also the questions asked to ensure appropriate data are collected. Information collected from recreational anglers, either in the field or via a variety of electronic survey mechanisms, is used to help agencies assess the success of management programs and determine what the fishing public expects from a particular lake or stream[20].

Other Gear and Techniques

A variety of other data-collection equipment and methods are used in fisheries. These include sonar, scuba, or snorkeling and telemetry. Detonation cord has been successfully utilized for fish collection in warm-water streams in some ecosystems, using block nets to aid in recovery.

Sonar is actually an acronym for "*so*und *n*avigation *a*nd *r*anging" and transmits sound waves through the water. They are the basis for fish finders widely used by anglers. As many anglers know, modern sonar units can show sound 'images' of fish that can often be identified to size and species. Review of the theory and effectiveness of sonar is beyond the scope of this treatment, but interested readers are referred to one or more of the excellent technique books published by the American Fisheries Society, including the most recent one[43].

Scuba (*s*elf-*c*ontained *u*nderwater *b*reathing *a*pparatus) gear can be used to make observations of both fish and habitat for extended periods of time. The reader is cautioned that even at shallow depths at sea level, serious injury or death can occur. Only fully trained and certified individuals should use scuba for any fisheries investigation. Snorkeling for habit evaluation and fish observations is much less dangerous than scuba and can be used for "in place" observations of fish behavior and species interactions.

Telemetry use for monitoring fish movement has become greatly more sophisticated and less expensive in recent years and provides highly specific data on individual fish movement. Body temperature sensors can provide useful information on temperatures selected by individual fish. Both radio and sonic tags have been used with great success in both large rivers such as the Colorado and the Columbia and in smaller streams across the continent. Both types of tags have inherent advantages and disadvantages. Advantages include the ability to track large numbers of fish simultaneously, inclusion of special sensors for temperature or depth, and lack of disturbance by humans during the study. Disadvantages include high initial equipment costs, high levels of equipment operation sophistication, potential signal disruption due to natural conditions, and the potential for disruption of normal fish behavior due to either trapping or placement of tags.

Fish Field and Laboratory Collection and Handling Procedures

Based on *Fishes of the Great Lakes Region* (revised 2004)[44] and *Recreational Fisheries: Management, Theory, and Application*[20].

Field and laboratory procedures are routinely performed in fisheries work. These tasks may vary from simple to complex, but should always be thoroughly documented for future reference, as sampling results are only as good as the information record[20,44].

One of the most important aspects of a fishery's fieldwork and fish collection is a field data sheet that should be made out for each fish or other field data collection as the occasion demands. Anglers interested in increasing the value of their fishing experience and in keeping long-term records of where and how they caught fish, are encouraged to maintain a field note book using forms similar to those used by biologists, described here.

Electronic data sheets have become popular in the past several years and can be completed on a variety of field data loggers. However, the same rules apply to electronic data collection as for paper field sheets[20,44].

As noted by some iconic North American ichthyologists, the pioneers in ichthyology were content to preserve only a few specimens of each kind of fish captured. Later, a goal was established to obtain statistically adequate series of all fishes encountered, common and uncommon forms alike, to provide materials for studies of geographic variation, ontogeny, life history, food habits, and relative abundance. Collection times may range from synoptic studies on riverine ecosystems to monthly sampling, which can be conducted during either daylight or nighttime hours[20,44].

Regardless of the collection mechanism, detailed notes should be made at the time of each collection and made a part of each collection. The form should describe the habitat, habits, ecological relationships, and breeding colors of all fishes encountered. Do not rely on your memory. Regardless of the initial information-collection technique, all copies of field notes should ultimately be filed electronically wherever the collections are to be housed for study. Examples of field information to include are listed below.

- Sampling event numbers, not site numbers, that are unique to the water body and location, date, equipment type, and collector should be recorded.

- Stream system and the exact geographical location (datum and coordinates) of the water area in which the collection was made should be noted.

- Bottom types, including a description of the nature of the bottom materials and relative abundance and location of types where pertinent, should be listed. Bottom types might include silt, sand, gravel, rubble, boulders, bedrock, clay, marl, and detritus. Bottom-type descriptions can be broken into as many as 27 categories, ranging from bedrock to debris or detritus although fewer categories often work well (bedrock, boulder, cobble, gravel, sand, detritus/mud). It is important that when a descriptive term is used, the term means the same thing in all reports.

- Temperature of the water during the collection should be recorded with either a pocket or an electronic thermometer; temperature gradients in lakes and ponds should be measured instrumentally.

- Turbidity, often expressed by the depth of the Secchi disk, should also be recorded for most ecosystems, as should color, which refers to the stain in the water itself, not to apparent color imparted to it by suspended matter.

- Hardness and other chemical features should also be listed. Dissolved oxygen concentration, pH, alkalinity, and other chemical parameters should be noted. All spaces on the form should be filled in or an NO (not observed) or NA (not applicable) included to indicate that the parameter was considered.

- In addition to the traditional state, county, drainage or watershed, and specific location, Global Positioning System (GPS) data should be noted for each individual gear set or capture method.

Information should be recorded for each type of capture gear used. For instance, net gear should be described as to type, length, depth, mesh size, and number of passes made. In the case of electrical equipment, voltage, amperage or watts, frequency, alternating or direct current, as well as continuous or pulsed flow and the time fishing occurs should be noted. Distances electrofished should also be recorded, and in streams upstream and downstream passes should be considered as distinct sampling units due to efficiency differences in fishing upstream versus fishing downstream. Measured stream width should be used whenever possible. Other ecological factors such as the amount of water in the ecosystem affected by vegetation, banks, and overhanging objects should also be recorded, as covered or another descriptive term. Weather conditions may be recorded, such as precipitation, percentage of overcast, and direction and velocity of wind.

All fish captured should be identified to species (or vouchered for later taxonomic analysis) and recorded in either absolute or relative numbers. Where hundreds or thousands of fish are taken in a single collection, it may be useful to record only relative abundance on a scale from 1 to 10 or 1 to 100. When individual age and growth or food habit data are taken, serial numbers of the specimens should be recorded on the data-collection form. Data taken on an individual fish should be complete enough to stand alone. When individuals are preserved, the preservative must be recorded so that subsequent treatment of the specimens can proceed from a known basis. This is of particular significance if the specimen comes to be used for histological, chemical, or molecular studies.

Weighing and measuring captured fish is an important first step to collecting size distribution, growth and condition data. Photo by JWS.

Weight mechanisms must contain the size fish collected. Photo by JWS.

Gillnetted fish. A work area like a dock is a plus for removing fish from nets. Photo by JWS.

Many hands make light work when removing fish from gillnets. Photo by JWS.

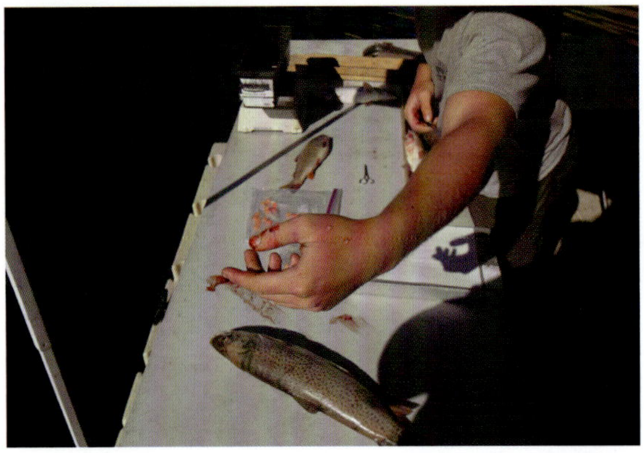

Special field collections. Note the removed otoliths on the researcher's left hand. Photo by JWS.

T-bar anchor tags may be a good option for an externally visible individual mark and allows collection of additional data. Photo courtesy IDFG.

Other data and information that can be included for each species in a "Notes" section should include life history stages recorded, such as larval, juvenile, adult, young of the year, and so forth; natural history notes such as reproductive readiness and habits as well as food and feeding activity; ecological notes, such as reference to pools, undercut banks, riffles, and the like; detailed color descriptions, particularly of breeding fish; and sources of pollution, which should be reported to the appropriate authorities. Specialized techniques such as removal of otoliths for specimen aging should be fully described. Techniques such as individual tagging can also be part of routine data collection.

Preserving Fish

Preservation of fish collected in the field can be done by a variety of means, depending on what the specimen will be used for. The easiest is often to throw it on ice (or keep it alive) and then freeze it later for preservation. The most commonly used field and laboratory fixative for fish is formalin. A variety of strengths have been used. Larger fish require stronger solutions, and small ones are preserved in more dilute solutions. Larger specimens more than 20 cm (7.9 in) long should have an incision made on the right side of the abdomen to facilitate penetration of the preservative. The right side is chosen for this and other operations, such as removal of scale samples, because the left side of the fish is ordinarily used for obtaining morphometric data and photographic images. Fixation of the average fish specimen tissue in 10 percent buffered formalin typically takes two days to one week. Formalin is removed from the specimens after fixing them by soaking in water up to two days, with the water being changed at least once. Fish specimens are then transferred to 70 percent ethyl alcohol to complete the removal of the formalin, and a change of alcohol is recommended prior to permanent storage. Other specialized solutions and techniques may be required for particular studies or protocols[20,44].

When measurements of weight and length are to be taken in the field, the use of an anesthetic greatly reduces handling stress for the fish and handling difficulty for the researcher. MS-222 has been previously used but has fallen out of favor with some researchers due to its carcinogenic nature and difficulties in proper disposal.

In a comprehensive project, three lengths of the fish-standard [SL], fork [FL], and total length [TL] may be measured to the nearest millimeter (See *Glossary*) depending on the study objectives. All three lenghts are often taken in a life history study where it is desirable to know the ratio between these three measurements or to deveop a conversion factor from one length to another. Taking weight measurements in the field is difficult for a variety of reasons, including the presence of water in or on the specimen. Weights are generally recorded in either grams or kilograms, and precision of the scale should be noted. Photographs of collected fish, with all the fins extended, can be used for subsequent idenification[20].

Introduction to Fish Life History Presentations

In the subsequent chapters, life history information is provided for all of the fish taxa known to exist in Idaho. In this text, fish life history information is organized into 10 subsections: Etymology, Importance, Description, Range and Distributional Status, Size and Longevity, Food and Feeding, Limiting Factors, Breeding Habits, Habits and Habitat, and Management and Conservation Status. Below we provide a summarization of what each of these subsections is intended to convey to the reader and, to some extent, why the information is provided for students of fishes.

Family, genus, and species names used in this text, as well as the order in which family, genus, and species lists are presented, are primarily based on the American Fisheries Society Special Publication 34[15]. Scientific names of species and higher taxa are those formed in accordance with the International Code of Zoological Nomenclature, a set of rules for naming animals[15].

Why are the terms "fish" or "fishes" used throughout the text? "Fish" is the general term for fish whether as individuals or collectively. "Fishes" is the term for individuals of more than one species. For instance, if a small aquarium contained only Goldfish, one could properly refer to the "fish" in the aquarium. However, if you added a Guppy to the aquarium there would now be two fishes, as well as two fish. If you were discussing all the species of fish in a lake, pond, or river, the appropriate term would "fishes": There are five fishes in the lake (meaning 5 species of fish). A species of fish is usually referred to in the plural, that is, "Rainbow Trout have been stocked in many waters in the western United States outside of their native range," rather than "Rainbow Trout has . . ."

A life history for a fish species is written based on information from literature, personal experience of the authors, research, and perhaps the authors professional or personal interests. Fish life histories may be brief or extensive depending on the amount of knowledge that is available and the intended audience[20]. Below we discuss the potential content of each of the life history sections and provide some rationale for inclusion of the selected information. Additional information can be found in a variety of current sources such as Moyle[3] and Moyle and Cech[2].

Etymology—Etymology, or the study of word origins, is included here because the genesis of fish scientific names, and to a lesser extent common names, provides interesting insights into the nature of the species. The etymology also provides connectivity from current usage to the original description and author.

Importance—Importance or status of a fish species may be evaluated economically or ecologically, and in some cases aesthetics may also play a role[20]. Species may be important regionally or across larger geographic areas.

A species may be important as an abundant game fish such as Bluegill or Black Bullhead or as trophy fish such as Muskellunge, Lake Trout, and Steelhead. Certain game species lend themselves to specialized and restrictive fishing opportunities, such as trout in fly-fishing-only streams. Some nongame species may be important as forage for sport-fishing species or as part of a food web in a particular ecosystem. Fish are often listed as nongame species if they do not take a hook readily or if they lack appeal to local anglers[20]. In decades past, these species were often referred to as "rough" or "trash" fish because they were all mistakenly thought to be harmful to gamefish populations or at least of no value. Current research reveals that many of these species are important parts of ecosystems supporting game species and provide useful insights into past geologic or ecosystem characteristics.

Piscivorous (fish eating) species can be important management tools for fisheries biologists. They can be used to reduce populations of less desirable species, such as the relationship between Northern Pike and a variety of sucker species[20]. Some species may have a negative impact on the ecosystems they inhabit and may, through all or portions of their range, be of no economic value. Common Carp are a good example of a negative ecological factor in most areas of North America[20]. Significant research and study have been devoted to mechanisms and educational approaches designed to make Common Carp a more desirable species in North America and to reduce populations of this species in ecosystems where they are a destructive influence[45]. Carp degrade the ecosystems in which they persist by increasing turbidity, reducing the functional depth of photosynthesis and associated primary production[20]. They are, however, now sought as game fish in some areas and in recent years have provided both fly-fishing and bow-and-arrow sporting opportunities. A publication no less prestigious than *Field and Stream* has devoted space to both techniques and popularity[46].

Fish that neither have measurable economic value nor exert undue observable influence on their resident ecosystem may nevertheless be of high ecological importance. These are the rare, sometimes indigenous, and often distinctive species that form an integral part of the vast aquatic ecosystem natural history complex or may be important links in the phylogenetic sequence, such as the Humpback Chub of the Colorado River basin[20].

In the United States, the federal Endangered Species Act, administered by the USFWS and the National Oceanic and Atmospheric Administration (NOAA Fisheries), provides protection to fish species under three categories: special concern, threatened, and endangered. Legal status of a species under the Endangered Species Act or parallel state regulations are discussed in the "Importance" section or in the "Management and Conservation Status" section[20].

Description [individual sections may also include taxonomy-morphology and genetics]—In life histories, fish are described by stating taxonomic characteristics in a narrative form. It is important to provide this "picture in words" of the fish as it is in its natural environment. Not only the general appearance but the color, spotting, stripe patterns, and other markings are included, partly because colors and markings are not included in taxonomic dichotomous keys and partly to address the high degree of variability some fish species exhibit from one life stage to another. For instance, many fish species are more colorful during the breeding season[20]. The classic lateral line stripe of the Rainbow Trout and the jaw slash of Cutthroat Trout are other examples.

If there are body form changes from larval to juvenile to adult form, these are presented. External morphological characteristics and the changes noted are included, at a minimum. Some characteristics of egg, embryo, larvae, subadult, and adult are discussed. Changes through each of these life forms are governed by age, species, and size[20]. Genetic data are becoming increasingly important and where available are included, particularly in species with numerous defined subspecies, such as Cutthroat Trout.

Range and Distributional Status [Range]—Local, regional, national, or international distribution of a species of fish is important to both fisheries scientists and anglers. Both the historic range (and reductions or losses) as well as the present range, including a discussion of where the species has been stocked and is considered a nonnative, are included. When historic range and present range are disparate, probable reasons for increased range, such as transplanting (stocking), constriction of range due to habitat loss, or construction of impassable dams on either major rivers or small streams, are be listed. Significant losses to disease are also presented. Range or distribution of a species is most appropriately discussed in terms of natural features such as stream drainages or lakes, rather than in terms of political subdivisions. If relative abundance across the range or in portions of the range is understood, that information can also be included. If a geological frame of reference for when the species first appeared is available, that is included as well[20].

Size and Longevity—Age composition of individual species is defined by the proportion of fish in each definable age group. Age-group breakdown ranges from numerous young to the few remaining old fish. This section is used to define both growth rates on an annual basis (TL at year end) and final maximum size as well as maximum age[20]. Age-group determination (age class) may be apparent from large numbers of individuals captured in field efforts. Aging fish can be conducted through the use of both nonlethal (scale) and lethal (otolith) examination.

One value of age and growth studies is how they are related to variables such as abundance and quality of food, spawning status, physical and chemical habitat characteristics, and presence and abundance of competitors and predators. Age at first spawning, number of times a species spawns, and age at senility are all reported[20].

Food and Feeding—Food habits (diet or items in the diet) and feeding habits (mechanisms for finding and taking food items) of fish are not always clearly separated in fisheries literature. In a food-habits study, dietary items are recorded in a systematic fashion, such as number, volume (by type), and frequency of occurrence, and are generally reported in comparison to what is available in that particular ecosystem in the time frame of sampling. These data allow development of selectivity indexes that can be useful in expressing a fish species' preference for the selected food items[20].

Feeding habits of a fish species are reported in relation to time of day, annual variations, place (within the ecosystem) of feeding, method of food capture, how food items are selected, frequency of feeding, abstention from feeding (as on spawning migrations in some species), and other variations[20].

Limiting Factors [Morbidity and Mortality Factors]—Limiting factors are those aspects of a species' life history that reduce its numbers or the general health of individuals and thus the entire population. Determining actual cause of death for individual fish, absent catastrophic events, may be difficult and impractical. However, long-term studies on individual populations can provide information on parasites and diseases that reduce population numbers. The ability of populations to rebound after decimation by natural events is addressed if known[20].

Reproduction or Breeding Habits—Studies of fish species' reproduction include the manner of spawning (deposition of eggs and sperm) or fertilization of the female, number, type, and size of eggs (or young) per female (preferably on a unit weight basis), and the age composition of the spawning population, both males and females. General spawning habitat, type of water or ecosystem area used, the type of redd (nest) constructed, and the manner of caring for young, if any, are included[20]. For species that lay eggs, a description of the spawn, either as laid eggs or fertilized eggs, are included, as are size, shape, color, opacity, texture, buoyancy, and adhesiveness[20].

Habits and Habitat [Ecology and Behavior]—Where fish spend their lives and the physical, chemical, and related ecosystem elements of those locations define a fish species' habitat. Water chemistry largely determines productivity of a particular ecosystem (high or low nutrient availability). Topography, latitude, altitude, and water temperatures determine both fish species and ecosystem productivity. Individual fish species may seek waters in which they can feed optimally but are too warm to facilitate reproduction. Individual fish species have differing habitat, temperature, water velocity, and depth requirements or preferences. Tolerances to ecosystem characteristics vary by life stage and time of year. Fish migrations, associated with either diel movements or annual feeding migrations or spawning migrations, are discussed in the context of timing, annual cycles, and frequency[20].

Management and Conservation Status [Conservation Issues]—This section discusses a species in the context of the socio-political environment in which it exists and has less to do with the fish itself than with how the species is managed. The Conservation status classification are defined in the Idaho Comprehensive Wildlife Conservation Strategy (IDFG 2005).

Fish management practices vary in relation to a wide variety of socioeconomic and ecological factors. In general, for both sport fishing and commercial fishing, management has as a primary goal an optimal sustained yield and some equitable division of the harvest[47].

Special protections, either under the federal Endangered Species Act or state-based special-concern status, are discussed for fish species that need protection from ecosystem degradation, fishing pressure, or other environmental factors.

Chapter Citations

1. Curet, 2012. 2. Moyle and Cech, 2004. 3. Moyle, 1995. 4. Helfman, et al., 2009. 5. Kohler and Hubert, 1999. 6. Sass and Allen, 2014. 7. Bonar, Hubert and Willis, 2009. 8. Guy and Brown, 2007. 9. Eades, et al., 2008. 10. Wallace and Zaroban, 2013. 11. Sigler and Sigler, 1996. 12. Sigler and Sigler, 1987. 13. Behnke, 2002. 14. Scott and Crossman, 1973. 15. Page et al., 2013. 16. Nelson, 2006. 17. Robbins et al, 1991. 18. Behnke, 1992. 19. Behnke, 2007. 20. Sigler and Sigler, 1990. 21. Hubert, Pope and Dettmers, 2012. 22. Williams, Horn and Punt, 2015. 23. Mach, 1875. 24. Hermann, 1885. 25. Hermann and Matthias, 1894. 26. McMichael, 1993. 27. Reynolds and Kolz, 2012. 28. Reynolds and Harlan, 2011. 29. Peterson, Thurow, and Guzevich, 2004. 30. Vibert, 1963. 31. Lamarque, 1990. 32. Sharber, et al., 1994. 33. Adams, Behmer and Weingarten, 1972. 34. Holliman, Reynolds, and Holmes, 2010. 35. Robb, et al., 2002. 36. Barton and Dwyer, 1997. 37. Bouck and Ball, 1966. 38. Burns and Lantz, 1978. 39. Dalbey, McMahon, and Fredenberg, 1996. 40. Swink, 1999. 41. Sparks et al., 2011. 42. Burger, et al., 2015. 43. Zale, Parrish and Sutton, 2012. 44. Hubbs and Lagler, 2004. 45. Sigler, 1958a. 46. Mervin, 2006. 47. Botsford, Castilla and Peterson, 1997.

Additional readings for this chapter are listed in the Supplmental References section.

© Link Jackson Art

3

BIOGEOGRAPHY OF THE NATIVE FISH FAUNA OF IDAHO

Gerald R. Smith, Ph.D., Professor Emeritus, University of Michigan, Ann Arbor
Paul K. Link, Ph.D. Professor, Geosciences Department, Idaho State University, Pocatello
Glenn D. Thackray, Ph.D. Professor, Geosciences Department, Idaho State University, Pocatello
Ernest R. Keeley, Ph.D. Professor, Biology Department, Idaho State University, Pocatello

Biogeography is one of the foundations of conservation. The distribution of fishes depends on where they can get to, and once there, how they are able to find permanent water, food, and sites for spawning. Looked at over the long haul of thousands of generations, favorable habitats are those that support survival and reproduction, while also being connected to other such habitats so that local extinctions are not permanent. The metapopulation (larger connected network of populations of a species) supplies exchanges of individuals and genes for recolonization and maintenance of genetic diversity. Conservation biology is applied biogeography in the sense that past events are the key to present ecology and the environmental conditions today determine the successes and failure of fishes for the future.

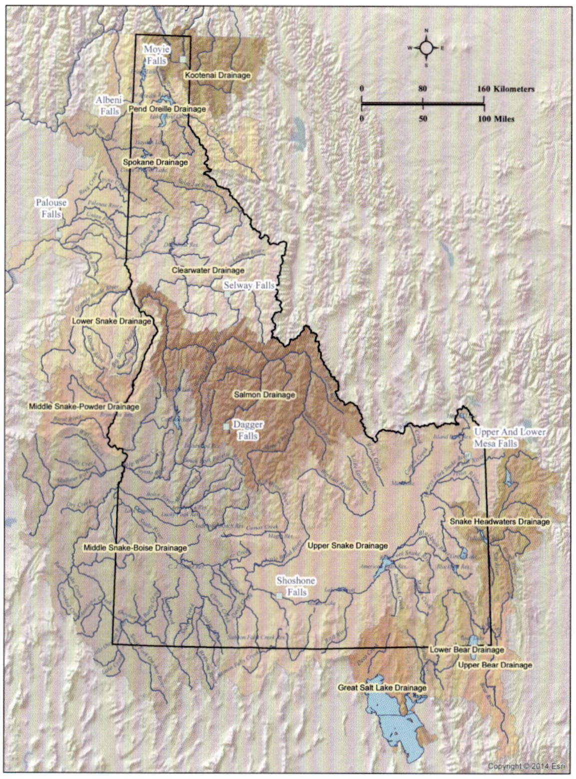

Figure 1 – Drainages, Major Waterbodies, and Major Waterfalls Affecting Fish Distribution.

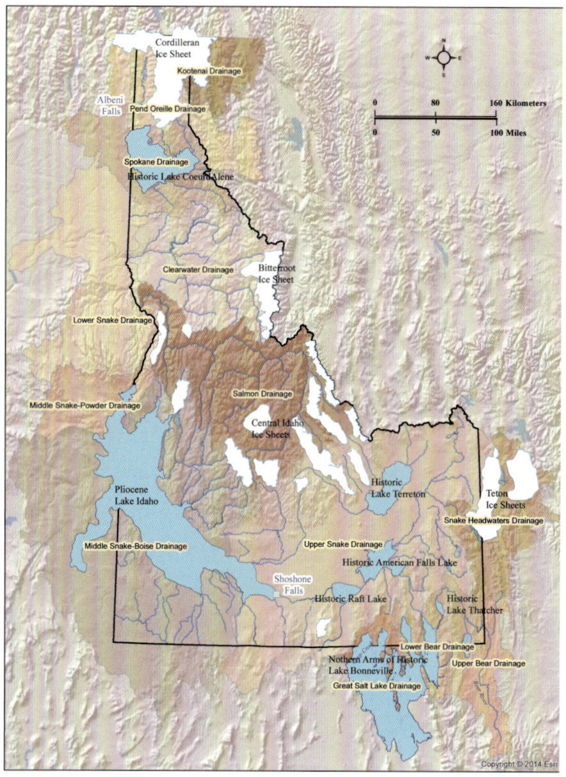

Figure 2 – Map of selected Pleistocene lakes, generalized extent of Pleistocene glaciers and selected geographic features, and Pliocene Lake Idaho.

Forty-eight native species or subspecies of freshwater and migratory fishes live all or part of their life histories in the waters of Idaho. These fishes belong to eight families, representing some of the major groups in North America, especially Cyprinidae (minnows), Catostomidae (suckers), Salmonidae (whitefishes, chars, trouts, and salmons), and Cottidae (sculpins). These family groups have inhabited this part of North America for millions of years, along with four smaller families which have one local species each—Petromyzontidae (Pacific Lamprey), Acipenseridae (White Sturgeon), Percopsidae (Sand Roller), and Gadidae (Burbot).

How all of these fishes came to be where they live now has been controlled by geological processes, such as mountain building, changing barriers, stream capture, lake spillovers, and dispersal routes. In the Snake River region of Idaho, we also need to understand the surprising ways by which volcanic activity and tectonics have influenced evolution and extinction in our aquatic ecosystems.

Southern Idaho has been drastically changed by the birth and passage of the Yellowstone Hot Spot beginning in the middle Miocene 17 millions of years ago (mya). The Eastern Snake River Plain was produced by the passage of the North American Plate (at a rate of about 4.5 mm [0.18 in] per year) over a stationary mantle plume or "hot spot" that is currently located beneath Yellowstone National Park[1–3]. The last major rhyolitic ignimbrite eruptions there were 600,000 years ago. This produced the Yellowstone Caldera where Yellowstone Lake currently sits[4]. This history of volcanic activity has reduced the diversity of minnows (Cyprinidae) and suckers (Catostomidae) in the Eastern Snake River Plain by almost 50 percent.

Over the past 10 million years, a drainage divide separating streams flowing west and east has migrated eastward across eastern Idaho, from the middle of the Idaho batholith to northwestern Wyoming (Yellowstone National Park). The continental divide, separating Pacific from Atlantic drainages and fishes, however, has been in the Yellowstone-Teton area for at least 6 m.y. Ironically, only one Atlantic drainage fish, the Lake Chub has colonized eastern Idaho in the last 10 m.y., according to the fossil record. Idaho's Atlantic drainage fishes—perch, Walleye, Sauger, two species of bass, five species of sunfish, two species of crappies, Brown Trout, shad, Golden Shiner, Spottail Shiner, Fathead Minnow, seven species of catfish, Brook Trout, Lake Whitefish, Alpine Char, Northern Pike, and six species of killifish and livebearers, were all introduced by agencies (with good records) in the past 145 years. The Idaho batholith forms the mountains of central Idaho, and has been and is a major barrier to fish colonization as well as human transportation and settlement. It is the roots of a large intrusive geological body some 85 million years old. Because this feature is deeply anchored within the continental crust, it has periodically been uplifted by isostatic forces, that continue to the present day.

As a result of subsidence of the Snake River Plain in the wake of active rhyolitic volcanism, streams which presently drain toward the Snake River Plain, from the north and south, formerly drained away from the rhyolitic volcanic plateau. Subsidence of the Eastern Snake River Plain in the wake of the Yellowstone Hot Spot allowed the Snake River to carry sand from the eastern Idaho thrust belt to the Western Snake River Plain and the Columbia River. The Holocene (less than 10,000 years ago) Snake River flows in a sandy and gravely path along the southern margin of the plain, pushed there by basalt eruptive centers. Ten million years ago, the western Snake River in southern Idaho and Oregon captured streams that formerly drained north, south and east, to become the major center of fish evolution in the West.

The modern Big Lost and Little Lost Rivers, as well as Birch Creek and several creeks to the east drain into areas of lakes and playas in the Northeastern Snake River Plain. This area includes the Big Lost Trough and the Lake Terreton area[5,6]. Westward drainage is blocked by the Arco-Big Southern Butte volcanic zone, so the streams drain internally and the water sinks into the porous Eastern Snake River Plain aquifer. Evidence for the path of pre-Pleistocene (before 2.5 Ma) Big Lost River has been located in a drillhole near Wendell (Figure 3) and Mountain Home, on the northern side of the Snake River Plain[7]. Isolation of Bull Trout, Mountain Whitefish, sculpins, and some minnows now extinct, was probably caused by these volcanic events.

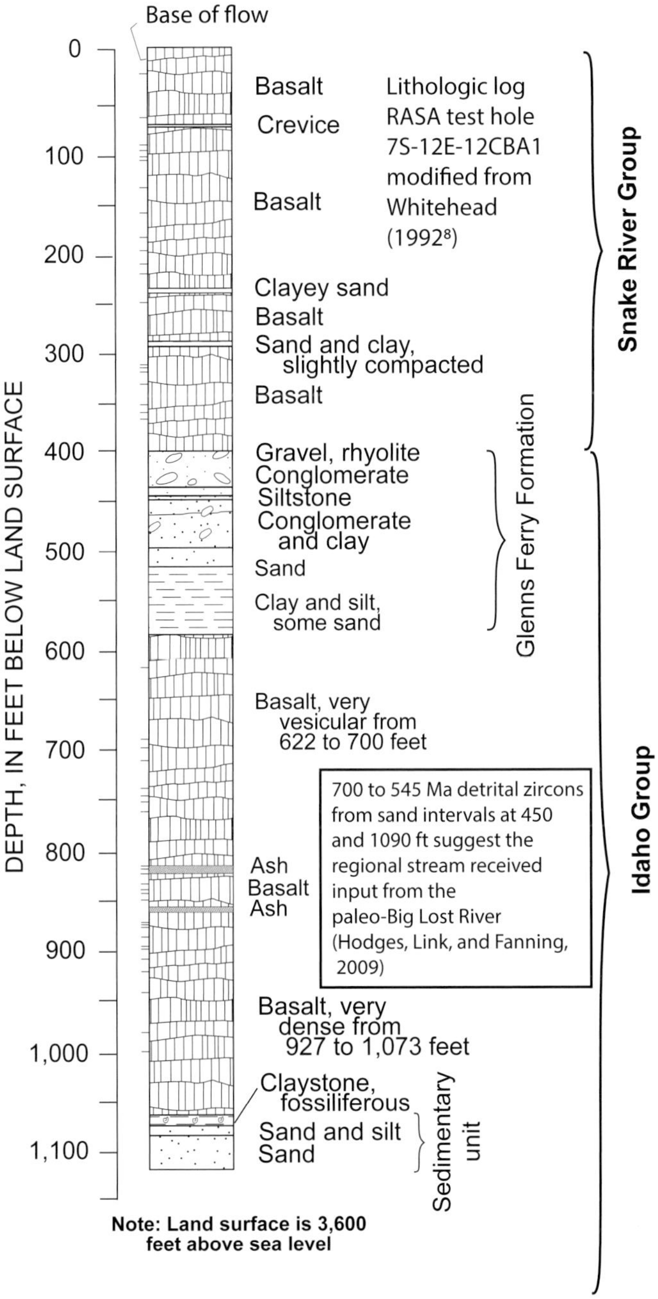

Figure 3 – Wendell Drill Hole[8]

As noted above, more than 50 other species of fishes have been introduced into Idaho waters over the past 145 years by people. Their history is an ecological problem, not a biogeographic problem.

Freshwater fish species vary in their success in different habitats and different drainages. Idaho's rivers and lakes are part of two major Pacific drainage basins—the Columbia River and Snake River Basins, and one other contributor—the Bonneville Basin, which drains internally to the continent and has no outlet to an ocean. The Columbia River Basin has three tributary systems in Idaho, all in the North—the Kootenai, Pend Oreille, and Spokane, with more than 20 shared with each other and with Snake River fishes to the south (See Figure 1 above).

The Snake River basin includes five sub-basins—the Hells Canyon area in the lower Snake River (with the Palouse River); above that, the Clearwater River, Salmon River, and the middle Snake River (including the Boise, Payette, Weiser, Owyhee, Bruneau, Big and Little Wood Rivers, and Salmon Falls Creek) drainages. The upper Snake River includes the Lost Rivers, Birch Creek, Camas Creek, Goose Creek, Raft River, and the many eastern headwaters of the Snake River, principally in Wyoming.

In Idaho, the middle Snake River is the largest drainage and has the richest fish fauna: it includes 35 different native fish species. The upper Snake River fish fauna consists of 14 species. For these fishes, the significant barrier between the middle and upper Snake Rivers is Shoshone Falls (Figures 4 and 5), 65 m (213 ft) high and passable by fish only in the downstream direction.

Figure 4 - 1894 Photo of Shoshone Falls (Courtesy of NOAA Database Modernization Program, National Climate Data Center Central Library, Silver Spring, MD)

Figure 5 - Shoshone Falls, 2015. Photo courtesy Ross Dean Mitchell.

The large Bonneville Basin, centered geographically in Utah and Nevada, but also present along the southern edge of Idaho, has fewer kinds of fishes for its size than the Columbia and Snake watersheds. It was at times a tributary to or connected to the upper Snake River, with which it shares 12 species. The Bonneville Basin also includes Bear Lake, (through its historical Bear River connections) a beautiful, long-lived lake in the southeast corner of the state, which is famous for possessing four endemic species (found nowhere else in the world), and some unique populations of Cutthroat Trout among its 17 species.

These drainages and subdrainages have fish resources whose abundance depends on how they are connected to each other now and how they were connected to each other in the past. Their future now depends on human communities and how faithfully they keep the waters clean and permanently flowing. Managers of the drainages below depend on the support of their human communities to maintain native fish diversity and the central role it plays in maintaining ecosystem health and functions. Perhaps the most useful function of these native fish species is to provide indicators of Idaho's special water quality.

A basic principle of fish diversity is that the number of species in a drainage is generally proportional to the drainage area, which is determined by its geological history. The middle and upper Snake River drainages are the largest in the state with the most diverse habitats, but the upper Snake River drainage fish fauna is only half as rich as the middle Snake River fauna because of its fiery volcanic history. Much of the upper Snake River landscape is buried in basalt and has experienced frequent extinctions over past millennia, with no possibility of recolonization from below because of the barrier presented by the great Shoshone Falls. examination of each of Idaho's drainage basins reveals different geological controls on habitat diversity, barriers, and dispersal routes from connected drainages and fishes.

Idaho Drainage Basins, Their Histories and Their Native Fish Faunas

Northern Idaho

The west-flowing drainages of northern Idaho were affected by Pleistocene glaciation, (Figure 2) but there is no evidence of major change in drainage basins through this time of glaciation and deglaciation. The Quaternary Period, spanning the last 2.5 million years and incorporating the Pleistocene and Holocene Epochs (2.5 mya to 10,000 years, and 10,000 years to present, respectively), has been a period of geologically rapid climate shifts that have impacted drainage basins in many ways. The marine oxygen isotopic record of global ice volume documents more than 30 major glacial-interglacial cycles of 41,000 and 100,000 year durations during the Quaternary period, with many shorter-term fluctuations occurring within the major cycles. The associated regional climatic fluctuations have imposed effects on Idaho river basins through the direct action of glacial erosion, mainly in central Idaho ranges and drainage basins, and more broadly on most drainage basins through variations in gravel production and transport, lake levels, river discharge, and other factors. Additionally, the cooler summers and higher discharges during glacial periods likely influenced stream water temperatures, with linked effects on fish biota.

The Cordilleran ice sheet formed in the mountains of British Columbia and Alberta and extended south into the northern United States in several areas[9]. In northern Idaho, the southward advance of the Pend Oreille lobe of the ice sheet set into motion a series of events leading to the spectacular Missoula Floods. The Pend Oreille ice lobe extended across the valley of the Clark Fork of the Columbia River and created a high ice dam. The dam blocked the Clark Fork, and lake sediments and shorelines indicate that a large lake filled the drainages upstream as far as present-day Missoula and the Bitterroot Valley, Montana. Widespread evidence of major outburst floods downstream of the ice dam in northern Idaho and eastern Washington documents several to dozens of outburst floods[10–12].

In northern Idaho, the Missoula Floods strongly affected the Rathdrum Prairie region and the Spokane River. The floods deposited thick gravel that now forms the Rathdrum Prairie Aquifer, and changed the character and course of the Spokane River. The Cordilleran ice sheet itself directly influenced the Priest River and other drainages to the north of the ice sheet limit.

In the Kootenai River drainage there are 19 native fish species. The Kootenai River is a large watershed but only a relatively small portion of it passes through the northeast corner of the Panhandle of Idaho. The Kootenai River originates in southeast British Columbia and flows south into western Montana before entering Idaho, and then flowing again north back into British Columbia and entering Kootenay Lake. Downstream of Kootenay Lake, Bonnington Falls was a major fish passage barrier but was inundated by hydroelectric development. Historically, however, the falls prevented upstream movement of migratory salmon and Steelhead. Fluvial and adfluvial populations of Bull Trout, Rainbow Trout, and Cutthroat Trout as well as Burbot, minnows, suckers, and sculpins are found in various areas of the watershed, and are examples of the rich fish species diversity because of proximity to the Columbia River—an example of the importance of dispersal routes into areas that were recently glaciated.

In the Lake Pend Oreille drainage there are 16 native fish species. This is an area of abundant water in the northeast corner of the Idaho Panhandle. It is thought to be less rich in fish species because Albeni Falls on the Pend Oreille River blocked access to immigration from the Columbia River. There are, therefore, no anadromous salmon or Steelhead populations naturally found in this area. However, early biologists that investigated the Pend Oreille watershed in the late 1800s as a salmon and Steelhead producer, questioned whether the falls could be considered a movement barrier to anadromous fishes, and wondered whether the apparent absence of migratory fishes was due to overharvest from downstream Columbia River salmon fisheries, even as early as the 1890s[14,15]. With the closure of dams on the Pend Oreille and Clark Fork Rivers, and especially with the completion of the Grand Coulee Dam on the main stem Columbia River in 1942, the question of whether anadromous fish populations were naturally found in the area was largely forgotten.

In the Spokane River drainage there are 14 native fish species. Prior to the construction of dams, anadromous salmon and Steelhead were plentiful in the lower sections of the river in Washington State, but the falls at Spokane have long been regarded as a barrier[14]. Hence, anadromous species are unknown upstream in the Coeur d'Alene, Saint Joe, and Saint Maries watersheds of Idaho. This area is better known for Westslope Cutthroat Trout (Figure 6), Bull Trout, and Mountain Whitefish as well as minnows, suckers and sculpin, given the benefits from abundant upstream water from the Idaho-Montana Rocky Mountains.

The lower Snake River drainage, including the Palouse River has 16 native fish species. Palouse Falls has isolated many fish populations for enough thousands of years to cause minor population differentiation. The Clearwater River drainage has 22 native fish species. Clearwater River tributaries are some of Idaho's most wonderfully large rivers; they are almost equally connected to the Snake and Columbia Rivers biogeographically, and are therefore rich in fish species, including eight salmonids. Upstream passage of migratory fishes has been greatly impeded by the presence of dams and water diversions in the watershed, but despite their impact, the river continues to support populations of salmon and Steelhead.

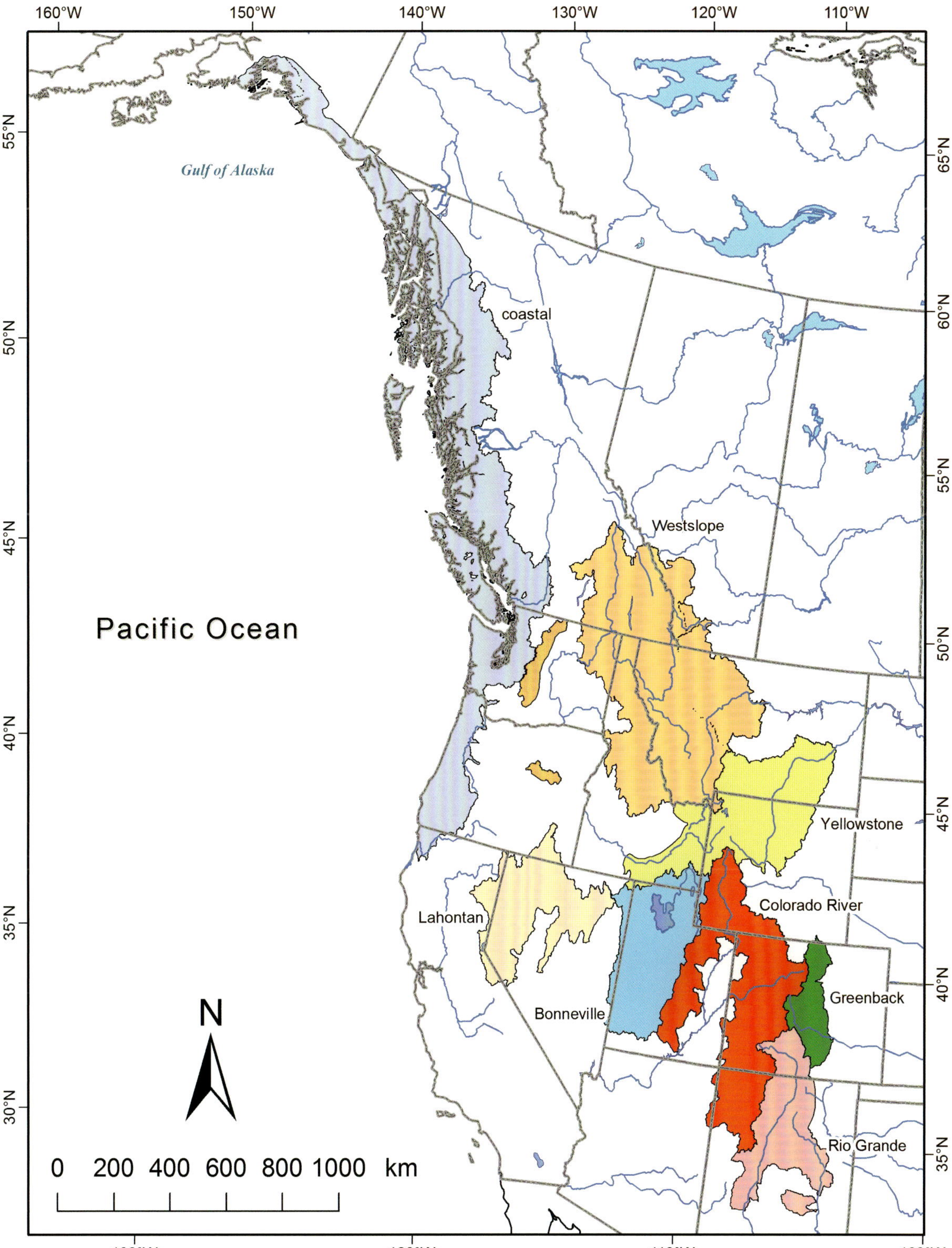

Figure 6 –Estimated historical distribution of Cutthroat Trout in western North America. Colored polygons indicate the extent of major subspecies of Cutthroat Trout, including three subspecies native to Idaho.[13]

The Salmon River drainage has 27 native fish species, including nine salmonids. The Salmon, one of our large, central rivers, is connected to both the Snake and Columbia. Famous for spawning migrations of Chinook Salmon, Sockeye Salmon, and Steelhead that have been migrating here for thousands, perhaps millions, of years, the Salmon River is a world class water in many regards, even today it is referred to as the "River of No Return" because of much of the watershed in unimpacted by civilization and the river still flows unimpeded by human development.

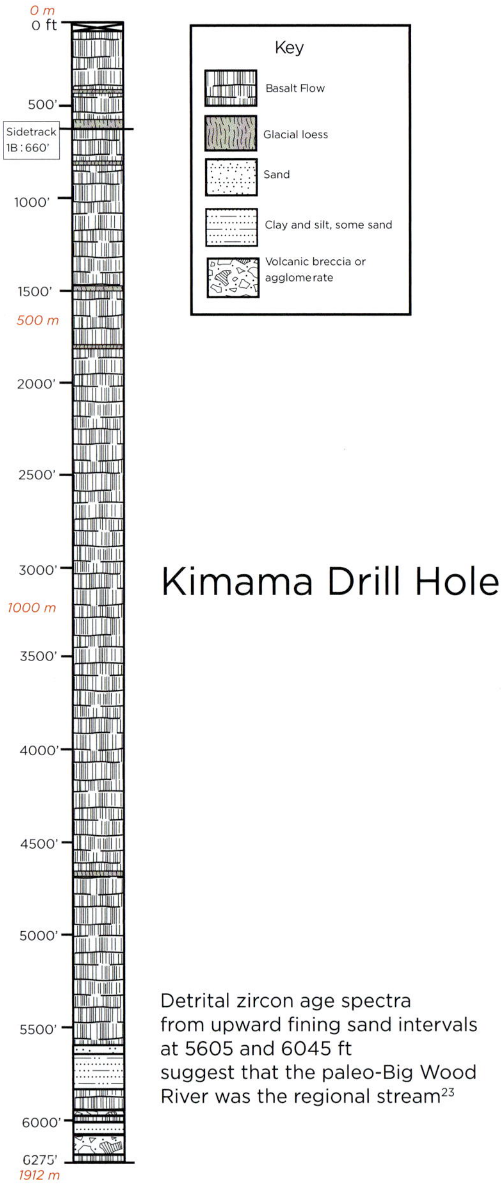

Figure 7 - Kimama Borehole[23]

Central Idaho and the Middle Snake River Drainage

Another active geological area that has changed dramatically over time and has influenced the organization of watersheds and fish species present is the Snake River and its volcanic plain. The Western Snake River Plain is structurally different—a rift valley—from the Eastern Snake River Plain and has more fish species because of this difference. Subsidence of the eastern Snake River Plain began about 10 mya and progressed northeastward, caused by crustal loading of a lower-crustal mafic sill[16], as well as thermal subsidence from detumescence of the topographic bulge, superimposed on early Basin and Range uplifts and basins[17,18], but this history should have contributed to fish diversity. It did not, probably because of the passage over the Yellowstone Hot Spot, discussed above.

Fluvial sediments contain detrital zircon grains, whose age of crystallization is preserved through multiple phases of recycling[19]. Detrital zircons are a powerful tool in deciphering paleo-drainage direction, since they fingerprint provenance[5]. Multiple-recycled Proterozoic zircons from the Cordilleran orogenic belt are not present in the Western Snake River Plain until after 7 Ma (late Miocene); this suggests that the central Idaho and Idaho-Wyoming thrust belts drained northward over eastward until late in the Miocene[20,21].

Before about 7 mya, the eastern Snake River drained eastward into the Green/Colorado River drainage[22]. Lake Idaho (Figure 2) on the Western Snake River Plain was fed by streams flowing west from the Idaho batholith. By 5 Ma the location of the Kimama borehole (Figure 7) was fed by the south-draining paleo-Wood River[23].

During much of the Late Miocene the western Snake River had southern headwaters in the Humboldt River basin and an outlet through northern California into the Klamath and Sacramento drainages. However, an eastern connection to the Cache Valley Formation of the Salt Lake Group at 10.3 Ma is also suggested.

By the late Miocene (ca. 7 Ma), drainage in north-central Nevada had changed direction, from south-southeastward to north-northwestward, as 42-35 Ma zircon grains from northern Nevada are present on the Western Snake River Plain of eastern Oregon[21]. An integrated Snake River is suggested by Pliocene to Recent (4-0 Ma) sediments of the central and western Snake River Plain. These sands contain abundant recycled Early to Middle Proterozoic grains proximally derived from thrust sheets of the Cordilleran orogenic belt[19].

There are more than 30 native fish species in middle Snake River drainage in Idaho. The middle Snake River and its tributaries has such a rich fauna because of its inheritance of fish diversity from large, long-standing and diverse Miocene and Pliocene lakes. These lakes filled the basin of the Western Snake River Plain between Hagerman, Idaho, and the Powder River of Oregon for several million years. The Miocene Chalk Hills Lake lasted almost 2 million years before it was desiccated partly by climate change. It had received its fishes through connections to the Sacramento River and Klamath River in California and Oregon. It had 18 species, including all of the families and most of the genera of fishes that now live in Idaho, plus sunfish and catfish, with ancient relationships to the Mississippi Basin about 20 m.y. before, but which were widespread in the west 12 to 6 million years ago. Although this lake was not especially deep or cold, compared to the lake that came later, it had five kinds of salmon, trout, and char, more than any other place in the known fossil record. Most remarkably, they were connected not to the Columbia River, which had only one salmon (known only from three bones) at that time, but to the Sacramento River, which shared with it many distinctive minnows and the sunfish. Hells Canyon did not yet exist.

We have no surface sedimentary record of a lake on the Snake River Plain between ~6-4.3 mya, but there must have been continuous aquatic habitat in the Boise-Payette area, because when tectonic and volcanic activity created a new natural impoundment, the Glenns Ferry Lake, it had most of the same lineages of fishes, although somewhat evolved. The Glenns Ferry Lake was about as large and as deep as Lake Ontario in the Laurentian Great Lakes. The new lake on the Western Snake River Plain was larger and deeper than the Chalk Hills Lake and it obviously had connections to the north because the Glenns Ferry Lake gained two new species of cold-water whitefish, a more modernized Redband Trout, and two new genera of cold-water sculpins that gradually evolved into eight new species. This was the richest fish fauna, 32 species, ever recorded in the Pacific drainage of North America, but as the ice age climate began and the lake was captured and drained through Hells Canyon, the Idaho sunfish, catfish, and warm-water fishes from California were extirpated. The capture of the northern corner of the lake, at the oxbow, 3.1 million years ago, created the familiar middle Snake River drainage species of today—sturgeon, lamprey, whitefish, Redband Trout, Northern Pikeminnow, Peamouth, Chiselmouth, dace, river suckers, mountain suckers, and river sculpins, with three species of anadromous salmon.

Surprisingly, we have found no bones of Cutthroat Trout in the rich fossil record of the Western Snake River Plain. The thousands of Redband Trout bones there, however, are testimony to that species' ability to exclude Cutthroat Trout. Salmon and Steelhead were abundant in many of the streams of the middle Snake River drainages prior to the early 20th century[14,24,25]. The Payette, Owyhee, Bruneau, and Boise rivers as well as Salmon Falls Creek and Rock Creek all had significant runs of anadromous fish. Relatively small dams and water diversions depleted many of these salmon and Steelhead populations during the early 1900s and they were completely extirpated from the Snake River above Hells Canyon with the completion of the Hells Canyon complex of hydroelectric dams by 1968.

Upper Snake River drainage, above Shoshone Falls

The Shoshone Falls rhyolite is one of the last, post-caldera, eruptive phases of the Twin Falls volcanic field, which was mainly active from 10 to 6 million years ago. It was dated at 6.53 Ma using the 40Ar/39Ar step heating method[26]. When the Shoshone Falls themselves became a knickpoint and obstacle to fish movement is surely Pleistocene (less than 2.5 Ma) but details are controversial.

This area has only 15 native fish species, fewer than expected, given abundant water both in the past and the present. The upper Snake River is isolated above Shoshone Falls, and so is cut off from colonization by fishes from downstream areas following Pleistocene extinctions by volcanic activity. Lava flows caused initial extinction in the contact area, but nutrient enrichment occurred shortly afterward, leading to ultimately higher diversity. Barrier falls prevent recolonization of upstream drainages. The upper Snake River has a history of frequent aquatic connections with the Bonneville Basin to the south. Its native fish fauna is dominated by minnows, suckers, sculpins, whitefish, and trout shared with the Bonneville Basin. Formerly, chars were also present in both drainages[27].

Southeastern Idaho Including Bear River, Bear Lake, and Lake Bonneville

Pleistocene Lake Thatcher occupied a small basin in the southern end of Gem Valley, and the history of the basin influences our understanding of the final diversion of the Bear River into the Bonneville Basin, the rise of Lake Bonneville to the Bonneville shoreline, the Bonneville Flood, and a temporary river connection between the Bonneville Basin and the Snake River. The Thatcher Basin hosted a series of lakes during the Quaternary period, collectively known as Lake Thatcher but revealing a rather complicated history[28–30]. Lake Sucker, *Chasmistes*, of Lake Thatcher is the key fish involved in this history. This is the likely route by which Lake Suckers gained access from the Snake River to the Bonneville Basin.

The Bear River currently flows northwest from the western Uinta Mountains into southeastern most Idaho, flows north to Soda Springs, and there turns south through southern Gem Valley, Thatcher basin, and Oneida Narrows to enter Cache Valley and the Great Salt Lake Basin. A low-elevation, low-relief divide (1670 m, 5480 ft) separated the Thatcher Basin from Cache Valley and the Bonneville Basin to the south and restrained its drainage into the Bonneville Basin; the divide is of similar elevation to the highest Lake Thatcher shoreline, suggesting that it formed the low barrier between the basins before the breaching of Oneida Narrows (see figure 2 above).

Recent research used amino-acid racemization on gastropod and bivalve shells to constrain the ages of Lake Thatcher sediments, and used strontium-isotopic compositions of the same shells to identify the input, or not, of the Bear River into the basin at various times[28]. The oldest known Bear River input to Thatcher Basin was around 140,000 years ago, though it could have occurred earlier. That appears to have been only a temporary Bear River connection, though, as the Thatcher basin was fed only by local streams, with no evidence of Bear River input, from 140,000 years to 50,000 years ago. Starting around 50,000 years ago, the Bear River was permanently diverted southward, and eventually the cutting of Oneida Narrows connected the Bear River and Thatcher Basin fully into the Bonneville Basin. By 20,000 years ago, the systems were clearly connected and Lake Bonneville sediments of that age were deposited in Thatcher Basin[31]. New geochronological and sedimentological work may soon clarify this complicated history[32].

The "flickering switch" (Figure 8) of Bear River input into Thatcher Basin and, ultimately, the Bonneville Basin, left a strong imprint on regional drainage development. The causes of the "flickering switch" are yet unknown, but are probably related to volcanic activity in the Blackfoot lava field (see figure 8) or the Gem Valley volcanic field (or both) diverting the Bear River flow from northward paths via the Blackfoot and/or Portneuf drainages (or both)[32]. Regardless of the cause, the addition of Bear River flow increased inflow into the Bonneville Basin by roughly one-third and contributed to the rise of the Bonneville Basin to its sill-controlled highstand and, ultimately, its outburst flood into the lower Portneuf drainage and the Snake River.

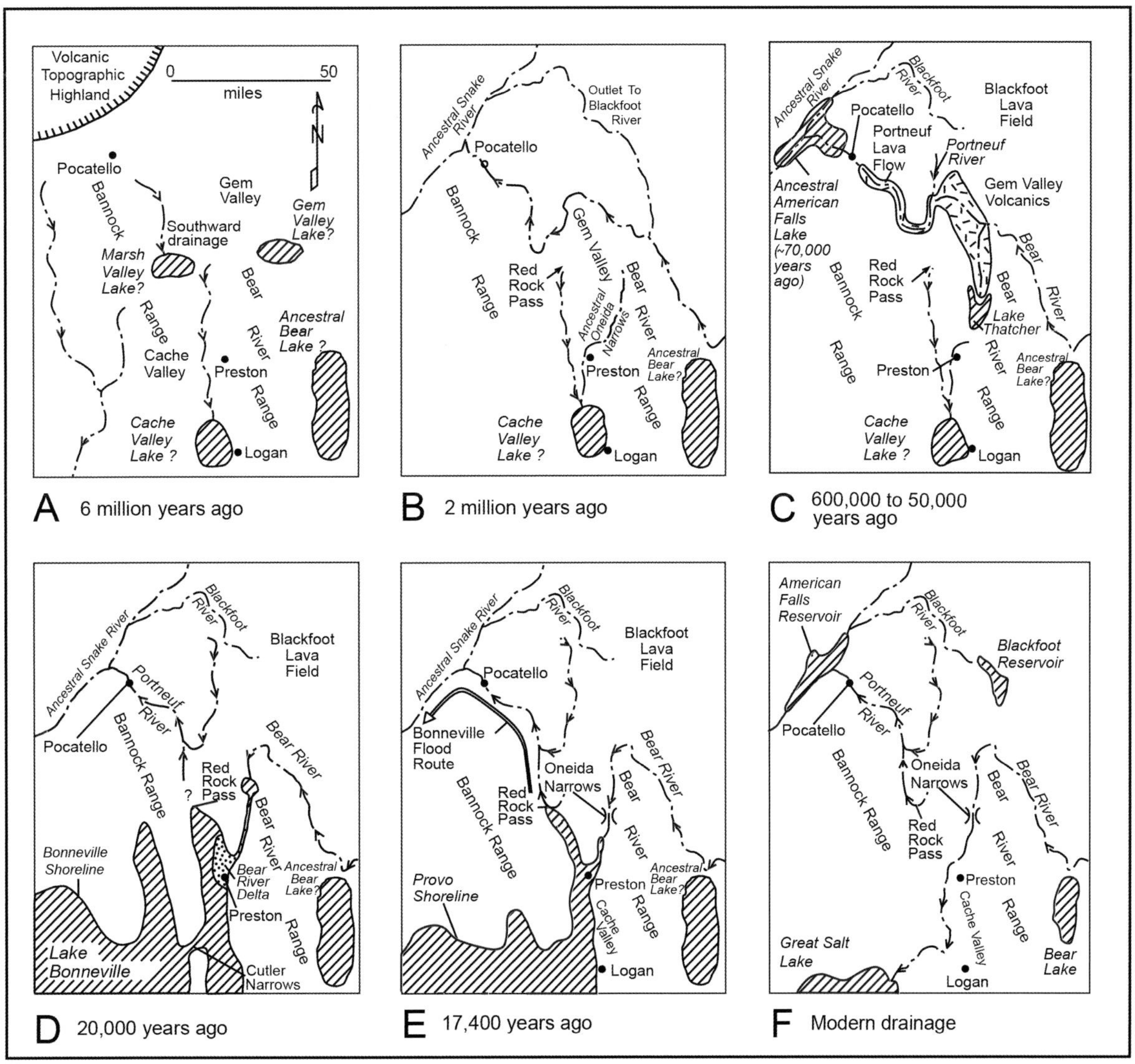

Figure 8 - Bear River "Flickering Switch"[33]

Lake Bonneville-Snake River Connection

The filling of Lake Bonneville to its highest (Bonneville) shoreline (about 20,000 years ago) set in motion a sequence of events that probably connected the Bonneville Basin with the Snake River for several hundred to more than two thousand years. This connection, in particular an inferred Bonneville outflow river that persisted 1,800-2,500 years after the catastrophic Bonneville Flood, would have permitted fish passage between the two systems, as detailed below, but fossil chubs, *Gila*, from Pleistocene deposits at American Falls (see below), and Lake Suckers, *Chasmistes*, from Lake Thatcher, as well as DNA dating, suggest earlier connections as well[34].

Red Rock Pass, lying between Cache Valley and Marsh Valley in far southeastern Idaho, is the lowest divide on the margins of the Bonneville Basin[35]. In 1890, Gilbert also first inferred that Lake Bonneville had reached the sill near Zenda, Idaho (at the Bonneville shoreline level), and, ultimately, generated an outburst flood via sill failure. He did not investigate that outflow further, but subsequent researchers[36,37], studied the features of the Bonneville Flood along its path from Red Rock Pass, through the Portneuf drainage and, ultimately, along the entire Snake River drainage downstream of the Portneuf confluence.

The detailed timing of Lake Bonneville's rise and fall and of the Bonneville Flood outbreak is the subject of continued study, and the general sequence of events is fairly well established. The Late Pleistocene rise of the lake began around 30,000 years ago[38,39]. The revised radiocarbon age of the Lake Bonneville Flood, which scoured the Devils Corral of the Wood River watershed and most recently excavated the present Shoshone Falls, is 17,400 years ago[40].

By that time, the lake had reached its sill level at Zenda, near modern-day Red Rock Pass. Within the coarse resolution of dates on lake deposits and Bonneville Flood deposits, it appears that Lake Bonneville spilled over the Zenda sill for as long as 500 years[41], and water may also have flowed through the limestone bedrock and gravels that formed the sill. This overflow and/or flow through the sill would likely have created a sizeable river flowing through Marsh Valley and the lower Portneuf drainage.

By best estimates, the Bonneville Flood outbreak at Red Rock Pass occurred around 17,000 years ago, although direct radiometric dates on flood deposits are few. Earthquakes on the nearby Riverdale Fault[42], landslides in the sill, erosion through the gravels of the sill, or any combination of these three processes, released Lake Bonneville through Red Rock Pass[43,44]. The catastrophic release produced a peak discharge around 1 million m^3/sec (35, 314, 667 ft^3/sec)[36], and eroded 100 m (324 ft) of the sill until stabilizing at a new bedrock sill. The outlet area then evolved through a series of sills inferred from paleochannels and related features in northernmost Cache Valley[42].

Once the lake level stabilized at the Provo level (about 16,000 to 12,000 years ago), controlled by the new sills, the excess water entering Lake Bonneville flowed through Red Rock Pass in non catastrophic fashion. Lake Bonneville stabilized at or near the Provo level for 1,800 to 2,500 years[40,45]. Alluvial features along extremely underfit Marsh Creek in southern Marsh Valley are likely a result of that inferred river[46]. This inferred river has important implications for fish distributions. Throughout this period, Lake Bonneville was connected to the Snake River drainage by a river, whose flow would have been modulated by the water balance of Lake Bonneville. The inferred river would have allowed fish passage between the two basins. The Idaho fish species shared with the Bonneville Basin as a result of this history include Utah Chub, Northern Leatherside Chub, Longnose Dace, Speckled Dace, Redside Shiner, Snake River Sucker (*Chasmistes*), Utah

Sucker, Mountain Sucker, Green Sucker, Mountain Whitefish, Cutthroat Trout, Mottled Sculpin complex, and Paiute Sculpin. Bear River and the Snake River minnows, suckers, trouts, whitefish, and sculpins show strong relationships with all but one of the Bonneville species (the Least Minnow, *Iotichthys*). The upper and middle Snake drainages show no such close genetic relationships with the neighboring Lahontan Basin to the south, in Nevada.

The Bonneville Flood modified many sections of its 1,100 km (684 mi) flow path between Red Rock Pass and Lewiston, Idaho. Its influences on Marsh Valley, the Portneuf Gap, and the lower Portneuf Valley at Pocatello were the most dramatic and include scoured basalt, extensive and very coarse gravel deposits, and scoured alluvial fan edges. Once the floodwaters entered the Snake River Plain, they were alternately concentrated in canyons, creating scour features and enhancing cataracts, and allowed to expand, creating dramatic gravel bars.

The Bear River and Bear Lake have 17 native fish species or subspecies in Idaho. Bear River was formerly tributary to Snake River before it was diverted to the Bonneville Basin near Soda Springs. Bear Lake, Idaho and Utah, was the source for Bonneville Basin fishes before desiccation. Lake Bonneville acquired 6 to 9 fish species from Bear Lake, which is the much older and more permanent lake. Lake Bonneville had a relatively short history before it shrunk and became too salty to support fish life, a process that had been repeated several times in its history. All but a few of its species were shared with the Snake River, to which it had been connected by the Bear River at times, but also by direct overflow at Red Rock Pass, about 17,400 years ago. The dramatic Bonneville Flood (briefly it was probably the largest volume river in the world), distributed some forms of fish all across Idaho and into Oregon and Washington to the Pacific Coast, but there is little evidence for dispersal in the upstream direction.

The Snake River Sucker, *Chasmistes muriei*[47], is closely related to the fossil *Chasmistes* sucker from Idaho's middle Snake River drainages and to the Utah Lake Sucker, *Chasmistes liorus*. These in turn are related to lake suckers in Klamath Lake, Oregon, and Pyramid Lake, Nevada. The biogeographic connections among the Lake Suckers in the Snake River and Bonneville Basin date back a few tens of thousands of years. The relationships of these derived from Idaho to the Klamath and Pyramid Lake Suckers date back several million years. They are not closely related to the River Suckers, *Catostomus*, that often occupy the same drainages, but they frequently hybridize when water withdrawal for irrigation forces them onto the same spawning grounds. Lake Suckers are known from a restricted northern Great Basin area in Idaho, Oregon, Nevada, and California over the past 6 million years. They are phylogenetically early branches in the sucker family, based on traits shared with the primitive buffalo fishes and carp suckers of the Mississippi River basin. River Suckers, *Catostomus*, are more advanced and more diverse in Idaho and surrounding states, along with Mountain Suckers and Green Suckers *(Pantosteus)*[48]. Three species of *Catostomus and Pantosteus* live in Idaho, there are others in the fossil record on the Snake River Plain.

Evidence of pluvial connections between the upper Snake River and Bear River are also found in the genetic relationships among populations of several fish species in southeastern Idaho and neighboring states. Populations of Cutthroat Trout, Northern Leatherside Chub, Redside Shiner, Utah Chub, and Utah Sucker all show greater similarity between the upper Snake River and Bear River than from areas west and south of the Wasatch Mountains, but still within the Bonneville Basin[13,49–51]

The form of Utah Chub of the Bonneville Basin *(Gila atraria)*, is replaced by a unique form, *Gila domninus* (Cope) in the Bear River and upper Snake River. It inhabited the American Falls lake in the Pleistocene (ISU Museum fossils) and its antiquity was identified from its divergent DNA[34]. Johnson, however, reports them all as Utah Chub[52].

There are 14 independent drainages on the Eastern Snake River Plain. In many of these geologically active drainages, fish populations have been affected by volcanic destruction and the formation of other barriers such as waterfalls and streamflow subsidence. Despite the presence of such barriers, only two, the lava plains and Shoshone Falls, have had a significant effect: 11 species are present, including species of whitefish, trout, char, dace, suckers, shiners, and sculpins. Others were present in the Eastern Snake River Plain in the Plio-Pleistocene, but are now only in the Western Snake River Plain[53].

An enhanced understanding of the relationships between Idaho geologic features and fish distribution throughout Idaho is being rapidly developed as new techniques and research protocols come online. These techniques include: detrital zircons, whole genome sequencing, transcriptome (RNA) analysis, isotopes in otoliths, fossil-calibrated DNA dating, radiotelemetry, and modern paleontology.

Chapter Citations

1. Pierce and Morgan, 1992. 2. Pierce and Morgan, 2009. 3. Rodgers, Hackett, and Ore, 1990. 4. Christiansen, 2001. 5. Geslin, Link, and Fanning, 1999. 6. Bestland, et al., 2002. 7. Hodges, Link, and Fanning, 2009. 8. Whitehead, 1992. 9. Waitt and Thorson, 1983. 10. Bretz, 1923b. 11. Waitt, 1985. 12. Bretz, 1923a. 13. Loxterman and Keeley, 2012. 14. Gilbert and Evermann, 1895. 15. McDonald, 1895. 16. McQuarrie and Rodgers, 1998. 17. Ore, Reid, and Link, 1996. 18. Rodgers, et al., 2002. 19. Link, Fanning, and Beranek, 2005. 20. Link, et al., 2002, 21. Beranek, Link, and Fanning, 2006. 22. Spencer, Smith, and Dowling, 2008. 23. Potter, 2014. 24. Fulton, 1968. 25. Chapman and Chandler, 2003. 26. Othberg, et al., 2012. 27. Broughton, Madsen, and Quade, 2000. 28. Bouchard, et al., 1998. 29. Bright, 1963. 30. Bouchard, 1997. 31. Oviatt, Curry, and Sack, 1992. 32. Pederson, et al., 2016. 33. Link, Kaufman, and Thackray, 1999. 34. Johnson and Belk, 1999. 35. Gilbert, 1890. 36. O'Connor, 1993. 37. Malde, 1968. 38. Scott, et al., 1983. 39. Oviatt, 1997. 40. Godsey, Currey and Chan, 2005. 41. Oviatt and Miller, 1997. 42. Janecke and Oaks, 2011. 43. Bright and Ore, 1987. 44. Shroder and Lowndes, 1989. 45. Benson, et al., 2011. 46. Thackray, Rodgers and Drabick, 2011. 47. Miller and Smith, 1981. 48. Unmack, et al., 2014. 49. Johnson, Dowling, and Belk. 2004. 50. Mock et al., 2006. 51. Houston, et al., 2014. 52. Johnson, 2002. 53. Hearst and Smith, 2002.

Additional readings for this chapter are listed in the Supplemental References section.

"Small acts, when multiplied by millions of people, can transform the world."

Howard Zinn

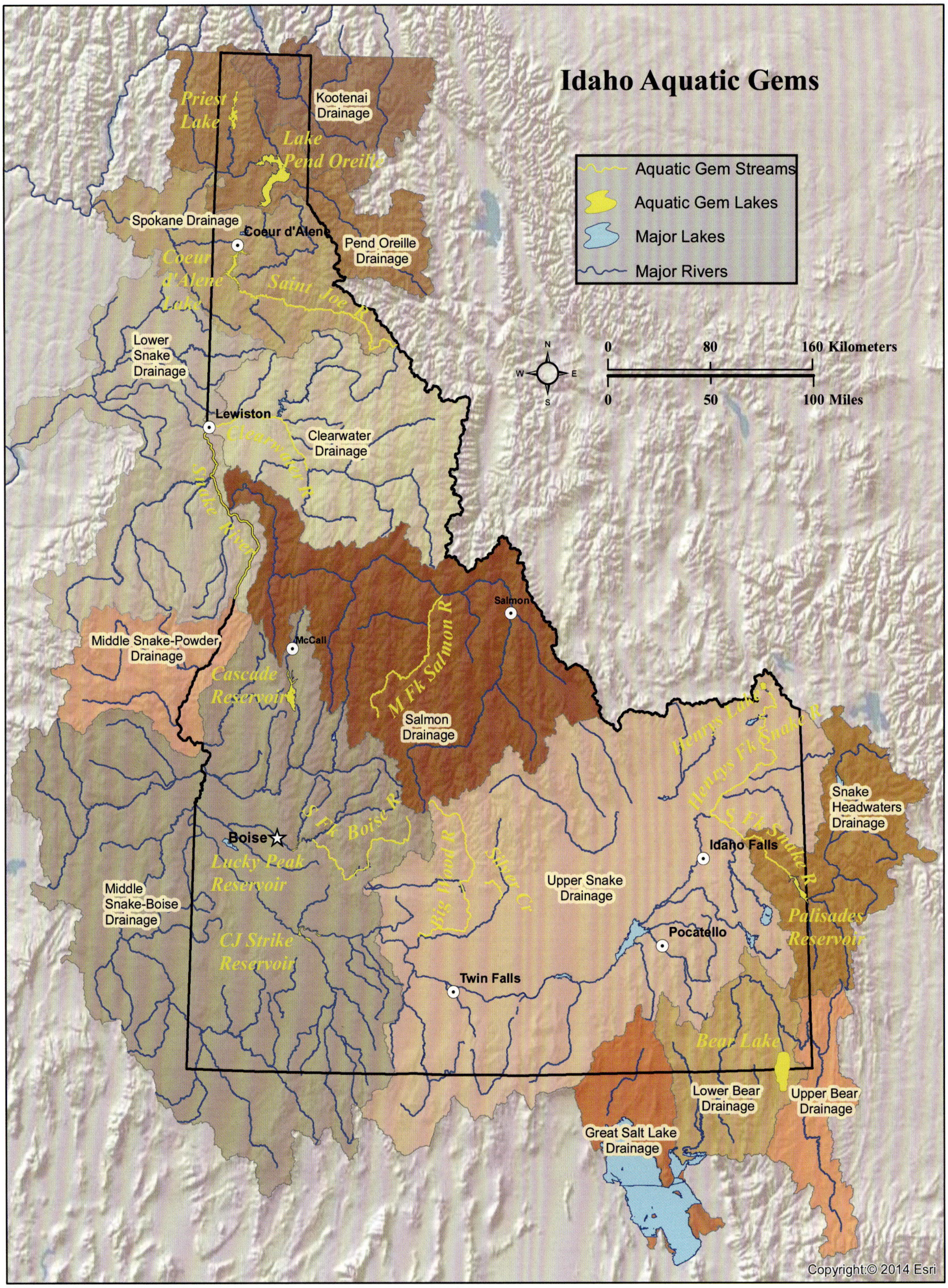
Idaho Aquatic Gems
Aquatic Gem Streams
Aquatic Gem Lakes
Major Lakes
Major Rivers
0 80 160 Kilometers
0 50 100 Miles
N
S
E
W
Priest Lake
Kootenai Drainage
Lake Pend Oreille
Spokane Drainage
Coeur d'Alene
Pend Oreille Drainage
Coeur d'Alene Lake
Saint Joe R.
Lower Snake Drainage
Lewiston
Clearwater R.
Clearwater Drainage
Snake River
Salmon
M Fk Salmon R
Middle Snake-Powder Drainage
McCall
Cascade Reservoir
Salmon Drainage
Henrys Lake
Henrys Fk Snake R
Snake Headwaters Drainage
S Fk Boise R
Boise
Lucky Peak Reservoir
Big Wood R
Silver Cr
Idaho Falls
S Fk Snake R
Upper Snake Drainage
Middle Snake-Boise Drainage
CJ Strike Reservoir
Pocatello
Palisades Reservoir
Twin Falls
Bear Lake
Lower Bear Drainage
Upper Bear Drainage
Great Salt Lake Drainage
Copyright:© 2014 Esri

4

IDAHO'S AQUATIC GEMS

Idaho is blessed with an amazing amount of surface water in the form of streams, rivers, lakes, and reservoirs. In the recent editions of *McClane's Fishing Encyclopedia*[1], Idaho waters are described as providing *"an almost limitless variety of fishing waters."* McClane goes on to extol the wonders of fishing in Idaho and then provides specific examples of where one can find great fishing in the state. Other publications have provided hints and tips on how to fish in Idaho.

Below we provide 17 "Idaho Aquatic Gems" presentations, written by the people who manage and fish them, and these are just a few of the most spectacular fishing waters in Idaho. This is not a complete list of great places to fish in Idaho, but rather presents the reader with some choices and helpful hints on a few of the best places to fish. Numerous publications exist that present various views on fishing in Idaho, and it is suggested that interested readers search for fishing information regarding both the locations and the species they wish to pursue. One good starting point is *Go Fish Idaho! An Angler's Guide to the Gem State*[2]. A great deal of useful information on places, species, seasons, regulations, and other materials can also be found on the IDFG website, idfg.idaho.gov.

Bear Lake

David Teuscher, Regional Fisheries Manager, Southeast Region, Idaho Department of Fish and Game

Location, Directions, and Access

Bear Lake straddles the Idaho-Utah state line, with about half of its surface area in Idaho. Lakeshore communities include Saint Charles, Fish Haven, Garden City, and Lake Town. Montpelier and Preston are larger communities within an hour's drive of the lake and provide a variety of services and accommodations. Bear Lake can be reached on US Highway 89 from Logan, Utah, or Montpelier, Idaho, or from the south on Utah Highway 30. Many boat ramps, state parks, and beach-access areas make Bear Lake a true outdoor recreation wonderland.

Physical Characteristics and Attributes

Bear Lake is the second-largest lake in Idaho, one of the oldest lakes in North America, and known for its striking turquoise-blue color. Because of its beautiful color—derived from suspended calcium carbonate prevalent across the Bear River Range—Bear Lake is known as the "Caribbean of the Rockies."

This oval-shaped lake is about 32 km (20 mi) long and 13 km (8 mi) wide and has 77 km (48 mi) of shoreline. It has a full pool elevation of 1,805 m (5,923 ft) with a maximum depth on the east shore of 63 m (208 ft). When full, Bear Lake holds more than 801,768 ha-m (6.5 million ac-ft) of water. Up to 20 percent of Bear Lake's volume supports irrigation needs for downstream farmers in Idaho and Utah.

St. Charles, Fish Haven, and Swan Creeks are the three primary tributaries that flow naturally into Bear Lake. Historically, Bear Lake was much larger in size and captured the Bear River at its northern end[3]. The Bear River is currently diverted to Bear Lake at Stewart Dam.

Bear Lake Fishery

Bear Lake supports 16 species of fish and is the only lake in North America with four endemic fishes: Bear Lake Whitefish, Bonneville Cisco, Bonneville Whitefish, and Bear Lake Sculpin. All four species originated some 10,000 to 35,000 years ago. Bear Lake also provides habitat for one of two remaining native adfluvial stocks of Bonneville Cutthroat Trout.

During the last century, significant efforts were made to enhance the fishery by introducing new species. In the 1890s, the first releases were American Shad and whitefish that were transported from the Great Lakes. The objective of those introductions was to create commercial fishing opportunities, but those efforts failed to establish reproducing populations. Experimenting with other species of exotic fish introductions continued over the next 75 years and included Chinook Salmon, Atlantic Salmon, Kokanee, Largemouth Bass, Rainbow Trout, Brook Trout, Lake Trout, and Yellowstone Cutthroat Trout. The unique water chemistry or other habitat limitations in Bear Lake resulted in failure of nearly all those introductions. The only species persisting from those stocking efforts are Rainbow Trout and Brook Trout that ascended the tributary streams.

To support the current sport fisheries of Bear Lake, about 200,000 native Bonneville Cutthroat Trout and 17,000 sterile Lake Trout are stocked annually. Both of these species of trout in Bear Lake tend to concentrate in the few areas of the lake where the bottom contains significant rocky substrate (e.g., Cisco Beach or the Rock Pile located off Rendezvous Beach). If you're a new angler to Bear Lake, find those areas on a map and try there first! Fishing for Bonneville Cisco and the native whitefish adds to the variety and unique approaches of fishing at Bear Lake.

Fishing Tips and Techniques

Bear Lake is open to fishing all year. Catch rates are generally highest in the spring and fall and during the ice-fishing season. The lake is famous for the Bonneville Cisco dipnet fishery and for the opportunity of catching large native Bonneville Cutthroat Trout. Native Cutthroat Trout live in excess of 10 years and can tip the scales at more than 4.5 kg (9 lb, 14 oz). The following paragraphs provide specific details on how, when, and where to fish Bear Lake.

Cisco

The dipnet fishery for Cisco is a fascinating and surprisingly entertaining endeavor. The fishing rules for Bear Lake were established so that anglers can pursue Bonneville Cisco during their spawning period with dip nets. Dipnetting is a successful method because of the large schools that form in shallow water during their spawning period. In most years, Bonneville Cisco spawning peaks in mid- to late January. Dipnetting requires a few exceptions to general fishing rules. Chief among them is that no restrictions are made for the size of hole that can be cut in the ice. Therefore, many anglers abandon normal ice augers in favor of chain saws.

Anglers either crouch down or lay on the ice to improve seeing the schools as they swim past. When a school is directly below the opening in the ice, a quick dip of the net targeted at the area slightly in front of the leading fish works best. If the school is large enough, just dipping the net in any manner can be successful. During the peak spawning period, limits of 30 Bonneville Cisco can be attained in 10 minutes or with just a few dips of the net.

The prized 300 mm (11.8 in) Bonneville Cisco make great bait for catching bigger Cutthroat Trout and Lake Trout and also make good table fare. Anglers travel long distances to participate in this unique experience. During peak netting, hundreds of anglers gather on the ice and have nicknamed the experience the "Cisco Disco." This unique dip-net fishery lasts only a few weeks and has no impact on the Cisco population that ranges between 2 and 6 million[4].

Cutthroat Trout

Trolling and jigging are the two most common methods for catching Cutthroat Trout. Less common methods include casting from shore and fly fishing. Trolling is generally best on the east shore, and common areas include First Point, Rainbow Cove, and Cisco Beach. Jigging and trolling near Rendezvous Beach are also productive. A silver-hammered spoon with red or orange stripes is a very popular lure.

During summer months, Cutthroat Trout move deeper, so trolling with downriggers, leaded line, or divers or jigging below 14 m (45 ft) is advisable. During the peak Cisco spawning period, fishing can slow down a little because of the abundant concentration of natural forage available. However, fishing for Cutthroat Trout a few weeks before and after the Cisco spawning period can be fantastic.

Harvest of Cutthroat Trout on Bear Lake is restricted to fish with a clipped adipose fin, as evidenced by a healed scar. Harvesting only hatchery stocked fish helps maintain the natural population, which is on the rise from successful habitat projects targeted at improving survival in the three largest spawning tributaries (Fish Haven, St. Charles, and Swan Creeks).

Lake Trout

There are three primary methods used for pursuing Lake Trout. The most common method is trolling along the east shore in water depths ranging from 18 to 24 m (60 to 80 ft). The same spoons used for Cutthroat Trout are successful as well as Rapalas. Jigging can also be productive at the same locations mentioned above for Cutthroat Trout.

Another method for catching Lake Trout is by long-lining. Long-lining is a method of fishing used by shore anglers. Long-lining is best along the east shore of the lake, as it has a steep drop off. Long-liners use small boats or other watercraft (e.g., canoes, kayaks, float tubes) to transport a baited Bonneville Cisco or other cutbait into deep water, and then they return to shore and wait for a strike. Long-liners are generally targeting Lake Trout, but can also catch a few Cutthroat Trout.

Fishing along the Utah State Marina in mid- to late October can produce some of the best Lake Trout catch rates. During that time, Lake Trout are concentrating to spawn along the rocks that protect the marina from wind and wave action. This is the only time and place where anglers can successfully target Lake Trout casting from shore. The Lake Trout are more aggressive than usual and can be caught on a variety of spoons, Rapalas, or plastic jigs.

Lake Trout in Bear Lake can be very long lived (30+ years) and grow to sizes over 9 kg (19 lb, 13 oz). However, most of the Lake Trout caught in the lake are between 2 and 5 kg (4 lb 7 oz and 11 lb, 0 oz). Hatchery-reared Lake Trout stocked in Bear Lake are sterile to prevent natural populations from establishing. Stocking sterile trout allows managers to control Lake Trout numbers. Population control for Lake Trout is important to meet fishery management goals that prioritize protection of the endemic species that thrive in Bear Lake.

Whitefish

Few anglers travel to Bear Lake specifically targeting whitefish. However, whitefish caught in Bear Lake are very good eating. Bonneville Whitefish are fish eaters and can be caught while jigging at the same places anglers target Lake Trout or Cutthroat Trout (see areas above). Harvest of whitefish is most common during the ice fishery. If you want to target whitefish, they have smaller mouths, so using slightly smaller gear increases the chances of hooking them.

Karen Teuscher with a Lake Trout. Bear Lake, Idaho. Photo by Dave Teuscher.

Panorama of Bear Lake, Idaho-Utah. Photo by John W. Sigler.

Casey Corrington with Bonneville Cutthroat Trout, Bear Lake, Idaho. Photo courtesy Dave Teuscher.

Big Wood River

Russell F. Thurow, Fisheries Research Biologist, US Forest Service: Rocky Mountain Research Station, Boise, ID
Fred E. Partridge, Regional Supervisor, Idaho Department of Fish and Game (retired), Jerome, ID

Location, Directions, and Access

US Highway 20 provides the primary access to southern portions of the Big Wood River immediately upstream from Magic Reservoir. Highway 75 provides primary access to northern portions of the Big Wood River through the towns of Bellevue, Hailey, and Ketchum and the headwaters near Galena Summit. Walk-in public access areas are found along the entire reach of the river between Bellevue and North Fork. For example, there are at least 13 public access park and walk-in sites along the river between Bellevue and Lake Creek. Contact the Blaine County Recreation District for more information and also visit http://www.bcrd.org/Assets/Wood-River-Trail-summer.pdf. Upstream from North Fork, the Big Wood River flows predominantly through public lands managed by the US Forest Service (USFS) so the public has full access except across isolated private inholdings. For information on trails and access to the abundant alpine lakes in the mountain ranges surrounding the Big Wood River, contact the USFS, Sawtooth National Forest, or visit their visitor center north of Ketchum. Driving times to Hailey are 2.2 hours from Boise; 4.1 hours from Salt Lake City; and 6.2 hours from Missoula, Ketchum is about 20 minutes north of Hailey. Hailey has direct air service, with major airports serving Boise and Salt Lake City.

Physical Characteristics and Attributes

Although most national and international trout anglers are familiar with Idaho's world-famous Silver Creek (addressed below in the Silver Creek section of this chapter), fewer are aware of the Big Wood River's exceptional wild trout fishing opportunities. The Big Wood River drains a large area encompassing more than 774,406 ha (2,990 mi^2) in south-central Idaho's Blaine County and feeds the aquifer that replenishes Silver Creek's upwelling springs. Other major Wood River drainage features include the Little Wood River and the Malad River, Silver Creek, and four earthen dams that store water in Magic, Little Wood, Fish Creek, and Mormon Reservoirs. Here we focus primarily on that portion of the Big Wood River upstream from Magic Reservoir, a river reach where anglers have the opportunity to fish for wild trout in a scenic valley bordered by multiple snowcapped peaks.

The Big Wood River drainage supports the largest area and most productive trout waters in south-central Idaho[5]. From its origin near the base of Galena Summit, the Big Wood River flows south-southwest for 93 km (58 mi) past the towns of Ketchum, Hailey, and Bellevue to its confluence with Magic Reservoir[6]. The upper Big Wood River is bordered by a series of scenic mountain ranges: to the north and east lie the Boulder and Pioneer Mountains, and to the west lie the Smoky Mountains. Principal upstream tributaries include the East and North Forks of the Big Wood River and Trail and Warm Springs Creeks. Many smaller streams are tributary to the Big Wood River, especially in headwater areas upstream from Ketchum. The upper Big Wood River and tributaries are free flowing except for a series of irrigation diversions. Substrates are primarily cobbles and gravels. Riparian habitats range from conifers in the headwaters transitioning to a mixture of willows, sage, cottonwoods, and river birches as the river flows toward Magic Reservoir.

Magic Dam, constructed in 1909, segregates trout populations above and below the dam. Downstream from Magic Dam, the Big Wood River flows south and west for 89 km (55 mi) and joins the Little Wood River near the town of Gooding, where it becomes the Malad River.

The Little Wood River originates in the Pioneer Mountains near 3,660 m (12,009 ft) Hyndman Peak and flows south-southwest for 145 km (90 mi) to its confluence with the Big Wood River. Silver Creek is a tributary to the Little Wood River and enters about 10 km (6 mi) downstream from Carey.

The Malad River flows southwest for 19 km (12 mi) before it enters the Snake River near Hagerman, Idaho. As an interesting historical note, the Malad River received its name in 1819 when Astoria fur trapper Donald Mackenzie camped along its banks[7]. It was reported that some of Mackenzie's men became sick and, believing the illness was caused by drinking water from the stream, named it "Malade," meaning "sick" or "bad" in French. Actually, the men had likely eaten flesh of beaver that fed on the poisonous roots of water hemlock. The beaver were immune to the poison, but the trappers suffered from their feast. Native tribes avoided this outcome by altering food preparation methods to include boiling, which apparently deactivated the poison[8].

Early inhabitants were Shoshoni and Bannock tribal members who hunted and fished in the area. In 1824, Alexander Ross and his expedition were likely the first Europeans to arrive in the Wood River Valley[9]. Ross was followed by fur trappers, miners, and ranchers who settled in the area.

Fish Species

The fish fauna of the Big Wood River reflect the geological isolation of the drainage from the Snake River by ancient lava flows[10]. Hubbs and Miller[11] had previously described the drainage as exhibiting partial isolation and disruption with faunal peculiarities. Wood River Sculpin are endemic and a State of Idaho Sensitive Species. Northern Leatherside Chub, a Lake Bonneville fauna, are present in Idaho only in the Wood River, Goose Creek, Raft River, and South Fork Snake River drainages[12]. Bridgelip Sucker found in the Wood River represent an isolated population that is "nearly 100 percent distinct" from other Bridgelip Sucker populations[13]. Smith reported that the distinctness of the Wood River Bridgelip Sucker population suggests it may be a unique species. Behnke believed Redband Trout were the indigenous trout in the drainage and suggested specimens represent an older, relict form of Redband Trout[14]. Behnke reported the presence of unique morphological characteristics in Wood River Redband Trout he examined[15].

Native fish fauna are represented by four families (Catostomidae, Cottidae, Cyprinidae, and Salmonidae), seven genera, and nine species[10]. Thurow lists Largescale Sucker and Bridgelip Sucker, Wood River Sculpin, Speckled Dace and Longnose Dace, Redside Shiner, Utah Chub and Northern Leatherside Chub, Mountain Whitefish, and Redband Trout[16].

Several introduced species are present in the basin. Nonnative Rainbow Trout as well as Brook Trout and Brown Trout are present. Although Redband Trout were native to the Big Wood River drainage, decades of stocking nonnative coastal-origin Rainbow Trout have resulted in genetic introgression of the native Redband Trout with introduced Rainbow Trout[17,18]. Current stocking of sterile hatchery Rainbow Trout occurs only in the uppermost reaches of the Big Wood where productivity is low and regulations allow more liberal stocked trout harvest. Brook Trout are most abundant in the tributaries and uncommon in the river. Brown Trout are most abundant in the lower Big Wood River below Bellevue. Mature Brown Trout migrate upstream from Magic Reservoir to spawn in lower reaches of the river. Lower reaches of the Little Wood River downstream from the confluence of Silver Creek also support a robust Brown Trout population.

Magic Reservoir supports populations of Yellow Perch and Smallmouth Bass. Other nonnative salmonids include exotic forms of Rainbow, Cutthroat, and Golden Trout and Arctic Grayling that have been introduced to formerly fishless, high-elevation headwater alpine lakes in the Boulder, Pioneer, and Smoky Mountain ranges.

Historical Angling

Historically, the Big Wood River was recognized as one of the premier wild trout fisheries in Idaho. Photographs in the Idaho State Historical Society archives document catches of native Redband Trout exceeding 5.4 kg (11 lb, 15 oz). Early trappers and explorers alternately called the river the Malade (see above) or the "Big Woody," likely as a result of the abundant large woody debris in the stream channel.

Anglers who had fished the Big Wood River since the 1940s were interviewed, and they described the pre-1960s river channel as being stuffed with a series of extensive log jams that often spanned the river channel[10]. Anglers reported a deep and narrow stream channel with debris jams that sorted spawning gravels and created pools that held the largest trout. Anecdotal accounts also suggest native trout formerly migrated very long distances within the drainage. It is likely juvenile trout reared in headwater stream reaches before migrating to more productive lower-elevation stream reaches with more favorable overwintering conditions. Prior to spawning, mature trout migrated back upstream to spawn. Although the construction of Magic Dam in 1909 blocked these extended migrations, it has been confirmed that, when possible, trout continued to migrate upstream from Magic Reservoir and lower reaches of the Big Wood River[19]. In the early 1960s, the IDFG treated Magic Reservoir with rotenone to remove introduced Utah Chub, and the project likely eliminated many of the large migratory Redband Trout in the lower river. However, by the 1980s, Thurow had confirmed that Redband Trout continued to migrate upstream from Magic Reservoir and lower reaches of the Big Wood River[10,20–22].

Established water rights in the Big Wood River drainage extend to the 1880s. Operation of the Baseline Bypass irrigation canal effectively blocks upstream migrations when stream flows are low and completely diverted into the canal. Hauck[19] and Thurow[20,21] reported that diverting water out of the river for irrigation adversely affected the fish population in four ways: fish enter unscreened diversions; the Baseline Canal dewaters a 6 km (4 mi) reach of river, killing all fish present; the Baseline Canal blocks fish migrations; and fish habitat is degraded by heavy equipment used to divert water.

Also beginning in the late 1800s, the Big Wood River drainage suffered from poor watershed practices and human developments that altered trout habitats and ultimately led to flooding conflicts in the 1950s. The conflicts led to excessive stream alterations in an attempt to protect infrastructures and property. Thurow[16] listed the most significant alterations as channel relocation, diking, clearance of large woody debris, rip-rapping, and removal of riparian vegetation. Most of these activities have been associated with attempts to control flooding, development of homes in floodplain areas, and road construction. Alterations of the stream channel and riparian habitat have adversely affected stream hydraulics, nutrient pathways, invertebrate production, and fish production. As a result of these habitat alterations, fish populations declined in the Big Wood River basin. Irizarry[23] reported trout densities in altered stream reaches were one-tenth of those in unaltered reaches. Thurow[21] reported eight to 10 times more trout in stream reaches with natural cover compared to altered reaches. Despite passage of Idaho stream-protection legislation in the 1970s, and improvements in streamside management practices in some areas, poor watershed practices continue, especially in response to high spring-flow events.

As trout populations declined, angling effort simultaneously increased; from 1960 to 1980, the population of the upper Big Wood River valley increased by 123 percent[24]. In response to declining trout numbers and increasing angler effort, hatchery-reared Rainbow Trout were stocked in areas of high angler effort to supplement the catch and buffer the wild trout fishery. In the 1980s, a fisheries research project was conducted to assess the status of the trout population, identify limiting factors, and develop recommendations for recovery. Research results documented trout growth rates comparable to those in the highly productive Henrys Fork Snake River and Silver Creek. Trout densities exceeded 1,000 fish per 1.6 km (1 mi), and more than 20 percent of the trout exceeded 304 mm (12.0 in)[10]. Research results were applied to phase out hatchery Rainbow Trout stocking in the most productive portions of the Big Wood River and develop angling regulations designed to improve trout abundance and size while maintaining diverse angler opportunities. With the implementation of restrictive harvest rules in the early 1990s, along with improved habitat protections, trout densities increased to more than 2,000 fish per 1.6 km (1 mi), and up to 26 percent of the trout exceeded 304 mm (12.0 in) in recent surveys[25,26].

Current Regulations

Current angling regulations provide diverse opportunities for fishing the Big Wood River. Three categories of fishing areas include those open to all angling gear (bait, lures, and flies) where some fish may be harvested, only lures and flies and only trout of certain lengths may be harvested, and only lures and flies and all trout must be released unharmed. Regulations allow summer and winter angling opportunities, and many consider the Big Wood River one of the finest winter trout angling destinations in Idaho.

There are 16 alpine lakes in the Wood River drainage that support populations of introduced trout and Arctic Grayling[6]. Rainbow and Cutthroat Trout populations are maintained by aerially stocking fingerling trout into the lakes every three years. Except for Baker Lake, which is managed for trophy trout, all have general fishing regulations. Access to these backcountry lakes varies from short hikes of less than 1.6 km (1 mi) to strenuous multiple-kilometer hikes and overnight trips.

Tackle and Techniques

After spring flows subside in late June, the Big Wood River and its tributaries may be waded and fished effectively. Fly, lure, and bait angling techniques are all effective for catching wild trout. Experienced bait anglers use a variety of natural baits, including worms, grasshoppers, crickets, and even cicadas to drift bait through areas where trout stage and feed. Most lure anglers use light spinning tackle and gold or silver spinners (size 1 or 2) with all but one hook removed and the barb on the remaining hook crimped. Lure anglers typically move their lures through likely holding waters by casting perpendicular to the current or quartering downstream and then retrieving the lure upstream against the current.

Most fly anglers use a rod approximately 2.4 to 2.7 m long (8 to 9 ft) with a 4- to 6-weight floating line, a tapered 3 to 4 m (9 to 13 ft) leader, and 7X (2 lb) to 3X (8 lb) tippet, depending on fly size. Anglers use dry flies, nymphs, and wet flies or streamers. Dry-fly anglers sight-cast to either rising trout or fish likely holding water along banks, near overhead cover, and in deeper pools and pockets. Fly anglers use a classic upstream presentation as well as a down-and-across presentation to avoid casting the fly directly over a feeding fish's head. Many anglers fish a dry fly with a small- to medium-size nymph suspended below the dry fly. Other anglers fish nymphs with a strike indicator usually positioned 1 to 2 m (3 to 6 ft) above the fly. In most situations, the nymph or leader is weighted. Often, a second nymph is fished as a dropper, about 0.5 to 1 m (1.5 to 3 ft) below the upper nymph. Upstream and across-stream presentations are most common. Streamer fishing can also be effective. An effective streamer technique is to cast down and across the current; allow the fly to sink; mend the line, enabling the fly to swing across the current; and then retrieve the fly by stripping the line.

Fly, lure, and bait angling techniques are also effective for catching trout in alpine lakes. A common and effective method for fishing alpine lakes with flies is to use spinning tackle with a weighted bobber (cast-a-bubble) and a fly attached 1 to 2 m (3 to 6 ft) behind the bubble with a clear leader. This approach helps avoid the need for space to accommodate a fly rod back-cast and allows anglers to cast farther.

Hatches and Fly Patterns

The Big Wood River has abundant insect hatches that provide year-round fly angling opportunities. Fly patterns primarily imitate three major aquatic insects: mayflies (Ephemeroptera), stoneflies (Plecoptera), and caddisflies (Trichoptera). Effective dry-fly mayfly patterns include the blue winged olive, pale morning dun, mahogany dun, green drake, and red quill. Effective surface stonefly patterns include the golden stone and yellow sally. Effective caddisfly imitations include the Elk-Hair Caddis, Stimulator, and Humpy. A host of terrestrial insect patterns are also effective. These include a variety of grasshopper patterns in hook sizes from 6 to 12, beetle patterns, and ant imitations. Nymph patterns including the Hare's Ear, Pheasant Tail, and Prince; imitations of larger stonefly nymphs are also effective. Streamer patterns that imitate sculpin, dace, and juvenile suckers can be effective for larger trout. During the winter, midge patterns may be very effective for both trout and whitefish.

Fly patterns that imitate mayflies, stoneflies, and caddisflies are also effective in alpine lakes. Streamers, Woolly Buggers, beetle, and black ant patterns may be effective when aquatic insect hatches are absent.

Other Tips

As described above, public fishing access within the Big Wood River basin is extensive, and most anglers are able to fish the river and its alpine lakes without the services of a guide. However, there are numerous local tackle shops and licensed outfitters and guides who have specific knowledge of the basin. Anglers seeking qualified guide services are encouraged to contact the Idaho Outfitters and Guides Association, PO Box 95, Boise, ID 83701; Office: (208) 342-1438; or visit http://ioga.org. IOGA also provides information on air transportation and trip shuttles; accommodations and dining; gear manufacturers, distributors, and suppliers; retail sports equipment and clothing rental or sales; and travel agents, tours, and tourist information.

Closing Thoughts

Although the trout population in the Big Wood River Basin has declined from its exceptional historical (pre-1950s) levels, it remains one of the premier wild trout fisheries in Idaho. Anglers on the Big Wood River continue to experience the thrill of catching large wild trout within scenic stream reaches lined with cottonwood trees and framed by snowcapped peaks. The future of the Big Wood River trout population will be dependent on efforts to successfully maintain and restore a foundation of diverse and high-quality in-stream and riparian habitats. These tasks will require broad public support, collection of data to fill information gaps, and enactment of protective legislation.

Wood River Rainbow. Photo by Susanne Connor.

Wood River Angler. Photo by Terry Ring.

CJ Strike Reservoir

Joe Kozfkay, Regional Fishery Manager, Southwest Region, Idaho Fish and Game

Location, Directions, and Access

CJ Strike Reservoir is located on the border of Elmore and Owyhee Counties in a rural portion of southwest Idaho. The surrounding uplands are mostly arid rangeland with some irrigated cropland. The closest communities are Grandview and Bruneau, where services such as restaurants, gas stations, and a few small tackle stores may be found. Lodging in these towns is limited. The closest lodging may be found in Mountain Home, about 32 km (20 mi) to the north. Although anglers visit CJ Strike from far-off places, most anglers travel from Mountain Home, other small surrounding communities, or the Treasure Valley. These anglers utilize several travel routes depending on the portion of reservoir they intend to target, including Highway 78, Highway 51, and Simco Road–Highway 67.

Fishing and boating access points as well as camping areas are numerous and well developed. Sites are maintained by IDFG, IPC, and Bureau of Land Management (BLM). Check these agencies' respective websites for more information on amenities and specific locations. The Snake River Arm may be accessed at three locations. Two boat ramps and a dock are located at Loveridge Bridge and the Highway 51 crossing of the Snake Arm. A third is located near the Cove Arm/Crane Falls Reservoir complex. Anglers should be forewarned that the road into this site is not easily found without a map, and it is bumpy. Furthermore, this ramp is rudimentary and should be used only by small boats. The Main Pool may be accessed from the north or south shores via the Air Force, Black Sands, or Cove Recreation sites. Of these, the Air Force Ramp is the most popular due to recent upgrades, spacious boat ramps, numerous docks, a fish-cleaning station, and nearby camping. The Bruneau Pool may be accessed from its south shore only at the Jacks Creek, Cottonwood, or Emigrant access sites, with Cottonwood being the most popular.

Physical Characteristics, Attributes, and History

CJ Strike is best described as four connected but drastically different sections due to the underlying topography, hydrology of inflowing rivers, and shape. The Main Pool, the area immediately upstream of the dam, is the most lake-like. This "pool" is approximately 6.4 km (4 mi) long by 2.4 km (1.5 mi) wide, with depths up to and exceeding 46 m (150 ft). Rocky points and mid-depth flats are found along its perimeter. This section reminds one of being on a large natural lake, especially when the winds are strong and out of the west. Upstream and to the north, the Snake Arm is heavily influenced by the Snake River and its changing flows. For most of its 26 km (16 mi) length, the Snake Arm is narrow, 0.4 to 1.2 km (0.25 to 0.75 mi) wide, and deep, with shorelines composed almost entirely of lava rock and boulders. In contrast, the extreme upstream end of the Snake Arm looks and acts differently from the rest of the section. Here, where flowing and flat waters collide and sediment and nutrients settle out, the Snake looks and acts like a giant freshwater marsh or delta. Water birds, cattails, and tules are abundant along with some submerged aquatic plant beds. Back downstream and east of the Main Pool sits the Narrows. Aptly named, the Narrows are a 3.2 km (2 mi) long by 0.2 km (.125 mi) wide channel that separates the Main and Bruneau Pools. Shorelines are steep and very rocky. Water depths in this section bottom out at about 15 m (50 ft). Farther east sits the Bruneau Pool. The Bruneau Pool is, 1.6 to 2.4 km (1 to 1.5 mi) wide, 5.6 km (3.5 mi) long, and relatively shallow. The deepest waters in this pool, about 12 m (40 ft), are found on the western side. The pool shallows to the east and becomes only a meter or so (couple of feet) deep and marsh-like on the easternmost edge where the Bruneau River and other small tributaries enter the pool. Inflows from the Bruneau River are much less than from the Snake, so water travel or retention time is much longer. Because of this and due to the depth, the Bruneau Pool warms more quickly and produces more phytoplankton and zooplankton—the bases of the aquatic food web. Last, because of the Bruneau Pool's shape, depth, and sediments, high winds cause bottom sediments to resuspend more easily, so the Bruneau Pool is usually more turbid than the rest of the reservoir.

CJ Strike Reservoir, commonly referred to as Strike by locals, is a tremendously popular and productive water body. The history of CJ Strike Dam and reservoir is a short one. CJ Strike just recently turned 65 years old. The intent of the project was, and is, power production. Because of this, IPC selected a high-gradient section of the Snake River and found a confined point in the valley. Here, they built the dam, with construction occurring from 1950 to 1952. When the dam closed and the 3,035 ha (7,500 ac) reservoir filled, about 32 km (20 mi) of the Snake River and 13 km (8 mi) of the Bruneau River were inundated. Due to the nature of operations and focus on power production, CJ Strike, unlike most other reservoirs, is held at essentially full-pool levels on a year-round basis.

CJ Strike Reservoir Fishery

CJ Strike Reservoir has very simple fishing rules. The reservoir is open all year. Anglers may harvest up to six trout of any size or six bass, none under 305 mm (12.0 in). White Sturgeon may not be retained at any time, nor may they be removed from the water. There are no bag or length limits on all other species likely to be caught by anglers.

Prior to dam construction, these portions of the Snake and Bruneau Rivers supported about a dozen native species, including members of the salmonid, minnow, and sucker families as well as White Sturgeon, and they acted as a migratory corridor for anadromous salmon and Steelhead. Most of these species are still present. Only the salmon and Steelhead have been extirpated. After reservoir formation, several species were introduced to increase and diversify fishing opportunities. The reservoir now contains at least 18 different species, including 12 game fishes. Game fishes include Black Crappie, Bluegill, Brown Bullhead, Channel Catfish, Largemouth Bass, Pumpkinseed, Rainbow Trout (hatchery), Smallmouth Bass, Warmouth, White Crappie, White Sturgeon, and Yellow Perch. Nongame fish present include Bridgelip Sucker, Chiselmouth, Common Carp, Largescale Sucker, Northern Pikeminnow, and Peamouth. Though not listed, a Walleye was caught by an angler during 2013, and an additional Walleye was collected by IDFG in the Snake River upstream of the reservoir during 2014. It is not known whether Walleye have established a self-sustaining population in the reservoir, but the following two instances are concerning. As of 2017, CJ Strike had produced two of the current state records. A 1.6 kg (3 lb, 8 oz) Bluegill was caught in 1966, and a 30.1 kg (66 lb, 5 oz) Common Carp was harvested by a pair of archers during 2011. Archery is a legal method of take only for unprotected nongame fish.

Largemouth and Smallmouth Bass

Bass are the most sought-after species in CJ Strike and were first introduced shortly after reservoir completion. Largemouth Bass were the dominant species in the system early on. Early reports from anglers and biologists alike indicated that the number of Largemouth Bass was high and that large individuals were present. Few, if any, Smallmouth Bass were present at this time. Fry and fingerling Smallmouth Bass were introduced in the mid-1970s and exploded in numbers. Today, Smallmouth outnumber Largemouth by at least 20 to 1. Although there is some overlap in habitat use, Largemouth Bass are more often found near some form of overhead cover such as cattails, tules, or Russian olive trees. Fishing tight to cover with soft plastics is a popular method for catching Largemouth Bass. Some trophy-size Largemouth Bass are still present. IDFG staff have collected more than a couple Largemouth Bass over 3.6 kg (8 lb) during recent fish survey efforts. Surprisingly, a high percentage of them had a tail of a 203 mm (8.0 in) or longer Rainbow Trout sticking out of their gullet. Don't be afraid to fish big baits if you have the desire to catch a trophy Largemouth Bass. Smallmouth Bass, on the other hand, are more often found farther offshore and adjacent to cobbles, boulders, and drop-offs. Smallmouth Bass, and other predators, have tremendously diverse and abundant food sources in Strike, including crayfish, Yellow Perch, crappie, and many other fish species. Despite these abundances, Strike's Smallmouth Bass grow slowly. Often a 305 mm (12.0 in) Smallmouth Bass is five or six years old, and Smallmouth Bass over 406 mm (16.0 in) usually exceed 10 years old. Just about anywhere in the reservoir, tube jigs, crankbaits, soft-plastic grubs, and top water plugs designed to mimic available forage are readily engulfed by eager Smallmouth Bass. Most Smallmouth Bass caught by anglers are in the 254 to 356 mm (10.0 to 14.0 in) range. However, anglers that fish early in the season or learn the offshore haunts of the bigger adults are rewarded by bass more than 406 mm (16.0 in) with an outside chance at a trophy in the 508 to 559 mm (20.0 to 22.0 in) and 2.3 to 2.7 kg (5.1 to 6.0 lb) range.

Black Crappie, White Crappie, and Yellow Perch

Black Crappie, White Crappie, and Yellow Perch, and to a lesser extent Bluegill and Pumpkinseed (i.e., panfish), draw anglers to Strike from far and wide. This is especially true when one of these species booms. During 2009–10, Black and White Crappie populations were booming, and word spread quickly. Since then, crappie production has dipped due to poor larval production or survival. In the interim, a strong Yellow Perch cohort was produced. These fish commonly reach 279 to 356 mm (11.0 to 14.0 in) and seem to provide much of the harvest opportunity in years when crappie abundance is down. Such is the nature of panfish in CJ Strike, with variable production and dramatic shifts in relative abundance. Regardless, anglers focus their panfishing efforts in the Narrows as well as on points and flats and along old channel breaks in the Main and Bruneau Pools. Jigs tipped with small chunks of garden worms catch most of the fish. Depending on the time of the year, jigs may be presented vertically, cast and retrieved, or fished under a slip bobber to catch suspended fish.

Rainbow Trout

Anglers who prefer cold-water species have the option of pursuing Rainbow Trout. This popular sportfish is stocked by IPC and IDFG to increase angling opportunity. Rainbow Trout feed on abundances of zooplankton, macro-invertebrates, and small fish, allowing them to grow at the remarkable rate of 25 mm (1.0 in) a month. Trout fishing is best during fall, winter, and early spring. Anglers fish bottom rigs baited with worms, power bait, or marshmallows from shore or troll small plugs, spoons, or spinners from boats. Catch rates often exceed a trout an hour, and it can be the hottest bite in the area during winter.

White Sturgeon

If trout and warm-water fish don't interest you, Idaho's largest fish, the White Sturgeon, can be found and caught in the Snake Arm and Main Pool. This is one of Idaho's strongest populations of White Sturgeon, with a population estimate of about 4,000 White Sturgeon greater than 0.9 m (3 ft). Sturgeon that frequent the reservoir make a long-distance migration into the Snake River to spawn in the Canyon Reach, near Bliss. After spawning, a portion of the adults seem to prefer spending time in the reservoir. Other than serving as habitat for adult sturgeon, the reservoir acts as a nursery for juvenile sturgeon. After hatching, larval sturgeon drift with the river's currents. After drifting and developing for several days, many settle out in the upper Snake River Arm. This area acts as a nursery, and the young sturgeon grow very quickly on a diet of chironomids (i.e., midge larvae) and other invertebrates. As they grow larger, they gradually disperse, with some electing to reside in the river, whereas others remain in the reservoir for future spawning migrations. Anglers targeting White Sturgeon generally pursue them from boats and drop large chunks of cutbait into deep holes, especially in the Snake River Arm. This is a tight-lipped angling group, so expect to do some exploring and spend a lot of time looking at a depth finder if you want to battle one of these living dinosaurs. No matter what your angling preference, it's hard not to find something to like and fish for in Strike.

Closing Thoughts

Anglers commonly report having large Yellow Perch engulf spinnerbaits, Channel Catfish eat rubber worms, and dinosaur-like White Sturgeon breach right next to the boat for no apparent reason. Expect the unexpected when fishing here.

CJ Strike Panorama and Dock Facilities near the Airforce Ramp facilities. Photo Courtesy IDFG.

Happy Anglers, CJ Strike Reservoir. Photo courtesy IDFG.

Successful Day, CJ Strike Reservoir. Photo courtesy IDFG.

The Smile Says It All, Happy Angler, CJ Strike Reservoir. Photo courtesy IDFG.

Clearwater River

Brett Bowersox, Fisheries Staff Biologist, Idaho Department of Fish and Game

Location, Directions, and Access

The heart of Idaho has three chambers, the Snake, the Salmon, and the Clearwater. The Clearwater River is the northernmost of these three iconic Idaho rivers. The Clearwater is also the smallest of the three but still sizable, with a drainage area of 24,980 km^2 (over 9,644 mi^2). The eastern end of the Clearwater is in the Bitterroot Mountains on the Montana state line and stretches all the way to the Washington state line, where it joins the Snake in Lewiston, Idaho, 224 km (140 mi) upstream of the Snake River confluence with the Columbia River. The drainage is bordered by the Palouse River and Saint Joe River to the north and the Salmon River to the south. The Clearwater River was the east-to-west travel route that carried Lewis and Clark toward the Pacific Ocean during the Voyage of Discovery[27]. This route continues to be the major artery through the lower portion of the Idaho Panhandle, with US Highway 12 following the Clearwater Drainage from Lolo Pass in the east to Lewiston in the west following the main-stem Clearwater River or its upper tributaries across the entire width of the state.

For main-stem Clearwater River access, US Highway 12 runs along the entire main stem of the Clearwater River and provides ample access to shore anglers. In addition, there are 17 improved boat ramps along US Highway 12 between Lewiston and Kooskia. This does not include a number of unimproved ramps that are still usable by drift-boat anglers. Just another wonderful aspect of the main stem Clearwater River: plenty of access points. For anglers looking for guided opportunities, a variety of options exist for guided trips on the main stem Clearwater River.

Diversity

Like the state of Idaho itself, the Clearwater River provides a diversity of fishing opportunities and fishing environments. The drainage and its tributaries begin in the Bitterroot Mountains surrounded by dense north Idaho forest and end as they enter the Snake River in dramatic steep-sided basalt canyonlands. Whether you enjoy watching the methodical ticking of a back-trolled plug rod waiting for that gracefully violent take of a fresh Steelhead or the ambling rise of a Westslope Cutthroat Trout emerging from an emerald-green abyss, the Clearwater River provides numerous world-class fishing opportunities.

The Tributaries

The crown jewel, or gem, as the case may be, of the Clearwater River tributary fishing involves the Idaho state fish. The main tributaries to the Clearwater River boast some of the finest Westslope Cutthroat Trout fishing in the world. Westslope Cutthroat Trout are gorgeous fish and just so happen to occupy equally gorgeous river reaches. Fishing is open year-round in most sections, with varying regulations with regards to harvest, so check prior to heading out fishing. Anglers target Westslope Cutthroat Trout using a variety of techniques, but many chose fly-fishing with dry flies. Westslope Cutthroat Trout are notorious for their willingness to take a dry fly; the fact that they aggressively rise to large dry flies is even better. Dry-fly fishing starts to heat up as the tributaries start to come down from their annual transport of snowmelt out of the high country, typically in mid- to late June.

Cutthroat Trout don't require an early wake-up call, as fishing often picks up as the sun warms the water. Fishing the heads of pools and runs where the water is deep enough to pick up a bit of green color is ideal for Westslope Cutthroat Trout. Fishing remains excellent throughout the summer and well into the fall. The fall season in the tributaries is especially coveted by anglers since the big October caddis (*Dicosmoecus* sp.) become plentiful and provide what is arguably the prime time to fish for Westslope Cutthroat Trout. Large dry flies such as the Stimulator, Sofa Pillow, Madame X, Foam Hopper, and variations on these themes are all productive flies. For those fly fishermen looking for some variety, nymphs and streamers can be very effective techniques for Westslope Cutthroat Trout as well. Streamers swung downstream across deep runs, pools, and pocket-water reaches often produce aggressive strikes by larger 400 mm (15.7 in) plus-size Westslope Cutthroat Trout. This style of fishing can be so enjoyable that you may forget your dry-fly box in the truck every once in a while. If fly-fishing isn't your thing, spinners and spoons cast within these pools and runs will also produce plenty of fish for anglers. Finally, using a spinning rod and clear bobber with your favorite dry fly attached is a great way to help young kids catch their first Westslope Cutthroat Trout.

For those of you who have spent some time in these tributaries to the Clearwater River fishing for Westslope Cutthroat Trout, it's likely you've been unexpectedly introduced to another inhabitant of the Clearwater River, the Bull Trout. These aggressive characters have a propensity to latch on to unsuspecting Westslope Cutthroat Trout that an even more unsuspecting angler is trying to land. The end result leaves the angler wondering what just happened and how a fish that big could be lurking down in the depths. Bull Trout are found throughout the Clearwater River drainage, but anyone wanting to target them should check out the North Fork of the Clearwater. Since the completion of Dworshak Dam and Reservoir in 1973, an adfluvial life history has emerged in Bull Trout that has boosted the size of Bull Trout occupying the drainage. Bull Trout in excess of 510 mm (20.1 in) are common. As the occurrence of Bull Trout grabbing smaller trout hooked by anglers would suggest, Bull Trout are piscivorous, meaning they like to eat other fishes. So if you plan on targeting Bull Trout, bring big streamers, spoons, or plugs such as Rapalas to try to catch them. Bull Trout spawning typically occurs in the month of September within the North Fork Clearwater River; however, they leave Dworshak Reservoir to start their upstream spawning migration in early summer. They can be caught throughout the main-stem river, but remember that as summer water temperatures increase, Bull Trout move into higher-elevation reaches within the main-stem river and tributaries. Also, remember that while Bull Trout may be targeted, they are closed to all harvest given their federal Endangered Species Act listed status.

Tributary Access

The Clearwater is blessed with three world-class Westslope Cutthroat Trout fishing tributaries, the Lochsa, Selway, and North Fork Clearwater Rivers. Each one is unique in its aesthetics and access. The Lochsa River and Selway River join at the town of Lowell to form the Middle Fork Clearwater River. The Lochsa River can be accessed along its entire course by US Highway 12. Numerous developed and undeveloped camping and fishing access sites are available to the angler. The route is breathtaking, and the 216 km (134 mi) route from Kooskia, Idaho, to Lolo, Montana, is a designated Scenic Byway, so please pull over and enjoy while scouting for an equally breathtaking fishing spot. The Selway River, on the other hand, can be accessed by road in two places, about 30 km (17 mi) from its mouth upstream to the edge of the Selway-Bitterroot Wilderness and along the Magruder Corridor Road in the upper headwaters of the drainage. The Magruder Corridor Road represents a unique and inspiring Idaho adventure in and of itself as the road snakes its way along the border ridge between the Clearwater River and Salmon River drainages. Taking this route, one should consider a few things prior to departure: how good is my spare tire, where is my camera, and how long do I want to stay in Paradise? Paradise is not only an appropriate descriptive term for this area but also the name of the upper access point to the Selway River. Paradise includes a USFS guard station and raft launch, a useful site, since drawing a coveted Selway River float permit is the best way to experience this wilderness area. However, anglers will find that even without a float permit, a backpack and good set of boots will provide a world of fishing opportunity. Moving downstream, the North Fork Clearwater River enters the main-stem Clearwater River just downstream of the town of Orofino. Dworshak Dam cut off the North Fork Clearwater River drainage from the ocean 3.1 km (2 mi) above its confluence with the main-stem Clearwater River. Main access to the North Fork Clearwater River from the west is through Orofino or Greer off US Highway 12. State Route 7 from Orofino and State Route 11 from Greer connect to Forest Service Road 247 that follows the North Fork Clearwater River all the way to Kelly Creek, its most famous tributary. At Kelly Creek, Forest Service Road 250 continues along the North Fork Clearwater River in the Black Canyon section and all the way out to Hoodoo Pass, which is the access point to the drainage from the east coming in from Montana. Numerous developed and undeveloped camping opportunities are available along the length of the North Fork Clearwater River.

Steelhead and Salmon Fisheries

The main-stem Clearwater River is a Steelhead- and salmon-fishing mecca. It is an Idaho anadromous gem in every sense for every type of Steelhead and salmon angler. However, the keystone of Clearwater River anadromous fisheries is the world-famous B-run Steelhead. Anglers travel from across the globe to target these brutes (sometimes exceeding 1,020 mm (40.2 in) in the classic Steelhead runs and gravel tailouts that make the Clearwater River famous. As a general rule, the B-run designation refers to upper Clearwater River wild fish and Dworshak National Fish Hatchery fish that spend another year in the ocean (2 to 3 years) and migrate back into the freshwater a month later than fish returning to the Snake River. These fish have evolved over time to endure the long migration to and from the ocean to maximize the chance of perpetuating by being *BIG*. It's an arduous journey, and the granitic geology of the Idaho batholith is not known for being very productive, so if you are going to go to the ocean and come back to spawn, make sure you are big so you have a lot of eggs to make offspring. The Clearwater River doesn't contain the only B-run populations in Idaho (they are also in the Middle Fork and South Fork Salmon Rivers), but they are the most famous and sought after.

The main-stem Clearwater is an angler's dream for targeting Steelhead. No single fishing technique is consistently more productive than others; rather, it's a matter of personal preference and the type and conditions of the water you are fishing that can dictate which technique to use. Drift fishing from shore and side-drifting from a boat have become popular techniques for good reason. These methods consist of using the current to slowly bounce baits near the bottom, where Steelhead tend to lie. Typically, anglers use a piece of pencil lead attached to a leadered hook tipped with their favorite combination of colored yarn, corkie, and bait and wait for that magic moment when something feels different in the drift. Rear back and let the games begin. Backtrolling, or "pulling plugs" as it's often referred to, is also popular and productive. This method typically requires a boat and consists of slowly backing plugs downstream through productive Steelhead water, waiting for a fish to become incensed and strike at the shiny rattling intruder. Popular plugs are Wiggle Warts, Hot Shots, and Kwikfish, with size and color depending on water conditions. Bobber-and-jig fishing is also very productive in the slower runs and pools. This method can work well from boats but is most often used from shore and is an excellent method for wading anglers. Most bobber-and-jig anglers tip their jigs with bait such as shrimp or add scent (or both) to help increase catch rates. Fly-fishing is also a tremendous technique for pursuing Clearwater River Steelhead. Late summer and fall are the peak time periods for targeting Steelhead on the fly. Anglers targeting the lower main-stem Clearwater River have predominantly moved toward Spey-style rods, given the size of the water. While a Spey rod is not a must, it makes covering water a lot easier in the lower Clearwater River. Most anglers choose to swing flies on floating or sink-tip lines during this time frame, although nymphing can also be productive. Popular fly patterns include classics such as the Green Butt Skunk, Purple Peril, and Freight Train as well as new patterns that include tube flies, Intruder patterns, and various Spey-pattern flies.

In addition to the main-stem Clearwater River, another sizable tributary, the South Fork Clearwater River, has become a very popular late-winter and springtime Steelhead destination for many anglers. Techniques outlined for the main-stem Clearwater River are also productive in the South Fork Clearwater River, minus techniques from a boat. Egg pattern flies and beads as well as drifting roe are very popular techniques in this later fishery. Fishing remains excellent in the South Fork Clearwater River until seasons close, and then it's time to gear up for spring Chinook Salmon.

Spring Chinook Salmon (a.k.a. Springers) were extirpated from the Clearwater River with the construction of Lewiston Dam in 1927. However, through a variety of efforts in the 1960s and 1970s, spring Chinook Salmon have been reestablished, with both hatchery and natural fish occupying the Clearwater River today. Annually, this fishery varies based on the size of the run coming back to the Clearwater River and its hatcheries. Season dates and harvest rules will change each year, so keep up to date on regulations as the season is set. As a general rule, fishing techniques that work for Steelhead can be sized up to work for Springers: bigger jigs under your bobber and bigger plugs trolled behind the boat for Springers. It's also good to have heavier-action rods and heavier-pound test line for doing battle with these aggressive critters in the high springtime river flow. Bringing a Springer to hand is truly a dietary celebration, as they are arguably the best table fare swimming around in Idaho waters.

In addition to Springers, fall Chinook Salmon, summer Chinook Salmon, and Coho Salmon are also found within the Clearwater River drainage. It is possible in the fall of the year to hit a Clearwater River triple and capture a Steelhead, fall Chinook Salmon, and Coho Salmon all in the same day—truly a unique opportunity in Idaho and almost anywhere you can cast a line. Given the diverse life history and run timing for these species, the angler is left with nearly year-round anadromous fishing on the Clearwater River. This leaves the true diehard Steelhead and salmon angler with only a few weeks to retie leaders, clean the roe stains off the inside of the boat, patch the leaks in their waders, and oil up the reels to get back at it when the next run comes in. What a wonderful problem to have!

Steelhead and Salmon Fishery Timing and Access

As a general rule, upstream migrating Steelhead start to enter the Clearwater River as early as July when the season begins and can be found in upper reaches of the river in April when the season closes. Adult Steelhead migrate through the fishery at different rates; recent studies by IDFG have shown that some fish move through the main-stem fishery to upstream reaches near hatchery release locations or springtime spawning grounds in a matter of weeks, and others remain in the lower river for months near Lewiston prior to continuing on their upstream migration. Early-season fishing centers on the lower river between Lewiston and the town of Orofino. Numerous boat ramps and shoreline access points are available to anglers along US

Highway 12 in this reach. Later in the season, the majority of angling pressure moves upstream, and anglers target fish returning to Dworshak National Fish Hatchery, located at the confluence of the main-stem Clearwater River and North Fork Clearwater River and higher on the main-stem Clearwater River between Orofino and the town of Kooskia. Main-stem river access continues along US Highway 12, and the North Fork Clearwater River below Dworshak Dam is accessed by taking State Highway 7 from Orofino to Ahsahka and driving up North Fork Drive along the northern side of the river. Late-season fishing is best on the South Fork Clearwater River, which is bordered by State Highway 14, and numerous shoreline access points are located along its course. The pattern of the fishery and access locations is similar for the spring Chinook fishery, though this fishery picks up when Steelhead fishing is coming to a close. Depending on the annual timing of the run, spring Chinook fishing in the Clearwater River starts up in the lower main stem in late April and early May near the town of Lewiston. Unlike Steelhead, which migrate through the main-stem river over the course of months, spring Chinook Salmon move quickly toward hatchery release sites and spawning areas, where they are collected at hatchery facilities or held in tributary reaches until spawning starts in late August and September. These fish move fast, so once anglers are catching spring Chinook Salmon in Lewiston, it's not long before fishing and catch rates increase in upstream fishing areas. Whether you are targeting Steelhead or spring Chinook Salmon, the Clearwater River provides ample access for both boat and shoreline anglers to enjoy these iconic Idaho fisheries.

For a main-stem Clearwater River access-point location reference, refer to "The Sportsman's Guide to the Clearwater River," available online at http://www.blm.gov/pgdata/etc/medialib/blm/id/recreation/rivers/clearwater.Par.0611.File.dat/Clearwater.pdf.

South Fork Clearwater River, Idaho, Sara Cassinelli Steelhead. Photo courtesy John Cassinelli.

Brian Knoth on the Clearwater River, Idaho. Photo by Brett Bowersox, courtesy of IDFG.

Ashore for a Rest, Winter on the Clearwater River, Idaho. Photo courtesy IDFG.

Coeur d'Alene Lake

Jim Fredericks, Chief of Fisheries, Idaho Department of Fish and Game

Location, Directions, and Access

Located in Kootenai County of the Idaho Panhandle and with State Highway 95 and Interstate 90 both passing briefly along its shores, Coeur d'Alene Lake is easily accessible at all times of the year. The city of Coeur d'Alene (population approximately 40,000) lies on the northern shore, at the outflow of the Spokane River. Coeur d'Alene is a popular vacation destination city that offers numerous services, including hotels, marinas, public boat ramps, restaurants, and a range of shopping opportunities. Elsewhere around the lake, there are several well-maintained public boat ramps as well as lakeside resorts and marinas.

The much smaller community of Harrison (population about 200) is located on the southeastern shore of the lake, near the mouth of the Coeur d'Alene River. Harrison also provides accommodations, a marina, and public boat ramp as well as a variety of services. Running along the eastern shoreline from Interstate 90 to Harrison, the Lake Coeur d'Alene Scenic Byway (Highway 97) offers a tremendous view of the lake and access to boat launches. Travelers should be aware, however, that the road winds its way along the shoreline, so it's not a route for those in a hurry.

Physical Characteristics and Attributes

Coeur d'Alene Lake is widely regarded as one of the most beautiful lakes in the western United States. At nearly 42 km (26 mi) long and generally 1.6 to 4.8 km (1 to 3 mi) wide, it is Idaho's second-largest natural lake (not including Bear Lake, which lies partly in Utah). In total, the lake is just over 12,545 ha (31,000 ac), with 217 km (135 mi) of shoreline. Looking for a moment beyond its beauty, Coeur d'Alene Lake, having produced several state-record fishes over the years, is one of the most popular and intriguing fisheries in Idaho. If one word can describe the Coeur d'Alene Lake fishery, it would have to be *diverse*. Whether an angler prefers warm-water or cold-water species, casting along the shore, or trolling the open water, there's something for everyone.

Coeur d'Alene Lake is not nearly as deep as its neighbor to the north, Lake Pend Oreille, with an average depth of about 31 m (100 ft) and a maximum depth just under 61 m (200 ft). Coeur d'Alene Lake is fed primarily by the Saint Joe River and the Coeur d'Alene River, both entering on the southern end of the lake. The lake then becomes the origin of the Spokane River, which flows westward into Washington from the northern end of the lake.

Though a natural lake, the elevation of Coeur d'Alene Lake is now controlled by Post Falls Dam for flood control and power generation. In the summer, the lake is held at an elevation of 648.6 m (2,128 ft) above sea level, about 1.8 m (6 ft) above the winter elevation. The summer pool elevation creates slack water in the lower Saint Joe and Coeur d'Alene Rivers and inundates a series of small lakes adjacent to the Coeur d'Alene River, collectively known as the "Chain Lakes." Most of these lakes are connected to the Coeur d'Alene River by boat channels, meaning anglers and other boaters are able to move freely between Coeur d'Alene Lake, the Coeur d'Alene River, and the Chain Lakes.

Aside from fishing opportunities, Coeur d'Alene Lake is well known among outdoor enthusiasts for its bird-watching opportunities—primarily raptors. The lake and associated slack-water habitats support the largest nesting population of Osprey (*Pandion haliaetus*) in the United States. Nests are located throughout the lake in trees, on utility poles, and on abandoned pilings. Osprey are easily seen throughout the spring, summer, and fall. From late November through early January, Bald Eagles (*Haliaeetus leucocephalus*) congregate by the dozens on the northern end of the lake, creating an immensely popular viewing and photographing opportunity. The eagles feast on spawned-out Kokanee for a few weeks prior to resuming their journey to wintering grounds farther south.

For those inclined to bike riding rather than fishing or bird-watching, there are numerous paved and unpaved bike trails all around Coeur d'Alene Lake. The Trail of the Hiawatha travels some 24 km (15 mi) through old railroad tunnels and across trestles high above numerous streams. This concessionaire-managed feature can be accessed from the Idaho Exit 0 on Interstate 90. The Trail of the Coeur d'Alenes provides more than 113 km (70 mi) of mostly paved gentle-gradient trail from Plummer near the south end of Coeur d'Alene Lake, along the lake, and up the Coeur d'Alene River. Other trails depart from the town of Coeur d'Alene and from various USFS trailheads throughout the area.

Coeur d'Alene Lake Fishery

Coeur d'Alene Lake and the adjoining waters are open all year, and fishing regulations are generally fairly simple, though rules do change from time to time, so it's important to consult the current rule book. Anglers should also be aware that while an Idaho fishing license is sufficient on the northern portion of the lake, the southern third (approximately) of the lake is owned and managed by the Coeur d'Alene Tribe, and a tribal fishing license is required.

Prior to the introduction of nonnative species, native game fish in Coeur d'Alene Lake consisted of Westslope Cutthroat Trout, Bull Trout, and Mountain Whitefish. With settlement of Idaho and the accompanying demand for additional food and sportfish, nonnative species began entering the picture as far back as the late 1800s. Initially, several spiny-ray and warm-water species were added to the system, including Largemouth Bass, Black Crappie, Yellow Perch, Pumpkinseed, and Brown and Black Bullhead, among others. All of these species are found throughout the main lake as well as the Chain Lakes. Much later, in the 1970s and 1980s, Northern Pike and Smallmouth Bass were illegally introduced. Both species have become firmly established throughout the system, and they now support popular fisheries, although that has come at the significant expense of other species.

Kokanee, which have become the cornerstone of the fishery, were first introduced in 1937. They quickly became a popular sportfish, supporting an annual harvest between a quarter- and a half-million fish through the 1970s. The high abundance of Kokanee in the late 1970s and the corresponding decrease in size led to the introduction of fall Chinook Salmon in 1982. The hope was that the aggressive predators would improve the Kokanee fishery by decreasing their abundance, while at the same time creating a trophy Chinook Salmon fishery. The program worked, and by 1987 a state-record fall Chinook Salmon of 19 kg (42 lb) was harvested. In the years since, the management strategy has been to strive for a balanced population of Chinook Salmon and Kokanee, with an abundance of Kokanee in the 229 to 279 mm (9.0 to 11.0 in) range, and a trophy Chinook Salmon fishery with catches of 5 to 9kg (10 to 20lb) fish being common. With occasional years of imbalance related to environmental conditions, the program has generally been very effective, and anglers interested in trolling can choose between fast-action Kokanee fishing, with liberal limits and high catch rates, or Chinook fishing, where anglers shouldn't expect multiple fish in an outing but have a chance at a real trophy.

For those anglers not so interested in trolling, the shallow vegetated bays and Chain Lakes offer excellent fishing for Largemouth Bass. The state-record Largemouth Bass, which was just under 5 kg (11.0 lb), was caught in Anderson Lake, one of the Chain Lakes. Largemouth Bass exceeding 2.7 kg (5 lb, 15 oz) are common in the system. In recent years, Smallmouth Bass have become established and are now widely dispersed throughout the system. Anglers are now able to find bass on a wide variety of shorelines around the lake. The abundance of both Largemouth and Smallmouth Bass, along with the size of the fish and the accessibility of abundant habitat, makes Coeur d'Alene and the Chain Lakes a favorite among tournament bass anglers. The Bassmasters organization has consistently ranked Coeur d'Alene Lake near the top of the it's list of the 100 best bass-fishing lakes in the country.

For the most part, where anglers find Largemouth Bass, they'll find Northern Pike as well. In terms of fish size, Coeur d'Alene Lake offers some of the best Northern Pike fishing in the country. Densities are not high, which means the aggressive predators haven't outrun their food supply, as often happens with Northern Pike. Instead growth rates in Coeur d'Alene Lake are among the best in the country. The previous state-record Northern Pike of 17.7 kg (39lb) was caught in Cougar Bay, on the north end of the lake, and fish exceeding 9 kg (20lb) are caught every year.

Kokanee, Chinook Salmon, Largemouth and Smallmouth Bass, and Northern Pike are the most sought-after species in the Coeur d'Alene Lake fishery, but other species attract angling attention as well. Panfish (primarily Yellow Perch, Bluegill, and Black Crappie) are found throughout the system, with greatest abundances in the Chain Lakes. While not as common as they once were, Westslope Cutthroat Trout are occasionally caught by both trollers and shoreline anglers. Westslope Cutthroat Trout and Bull Trout are both managed strictly for catch-and-release fishing.

Tips and Techniques

The diversity of the Coeur d'Alene Lake fishery provides a range of options to suit anglers of all levels of skill, experience, and available time. Though shoreline fishing is possible and even productive at times of the year, the vast majority of angling on Coeur d'Alene Lake is done with a boat. Given the size of the lake and the diversity of opportunities, most anglers use midsize boats in the 5 to 6.7 m (17 to 22 ft) range so that they may easily move around the lake. That doesn't mean smaller boats aren't suitable. Much of the lake is sheltered and narrow, and outside of the summer boating season, the lake is often quite calm, meaning 3.7 to 4.9 m (12 to 16 ft) boats are often adequate. A few anglers actually use flat-water kayaks trolling for Kokanee with great success, and others often use pontoon "kick boats" to fish the shoreline for bass and pike.

Kokanee

Kokanee are a popular target for anglers of all ages and levels of experience. Though not particularly large, Kokanee are excellent eating, and they are generally quite abundant, making for fairly fast action. Gear requirements are fairly simple. Typically, a wedding-ring spinner tipped with a maggot or corn and trolled behind "pop gear" (multiblade flashers) or mini-dodgers is the most common technique. Kokanee have soft mouths and can be difficult to keep hooked when reeling them in behind flashers. For this reason, a key component of a Kokanee rig is a "rubber snubber" between the flasher and the spinner that acts as a shock absorber and decreases the number of Kokanee that come unhooked. Specially designed rubber snubbers are available at tackle shops, but a rubber band works quite well.

Most anglers use leaded line or 28 to 57 g (1 to 2 oz) of lead to get the gear down to the appropriate depth, which generally ranges from 3 to 25 m (10 to 80 ft) depending on the season and water temperature. Though the fish are small, a medium or medium- to heavyweight rod is ideal because of the additional drag of the weight and flashers. Any reel will work, but many anglers prefer to use level-wind reels with line counters to make depth adjustment easy and consistent.

Kokanee angling typically begins in late spring and gets progressively better through the summer into early fall. All Kokanee die after spawning, and it's the adults that provide the fishery. The fishing action migrates in the lake along with the adult Kokanee. Early in the season, most of the fish are in the southern third of the lake, but they move northward as the summer progresses. Virtually all of the Kokanee in Coeur d'Alene Lake spawn in the northeast portion of the lake, in Wolf Lodge Bay in November and December. The migration to the spawning area means that highest Kokanee density (and fishing success) is in Wolf Lodge Bay from August through October. Though there are still plenty of Kokanee to be caught in November and December, the quality of their flesh deteriorates as they prepare to spawn.

Chinook Salmon

For anglers more interested in the challenge of a potential trophy, Chinook Salmon can be caught year-round. Anglers generally use a high-capacity trolling reel equipped with 15 to 20 lb test line and a medium-heavy- to heavy-action rod. Though effectiveness of assorted terminal tackle varies with the season, year in and year out herring are generally the most consistently effective bait. Most anglers use a helmet rig to provide the correct "roll" in the bait that will elicit a strike. In addition to herring, common gear includes hoochies, mini-squid, spoons, flies, and plugs. When trolling deep, flashers and dodgers are typically included in the setup.

Like Kokanee, where Chinook Salmon can be found depends on season and water temperatures. In the spring, fish are generally near the surface, and most anglers keep their gear in the top 9 m (30 ft) of water. Setting downriggers shallow or trolling directly behind the boat with an ounce of weight are both effective techniques, but many anglers use planer boards when fishing shallow to cover a broader swath of water. In the summer and early fall, fish tend to move deeper, and anglers generally target 12 to 24 m (40 to 80 ft) of water. During these months, downriggers are critical. Most Chinook Salmon from Coeur d'Alene Lake spawn in the Coeur d'Alene River in early October, so the largest fish are caught in the lake in late August and early September just prior to their spawning migration. From October through March, juvenile Chinook move into deeper water and associate closely with the bottom. Most anglers fish in 21 to 31 m (70 to 100 ft) of water and keep their gear right next to the bottom. Though this isn't the time to catch a monster, it's probably the best time in terms of catch rates. A 1 m (40 in) 14.5 kg (32 lb) land-locked Chinook Salmon was caught by Larry Masuda in 2016 from Coeur d' Alene Lake.

Chinook Salmon fishing can be as complex or simple as a person wants to make it, but anglers who have the most success are the ones who pay attention to detail. Dozens of books, articles, and websites provide advice on trolling for Chinook, and for an angler new to Coeur d'Alene Lake some advance research on techniques would be time well spent. Alternatively, there are a number of excellent guides on Coeur d'Alene Lake who have Chinook Salmon fishing down to a science.

Pike and Bass

For anglers who simply don't have access to a boat, Northern Pike provide a tremendously popular shoreline fishery in early spring. Shortly after ice out, large, mature Northern Pike move into the shallow bays in preparation for spawning. During this period, many anglers take advantage of the opportunity to catch a trophy fish. Water temperatures in March and April are quite cold and fish are relatively sluggish, so the most common technique is to suspend dead bait fish underneath a float and then wait for a cruising Northern Pike to find it. Wolf Lodge Bay, Cougar Bay, Blue Creek Bay, Chatcolet Lake, and all of the Chain Lakes are popular areas for springtime shoreline Northern Pike fishing.

As water temperatures warm in late May and June, Northern Pike become more active and fishing heats up, though a boat is essential. Bait is still effective, but artificial baits work well throughout the summer. Northern Pike are known for being aggressive, and there aren't many lures they *won't* hit. Spinnerbaits, buzz baits, jigs, Rapalas, spoons, topwater plugs, and softbaits will all catch Northern Pike. A medium-weight rod equipped with 8 to 12 lb test line is generally sufficient, but it's a good idea to use a steel leader to keep the sharp-toothed fish from biting off the lure.

Fishing for Largemouth and Smallmouth Bass typically begins in April, but doesn't really heat up until May and June, when water temperatures warm and fish move close to the shoreline in preparation for spawning. In early summer, the best bass fishing is found in the shallow vegetated bays of the main lake and around the shoreline of the Chain Lakes. Most anglers cast softbaits, such as rubber worms or crayfish, toward the shore, where bass are guarding nests. After spawning, in July and August, fish move to deeper water and associate with some sort of structure, such as fallen trees, docks, deep weed beds, and rocky humps. Anglers typically cast deep-water plugs, crankbaits, spinnerbaits, or drop-shot softbaits in the late-summer months. Though most bass fishing is done from a boat, it doesn't have to be. Smallmouth Bass can be caught from most rocky shorelines around the lake or from the many public docks.

Across the lake at Coeur d'Alene. Photo courtesy of IDFG.

Another beautiful day on Lake Coeur d'Alene. Photo courtesy IDFG.

A 32 lb (40" long) land-locked Chinook Salmon. Photo Courtesy Larry Masuda in 2016.

Hells Canyon

Joe DuPont, Regional Fishery Manager, Clearwater Region, Idaho Department of Fish and Game

Location, Directions, and Access

When Hells Canyon is mentioned, many thoughts often come to one's mind: the deepest river gorge in North America, world-renowned white-water rafting, Evel Knievel's failed "sky-cycle" jump (even though it actually didn't occur there), and breathtaking scenery. Interestingly, people often overlook just how amazing this place is to fish. It truly is one of Idaho's fishing gems. Nowhere else will you will be able to fish for Smallmouth Bass, giant White Sturgeon, Steelhead, and fall Chinook Salmon in a remote and rugged river canyon rich in history.

Hells Canyon is the 130 km (80 mi) stretch of the Snake River that spans from Heller Bar upstream to Hells Canyon Dam. This section of the Snake River was first referred to as "Hells Canyon" by a steamboat operator in 1895 as he began his voyage into this canyon. The name stuck, as many steamboats met their demise as they tried to navigate these treacherous waters.

Hells Canyon forms the border between Idaho and Oregon for all but the lower 13 km (8 mi), where it forms the border between Idaho and Washington. Access to Hells Canyon is very limited, as no roads cross or parallel the river. There are basically three entry points to this section of river: Heller Bar, Pittsburg Landing, and Hells Canyon Dam. A few other access sites occur but are not recommended unless one is willing to drive on rough and windy roads. To get to Heller Bar, travel Snake River Road about 34 km (21 mi) upriver from the town of Asotin, Washington. Heller Bar is about the most upstream point you can drive on this road; a public boat ramp and parking area serve as a launching spot to explore the lower canyon. The next easy access point is about 76 km (47 mi) upstream at Pittsburg Landing. To get to Pittsburg Landing, exit US 95 near White Bird onto the Old Highway 95 road, cross the Salmon River, and then take a left onto Deer Creek Road (NF 493). Travel Deer Creek Road about 27 km (17 mi), which will take you across the divide between the Salmon and Snake River Drainages and eventually to Pittsburg Landing, where public camping, parking, and a boat ramp are available. Finally, you can enter the most upstream extent of Hells Canyon by driving to Hells Canyon Dam. To do this, travel to Copperfield, Oregon, and then follow Hells Canyon Road (NF 454) downriver 35 km (22 mi) to Hells Canyon Dam. The road crosses the dam and continues another mile to the Hells Canyon Visitor Center on the Oregon side, where public parking and a boat ramp are present.

If one wants to rough it, a trail borders the Snake River on the Idaho side from Hells Canyon Dam 53 km (33 mi) downstream to Pittsburg Landing. A trail also borders about 80 km (50 mi) of the Oregon side of the river; however, one must access this trail from the mountains above, which entails a decent of more than 1,000 m (3,281 ft). Both trail systems tend to wind through the rocky canyon walls along the Snake River, with access to the river often being difficult. As a result, this is not the best strategy if one's main goal is to fish Hells Canyon, but if one has the desire, it is certainly doable.

If one truly wants to experience the amazing fishery this river has to offer, the best way to do it is by raft or jet boat. As a word of caution, this is not a river where one just jumps in a boat or raft and goes fishing. There are many rapids with Class III or IV ratings one must negotiate, and numerous boulders and shallow bars occur just under the surface that can cause serious damage to a jet boat. As such, if you are not experienced with white-water rafting or jet boating, it is best to go with someone who is. There are many guide services on this river that will be able to give you the experience you are looking for.

The Snake River in Hells Canyon is designated a Wild and Scenic River. The 51 km (32 mi) section of the river between Hells Canyon Dam and Pittsburg Landing is designated Wild, and the 63 km (39 mi) section of river immediately downstream from Pittsburg Landing is designated Scenic. In both these areas, permits are required to boat or float the river. Permitting varies between the two river secttions and times of the year. For more information about permitting, contact the USFS, which manages the Hells Canyon National Recreation Area. No permits are required to utilize the lower 14 km (9 mi) of Hells Canyon from Heller Bar upstream to Cache Creek (Forest Service Check-in Area). This section of river is probably the easiest to boat, although there are still rapids, gravel bars, and hidden boulders to negotiate. Fishing is excellent in this reach of river, so don't pass it up just because it's closest to civilization.

The Fishery and Techniques

Smallmouth Bass

There are many different fishes people like to target in Hells Canyon, but probably the easiest to catch and the most popular species is Smallmouth Bass. Smallmouth Bass are very abundant and can be caught throughout the entire reach of this river. Once water temperatures reach or exceed 10°C (50°F), Smallmouth Bass fishing gets good. Water temperatures often reach 10°C (50°F) in April and often last through November. Serious bass anglers often love fishing in the spring, as the fish tend to be uneducated and hungry and will typically congregate in areas with lower water velocities and cover (typically boulders). Even the larger fish will often move into these areas, making them more vulnerable to anglers in the spring. Most Smallmouth Bass you will catch in Hells Canyon range from 200 to 380 mm (7.9 to 15.0 in) in length, although during a Smallmouth Bass predation study, the US Geologic Survey recently captured a 560 mm (22.0 in) monster that topped 2.3 kg (5 lb, 1 oz)[28]. Many anglers fish for Smallmouth Bass in the summer when the water is warm and clear and high flows have subsided. Even though people love to be on the water during this time of year, fishing pressure is still relatively light due to the remoteness of this river. Fishing is often outstanding, and if you are serious about it, catching 30 or more fish in a day would not be unthinkable. The best areas to fish in the summer tend to be rocky areas that provide cover but still have noticeable flow. Fish can be right along the bank, but don't forget to fish deep, especially since in the summer this is often where those larger fish are lurking. Baits of choice tend to be jigs rigged with rubber crayfish or twister tails, although spinners, crankbaits, poppers, or worms can be effective. For those spending multiple days on the river, these fish can provide a delicious camp meal. Idaho allows harvest of these fish, but consult the fishing rules before heading out on the river.

White Sturgeon

White Sturgeon are the largest fish in North America, and their sheer size often leaves people gasping in amazement. When fishing Hells Canyon, one can expect to catch White Sturgeon of all sizes. Because this population of sturgeon grows so slowly, Idaho, Oregon, and Washington require all sturgeon to be released in this stretch of river. Based on recent work by the IDFG and IPC, many female sturgeon don't spawn for the first time until they exceed 50 years of age and 7 ft in length. Those truly giant fish that exceed 2.4 m (8 ft) are likely over 80 years old. When conducting sturgeon research in Hells Canyon, IDFG recently sampled a 3.3 m (10.7) monster that weighed more than 227 kg (500 lb). This fish was likely over 100 years old. If you are wondering, the Idaho state rod-and-reel record for White Sturgeon is 179 kg (394 lb, 11 oz), which was set back in 1956 when fish could be harvested.

If you want to fish for sturgeon in Hells Canyon, you need to come prepared. Recommended gear includes no less than 23 kg (50 lb) test line, and use something stronger if your reel can fit at least 137 m (150 yd) of it. When fighting, a big fish often hugs the bottom, where abrasion from boulders will easily break lighter line. If using braided line, use at least 69 kg (150 lb) test, as its thinner diameter makes it more prone to breaking from abrasion. Anglers use a wide variety of rods, from short stout ones that are easier to handle out of boats but are more difficult to cast to long surf rods that will allow a person to cast long distances but are more cumbersome to handle. It's really a personal choice, as long as it has enough backbone to withstand the pressure when one really puts their thumb down when trying to slow what seems like a freight train screaming downriver. When rigging up your weight, Idaho rules requires that they be attached to a slider with lighter test line than the main line (refer to Idaho's fishing rules for details). This rule was created to help reduce the number of hooks that are broken off on the bottom after the weight becomes stuck in the rocks. Recent research has found sturgeon will ingest these broken off hooks if still baited.

Although you can catch sturgeon year-round in Hells Canyon, the best times to fish tend to be when water temperatures range from 10 to 21°C (50 to 70°F). Typically, these temperatures occur from April through mid-July and mid-September through late November. Fishing for sturgeon is really quite simple. It basically entails putting your favorite bait on a large barbless hook (8/0 to 10/0 hooks depending on the brand), casting it into a good-looking spot, and waiting for the bite. When looking for areas to fish, focus on areas that are at least 6 m (20 ft) deep and near areas of calmer water. Often pools or runs with back eddies are excellent spots to fish. Try casting right into the calm spot in the middle of the back eddy. There are many choices of baits one can use when sturgeon fishing. Most tackle shops will sell pickled squid and herring, which will catch fish, but many of the better anglers tend to catch their own bait. I've seen sturgeon caught on Steelhead or salmon bellies, Northern Pikeminnow, Peamouth, Smallmouth Bass, Rainbow Trout, American Shad, crayfish, and even worms. If you use fish as bait, they must be dead, and it is not necessary to use large chunks to catch large sturgeon. Many 2.4 to 3.0 m (8 to 10 ft) White Sturgeon are caught on a 10 cm (4 in) piece of meat. The type of bait you use can make a big difference in your success. If something isn't working, try experimenting with other baits, or try curing your bait with various scents. If you have the right bait and water temperatures are ideal, it usually doesn't take long for a sturgeon to bite. As such, if you have fished a spot for a half hour without a bite, move to a new location.

Many sturgeon will jump within a couple minutes of being hooked. For those looking for a good picture, get ready once somebody hooks into a large fish. Watch the line, and as it starts to rise, focus just past it and get ready to shoot. If you're ready—and a little lucky—you'll have an incredible image and a great memento of your trip.

Steelhead

Many will argue that pound for pound, Steelhead are one of the best fighting freshwater fish there are. Combine that with their exquisite table fare, and it's no wonder that when Steelhead enter Hells Canyon, people flock to this river with the hopes of landing one of these hard-fighting ocean-going fish. This Steelhead harvest fishery is created from more than three million hatchery smolts that are released into rivers in and around Hells Canyon each year. Most of these smolts return after spending one year in the ocean as

2.3 to 3.6 kg (5.1 to 7.9 lb) Steelhead, with a smaller percentage returning two years later as 4.1 to 5.9 kg (9.0 to 13.0 lb) fish.

The Idaho Steelhead season in Hells Canyon is open from September 1 to May 31 (please consult Idaho's rules before fishing, as they can change from year to year). Typically, fishing doesn't pick up until mid- to late September, as water temperatures cool down and a surge of fish pushes upstream into Idaho. Early on (September through October), the best fishing tends to be from Heller Bar upstream 32 km (20 mi) to the confluence with the Salmon River. During good return years, it can be very crowded in certain areas, particularly around the Heller Bar Boat Ramp. Fish tend to stack up in this area, due to the fast shallow water and because large numbers of fish often return to the Grande Ronde River, which enters the Snake River just upstream of the boat ramp. When the fishing is good, it is not unusual to see multiple anglers fighting fish at the same time. For those who want to get away from the crowds, the farther upstream you go, the fewer people there are. Fishing between Heller Bar and the Salmon River tends to die down by Christmas as these fish push upstream.

By mid-October some Steelhead will have pushed all the way upstream to Hells Canyon Dam, where they are trapped for broodstock. Using the eggs collected from this broodstock, more than 500,000 smolts are released below Hells Canyon Dam every year to create this fishery. After October these fish tend to hold from the Salmon River upstream to Hells Canyon Dam. Typically, the later in the season it is, the farther upstream you will want to fish. The angling effort in this reach of river is very low except for within five miles of Hells Canyon Dam, where the fish stack up.

Catch rates for Steelhead in Hells Canyon average around 10 hours a fish but will vary considerably based on when, where, and how you are fishing, your level of experience, and the strength of the run. The most popular techniques anglers use to fish for Steelhead are drifting yarn or bait (or both) and back-trolling plugs. To have good success when drift fishing, try to keep the boat moving downstream at the same speed as your gear, thereby allowing it to bounce along without pulling it off the bottom. When your line tightens or you feel a strike, set the hook. Most anglers have their favorite yarn colors and bait types. If you are having trouble catching fish, experiment or talk to some of the more successful anglers. When back-trolling plugs, the key is to slowly work back through a run or pool (often zigzagging back and forth), trying to keep your plug as close to the bottom as possible. When you get toward the end of a pool or run and the river shallows, stay focused. At times, you can actually herd fish downstream until they reach this shallow water. There are a large variety of plug styles and colors to use when backtrolling. In addition, many people add scent or bait to their plugs to increase their catch rates. If new to the area, try visiting tackle shops and talking to experienced anglers to see what is working. When Steelhead fishing in Hells Canyon, it is not unusual to hook into a fall Chinook Salmon or Channel Catfish. Be sure to have the appropriate licenses and permits if you wish to harvest one of these magnificent fish.

Despite the large size of the Snake River in Hells Canyon, experienced fly anglers can be quite successful catching Steelhead. Most fly anglers in Hells Canyon use jet boats to access areas where they can wade and fish specific runs, riffles, or pool tailouts. Most anglers also use Spey rods because they can cast farther, don't require room for a back cast, and take less effort than single-handed rods. The preferred technique is to cast slightly downriver and let your fly swing toward shore. When you get a hit, don't set the hook; let the fish hook itself, or you will likely find yourself screaming obscenities. Many fly anglers claim that nothing beats a Steelhead slamming a fly on the surface, but as water temperatures cool and Steelhead become less active, you may want to try fishing deeper if you aren't having any luck. Another thing to remember is that in cooler water temperatures, Steelhead will often move into slower water to conserve energy. In these situations, be sure to fish the slower water closer to shore.

Fall Chinook Salmon

Fishing for fall Chinook Salmon is a relatively new thrill in Hells Canyon because their numbers have only recently increased to the point where anglers are allowed to fish for them. The resurgence of this run of fish was a collaborative effort that involved utilizing wild fish in hatcheries, regulating outflow from Hells Canyon Dam during critical spawning and incubation periods, and making efforts to increase the survival of juveniles as they migrate downstream. These actions have been so successful that the run of fall Chinook Salmon returning to Idaho increased from 78 fish in 1990 to around 80,000 adult fish in recent years. The year 2008 was the first time in more than 40 years that Idaho provided opportunities to fish for fall Chinook Salmon. The season for fall Chinook Salmon, in recent years, runs from September 1 to October 31, except for a short reach downstream of Hells Canyon Dam. Here it typically ends around mid-November. Please consult Idaho's fishing rules if you wish to fish for fall Chinook Salmon, as seasons and rules can change.

Because this fishery is so new, anglers are just learning how to catch them. In Hells Canyon most people initially caught them incidentally when fishing for Steelhead. Now some anglers are starting to target them using jet divers to fish some of the deeper holes with cut herring, superbaits, or a quick fish trailed behind. One of the exciting things about fall Chinook Salmon is they can get big. In recent years, the trapping operation at Lower Granite Dam (the first dam downstream of Hells Canyon) has captured fall Chinook Salmon that exceed 23 kg (50 lb, 11 oz). The Idaho State record Chinook Salmon is 24kg (54 lbs) caught back in 1956. Although most fish range between 2.3 and 6.8 kg (5.1 and 15.0 lb), the new state record is out there to be had.

Whatever you prefer fishing for, don't pass up an opportunity to fish Hells Canyon. When you combine its breathtaking scenery, remote setting, world-renowned white water, and amazing fishing, it is an experience that you won't soon forget.

Hells Canyon Scenic View, Snake River, Idaho. Photo by Joe DuPont.

Boat Fishing, Hells Canyon, Snake River, Idaho. Photo by Joe DuPont.

Henrys Fork

Rob Van Kirk, Senior Scientist, Henry's Fork Foundation

The Henrys Fork is a gem not only within Idaho; it is one of the premier trout streams anywhere. In 1998, in preparation of celebrating its 40th anniversary, Trout Unlimited polled its members to identify the top 100 trout streams in the country. The Henrys Fork ranked number one.

Location, Directions, and Access

US Highway 20 provides primary access to the Henrys Fork and parallels the river for its entire length. The most popular route to the Henrys Fork from the south and west is I-15, which intersects Highway 20 at Idaho Falls, Idaho. From the north, I-90 leads to US Highways 287 and 191, which both intersect Highway 20 at West Yellowstone, Montana. From the east, Highway 20 leads directly from Cody, Wyoming, through Yellowstone National Park to West Yellowstone, although this route is not open during the winter. From Jackson, Wyoming, take Wyoming State Route 22 west over Teton Pass to Idaho Route 33 and then north on Idaho Route 32 to Ashton, Idaho. From Ashton, the Henrys Fork can be accessed from county roads, from US Highway 20, and from Idaho Route 47. Ashton lies near the geographic center of the Henrys Fork watershed, providing the optimal location from which to base a fishing trip. Driving times to Ashton are one hour from West Yellowstone, one and a half hours from Jackson, four and a half hours from Salt Lake City, five hours from Missoula, and five and a half hours from Boise.

The watershed of the Henrys Fork, also known as the North Fork of the Snake River, lies in the northern corner of eastern Idaho and is bordered on the north and east by the Centennial Mountains, the Yellowstone Plateau, and the Teton Range. On a map, the Henrys Fork appears to originate as the outflow of Henrys Lake, nestled in a crook in the Continental Divide about 20 km (12 mi) from the west entrance of Yellowstone National Park. However, the hydrologic source of the Henrys Fork is about 15 km (9 mi) southeast of Henrys Lake at Big Springs, where a full-blown river emerges from the base of the Yellowstone Plateau at a constant flow of 5.7 m^3/sec (200 cfs).

Fish Species

The only trout native to the Henrys Fork watershed is the Yellowstone Cutthroat Trout. Non-native trout were introduced in the late 19th century, and by the 1960s Rainbow Trout had become the primary sport fish[30,31]. However, Cutthroat Trout are still found in some locations and are the dominant trout species in certain tributaries. Brook Trout, Brown Trout, Cutthroat × Rainbow hybrid trout, and Kokanee are also found in the Henrys Fork watershed. Mountain Whitefish, a native salmonid, are abundant. Although hatchery fish are stocked into a few water bodies, self-sustaining populations of Rainbow Trout and Brown Trout provide the vast majority of angling opportunities on the Henrys Fork. Recreational angling for these wild trout contributes tens of millions of dollars annually to the local economy[32,33].

The Allure of the Henrys Fork

The Henrys Fork is one of many rivers that form the heart of trout fishing in the western United States. Within a three-hour drive of the Henrys Fork are the Bighole, Madison, Gallatin, Yellowstone, Green, and Snake Rivers—as well as Silver Creek and numerous others. The Henrys Fork stands out even among these great trout streams because it offers easy public access to a large amount of water, incredible scenery, and unparalleled opportunities to cast to rising fish in a spring-creek setting.

The subtitle to a recently published guidebook to the Henrys Fork reads "80 Miles of World-Class Water"[34]. The upper half of the river and most of the tributary streams flow through the public lands of the Caribou-Targhee National Forest and Harriman State Park. Over the lower half of the river's course, the banks are mostly private farmland, but public access to the river is still plentiful along county roads and at public easements. Only a handful of rivers in the Rocky Mountains can boast 80 miles of world-class trout fishing, all readily accessible to the general public.

The Henrys Fork watershed is the product of volcanic activity that has occurred in the past two million years. The river emerges from giant springs at the base of the Yellowstone Plateau and flows through canyons cut into volcanic rock, across a flat caldera formed by collapse of a 600,000-year-old volcano, over sheer waterfalls, and onto the Snake River Plain. The resulting scenery is not only geologically unique but also spectacular. The iconic view of the Henrys Fork is in Harriman State Park and looks southwest across the caldera toward the Teton Mountain Range, about 80 km (50 mi) away. There are few angling experiences in the world that can equal that of standing in pastel evening light in a spring creek that is 150 m (492 ft) wide, sharing your stretch of water with a couple of moose, and watching the sun set on the Grand Teton. On the lower Henrys Fork, the riparian forest in autumn colors rivals that of any river in the region.

Although anglers who frequent the Henrys Fork value its accessibility and scenery, they come for the dry-fly fishing. Throughout its length, the river supports prolific hatches of a broad array of insects, ranging from the tiniest midges (Chironomidae) to the Giant Salmonfly (*Pteronarcys californica*). The most famous dry-fly fishing water is the 13 km (8 mi) that flow through Harriman State Park, also known as the "Railroad Ranch," which was donated to the State of Idaho by the Harriman family in 1977. "The Ranch" is flat spring-creek water on a grand scale and offers a unique opportunity for anglers to fish for large Rainbow Trout feeding on insects at or near the water's surface. Many modern techniques and fly patterns for this type of "match-the-hatch" fishing were developed and perfected on the Ranch, and there is an entire book devoted just to this stretch of river[35]. Many anglers consider landing a large Rainbow Trout on the Ranch to be the ultimate trout-fishing experience because of the challenges involved. The easiest part of catching one of these fish, which generally range from 360 to 600 mm (14.2 to 23.6 in) in length, is spotting its rise in the gentle current. After that, the successful angler must correctly identify the insect on which the trout is feeding, select an appropriate fly to imitate the insect, wade into a position that will allow an accurate cast to the fish without spooking it, execute a perfect cast and presentation, hook the fish if it takes the fly, and land it. Even if everything else goes well, these last two steps are more difficult on the Henrys Fork than they would be in most other trout streams, because more often than not, the appropriate fly is small and the required tippet is fine. Landing the fish is a fine balance between keeping the fish out of the lush vegetation on the river bottom and breaking the tippet. Even in other reaches of the Henrys Fork, where riffles help conceal both the stalking angler and flaws in fly presentation, trout can be extremely difficult to catch.

Reaches of the Henrys Fork

The river is closed to fishing from Big Springs downstream about 2.5 km (1.5 mi) to the confluence of Henrys Lake Outlet. However, Big Springs is worth a visit both to witness the birth of the river and to view large trout that often congregate there. Downstream of Henrys Lake Outlet, the river is open to fishing and contains wild Rainbow Trout, Brook Trout, hatchery Rainbow Trout, Kokanee, and large Rainbow Trout that migrate seasonally upstream from Island Park Reservoir, and Cutthroat Trout and hybrid trout that migrate downstream from Henrys Lake.

The most popular reach of the Henrys Fork is the 45 km (25 mi) from Island Park Dam downstream to Mesa Falls, the reach that flows across the floor of the Henrys Fork Caldera. Just downstream of the dam is Box Canyon, 5 km (3 mi) of riffles, rapids, and pocket water. This type of water is also found in the lower canyon, roughly from Riverside Campground downstream to Mesa Falls. The centerpiece of the caldera reach is the flat water in Harriman State Park. Immediately upstream and downstream of this flat water are the gentle riffles of Last Chance and Pinehaven, respectively.

The adventurous angler willing to slide a boat or raft down a steep slope can find some good fishing in the deep canyon between Lower Mesa Falls and Warm River, particularly during May and June. Between Warm River and Ashton Reservoir, the river is easily accessible to the more casual floater as well as to the wade-fishing angler and provides consistent fishing year-round. Downstream of Ashton Reservoir, the Henrys Fork leaves the lava canyons and flows out onto the Snake River plain, where it winds through riparian forests and wetlands. Gravel bars, side channels, islands, and cottonwood logs provide trout habitat and a wide variety of angling opportunities ranging from dry-fly fishing to rising Rainbow Trout to streamer fishing for large Brown Trout. Regularly spaced boat ramps allow floating access to most of the lower river. However, irrigation diversion dams constitute dangerous, and in some cases impassable, barriers to floaters, so seek local advice before floating the lower Henrys Fork.

Tributary streams not only provide spawning and rearing habitat for Henrys Fork trout but also offer additional angling opportunities. Henrys Lake Outlet contains Cutthroat, hybrid, Rainbow, and Brook Trout. In the spring and early summer, large Cutthroat Trout can be taken on dry flies. The Nature Conservancy's Flat Ranch provides public access to the outlet. Buffalo River and Warm River are both large spring creeks that provide good fishing for smaller trout, easy wading, and great places for beginners and kids to hone their skills before tackling the main river. Wild Rainbow Trout and Brook Trout provide most of the action, but Warm River is stocked with hatchery Rainbow Trout and also contains Brown Trout. The two tributaries in the lower watershed are both large rivers and high-quality fisheries in their own right. Falls River originates in the southwest corner of Yellowstone National Park and is confined to a narrow canyon over most of its length. Riffles and pocket water provide great dry-fly and nymph fishing for Rainbow Trout. The Teton River is born from a unique combination of snowmelt from the Teton Range and springs on the floor of Teton Valley. It is the only stream in the Henrys Fork watershed that is still dominated by native Yellowstone Cutthroat Trout and features a diversity of water ranging from the flat glides of the upper river to the rapids of Teton Canyon. Local advice is particularly important before floating the Teton River or Falls River.

Seasons

Most of the Henrys Fork and its tributaries are open to fishing all year, but consult regulations because closures, harvest limits, and gear restrictions vary by season and stream reach. Most anglers visit the Henrys Fork from mid-April through mid-October, when the majority of insect hatches occur. Because the Henrys Fork receives much of its inflow from groundwater springs, it remains fishable throughout the spring snowmelt, unlike many of its neighbors. Flow in the Henrys Fork is regulated by dams at Henrys Lake and Island Park Reservoir and by irrigation diversion. Flows in the lower reaches of Falls River, the Teton River, and the

Henrys Fork can become very low during July and August due to irrigation withdrawals, and fishing should be avoided in these areas when water temperatures are high. Streamer fishing for large Brown Trout peaks in mid-October but continues through November. Midge hatches continue to provide dry-fly fishing on most of the river from late fall through early spring, and fish will take nymphs all year. However, seek up-to-date information on weather and water conditions before planning an off-season trip.

Tackle and Techniques

The vast majority of anglers who visit the Henrys Fork use flies. A good all-around outfit is a 3 m (9 ft) rod with a 5- or 6-weight floating line, a tapered leader of 3 to 4 m (9 to 12 ft), and 3X to 5X tippet, depending on fly size. For the flat water of Harriman State Park, a 3 to 3.5 m (9 to 10.5 ft), 4-weight rod is preferred. Longer and finer leaders are also required. For the smaller streams, a shorter rod with a 5-weight line is preferred. Anglers coming specifically to fish the Salmonfly hatch or to cast streamers may want to bring a 7-weight rod.

Four standard fly-fishing techniques will cover 95 percent of the fishing situations one will encounter on the Henrys Fork and its tributaries. The first is dry-fly fishing, whether it be sight-casting to rising trout or blind-fishing banks, riffles, and pockets. In tributaries, small headwater reaches, braided channels, and pocket water, classic upstream presentation of the dry fly is preferred. However, when sight-casting to rising fish on flat water and open riffles, most experienced Henrys Fork anglers use a down-and-across presentation to avoid casting the fly directly over the fish's head. When fishing larger stream reaches from a boat, dry flies are usually cast perpendicular to the current.

The second method, which is most effective in the canyon reaches and on the lower river, is to fish a medium-size to large dry fly with a small- to medium-size nymph dropper. The nymph should be suspended about 0.5 to 1 m (1.5 to 3 ft) below the dry fly. Presentation methods are generally the same as described above for dry flies, except that the dry-dropper method is not widely used for sight-fishing to rising fish.

The third method, again most often used in the canyon reaches and on the lower river, is nymph fishing with a strike indicator. Anglers vary greatly on their preference for type of strike indicator, but the indicator is usually positioned 1 to 2 m (3 to 6 ft) above the fly. In most situations, the nymph or leader is weighted. Often, a second nymph is fished as a dropper, about 0.5 to 1 m (1.5 to 3 ft) below the top fly. Upstream and across-stream presentations are most common.

Streamer fishing can often yield the largest trout. Early in the fall, when water temperatures are still relatively warm and flow is low, streamers can be fished on a floating line, as long as the fly is weighted. As water temperatures become colder later in the fall, or when flows are high during the spring, a 3 m (9 ft) sink tip helps keep the fly closer to the fish. The most effective presentation method is to cast the streamer slightly upstream perpendicular to the current, let the fly sink for a few seconds, and then retrieve with a series of short, erratic strips.

Hatches and Fly Patterns

The four major aquatic insects of importance on the Henrys Fork are mayflies (Ephemeroptera), stoneflies (Plecoptera), caddisflies (Trichoptera), and midges. Extensive technical information on taxonomy, ecology, and imitation of these insects is available online and in books and magazines. Here, we provide only basic information, with insects and flies referred to only by their common names.

The iconic Henrys Fork mayfly is the Green Drake, a dark-winged size 10 to 12 insect that hatches on most reaches of the river during June. The Gray Drake (size 10 to 12) hatches from mid-June to early July, primarily downstream of Ashton, and the Brown Drake (size 6 to 8) hatches in the flat water of Harriman State Park during late June and early July. Important medium-size mayflies (size 14) include the March Brown (April–May), Small Western Drake (July), and Mahogany Dun (September). The important small mayflies (sizes 16 through 24) include the Blue-Winged Olive, Pale Morning Dun, and Trico. Tricos hatch in flat-water sections of Harriman State Park and the lower river during August and September. The other two are common names for several different taxonomic groups of insects that hatch on nearly every stream in the watershed. Although they can hatch at any time from April through October, they tend to be most prolific during spring and fall in the lower watershed and during the summer in the upper watershed. In many situations where fish are feeding on one or more of these mayflies, specific imitations are required, but it is also possible to catch finicky Henrys Fork trout on standard dry-fly patterns such as the Adams and Royal Wulff, if they are the correct size and are presented well. Hare's Ear, Pheasant Tail, and Prince Nymph are popular mayfly nymph patterns.

The major stonefly on the Henrys Fork is the Giant Salmonfly, which hatches in almost all reaches of the river upstream of Saint Anthony, except for the flat water of Harriman State Park. The Salmonfly also hatches in many tributaries. The hatch starts in early to mid-May and continues through late June. Timing of the hatch varies across river reaches, dependent on water temperature. The Henrys Fork Salmonfly in a size 4 or 6 is an effective pattern. Several smaller stoneflies, including the Golden Stone (sizes 6 to 8) and Yellow Sally (size 14), hatch in most of the same stream reaches as the Salmonfly from late spring through midsummer. Standard stonefly patterns and down-wing attractors such as the Stimulator and Madam X are effective imitations of these smaller stoneflies. A wide variety of stonefly nymph patterns are available; the standard Black Rubberlegs nymph in sizes 4 to 10 is one of the very best fly patterns on the river, period.

A large number of caddisfly species hatch throughout the watershed, from early April through mid-September. The Elk-Hair Caddis and Spent Partridge Caddis in sizes 12 to 18 are two effective patterns, although there are times and places where more technical caddisfly patterns, including emergers and pupae, are required.

Midges are most important from November through March. Prolific swarms can occur on nearly every river reach, usually on calm afternoons and evenings. Griffith's Gnat, Parachute Adams, and Zebra Midge Nymph in sizes 18 or 20 are popular patterns.

Other aquatic and terrestrial organisms are important to the trout diet on the Henrys Fork. Grasshopper, ant, and beetle imitations are effective on most river reaches from late July through September. Streamers in sizes 2 to 8 that imitate leeches and small fish are effective year-round.

Other Tips

Although most of the Henrys Fork is easily accessible and can be waded effectively under normal water conditions, floating the river is necessary on some reaches, especially during high water. Even during low water, floating can allow the angler to cover more water and experience more of the river's scenery and wildlife. Most sections of the Henrys Fork can be floated in a standard low-sided drift boat, skiff, or pram, but many anglers use rafts. If you are not familiar with the river, seek local advice before floating. Of course, another option for your first trip to the Henrys Fork is to hire a guide. Whether you take a wade trip or a float trip, there is no substitute for the experience and knowledge of a professional guide, especially if you want to land a large Rainbow Trout on a dry fly. There are several outfitters that are licensed to guide on the Henrys Fork.

Henrys Fork Rainbow. © William Teuscher, provided courtesy of Henry's Fork Foundation.

Henrys Fork "fish tail". © Kelly Glissmeyer, provided courtesy of Henry's Fork Foundation.

Big Bend on Henrys Fork in Harriman State Park. © Marine Ventures Foundation, provided courtesy of Henry's Fork Foundation.

Sunset on Henrys Fork downstream of Ashton. © William Teuscher, provided courtesy of Henry's Fork Foundation.

Osborne Bridge, Henrys Fork from the Railroad Ranch. Photo courtesy of John Sigler.

Sheep Creek Falls on Henrys Fork. © William Teuscher, provided courtesy of Henry's Fork Foundation.

Rainbow release, Henrys Fork. © Mike Lawson, provided courtesy of Henry's Fork Foundation.

Henrys Lake

Jessica Buelow, Regional Fisheries Biologist, Henrys Lake, Idaho Department of Fish and Game

Location, Directions, and Access

Henrys Lake is located in the mountains of southeastern Idaho in northern Fremont County. The lake is just two miles south of the Continental Divide along the Idaho-Montana state line and provides beautiful panoramic views of the Centennial Mountain Range and the Henrys Lake Range. Henrys Lake is part of the Greater Yellowstone Ecosystem and is 24 km (15 mi) from the west entrance to Yellowstone National Park in West Yellowstone, Montana, and about 480 km (300 mi) north of Salt Lake City, Utah. To reach the lake from West Yellowstone, take Highway 20 west; to reach the lake from Salt Lake City, take I-15 north and then US 20 east toward West Yellowstone.

Henrys Lake is managed as a trophy fishery, and anglers value the large trout that can be caught. As a bonus to the trout fishery, the lake is surrounded by beautiful mountain ranges and many opportunities to see wildlife. Most of the shoreline is private, but with the exception of a few areas, much of the shoreline remains undeveloped. Easements are in place in many areas to ensure development is minimized and the scenic beauty surrounding the lake remains intact for years to come. There are several locations where anglers can fish from shore. They include Henrys Lake State Park, Frome County Park, IDFG hatchery, BLM South Shore access, and Pintail Point access. There is also some BLM access along the west shoreline, though it does not have signs. Float tubes are frequently launched from these locations to access more areas of the lake. There is a public boat launch at Henrys Lake State Park and at Frome County Park and private launches at Wild Rose and Staley Springs where anglers can launch for a fee. If you are out in a float tube or boat, make sure to keep an eye on the weather. Henrys Lake can get very rough in windy conditions, and the weather can change quickly.

Physical Characteristics and Attributes

Henrys Lake is a natural alpine lake fed by numerous tributary streams and springs. It is located between two mountain ranges at 1,973 m (6,470 ft) above sea level. In 1924 a dam was installed on the outlet of the lake, increasing its storage capacity to meet irrigation demands downstream. In part because of the importance of the fishery, and because of the limited capacity to refill water released for irrigation, Henrys Lake is drawn down significantly only during drought years. The lake is 21 km^2 (8 mi^2) in area, 6.4 km (4 mi) long, and 3.2 km (2 mi) wide. Maximum depth is around 6.6 m (20 ft), with a mean depth of 4 m (13 ft). The lake is very productive, and native aquatic vegetation provides cover for fish and plenty of aquatic organisms needed for producing large trout.

The Fishery

In Henrys Lake, the fishery is supported by native Yellowstone Cutthroat Trout; sterile hybrid trout, which are a cross between Rainbow Trout and Yellowstone Cutthroat Trout; and sterile Brook Trout. All of these species grow very fast and large, with Yellowstone Cutthroat Trout being 310 to 360 mm (12.2 to 14.2 in) by age two. Hybrids can grow to weights in excess of 6.4 kg (14 lb, 1 oz): the state record Brook Trout was caught in Henrys Lake and weighed 3.7 kg (8 lb, 2 oz). These trophy fish attract anglers from all over the nation for the opportunity to hook and land a Henrys Lake monster.

Yellowstone Cutthroat Trout

Yellowstone Cutthroat Trout are the only trout native to eastern Idaho and are considered a "species of special concern" or a "sensitive species" by many federal and state agencies. However, the Yellowstone Cutthroat Trout population in Henrys Lake is very healthy and is the stronghold for the entire Henrys Fork watershed. Yellowstone Cutthroat Trout can be identified by their golden color, red slashes under their jaw, and round spots occurring mostly behind the dorsal fin. Occasionally, these trout will have spots in front of the dorsal fin, but they are always circular, as opposed to the crescent- or irregularly shaped spots of a hybrid trout. IDFG runs a hatchery on Hatchery Creek (on the north shore) that collects Yellowstone Cutthroat Trout as they swim up the fish ladder to spawn; this operation supplements natural reproduction in Henrys Lake. Yellowstone Cutthroat Trout are present in the ladder from March through mid-May, and many spectators come to watch the fish swim up the ladder and into the spawning shed. It's a sight worth seeing, with hundreds of trout jumping the rungs of the fish ladder on a good day. Hatchery staff generally take eggs twice a week during this time, and volunteering or watching the operation is a popular activity and encouraged. Yellowstone Cutthroat Trout eggs are incubated for 30 days until they reach eye-up stage, which is when the egg has developed enough that you can see the black eyes inside the orange egg. At this point, the eggs are very durable and are shipped to the IDFG's Mackay Hatchery, where they are raised until fall.

Each fall the lake is stocked with 750,000 to 1.3 million 75 to 100 mm (3.0 to 3.9 in) Yellowstone Cutthroat Trout, 200,000 75 to 100 mm (3.0 to 3.9 in) sterile hybrids, and 100,000 125 to 150 mm (4.9 to 5.9 in) Brook Trout. Numbers of Yellowstone Cutthroat Trout vary from year to year based on estimations of what the natural spawning population is contributing to the fishery. Habitat restoration projects on tributary streams have improved spawning habitat and fish passage, increasing the number of naturally produced fish. When trout abundance gets too high, food resources become limited and growth slows, resulting in smaller fish on average and fewer large trophy fish. However, angler catch rates increase when fish numbers are up. In order to balance the number of fish in the lake and maintain the quality size anglers desire, stocking rates are adjusted upward or downward each year.

Other Fish Species

Hybrid trout have been present in Henrys Lake since the early 1900s after Rainbow Trout escaped from an adjacent pond. Hybrids became a popular fish because they attain a larger size and fight more aggressively than Yellowstone Cutthroat Trout. In the 1950s IDFG began stocking them in Henrys, with intensive stocking between 1961 and 1972. Hybrids are produced at the hatchery using eggs from Yellowstone Cutthroat Trout and milt from Rainbow Trout, which comes from a different hatchery. In 1972 stocking of hybrid trout was discontinued, and management focused on conservation of the native Yellowstone Cutthroat Trout. However, public interest and demand for the hybrid fishery remained strong. In 1976 the department started stocking hybrids again, with the added objective of developing a sterile hybrid. Early on, experiments focused on heat shocking fertilized hybrid eggs to create sterile triploids. This method met with varied success and was never implemented on a large scale. By the late 1990s, advancements in this technology allowed biologists to heat shock all of the hybrids for Henrys Lake and stock fish that were 90 percent or more triploids. In 2004 the sterilization method was changed to a pressurization technique, which has a triploid induction rate of more than 99 percent. The overall attention to managing genetic risk to Yellowstone Cutthroat Trout has paid off, and the most recent assessment of Yellowstone Cutthroat Trout in Henrys Lake indicates very low levels of Rainbow Trout introgression despite the long history of hybrid stocking.

Brook Trout were introduced into Henrys Lake by early settlers and adapted to the conditions of the lake very well, growing very large. They were later stocked in the early 1900s in order to supplement the natural recruitment and increase angler catch rates. Stocking programs have varied over the years, but started on a regular basis in 1974. In 1999 IDFG stopped stocking Brook Trout, and the population dramatically decreased. Stocking resumed in 2004 after IDFG developed techniques to produce sterile triploid Brook Trout. Eggs for use in the hatchery Brook Trout program cannot be collected from Henrys Lake because most of the population in the lake are sterile and don't produce eggs or sperm. Brook Trout eggs are acquired or purchased from other hatcheries and reared at American Falls Hatchery until they are stocked in the fall. Brook Trout are extremely popular in Henrys Lake due to their size, taste, and great fighting abilities when hooked.

Facilities and Amenities

There are camping facilities available at Henrys Lake State Park and Frome County Park. There is also dispersed camping available around the lake on BLM land and USFS property. If you are looking for something a little less rustic, there are cabins available for rent as well as hotels available in West Yellowstone or farther south in Island Park.

Seasons and Fishing Options

Fishing season begins on the Saturday of Memorial Day weekend and runs through January 1. There are no bait restrictions or trout-size restrictions, but there is a bag limit of two fish and you must stop fishing after you harvest your second fish. Unlike many areas in Idaho, at Henrys Lake Brook Trout are counted in the two-fish bag limit. There are two areas closed to fishing: Staley's Spring and near the mouth of Hatchery Creek. These areas are marked by buoys.

There are many options on how to fish the lake, and they vary by time of year. Trout cruise the shoreline for the first couple of weeks after the ice comes off, which can vary from the end of April to the end of May. This time of the year you can see large fish in very shallow water, and fishing from shore is usually a great option for the Memorial Day weekend opener. Anglers will use bait (worms), spinners, and flies from shore during this early season. Flies most commonly used are Woolybuggers, leech patterns, scuds, and small nymphs. Boat fishing is also popular this time of year. Anglers troll Rapalas, spinners, or wet flies.

As spring and summer advance, the fish tend to move away from the shoreline into deeper water. Fishing from shore is less productive in most of the lake, but fishing along the cliffs at the South Shore access can remain good well into summer. A float tube or boat is your best option for finding fish. Native aquatic vegetation will start to grow tall by midsummer, making trolling difficult. Many anglers will select an area and cast flies or lures or sink a worm. Search for channels in the vegetation; the fish use them as highways searching for food. Through late June, the fishing is very good in water depths around 3 to 4 m (10 to 14 ft). Later in July, the water starts to warm up, and the fish seek thermal refuge at the mouths of tributary streams and springs around the lake. This time of the year, popular fishing spots are Targhee Creek, Howard Creek, Duck Creek, South Shore Access, and Staley's Spring. Fishing is best in early morning and tapers off as the heat of the day increases. You can start the morning fishing shallow near the tributaries, then move into deeper water as the day progresses. A variety of flies will work well this time of year, including Woolybuggers, leech patterns, small midge, Pheasant Tail, or scud patterns. A *Callibaetis* hatch will start in the summer, and these are a good option as a nymph, or on a calm day tie on a dry fly. Henrys Lake is also known for its damselfly hatch early in July. This can be a spectacular site, with thousands of them flying around. They are also a great option to fish as a nymph.

The lake starts to cool down in mid-September, and the lake vegetation will also start to die back. If you are nymph fishing this time of year, plan on using a sink tip or full sinking line to get down to where the fish are. During this time, the trout start to move back into the shallow water and cruise the shoreline. Trolling with lures or flies can be a very good option this time of year. A lure that mimics a minnow can also be a very good option, as larger trout key in on the young trout stocked each fall. Woolybuggers, nymphs, or anything with some sparkle seems to work well. Popular fall fishing locations are around Howard Creek, Targhee Creek, and Duck Creek. You don't need to be very close to the mouths of the tributaries in the fall, as temperatures have dropped, allowing trout to spread back out through the lake. This can be a great time of the year to fish, so if you aren't catching anything, change your location, lure, and technique a bit. Weather can also be a bit unpredictable during the fall. Be prepared to be snowed on, to be rained on, or to enjoy a wonderful sunny fall day on the lake.

Henrys Lake will freeze sometime in November. In some years, that can be as early as the first week of November, while in other years the lake caps over late in the month. Fishing can be a bit difficult during the freeze-thaw days; ice forms, causing access issues along the shore. Under some late-fall conditions, it's possible to find anglers ice fishing, fishing from shore, and fishing from boats on the lake at the same time. Ice fishing has been allowed on Henrys Lake since 2005, initially only until November 30, which limited the amount of days for fishing. In 2011 the fishing season was extended through January 1, which provided much more opportunity for those willing to brave the elements. Popular methods for ice fishing are jigging, dead sticking, or using a tip up. Popular lures are jigs, jigs with grubs, Swedish pimples, ice-fishing Rapalas, or a hook and worm. Fishing can be very good in shallow water at the beginning of the season, but it is better in deeper water later on in the season. Fish at a variety of depths, and use a variety of bait and colors.

There are several great books, websites, videos, and other resources for fishing advice at Henrys Lake, and guide services are also available. Search the Internet to find general fishing advice about the lake, how and where people are fishing throughout the year, and photos of some huge fish! For technical information, feel free to contact the IDFG Upper Snake Region office in Idaho Falls.

Late summer storm on Henrys Lake. Photo Courtesy John Sigler.

View of Henrys Lake from Sawtell Peak, Summer 2016. Photo courtesy John Sigler.

Bill Garren with a Henrys Lake beauty. Photo courtesy of Dan Garren.

Lake Cascade

Dale Allen, Regional Fisheries Manager, Southwest Region, Idaho Department of Fish and Game

Location, Directions, and Access

Lake Cascade is a federal reservoir constructed on the North Fork Payette River by the US Bureau of Reclamation (BOR) after World War II. The reservoir is one of the larger water bodies in Idaho and quickly became a popular fishery. The reservoir lies in Long Valley, with the dam in the small town of Cascade. Located just off Highway 55, Lake Cascade is approximately 130 km (80 mi) north of the Treasure Valley, where about 40 percent of Idaho's entire population lives.

Physical Characteristics and Attributes

The reservoir is nearly 34 km (21 mi) long and 7 km (4.5 mi) wide at the widest point. Water levels fluctuate about 3.6 m (12 ft) per year because of irrigation deliveries. The reservoir is fairly shallow, with a mean depth of 8 m (26 ft). At full pool the reservoir is about 10,650 surface ha (26,307 ac), has 138 km (86 mi) of shoreline, and contains 85,500 ha-m (693,000 ac-ft) of water.

A small ribbon of land around the reservoir is owned by BOR, with small holdings by the USFS and the State of Idaho; the remainder of the shoreline is in private hands. Overall access for campers, boaters, and anglers is excellent. Idaho Parks and Recreation operates Lake Cascade State Park, which includes seven campgrounds and other access and boat-launch facilities. The USFS operates two public campgrounds, and there is one privately owned campground on the reservoir. There are 10 boat ramps offering access to different parts of the reservoir. Access to the west side of the lake is via Roseberry Road in the town of Donnelly and ties into a paved road, West Mountain Road, that lies on the west shore and loops around the southern end of the reservoir and back into the town of Cascade. These roads are open all year and have good winter snow maintenance. The campgrounds are heavily used from late May until Labor Day, when a large drop in use occurs. There are many other areas for camping away from the reservoir on both private and public lands.

Surrounding Lake Cascade, the west-central mountains of Idaho present many diverse recreational opportunities, and the region is a popular destination for residents and tourists alike. The area economy has slowly transformed itself from reliance on the timber industry and wood products to a four-season recreation area. The area is becoming well known and is seeing more out-of-state tourism. In particular, the fishery in Lake Cascade is again on the radar of the traveling angler.

The state fishing rules pertaining to Lake Cascade are simple, with no gear restrictions and a lot of harvest opportunity. The general Southwest Region fishing rules apply, which means daily bag limits can include six trout, six Kokanee, six bass with none under 310 mm (12.2 in) allowed, and two Tiger Muskie with none under 1,020 mm (40.2 in). There are no bag limits or size limits for other species. Anglers are drawn to Lake Cascade primarily for the Yellow Perch and Rainbow Trout fisheries. Kokanee fishing also gets considerable attention, though it's less consistent from year to year. And last, Lake Cascade has recently attracted the interest of tournament bass-club anglers for a try at a growing Smallmouth Bass population.

The Fishery

Yellow Perch

Yellow Perch fishing is back at Lake Cascade. The reservoir is well known for producing large catches of good-size Yellow Perch. A little history on the Yellow Perch fishery is worthwhile to learn. Yellow Perch established quickly after creation of the reservoir, most likely with individuals drifting down from Payette Lake upstream. The population expanded rapidly and created a fishery in the 1950s. By the late 1960s, Northern Pikeminnow (formally called northern squawfish) began to impact the Yellow Perch and trout fishery. Control efforts on the Northern Pikeminnow were conducted for several years and resulted in resurgence of the trout and Yellow Perch. The Yellow Perch fishery increased in popularity into the early 1990s and then went into a steep decline. Fishery managers determined that predation by Northern Pikeminnow was again contributing to the decline of the Yellow Perch fishery. Actions were taken to reduce predators and increase numbers of spawning Yellow Perch in the early 2000s. As the population of Yellow Perch expanded, the anglers have returned to Lake Cascade. In March 2014 a 1.2 kg (2 lb, 10 oz) Yellow Perch was certified as a state-record fish and later as a world-record Yellow Perch taken on ice-fishing tip-up gear.

Ice fishing for Yellow Perch on Lake Cascade is extremely popular and increasingly drawing new anglers to the reservoir. The reservoir generally has safe-fishing ice from the end of December till late March. Access to the ice on the east side is at the Cascade City Boat Ramp, Sugarloaf Boat Ramp, and Boulder Creek Day Use Area and on the west side the Poison Creek Boat Ramp. Anglers do best by being mobile and searching for larger-size Yellow Perch. Many anglers use snowmobiles or all-terrain vehicles to be able to try many areas. The trend in ice fishing is to use lighter jigs and lighter lines than in the past. More varieties of ice-fishing jigs and small lures are on the market now than ever before. Using worms or cutbait tipping a jig or a bare hook is still the most popular method of ice fishing.

In Lake Cascade the open-water Yellow Perch fishery begins when the ice recedes each April, with anglers happy to get on the lake after a long winter. Fishing is generally best right after ice out, with a drop in success after the spawn as the fish recover and move a little deeper into the lake. Again like ice fishing, the Yellow Perch fishery is predominantly a bait fishery, with worms being the preferred bait. The most consistent fishing is by boat, and all ramps and access areas are open when the ice comes off the lake. As the season progresses into summer, the Yellow Perch move farther offshore into 4.5 to 7.6 m (15 to 25 ft) of water. Successful anglers search humps and deep weed lines and move until they locate larger active fish. A good technique is to drift a night crawler on the bottom with the wind until a hookup. A note of caution: Lake Cascade can get rough quickly when a front comes in over the western ridge.

Some of the best fishing is mid-September to November, with a lot of larger fish caught as they go on a bit of a feeding binge. By fall the reservoir is drawn down to meet irrigation demands, leaving most boat ramps and docks inaccessible. The exception is the Blue Heron Boat Ramp on the south side, which is accessible year-round.

Rainbow Trout

Rainbow Trout are the preferred target for many anglers on Lake Cascade. At ice out or even before, as the ice cover recedes from shore, anglers have good success fishing the first open water of the year. Anglers generally fish with bait, either night crawlers or synthetic scented baits, to catch larger trout cruising the shoreline. Lake Cascade grows nice Rainbow Trout from the 254 mm (10.0 in) fish stocked by the state. Just 18 months after stocking, these Rainbow Trout average about 432 to 508 mm (17.0 to 20.0 in), with good full bodies. The trout fishery shifts to trolling as the season progresses. Anglers find that trolling off the west-side golf course and west and north of Sugarloaf Island to be the most productive. There seem to be two types of trollers: anglers who use the large gang spinners (beer can) with a small spoon or hook and worm, and those who use downriggers and smaller trolls or plugs. A good bet is a smaller flasher with a hootchie or small plug. These trout are eating small Yellow Perch to attain their size, so a plug mimicking the bait is worthwhile. As late fall comes along and the water cools, the shoreline bait fishery really picks up. Anglers tend to fish from the beaches near the golf course and south.

Ice fishing for trout can be productive and is usually done while looking for Yellow Perch. Many nice trout are captured on worms that are fished on the bottom. A smart bet is to set one of the ice-fishing rods to fish about five feet under the ice for cruising trout. Large trout have stolen many ice-fishing poles from unwary Yellow Perch anglers.

Smallmouth Bass

Smallmouth Bass are the third fish species that have their own following in Lake Cascade. The population has expanded in the past decade and has some larger fish that anglers enjoy. The size of the Smallmouth Bass population has attracted tournament bass clubs from all over western Idaho. Currently, there are several year classes of bass greater than 410 mm (16.1 in) in length. More bass are probably caught and harvested by open-water perch anglers than are harvested by anglers targeting the bass. The heavier-used areas for Smallmouth Bass fishing tend to be on the east side of the lake from the dam to Sugarloaf Island, and prime time is typically May to early October. Anglers follow the bass into deeper water as the summer arrives, concentrating on the points and submerged humps. The main river channel is also a good bet to find some active bass in the middle to northern part of the reservoir. Most anglers use a variety of soft plastic baits and smaller crankbaits.

Lake Cascade is an Idaho gem that is easily accessible for family fishing throughout the year. The area around Lake Cascade has great camping, resort accommodations, access to federal lands of all types, and great fishing. Excellent family trips are available all summer, especially if you have a boat to get around. Chase large trout in the morning, and then go hunt for schools of Yellow Perch. Visit the area in the fall for fine fishing; it's a great time to pursue large Yellow Perch on their fall feeding binge, see some beautiful changing fall colors, and have a huge reservoir all to yourself. As winter sets in, Lake Cascade is a great ice fishery for Yellow Perch and big trout. No matter the season, Lake Cascade has become one of the destination fisheries in southern Idaho.

Yellow Perch, Lake Cascade, Idaho. Photo by Dale Allen.

Yellow Perch, Lake Cascade, Idaho. Photo by Dale Allen.

Lake Pend Oreille

Andy Dux, Regional Fishery Manager, Panhandle Region, Idaho Department of Fish and Game

Location, Directions, and Access

Lake Pend Oreille is located in the panhandle region of northern Idaho. Most of the lake lies within Bonner County, although the southern tip of the lake is in Kootenai County. Sandpoint is the largest lakeshore community and provides many services and accommodations. There are numerous other small communities around the lake, including Bayview, Garfield Bay, Hope, and Clark Fork. Additionally, the cities of Coeur d'Alene and Spokane are located 48 to 97 km (30 to 60 mi) south of the lake and offer the greatest variety of services and accommodations, including the Spokane International Airport. There are several primary highways that provide access to areas around Lake Pend Oreille. Highway 95 is the main travel route from Coeur d'Alene or from Bonners Ferry or Canada. Highway 2 provides access from northeastern Washington, and Highway 200 can be used to access the lake from northwestern Montana.

Physical Characteristics, Attributes, and Geologic History

Lake Pend Oreille's grand size and beauty are part of what makes it one of Idaho's aquatic gems. It is the largest lake in Idaho, with a surface area of approximately 34,803 ha (86,000 ac). The lake is 69 km (43 mi) long and 9.7 km (6 mi) wide at its widest point, and it has 179 km (111 mi) of shoreline. Maximum depth of Lake Pend Oreille is 353 m (1,158 ft), making it the fifth-deepest lake in the United States. Only Crater Lake at 599 m (1,964 ft), Lake Tahoe at 501 m (1,644 ft), Lake Chelan at 453 m (1,486 ft), and Lake Superior 406 m (1,332 ft) are deeper. Lake Pend Oreille is situated in the southern end of the long, steep-sided Purcell Valley. The Cabinet, Selkirk, and Coeur d'Alene mountain ranges surround the lake, resulting in spectacular scenery in all directions. Northern Idaho receives more annual precipitation than most other regions of Idaho, supporting dense coniferous forests and lush vegetation around the lake. The Clark Fork River provides most of the inflow to the lake, while the Pend Oreille River is the only outflow. Shoreline development has occurred in many areas around the lake, but the majority of land around the lake is public and undeveloped.

Remarkable geologic events led to the formation of Lake Pend Oreille. Lake Pend Oreille was created following a series of floods that are among the largest known floods in geologic history. Near the end of the last Ice Age, much of northern Idaho was covered by ice sheets. An ice mass blocked water from Montana's Clark Fork Drainage from entering Idaho's Purcell Valley. Behind this ice dam was Glacial Lake Missoula, which covered a large portion of western Montana. As the ice sheets advanced and retreated over time, a series of giant floods occurred, and eventually Glacial Lake Missoula was drained. Lake Pend Oreille was formed during this process. This geologic history is well documented, and numerous books and other resources can be found with detailed information (including the Biogeography chapter of this book).

History and Dynamics of the Lake Pend Oreille Fishery

The Lake Pend Oreille fishery was supported almost exclusively by native species until the 1930s. Historically, the fishery was fairly simple. Three native sport fishes, Westslope Cutthroat Trout, Bull Trout, and Mountain Whitefish, were predominant. Several native nongame fishes were also present. Westslope Cutthroat Trout were abundant and readily caught by anglers. Bull Trout were also targeted by anglers, both in the lake and in spawning tributaries during the fall. Similarly, spawning runs of Mountain Whitefish were targeted in tributaries and supported a commercial fishery for many years.

All of the native fish species are still present in the lake, but a multitude of human-related impacts have created an altered and dynamic fishery. Construction of hydroelectric dams, overharvest, land-use activities (i.e., logging, road building, and residential development), and introduction of nonnative fish species are some of the major activities that have influenced the fishery.

Many human impacts have been negative, but some resulted in changes to the fishery that quickly became desirable to anglers. Most notable was the introduction of Kokanee. During the winter flood of 1933, Kokanee were washed downstream from Flathead Lake, Montana. They had been stocked in Flathead Lake from Lake Whatcom, Washington, in 1916. Kokanee quickly became established in Lake Pend Oreille. They were appealing to anglers for several reasons: they were exceptionally good to eat, abundant, and easy to catch with simple angling techniques. In fact, one of the most popular historical angling techniques was hand-lining (vertical jigging without a fishing rod). A popular Kokanee fishery developed almost immediately and thrived until the late 1960s. Kokanee eventually supported both sport and commercial fisheries, which together accounted for an average annual harvest of about one million fish during the 1950s and early 1960s.

Soon after Kokanee were established, Rainbow Trout were stocked in Lake Pend Oreille. The Gerrard strain of Kamloops Rainbow Trout, native to Kootenay Lake, British Columbia, was stocked in 1941 and 1942. Hereafter referred to as Rainbow Trout, this particular strain grows to extraordinary size when Kokanee, their preferred food, are abundant[36]. Like Rainbow Trout, native Bull Trout also prefer to eat Kokanee when they are available. With a newly established and super-abundant Kokanee population, Lake Pend Oreille was soon producing huge fish. In 1947 a world-record Rainbow Trout of 16.8 kg (37.0 lb) was caught, followed by what still stands as the world-record Bull Trout at 14.5 kg (31 lb, 15 oz) in 1949. The historical fishery quickly transformed after Kokanee introduction, and Lake Pend Oreille developed a reputation as a world-class trophy fishery for Rainbow Trout and Bull Trout.

Great fishing lasted for several decades after Kokanee were established. Kokanee were abundant and relatively easy to catch, thus offering a high-yield and low-tech fishery. Bull Trout and Rainbow Trout provided an unrivaled trophy fishery for anglers who were willing to use more specialized angling methods. In combination, this resulted in Lake Pend Oreille being the most popular fishery in the state of Idaho in the 1950s and early 1960s. Unfortunately, a combination of factors led to a dramatic decline and eventually a near collapse of the Kokanee population. Construction and operation of Albeni Falls Dam downstream on the Pend Oreille River and Cabinet Gorge Dam upstream on the Clark Fork River began in the 1950s, ultimately having negative effects on the Kokanee population. Additionally, small freshwater shrimp known as mysid shrimp (*Mysis diluviana*) were introduced to Lake Pend Oreille in the late 1960s. These shrimp were intended to provide additional food for Kokanee, as had occurred in Kootenay Lake. However, the extreme depths of Lake Pend Oreille allowed the mysid shrimp to migrate deep enough during the day to avoid Kokanee predation. Compounding the problem, they competed with Kokanee for the same zooplankton food source by feeding in shallower depths at night.

Mysid shrimp eventually had an even greater negative effect on the Kokanee population: they provided an abundant food supply for juvenile Lake Trout. Lake Trout were stocked in 1925 but existed at low density and were rarely caught by anglers until the 1990s. The Lake Trout population exploded in the 1990s, after an initial time lag, because of increased juvenile survival from the newly available mysid shrimp food source. Kokanee are the preferred prey of Lake Trout, and added predation quickly threatened an already depressed Kokanee population. By 2000 the combined predation from Rainbow Trout, Bull Trout, and Lake Trout had caused what is known as a predator trap. There were simply too many predators to sustain a Kokanee population, and the Kokanee fishery was closed to harvest. In addition, Lake Trout threatened the native Bull Trout population. Bull Trout have declined or been extirpated in every instance where Lake Trout have been introduced in the western United States. Suddenly, the once-storied Lake Pend Oreille fishery was at risk of collapse.

In an effort to recover the fishery, IDFG initiated an aggressive predator management program in 2006. A commercial fishing company with Lake Trout netting experience from the Great Lakes was contracted to remove Lake Trout. Additionally, anglers were enlisted to help remove predators through an incentivized harvest program that paid $15 for each Lake Trout and Rainbow Trout turned in. The Lake Trout population was dramatically reduced, and the Kokanee population rebounded in response to less predation. The Lake Trout netting program and bounty program are ongoing to maintain a low-density Lake Trout population. The Rainbow Trout bounty has since been removed, and trophy management regulations are again in place. Rainbow Trout over 9.1 kg (20 lb, 1 oz) are becoming increasingly common as Kokanee become more abundant. Additionally, the Kokanee fishery was reopened in 2013 after a 13-year closure. Native Bull Trout and Westslope Cutthroat Trout are also benefiting from reduced Lake Trout abundance. Both species continue to be managed under catch-and-release regulations and are regularly caught by anglers. A resurgence of the Lake Pend Oreille fishery appears to be under way. Time will tell if the fishery recovery process will continue to gain momentum and be sustainable, but right now the future for angling on Lake Pend Oreille looks brighter than it has in years.

While much of the attention is generally focused on the lake's renowned traditional fishery, other peripheral fisheries have developed over time. Numerous other nonnative sport fishes have become established in Lake Pend Oreille that now offer angling opportunity. Several of the most notable species that are now pursued by anglers are Smallmouth Bass, Largemouth Bass, Brown Trout, Black Crappie, and Walleye. Smallmouth Bass are widespread along the entire shoreline of the lake, while the other aforementioned species are more common on the lake's shallower northern end and in the Pend Oreille River. Also, Lake Whitefish are sometimes targeted by anglers.

Fishing Tips and Techniques

Fishing Lake Pend Oreille can range from fairly easy to extremely difficult depending on your level of angling experience, the species you wish to catch, the time of year, and weather conditions. Many anglers, both local and visiting, have successful self-guided trips. For those less familiar with the lake, numerous guide services are available in the area. There are also boat rentals available at several locations around the lake. Because the Lake Pend Oreille fishery is diverse, anglers can easily find a species to target that is the best fit for them. Some fishing tips for the most commonly targeted species follow.

Kokanee

Expect Kokanee to average 229 to 305 mm (9.0 to 12.0 in) in size, although some larger fish exist. Fishing for Kokanee is best from May through October. They can typically be caught at all times of day, although their depth distribution may shift deeper during the day. Kokanee are widespread throughout the entire offshore portion of lake when they are abundant. This makes them fairly easy to fish for because they can be found without prior knowledge about specific fishing locations. Since Kokanee live offshore, a boat is necessary to effectively target them. Trolling at slow speeds, usually 1.6 to 2.4 kph (1 to 1.5 mph), is the primary angling technique. Kokanee can be caught near the surface during the spring, but they move deeper as the summer progresses. An electronic fish finder is helpful for locating the depth where Kokanee are most numerous. Trolling in depths of 6 to 18 m (20 to 60 ft) is most common during summer. Downriggers are helpful for reaching these depths, but leaded lines or 1-3 trolling weights are most commonly used by anglers. A variety of fishing rods can be used, but 2.1 to 2.3 m (7 to 7.5 ft) rods are common. A level-wind reel and 6- to 10-lb test line are preferred by many anglers. Dodgers or multiblade attractors (i.e., Jack Lloyd, Hildebrandt, and Ford Fender) fished ahead of a lure are the most common tackle used. Numerous lures are available, but popular lures include wedding-ring spinners, Apex Kokanee Killers, or small hoochies. Lures are usually tipped with white corn or maggots. Kokanee have very soft mouths, so some anglers use a rubber snubber to improve hooking and landing rates. Having a landing net also greatly increases the number of fish that make it into the boat.

Rainbow Trout

Most Rainbow Trout are 0.91 to 3.63 kg (2.0 to 8.0 lb), and fish up to 9.1 kg (20 lb, 1 oz) are caught regularly. Fish in the 9.1 to 13.6 (20.1 to 30.0 lb) range are less common, but some are caught each year. Rainbow Trout fishing is best in the months of May, June, October, and November. These are the months when water temperatures are cooler and fish are found near the surface. They can also be caught in the summer, but they move deeper when water temperature warms and are more difficult to target. Rainbow Trout feed on Kokanee, so they are widely distributed throughout the offshore portion of the lake. A boat is necessary to effectively target them, but much like Kokanee, knowledge of specific fishing locations is not critical because they are so widespread. The key is to start trolling, cover a lot of water, and spend more time where you see schools of Kokanee. Rainbow Trout exist at a much lower density than Kokanee and require more specialized angling techniques. When they are located near the surface, the presence of a boat will spook them. As a result, most anglers fish with planer boards that spread their fishing lines as much as 46 m (150 ft) to the sides of the boat. In addition, the lure or fly is commonly run 30.5 to 91.4 m (100 to 300 ft) behind the planer board line, getting it well away from the boat. In the summer, fish are deeper, and downriggers are generally used instead of planer boards. The potential to catch a fish over 9.1 kg (20 lb, 1 oz) that is capable of making long runs is the big appeal to this fishery. This means a heavy trolling rod (often about 2.7 m or 9 ft), a level-wind reel with a good drag and high-line capacity of 244 to 305 m (800 to 1,000 ft), strong line (16- to 20-lb test), and strong leaders (10- to 14-lb test) are needed. Trolling speed is something anglers pay close attention to, often running 4.0 to 4.8 kph (2.5 to 3 mph). Dedicated Rainbow Trout anglers often have boats with an autopilot system that allows them to spend time on the deck tending to rods without having to manually drive the boat. Terminal tackle choices vary, with both flies and lures being popular. Frisky Jenny flies are streamers that many anglers use when fish are near the surface. Apex lures and Rapala lures are very popular, whether fished at the surface or deeper.

Lake Trout

A good Lake Trout fishery existed from the mid-1990s until about 2010. As intended, the aggressive netting and angler bounty program started in 2006 has reduced the population so much that fishing for them is now fairly difficult. Some anglers still target Lake Trout, and those who do can claim a $15 reward for each fish they harvest. Fish up to 11.3 kg (24 lb, 14 oz) are still caught on occasion. Lake Trout can be caught throughout the year, but April through October is best. They are found throughout the lake, both suspended offshore and near the bottom closer to shore. Traditionally good areas include Whiskey Rock, Lee's Point, and along the north shore from the Clark Fork River to Hope. Trolling is the most common angling technique. In the spring, Lake Trout are commonly caught near the surface, but during the summer and fall the use of downriggers is important to fish deeper. Fishing depths of 15.2 to 38 m (50 to 125 ft) are most common. Trolling rods, reels, and lines similar to those described for Rainbow Trout fishing work well. Popular lure choices include Apex lures, Rapalas, Lyman plugs, and hoochies. Running a dodger ahead of the lure is done by many anglers.

Bull Trout

Lake Pend Oreille supports one of the strongest Bull Trout populations throughout the range of the species. They reach large sizes, with fish over 4.5 kg (9 lb, 14 oz) fairly common. It is a catch-and-release fishery, but anglers are allowed to target them. Bull Trout are strong fighters, and Lake Pend Oreille offers one of the best opportunities to catch one of these beautiful native fish. The best fishing occurs from April through November, although many of the larger fish leave the lake during the fall to spawn in streams. Effective fishing techniques are similar to those described for Lake Trout. Bull Trout are often caught by anglers fishing for both Rainbow Trout and Lake Trout. Make sure you know how to identify a Bull Trout so that you don't accidentally harvest one!

Westslope Cutthroat Trout

There are not as many Westslope Cutthroat Trout in Lake Pend Oreille as there were historically, but they are still commonly caught by anglers. Fish size is typically 356 to 510 mm (14.0 to 20.1 in). The best fishing is from May through October. Westslope Cutthroat Trout are commonly distributed closer to the shoreline. They can be caught from shore in places, but the most effective strategy is to troll along the shorelines of the lake. They are generally located in fairly shallow water, so downriggers aren't necessary. A wide variety of spinners and spoons are effective. This is a catch-and-release fishery, but it offers anglers the opportunity to catch one of Idaho's most popular native sport fish.

Smallmouth Bass

Lake Pend Oreille offers good Smallmouth Bass fishing. They are distributed around the entire shoreline of the lake. Fish up to 2.3 kg (5 lb, 1 oz) are commonly caught. Smaller fish are abundant and fairly easy to catch in shallow waters around the lake, while larger fish often occupy slightly deeper waters along the shoreline. The somewhat shallower northern end of the lake, including around the islands, is the most popular area for Smallmouth Bass fishing. Focus on areas with rocky shorelines, docks, pilings, and other forms of structure that Smallmouth Bass like to use as cover. The best fishing occurs from May through October. In the spring, warmer water temperatures over 10°C (50°F) trigger the start of good Smallmouth Bass fishing. They spawn along the lakeshore in the spring, which is an effective time to target them. Smallmouth

Bass can be caught using a wide variety of rods and tackle. Rods are commonly 1.8 to 2.1 m (6 to 7 ft) long and geared with a spinning or bait-casting reel and 3.6 to 5.4 kg (8 to 12 lb) test line. Tackle also varies widely, ranging from plastic jigs to crankbaits. Fish can be caught from shore, but working along the shoreline in a boat is most effective.

Lake Pend Oreille, Idaho. Photo by Glenn Oakley, courtesy of IDFG.

One happy angler, Alexander Fremier Lake Pend Oreille, Idaho. Photo by Carson Jeffres.

Kameron Roslak with a 13lb. Rainbow Trout caught in Lake Pend Oreille. Photo by Ryan Roslak.

Lake Pend Oreille, Idaho. Photo by Glenn Oakley, courtesy of IDFG.

Lucky Peak Reservoir

Joe Kozfkay, Regional Fishery Manager, Southwest Region, Idaho Fish and Game

Location, Directions, and Access

Located just outside of the Treasure Valley, near Idaho's western border, Lucky Peak Reservoir is one of the most heavily fished waters in the state. Lucky Peak Reservoir is just between Boise and Idaho City, the closest neighboring communities. About 64 km (40 mi) of Highway 21, the main access road to Lucky Peak, separate the communities. Despite the close proximity, these communities couldn't differ more. Boise is a popular and growing urban area that offers a wide variety of outdoor-based recreational opportunities along the Boise River and in the surrounding foothills, as well as all the services that can be expected from a large city. On the opposite side of the spectrum, Idaho City is much smaller. Idaho City was founded in the 1860s, during the gold rush days, and by all accounts was a bustling place. Mining activity has declined, and the human population shrank gradually from its high of 7,000 residents during the 1870s to its current population of just less than 500. Many relics and historical sites from the glory years still remain. Tourism and outdoor recreation are important to the local economy.

Lucky Peak Reservoir has several access areas. Two of the access areas are located on the south end of the reservoir, near or on the dam. Barclay Bay and Turner Gulch access sites offer ramps, restrooms, and parking areas. The other access areas are located farther northeast. From Boise, travel farther north on Highway 21 until you cross the high bridge, and immediately afterward turn east on Arrowrock Road (East Spring Shores Rd). Spring Shores Marina and Macks Creek access areas are found between the reservoir and the road. Spring Shores Marina is one of the most popular launching areas on Lucky Peak Reservoir, as it is spacious, possesses a long ramp that is accessible at most water levels, is centrally located, and possesses good amenities. Access areas are managed by US Army Corps of Engineers or by Idaho Department of Parks and Recreation (IDPR), and their respective fee schedules apply. It is important that anglers check reservoir levels before selecting a boat ramp to ensure access to the water. In addition, IDPR maintains a series of docks and camping areas around the reservoir at which anglers may stay and fish for a couple of days.

Physical Characteristics and Attributes

Lucky Peak's most popular forms of recreation include pleasure boating, angling, and sightseeing. Most people visit Lucky Peak Reservoir just to be on the water, to swim, to bask in the sun, to enjoy its small beaches. Others visit for more active hobbies, such as water skiing or wakeboarding. Wildlife viewing is also a popular activity. Surrounded by the Boise River Wildlife Management Area and other public lands, Lucky Peak can be a great wildlife viewing area, especially during the late fall and winter months when deer and elk congregate on the reservoir's slopes to spend the winter. During a harsh winter or late spring, it's not uncommon to see several hundred deer and elk while on the reservoir. Osprey, Bald Eagle, Common Loon, and other birds frequent the reservoir, especially in the spring months, to feed on fish during their seasonal migrations. At just about the same time, anglers begin targeting the reservoir's most prized fish, Kokanee. Sure, other anglers visit to target Rainbow Trout or Smallmouth Bass, but without a doubt Kokanee is king in Lucky Peak Reservoir.

Lucky Peak Reservoir is one of four reservoirs in the Boise Project: a system of dams and reservoirs designed and operated to provide irrigation water, flood control, power production, fish and wildlife habitat, and recreation. Lucky Peak was the last dam built within the project, and construction occurred from 1949 to 1955. Since its primary purpose is flood control, Lucky Peak is operated by the USACE, as opposed to the other reservoirs in the system that are managed by the BOR. When filled to full pool levels, Lucky Peak Reservoir is about 1,222 ha (3,019 ac) and contains 377, 445, 442 cu m (306,000 ac-ft) of water. Close to Idaho's largest metropolitan area, the area sees heavy recreational use. A 2011 estimate indicated that almost 800,000 people visited Lucky Peak Reservoir. According to a different but similarly timed survey, fewer than 10 percent of the people visiting Lucky Peak were there for the fishing.

Lucky Peak Reservoir is deep, cold, clear, and relatively infertile. The earthen-filled dam is about 76 m (250 ft) tall and is located in what was a relatively narrow U-shaped or incised canyon. Much of the reservoir's shorelines is rocky and steep, as is the bottom of the reservoir. Lucky Peak is also located directly downstream of Arrowrock Dam from which water is doled out low in the water column. This water-release pattern contributes to cold water temperatures in Lucky Peak for much of the year. Furthermore, Arrowrock Dam and Reservoir trap much sediment and nutrient material from the North, Middle, and South Forks of the Boise River, causing Lucky Peak to be relatively sterile and clear.

Lucky Peak Fishery

Kokanee

Lucky Peak Reservoir is first and foremost a Kokanee fishery created by the annual stocking of early-run Kokanee from IDFG hatcheries. Kokanee eggs are collected at Deadwood Reservoir and reared at Cabinet Gorge Hatchery and Mackay Fish Hatchery. About 250,000 fingerlings are transported by semitruck and transport trailer and stocked in late May or early June. By this time, the young Kokanee have grown to lengths of approximately 76 mm (3.0 in) on a diet of hatchery pellets. The timing of stocking is set to correspond to upswings in zooplankton abundance, Kokanee's primary food. Young Kokanee gorge themselves on *Daphnia*, copepods, and other zooplankton. By June of the following year, they will have grown to the 254 to 305 mm (10.0 to 12.0 in) range and start to be caught by anglers. By the following June, the now two-year-old fish will have reached or exceeded 356 to 406 mm (14.0 to 16.0 in) and will be the most frequently encountered age class in the creel. Most Kokanee in Lucky Peak mature and die later in the fall as age-two 406 to 457 mm (16.0 to 18.0 in) adults. However, a small portion of the population delays maturation another year and continues to grow. These fish occasionally reach or even slightly exceed 508 mm (20.0 in).

Kokanee fishing is primarily a trolling game. Anglers use several methods to ensure their offerings reach the proper depth where Kokanee reside. Anglers tailor their methods to the time of year. In the spring, Kokanee are spread throughout the water column, with some right on the surface and others as deep as 15 or 19 m (50 or 60 ft). As spring progresses into early summer, most Kokanee are caught in the 4.5 to 9 m (15 to 30 ft) range. By mid- to late summer, it's not uncommon for most Kokanee to be caught in the 7.6 to 12 m (25 to 40 ft) range. Leaded line, trolling weights, and downriggers are the messengers that send baits to the proper depth. Terminal gear usually includes some sort of flasher or dodger followed by a bead and spinner assortment with a hook or a pair of hooks. Alternatively, small squids (a.k.a. hoochies) with paired hooks have become popular recently. Nearly all Kokanee anglers tip these offerings with shoepeg corn. Not stopping there, a variety of scents are added to the corn. The number of combinations is almost limitless. Rigs are trolled at slow speeds, usually in the neighborhood of 0.8 to 2.9 kph (0.6 to 1.8 mph).

Rainbow Trout

Lucky Peak Reservoir is also stocked with about 11,500 catchable-size Rainbow Trout annually. These fish are usually stocked during spring, but can be caught throughout the year. Rainbow Trout are commonly caught while trolling for Kokanee. Others are caught from shore. Shore fishing can be outstanding in late winter, when anglers utilize bottom rigs and present baited hooks on gently tapered sand beaches. Also, at this time of year, Rainbow Trout can be caught immediately downstream of the safety exclosure area of Arrowrock Dam. Anglers are reminded to be careful in this area, as footing can be treacherous on the steep banks used to access this area. Plus, Lucky Peak harbors some wild trout. The origins and genetic composition of these fish are not well understood. It is more than likely that some come from small tributaries in the Mores Creek drainage, whereas others are produced upstream of Arrowrock Dam and are entrained.

Smallmouth Bass

Lucky Peak possesses an overlooked Smallmouth Bass population and fishery. Though this fishery will never compare to CJ Strike Reservoir or Brownlee Reservoir from a numbers perspective, it makes up for its low numbers by having solid average sizes. During the early season, anglers focus their efforts in the back ends of coves to take advantage of warmer waters found there. These slightly warmer waters attract Smallmouth Bass and their prey and make bass more likely to bite. Later in the season, anglers target rocky outcroppings, specifically those that have more moderate slopes. Due to the extremely clear waters, Smallmouth Bass hold deeper than in most other waters. Heavy jigs, drop-shot rigs, or deep-diving crankbaits capture most of the fish.

Kokanee in Hand, Martin Koenig, Lucky Peak Reservoir, Idaho. Photo courtesy IDFG.

Lucky Peak Reservoir, Idaho. Photo courtesy IDFG.

Early morning trolling on Lucky Peak Reservoir, Idaho, photo courtesy of IDFG.

Middle Fork Salmon River
Frank Church–River of No Return Wilderness

Russell F. Thurow, Fisheries Research Biologist, US Forest Service: Rocky Mountain Research Station, Boise, ID
Bart L. Gamett, Fisheries Biologist, US Forest Service: Salmon-Challis National Forest, Mackay, ID
Richard N. Williams, Fisheries Geneticist, Department of Biology, The College of Idaho, Caldwell, ID

Location, Directions, and Access

The Middle Fork Salmon River (MFSR) flows through the central Idaho wilderness. US Highway 21 provides primary access to southern portions of the basin near Stanley. Highway 93 provides access to eastern portions near the towns of Challis and Salmon. Highway 55 provides access to western portions near McCall. Each of these towns supports air-taxi and guide services. Driving times to Salmon are 2.5 hours from Missoula, 5 hours from Boise, and 5.5 from Salt Lake City.

Physical Description and Attributes

Idaho's Middle Fork Salmon River flows through the heart of the 906,136 ha (2.2 million ac) Frank Church–River of No Return Wilderness (the Frank), the largest contiguous wilderness in the lower 48 states. In 1968 the MFSR was one of the original eight rivers designated during creation of the National Wild and Scenic Rivers System. With more than 100 named rapids and a 900 m (2,950 ft) drop in elevation along its 171 km (106 mi) length, the MFSR is one of the world's premium fishing and rafting destinations. As a result of its exceptional native trout fishery, spectacular scenery, abundant wildlife, archaeological and historical sites, challenging white water, an opportunities for solitude, many consider a float trip down the MFSR an adventure unparalleled in the United States. The MFSR basin also contains hundreds of mountain lakes that offer scenic and spectacular alpine fishing opportunities.

Wilderness Values

From its origin at the confluence of Bear Valley and Marsh Creeks, the MFSR flows north-northwest through the Salmon River mountains and joins the Salmon River 92 km (57 mi) downstream from Salmon, Idaho. Twelve major streams and hundreds of smaller ones are tributary to the river. Human access in the lower 154 km (96 mi) of the main stem and most tributaries is limited to backcountry airstrips, float craft, or by trail. From 1930 to 1980, a majority of the region was managed in "Primitive Area"[37] status. In 1980 the Central Idaho Wilderness Act established this wilderness that remains the largest in the national forest system[37]. In 1984 its name was adopted in honor of the late Senator Frank Church's efforts to secure wilderness designation.

Most of the MFSR drainage has been relatively undisturbed by anthropogenic activities, so habitat quality is good to excellent. Its protected status also enables natural processes to function relatively unimpeded across this extensive wilderness landscape. As a result, fires, floods, debris flows, and snow avalanches recruit wood and sediments to streams. These processes create and maintain a dynamic mosaic of landscape conditions that support diverse, high-quality, and connected habitats[38]. These habitats provide a physical template essential to the expression of native species' life history and genetic diversity. The MFSR basin contains some of the highest-quality, most diverse, and most functional natal, wild salmon, Steelhead, and trout habitats in the Columbia River basin. Although past land use degraded habitat in isolated areas, since wilderness designation past perturbations have largely been eliminated. With additional land-use constraints to protect species listed under the Endangered Species Act (ESA), most degraded habitats are recovering.

As a result of its large size, functioning natural processes, and the diverse, high-quality, connected habitats they create, a nearly complete native species assemblage persists in the Frank; only grizzly bears (*Ursus arctos horribilis*) and indigenous people are absent. Today a host of federally listed species (Chinook Salmon, Steelhead, Bull Trout, and lynx [*Lynx canadensis*]), federal candidate species (white-bark pine [*Pinus albicaulis*]); recently federally delisted species (peregrine falcon [*Falco peregrinus*], bald eagle [*Haliaeetus leucocephalus*], and gray wolf [*Canis lupus*]), USFS Regions 1,4-Sensitive Species (Pacific Lamprey; Westslope Cutthroat Trout; inland Redband Trout; Columbia spotted frog [*Rana luteiventris*]; black-backed [*Picoides arcticus*] and American three-toed [*Picoides dorsalis*] woodpeckers; boreal [*Aegolius funereus*], flammulated [*Psiloscops flammeolus*], and great gray [*Strix nebulosa*] owls; spotted [*Euderma maculatum*] and Townsends big-eared [*Corynorhinus townsendii*] bats and fringed myotis [*Myotis thysanodes*]; common loon [*Gavia immer*]; harlequin duck [*Histrionicus histrionicus*]; northern goshawk [*Accipiter gentilis*]; wolverine [*Gulo gulo*]; and fisher [*Martes pennanti*]), and several sensitive native plants depend on this wilderness and its natural processes for essential habitats[38].

Fish Species

The MFSR supports 15 native fishes, including seven salmonid forms: Westslope Cutthroat Trout; Bull Trout; Mountain Whitefish; Redband Trout, which include both resident and anadromous (summer Steelhead) stocks; and spring and summer Chinook Salmon[39]. Many Redband Trout less than 230 mm (9.1 in) are actually juvenile Steelhead that will migrate to the ocean after physiologically transforming to a smolt. Healey[40] categorized juvenile Chinook Salmon that migrate seaward after one or more years as stream type. MFSR stream-type Chinook Salmon include both spring- and summer-run fish[41–44]. Spring Chinook Salmon cross Bonneville Dam on the Columbia River from March to May, summers from June to July, and falls from August to September[45,46]. Wild, indigenous stream-type Chinook Salmon and summer Steelhead populations like those in the MFSR are rare. Thurow et al.[47] reported their presence in 4 percent and 10 percent of the potential historical range, respectively, and 15 percent and 22 percent of the current range, respectively, in the Columbia River and portions of the Klamath River basins. Most wild populations in the Columbia River and Klamath River basins were either extirpated or have been supplemented with hatchery-reared fish. MFSR Chinook Salmon are especially unique because they spawn at the highest elevations of any spring/summer Chinook Salmon population in the world[48]. Other native fishes include three sculpin species, Torrent Sculpin, Mottled Sculpin (now provisionally described as Columbia Sculpin *Cottus hubbsi*), and Shorthead Sculpin; Speckled Dace and Longnose Dace; Largescale Sucker and Bridgelip Sucker; Redside Shiner; Pacific Lamprey and Northern Pikeminnow[39].

Introduced fishes are uncommon in the MFSR and most tributaries. Only Brook Trout are locally abundant in a few streams, and they tend to be patchily distributed[49] primarily in road-accessible meadow streams. Five other species, including exotic forms of Rainbow Trout and Westslope Cutthroat Trout, Yellowstone Cutthroat Trout, Golden Trout, and Arctic Grayling, have been primarily introduced to alpine lakes. Hybrids between Yellowstone Cutthroat Trout and Westslope Cutthroat Trout and natural hybrids between native Redband and Westslope Cutthroat Trout have been detected in the MFSR basin[18].

Nearly every major MFSR tributary has headwater alpine lakes: the Bighorn Crags, Roughneck Peak, Sleeping Deer, and Soldier Creek regions all have numerous alpine lakes and are great places to begin a MFSR mountain lake–fishing adventure[50]. Although fish currently occupy many alpine lakes, few, if any, naturally supported fish. Most of the lakes were barren because their outlet streams are too steep for fish to ascend or they lack an outlet stream connected to downstream areas. As Euro-Americans settled central Idaho, they recognized the potential of these lakes and began stocking fish. Over many decades, exotic forms of the six species listed above were introduced. Today, fisheries in some lakes have self-sustaining populations established through earlier introductions, while fisheries in other lakes are maintained by regular aerial or horse-supported stocking. Other lakes are managed for the protection of native aquatic communities (i.e., insects and amphibians), so remain fishless.

Historical Angling

Native Americans inhabited the MFSR and utilized its fisheries resources for at least 10,000 years[51]. Mountain Shoshoni (Sheepeaters) were the principal inhabitants, and they shared fishing grounds and traded with neighboring Flathead and Nez Perce tribes[52,53]. Historical accounts confirm the exceptional trout fishing in the MFSR. Carrey and Conley[54] report 1.4 to 1.8 kg (3.1 to 4.0 lb) trout were common, as were catches of 100 trout per day. Trout populations began to decline in the 1950s, as human populations increased and citizens enjoyed more leisure time and better gear[55]. Although Westslope Cutthroat Trout have a life history that is uniquely adapted to the MFSR, excessive angler harvest threatened their persistence. Mallet[56] used tag recoveries in 1960 to document the migratory nature of Westslope Cutthroat Trout that spent about two years in tributaries before migrating distances of often more than 150 km (93 mi) to overwintering habitats in the lower MFSR and Salmon River. More recently, researchers[57] used radio telemetry to track movements of 30 adult Westslope Cutthroat Trout that were collected in Bear Valley Creek. The authors reported that by February, tagged fished were distributed over more than 150 km (93 mi) of habitat, from Bear Valley Creek to the Salmon River. Trout began returning upstream in March as water temperatures and discharge increased, and movements peaked in May. The longest total and downstream movements were 475 km (295 mi) and 194 km (120 mi), respectively[57]. When unconstrained by barriers, Westslope Cutthroat Trout migrate long distances and exhibit variable seasonal habitat use.

In response to the MFSR's cold water temperatures and relatively low productivity, Westslope Cutthroat Trout tend to feed very actively, aggressively rising to terrestrial and aquatic insects. As a result of their slow growth rates, most Westslope Cutthroat Trout in the MFSR do not spawn until age six[55]. This combination of slow growth, aggressive feeding, and vulnerability to anglers in both the MFSR and the Salmon River led to very high angler exploitation. Their vulnerability was exacerbated by increasing numbers of floaters and anglers. In 1949 about 25 people floated the MFSR; twenty years later, in 1969, there were about 1,624 floaters[53]. Sixty years later, in 2009, there were 10,003 floaters; more recently, in 2013, 11,360 people floated the MFSR[58]. With increasing angler effort, most Cutthroat Trout were being harvested before reaching reproductive maturity. Declines in the Cutthroat Trout populations led to a series of more restrictive angling regulations: in 1963 the Cutthroat Trout limit was reduced from 15 fish per day to three fish per day; in 1969 the limit was reduced to two trout per day, bait fishing was prohibited, and barbless hooks were required; in 1972 catch-and-release was implemented[55]. The response of MFSR Westslope Cutthroat Trout to restrictive angling regulations was dramatic. In 1972 13 percent of the Cutthroat Trout observed in snorkel transects exceeded 304 mm (12.0 in)[59]; by 1977 this percentage increased to 31 percent[60], and by 1982 62 percent of the Cutthroat Trout observed in transects exceeded 304 mm (12.0 in)[61]. In 2012 46 percent of observed Cutthroat Trout exceeded 304 mm (12.0 in)[61,62]. Reduced harvest has enabled more trout to survive to an older age and successfully spawn. Kennedy and Meyer[63] summarized long-term snorkel survey data and reported statistically significant Westslope Cutthroat Trout population increases in the mainstem MFSR and tributaries from 1985 to 2012. Despite more than 10,000 people annually floating the MFSR, restrictive angling regulations coupled with diverse, connected, high quality habitats have contributed to a secure native Westslope Cutthroat Trout population.

The MFSR and its tributaries also contained robust populations of Chinook Salmon and Steelhead, which formerly supported popular sport fisheries. In the 1880s, an estimated 1.5 million Chinook Salmon returned to the Snake River basin, including the Salmon River basin[64]. Salmon and Steelhead populations plummeted as the Northwest was developed, and by 1995 fewer than 1,200 wild Chinook Salmon returned to the Snake River basin[65]. Federal fisheries management agencies identified "four *H*s" (habitat degradation, harvest, hatchery practices, and hydrosystem operation) as the primary causes of anadromous fish declines[66]. By the early 1990s, all wild Snake River Chinook Salmon and Steelhead populations were federally listed under the ESA. Despite abundant, high-quality natal habitat, absence of hatchery fish, and low ocean harvest rates verified by tag returns, MFSR Chinook Salmon and Steelhead remain at risk of extirpation, primarily as a result of outside basin factors in the Columbia River and Snake River migration corridor, estuary, and ocean. Today, all anadromous fish in the Salmon River basin must navigate eight dams (four in the Columbia River and four in the lower Snake River) to reach the ocean as smolts and then ascend those same eight dams as adults returning to spawn in MFSR natal areas. Researchers documented the adverse effects of these dams and impoundments on salmon and Steelhead survival[67]. As a result of the severe MFSR population declines, angling has been closed for Chinook Salmon and Steelhead since 1978. All salmon and Steelhead must be immediately released unharmed if accidentally hooked. Salmon are keystone species that provide ocean-derived nutrients to aquatic and terrestrial ecosystems[68]. The severe declines in anadromous fish within the Columbia River basin (current wild populations are about 2 percent of historical numbers)[69] have reduced the productivity and adversely affected the functioning of the entire MFSR ecosystem.

Current Regulations

Current MFSR angling regulations require catch-and-release fishing with barbless hooks, no bait is allowed, and harvest of all native trout is prohibited. Only Mountain Whitefish and unprotected species such as Northern Pikeminnow and nonnative Brook Trout (considered a nuisance species and a hybridization threat with native Bull Trout)[49] may be harvested. Alpine lakes are managed under a general regulation that allows harvest and fewer gear restrictions relative to MFSR streams. In alpine lakes, bait and barbed hooks are allowed, and anglers may harvest six fish of all species (Golden, Rainbow, and Cutthroat Trout forms; hybrid trout; and Arctic Grayling) combined. Regulations allow separate harvest of 25 alpine-lake Brook Trout.

Tackle and Techniques

Both fly and lure angling techniques are effective for native trout in the MFSR and its tributaries. Most lure anglers use light spinning tackle and gold or silver spinners (size 1 or 2) with all but one hook removed and the barb on the remaining hook crimped. Lure anglers typically move their lures through likely holding waters by casting perpendicular to the current or quartering downstream and then retrieving the lure upstream against the current.

Most fly anglers use a rod approximately 2.4 to 2.7 m long (8 to 9 ft) with a 4- or 5-weight floating line, a tapered 3 to 4 m (9 to 13 ft) leader, and 6X (3 lb) to 3X (8 lb) tippet, depending on fly size. Anglers use dry flies, nymphs, and wet flies or streamers. Dry-fly anglers either sight-cast to rising trout or fish likely holding water along banks, near overhead cover, and in deeper pools and pockets. Wading fly anglers use a classic upstream as well as a down-and-across presentation to avoid casting directly over feeding fish. When fishing from a float craft, dry flies are usually cast downstream and across or perpendicular to the current. Many anglers fish a dry fly with a small- to medium-size nymph suspended below the dry fly. Other anglers fish nymphs with a strike indicator, which is usually positioned 1 to 2 m (3 to 6 ft) above the nymph. Often, a second nymph is fished as a dropper, about 0.5 to 1 m (1.5 to 3 ft) below the upper nymph. Upstream and across-stream nymph presentations are most common. Streamer fishing can also be effective, typically casting down and across the current, allowing the fly to sink; mending the line, enabling the fly to swing across the current; and then retrieving the fly by stripping.

After spring flows subside in July, most MFSR tributaries may be fished effectively. The lower reaches of some of the larger tributaries offer exciting dry-fly fishing opportunities for wading anglers. Although the main-stem MFSR may be waded along its margins, most anglers agree that floating provides the best option for fishing and accessing holding waters in the main stem. Catch rates for anglers floating the length of the MFSR sometimes decrease in the lower Impassable Canyon reach, where there are fewer major tributaries. Here, fishing deeper holding waters along canyon walls and amid large boulders can be very effective for larger fish.

Fishing MFSR alpine lakes can be a very rewarding experience for anglers of all ages and skill levels. Lake fish often feed aggressively and will typically respond to a wide range of lures, flies, and baits. An effective alpine-lake technique is to use spinning tackle with a weighted bobber (cast-a-bubble) and fly attached 1 to 2 m (3 to 6 ft) beneath the bubble. This approach allows anglers to cast farther and avoids the need for back-casting space.

Hatches and Fly Patterns

Effective MFSR fly patterns primarily imitate three major aquatic insects, mayflies (Ephemeroptera), stoneflies (Plecoptera), and caddisflies (Trichoptera), and one terrestrial insect, grasshoppers (Orthoptera). Effective dry-fly patterns are easily cast, float high in the water, and are readily seen. Effective dry patterns include the Elk-Hair Caddis, Stimulator, Renegade, Humpy, Adams, and Royal Wulff. A variety of grasshopper patterns such as the foam-bodied Chernobyl Ant (and its variations) are durable and effective for locating feeding fish. Nymph patterns including the Hare's Ear, Pheasant Tail, Prince, and imitations of larger stonefly nymphs are also effective.

Fly patterns imitating mayflies, stoneflies, and caddisflies are also effective in alpine lakes. Streamers, Woolly Bugger, beetle, and black ant patterns may be effective when aquatic insect hatches are absent.

Other Tips

With the exception of isolated private inholdings, the vast majority of lands and waters within the MFSR are publicly owned, so they are accessible to any licensed angler willing to exert the effort. Upper reaches of several tributaries are road accessible outside the wilderness boundary. Within wilderness, access is gained via trail, float craft, or backcountry airstrips. Anglers seeking qualified backcountry trail, float, or air services are encouraged to contact the Idaho Outfitters and Guides Association (IOGA), PO Box 95, Boise, ID 83701; Office: (208) 342-1438; or visit http://ioga.org. IOGA provides information on air transportation and trip shuttles; accommodations and dining; gear manufacturers, distributors, and suppliers; retail sports equipment and clothing rental or sales; and travel agents, tours, and tourist information.

Trail access provides many options within the MFSR; anglers may drive to a trailhead, hike in, and return to the same trailhead. Or anglers may hike into one drainage and loop back out a different drainage to a new trailhead. Anglers who prefer assistance in accessing angling areas via trail may hire experienced guides to carry their gear on horseback. Backcountry airstrips enable anglers to hike from a trailhead to an airstrip or hike from an airstrip to a trailhead. In either case, anglers may hire experienced pilots through a local flight service. Those rare anglers who are experienced backcountry pilots may use public airstrips and, for a nominal landing fee, private landing strips. Aspiring do-it-yourself pilots should take note, however, that many landing strips are short (<240 m [787 ft]), and the terrain, weather, and local airstrip conditions quickly create hazardous conditions for the inexperienced.

As a result of the very high demand (11,360 floaters in 2013), MFSR float-boat access is regulated with a lottery and permit system. Anglers who wish to float and fish and who have access to float gear may apply for launch permits through the Salmon-Challis National Forest (see http://www.fs.usda.gov/detail/scnf/passes-permits/recreation/?cid=fsbdev3_029568). Anglers who need gear or assistance may hire experienced guides to completely outfit their trip.

Closing Thoughts

The MFSR is one of few North American rivers where anglers may experience the thrill of catching native trout in exceptionally beautiful streams (or introduced trout in spectacular alpine lakes) within a large wilderness area that supports abundant native wildlife and still functions very much as it did in the era of Lewis and Clark. Even the most experienced and world-traveled anglers may be spontaneously overcome with wonder as they witness a native Westslope Cutthroat Trout rising from several meters below the surface to suddenly flash through crystal waters and strike their fly.

The Middle Fork Salmon River within the Frank Church River of No Return Wilderness. Photo by Alex Sholes, Alex Sholes Photography.

Westslope Cutthroat Trout, Middle Fork Salmon River, Idaho. Photo by Willi Cannelli, Solitude River Trips.

Priest Lake

Jim Fredericks, Chief of Fisheries, Idaho Fish and Game

Location, Directions, and Access

In the northwestern corner of the Idaho Panhandle lies one of the most precious of Idaho's aquatic gems. While it might be possible to visit Priest Lake and leave without a lasting impression, it wouldn't be easy. At 31 km (19 mi) long, up to 7 km (4.4 mi) wide, and just over 110 m (360 ft) deep, the 9,310 ha (23,000 ac) lake is the state's fourth-largest natural lake and without a doubt one of the most beautiful. Located at the base of the Selkirk Mountains, Priest Lake is surrounded by vast amounts of state and federal public lands, limiting development and providing a wide range of recreational opportunities, with fishing, boating, camping, huckleberry picking, and hunting being prominent among them.

Aside from the unincorporated communities of Priest Lake, Coolin, and Nordman (combined population fewer than 1,000 people), Priest Lake is relatively isolated, essentially lying at the "end of the road." The nearest town is Priest River, about 48 km (30 mi) away, and the nearest cities, Coeur d'Alene and Spokane, are about 120 km (75 mi) to the south. Getting to Priest Lake is fairly simple. Whether coming from the east or west, take Highway 2 along the Pend Oreille River to the intersection of Highway 57 at the town of Priest River. Highway 57 provides access to destinations along the western side of Priest Lake, including three lakeside resorts, private marinas, and USFS campgrounds. The southern shore (including the town of Coolin) and the eastern side of the lake are accessed via the Dickensheet Highway, which intersects Highway 57 just south of Priest Lake. Features on the eastern side of the lake include two state parks and access to roads and hiking trails in the Selkirk Mountains.

Several state and federal campgrounds offer mature timber, boat ramps, swimming areas, and excellent facilities for tent and RV camping. For many campers, however, the biggest draw is the ability to camp away from the roads and vehicles at one of the lake's many boat-access campsites. The USFS manages camping on both Bartoo and Kalispell Islands, as well as on several secluded beaches along the western shore of the lake. In addition to camping opportunities, several resorts provide a range of accommodations, and many of the privately owned lake-front homes and cabins are available as vacation rentals.

Even farther to the north lies Upper Priest Lake, a much smaller but entirely undeveloped version of Priest Lake. There is no road access to Upper Priest Lake, so visitors have to hike up a trail on the west side of the lake or boat through the Thorofare, which is a 4 km (2.5 mi) low-gradient stream channel connecting the two lakes. The isolation of Upper Priest Lake makes it an appealing destination for many campers and boaters. The smaller size of the lake and the scenic trip through the Thorofare make it particularly popular with canoeists and kayakers.

The geology of the Priest Lake basin is primarily granitic, which means highly oligotrophic (nutrient-poor) water, not unlike an alpine lake. The low nutrients translate to a low abundance of phytoplankton (microscopic plants) in the tributary streams and the lake itself, resulting in unusually clear water. It is not uncommon for visibility in Priest Lake to exceed 9 to 12 m (30 to 40 ft).

Priest Lake Fishery

Given that sunlight and nutrients form the base of an aquatic food chain, it follows that Priest Lake supports less aquatic life, from plankton to fish, than a more nutrient-rich system. That's not, however, to say that Priest Lake doesn't support a popular and valued fishery. Historically, Priest and Upper Priest Lakes contained three native sport fishes, Westslope Cutthroat Trout, Bull Trout, and Mountain Whitefish, with the Cutthroat fishery being the most popular throughout the early part of the 20th century. Lake Trout (Mackinaw) were introduced in 1925 but remained at low abundance and provided the occasional catch of very large fish. It was Kokanee, introduced in the 1940s, that created one of the region's most popular fisheries. Kokanee not only provided a high-yield sport fish for anglers but also served as an abundant food source that allowed Bull Trout and Lake Trout to reach trophy sizes. This popular fishery based on a balance between large trophy fish and an abundant Kokanee population lasted through the 1970s. During this period, the state-record Lake Trout of 26 kg (57 lb, 5 oz) was caught in Priest Lake.

Unfortunately, the delicate balance didn't last. Though declines in Westslope Cutthroat Trout and Bull Trout populations were likely exacerbated by overharvest and loss of spawning habitat, there is little doubt that the ultimate collapse was related to the introduction of *Mysis* shrimp. *Mysis* shrimp were introduced into Priest Lake in 1965 to provide a supplemental food item for Kokanee. Though some Kokanee fed on the shrimp and reached trophy size, as evidenced by the state record of 3 kg (6 lb, 9 oz) caught in 1975, *Mysis* shrimp were an even greater boon to Lake Trout, greatly enhancing juvenile survival and reproductive success. Increasing Lake Trout predation soon caused a collapse of the Kokanee population and major declines in Cutthroat and Bull Trout numbers. Bull Trout were closed to harvest in 1984, and by 1985 Bull Trout spawning runs into Priest Lake tributaries were essentially gone. Westslope Cutthroat Trout were closed to harvest in 1992. Through the 1980s millions of Kokanee fry and hundreds of thousands of Cutthroat Trout fingerlings were stocked in Priest Lake and its tributaries in an effort to overcome the predation impact of Lake Trout. The efforts were unsuccessful, and the fishery shifted from a diverse yield and trophy fishery to one dominated by Lake Trout.

Since the 1980s when Lake Trout became dominant in Priest Lake, the IDFG has managed Priest Lake as a Lake Trout fishery, while focusing native-fish conservation efforts on Upper Priest Lake. Lake Trout were not known to be present in Upper Priest Lake until the mid-1980s but became well established in the 1990s through immigration from Priest Lake through the Thorofare. Since 1998 IDFG has implemented a Lake Trout gillnetting program in Upper Priest Lake, generally removing one to four thousand Lake Trout annually. Because Lake Trout have continually repopulated the upper lake each year through immigration and reproduction, the removal efforts amount to an annual maintenance program.

Despite the fact the Lake Trout angling has constituted the majority of the recreational fishing in recent years, Kokanee have provided good fishing opportunity in some years. The population fluctuates around a low density, but in some years offers decent catch rates for Kokanee that are 14-18" long. Even though Kokanee effort and harvest are minor in comparison to previous decades, they have been sufficiently abundant to add angling diversity to the fishery.

In addition to Kokanee, Cutthroat Trout have also continued to provide angling opportunity, despite being less abundant than they were historically. The Cutthroat Trout fishery, which continues to be managed under catch-and-release rules, can be quite prolific at times and supports a loyal following of anglers. Finally, whether by illegal introduction or migration from the Pend Oreille River, Smallmouth Bass have found their way into the Priest Lake system in recent years and are now widely distributed throughout the lake. Many regard Smallmouth Bass as an unwelcome addition to the lake and are concerned that they may have negative effects on an aquatic ecosystem that has already been severely impacted. Nevertheless, they have provided an additional angling opportunity in a system that has been lacking in diversity.

Tips and Techniques

Lake Trout

Lake Trout fishing on Priest Lake tends to be fairly predictable. Regardless of season, most successful anglers use one of two techniques—trolling or jigging. Lake Trout are usually closely associated with the bottom and are typically found in depths between 31 and 61 m (100 and 200 ft). For that reason, most trolling is done with downriggers, though leaded or steel line is sometimes used. Lake Trout are not particularly finicky and will hit a variety of trolled gear, including plugs, spoons, or flies and hoochies pulled behind a flasher. A small strip of fresh-cut bait or night crawler will greatly improve the chances of a hookup. Natural color combinations of black or blue over silver or white are good choices for lures, as are lighter-colored, fluorescent, and glow-in-the-dark colors. A good depth finder with a zoom feature will help locate fish near the bottom.

When the weather conditions are conducive, meaning the winds are calm, jigging can be an exciting and less gear-intensive way to catch Lake Trout. As with trolling, Lake Trout will hit a variety of terminal tackle, and the addition of cutbait or night crawlers helps entice a hit. Most anglers use 0.5 to 2.0 oz lead-headed buck-tail or rubber-tail jigs, jigging spoons, or buzz baits. Drop-shotting is very effective and is becoming a favorite technique of many jiggers. The keys to successful jigging are finding aggregations of fish around underwater ridges, shelves, and drop-offs, typically between 31 and 55 m (100 and 180 ft), and then keeping the boat directly over the top of them. Once there, it's critical to use nonstretch, multifilament line, as detecting a subtle bite in 43+ m (140+ ft) of water is nearly impossible with monofilament line. A strong hook set is also important.

Though conditions don't allow for it every year, when winter provides a sufficient freeze on Priest Lake, jigging for Lake Trout through the ice can be productive and exciting. The same methods used when jigging from a boat in the summer apply when fishing through the ice in the winter, without the added challenge of keeping the boat from drifting away from the fish! Cavanaugh Bay is typically one of the first areas of the lake to freeze and is a popular area for ice fishing.

Kokanee

Although the Kokanee population is a far cry from what it was in the 1950s and '60s when more than 100,000 Kokanee were harvested each year, recent years have seen numbers sufficient to maintain fishing opportunity. The lack of Kokanee abundance is offset by their large size, with anglers commonly catching 355 to 457 mm (14.0 to 18.0 in) fish. As Kokanee fishing almost exclusively involves trolling, it requires a boat, though not necessarily one equipped with downriggers. Unlike Lake Trout, which live in the deeper portions of the lake, Kokanee move throughout the water column, and for much of the year they can be found in the top 12 to 15 m (40 to 50 ft) of water. The most common method of fishing for Kokanee is to troll attractors, often referred to as "pop gear" or "cow bells," followed by a "wedding ring"–style spinner.

Depending on the depth of the Kokanee, leaded line or the use of a 1 to 3 oz of trolling weight will help get the gear to fish. Though not necessary, some anglers prefer to fish with downriggers and use small dodgers trailed by mini-hoochies or Kokanee flies. Tipping the hook with maggots, white corn, or even a piece of worm is helpful.

Cutthroat Trout

Westslope Cutthroat Trout angling is at its best during the "shoulder-season months" of May, June, September, and October when water temperatures are relatively cool and hatches of insects have the fish feeding near the surface. Calm mornings and evenings on both Priest and Upper Priest Lake can provide some very good fly-fishing or casting with small spoons and spinners. Though most Priest Lake Cutthroat Trout are in the 305 to 356 mm (12.0 to 14.0 in) range, it's not uncommon to catch fish up to 457 mm (18.0 in).

Smallmouth Bass

Smallmouth Bass provide another option for anglers more interested in casting around the shoreline than trolling the open water. Smallmouth Bass have expanded rapidly in Priest Lake and are now abundant and relatively easy to find near rocky shoals, outcrops, and docks around the shoreline. The cool water temperatures and lack of abundant forage fish species translate to relatively slow growth rates for Smallmouth Bass, and 254 to 305 mm (10.0 to 12.0 in) fish are the most abundant, though fish up to 508 mm (20.0 in) are occasionally caught. In the early summer, Smallmouth Bass are typically caught casting rubber-tail jigs, crankbaits, or softbaits toward the shoreline and retrieving them slowly into deeper water. As summer progresses and spawning season ends, Smallmouth Bass move into deeper water, and many anglers target 9 to 15 m (30 to 50 ft) depths and use a drop-shot setup.

Visitors who find themselves in the Priest Lake area without a boat or fishing experience need not despair. There are some very experienced and capable Priest Lake fishing guides. Though guided trips most commonly target Lake Trout, anglers can arrange for any one of the many fishing opportunities Priest Lake has to offer.

Priest Lake, Idaho. Photo by Glenn Oakley, courtesy of IDFG.

Priest Lake, Idaho. Photo by Glenn Oakley, courtesy of IDFG.

Priest Lake, Idaho. Photo by Glenn Oakley, courtesy of IDFG.

Saint Joe River

Chip Corsi, Regional Supervisor, Panhandle Region, and Matt Corsi, Principal Fishery Research Biologist, Idaho Department of Fish and Game

Location, Directions, and Access

On the western shoulder of the Bitterroot Divide, at 1,962 m (6,475 ft) elevation, the outlet of Saint Joe Lake marks the head of the Saint Joe River. From Saint Joe Lake, the river flows some 233 km (140 mi) and drops 1319 m (4354 ft) to its confluence with Coeur d'Alene Lake near the town of Saint Maries. Along the way, the Saint Joe flows past the hamlets of Avery, Calder, and Saint Joe City. Between Calder and the old railroad stop at Saint Joe City, the river loses gradient, making the transition from "swift water" to a slow, meandering channel that winds through the lower Saint Joe Valley. Near its mouth, the channel flows between Benewah and Chatcolet Lakes near Heyburn State Park, reaching Coeur d'Alene Lake by the old railroad bridge, which is now a part of the Trail of the Coeur d'Alenes bike path.

Often referred to as the "Shadowy Saint Joe," the river flows from the dense coniferous woodlands of the Saint Joe National Forest (now a part of the Idaho Panhandle National Forests) and through the cedar- and cottonwood-lined valley bottoms. The Saint Joe also has the distinction of being the highest navigable river in the world. Steamboats regularly traveled upstream to Saint Joe City and beyond in the late 19th and early 20th centuries. In 1976 more than 110 km (68 mi) of the Saint Joe upstream from Avery received designation by Congress under the National Wild and Scenic Rivers Act, with the upper 38 km (24 mi) being designated as "wild" and the remainder "recreational." Some 28 km (17 mi) of the upper Joe are in a roadless canyon, in some of the wildest country in the Idaho Panhandle.

Access

Early access into the swift water reaches of the Saint Joe was by trail, followed by rough two-track roads and the Chicago, Milwaukee, Saint Paul, and Pacific Railway that came along in the early 20th century. The Milwaukee went bankrupt in the early 1980s, and a paved two-lane road (Forest Highway 50) from Saint Maries to the top of Gold Summit on the Idaho-Montana border has replaced the rail bed and early road system, following the river eastward for 125 km (75 mi) upstream to the mouth of Gold Creek. At Gold Creek, a paved one-lane road follows the river upstream another 16 km (10 mi) to the Red Ives Ranger Station. For another 3.5 km (2 mi), a one-lane dirt road follows the river upstream to the Spruce Tree Campground, where it dead-ends. Another one-lane dirt road leaves the river at Red Ives, crosses over a ridge, and then drops back down to the Saint Joe at Heller Creek 28 km (17 mi) upstream of Spruce Tree.

Along with the river road from Saint Maries, vehicle access to the Saint Joe is commonly gained by traveling along Interstate 90 to the town of Wallace and then taking the Moon Pass Road from the southwest corner of Wallace up and over Moon Pass and then down the North Fork of the Saint Joe to Avery, a distance of about 50 km (31 mi). Other popular routes in are from the I-90 towns of Saint Regis and Superior, both in Montana. From Saint Regis, head up the Little Joe Road southwest of town and then up and over Gold Summit, about a 45 km (27 mi) drive to the Saint Joe River at the mouth of Gold Creek. Less traveled is the Cedar Creek road south of I-90 from Superior over the Bitterroot Divide to the Saint Joe River headwaters, approximately 42 km (26 mi).

Halfway Hill, Saint Joe River, Idaho. Photo by Clay Hayes.

Westslope Cutthroat Trout, Saint Joe River, Idaho. Photo by Clay Hayes.

Stonefly, Saint Joe River, Idaho. Photo by Clay Hayes.

Angler, Saint Joe River, Idaho. Photo by Clay Hayes.

Fishing Access

Virtually all of the Saint Joe between Falls Creek at the lower end of the swift water and the end of the road at Spruce Tree is accessible to anglers by driving the Saint Joe Road (FH 50). Upstream from Spruce Tree, the river is accessible by foot or horseback in the 27 km (17 mi) between Spruce Tree and Heller Creek. A short reach is accessible by road between Heller Creek and Medicine Creek. Depending on flows, much of the river is floatable in rafts and kayaks. The Saint Joe includes some Class III white water in several stretches upstream from Avery, so anglers wanting to float and fish should be experienced boaters if they attempt white-water reaches such as Skookum Canyon, or the Tumbledown Reach between Gold Flats and Bluff Creek. Float season in the river from Spruce Tree down to Avery typically ends in early to mid-July. Large debris jams and the smaller size of the river in the roadless reach upstream from Spruce Tree prevent use of that stretch for float fishing.

Downstream from Avery the gradient is gentler and the river larger, with inflow from the North Fork and Marble Creek, and can generally be floated year-round, although ice may prevent winter floats. Drift boats and canoes can be used to effectively fish in the Avery to Falls Creek reach.

Because the Saint Joe is a relatively small river, there is substantial potential for floaters and wading anglers to come in contact, although much of the upper river is unfloatable by the time it becomes wadable. Where there is overlap, floaters are advised to observe good fishing etiquette and not fish water occupied by wade fishers, moving through those sites quickly and behind the wading anglers, when possible.

Saint Joe Area History

The Saint Joe has a rich history. Named for Saint Joseph by the Jesuit missionary Father DeSmet, the river corridor included a trail used by the Coeur d'Alene Tribe as a route to trading and hunting areas in what is now western Montana. The Milwaukee Railroad was the last of the transcontinental railways built across the United States and was constructed through the Saint Joe country at the end of the first decade of the 20th century. Along with a "perfect storm" of weather conditions, sparks from the railroad's steam locomotives were credited with igniting many of the starts of the Great Fire of 1910. The Saint Joe was in the very heart of the conflagration, which cost more than 80 people their lives, resulted in the loss of several communities, and was the catalyst for establishing wildland firefighting policies and methods for the United States. The timber industry has a prominent history in the Joe, and timber and forest products remain a mainstay of the Saint Maries and neighboring communities.

The Trail of the Hiawatha bike path now occupies many of the old trestles and tunnels of the Milwaukee railway, and the old station house in Avery is used as a community center. Remnants of the "Big Burn" are still readily apparent along the North Fork of the Saint Joe, in the form of blackened, hulking cedar "ghosts"—the burned-out shells of cedar trees, some more than eight feet in diameter and 40 feet tall[71]. Spring board notches can still be seen in cedar stumps in the Saint Joe, where loggers would insert a board to stand on, so they could use axes or crosscut saws to cut down the tree above the boll.

The Fishery

The Saint Joe also has a rich fishing history. Native Americans, principally the Schitsu'umsch (Coeur d'Alene) Tribe, regularly fished the Saint Joe and associated waters for subsistence purposes. In the mid-19th century,

Captain John Mullan of the US Army commanded an engineering crew charged with building a road from Fort Walla Walla in Washington Territory to Fort Benton in Montana Territory. Along the way they crossed the lower Saint Joe and observed tribal fishing. Mullan described the Saint Joe's receiving water, Lake Coeur d'Alene, as "a noble sheet of water, teeming with an abundance of delicious salmon-trout." Many of those fish were migratory Westslope Cutthroat Trout originating in the Saint Joe River. Later, in his personal account of European settlement of the Saint Joe Valley (*Pioneer Days on the Shadowy Saint Joe River*), author Orland B. Scott described catching and observing Cutthroat Trout and Dolly Varden (Bull Trout) in the Saint Joe River[72]. Steamboat expeditions provided access up the river where a major tourist activity was trout fishing. By the early 20th century, the Saint Joe had a reputation as being a fine trout fishery, and anglers used the newly constructed railway to gain access to the river between Saint Maries and Avery.

Eventually, a road system was developed that provided automobile access up and down the Saint Joe River from Saint Maries to Red Ives, more than 125 km (78 mi) upstream. As late as the early 1980s the roads were rough dirt roads, often only one lane, and busy with log-truck traffic. In those days, the old bumper sticker "Hell yes I'm dirty, I drive the Joe!" actually meant something. Nonetheless, due to liberal bag limits and high catch rates, anglers depleted the fishery to the point where special fishing regulations limiting harvest and requiring the release of fish less than 330 mm (13.0 in) long were adopted by the Idaho Fish and Game Commission. The rules were based on pioneering research led by IDFG and the University of Idaho in the late 1960s and early 1970s, with the length limit established on the pretense that fish should be protected until they were old enough to have spawned at least once. Idaho was at the forefront of using specialized fishing regulations requiring the release of some or all fish to improve population strength and ultimately to improve catch rates, and the Saint Joe was among the first waters anywhere to be managed with special regulations.

The fishery responded to the new rules, and catch rates recovered. With the bankruptcy of the Milwaukee Railroad, though, the tracks were replaced with a paved highway up the river up to and well past Avery (today the road is paved all the way to Red Ives). With better access came more anglers. Biologists and anglers alike recognized that Cutthroat Trout larger than 330 mm (13.0 in) were a rarity, because they were rapidly being cropped off when they reached a legal size. With changes in Coeur d'Alene Lake, including the advent of nonnative predators such as illegally introduced Northern Pike, and impairments to lower-river tributary streams due to roads and timber harvest prior to implementation of rules such as the Forest Practices Act, the adfluvial (lake migratory) component of the fishery had declined substantially by the 1980s. Anglers, still pining for "the good ol' days" when large Westslope Cutthroat Trout were common, demanded changes to improve the fishery. Rules were gradually tightened and the size limit increased, and catch-and-release rules were applied to the fishery upstream from Prospect Creek. By the end of the 1990s, with strong angler support, the entire river was designated as a catch-and-release fishery for native Cutthroat Trout and Bull Trout.

Like most trout fisheries across Idaho and the rest of the United States, fishery management in the early 20th century and into the 1960s was primarily an exercise in stocking hatchery-reared fish of different species and sizes to try to keep up with increasing harvest. Despite the advent of special regulations to improve the fishery, fish stocking continued until the early 2000s in portions of the lower river in an effort to provide people opportunities to harvest fish. Species stocked in the Saint Joe watershed included Brook Trout, Coho Salmon and Kokanee, Cutthroat Trout (to include what we now know as Yellowstone and Westslope Cutthroat Trout), and, of course, a variety of Rainbow Trout strains. In the lower slack-water reaches of the

Saint Joe, Channel Catfish and Tiger Muskies were stocked in an effort to provide a diversified sport fishery for a few years in the late 1980s and early 1990s.

Brook Trout are established in some of the Saint Joe tributaries, and there is evidence to indicate they have outcompeted native Cutthroat Trout in some of the lower-elevation streams, under some circumstances[73]. There is little to no evidence documenting any hybridization of Brook Trout with native Saint Joe Bull Trout, a serious problem for Bull Trout conservation elsewhere. Today, Brook Trout are an anomaly in most of the drainage.

Coho Salmon never established, and while early-fall spawning runs of Kokanee from Coeur d'Alene Lake were somewhat notable as far upstream as the Big Eddy near Calder into the 1970s, Kokanee runs in the Saint Joe are now just a memory. A few Chinook Salmon, stocked in the lake during the 1980s and 1990s, do show up in the lower end of the swift-water reach every fall to spawn.

With increasing concerns about the potential for stocked Rainbow Trout to hybridize with native Westslope Cutthroat Trout, the Rainbow Trout stocking program shifted to the use of sterile fish in 2000. This change essentially eliminated any concerns about continued hybridization. Furthermore, an assessment of angler catch of stocked fish showed a very low return to the creel, even with the stocking program targeting high-use areas such as campgrounds. Raising and stocking hatchery fish is an expensive proposition, and low return rates, coupled with a flourishing Westslope Cutthroat Trout fishery, begged the question of the value of the program. With a very high level of support from anglers, the decision was made to put the hatchery fish to better use elsewhere, and the stocking program was discontinued after the 2002 fishing season. A few wild Rainbow Trout can still be encountered in the lower swift-water reaches, but their numbers appear to be declining and genetic studies indicate hybridization with native Westslope Cutthroat Trout is virtually nonexistent.

Tiger Muskie, a sterile cross of Northern Pike and Muskellunge, and Channel Catfish, which do not reproduce in the colder waters of the Saint Joe, have not provided any significant fishery in the lower river. These stocking programs have been discontinued, and it is improbable any of these fish remain from the original stockings. On the other hand, illegally introduced Northern Pike have taken up residence in the slack-water reach and are periodically taken by anglers. Also present are Smallmouth Bass, Yellow Perch, Bullhead, and Pumpkinseed. Concerns have been raised about the potential for pike and Smallmouth Bass to impact juvenile Bull Trout and Cutthroat Trout migrating to Coeur d'Alene Lake, but whether they actually impact these Saint Joe natives to any significant degree in the river is unknown. Studies in Coeur d'Alene Lake have demonstrated that Northern Pike do consume substantial numbers of Westslope Cutthroat Trout in several bays and may be affecting adfluvial populations. [75]

Despite a litany of human-caused disturbances to the landscape and exotic additions to the fish community—circumstances that have resulted in the loss of native fishes and declines in sport fisheries elsewhere—the Saint Joe River harbors one of the finest native trout fisheries anywhere. The Saint Joe's Westslope Cutthroat Trout population has been estimated at more than 1,280 catchable-size trout per km (800 per mi). They appear to have dodged the bullet of hybridization with Rainbow Trout that has impacted a number of other populations across the northern Rockies. Today the Saint Joe is considered a stronghold for genetically pure Westslope Cutthroat Trout[74] and is a destination fishery for anglers from all over the United States.

Fishing Gear Options

Fly-fishing is the most popular method for catching Saint Joe Cutthroat Trout, with popular patterns including Stimulators, Elk-Hair Caddis, Renegades, Hoppers, Muddler Minnows, and a variety of attractors such as Humpies and the Wulff patterns. Later in the summer Saint Joe Cutthroat Trout can get a little bit picky; they see a lot of flies by early September. Anglers will do well to try to match hatches and use light tippets when conditions are low and clear. Oftentimes what appear to be fish rising to dry flies are actually fish feeding on emergers. Tying on an emerger of the right pattern and fishing it behind a dry fly can be very effective under those circumstances. Bull Trout are not abundant in the Saint Joe but are well distributed and may be caught (and released) using streamers. Well-prepared anglers will also have a supply of nymphs and be ready to "dredge" the bottom for stubborn fish. That's also a good method for catching the Saint Jo-e's other native salmonid, the Mountain Whitefish. Also known as "Rocky Mountain bonefish" to aficionados, Mountain Whitefish provide good sport, and because they are abundant and prolific, they may also be kept for excellent table fare. Hares Ear, Pheasant Tail, Copper John, and stonefly patterns such as the Prince Nymph and the Bitch Creek are useful to have in the fly box. Spin fishers can also enjoy good success using spoons and spinners or a fly-and-bubble rig. Bait fishing is allowed downstream of the North Fork near Avery.

Fishing Seasons

Fishing is open year-round, but the road upstream of Avery is not maintained for winter travel by automobile, and the east-west orientation of the canyon means very little direct sunlight reaches the river. Thus, the upper Saint Joe is typically iced over most of the winter. For those anglers hardy enough to try, nymphing the larger pools and slowest runs is apt to produce the best results during the winter.

More anglers are discovering good late-spring fishing prior to runoff and may encounter hatches that warrant breaking out the dry-fly box. Runoff typically occurs in May and June, with the river typically becoming fishable again by mid- to late June.

Summer and fall are when the Joe really comes into its own as a fishery. The Saint Joe's Westslope Cutthroat Trout are generally "polite" fish, in that they don't necessitate early-morning or late-evening fishing to be successful, and in fact midday to early evening is often the best time to be fishing.

On a warm summer day, the angler may be standing in the shade of Idaho's state tree, the white pine, enjoying the fragrance of a syringa, the Idaho state flower. If it's a little later in the summer, there's even the opportunity to snack on a fresh-picked huckleberry, the Idaho state fruit, from a nearby shrub, all the while listening to the call of Idaho's state bird, the mountain bluebird. The curious angler may note the dark-red hue of some of the sand along the river's edge while wading the upper Joe and quickly determine that they are standing on garnets—the officially designated gem of the "Gem State." With a little bit of luck, he or she will cast true and catch Idaho's iconic state fish and one of its living gems, the Cutthroat Trout. No other gem in the Gem State affords the angler all of these opportunities to come into contactwith these iconic state symbols.

Silver Creek

Chad Chorney, Big Wood Project Manager, Trout Unlimited

Location, Directions, and Access

Silver Creek, nestled near the small central Idaho ranching community of Picabo along Highway 20, is likely America's most famous spring creek. Located approximately 225 km (140 mi) southeast of Boise in Blaine County, Silver Creek (described by Native Americans as the "shining water") has lured trout anglers from all 50 states and more than a dozen countries. Large and finicky Brown Trout and Rainbow Trout, prolific mayfly hatches, and stunningly clear water all combine to make Silver Creek one of trout fishing's premier destination waters.

Centrally located in Idaho just 65 km (40 mi) southeast from the famous resort town of Sun Valley, Silver Creek is surrounded by sagebrush-covered hills, pasture, and alfalfa and barley fields. Few full-time residents call the small town of Picabo and surrounding areas home, adding to the quiet and solitude of a day on the "creek." As mentioned, Silver Creek is one of the country's premier trout-fishing destinations, and for good reason. In addition to mouse-eating Brown Trout, a healthy population of Rainbow Trout calls Silver Creek home. It's these Rainbow Trout and their affinity for eating a variety of mayflies on Silver Creek's placid surface that bring anglers to the creek's fabled riffles and runs. In fact, Silver Creek boasts some of the highest densities of fish per mile of any of Idaho's many world-class trout streams. Silver Creek is a high-desert spring-fed ecosystem, with water levels and temperatures that remain relatively stable year-round. These conditions, coupled with nutrient-rich water, translate to prolific aquatic life. The fertile waters of Silver Creek are a literal "insect factory," providing trout a multitude of food sources—numerous species of mayflies, caddis, midges, stoneflies, and damselflies are all on a trout's menu. On a typical summer day of fishing Silver Creek, anglers often see caddis, PMDs, BWOs, *Callibaetis*, damselflies, and terrestrial insects like beetles, ants, and grasshoppers. This abundance of aquatic insect life often confounds anglers on Silver Creek and provides true "match the hatch" fishing.

Silver Creek can be roughly broken down into three sections. Perhaps the most famous portion of Silver Creek lies upstream, not far from the area where Silver Creek gets its start. This portion is owned by the Nature Conservancy and is called the Silver Creek Preserve. Approximately 4.8 km (3 mi) of stream run through the preserve, which encompasses more than 344 ha (850 ac). An additional 5,100 ha (12,600 ac) are adjacent and protected through conservation easements, translating to one of the most successful stream-conservation efforts undertaken for public benefit. In addition to angling, the preserve is popular among canoeists, hikers, birders, and photographers. An abundance of wildlife call the preserve home, and visitors are often treated to sightings of eagles, hawks, songbirds, waterfowl, coyotes, bobcats, mountain lions, otter, moose, mule deer, and elk. More than 150 species of birds have been identified on the preserve, and during the winter when angling is closed, Silver Creek becomes a bird-watcher's paradise.

Fishing Techniques

On the preserve, Silver Creek is slow moving, crystal clear, and characterized by silt, sand, and fine gravel substrate. Aquatic vegetation is prevalent, with late summer typically exhibiting the highest growth and density. To say that fly-fishing on the Silver Creek Preserve is challenging is an understatement. In fact, it's been said that Silver Creek's trout aren't just college graduates, but have their PhDs as well! To be successful on the Preserve (or anywhere on Silver Creek), anglers have to become hunters and adopt many of a hunter's traits. The ability to be quiet, move slowly and surely, and exhibit patience leads to successful fishing. To continue with the hunting analogy, anglers should never "flock shoot," but instead concentrate their efforts on single fish that have fallen into a distinct feeding pattern.

Anglers should minimize wading; some of Silver Creek's most successful anglers barely get their feet wet during a day's fishing. Since the creek is slow moving and calm, heavy wading spooks trout and can send silt and sediment downstream, alerting more fish to an angler's presence. It's also a good idea to limit false casting; avoid bright, gaudy clothing; and move "low and slow" along the stream bank. Silver Creek trout are always looking up, and any foreign movement or action alerts them to predators and effectively puts those fish off the bite for the short term.

Downstream, "fly first" presentations are by far the most effective on Silver Creek. If a feeding trout sees an angler's leader or line before it sees the fly, chances are that it will refuse your offering. Before a visit to Silver Creek, fly fishermen should brush up on their casting and mending skills; the creek is not a forgiving place. A note of caution: when attempting to set the hook on a straight-downstream drift, be sure to adjust for the angle and allow a bit more time on the set. This is especially true considering the fine tippets and small flies that are often used on Silver Creek. Successful Silver Creek anglers have a saying: "First let 'em have it, and *then* let 'em have it!"

Fishing on the Nature Conservancy's Silver Creek Preserve is split into two sections—the walk-and-wade section and the float-tube section. The walk-and-wade portion is located on the upstream area of the preserve and essentially covers Silver Creek from Stalker Creek down to Loving Creek. Below that, wading is not permitted, and fishing is allowed from float tubes only (this section ends at Kilpatrick Road Bridge). Keep in mind that fishing on the float section is downstream only (you may not use your float tube to move upstream). Access maps are available at most of the fly shops serving Silver Creek, as well as at the Nature Conservancy Visitor Center cabin located on the preserve.

Moving downstream, the next section of Silver Creek is known by locals as "the Double R." This section starts at the Kilpatrick Road Bridge, ends downstream at the Highway 20 crossing, and runs entirely through a private fishing club—the RR Ranch. However, access to float-tube angling is allowed immediately adjacent to and downstream of the Kilpatrick Road Bridge.

This section of Silver Creek fishes well throughout the season, with excellent hatches of a variety of mayflies. Anglers often cast to actively cruising and feeding fish, experiencing some of Silver Creek's most exciting and consistent fishing. As mentioned, fishing in this area is done out of float tubes, and wading is not permitted. This section of Silver Creek is fairly wide, deep, and similar to fishing still water. Long-leader dry-fly presentations are the standard here, although at select times, streamers and even dropper nymphs can be productive.

The last section of Silver Creek of note to anglers lies downstream of the Highway 20 crossing and covers the remainder of the creek before its confluence with the Little Wood River near Carey, Idaho. There are several access areas on this section of Silver Creek. They include (moving downstream) Highway 20, George Martin Bridge, Point of Rocks, North Picabo Road, and Priest Rapids. From Highway 20 down to North Picabo Road, the character of Silver Creek exhibits a fair amount of variety. Anglers find deep pools, meandering meadow water, and extensive undercut banks. Hatches throughout these sections are less abundant than those found on the Silver Creek Preserve, but still prolific at times. Since these sections run through grasslands, pasture, and some cultivated fields, terrestrial insects such as grasshoppers are an important food source, and anglers eagerly await late-summer "hopper" fishing here. During the winter fishing season, these areas can fish very well with streamers on calm, overcast days, and some of Silver Creek's largest trout are caught here each year.

Selection of PMDs and Caddis for a morning on Silver Creek. Photo by Chad Chorney.

Silver Creek Rainbow Trout Ready for Release, Silver Creek, Idaho. Photo by Chad Chorney.

Opening Weekend on Silver Creek, Idaho. Photo by Chad Chorney.

Silver Creek, Photo courtesy of IDFG.

Heading farther downstream, Silver Creek takes on a very different character before it eventually runs into the Little Wood River. The stream substrate here is characterized by lava rock, and riffles, runs, and pocket water replace Silver Creek's typical placid flows. These areas of Silver Creek typically fish well early and late in the season, since mid-summer flows can be fairly warm and intolerable to trout.

Hatches and Gear

As mentioned previously, hatches on Silver Creek are varied and abundant. The incredible insect life and world-class dry-fly fishing are what bring anglers back year after year. Two hatches in particular define Silver Creek and set it apart from other waters. The first is the famous, or perhaps infamous, brown-drake hatch. Occurring in late May or early June, the brown-drake hatch is truly a wonder of nature. These mayflies are large (size 8 to 12), appear for a very short time, and hatch in incredible numbers. Trout often become so engorged on brown drakes that they simply can't eat another bug, even when Silver Creek is covered by thousands of these mayflies. But if an angler is patient, willing to stay late into the evening, and possessed with a fair amount of luck, he or she can experience one of trout fishing's most incredible hatches and have fishing that they will never forget. Since timing of the brown-drake hatch is critical, it's suggested that visiting anglers pay attention to the local fly shops and the fishing reports that they provide. Unlike most dry-fly fishing on Silver Creek, leaders and tippets used when fishing the brown drake are on the heavier side—4X and 3X are typical. A stout leader is needed to turn over the large brown drake imitations and land large Brown Trout and Rainbow Trout during late nights on the water.

The second big hatch on Silver Creek is the Trico, a diminutive mayfly (size 20 to 24) that hatches during the hustle and bustle of summer. The Trico is an early-morning hatch that occurs during late July and often lasts until early fall. In contrast to the brown drake, when anglers get on the water in the evening and stay well past dark, Tricos appear on warm mornings not long after sunrise, and the action is usually over by late morning. When fishing the Trico hatch, anglers can increase their odds of success by using long, light leaders and tippets (12 to 15 ft, in 6X and 7X), fishing a straight-downstream drift, and looking for pods of actively feeding trout.

Specific tackle, gear, and flies can be the difference between a successful or frustrating outing on Silver Creek. When selecting fly rods for Silver Creek, anglers should lean toward softer, slower-action rods. A softer, slower fly rod is required when fishing long, light leaders and small flies, both the norm on Silver Creek. With the exception of fishing streamers or grasshopper imitations, leaders and tippets used on Silver Creek will rarely be heavier than 5X. Today, 6X leaders and tippets are standard on Silver Creek, and many anglers fish 7X during the Trico hatch. For leaders, the spring-creek variety in a minimum length of 3 m (10 ft) is a good bet. Many anglers fish 3.7, 4.3, and 4.9 m (12, 14, and even 16 ft) leaders, which put more distance between the fly and fly line, reducing the chance of spooking trout. Soft, slow rods are much more effective at handling these fine leaders and tippets and provide a delicate presentation to wary trout. Also, slow- or moderate-action fly rods are better for playing and landing large trout on delicate tippets and small flies. These rods have more flex and act as a shock absorber when a 508 mm (20.0 in) Brown Trout heads for an undercut bank. In fact, many fly fishers are using fiberglass or bamboo rods when dry fly-fishing on Silver Creek, as these rod materials are much more forgiving than standard graphite fly rods. For dry fly fishing, rods of three, four, and five weight work well, with heavier rods better for larger flies, stouter leaders, and windy days.

When fishing large, heavy flies like streamers, or wind-resistant flies such as grasshopper imitations, shorter and heavier leaders work well. Anglers should consider leaders in 1.8 to 2.7 m (6 to 9 ft) lengths, tapering to 2X or 3X. Fluorocarbon leaders for this type of fishing are a good bet, particularly on bright, sunny days. A medium- or fast-action fly rod in a 5 or 6 weight is ideal.

Fly selection is critical to success on Silver Creek. Flies should be more imitative than suggestive and be tied sparsely. Leave the bushy attractor flies at home, and instead use spring creek–style flies that are more realistic. Effective dry flies on Silver Creek tend to float low in the surface film, incorporate natural materials, and mimic the natural insects present. Fly patterns designed by Bob Quigley, Shane Stalcup, and Rene Harrop are successful on Silver Creek and are carried by most of the local fly shops. Visiting anglers would do well to stop in these fly shops and inquire about flies for Silver Creek, as many of the shop employees and guides are on the water every day and know which patterns are most effective.

It's common to experience several hatches during a single day on Silver Creek, and fish often prefer one insect over another. To complicate matters, trout often key in on a particular stage of an insect (emerger instead of dun, dun instead of spinner, etc.), so it pays to carry a fairly extensive selection of flies that imitate a variety of insect species, life stages, and insect sizes.

Accommodations

Visiting anglers can find a variety of accommodations when planning a trip to Silver Creek. Simple tent camping is available at the George Martin Bridge and Point of Rocks fishing access sites, while the IDFG Hayspur Fish Hatchery and Picabo RV Park provide more services. For anglers who don't want to rough it, numerous hotels, condos, and vacation rentals are available in nearby Hailey, Ketchum, and Sun Valley. There's also a great selection of restaurants to choose from as well.

Five fly shops serve Silver Creek and surrounding areas. Four of the five are located in Hailey and Ketchum. The newest shop is located in Picabo and specializes in fishing Silver Creek. All the shops offer guide services, and for anglers looking for on-stream instruction, a day of guided fishing can pay dividends. Angling regulations and seasons vary on different sections of Silver Creek, so be sure to consult current IDFG regulations and be in possession of a valid Idaho fishing license.

While fishing Silver Creek should be at the top of every angler's list, visitors should spend some time fishing other local waters such as the Big Wood River, Big Lost River, and Magic Reservoir. The Wood River valley of south-central Idaho is blessed with a host of world-class trout waters, and after a tough outing on Silver Creek, one of these other waters can really save the day!

When fishing Silver Creek, be sure to take the time to "stop and smell the roses." Few places are as peaceful and serene, while at the same time providing one of fly-fishing's greatest challenges. Silver Creek is a special place and a shining example of the Gem State's incredible trout fishing.

South Fork Boise River

Jeff Dillon, State Fishery Manager, and Art Butts, Regional Fisheries Biologist, Southwest Region, Idaho Department of Fish and Game

The South Fork Boise River in southwestern Idaho includes two very different reaches divided by Anderson Ranch Reservoir. Anderson Ranch Dam was completed in 1951, the second of three major reservoir projects on the Boise River drainage following Arrowrock Dam (1915) and just ahead of Lucky Peak Dam (1955). At 142 m (465 ft), the dam was the tallest embankment dam in the world at that time and created a 1,940 ha (4,800 ac) reservoir pool. The hydrologic result is a river basin above the reservoir with a relatively large and complex drainage and natural hydrograph and temperatures, while the river below the dam has highly regulated flows and consistent cold water temperatures. The access, amenities, and the fisheries themselves differ markedly between reaches, but both offer something special for anglers and other recreationists.

Location, Directions, and Access

The headwaters of the South Fork Boise River tumble down the southern slopes of the Sawtooth Mountains, flowing generally southwest through the small communities of Featherville and Pine before flowing into Anderson Ranch Reservoir. Coming from the east or west, most visitors reach the upper South Fork off of Highway 20, taking County Road 61 (Louse Creek Road) to Anderson Ranch Reservoir and upstream to Pine. Turn right (north) on the Pine-Featherville Highway. Between Pine and Featherville is a mix of private and public land with several access areas to reach the river. Above Featherville, bear right onto FS Road

Some of the miles of gorgeous scenery on the South Fork Boise River, Idaho. Photo courtesy IDFG.

Geared up for a South Fork Boise River, Idaho float. Photo courtesy IDFG.

Rainbow Trout beauty release, South Fork Boise River, Idaho. Photo courtesy IDFG.

227, which parallels the South Fork all the way to the headwaters. This stretch of freestone trout stream is bordered mostly by public land, with a number of developed and dispersed camping opportunities and very good access to the river. As a scenic alternative, one can also reach the drainage from the top by taking the Soldier Creek Road (FS Road 94) north out of Fairfield, over Couch Summit, and down Little Smoky Creek.

Below Anderson Ranch Dam, the river turns to the northwest and flows another 48 km (30 mi) or so before joining the Middle Fork Boise at Arrowrock Reservoir. Access to the tailwater is off of Highway 20 via the Cow Creek Road (FS Road 131) or the Anderson Dam Road (FS Road 134). The first 20 km (12 mi) below the dam are accessible along Forest Service Road 121 (Anderson Ranch Dam Road) as the river winds through a narrow valley. Below Cow Creek the floodplain widens, and the river becomes braided with occasional side channels. Throughout the roaded reach, some limited dispersed campsites and restrooms are available among the cottonwoods, ponderosa pines, and willows. There are numerous developed and undeveloped boat launches available from the Tailwaters and Reclamation Village Boat Ramps just below the dam, down to the Danskin ramp, the last ramp in the roaded section. Prior to 2013 this reach was a relatively easy float, with a few sections of rock gardens and small (Class II) rapids. This section is popular with drift boaters but is not recommended for novice rowers. In late summer 2013 the Pony/Elk Complex wildfire swept through the canyon, burning much of the surrounding hillsides and riparian area. Intense thunderstorms following the fire triggered several landslides into the river. As of this writing, the riparian zone in the roaded section is recovering, but the river channel itself is still changing as the new material gets sorted. New rapids and debris have made this run more difficult. Float anglers are advised to check conditions before launching.

From the last developed boat ramp at Danskin downstream to the takeout at Neal Bridge (FS Road 189, also known as Blacks Creek Road), the river's character changes considerably. About 6 km (4 mi) downstream of the Danskin Bridge, the valley narrows, gradient increases, and the river enters a steep, narrow canyon bordered by basalt cliffs. For the next 20 km (12 mi), this is a remote and roadless stretch of fast-moving riffles and white water where the trout fishing and the scenery compete for your attention. Mule deer, river otters, ospreys, and Bald Eagles are just some of the wildlife common to the canyon. At summer and early fall flows, floating the canyon is a long day trip; there are also a few beaches and primitive camping sites for a shore lunch or an overnight stay. Even before the fire and debris flows in 2013 and 2014, floating this canyon reach was not recommended for inexperienced boaters. It was a moderately technical float with a lot of rock dodging and some Class III–IV water. As of this writing, numerous new rapids were created from recent debris flows, some considered Class IV+. Only advanced white-water boaters should attempt to float this reach and should be prepared to portage if needed. Check river conditions before your trip.

The Fisheries

While portions of the upper South Fork Boise River drainage have been significantly altered by mining, timber harvest, and road development, the area remains a very popular and accessible trout fishery. This section is a typical freestone trout stream, flowing through a valley of rolling hills and mixed pines. The main stem has a healthy population of wild Rainbow Trout, supplemented in high-use areas with hatchery trout; Mountain Whitefish are abundant; and both resident and adfluvial Bull Trout are found in much of the drainage. Brook Trout are common in some tributaries, and from late August to early October both anglers and nonanglers alike can enjoy the sight of thousands of bright-red Kokanee filling the river as they migrate from Anderson Ranch Reservoir to upstream spawning grounds.

Although the upper South Fork Boise has much to offer, most anglers regard the lower river as the true gem of southwest Idaho wild trout fishing. As in many tailwater fisheries, fish populations below Anderson Ranch Dam benefit from the boost in nutrients and consistent water temperatures provided by deep-water reservoir releases. Water releases are managed primarily to provide flood control and irrigation benefits, but are also generally within sideboards that ensure adequate winter habitat and spawning conditions for wild trout. Under these prime conditions, and with restrictive harvest rules in place for more than thirty years, the wild Rainbow Trout population has flourished. Wild Rainbow Trout densities are considerably higher in the tailwater, with the most recent surveys indicating an average of 1,079 fish per km (1,739 fish/mi)[70]. Above Anderson Ranch Reservoir, Rainbow Trout densities are typically 450 fish per km (724 fish/mi)[26]. Growth rates and average size are also much higher in the tailwater, where much of the catch is composed of 406 to 508 mm (16.0 to 20.0 in) fish. Mountain Whitefish are also abundant, and anglers may encounter Bull Trout seasonally from about October through spring.

From Memorial Day weekend until late summer, water releases from Anderson Ranch Dam are typically held stable at around 45 to 51 cms (1600 to 1800 cfs). This is the prime float-fishing season. Shore or wade fishing is a little more challenging at these flows but can still be productive. Flows typically decrease to around 17 to 23 cms (600 to 800 cfs) by the end of August, stay at this level until October, and then decrease again to around 8.5 cms (300 cfs) for the winter. The fall and winter flows provide excellent wade-fishing conditions throughout the roaded section. Spring flows typically increase to 17 cms (600 cfs) in early April, in part to increase availability of side-channel spawning habitat.

Fishing Regulations

The upper South Fork Boise River drainage (including tributaries) is open to angling year-round and is generally managed with simple fishing rules that provide harvest opportunity for wild and hatchery trout and Mountain Whitefish. For anglers who want to focus on wild Rainbow Trout, the main-stem South Fork from Beaver Creek upstream to the confluence with Big Smoky Creek (about 18 km [11 mi]) is managed with a two-trout limit, none under 356 mm (14.0 in), and gear is restricted to artificial lures and flies with single barbless hooks.

Fishing regulations on the South Fork Boise tailwater from Anderson Ranch Dam downstream to Neal Bridge, about 45 km (28 mi), are designed to conserve the trophy wild Rainbow Trout population. Anglers are restricted to artificial lures and flies only with single barbless hooks. The fishing season opens the Saturday of Memorial Day weekend. Through November daily harvest is restricted to two Rainbow Trout, none under 508 mm (20.0 in). From December 1 to March 31, Rainbow Trout fishing is catch-and-release only. Throughout the open season, Mountain Whitefish and Brook Trout bag limit is 25 fish per day. Bull Trout are fairly common, especially in fall and winter, and must be released if caught. Fishing is closed from April 1 through the Friday before Memorial Day weekend, the time frame when most Rainbow Trout spawning occurs.

South Fork Snake River/Palisades Reservoir

Dan Garren, Regional Fisheries Manager, Upper Snake Region,Idaho Department of Fish and Game

Location, Directions, and Access

The headwaters of the Snake River originate north of Jackson, Wyoming, in Yellowstone National Park and are regulated by Jackson Lake Dam in Wyoming and Palisades Dam in Idaho. Flowing from the border between western Wyoming and southeastern Idaho westward to the small town of Menan, the Snake River (typically referred to as the South Fork in Idaho) joins the Henrys Fork. Palisades Reservoir straddles the border between Wyoming and Idaho and creates the second-largest reservoir in Idaho. When full, Palisades Reservoir stores enough water to cover the entire surface area of nearby Yellowstone National Park in more than 15 cm (6 in) of water (148,000 ha m/1.2 million ac ft). The reservoir provides flood control for downstream residents, and the stored water is metered out through the summer and fall to meet agricultural demands in the heavily irrigated Snake River Plain and points downstream. The variation in seasonal flows resulting from irrigation storage and delivery can affect fishing opportunity at times. Anglers can reach the South Fork by traveling to Idaho Falls, which is about 345 km (215 mi) north of Salt Lake City, Utah. Palisades Dam is approximately 80 km (50 mi) east of Idaho Falls and marks the start of the South Fork. The city of Menan lies approximately 30 km (18 mi) north of Idaho Falls and marks the terminus of the South Fork.

Fishery

The South Fork provides the largest, most robust population of Yellowstone Cutthroat Trout in Idaho and one of the most abundant populations of Yellowstone Cutthroat Trout remaining in their native range. Densities of Cutthroat Trout range upward of 1,500 fish per km (2,200 fish per mi) in the upper river near Palisades Dam. Cutthroat are popular for their eagerness to bite and willingness to rise to dry flies, making for exciting fishing on the South Fork. Combined with the prolific insect hatches common along this river, the allure of the South Fork and the chance to catch a large native Yellowstone Cutthroat Trout in its native range draw anglers from across the country and worldwide. Although Cutthroat Trout provide the backbone of anglers' catch, Rainbow Trout and Brown Trout are also common and increase fish densities to more than 3,300 fish per km (5,300 fish per mi).

The Fishery and Its Segments

Many anglers view the South Fork as a single river, but it can more aptly be broken into three distinct segments, each with a unique character. The uppermost segment runs from Palisades Dam downstream to Pine Creek and flows through Swan Valley, a relatively unconfined stretch of river. The first 13 km (8 mi) of this river reach are a single channel, but afterward the river braids into numerous side channels and islands that add to the character of the river. South Fork Cutthroat Trout generally spawn in four major tributaries, three of which (Palisades, Pine, and Rainey Creeks) can be found in this upper stretch of the river. These tributaries can also provide outstanding fishing during the summer months. Angler access is plentiful in this section of river, with developed boat ramps found at Palisades Dam, Palisades Creek, Spring Creek, and Conant Creek. There is also public ground along the south side of the river, providing access for anglers looking to wade as opposed to fishing from a boat. The upper section of the South Fork is dominated by Yellowstone Cutthroat Trout followed by Rainbow Trout and increasing densities of Brown Trout.

A South Fork Snake River beauty. Photo Courtesy BLM.

South Fork of the Snake River, Idaho. Photo by Roger Phillips.

Drift boat action on the South Fork Snake River. Photo Courtesy BLM.

Palisades Reservoir July 2016. View of Blowout Boat Launch from the Edwards Creek dock. Photo courtesy John Sigler.

Falls on the South Fork Snake River. Photo Courtesy BLM.

The next stretch of river runs from Pine Creek downstream to the Heise area and is commonly referred to as the canyon stretch. Burns Creek, which is the final major spawning tributary for Cutthroat Trout, is located here. South Fork Canyon represents the classic western trout river, with crystal-clear water, exceptional scenery, and outstanding fishing. This reach, along with portions of the reaches above and below here, contains one of the largest riparian cottonwood galleries outside of Alaska. This cottonwood gallery provides habitat for a variety of species, including Shiras moose, elk, white-tailed and mule deer, bobcat, mountain lion, eagles and osprey, beaver, and mink. Some of these species can be seen on almost all trips through the canyon, and all have been seen at one time or another by anglers in this stretch. The BLM is responsible for managing numerous designated camping areas throughout the canyon, making overnight stays and multiday trips common during the warmer months. Although there are portions of this reach that are braided and have multiple channels and islands, the majority of the river is contained in a single channel. Road access to large portions of this reach is lacking, and anglers who are lucky enough to venture here need to be prepared with the right gear to accommodate changing weather conditions as well as changing fishing conditions. The nearest boat ramp on the upper end is the Conant Ramp, and the nearest public takeout is at Fullmer (Cottonwood) Boat Ramp, followed by Wolf Flats. Some anglers will use these two ramps, but most anglers choose to take their boat out at the Byington Boat Ramp since it's easier to run a shuttle from Conant to Byington. This reach has more native Cutthroat Trout than any of the other species, with Brown Trout second in abundance and Rainbow Trout found in lower numbers.

The final section of river runs from Heise downstream to the confluence with the Henrys Fork and is commonly referred to as the lower section. Once again, roads connect and cross the river at numerous locations, making access to the river much easier than the canyon section above. Brown Trout dominate

Panoramic view of the South Fork Snake River. Photo Courtesy BLM.

this lower end of the river, although Cutthroat Trout can still be found in good numbers. Rainbow Trout are rare, but still present in small numbers. Numerous irrigation diversions begin to appear in this lower reach, delivering water from the river to fields to grow crops like alfalfa, potatoes, and grains that help support the local economy. Unlike the canyon, the lower river becomes more braided as it spreads across the valley floor, no longer hemmed in by the Big Hole Mountains. Side channels and islands are the norm here, with high flows rearranging the river channel annually. This shifting of the river channel adds to the high-quality riparian and aquatic habitat found in the South Fork. Throughout the river channel, the river hits places where downwelling occurs, often noted as "losing reaches," where water disappears from the river channel. The subterranean journey of the water results in filtering fine particles and impurities, and when the water reemerges in a "gaining reach," it's cleaner and cooler than when it first entered the gravel. Over time, these gravel areas plug up with detritus and fine sediments, causing the water to remain above ground and increase in velocity. Eventually, this increased velocity carves out new channels and re-sorts and cleans the substrate, and the cycle starts again. Luckily, the South Fork is not overly confined by development and remains as wild as any western river; these natural processes are allowed to continue and improve aquatic habitat. The adage that a man can never fish the same river twice is very accurate for the South Fork Snake River.

Fishing Regulations

Unlike regulations on many wild trout rivers across the West, the South Fork still welcomes anglers of all types and does not exclude anyone from fishing these highly productive waters. Fly-fishing tends to be the dominant gear of choice for anglers from near and far, mainly because it can be used to successfully imitate the bulk of trout diets on the river. But spin and even bait anglers are still welcome to ply their craft on the South Fork's waters. Similarly, many anglers prefer to fish from a drift boat, stopping periodically to target a particular pod of rising fish. Still other anglers choose to wade, while some fish from motorized craft, mainly jet boats. Common courtesy, tolerance, and respect for other anglers make these sometimes opposite recreational pursuits possible on the South Fork. Anglers should review current regulations prior to fishing this area.

Seasonal Fishing

Changing seasons bring different opportunities to anglers who choose to fish the South Fork. During the winter, flow releases from Palisades Dam are low, generally ranging around 25 to 34 cms (900 to 1,200 cfs). This provides the best opportunity for wading anglers to stalk trout and can at times make boating difficult. Fish tend to move into winter holding patterns and congregate in deeper holes and pockets. Insect hatches are not common during the winter, although midge hatches can occur at any time of the year. Fly anglers will typically choose to nymph-fish slow runs or, on rare days, fish small dry flies to casually rising trout. Streamers worked slowly on sinking line can entice big browns into becoming active as well. Spin anglers can slowly roll big spinners like a Vibrax Blue Fox #4 through deep holes and elicit strikes. And anglers who choose to use bait can drift a night crawler through deep holes and runs and find active fish too. When the action gets slow for trout, patient anglers can sometimes find pods of Mountain Whitefish, another Idaho native, active at the tail end of runs. Although not as glamorous as a feisty Cutthroat Trout in some anglers' view, Mountain Whitefish have saved many trips on the South Fork.

Spring can be a great time to get out on the river. Many anglers don't enjoy Idaho's variable spring weather and choose to pursue other activities as opposed to getting out on the water. That leaves an increasingly active fish population available to the minority of folks who are willing to roll the dice on the weather and go chase trout. Flows generally remain fairly low until April, when increases of several hundred cubic meters per second (4,000+ cfs) can make wading more challenging. Many areas are still very accessible to wade fishing, but anglers will need to be more cautious with these higher flows. Rainbow Trout in the upper river begin to stage for spawning in March, becoming more active and moving to spawning grounds below Palisades Dam and in slower water along the river's edge where smaller gravel is found. Fish staging to spawn will willingly take Prince Nymphs, Hares Ears, Prince and similar patterns. Blue Winged Olives start to emerge in the spring, creating the first consistent dry-fly fishing of the year. Midges will continue to hatch and diversify the dry-fly action. As spring water temperatures increase, Brown Trout in the lower river become more active as well and will start aggressively chasing down streamers and Woolybuggers. Spin anglers can start working the bank habitat around logs and brush with Rapalas and similar minnow imitations and find fast action. Bait anglers will find success with worms, particularly as flows increase and erosion of the banks begins. This erosion naturally washes invertebrates into the water, making a night crawler more of a naturally occurring part of the diet.

Summer is when the South Fork really shines, but expect to see other anglers out enjoying the fantastic weather and great fishing that the river offers at this time. Flows generally peak in late May through June, making wading more difficult. By July flows settle down to a more modest 280 cms (10,000 cfs), and fishing becomes excellent. Most anglers prefer to fish from a boat during the summer, stopping on riffles and side channels to get out and wade around. Water temperatures are still too cool to comfortably swim due to releases from the dam upstream. However, this cool water keeps trout very active through the summer months. Insect hatches are diverse, with Pale Morning Duns, Caddis, and Yellow Sallies providing fast action when conditions favor abundant hatches. Of particular interest to many anglers is the hatch of both Golden Stoneflies and Salmonflies that occurs in late June and July. These large protein packages get the trout looking toward the shoreline in an attempt to catch a careless adult that falls from nearby vegetation and provide for some of the most memorable dry-fly fishing that can be found anywhere. Late summer sees grasshoppers, ants, and other terrestrial insects becoming more common. The grasshopper bite can be another great way to catch a lot of trout in August. Kids and new anglers may enjoy perfecting their dry-fly fishing on one of the many pods of Mountain Whitefish that can be found rising in slower runs. Whitefish can be more susceptible to smaller flies like a Parachute PMD or Light Cahill, as they have a smaller mouth than trout do. Spin anglers will continue to be successful with spinners and Rapalas and may try using smaller-size spinners such as Rooster Tails, Panther Martins, and Mepps to up the catch rates. Bait anglers should look to the lower river during summer and drift dead sculpin through deep holes to find lurking Brown Trout.

Fall brings cooler temperatures to the region and a departure from the more crowded conditions of summer. Flows generally recede once again during the fall, as irrigation demand wanes and water is stored behind Palisades Dam for delivery the following season. Foliage starts to change color, starting with aspens in September, which turn a bright yellow, painting the valleys in stark contrast to the dry sagebrush nearby. Changes to the cottonwood gallery that provides shade during the summer follow the aspens, turning vivid yellow and orange as fall progresses. Cooler morning temperatures require a couple layers of clothing or a warm jacket, which can be shed by early afternoon with some warming sunshine. Trout respond to the shortening day length and cooler temperatures by putting on the feed bag to prepare for the slower wintertime.

Fly anglers can catch the last portion of the ants, beetles, and grasshoppers into October and will see a reemergence of Blue Winged Olives until late November. As fall progresses, midges once again take center stage, with patterns like the Griffith Gnat working on the surface and nymphs such as the Brassie working under an indicator. Fall also brings more activity to Brown Trout, as they stage for spawning and try to put on some pounds for the colder months ahead. Streamers once again shine for big Brown Trout, although Yellowstone Cutthroat Trout and even Rainbow Trout can be picked up on patterns like Double Bunnies and Woolly buggers. Watching a streamer twitching beneath the water's surface and then be engulfed by an unseen mouth can be a memorable experience during the fall. Spin fishing techniques remain similar to early-season patterns—big spinners and minnow-imitating plugs and crankbaits. Similarly, bait anglers can drift dead minnows through deep holes to connect with Brown Trout and Yellowstone Cutthroat Trout.

While the likelihood of a visiting angler encountering a fish on the South Fork on his own is good, the odds of success go up when tapping into a local knowledge source like a guide or outfitter. There are several high-quality, reputable guide shops that operate on the South Fork. Having access to and fishing with these local guides can help a visiting angler quickly get up and over the learning curve and on to more successful fishing. Guides also have access to the best equipment and boats, which can get you to the fish faster and more effectively than a do-it-yourself trip. A quick Google search can get you in contact with an appropriate outfitter and help get you to a picture-worthy moment on the South Fork.

Management

While the South Fork offers one of the West's best trout-fishing opportunities for native Yellowstone Cutthroat Trout, their future is uncertain. Our native trout are threatened by an increasing abundance of Rainbow Trout, which can interbreed and compete with our native Cutthroat Trout. Several management actions are ongoing on the South Fork, including managing flows from Palisades Dam to benefit Cutthroat Trout when possible and managing spawning tributaries to keep Rainbow Trout from invading. But possibly the most effective management tool available is angler harvest. Biologists are encouraging anglers to keep all the Rainbow Trout they catch on the South Fork to help suppress the population. Removing Rainbow Trout entirely will never be possible and isn't necessary for Yellowstone Cutthroat Trout persistence. But reducing Rainbow Trout abundance will help keep Cutthroat Trout viable into the future and keep the South Fork unique. Anglers can do their part to help manage these rare Idaho gems by enjoying a fresh Rainbow Trout dinner while fishing the South Fork or donating their fish to the IDFG.

Idaho's Urban Fishing Opportunities

Time for recreation seems to becoming increasingly scarce these days. The number of outdoor opportunities competing with fishing has never been higher. With limited time and so many recreational possibilities, fishing opportunity close to home is increasingly important to anglers. Community ponds are becoming popular fishing destinations across Idaho. These ponds are important recreational amenities and provide easily accessible fishing options to anglers of all ages and skill levels.

Community ponds are present throughout Idaho and are located in a wide variety of settings, from small ponds in urban city parks to larger lakes in rural settings. Around Idaho, community ponds typically offer angling for hatchery Rainbow Trout, and several warm water species including Largemouth Bass, Bluegill, Pumpkinseed and Channel Catfish. While Rainbow Trout are provided by IDFG hatchery programs, warm water fish populations depend on natural reproduction or IDFG transfers from other waters. Trout fishing is usually best during the cool months of the year. Summer water temperatures in some ponds may get too high to allow trout stocking, which is a better time to focus on bass and panfish.

Fishing rules for community ponds are usually simple to encourage new anglers to try their luck, but be sure to check the local rules before you go. Fishing rules are provided annually in a printed booklet available where licenses are sold. Rules are also available on the IDFG website or try using the mobile-friendly Idaho Fishing Planner website.

There are a huge number of community ponds available to anglers throughout Idaho. The following list represents just a sample of those available and which species you might expect to catch. More complete information is available through the IDFG website. From the IDFG homepage, scroll to the bottom and find the Idaho Fishing Planner logo for the most complete information on access, stocking, and maps!

Panhandle Region

Antelope Lake

Directions: This is a scenic little lake near the town of Clark Fork, north of the Clark Fork River. Travel east of Clark Fork on State Hwy 200 approximately 2.4 km (1.5 mi) to a dirt road on the north side of the highway. Road winds approximately 1.3 km (0.8 mi) up the hill to Antelope Lake. Road may be seasonally impassable depending on snow conditions.

Species: Westslope Cutthroat Trout, Rainbow Trout

Fernan Lake

This popular fishing spot is only 10 minutes from downtown Coeur d'Alene. Anglers of all abilities can enjoy fishing from floating docks, miles of shoreline, or from boats. From Coeur d'Alene, Fernan Lake is located just east of Interstate 90 off the Sherman Street exit.

Species: Rainbow Trout, Brook Trout, Cutthroat Trout, Bluegill, Largemouth Bass, Northern Pike, Channel Catfish

Jewel Lake

Jewel Lake is a peaceful lake stocked with Rainbow Trout and Channel Catfish. Also, lots of small Bluegill will thrill young anglers. From US Hwy 95, 4.8 km (3.0 mi) south of Sagle or 3.2 km (2.0 mi) north of Westmond, turn onto Dufort Road (by the dumpsters on the west side of the highway.) Go past Round Lake State Park, cross Cocolalla Slough and about 1.3 km (0.8 mi) after crossing the slough watch for a small sign on the left that marks the dirt road to Jewel Lake. Follow this road up and over the hill approximately 4.8 km (3.0 mi) to Jewel Lake.

Species: Rainbow Trout, Bluegill, Channel Catfish

Post Falls Park Pond

Post Falls Park Pond is a great fishing spot for kids and anglers with special needs. Anglers can easily fish the small pond from a fishing bridge, shore, or fishing platforms, and enjoy the adjacent park. From Interstate 90 take the Spokane Street Exit at Post Falls and travel south one-half block to Fourth Street. Go west on Fourth Street to a large parking lot and walk a short distance to the pond.

Species: Rainbow Trout, Largemouth Bass, Bluegill

Spicer Pond

Spicer Pond is a small, easily accessed pond next to the Saint Maries River. From the east end of Saint Maries, cross the Saint Maries River on US Hwy 3 (E College Ave) and continue south about 1.3 km (0.8 mi). The pond and parking lot are on the right.

Species: Rainbow Trout, bullhead, Pumpkinseed, Yellow Perch

Clearwater Region

Kiwanis Park Pond

This pond is conveniently located in the heart of Lewiston along the Snake River inside Kiwanis Park. The park is located on the west side of downtown at the intersection of Snake River Ave and Hwy 12 (Main Street).

Species: Rainbow Trout

Robinson Pond

This pond is located about 4.8 km (3.0 mi) west of Kamiah up the Lawyer Creek canyon. From Kamiah, take Pine Street, which turns into 7 Mile Road (ID-162). Follow 7 Mile Road and veer right onto Lawyer Creek Road. Travel another 1.1 km (0.7 mi) and the pond is on the left.

Species: Rainbow Trout, Largemouth Bass

Mann Lake

This is a popular small lake for trout and panfish just a short drive east of Lewiston. From Thain Road, turn east on Powers Ave. Travel 7.8 km (4.85 mi) to the lake. This location has excellent facilities including a boat ramp, toilette and docks.

Species: Rainbow Trout, Largemouth Bass, White Crappie, Channel Catfish, Bluegill, bullhead

Southwest Region

Boise River

This stretch of the Boise River offers miles of fishing in one of the finest urban trout rivers in the country. Wild trout are abundant along with stocked hatchery Rainbow Trout and Brown Trout. Access is easy along the many paths of the Boise River Greenbelt. This section of the Boise River is accessible from multiple locations along the Boise River Greenbelt from Lucky Peak Reservoir through Boise to Eagle, as well as other access points farther west. Easy access points include bridges at Star Rd, Linder Rd, Eagle Rd, Glennwood Rd, Veterans Blvd, as well as multiple city parks.

Species: Rainbow Trout, Brown Trout, Mountain Whitefish

Caldwell Rotary Pond

Rotary Pond is located in a city park a park-like setting with plenty of trees to offer shade in the summer. Lots of bank fishing offers young anglers a safe place to fish for Bluegill and bass. This pond is part of Canyon County's park system and is west of Whittenberger Park on West Chicago Street. More information may be found on the Canyon County website. From Interstate 84 take Exit 27 (Greenleaf/Wilder). Turn south on Centennial Way. Travel 0.6 km (0.4 mi) to Chicago Street and turn west. Go 0.32 km (0.2 mi) to Rotary Pond.

Species: Rainbow Trout, Largemouth Bass, Bluegill, Channel Catfish

Kleiner Pond

Kleiner Pond is two connected waters located within the City of Meridian's Julius M. Kleiner Memorial Park. A large fishing platform and plenty of shoreline access make this a family-friendly water with city park activities nearby. Contact the City of Meridian for more information.

Species: Rainbow Trout, Largemouth Bass, Bluegill

Park Center Pond

Park Center Pond is a popular fishing pond within a City of Boise Park. Amenities include docks, tables, and shade. This convenient pond is located near the greenbelt and a variety of restaurants. Located near downtown Boise, this pond is 0.6 km (0.4 mi) east of the West Park Center Bridge between Beacon Street and Mallard Drive.

Species: Rainbow Trout, Largemouth Bass, Bluegill, Channel Catfish

Riverside Pond

Riverside Pond is tucked into the trees off Glenwood Road and offers good bank fishing access. This neighborhood pond is right off the Boise Greenbelt next to the Boise River. Located near the fairgrounds in Garden City on the west side of Glennwood Road alongside the Boise River. Go north across the river on Glenwood Road, turn west on Riverside Drive and travel 0.32 km (0.2 mi) to a small parking area on the left.

Species: Rainbow Trout, Largemouth Bass, Bluegill

Sego Prairie Pond

This small neighborhood pond offers easily accessible fishing for trout and panfish within Nicholson Park next to Indian Creek in Kuna. From Interstate 84 in Meridian, travel south on State Hwy 69 about 10 km (6.2 mi), then go west on Deer Flat Road for 3.2 km (2 mi). Turn south on Ten Mile Road and travel 0.6 km (0.4 mi), turning west onto W. Sego Prairie Street. The road ends near the pond in Nicholson Park.

Species: Rainbow Trout

Weiser Community Pond

This community pond offers an easily accessible shoreline and dock fishing near the Weiser River. From US Hwy 95 in Weiser, travel east on Commercial Street. Turn south on E 4th Street, which ends at the pond.

Species: Rainbow Trout

Wilson Springs Pond complex

Wilson Springs Ponds offers a rich natural environment that is very accessible. Paved pathways provide wheelchair access to several fishing areas, while unimproved trails let kids fish from the shore and explore nature, making this a popular family destination. From Interstate 84, take Exit 38. Travel south on Garrity to the first stoplight. Turn left onto Flamingo Avenue and continue 0.32 km (0.2 mi). The road turns south and becomes Happy Valley Road. Travel nearly 6.4 km (4 mi) to Greenhurst. Turn west on Greenhurst and travel 3.2 km (2 mi) to South Powerline Road. Turn south and travel 1.1 km (0.7 mi) to the Wilson Ponds parking area.

Species: Rainbow Trout, Largemouth Bass, Bluegill

Magic Valley Region

Dierke's Lake

This is a conveniently located pond inside Shoshone Falls Park near Twin Falls. This scenic county park also features hiking, swimming docks, picnic areas, swim beach and other amenities nearby. From US Hwy 93/ Blue Lakes Boulevard in Twin Falls, travel 4.8 km (3.0 mi) east on Falls Ave E. Turn north on N 3300 E and go 1.6 km (1.0 mi) to the park entrance. County park fee applies. From the kiosk continue for about 1.0 km (0.6 mi) and veer right on Dierke's Lake Road to the pond.

Species: Rainbow Trout, Largemouth Bass, Bluegill

Dollar Lake

Dollar Lake offers great rustic fishing on a small pond that is also close to stream fishing, camping, hiking, biking and even shopping in Wood River Valley. About 0.4 km (0.25 mi) past Dollar Lake, Penny Lake is on the opposite side of the road. In Ketchum at the intersection of Sun Valley Road and Main Street go 0.32 km (0.2 mi) north on Main Street. Take soft left onto Warm Springs Road and go 6.1 km (3.8 mi) west. The parking lot is on the right side of the road and it's a short walk up the hill to the pond.

Species: Rainbow Trout

Emerald Lake (near Heyburn)

This pond is located just outside Heyburn and is also close to the cities of Burley and Paul. Anglers can access the pond off S 400 W near the intersection of I-84 and Hwy 30/24. The pond offers seasonal trout fishing with bass and panfish during warmer months.

Species: Rainbow Trout, Largemouth Bass, Bluegill

Rock Creek in Rock Creek City Park

Rock Creek is an easily accessed stream in a park setting with park amenities. The creek is stocked regularly with Rainbow Trout and once a year with Brown Trout fingerlings. Best fishing is before and after spring runoff. The park is located on Rock Creek County Park in Twin Falls. From Blue Lakes Boulevard go 3.5 km (2.2 mi) west on W Addison Avenue. Turn right (north) at the sign for the County Park.

Species: Rainbow Trout, Brown Trout

Chapter Citations

1. McClane, 1998. 2. Idaho Department of Fish and Game, (undated). 3. Rosenbaum and Kaufman, 2009. 4. Tolentino and Teuscher, 2010. 5. Idaho Department of Fish and Game, 1996. 6. Idaho Department of Fish and Game, 2013. 7. Matthews, 2013. 8. McDevitt, 2001. 9. Idaho State Historical Society, 1964. 10. Thurow, 1990. 11. Hubbs and Miller, 1948. 12. Wallace and Zaroban, 2013. 13. Smith, 1966. 14. Behnke, 1979. 15. Behnke, 1988. 16. Thurow, 1987. 17. Williams and Shiozawa, 1993. 18. Kozfkay, et al., 2007. 19. Hauck, 1949. 20. Thurow, 1988a. 21. Thurow, 1988b. 22. Bell, 1988. 23. Irizarry, 1969. 24. US Department of Commerce, 1984. 25. Ryan, Gutknecht and Megargle, 2008. 26. Stanton and Megargle, 2014. 27. Ambrose and Abell, 2002. 28. Saint John, 2014. 29. Bentz, 2014. 30. Van Kirk and Gamblin, 2000. 31. Van Kirk and Griffin, 1997. 32. Loomis, 2006. 33. Grunder, et al. 2008. 34. Lawson, 2012. 35. McDaniel, 2012. 36. Behnke, 2002. 37. US Forest Service, 1998. 38. Thurow, 2015. 39. Thurow, 1985; 40. Healey, 1991. 41. Fulton, 1968. 42. Gebhards, 1959. 43. Idaho Department of Fish and Game, 1992. 44. Parkhurst, Bryant and Nielson, 1950. 45. Burner. 1951. 46. Matthews and Waples, 1991. 47. Thurow, Lee, and Rieman, 2000. 48. Crozier, et al., 2008. 49. Gamett and Bartell, 2011. 50. Idaho Department of Fish and Game, 1973. 51. Knudsen, et al., 1982. 52. Liljeblad, 1957. 53. Smith, 1973. 54. Carrey and Conley, 1977. 55. Mallet, 2013. 56. Mallet 1960; 57. Zurstadt and Stephans, 2004. 58. Leuzinger, 2015. 59. Corley, 1972. 60. Jeppson and Ball, 1979. 61. Thurow, 1982. 62. Flinders, et al., 2013. 63. Kennedy and Meyer, 2015. 64. Bevan, Harville and Bergman, 1994. 65. US vs Oregon Technical Advisory Committee, 2008. 66. National Marine Fisheries Service, 2000. 67. Raymond, 1979. 68. Willson and Halupka, 1995. 69. Williams, 2006. 70. Butts et al., 2016. 71. Egan, 2010. 72. Scott, 1968. 73. MacPhee, 1966. 74. Shepard, May and Urie, 2003. 75. Walrath, Quist, and Firehammer. 2015.

"A true conservationist is a man who knows that the world is not given by his fathers, but borrowed from his children."

John James Audubon

5

KEY TO THE FAMILIES AND TAXA OF FISHES IN IDAHO

Gerald R. Smith

The use of keys for identification of fishes is sometimes frustrating. Fish colors, sizes, shapes, and numbers of scales and fin rays are naturally variable for genetic, age, and environmental reasons. Keys are intended to pick out the most diagnostic traits of species or other groups and arrange them in contrasting pairs so as to present clear choices that will lead to correct identification of the fish in hand, but juveniles often fail to exhibit clearly the most diagnostic traits, and some individuals may be injured or genetically near the limit of species variation. In the worst cases, the individual's parents might have been careless in their mate choice and produced a hybrid with mixed species characters. Some times hybridization and gene flow have caused species characters to overlap. In case of difficulty in using the keys, consider these possibilities, study the figures and descriptions associated with the couplet in which they appear, and try again. In the most frustrating case, you might be trying to identify a newly invasive or introduced species!

Key to Idaho Fish Families

1a. Mouth a round or oval sucking wreath with no jaws; no paired fins; a line of 7 small gill openings on each side (Fig. 1a). Lampreys, **Petromyzontidae** (Chapter 6), 1 species,.. Pacific Lamprey, *Entosphenus tridentatus*

Fig. 1a

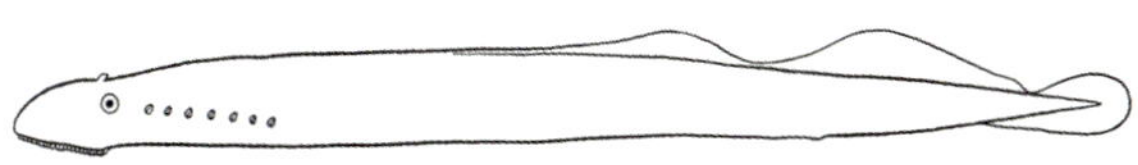

1b. Mouth with distinct upper and lower jaws, paired pectoral and pelvic fins present; 1 large gill opening on each side.. ray-finned fishes, 2

2a. Head triangular from above, narrow in front and flat below; 4 barbels in a transverse row ahead of mouth; tail with much larger upper lobe; 5 rows of bony plates; no scales (Fig. 2a). Sturgeons, **Acipenseridae** (Chapter 7), 1 species .... White Sturgeon, *Acipenser transmontanus*

Fig. 2a

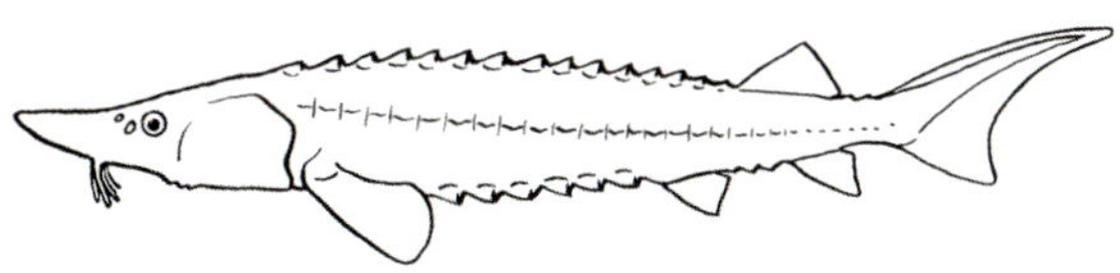

2b. Head rounded as seen from above, not a pointed triangle; lower jaw usually prominent; tail with upper and lower lobes about equal; barbels absent or if present, around mouth; scales present or absent. Bony Fishes, Teleostei 3

3a. Small adipose fin (no rays or spines) present on back between dorsal and caudal fins. CATFISH, SMELT, GRAYLING, WHITEFISH, SALMON, TROUT, CHAR, and TROUT-PERCH 4

3b. No adipose fin; 1 or 2 dorsal fins, always with rays or spines 7

4a. Scales absent, 4 pairs of long barbels on front of head; stout spines at front of dorsal and pectoral fins but not anal fin (Fig. 4a). CATFISHES, **Ictaluridae** (Chapter 10) ....7 species

Fig. 4a

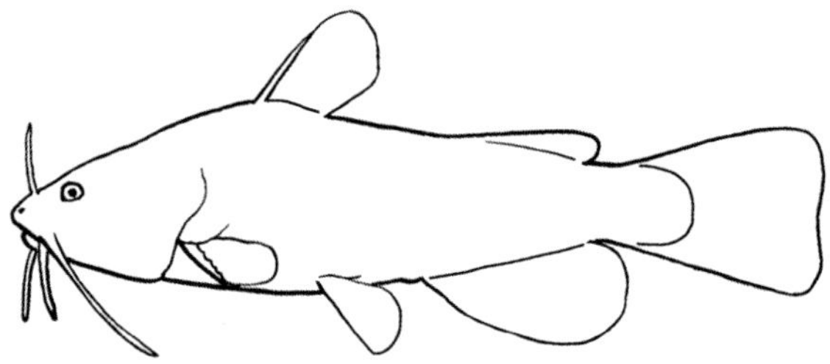

4b. Scales present; barbels absent; no spines in pectoral fin (Figs. 6a, 6b, 5b) ...... SMELTS, SALMONIDS, AND TROUT-PERCH 5

5a. No fin spines; cycloid scales; pelvic axillary process often present. SMELTS, GRAYLING, WHITEFISHES, SALMONS, TROUTS, CHARS 6

5b. Spines at front of dorsal and anal fins, ctenoid scales; no pelvic axillary process (Fig. 5b). TROUT-PERCHES, **Percopsidae**, (Chapter 14) 1 species..................................SAND ROLLER, *Percopsis transmontana*

Fig. 5b

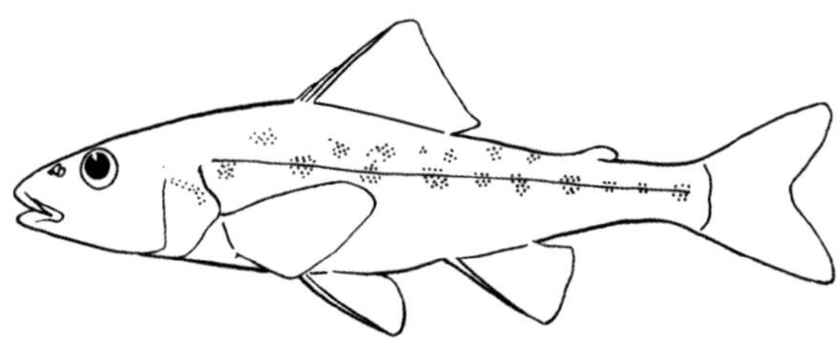

6a. No pelvic axillary process; scales large; teeth large (Fig. 6a); pyloric caeca few (Fig. 6a; see Salmoninae figure 2a for illustration of pyloric caeca) SMELTS, **Osmeridae** (Chapter 11), 1 species ....
.... RAINBOW SMELT, *Osmerus mordax*

Fig. 6a

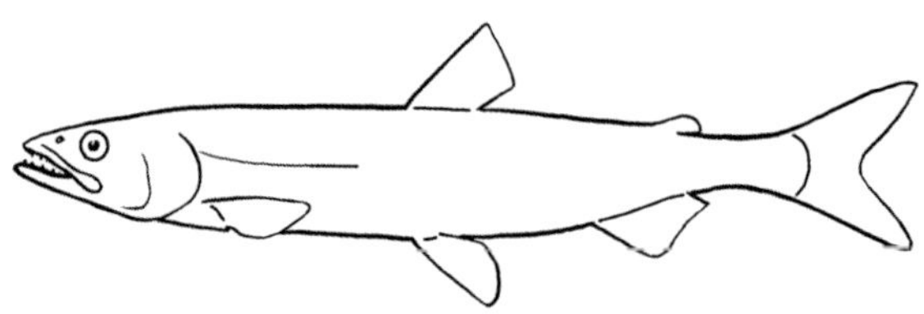

6b. Pelvic axillary process present; scales large or small, but if scales are large, teeth are small (Fig. 6b). **Salmonidae** (Chapter 12): Whitefishes, Coregoninae; Grayling, Thymallinae; Chars, Salmons, Trouts, Salmoninae, ... 26 species and subspecies

Fig. 6b

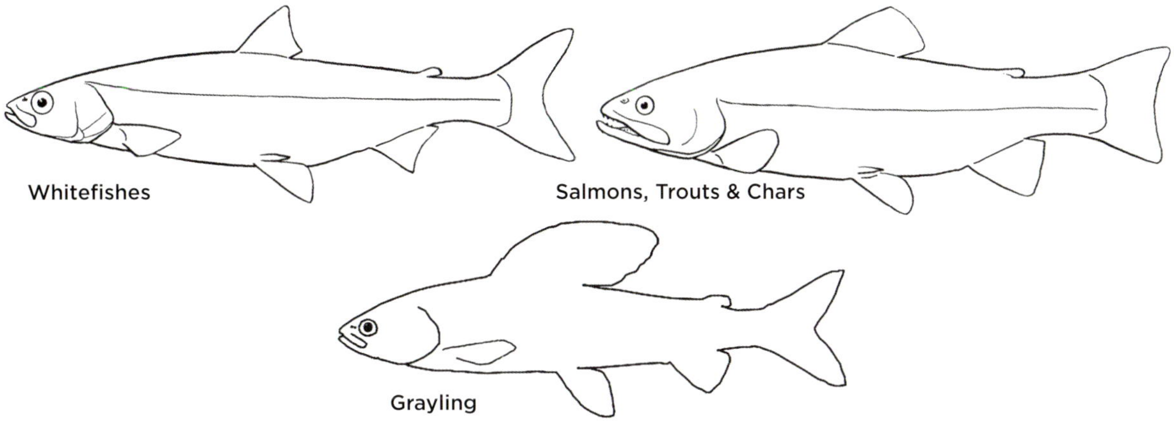

Trout Anatomy*
*Total Length omitted due to its high variability and limited usefulness in research.

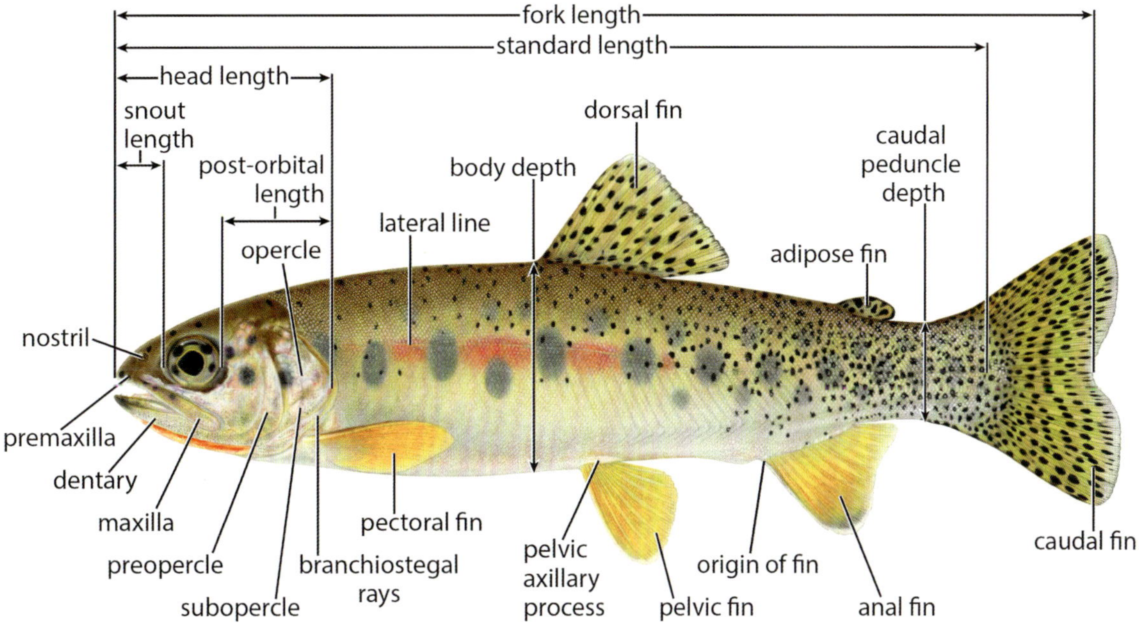

7a. A row of sharp scales along midline of belly; no lateral line; body laterally compressed; thin cycloid scales; adipose eyelids present (Fig. 7a). Herrings, **Clupeidae** (Chapter 8), 1 species .... American Shad, *Alosa sapidissima*

Fig. 7a

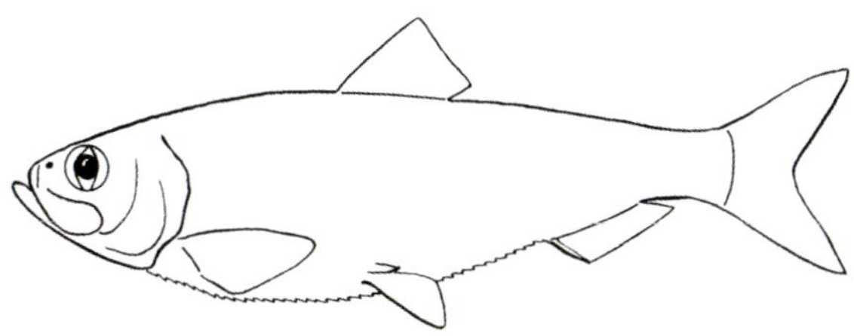

7b. No row of sharp scales along midline of belly; no adipose eyelids; many soft-rayed and spiny-rayed families........ 8

8a. Branchiostegal rays 11–19; long, sharp triangular teeth in "duckbill-shaped" jaws; dorsal and anal fins small, far back, and opposite each other; pelvic fins at midbelly (Fig. 8a). PIKES, **Esocidae** (Chapter 13),1 species........ NORTHERN PIKE, *Esox lucius*

Fig. 8a

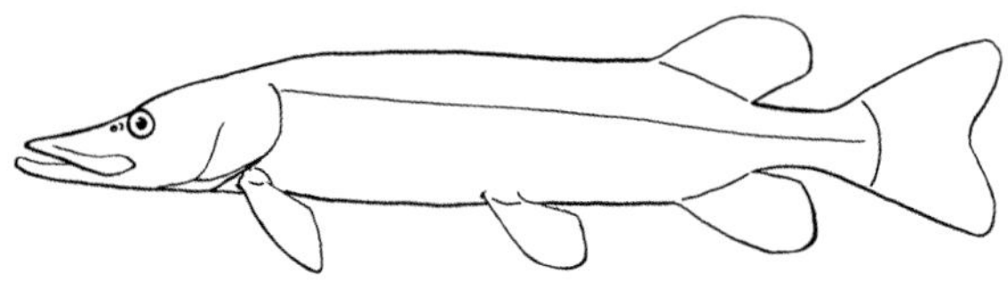

8b. Branchiostegal rays 3–7; teeth not large and triangular; jaws shorter; dorsal and anal fins not far back and opposite each other; pelvic fins abdominal or thoracic (Fig. 17b)........ 9

9a. Dorsal and anal fins long, with 60 or more rays; a small dorsal fin ahead of the long fin; body long; 1 barbel on chin (Fig. 9a). CODS, **Gadidae** (Chapter 15), 1 species........ BURBOT, *Lota lota*

Fig. 9a

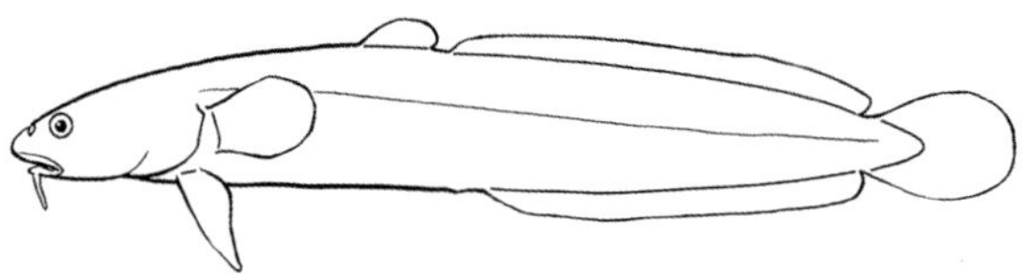

9b. Dorsal and anal fins much shorter than 60 rays; no barbel on chin; body shape variable........ 10

10a. Jaws with no teeth; 1-3 rows of well-developed pharyngeal teeth in throat (Figs. 12a, 12b, 11a). SUCKERS, MINNOWS, LOACHES........ 11

10b. Jaws upturned with small teeth; patches of small teeth in throat; size less than 80 mm (3.1 in) long... 13

11a. Body small and eel-like; mouth surrounded by 5 pairs of small barbels; tail rounded (Fig. 11a). LOACHES, **Cobitidae** (Chapter 9), 1 species........ ORIENTAL WEATHERFISH, *Misgurnus anguillicaudatus*

Fig. 11a

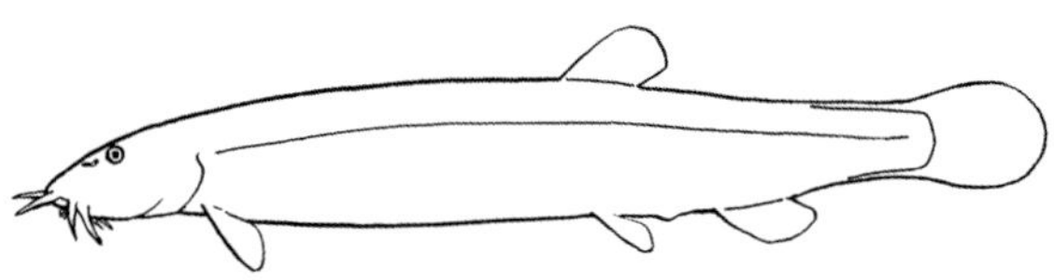

11b. Body never eel-like; mouth with no more than 2 pairs of small barbels; tail forked MINNOWS and SUCKERS (Figs. 12a, 12b)........ 12

12a. Mouth with fleshy lips, usually subterminal or ventral; pharyngeal teeth in 1 long, curved row of 20 or more; anal fin usually closer to caudal fin than to pelvic fins; no spines; no barbels (Fig. 12a). Suckers, **Catostomidae** (Chapter 9)....................8 species

Fig. 12a

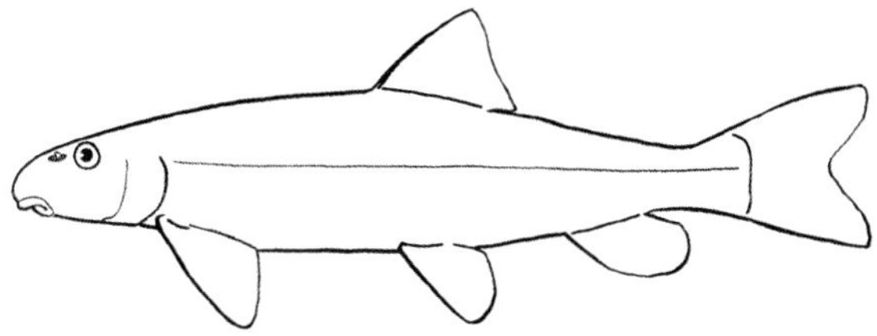

12b. Mouth with thin lips, usually terminal or subterminal; pharyngeal teeth in 1–3 rows of 4 or 5 teeth; some species with small barbels; anal fin variable (Fig. 12b). Minnows, **Cyprinidae** (Chapter 9)........... 19 species

Fig. 12b

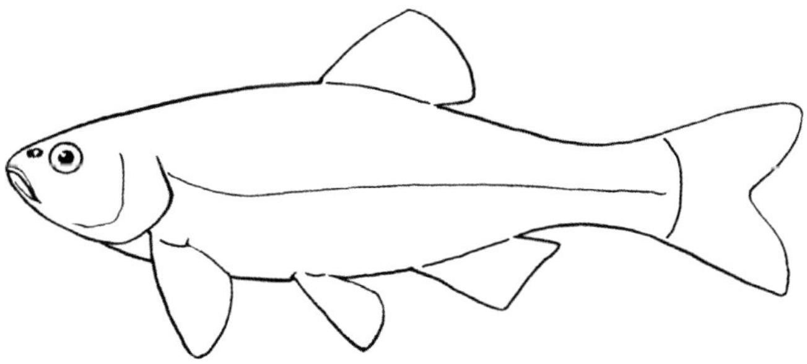

13a. Small, upturned mouth; head small and somewhat flattened above; body slender; dorsal and anal fins small; pelvic fins more or less abdominal (Figs. 14a, b). Topminnows and Livebearers......................... 14

13b. Large terminal mouth; head large and wide; body variable; dorsal fins large and sometimes 2; pelvic fins thoracic (Figs. 15a, 16b, 17a, 17b). Sculpins, Perch, Bass, Sunfish, Walleyes, Crappies, Cichlids...... 15

14a. Anal fins always small and rounded (Fig. 14a). Topminnows, **Fundulidae** (Chapter 16), 1 species.
....................Banded Killifish, *Fundulus diaphanus*

Fig. 14a

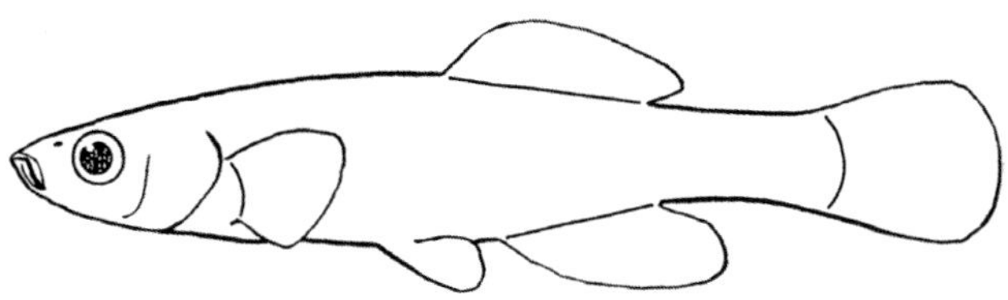

14b. Anal fins of males elongated into a sperm-transfer organ (Gonopodium); female abdomens often swollen with live young (Fig. 14b). LIVEBEARERS, **Poeciliidae** (Chapter 16) ....5 species

Fig. 14b

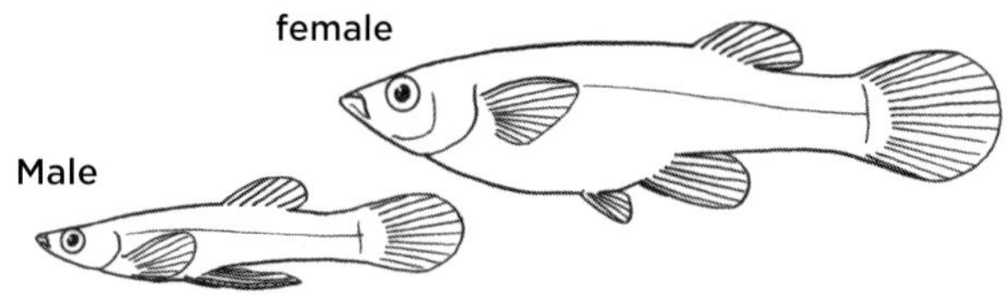

15a. Skin scaleless or with small, sharp prickles; dorsal and anal fins with short, unbranched rays and no spines; pectoral fins large, rounded, and tilted forward; head wider than deep with eyes directed slightly upward; color mostly brown or grayish (Fig. 15a). SCULPINS, **Cottidae** (Chapter 17) ....10 species

Fig. 15a

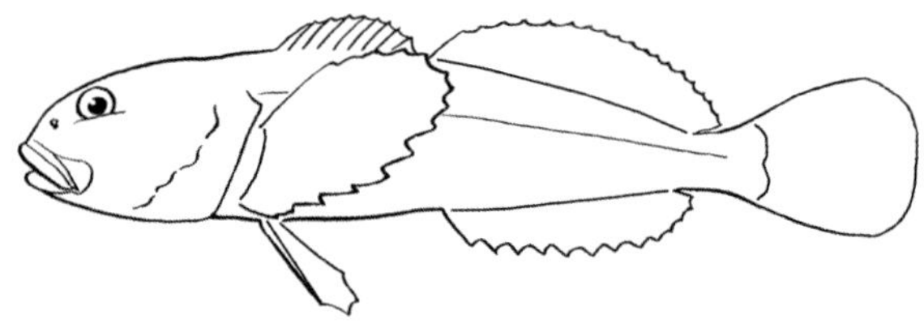

15b. Ctenoid scales; dorsal fin with sharp spines in front; branched rays behind; anal fins with 1–6 spines and up to 19 branched rays; pectoral fins high on sides; heads laterally compressed or as wide as deep; skin pigment patterns often colorful (Figs. 16b, 17a, b). PERCH, BASS, SUNFISH, CRAPPIES, CICHLIDS 16

16a. Nostrils with separate incurrent and excurrent openings on each side; lateral line continuous (compare color illustrations). PERCHES and SUNFISHES 17

16b. Nostrils with 1 opening on each side; lateral line divided and offset below second dorsal fin (Fig. 16b). CICHLIDS, **Cichlidae** (Chapter 18) ....5 species

Fig. 16b

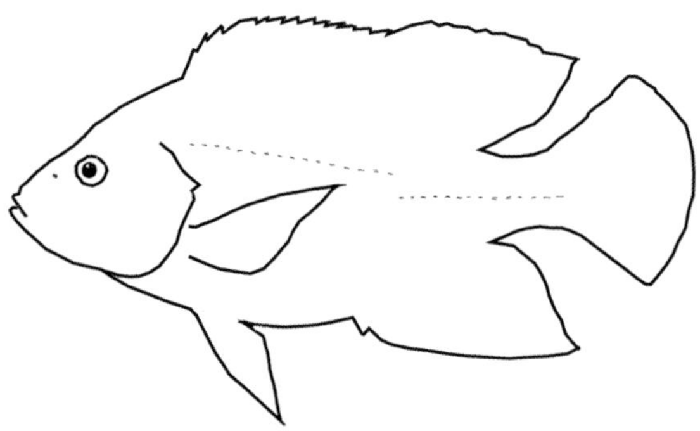

17a. Anal spines 1 or 2 (Fig. 17a). PERCH, WALLEYE, SAUGER, **Percidae** (Chapter 18)....................3 species

Fig. 17a

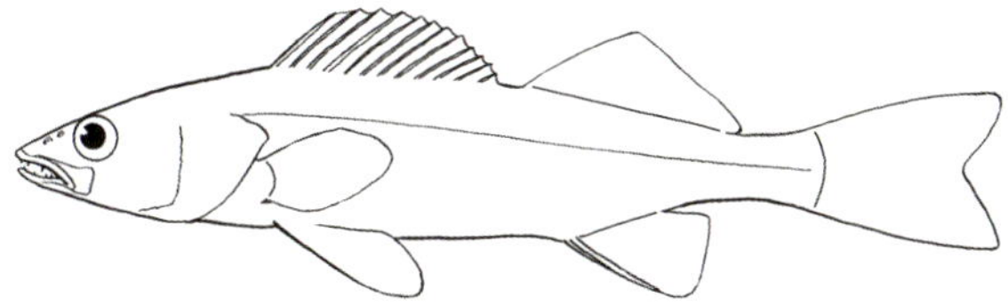

17b. Anal spines 3–6 (Fig. 17b). BASS, SUNFISH, CRAPPIES, **Centrarchidae** (Chapter 18) ....10 species

Fig. 17b

Bass Anatomy*

*Total Length omitted due to its high variability and limited usefulness in research.

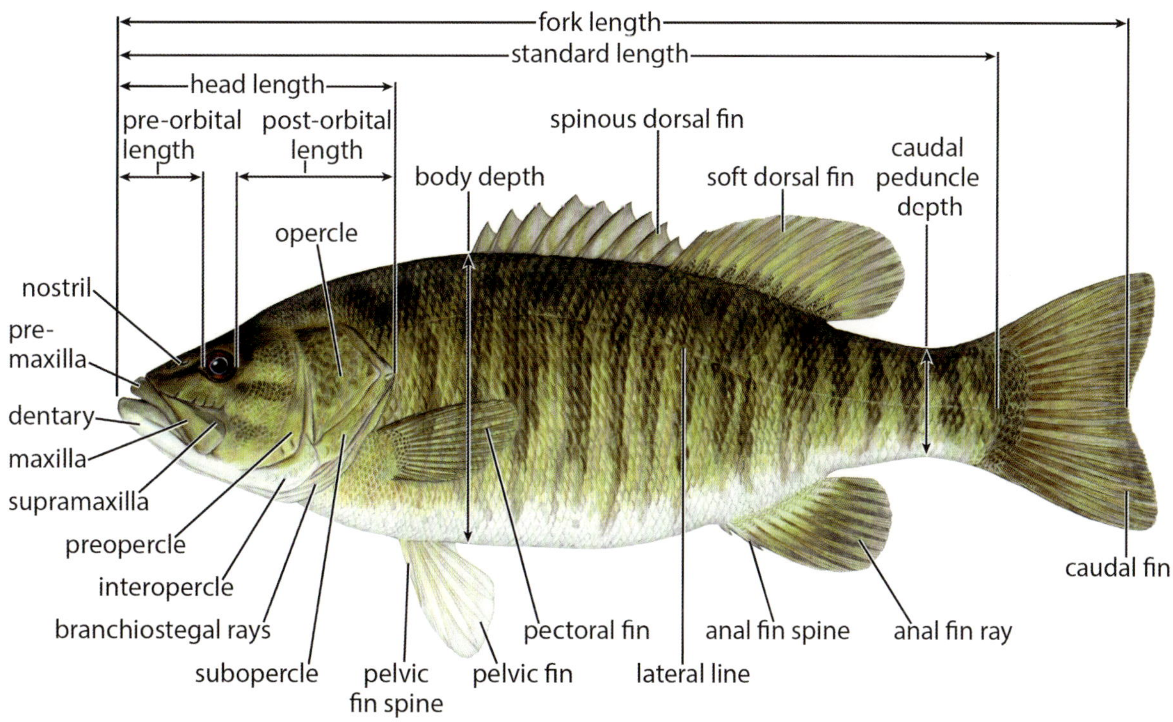

Key to Idaho Carps and Minnows
Family Cyprinidae (Chapter 9)

1a. Dorsal and anal fins with spines at the leading edge; dorsal fin with more than 15 rays CARP and GOLDFISH.. 2

1b. Dorsal and anal fins with no spines; dorsal fin with fewer than 13 rays.. 3

2a. Upper jaw with 2 prominent barbels; gill rakers 21–27; sometimes with large scattered scales; pharyngeal teeth are molariform (Fig. 2a).. COMMON CARP, *Cyprinus carpio*

Fig. 2a

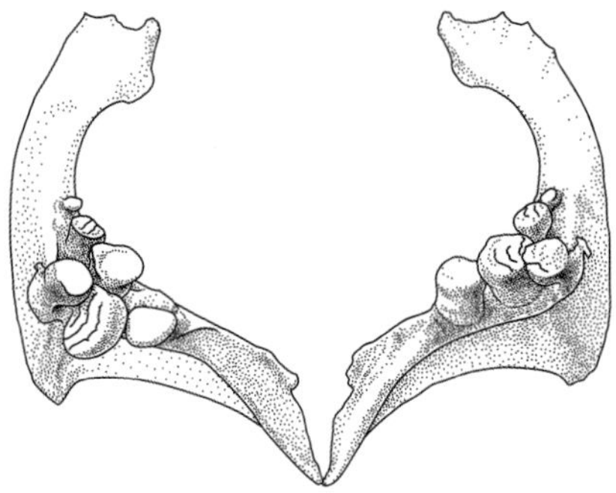

2b. Upper jaw with no barbels; gill rakers 37–43; color orange or olive green .... GOLDFISH, *Carassius auratus*

3a. Eye low on side of head, even with tip of upper lip, which is noticeably thickened; large pharyngeal teeth serrated (Fig. 3a)... GRASS CARP, *Ctenopharyngodon idella*

Fig. 3a

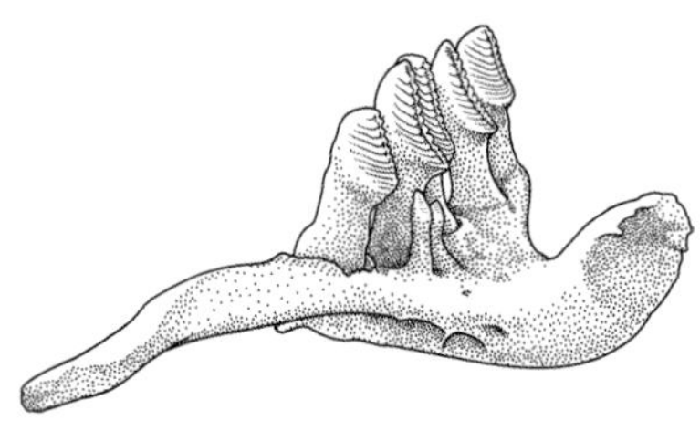

3b. Eye above midline; upper lip not thickened; pharyngeal teeth not serrated (Figs. 2a, 4a, 5a3, 7b3, 14a, 16a, 18a, 19a, 19b).. 4

4a. Caudal peduncle deep, its depth goes into standard length 5–6 times; scales small in more than 80 diagonal rows along sides; pharyngeal teeth are molariform or with grinding surfaces (Fig. 4a) .... TENCH, *Tinca tinca*

Fig. 4a

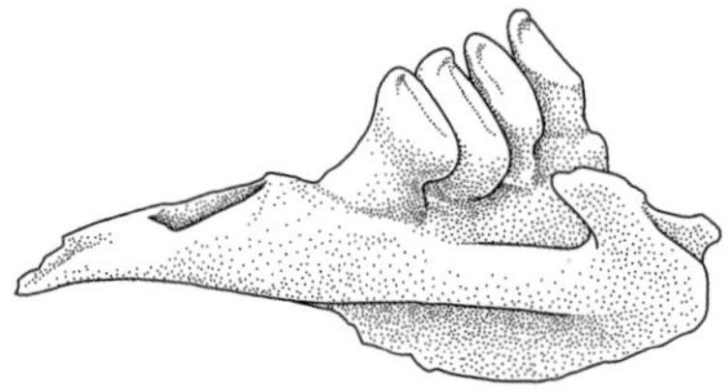

4b. Caudal peduncle depth goes into standard length more than 7 times; scales and teeth variable.......... 5

5a. Lower jaw ventral, with a straight cartilaginous scraping sheath; mandible bends at right angles toward the midline; pharyngeal teeth with a long scraping surface on the upper side (Fig. 5a); scales small, caudal peduncle narrow; fins large.. Chiselmouth, *Acrocheilus alutaceus*

Fig. 5a

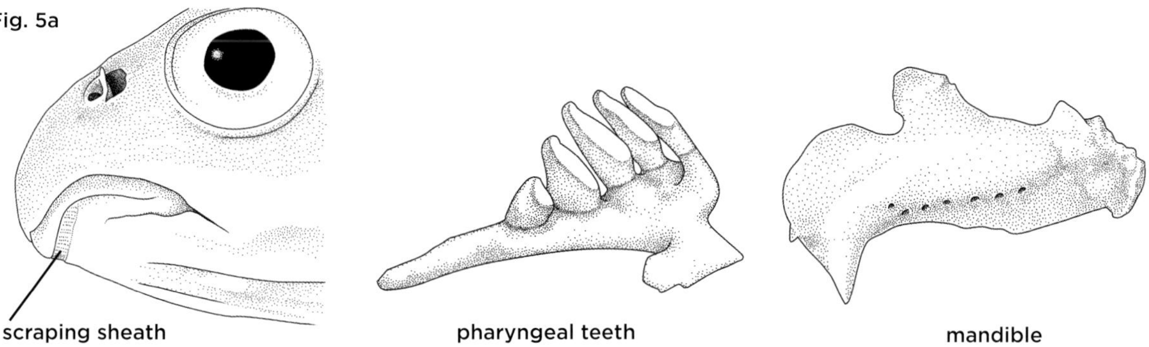

5b. Lower jaw subterminal or terminal but always curved and convex forward (Fig. 5b); other characters variable.. 6

Fig. 5b

6a. Barbels typically present at corners of mouth (sometimes absent from 1 side); snout overhangs or projects slightly ahead of upper lip (Fig. 6a) .... Dace and Lake Chub, 7

6b. Barbels absent; snout even with upper lip (Fig. 6b) 11

Fig. 6a

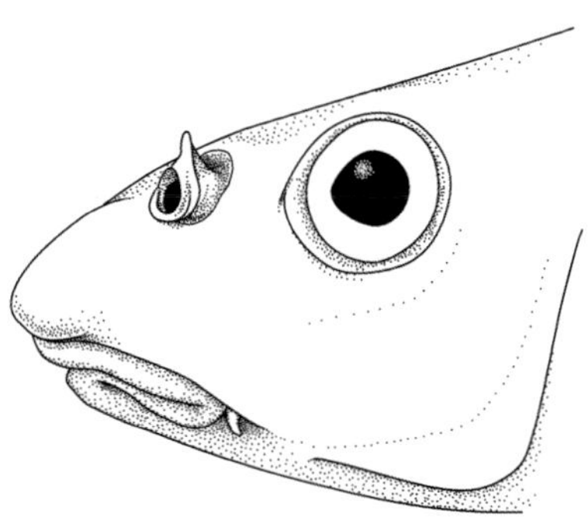

Fig. 6b

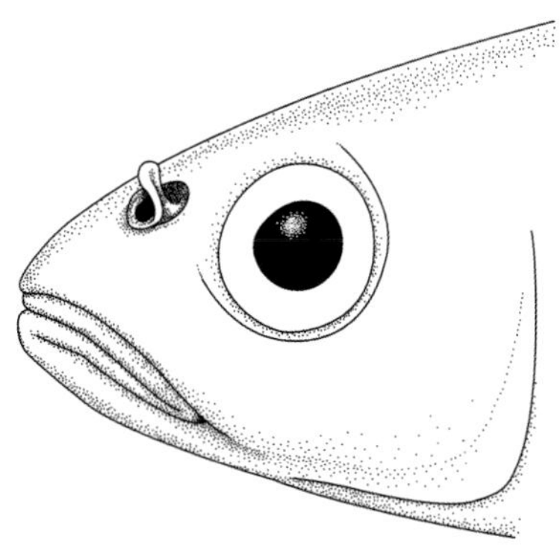

7a. Color pattern usually blotched or speckled; width of isthmus (distance between lower ends of gill clefts) less than half the distance between the eyes (interorbital width); scale radii present in all fields; pharyngeal teeth tapered with a rounded curved tip (Fig. 7a) .... DACE, *Rhinichthys*, 8

Fig. 7a

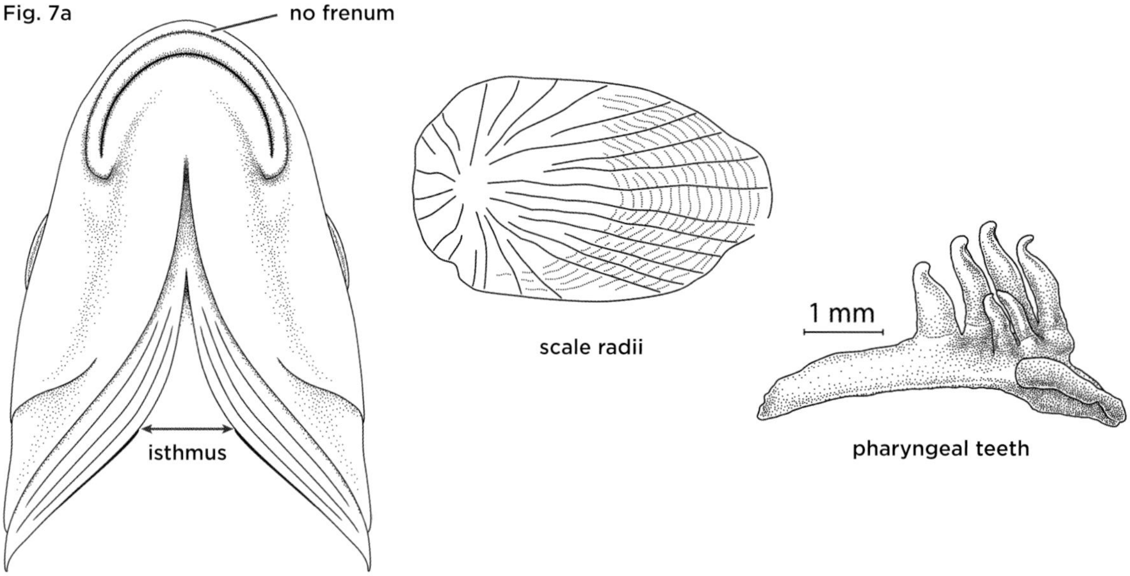

7b. Color pattern gray above, white below, with gray lateral band (sometimes with few darkened scales below) isthmus wider than half interorbital width; scales with radii absent in posterior field; pharyngeal teeth tapered with curved tips (Fig. 7b) .... LAKE CHUB, *Couesius plumbeus*

Fig. 7b

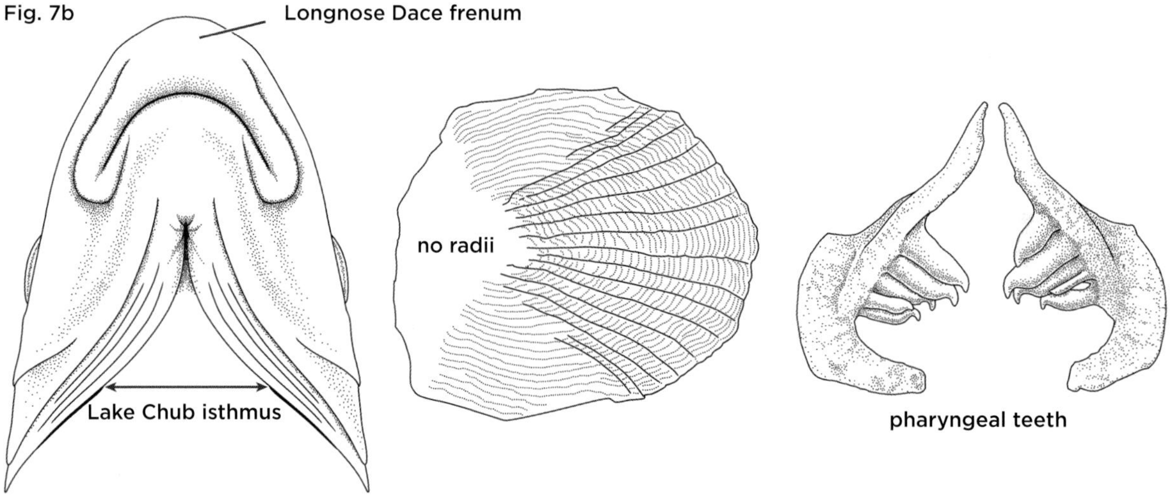

8a. Bridge of skin (frenum) joins snout to upper lip (Fig. 7b) ....LONGNOSE DACE, *Rhinichthys cataractae*

8b. A groove separates snout and upper lip, no frenum (Fig. 7a) 9

9a. Snout long; dorsal and anal fins falcate; caudal peduncle slender; color pattern of large dark blotches .. .. LEOPARD DACE, *Rhinichthys falcatus*

9b. Snout blunt; dorsal and anal fin margins straight or slightly falcate; caudal peduncle thicker; color pattern of small speckles or darkened individual scales 10

10a. Color pattern dark, speckled with small dark pigment cells.......... Speckled Dace, *Rhinichthys osculus*

10b. Color pattern pale, slightly darker above, with random dark scales .... Umatilla Dace, *Rhinichthys umatilla*

11a. Mouth small, about as long as eye diameter in side and face views and angled upward; scales large, with fewer than 52 diagonal scale rows along sides; dorsal fin small, usually with 8 rays, standard length usually less than 100–200 mm (4–8 in). .... Fathead Minnow, Spottail Shiner, and Golden Shiner, 12

11b. Mouth larger than eye in face or ventral view; scales small, usually more than 45 diagonal rows along sides; dorsal fin larger, with 9 or more rays; adult length variable.. 14

12a. Anal rays usually 10–13; lateral line decurved; golden color. Golden Shiner, *Notemigonus crysoleucas*

12b. Anal rays 7 or 8; lateral line slightly curved; color variable... 13

13a. Scales in front of dorsal fin small and crowded; anal rays 7; dorsal fin with first ray separated with a membrane of skin and approximately one half the length of the second ray; color rather dark, with faint, slightly oblique line at base of tail..Fathead Minnow, *Pimephales promelas*

13b. Scales in front of dorsal fin not small or crowded; anal rays 8; first dorsal rays not separated by membrane; silvery with dark spot at base of tail................................ Spottail Shiner, *Notropis hudsonius*

14a. Face elongate; mouth large, reaching under front of eye; caudal peduncle deep relative to body depth, its depth going into body depth less than 2.2 times; pharyngeal teeth tapered with curved tip (Fig. 14a) .. .. Northern Pikeminnow, *Ptychocheilus oregonensis*

Fig. 14a

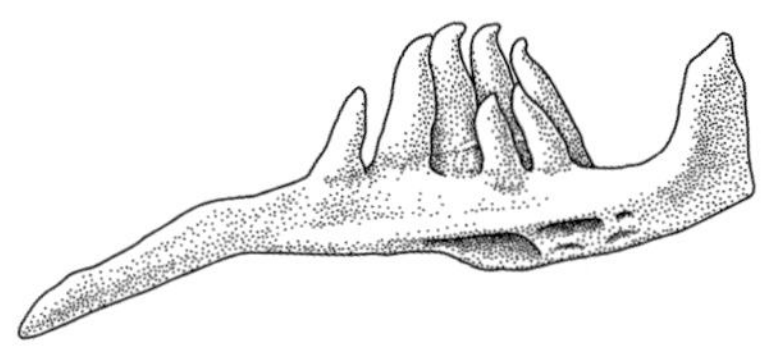

14b. Face conical; mouth smaller, not reaching under front of eye; caudal peduncle not as deep relative to body depth, its depth going into body depth more than 2.3 times 15

15a. Caudal peduncle slender, its depth goes into standard length more than 5 times................................ ....Peamouth, Redside Shiner, Leatherside Chub, 16

15b. Caudal peduncle deeper, its depth goes into standard length fewer than 5 times................................ .... Utah Chub, Tui Chub, 18

16a. Mouth small, reaching to below nostrils; pharyngeal teeth molariform (Fig. 16a) .... Peamouth, *Mylocheilus caurinus*

16b. Mouth larger, reaching nearly to below front of eye; pharyngeal teeth not molariform but tapered, with a rounded, curved tip (Fig. 16b)...................... Redside Shiner, Northern Leatherside Chub, 17

Fig. 16a

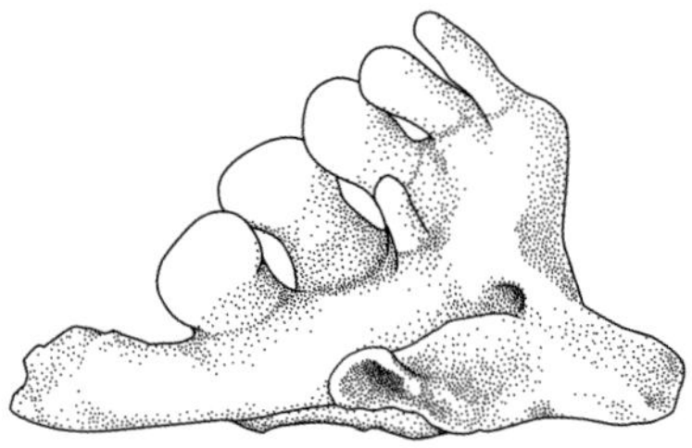

Fig. 16b

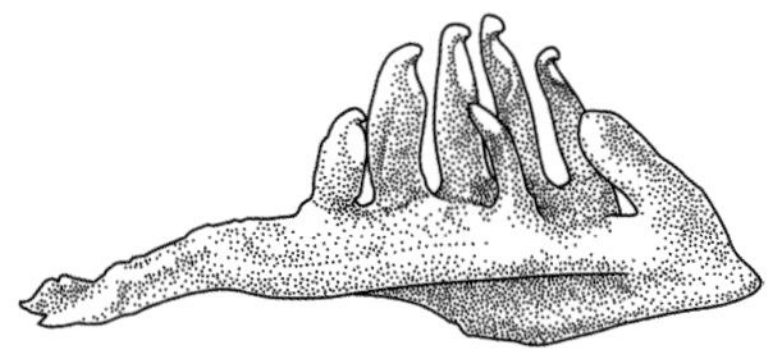

17a. Anal rays 10–13; lateral line scales 52–70; pharyngeal teeth tapered with a rounded, curved tip (Fig. 16b)..Redside Shiner, *Richardsonius balteatus*

17b. Anal rays 8; lateral line scales 75–90..................... Northern Leatherside Chub, *Lepidomeda copei*

18a. Pharyngeal arch with 4 or 5 teeth* in the major row; no minor teeth; pharyngeal tooth formula usually 0,5–4,0 (Fig. 18a), caudal peduncle deep ....Tui Chub, *Siphateles bicolor*

18b. Pharyngeal arch with 4 or 5 teeth* in the major row and one 1 or two 2 teeth in the minor row, pharyngeal tooth formula usually 1,5–4,1 (Fig. 18b)...Utah Chub, *Gila atraria*

Fig. 18a

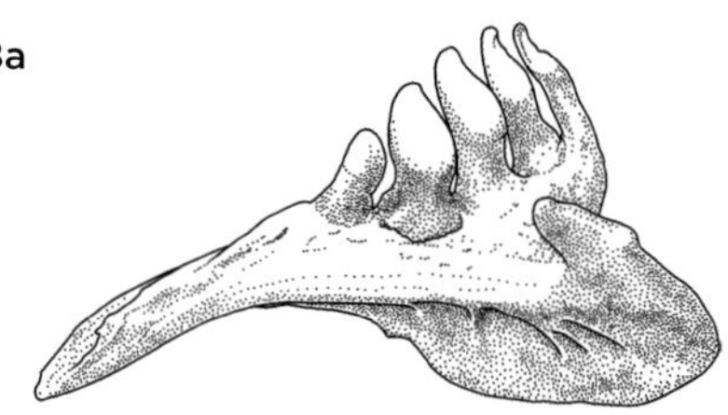

Fig. 18b

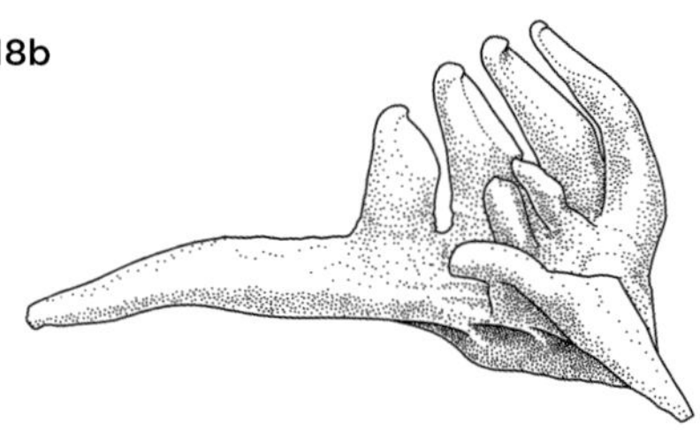

*Note: These two 2 species, despite being outwardly similar, have different DNA, cranial bones, and pharyngeal teeth. Dissection of the pharyngeal bones requires cutting through the muscle between the shoulder girdle (cleithrum) and fifth pharyngeal arch (Dissection Fig.), then slicing deeply upward and then deeply antero-ventrally to sever other muscles (Dissection Fig.), and then pulling the arch out of the head with small forceps. Unfortunately, this is the easiest way. Master this and you can become a brain surgeon or an ichthyologist.

Dissection Fig.

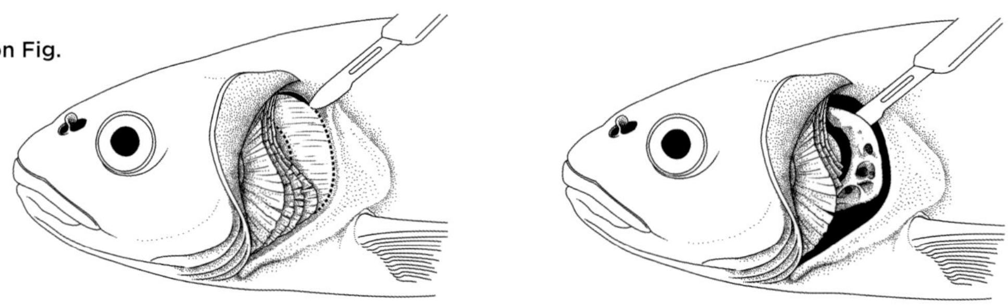

Key to Idaho Suckers
Family Catostomidae (Chapter 9)

1a. A gap between bases of lower lips (Fig. 1a); mouth opens subterminally to ventrally (see color illustration); gill rakers finely branched (Fig. 1a). Lake Suckers, *Chasmistes*....... SNAKE RIVER SUCKER, *Chasmistes muriei*

1b. No gap (Figs. 5a, 6a, 6b), left and right lower lips touching at their bases; mouth opens ventrally; gill rakers knobby, not branched (Fig. 1b) .... RIVER SUCKERS, *Catostomus* and MOUNTAIN SUCKERS, *Pantosteus*, 2

2a. Lower lip divided by a cleft to its base; usually no row of papillae crossing between the notch and the cartilaginous edge of the jaw (Fig. 2a); upper and lower lips joined and rounded at the corners (Fig. 2a); weak cartilaginous ridges at edges of jaws (Fig. 2a) .... RIVER SUCKERS, *Catostomus* spp. 3

2b. Lower lip notch shallow, with a bridge of 2 or more rows of papillae crossing between the cleft and the cartilaginous edge of the lower jaw (Fig. 5a, 6a, 6b); upper and lower lips strongly or weakly notched at the corners (Figs. 5a, 6a, 6b) (notches on 1 or both corners may be absent in *P. columbianus*); jaws with strong cartilaginous edges (Figs. 5a, 6a, 6b)... MOUNTAIN SUCKERS, *Pantosteus* spp. 5

3a. Scales small; diagonal scale rows along sides more than 90 (see color illustration) .... LONGNOSE SUCKER, *Catostomus catostomus*

3b. Scales large; fewer than 75 diagonal rows along sides or in lateral line (see color illustration)............... *Catostomus* spp. 4

4a. Scales around caudal peduncle small (see color illustration), usually 18–20 around caudal peduncle (counted diagonally)... UTAH SUCKER, *Catostomus ardens*

4b. Scales around caudal peduncle large (see color illustration), usually 16 or 17 around caudal peduncle (counted diagonally)... LARGESCALE SUCKER, *Catostomus macrocheilus*

5a. Two to 4 rows of papillae bridging the lower lip between the notch and the cartilage of the lower jaw (Fig. 5a); jaw cartilages less truncate (not square across); upper lip may have papillae on its anterior face; isthmus narrow, its width going 3.5–4 times in head length; gill rakers with spines in clusters .... BRIDGELIP SUCKER, *Pantosteus columbianus*

5b. Four to 6 rows of papillae bridging the lower lip between the notch and the cartilage of the lower jaw, jaw cartilages more or less straight across; upper lip usually smooth, with no strong papillae on its anterior face (Figs. 6a, 6b); isthmus broad, its width going 2.5–3 times in head length; gill rakers with 2 rows of spines... MOUNTAIN SUCKERS and GREEN SUCKER, 6

6a. Lower lip with a convex-forward rosette of large papillae in the anterior center, with small spaces on each side, between the rosette and the fold between the upper and lower lips; about 5 rows of papillae between lower lip notch and edge of jaw (Fig. 6a); gill rakers in outer row of first arch 23–31 and inner row of first arch 33–39...MOUNTAIN SUCKERS, 7

6b. Lower lip papillae small, the anterior row follows the shape of the edge of the jaw and extends to the folds between the upper and lower lips; about 6 rows of papillae between lower lip notch and edge of jaw (Fig. 6b); gill rakers in outer row of first arch 33–43 and inner row of first arch 44–54 .... Green Sucker, *Pantosteus virescens*

7a. Gill rakers on the external row of the first arch 29–37; gill rakers on the internal row of the first arch 35–51; lateral line scales 75–85 (means 77–82); predorsal scales 41–58 (means 43–56) .... Cascadian Sucker, *Pantosteus bondi*

7b. Gill rakers on the external row of the first arch 23–31; gill rakers on the internal row of the first arch 33–39; lateral line scales 80–90 (means 83–86); predorsal scales 40–45 (means 42–44) ....Mountain Sucker, *Pantosteus platyrhynchus*

Fig. 1a

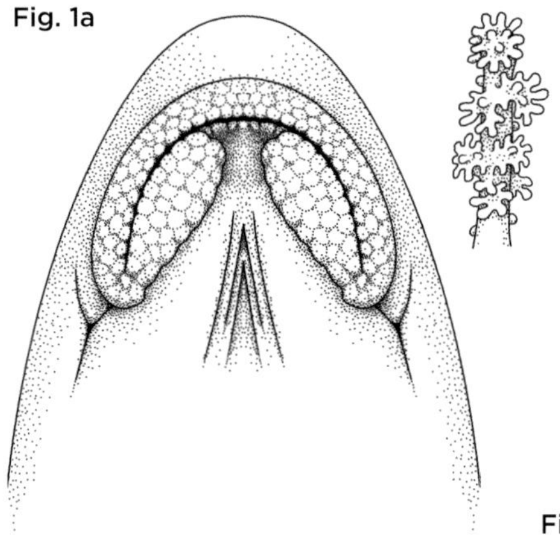

Fig. 1b

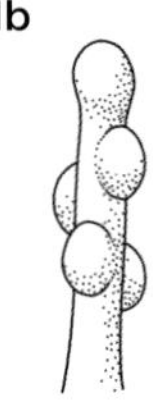

Fig. 2a

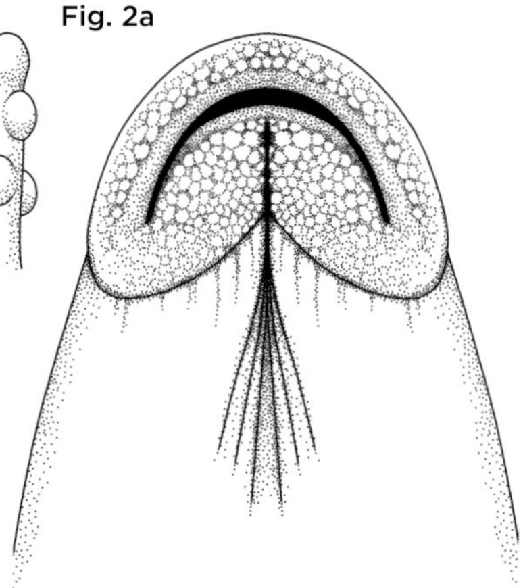

Fig. 5a

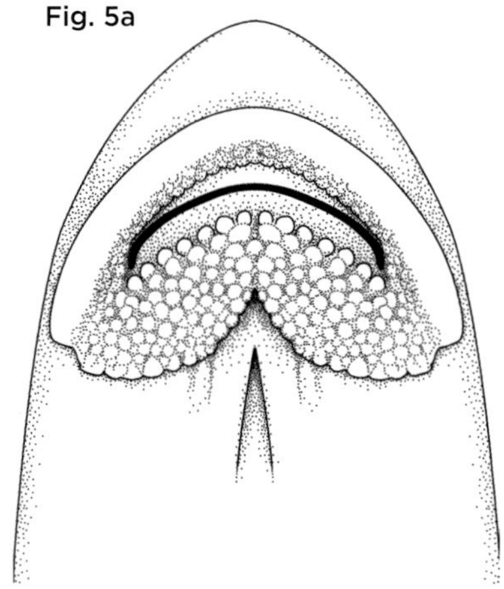

Fig. 6a

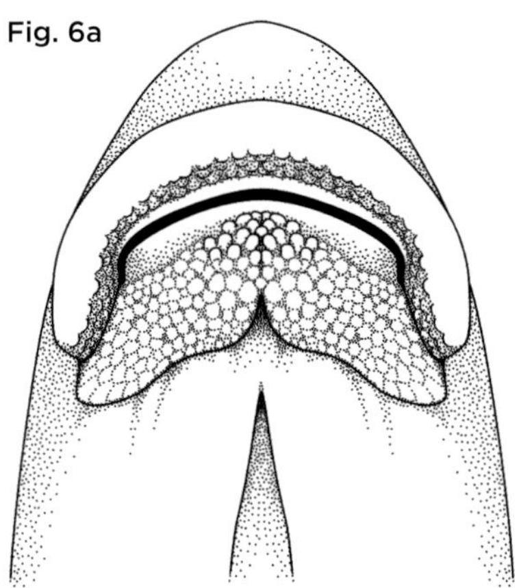

Fig. 6b

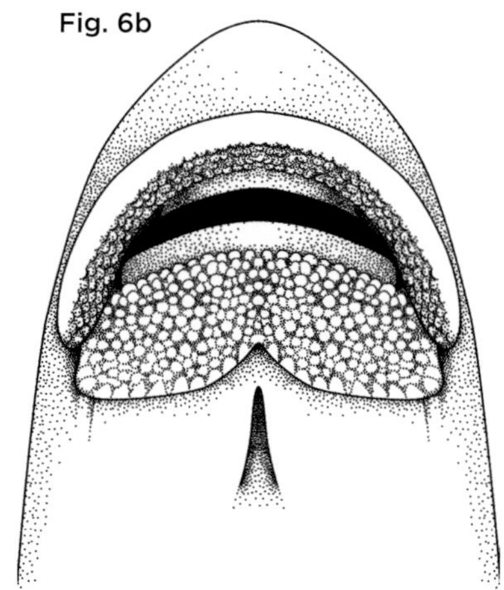

Key to Idaho Catfishes
Family Ictaluridae (Chapter 10)

1a. Caudal fin deeply forked (compare color illustrations); upper jaw projects slightly ahead of lower jaw.. ....Channel Catfish and Blue Catfish, 2

1b. Caudal fin rounded or truncate (compare color illustrations), not forked; upper and lower jaws about even or lower jaw projects forward 3

2a. Anal fin rounded, with 24-30 (counting all rays, including small ones); color pattern often with black spots on silver background, eyes high on sides of head (see color illustration)... ....Channel Catfish, *Ictalurus punctatus*

2b. Anal fin longer and straight edged; with 30-36 rays (counting all rays including small ones); color pattern never spotted; eyes nearly midway down on the sides of head (see color illustration) .... Blue Catfish, *Ictalurus furcatus*

3a. Adipose fin flattened to a low ridge which is mostly continuous with the caudal fin (see color illustration); no posterior serrations on pectoral spines (Fig. 3a); caudal fin often longer than head ....Tadpole Madtom, *Noturus gyrinus*

Fig. 3a

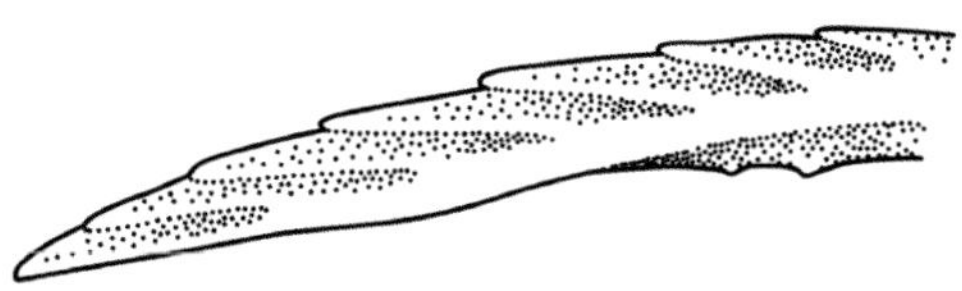

3b. Adipose fin is a flap with a free edge not connected to caudal fin (see color illustration); pectoral spines at least weakly serrated; caudal fin shorter or about equal to head length... 4

4a. Lower jaw projects ahead of upper jaw; premaxillary tooth band at front of roof of mouth with backward lateral projections (Fig. 4a); head elongate and flattened ahead of the eyes... ....Flathead Catfish, *Pylodictis olivaris*

4b. Lower jaw about even with upper jaw; premaxillary tooth band at front of roof of mouth lacking backward lateral projections (Fig. 4b); head not elongate or flattened ahead of the eyes............................. .... Bullheads, *Ameiurus*, 5

Fig. 4a

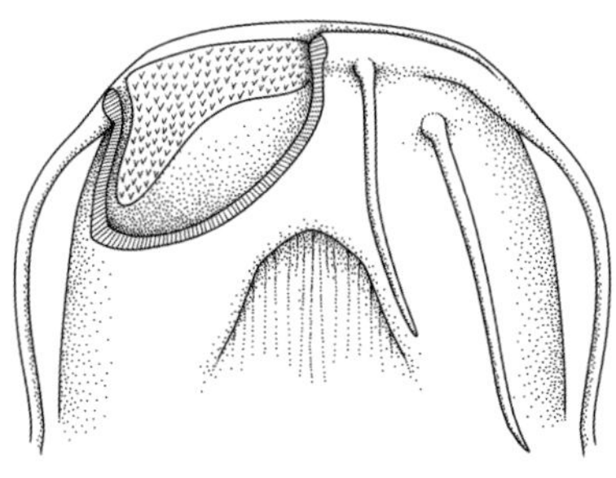

Fig. 4b

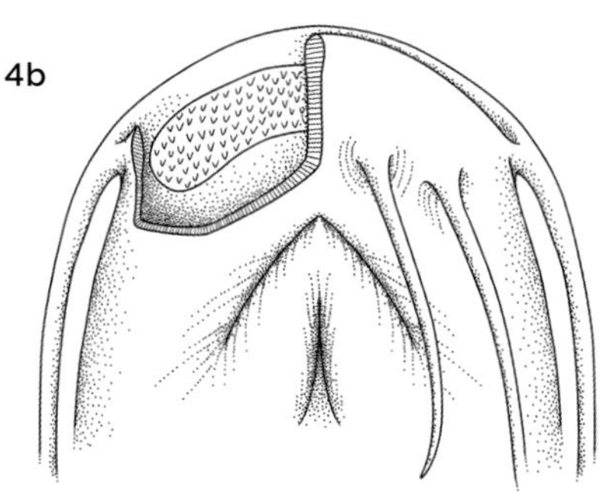

5a. Chin barbels pale yellow; anal fin with at least 24 rays (counting all rays); pectoral spine with anterior barbs and posterior serrations (Fig. 5a) .... Yellow Bullhead, *Ameiurus natalis*

Fig. 5a

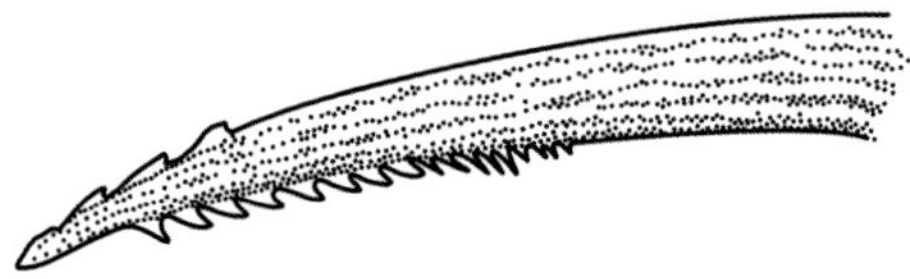

5b. Chin barbels with dark pigment; anal fin with fewer than 24 rays; pectoral spine with posterior serrations or roughness 6

6a. Pectoral spines with strong posterior serrations; gill rakers 11–14; supraethmoid bone (anterior part of roof of cranium) with a narrow anterior notch (Fig. 6a) .... Brown Bullhead, *Ameiurus nebulosus*

6b. Pectoral spine feels rough but does not show strong serrations; gill rakers 15–19; supraethmoid bone with wide convexity, not a narrow notch (Fig. 6b) .... Black Bullhead, *Ameiurus melas*

Fig. 6a

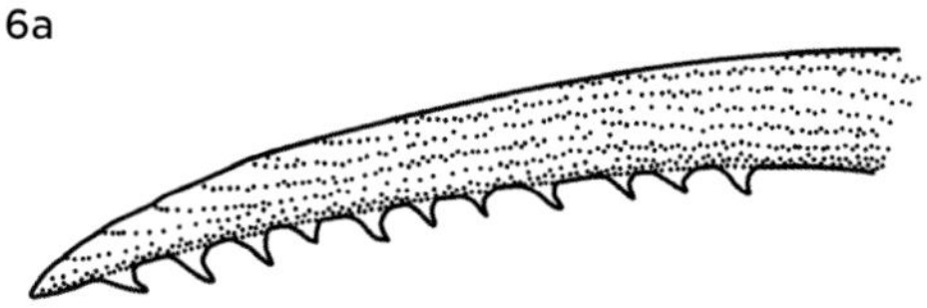

pectoral spine

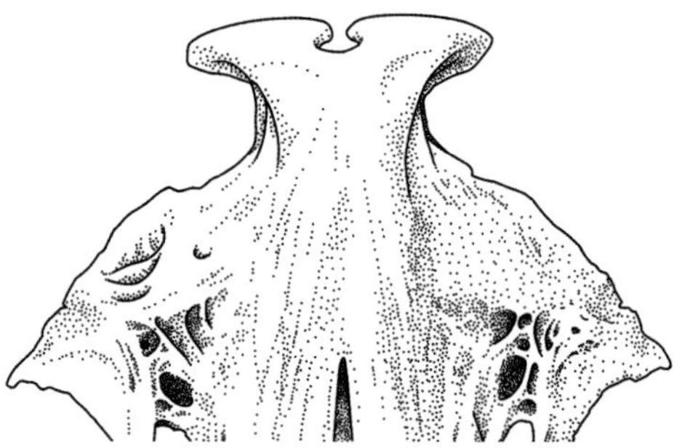

supraethmoid bone

Fig. 6b

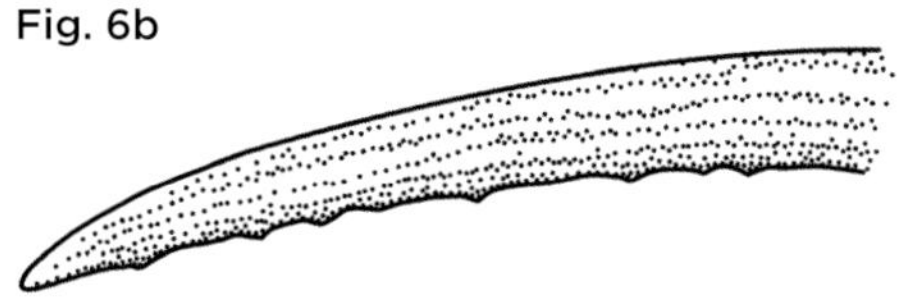

pectoral spine

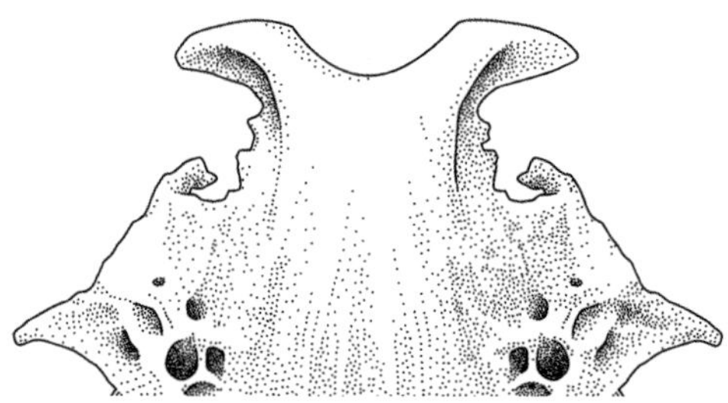

supraethmoid bone

Key to Idaho Whitefishes
Subfamily Coregoninae (Chapter 12)

1a. Two flaps between incurrent and excurrent nostrils (Fig. 1a); transparent eyelid without a small notch below back of pupil (see color illustration) .... LAKE WHITEFISH, *Coregonus clupeaformis*

1b. One flap between incurrent and excurrent nostrils (Fig. 1b); transparent eyelid with a small notch below back of pupil ....*Prosopium*, 2

Fig. 1a

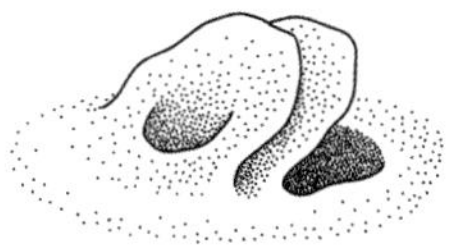

Fig. 1b

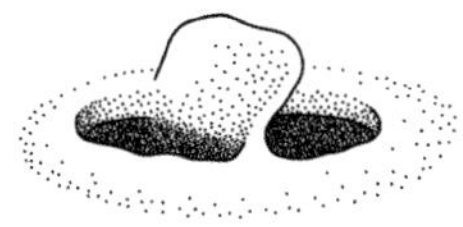

2a. Gill rakers 37–45 (Fig. 2a); mouth extends sharply forward; standard length of adults less than 200 mm (8 in) (see color illustration) ....BONNEVILLE CISCO, *Prosopium gemmifer*

2b. Gill rakers fewer than 30 (Fig. 2b); front of head rounded; length variable.. 3

3a. Gill rakers 13–20 (Fig. 3a); lateral line scales 54–70; adult size usually less than 150 mm (6 in); pyloric caeca 13–33 (see Salmoninae figure 2a for illustration of pyloric caeca)....PYGMY WHITEFISH, *Prosopium coulterii*

Fig. 2a

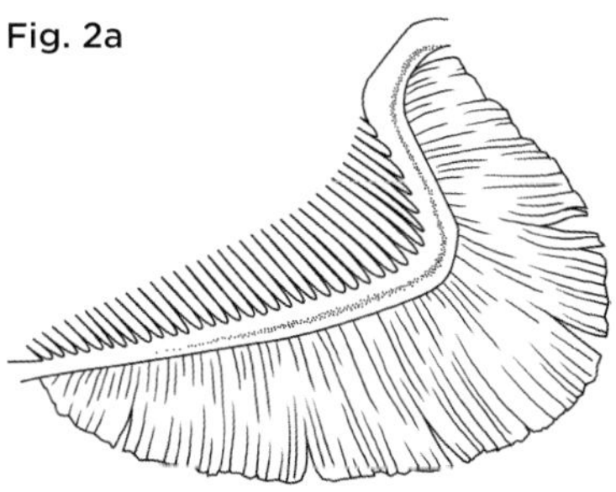

Fig. 2b

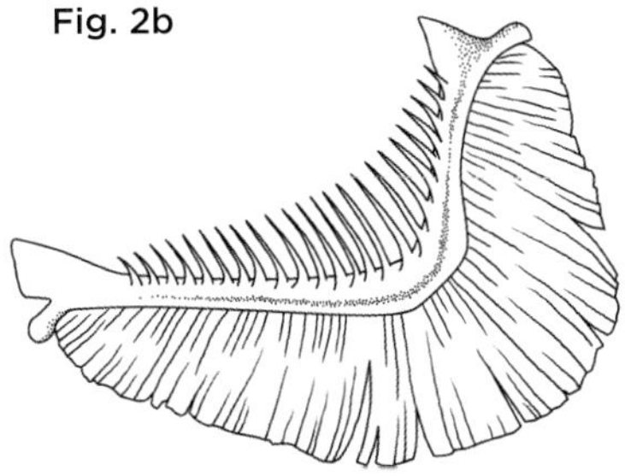

Fig. 3a

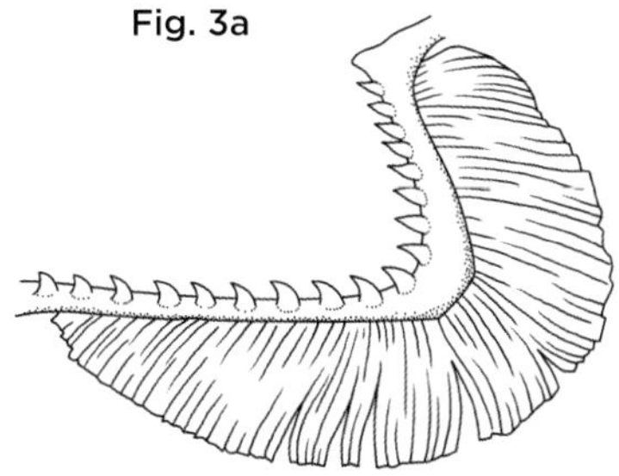

3b. Gill rakers 19–26; lateral line scales 67–89; adult size over 200 mm (8 in); pyloric caeca over 33 (see Salmoninae figure 2a for illustration of pyloric caeca).. 4

4a. Dorsal fin rays 12–13; anal fin rays 11–13 (see color illustration) .... MOUNTAIN WHITEFISH, *Prosopium williamsoni*

4b. Dorsal fin rays 10–12; anal fin rays 9–11 (see color illustrations)... .... BEAR LAKE WHITEFISH and BONNEVILLE WHITEFISH, 5

5a. Lateral line scales 67–78; maxilla about equal to eye diameter (see color illustration); adult size about 200 mm (8 in); no teeth on tongue .... BEAR LAKE WHITEFISH, *Prosopium abyssicola*

5b. Lateral line scales 76–86; maxilla longer than eye diameter (see color illustration); adult size over 300 mm (12 in) long; a few teeth on tongue (requires preparation to see) .... BONNEVILLE WHITEFISH, *Prosopium spilonotus*

Key to Idaho Grayling
Subfamily Thymallinae (Chapter 12)

1a. Scales large, fewer than 100 (usually 80-90) in lateral line; caudal fin distinctly forked; teeth small; 17 or more dorsal fin rays (see color illustration) .... Arctic Grayling, *Thymallus arcticus*

1b. Scales small, 115-200 in lateral line; teeth relatively large; caudal fin truncate or forked; fewer than 17 dorsal rays .... Salmoninae

Key to Idaho Chars, Salmons, and Trouts
Subfamily Salmoninae (Chapter 12)

1a. Color pattern of pale markings (excluding parr marks in juveniles) on a dark background on back and sides (compare color illustrations); vomer with cluster of 4–6 teeth at the front of the roof of the mouth only about 3 teeth in midline of roof of mouth (Fig. 1a) .... CHARS, *Salvelinus*, 2

Fig. 1a

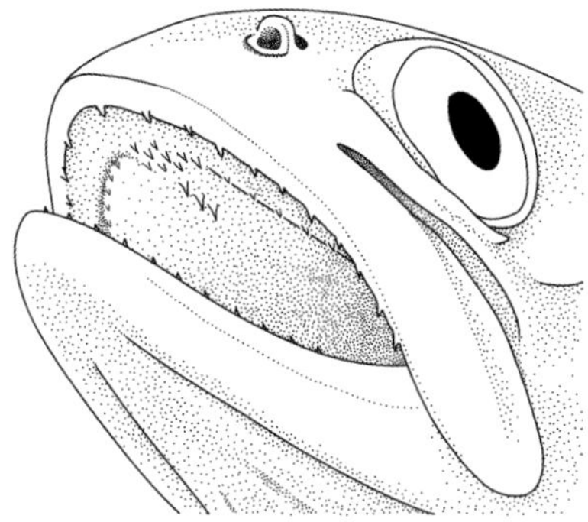

1b. Color pattern of dark spots (excluding parr marks in juveniles) on a pale background on the back and sides (compare color illustrations); vomer with cluster of 4-6 teeth at the front of the roof of the mouth and a longer row of teeth in midline of roof of the mouth (Fig. 1b) .... SALMON and TROUTS, 5

Fig. 1b

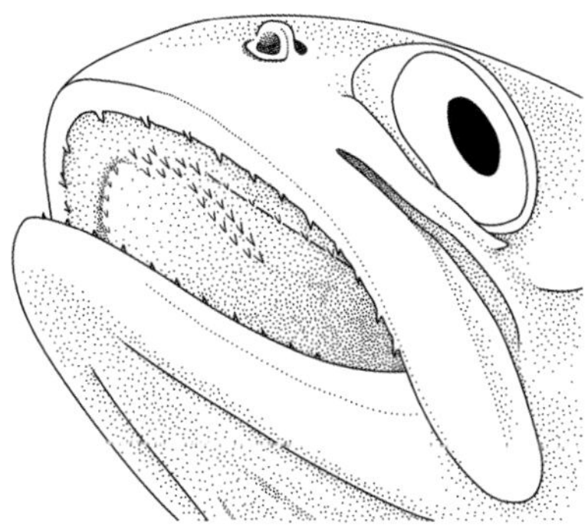

2a. Caudal fin deeply forked (compare color illustrations); pyloric caeca (Fig. 2a) more than 90; no orange or red pigment; body with small, irregular pale spots closely spaced on gray to greenish background.......... .... LAKE TROUT, *Salvelinus namaycush*

Fig. 2a

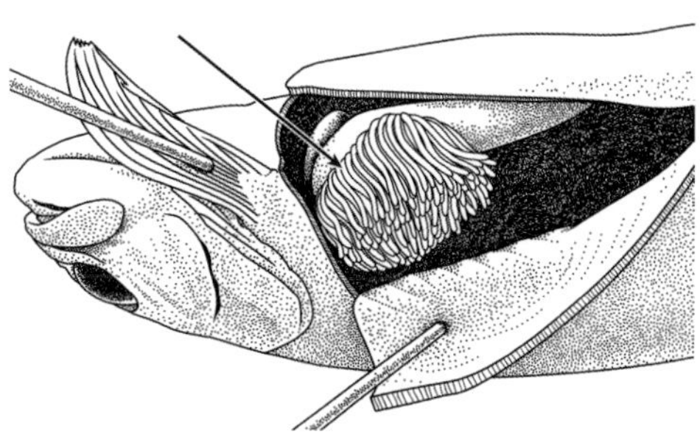

2b. Caudal fin moderately or not forked (compare color illustrations); pyloric caeca (Fig. 2a) fewer than 75; body often with red or pink spots and (or) orange belly in breeding season 3

3a. Dorsal and caudal fins with distinct black spots or wavy markings; caudal fin not forked; sides with pink or red spots, some surrounded by blue; lower fins with white edges ahead of a distinct black line; fins often reddish (compare color illustrations) .... Brook Trout, *Salvelinus fontinalis*

3b. No wavy black markings in dorsal fin; caudal fin slightly or moderately forked; sides with pale off-white or pink round spots; lower fins with thin white edges often followed by thin black line (compare color illustrations) 4

4a. Head and jaws large, head goes into standard length about four times; snout about 1/3 of head length; eyes back of long, tapered snout; eyes high on sides of somewhat flattened head (compare color illustrations) ....Bull Trout, *Salvelinus confluentus*

4b. Head and jaws shorter, head length goes into standard length about 4.5 times; snout goes into head length about 4 times; eyes not high on sides of head; top of head rounded (compare color illustrations) ... . Sunapee Trout, *Salvelinus alpinus oquassa*

5a. Caudal fin without black spots except sometimes a few at upper and lower edges; postorbital bones reach only halfway to preopercle bone (probe with sharp point to find edges of bones; Fig. 5a); often with red spots on sides; pelvic fin rays usually 8 or 9; anal fin rays always 12 or fewer (compare color illustrations) ....Atlantic Trouts and Salmons, *Salmo*, 6

Fig. 5a

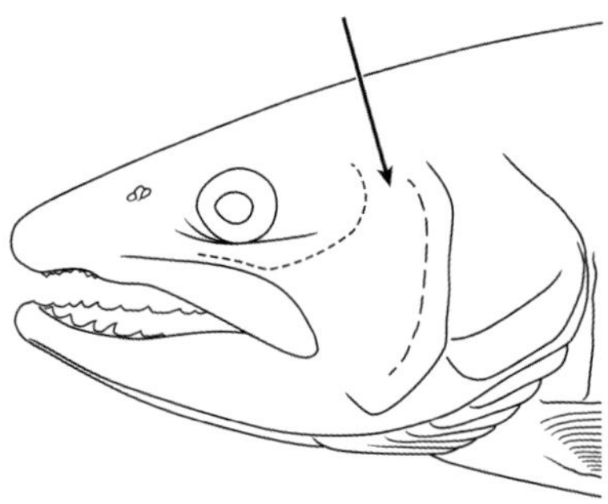

5b. Caudal fin with black spots (except in Sockeye and only on upper lobe in Coho Salmon); postorbital bones reach almost to preopercle (Fig. 5b); no red spots on sides; pelvic fin rays usually 10; anal fin rays 13 or more in salmon, fewer in trouts (compare color illustrations).. ....Pacific Trouts and Salmons, *Oncorhynchus*, 7

Fig. 5b

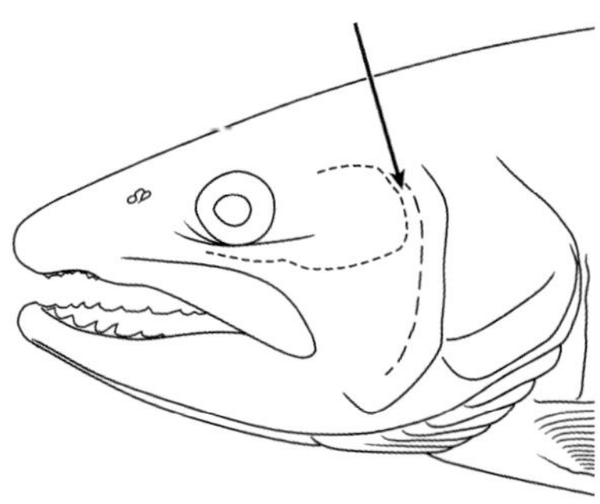

6a. Scale rows around the caudal peduncle 53–61; lateral line scales 120–130; caudal fin not as forked as salmon of same size; mouth large; maxilla extends to back margin of orbit or beyond (compare color illustrations) .... Brown Trout, *Salmo trutta*

6b. Scale rows around the caudal peduncle 43–49; lateral line scales 109–121; caudal fin forked in juveniles; mouth smaller; maxilla extends to below eye except in some large adults (compare color illustrations) .... Atlantic Salmon, *Salmo salar*

7a. Principal anal fin rays 12 or fewer; length of anal fin base slightly greater than depth of caudal peduncle (compare color illustrations) .... Trouts, 8

7b. Principal anal fin rays 13 or more; length of anal fin base much greater than depth of caudal peduncle (compare color illustrations) .... Salmon, 15

8a. No basibranchial teeth in throat behind tongue (Fig. 8a); adults usually with red lateral stripe (except in some silver Steelhead); snout often blunt and maxilla shorter except in most Steelhead; black spots on sides usually smaller (except in Golden Trout) (compare color illustrations) .... Rainbow Trout series, 9

8b. A row of small basibranchial teeth present in throat behind tongue (Fig. 8b); adults usually with no red lateral stripe; snout usually long and tapered; maxilla longer; black spots on sides usually large and concentrated posteriorly in some subspecies (compare color illustrations) .... Cutthroat Trout Series, 11

Fig. 8a

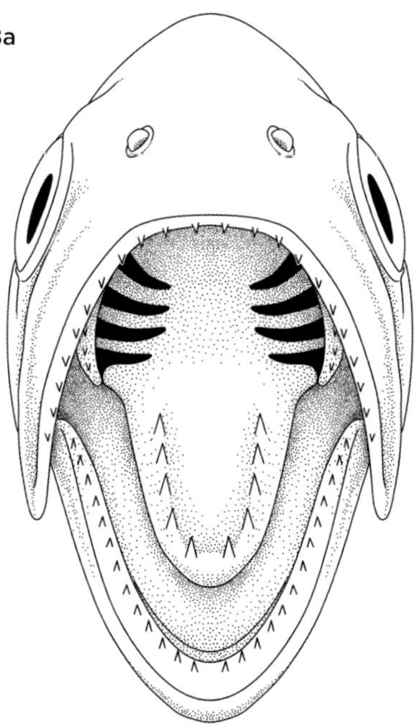

Fig. 8b

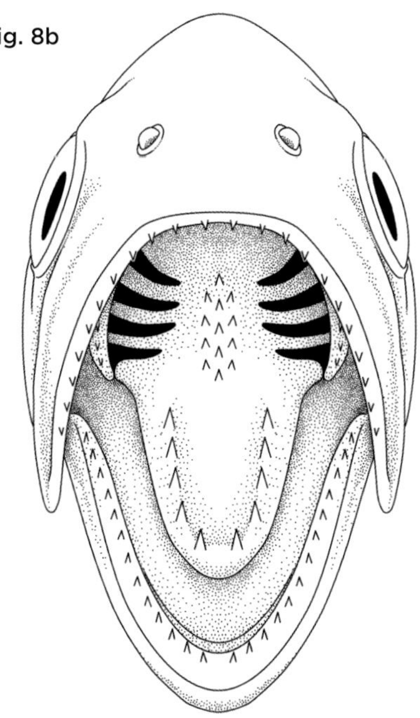

9a. Red throat stripe present; red lateral band usually not intense (compare color illustrations) .... Golden and Redband Trouts, 10

9b. Red throat stripe not present; red lateral stripe well developed (compare color illustrations) .... Rainbow Trout, *Oncorhynchus mykiss irideus*

10a. Sides golden with strong red band; lower fins reddish (compare color illustrations) .... Golden Trout, *Oncorhynchus aguabonita*

10b. Sides pale bluish or silvery, usually with dispersed red lateral band; lower fins sometimes orange or yellow (compare color illustrations) .... Redband Trout, *Oncorhynchus mykiss gairdneri*

11a. Dense, fine spots on back, sides, dorsal fin, and caudal fin (compare color illustrations) .... Snake River Fine-Spotted Cutthroat Trout, *Oncorhynchus clarkii behnkei*

11b. Spots not as fine, medium to large, not densely spaced, sometimes concentrated posteriorly (compare color illustrations) 12

12a. Spots medium to large and only slightly denser posteriorly, dispersed over back and sides (compare color illustrations) .... Bonneville and Lahontan Cutthroat Trouts, 13

12b. Spots largely concentrated posteriorly (compare color illustrations) 14

13a. Spots over most of body, but not head; 2 forms: those in Snake Valley have 20–22 gill rakers and 20–28 basibranchial teeth, those in most of the Bonneville drainage are variable but with 18–19 gill rakers and 5–10 basibranchial teeth; Bear River fish with small spots, sometimes with bluish color (compare color illustrations) .... Bonneville Cutthroat Trout, *Oncorhynchus clarkii utah*

13b. Spots sparsely dispersed over back and sides, including head; often least spotted at midbody (compare color illustrations) .... Lahontan Cutthroat Trout, *Oncorhynchus clarkii henshawi*

14a. Large, nearly round black spots dispersed on back and posterior third of body, not crowded except on caudal peduncle (compare color illustrations) ....Yellowstone Cutthroat Trout, *Oncorhynchus clarkii bouvieri*

14b. Irregular black spots on back and posterior third of body, especially crowded on caudal peduncle (compare color illustrations) .... Westslope Cutthroat Trout, *Oncorhynchus clarkii lewisi*

15a. No spots; gill rakers more than 30; includes landlocked and anadromous forms 16

15b. Spots on back and part of caudal fin (compare color illustrations); gill rakers fewer than 30.......... 17

16a. Large, anadromous, migrate to lakes; 550–650 mm (21.6–25.6 in) long and weigh 1.8–3.2 kg (4.0–7.0 lb) when they reach our waters .... Sockeye Salmon, *Oncorhynchus nerka*

16b. Small, landlocked in lakes; adults 300–400 mm (11.8–15.8 in) long and weigh 1.0–1.5 kg (2.2–3.3 lb) .... Kokanee, *Oncorhynchus nerka kennerlyi*

17a. No distinct black spots, only small speckles on back and caudal fin (compare color illustrations)........ .... Chum Salmon, *Oncorhynchus keta*

17b. Distinct black spots on back and all or part of caudal fin (compare color illustrations) .... Coho and Chinook Salmon, 18

18a. Small spots restricted to upper lobe of caudal fin; gums at base of lower teeth pale (compare color illustrations); pyloric caeca fewer than 80 .... Coho Salmon, *Oncorhynchus kisutch*

18b. Small spots on both lobes of caudal fin; gums at base of lower teeth black (compare color illustrations); pyloric caeca more than 140 .... Chinook Salmon, *Oncorhynchus tshawytscha*

Key to Idaho Sculpins
Family Cottidae (Chapter 17)

1a. No palatine teeth in roof of mouth; prickles absent or restricted to a few under pectoral fin; incomplete lateral line, no lateral line pores on caudal peduncle (Fig. 1a) .... *C. beldingii, leiopomus, cognatus,* 2

Fig. 1a

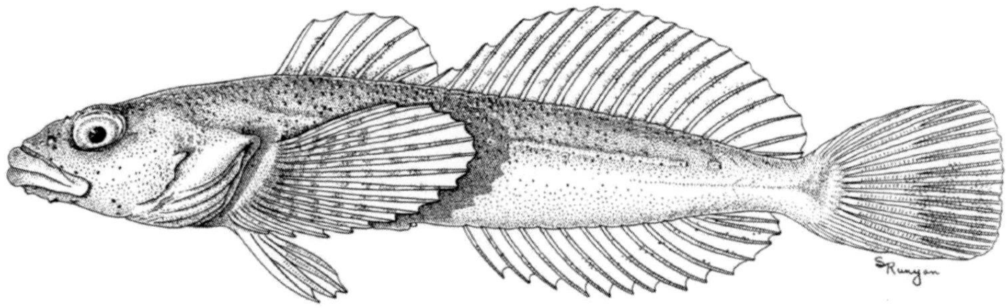

1b. One or more rows of teeth in roof of mouth; prickles present under pectoral fin and sometimes elsewhere; lateral line variable (Fig. 1b). (*C. bairdii* complex, *confusus, extensus, greenei, hubbsi, rhotheus, schitsuumsh*) 4

Fig. 1b

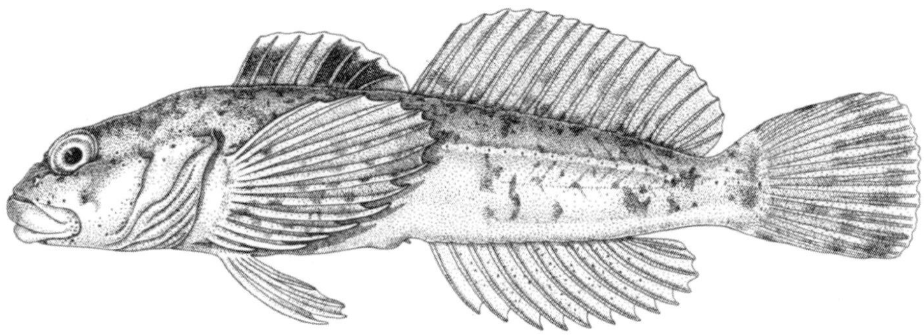

2a. No prickles on body; eye diameter goes into depth of caudal peduncle 1.5 to 2 times (see color illustration); caudal peduncle depth goes into head length 2.9–3 times (see color illustrations) .... *C. beldingii and leiopomus,* 3

2b. Few prickles under pectoral fin; eye diameter goes into depth of caudal peduncle 1.3 times; caudal peduncle depth goes into head length 4.2 times (see color illustrations); 2 or 3 preopercular spines (Fig. 2b) ....Slimy Sculpin, *Cottus cognatus*

3a. One or 2 spines on preopercle (Fig. 2b) ....Paiute Sculpin, *Cottus beldingii*

3b. No spines but possibly a small nub on preopercle (Fig. 2b) .... Wood River Sculpin, *Cottus leiopomus*

4a. Caudal peduncle slender to medium, less than 1.4 eye diameters deep, and depth goes into head length more than 3.7 times (see color illustrations) .... (*C. hubbsi, rhotheus, extensus, bairdii* complex), 5

4b. Caudal peduncle deeper, 1.5 or more eye diameters deep, and depth goes into head length fewer than 3.7 times (see color illustrations) .... (*C. schitsuumsh, greenei, confusus*), 8

Fig. 2b

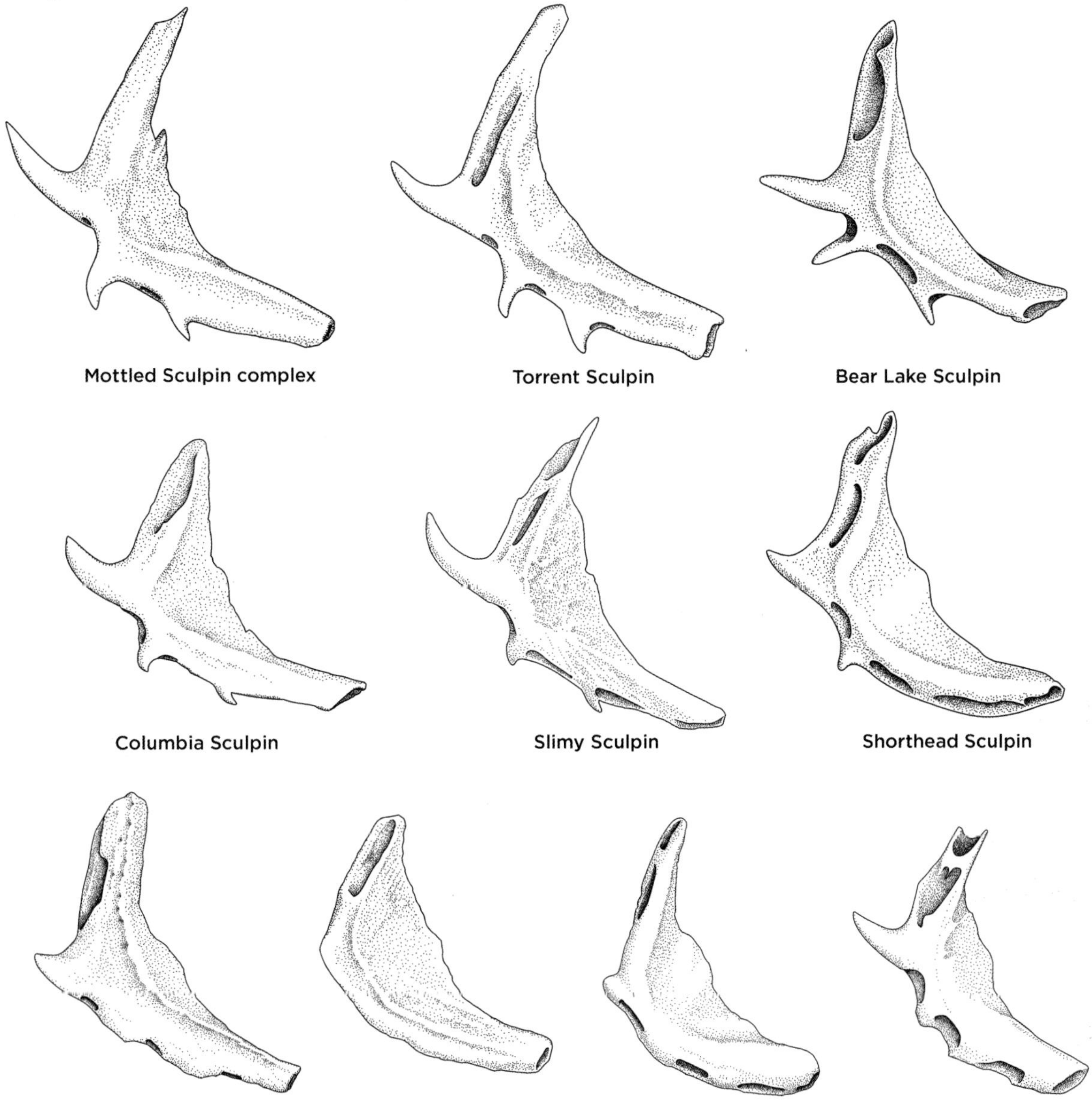

Mottled Sculpin complex / Torrent Sculpin / Bear Lake Sculpin / Columbia Sculpin / Slimy Sculpin / Shorthead Sculpin / Shoshone Sculpin / Wood River Sculpin / Cedar Sculpin / Paiute Sculpin

5a. Lateral line complete (or nearly so in juveniles) .... (*C. hubbsi, rhotheus*), 6

5b. Lateral line does not extend beyond posterior end of second dorsal fin .... (*C. bairdii* complex, *extensus*), 7

6a. Prickles dispersed over much of body; caudal peduncle slender, about 1.1 eye diameters deep, its depth goes into head length 4.8–6.4 times; dark crossbars slant forward and downward almost to anal fin (see color illustration); 3 or 4 preopercular spines (Fig. 2b) .... Torrent Sculpin, *Cottus rhotheus*

6b. Prickles dense in patch under pectoral fins, sparse elsewhere; caudal peduncle medium slender, about 1.3 eye diameters deep, and depth goes into head length 3.7–4.6 times; 3 dark crossbars beneath second dorsal fin extend to or below the lateral line (see color illustration); 3 preopercular spines (Fig. 2b)............ .... Columbia Sculpin, *Cottus hubbsi*

7a. Body with prickles restricted to patch under pectoral fin; body robust anteriorly (see color illustration); .... Mottled Sculpin complex, *Cottus bairdii* complex

7b. Body with prickles abundant on back and sides (Fig. 1a); body slender, not robust anteriorly (see color illustration) .... Bear Lake Sculpin, *Cottus extensus*

8a. Preopercle with 1 spine, perhaps only a rounded nub (Fig. 2b)...... Cedar Sculpin, *Cottus schitsuumsh*

8b. Preopercle with 1 or more sharp spines 9

9a. One preopercular spine (Fig. 2b); caudal peduncle deep, about 1.8 eye diameters deep, and depth goes into head length about 2.1 times (see color illustration); no lateral line pores on the caudal peduncle .... Shoshone Sculpin, *Cottus greenei*

9b. Two or 3 preopercular spines (Fig. 2b); caudal peduncle intermediate, about 1.5 eye diameters deep, and depth goes into head length 3.5 times (see color illustration); lateral line pores usually present on the caudal peduncle (Fig. 1b) ....Shorthead Sculpin, *Cottus confusus*

Table of Sculpin Characteristics.

Taxa	Palatine Teeth	Preopercle spines	LL pores on CP	CP in Eye	CP in Head	Prickles
C. bairdii	Typically 2 or 3 rows	2 or 3	no	slender 1.2	4.2	few under P1
C. beldingii	no	1 or 2	no	deep 2	3	none
C. cognatus	no	2 or 3	no	medium slender 1.3	4.2	few under P1
C. confusus	Typically 1 or 2 rows	2 or 3	Typically yes	medium 1.5	3.5	under P1+along LL
C extensus	yes	3	no	1–1.1	4.2	back and sides
C. greenei	yes	1	no	deep 1.8	2.1	under P1
C. hubbsi	at least 2 rows	3	yes	medium slender 1.3	3.7–4.6	under P1+ sparse elsewhere
C. leiopomus	no	none or nub	no	medium 1.5	2.9	none
C. rhotheus	at least 2 rows	3 or 4	yes	slender 1.1	4.8–6.4	over most of body
C. schitsuumsh	Typically 1 or 2 rows	1 or nub	Typically no	medium 1.6	3.5	under P1 + along LL

LL = lateral line; CP = caudal peduncle; P1 = pectoral fin.

Key to Idaho Sunfishes, Bass, and Crappies
Family Centrarchidae (Chapter 18)

1a. Anal fin spines usually 6 or 7 (compare color illustrations) Rock Bass and Crappies 2

1b. Anal fin spines usually 3 (compare color illustrations) Bass and Sunfish 4

2a. Dorsal fin spines 11 or 12; dorsal soft rays 10 or 11 (compare color illustrations) ....Rock Bass, *Ambloplites rupestris*

2b. Dorsal fin spines 6–8; dorsal soft rays 14–17 (compare color illustrations) .... Crappies, *Pomoxis* 3

3a. Dorsal fin spines usually 6; dorsal fin base much shorter than distance from front of dorsal fin to eye; dark markings on sides arranged in irregular vertical bands (compare color illustrations)........................... .... White Crappie, *Pomoxis annularis*

3b. Dorsal fin spines usually 7 or 8; dorsal fin base as long as distance from front of dorsal fin to eye; dark markings on sides not arranged in irregular vertical bands but mottled (compare color illustrations).......... .... Black Crappie, *Pomoxis nigromaculatus*

4a. Body not deep, depth about 1/3 times as deep as standard length; scales small, more than 60 in lateral line; notch between spiny and soft-rayed dorsal fins (compare color illustrations) Bass 5

4b. Body deep, about 1/2 as deep as standard length; scales large, fewer than 60 in lateral line; no deep notch between spiny and soft-rayed dorsal fins (compare color illustrations) Sunfish 6

5a. Upper jaw reaches past eye in adult; spinous dorsal fin peaked; next-to-last dorsal spine short, at the bottom of a deep notch; dorsal and anal fins lack scaled membranes along base (compare color illustrations) .... Largemouth Bass, *Micropterus salmoides*

5b. Upper jaw does not reach past eye in adult; spinous dorsal fin a low curve; next-to-last spine about half the length of the longest dorsal fin spine (compare color illustrations)... .... Smallmouth Bass, *Micropterus dolomieu*

6a. Teeth on tongue (Fig. 6a); supramaxilla longer than width of maxilla (Fig. 6a) .... Warmouth, *Lepomis gulosus*

Fig. 6a

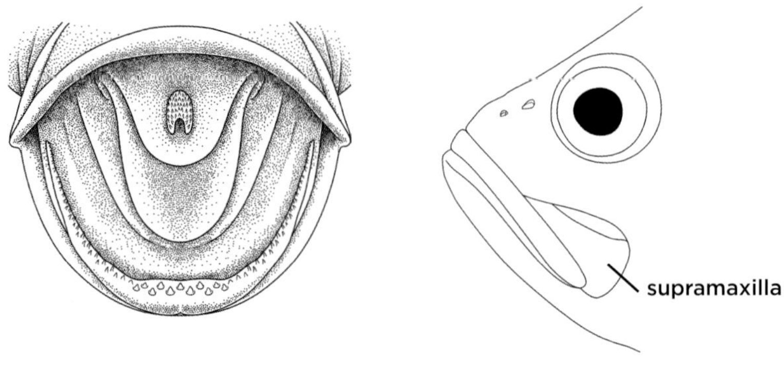

6b. No teeth on tongue; supramaxilla much shorter than width of maxilla................................ SUNFISH, 7

7a. Mouth longer than one-half length of short, rounded pectoral fin and short dorsal fin spines (compare color illustrations) .... GREEN SUNFISH, *Lepomis cyanellus*

7b. Mouth shorter than one-half length of long pectoral fin (compare color illustrations) 8

8a. Gill flap with red spot at edge (compare color illustrations); gill rakers short and knobby (Fig. 8a); pharyngeal teeth molariform (Fig. 8a) .... PUMPKINSEED, *Lepomis gibbosus*

Fig. 8a

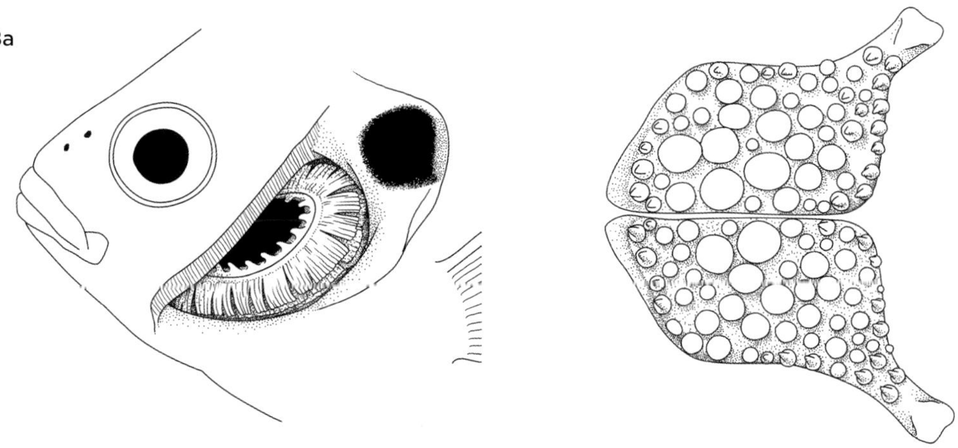

8b. Gill flap with no red spot (compare color illustrations); gill rakers long and slender (Fig. 8b); pharyngeal teeth sharp (Fig. 8b) 9

Fig. 8b

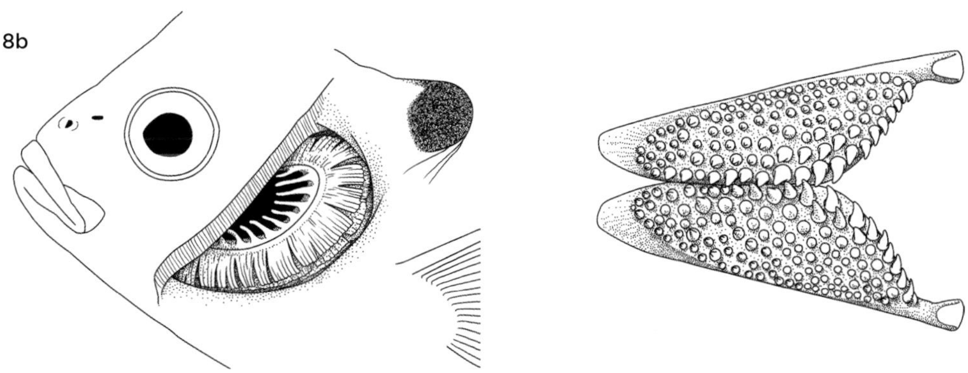

9a. Gill flap black with no white edge; pectoral fin pointed; no bright red or brown spots in adult; anal fin with 3 spines and 10–12 branched rays; black blotch in posterior of soft dorsal fin (compare color illustrations) .... BLUEGILL, *Lepomis macrochirus*

9b. Gill flap with white edge; pectoral fin rounded; bright red or brown spots on sides of adults; anal fin with 3 spines and 7–9 branched rays; no black blotch in posterior of soft dorsal fin (compare color illustrations) .... ORANGESPOTTED SUNFISH, *Lepomis humilis*

Key to Idaho Perches
Family Percidae (Chapter 18)

1a. Anal rays 6–8; jaw teeth small, with no canines; body deep and laterally compressed; 6–9 definite dark cross bands (compare color illustrations) on sides; pelvic fin bases close together .... Yellow Perch, *Perca flavescens*

1b. Anal rays 12–13; jaw teeth include large canines on jaws and palatines; body slender; dark cross bands indefinite (compare color illustrations); pelvic fins widely separated .... *Sander*, 2

2a. Lower lobe of caudal fin with milk-white tip; spiny dorsal fin without prominent dark spots; large black spot (compare color illustrations) at base of posterior dorsal spines .... Walleye, *Sander vitreus*

2b. Lower lobe of caudal fin without milk-white tip; spiny dorsal fin with prominent dark spots; no large black spot (compare color illustrations) at base of posterior dorsal spines .... Sauger, *Sander canadensis*

"The idea of wilderness needs no defense, it only needs defenders."

Edward Abbey

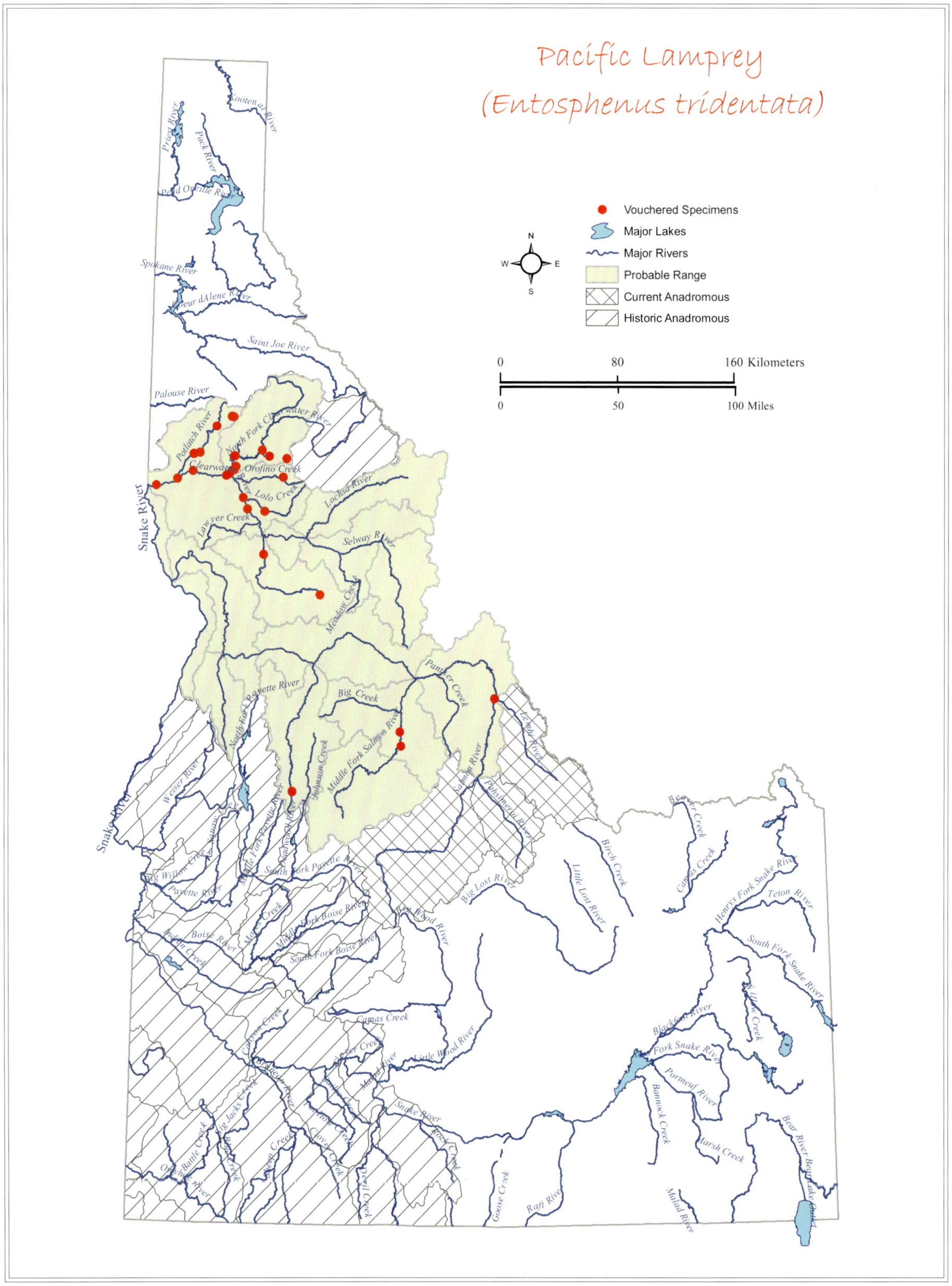
Pacific Lamprey
(Entosphenus tridentata)
Vouchered Specimens
Major Lakes
Major Rivers
Probable Range
Current Anadromous
Historic Anadromous
N
W
E
S
0
80
160 Kilometers
0
50
100 Miles
Snake River
Palouse River
Saint Joe River
Potlatch River
Orofino Creek
Lolo Creek
Lochsa River
Lawyer Creek
Selway River
Meadow Creek
Big Creek
Middle Fork Salmon River
Salmon River
Pahsimeroi River
Birch Creek
Little Lost River
Big Lost River
Teton River
South Fork Snake River
Camas Creek
Portneuf River
Bannock Creek
Marsh Creek
Malad River
Goose Creek
Raft River
Weiser River
Payette River
Boise River
South Fork Boise River
Spokane River
Pack River
Priest River

6

PETROMYZONTIFORMES

Lampreys

Order Petromyzontiformes

The body is naked and eel-like with one or two dorsal fins and no paired fins; barbels are absent. Lampreys are either parasitic or nonparasitic. Lampreys are characterized by metamorphosis from blind, toothless larva (ammocoetes), to immature juveniles (macropthalmia), to toothed, sighted adults in one to four years. This species has a round sucker mouth (jawless) armed with horny teeth that are often sharp. They occur in a single living family, Petromyzontidae. Four families are noted, with 10 genera and 38 extant species[1].

Family Petromyzontidae—Lampreys

All described lampreys exist in this single family of extant taxa. There are about 30 species in 10 genera. All larvae or ammocoetes live in freshwater, generally in the soft bottom of lakes and streams. Many species are externally parasitic, attaching themselves to a host with the suctorial mouth and using the rasping tongue to break the surface of the host and consume blood slowly[2]. In the genus *Entosphenus,* there are both parasitic and nonparasitic forms in seven species[1]. One species occurs in Idaho.

Pacific Lamprey

Entosphenus tridentatus

(Gairdner, 1836)

Etymology

Entosphenus = within a wedge, after the wedge-shaped tongue tooth; *tridentatus* = three-toothed, since many of the teeth have three cusps.

Importance

Pacific Lamprey are valued by Native American tribes[3], but until recently state fish and wildlife agencies treated them as threats to commercial and recreational salmonid fisheries. Within the Columbia River Basin, for example, the Umatilla River was once a preferred harvest location for Pacific Lamprey and salmonids by many Native American tribal members[3]. Native Americans in the Pacific Northwest used lamprey for food for centuries, preserving it by smoking or drying. This and other lamprey often have only an indirect relationship to man, resulting from the potential destruction of host species that were considered to be of greater value to sport or commercial anglers[2,4].

Protecting native anadromous lamprey populations has not been a management priority until recently, when states, tribes, and federal agencies signed the first Conservation Agreement for Pacific Lamprey Conservation[5]. Even though native lamprey populations are an important element in freshwater inland ecosystems, widely publicized predation by nonindigenous Sea Lamprey on game species in the Great Lakes has resulted in a generally negative public perception of all lamprey species. In reality, native lamprey in the Pacific Northwest are susceptible to many of the same threats facing anadromous species managed for sportfishing in Washington, Oregon, and Idaho[6].

Larval Pacific Lamprey can represent a large portion of the biomass in streams where they are abundant, making them an important component, along with aquatic insects, in processing nutrients, nutrient storage, and nutrient cycling. Larvae facilitate the conversion of the nutrients derived from detritus and algal biomass, while the undigested material is processed into fine particulate matter. The reduced matter is then exported from the system or taken up by other organisms such as filter-feeding insects. Larval Pacific Lamprey constitute a food source for other animals during two periods. Larvae are subjected to predation during emergence from nests and during stream-scouring events that dislodge larvae from their burrows. Recent reports document that Coho Salmon fry ate emergent larval lampreys[7].

Description

All members of this family have an elongate and cylindrical body form. They have no scales, no paired fins, and no jaws. The mouth is surrounded by a circular sucking disc that is armed with horny, multi-cuspid plates. A single median nostril occurs on the dorsal surface of the head; a series of seven circular gill openings occurs on each side of the body. No lateral line is present. Transformed adults are blue-black to dark brown and are lighter ventrally. No color patterns are generally present. Adult teeth are usually yellow to orange, many with three cusps. Adult lampreys can be identified by the arrangement of the teeth in the sucker-like mouth[2,7,8].

Outside the breeding season, sexes are difficult to distinguish, but their morphologies diverge as they approach spawning. Males develop a slender, short urogenital papilla, while the female's urogenital papilla is inconspicuous; however, females develop a single anal fin (a prominent swelling behind the vent) that is absent in males. There are also sex differences in dorsal fin height and body proportions[8].

Range and Distributional Status

This species is native to Idaho. Pacific Lamprey are found on the Pacific Coast of the United States and Canada from Baja California, Mexico to the Aleutian Islands and to Hokkaido, Japan. In Idaho, this species was originally distributed in tributaries of the Snake River below Shoshone Falls, and original range overlapped most of the Chinook Salmon and Steelhead habitat in Idaho. They were thought to be restricted to the Clearwater and Salmon River drainages and tributaries to the Snake River below the confluence with the Salmon River[6]. Recently they have been documented in portions of the Snake River, Clearwater River, Salmon River, Imnaha River and the Lochsa River[9]. It may have become landlocked in lakes and reservoirs and has been reported in both Pend Oreille Lake and Dworshak Reservoir. Long-term persistence of the species in these habitats has not been documented[2,7].

Size and Longevity

Although minute when hatched, lamprey ammocoetes grow rapidly, usually reaching 122 to 303 mm (4.8 to 12.0 in) at transformation. Adult Pacific Lamprey may reach a total length of 762 mm (30 in). They reach an average of 537 mm (21 in) total length when migrating inward from the sea[2,7,10]. Ammocoetes live in freshwater for four to seven years[7]. Adult Pacific Lamprey spend up to three to five years in the sea and reach lengths up to 720 mm (28 in) before returning to freshwater; there is, however, considerable variation in adult length[8].

Food and Feeding

During its freshwater-life stage, Pacific Lamprey are generally a filter feeder, with desmids and diatoms constituting a notable portion of the total diet. As an adult it becomes parasitic and attaches to a variety of hosts. Using the toothed tongue, a hole is rasped through the scales and skin of the host, and blood and body fluids are extracted[2,7,10].

Limiting Factors

A growing body of studies suggest that hydroelectric dams, diversions, and low-head irrigation dams may be a principal cause of sharp declines in abundance of lamprey into the interior Columbia and Snake River basins. Large hydroelectric dams were constructed on the main-stem Columbia and Snake Rivers from the 1930s through the 1970s. These caused severe declines in anadromous salmonids and also likely affected the migration, distribution, and abundance of Pacific Lamprey. Recent evidence showed that fewer than 50 percent of adult Pacific Lamprey encountering the lower Columbia River dams were able to pass above the dams using the fishways, and they experienced longer delays in migration than those experienced by salmon. Low-head irrigation dams may have had similar impacts on migrating adults in the tributaries as the major hydroelectric dams do in the main stem of the Columbia River. Recent studies show that adult lamprey passage efficiency was less than 50 percent at these dams on the Umatilla River. In the Columbia River basin, the construction of irrigation diversion dams in the tributary rivers began in the early 1900s, and hundreds of these dams now exist. Cumulative effects of these dams may have been significant. Larval recruitment of Pacific Lamprey in the tributaries of the Columbia and Snake Rivers where these dams occurred, for example, appears reduced from historical levels. Ignorance of the ecological role of lamprey and conflicting social values associated with different species also contributed to the decline[3].

Range reductions have occurred for Pacific Lamprey, principally due to migration barriers and other adverse ecosystem modifications. Both ammocoetes and adults (primarily post-spawning carcasses) are food for other species[10].

Threats to Pacific Lamprey in Idaho include some of the same factors adversely affecting other native anadromous species such as Chinook Salmon and Steelhead, that is, loss of spawning and rearing areas, ecosystem alterations associated with stream modifications and losses (mortalities) at main-stem Snake and Columbia River dams. In addition, with their protracted residence in freshwater, Pacific Lamprey are subject to a wide variety of pollution sources and sediment dredging. These factors all combine to make lampreys highly subject to extirpation[6,11]. Lamprey were extirpated from the Umatilla River by poisoning with rotenone treatments in 1967 and 1974[3].

Negative effects on downstream migrants have also been noted. Lamprey migrate downstream as macropthalmia and ammocoetes. They may pass under fish screens on dams or may pass through bypass screens that have openings larger than the diameter of lamprey. It has been suggested that the highest potential for mortality comes from the fish screens, and anecdotal observations by biologists working on main-stem dams on the Columbia and Snake Rivers during the 1970s and 1980s indicated that juvenile lamprey impinged on the perforated plates that blocked various openings across the forebay faces of the dams and on the juvenile bypass screens in huge but undocumented numbers. Such observations have been documented to some extent by recent research[5].

Increased mean water temperatures from land-use practices and hydro modifications may also affect the survival of juvenile Pacific Lamprey. It has been reported that lamprey mortality may occur at temperatures in the 22 to 28°C (72 to 82.4°F) range. Elevated temperatures may adversely affect lamprey survival[5].

Breeding Habits

Adult Pacific Lamprey build a nest and deposit and fertilize 30,000 to 230,000 eggs. Adults enter freshwater in late spring and early summer to spawn (April to July) when water temperatures are in the 10 to 15°C range (50 to 59°F). Males go through rather elaborate courting rituals, releasing sperm, while the female is attached to a rock and deposits eggs. After two to three weeks, the elliptical eggs hatch, and the ammocoetes burrow into the mud, where they remain for up to five years. During transformation, macropthalmia and adults migrate downstream to the ocean. Adults may remain in the parasitic phase in the ocean for one to two years before returning to freshwater to spawn[4,7,10].

Pacific Lamprey mature at about eight to 13 years (three to eight years as ammocoetes, four years as parasitic adults, and one year during the spawning migration) and typically have a single lifetime spawning event and incredibly high female fecundity, ranging from 30,000-238,400 eggs[14].

Pacific Lamprey spawn in the spring, generally April through June, and typically die shortly after spawning. However, in some coastal ecosystems, substantial numbers of adults survive spawning and migrate back downstream toward the ocean. Some of these downstream migrants may return to spawn a second time. Typically, spawning is initiated when water temperatures rise to about 10°C (50°F). Behavior during nest construction is similar to that described for other lampreys, with males moving most of the gravel but with some participation by females. Fecundity increases with female body size and ranges from 34,000 to 238,400 eggs[8].

Habits and Habitat

Pacific Lamprey ammocoetes inhabit fine silt deposits in backwaters and other quiet portions of small to large streams. Macropthalmia and adults migrate to the ocean, where they become parasitic, although some landlocked populations exist[4,7,10].

Upon entering the sea, Pacific Lamprey move offshore into relatively deep water. Young adults have been caught at depths of 100 to 250 meters (328 to 820 ft) in the Strait of Georgia and off the west coast of Vancouver Island[8].

Management and Conservation Status

Although previously abundant in Idaho waters, Pacific Lamprey range has been reduced due to migration barriers and other adverse ecosystem factors. They are considered a critically imperiled species by the State of Idaho[10,12-14]

The IDFG objectives for distribution, abundance, and connectivity are largely unmet. Improvements in upstream passage are necessary for any chance of recovering Pacific Lamprey populations in Idaho. Direct actions to facilitate improvement of downstream and upstream passage at main-stem lower Snake and Columbia River hydroelectric facilities are outside the regulatory purview of the IDFG. However, the IDFG can support and promote modifications and improvements at the Federal Columbia River Power System hydroelectric facilities so that objectives for Idaho populations of Pacific Lamprey can be realized. More directly, the IDFG is actively pursuing and implementing fish habitat restoration throughout the Clearwater and Salmon River basins for the benefit of salmon and Steelhead. The IDFG and its partners will need to give additional consideration to accommodate Pacific Lamprey passage at these project sites.

Chapter Citations

1. Nelson, 2006. 2. Scott and Crossman, 1973. 3. Close et al., 2009. 4. Simpson and Wallace, 1978. 5. Columbia River Basin Lamprey Technical Workgroup, 2004. 6. Idaho Department of Fish and Game, 2011. 7. Wydoski and Whitney, 2003. 8. McPhail, 2007. 9. Mcllraith, et al., 2015. 10. Wallace and Zaroban, 2013. 11. Clemens, 2011. 12. Moser and Close, 2003. 13. Idaho Department of Fish and Game, 2005. 14. Luzier, et al., 2011.

Additional readings for this chapter are listed in the Supplemental References section.

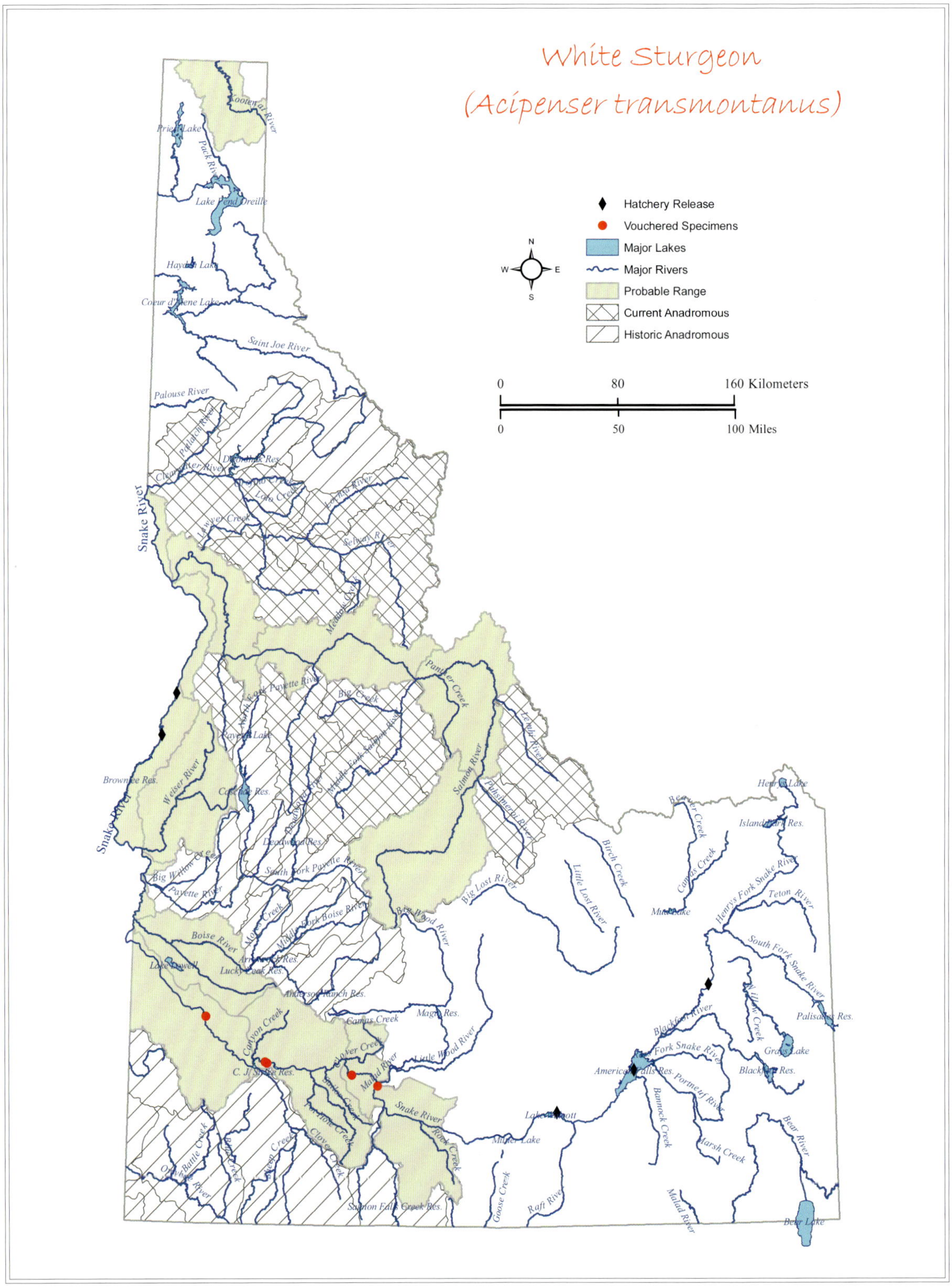
White Sturgeon
(Acipenser transmontanus)
Hatchery Release
Vouchered Specimens
Major Lakes
Major Rivers
Probable Range
Current Anadromous
Historic Anadromous
N
W
E
S
0
80
160 Kilometers
0
50
100 Miles
Kootenai River
Priest Lake
Pack River
Lake Pend Oreille
Hayden Lake
Coeur d'Alene Lake
Saint Joe River
Palouse River
Snake River
Dworshak Res.
Lawyer Creek
Selway River
Panther Creek
Big Creek
Payette Lake
Weiser River
Brownlee Res.
Cascade Res.
Deadwood Res.
Salmon River
Lemhi River
Pahsimeroi River
Birch Creek
Little Lost River
Big Lost River
Big Wood River
Camas Creek
Mud Lake
Henrys Lake
Island Park Res.
Henrys Fork Snake River
Teton River
South Fork Snake River
Palisades Res.
Willow Creek
Grays Lake
Blackfoot River
Blackfoot Res.
American Falls Res.
Portneuf River
Bannock Creek
Marsh Creek
Bear River
Bear Lake
Malad River
Raft River
Goose Creek
Milner Lake
Lake Walcott
Snake River
Rock Creek
Salmon Falls Creek Res.
C. J. Strike Res.
Canyon Creek
Big Willow Creek
Payette River
South Fork Payette River
Boise River
Lake Lowell
Lucky Peak Res.
Arrowrock Res.
Anderson Ranch Res.
Magic Res.
Little Wood River
Camas Creek
Clover Creek
Malad River

7

ACIPENSERIFORMES

Sturgeon and Paddlefish

Order Acipenseriformes

Sturgeons and Paddlefishes, within the class Actinopterygii, are heavy bodied and almost cylindrical, with a heterocercal tail. Paired fins are well developed, and the pelvic fins are abdominal with true spines. These fish inhabit temperate marine and fresh water; marine forms spawn in freshwater. This order consists of two living families, Acipenseridae and Polyodontidae, with six genera and 27 species[1,2].

Family Acipenseridae—Sturgeons

Sturgeons are heavy sub-cylindrical fishes that have hard, extended snouts and a ventral, protrusible mouth that is preceded by four barbels. The body is covered in large bony scutes, in five rows, and the head is covered in bony plates. Sturgeons are either amphidromous or freshwater fishes. Four genera contain 25 species[1,2.] One species is found in Idaho.

White Sturgeon

Acipenser transmontanus
(Richardson, 1836)

Etymology

Acipenser = Old World name for Sturgeon; *transmontanus* = beyond the mountains. Richardson, who had come from the East, was alluding to the Pacific Coast distribution.

Importance

As a food fish, White Sturgeon are highly acceptable, both for the flesh and for the eggs. As many as 113 kg (250 lb) of eggs can be obtained from a single female[2]. Sportfishing for this large species is popular in a variety of locations in Idaho, including below American Falls Reservoir, CJ Strike Dam, and Hells Canyon Dam.

Description

White Sturgeon are grayish to gray-brown above, slightly lighter to white below. An elongated snout protrudes beyond the completely ventral mouth. Eyes are relatively small, and the skin has a rough texture to it; there is no lateral line. Typical of this family, the tail is heterocercal, with the upper lobe longer and more curved than the lower lobe. Juveniles are not distinctly different in color from adults, but generally lighter gray[2,3].

Range and Distributional Status

This species is native to Idaho. The range of White Sturgeon includes Pacific Northwest rivers, estuaries, and the near-shore ocean from Baja California, Mexico, and to the Aleutian Islands, Alaska. Significant populations extend into major river basins for hundreds of miles, including the Sacramento–San Joaquin, Columbia, and Fraser River systems[3,4].

In Idaho, White Sturgeon occur in the Snake River upstream to Shoshone Falls, the Salmon River and as far upstream as McKim Creek, and the Kootenai River drainage in northern Idaho[5]. White Sturgeon have been introduced into the Snake River above Shoshone Falls upstream as far as the city of Idaho Falls[3,4]. The Kootenai River, which flows into Kootenay Lake, British Columbia, Canada, supports the only naturally landlocked White Sturgeon population in North America.

Size and Longevity

White Sturgeon are the largest North American sturgeon and the largest fish in North American fresh waters. Adults may reach lengths of 6.1 m (20.0 ft) and weigh up to 816 kg (1,799.0 lb)[2]. Individual growth rates are highly variable. The following age and length measurements (mean fork length) in mm (in) have been reported for individuals in the Fraser River: age 1, 221 (8.7); age 5, 493 (19.4); age 10, 772 (30.4); age 20, 1,282 (50.5); age 35, 2,037 (80.2). White Sturgeon found in more southern latitudes exhibit faster growth[6]. Ages in excess of 100 years have been reported[2], but fish over 80 years of age are rare. The all-tackle record for this species is 212.3 kg (468.0 lb) caught near Benicia, California, in July 1983[7].

White Sturgeon free embryos and larvae collected in the Fraser River, British Columbia, Canada, ranged from 9 to 20.5 mm (0.4 to 0.8 in) in length, the majority being less than 16 mm (0.6 in)[8]. In general, White Sturgeon growth tends to slow with increased latitude and distance inland due to cooler temperatures, lower system productivity, and reduced food availability[9,10]. The upper reaches of the Snake River below Shoshone

Falls are an exception because of the moderated thermal regime influenced by a large volume of groundwater inflow to the Snake River in this region. Longevity of the species results from low natural mortality beyond the first few years of age[10], with annual survival rates often exceeding 90 percent[11–13].

In the Snake River in Idaho, historical captures of large sturgeon have ranged from 231 kg (509.3 lb) near Glenns Ferry to 306 kg (675 lb) and 3.3 m (10.8 ft) below Shoshone Falls in 1908, 311 kg (686 lb) near the mouth of the Salmon River, and 680 kg (1,499.1 lb) near Weiser in 1898[14]. Present-day maximum size approaches 3.4 m (11.2 ft), but individuals of this size are uncommon.

The wild Kootenai River White Sturgeon population is currently dominated by large older fish. Wild juveniles in the 1,000 mm (39.4 in) FL virtually disappeared from length distributions after 1983. Hatchery fish contributed 94 percent of the 659 juveniles that have been captured since hatchery fish were first released. The age composition of wild fish has shifted from a population dominated by fish of age 25 and younger to one comprising primarily fish older than 25. Kootenai River median female maturation at 1,400 mm (55.1 in) FL was smaller than the medians of 1,580 to 1,940 mm (62.2 to 76.4 in) reported for lower Columbia River populations, perhaps reflecting the slow growth rates of the Kootenai River population[15].

Food and Feeding

White Sturgeon are invertivore-piscivores, feeding on the bottom for any available organisms. Adult diet consists mainly of fish, dead or alive. Juveniles feed largely on larval forms of aquatic insects, crustaceans, and mollusks[3,4]. Food items are detected with chemo- and electroreceptors located on four sensory barbels and the snout rather than by sight alone[16,17].

Limiting Factors

Liberal fishing regulations can decimate populations of long-lived species such as White Sturgeon. The first significant human exploitation of White Sturgeon probably began in the late 1880s, when demand for caviar ballooned on the East Coast of the United States. During that era, the harvest of White Sturgeon in the lower Columbia River increased to more than 453,592 kg (1 million lb) annually. This high rate of exploitation was not sustainable, and White Sturgeon populations became precipitously low. Although historical information for the Snake River is lacking, it is likely that White Sturgeon in Idaho were also significantly exploited and that populations were impacted by overharvest[5,14]. Additional anthropogenic effects include pollution and habitat fragmentation and destruction through dam construction for power, irrigation, and flood control[3,10,12]

White Sturgeon have declined in Idaho for a variety of reasons but primarily due to the construction of dams that began in the early 1900s through the late 1960s. These dams fragmented populations and eliminated or severely altered diverse spawning, rearing, and feeding habitats[5]. The Snake River is extensively managed to provide water for agriculture, hydropower, municipalities, and flood control, with almost half of its estimated volume diverted for agriculture. This has resulted in alterations in the natural hydrograph and significant reductions in the natural flow and water quality of the Snake River. Recruitment monitoring in the middle Snake River shows that increased production occurs during years with significant spring freshets[18,19]. Of the Snake River's nine reach populations in Idaho, only two (Bliss to CJ Strike and Hells

Canyon to Lower Granite) support viable populations characterized by self-sustaining natural recruitment. Remaining reaches show little or no detectable reproduction[18].

White Sturgeon in the Kootenai River are significantly impacted by flow releases from Libby Dam and the resultant changes in habitat quality[5]. There has been little or no annual natural recruitment since 1974, corresponding with the closure of Libby Dam, though natural recruitment was depressed or absent prior to 1974. Apparent causes of this decline are spawning over low-quality sand-silt substrates and loss of ecosystem function (riparian, diking, floodplain connection, food web)[20–23].

One of the most common causes of mortality of fish during early life stages is predation, and predator-prey interactions are affected by environmental conditions[26]. Habitat conditions in many rivers have been altered by dams and other human activities, which may have shifted White Sturgeon predator-prey dynamics and led to decreased survival. Three factors are often implicated in changes in predator-prey interactions in altered ecosystems: decreased turbidity levels, higher light levels, and altered substrates (i.e., cover availability). The effect of turbidity on predator-prey interactions is complex. Turbidity reduces the distance at which a predator can detect prey and also reduces prey-reaction distance[26].

Significantly more White Sturgeon yolk-sac larvae were eaten at lower turbidity levels, although some larvae were still eaten at the highest turbidity level tested, 360 NTU. This suggests that lower turbidity levels now present in many river systems throughout the Pacific Northwest because of impoundment may result in increased predation on White Sturgeon larvae. For example, Columbia River reservoirs typically have very low turbidity levels (mean, 6 NTU; maximum, 23 NTU) during May to July, when White Sturgeon larvae would be expected to be present. Light levels in nonturbid water, cover, and prey size also significantly affected the outcome of predator-prey interactions between Prickly Sculpin *(Cottus asper)* and White Sturgeon larvae. The lowest mean proportion of White Sturgeon eaten was during trials with the lowest light level, the presence of cover, and newly hatched larvae. This could be due to the behavior of the predator or that of the prey, or a combination of the two. Prickly Sculpin ate more White Sturgeon larvae at lower turbidity levels; higher light levels may similarly aid their predatory efforts[26].

Five of the nine populations of White Sturgeon, between dams on the middle Snake River, have declined from historical levels, and one explanation is associated with reduced recruitment. Ranking factors that influence recruitment addresses some of these concerns. Seven mechanistic factors were evaluated that contribute to lost recruitment in each river segment: (1) temperature-related mortality during incubation, (2) flow-related mortality during incubation, (3) downstream export of larvae, (4) limitation of juvenile and adult habitat, (5) mortality of all ages during summer episodes of poor water quality in reservoirs, (6) entrainment mortality of juveniles and adults, and (7) angling mortality. Model predictors that influence recruitment differed among river segments, but there was a clear distinction between river segments limited by episodic poor water quality and those with adequate water quality. Among river segments with better water quality, short river segments were regulated by different factors than longer segments. Additional studies are required[27].

Breeding Habits

Water temperatures suitable for spawning range from 10 to 18°C (50 to 64°F)[28]. White Sturgeon are broadcast spawners, releasing their eggs and sperm into the water column over boulder and cobble substrates, where fertilization occurs[10,28]. Fish gather in aggregations to spawn, and several males typically spawn with each female. Spawning is accompanied by darting, rolling, and breaching activity, often associated in turbulent waters with high habitat complexity, which removes fine sediments from spawning areas that reduce embryo survival[28,29-31]. Spawning periodicity for lower Columbia River populations, three to five years at full maturation, is close to the physiological minimum of two to three years observed in California commercial aquaculture operations[15].

Spawning populations are found in three major river basins: the Sacramento–San Joaquin, Columbia (including Kootenai and Snake Rivers), and Fraser River systems. Across the species range, spawning occurs between February and August, with late-winter spawning occurring at the southern areas of their range (Sacramento–San Joaquin River)[32] and continuing into late summer with increasing latitude (upper Columbia and Fraser Rivers)[8,33]. In the Snake River, spawning activity occurs between March and June[18], with Kootenai White Sturgeon spawning from May to July[31]. Outside of Idaho, this species is considered amphidromous, although marine environments are not obligatory to complete their life cycle, and both amphidromous and resident freshwater populations occur.

Males in Idaho populations may mature as early as seven years of age[3], while females may not develop eggs and spawn until 16 to 18 years of age[34]. Females may spawn on a three- to 11-year basis. Mature females can produce more than one million eggs that are released and fertilized in the water column[2,3]. Optimal water temperature for egg incubation is from 14 to 16°C (57.2 to 60.8°F), with increasing egg mortality at <8°C (46.4°F) and >18°C (64.4°F)[35]. Embryos hatch within seven to 14 days, depending on water temperature, and metamorphosis into juveniles occurs within about 30 days[17]. In the Kootenai River, spawning movements of White Sturgeon occur each autumn and spring when adults migrate from the lower Kootenai River and Kootenay Lake, British Columbia, to prespawn staging reaches in Idaho. In spring, individuals migrate farther upriver to a spawning reach near Bonners Ferry, Idaho. In recent studies, White Sturgeon responded to augmented flows, migrating substantial distances to a spawning reach soon after the onset of local runoff and rising water temperatures. Males migrated at temperatures of 5.5 to 12.1°C (41.9 to 53.8°F) two weeks before spawning; females followed about a week later, at slightly warmer temperatures. Females stayed in the spawning reach one to 28 days, averaging 10.5 days. Males spent seven days to two months in the spawning reach, averaging 30 days. After spawning, 63 percent of the females moved immediately to Kootenay Lake; the remainder spent a longer time in the river downstream of the spawning reach. Some (52 percent) males remained in the river, and the remainder migrated to Kootenay Lake. Female behavior and migration were more attuned to environmental conditions than was male behavior. Changes in temperature and river stage were the best predictors of the probability that females would migrate to the spawning reach[31].

Channel substrate at spawning sites in Fraser River, Canada side channels was similar and relatively heterogeneous, and all size classes of sediment were represented except boulder. One side-channel spawning area consisted mostly of sand with some gravel-size sediment; one side-channel spawning area was predominantly cobbles with no sediment less than 2 mm (0.1 in); and one main-channel spawning area consisted of boulders and cobbles. The majority of White Sturgeon eggs and larvae were collected at temperatures ranging from 13 to 15°C (55 to 59°F). A small number of larvae were found at temperatures exceeding 18°C (64°F)[8].

Turbidity as a factor affecting spawning habitat has not been examined in previous studies, although it is recognized as a potentially important factor in the Kootenai River. Hydroelectric dams on most rivers with White Sturgeon populations trap sediment loads and substantially reduce turbidity. In contrast, the Fraser River has naturally high turbidity, particularly during freshets; light attenuation based on Secchi disk readings is often less than 30 cm (11.9 in). Seasonally high concentrations of suspended sediment may provide sufficient cover, allowing habitats of relatively shallow depth to become suitable for spawning. Before hydroelectric development, rivers such as the Columbia and Kootenai would have been turbid during spring snowmelt. If White Sturgeon are generally photophobic because of this evolutionary history, light attenuation may be an important factor affecting spawning depth[8].

White Sturgeon in the unregulated Fraser River used a range of substrate types for spawning, and differences among spawning sites were influenced by channel type and morphology. The main-channel site was similar to Columbia River spawning areas, consisting of cobble and boulder substrates. Spawning areas in side channels were composed mostly of gravel and some sand, which is similar to observations reported for the Sacramento and Kootenai Rivers. The fact that White Sturgeon have been found to spawn over this range of substrates suggests that sediment size may be of less importance in determining habitat suitability compared with factors such as water depth and velocity. Additionally, adult White Sturgeon were seen surfacing, breaching, and rolling in a side channel during spawning, which supports the side-channel spawning evidence that is identical to behavior reported for the Columbia River and was attributed to spawning activity because adhesive eggs were collected downstream of the site immediately following those observations[8].

In a 1998–99 Fraser River study, water temperatures during White Sturgeon spawning ranged from 13 to 19°C (55.4 to 66.2°F) and increased during the study period of each year. Differences in water temperature were not significant between main and side channels[8]

A positive correlation has been reported between median length of female maturation and growth rate. No evidence of reproductive senescence or changes in gonad appearance was obvious in the reduced maturation frequency of large old fish[15].

Habits and Habitat

White Sturgeon are a benthic (bottom-dwelling) species using a wide variety of habitats, including reservoirs and lakes and deeper holes of large cool rivers. Individuals may range widely to take advantage of scattered and seasonally abundant resources associated with feeding, spawning, and overwintering[36–38]. It is believed that White Sturgeon migrated freely between the Snake and Columbia River drainages prior to the closure of the main-stem Snake and Columbia River dams[2–4]. During early life stages, age 0 White Sturgeon in the lower Columbia River prefer deep water of 9 to 31 m (29.5 to 101.7 ft) and low-velocity areas, where substrate particle sizes are small (e.g., sand)[28], and seem to prefer the main river channel. Juvenile and subadult White Sturgeon are known to use both main- and off-channel habitats in the lower Columbia River[39,40], preferring those habitats with moderate riverbed roughness and slope[41].

Habitat use of subadults and adults varies with habitat availability. Many White Sturgeon movement and migration patterns appear to be associated with feeding. Where habitat is relatively homogenous, such as in marine waters, estuaries, low-gradient main-stem areas of the lower basin, and reservoirs, White Sturgeon move frequently and range widely, presumably in search of scattered or mobile food resources[40,42,43].

In riverine environments, White Sturgeon seem to prefer free-flowing stretches[42,43], though they are also known to inhabit lower-velocity lacustrine habitats[43]. Fish are more typically sedentary in the upper basin where the river consists of interspersed rapids and pools where fish can hold and feed on prey delivered by the river or take advantage of the patchy distribution of vertebrate and invertebrate prey items. High-use areas by White Sturgeon in the Canadian portion of the upper Columbia River and the river–Lake Roosevelt transition zone are all deep depositional environments where food items settle and become available to benthic-oriented feeders such as White Sturgeon. These low-velocity areas adjacent to fast water allow sturgeon to optimize energetic benefits[10].

Management and Conservation Status

White Sturgeon in the Kootenai River are listed as endangered under the federal Endangered Species Act. Snake River and Salmon River populations are an Idaho state species of special concern. Populations that exist in isolation between dams have little or no opportunity for movement above or below the impoundments. No angling is allowed for the Kootenai River population, and the Snake River and Salmon River populations have a catch-and-release-only fishery.

While significant numbers of reproductive White Sturgeon remain in the Kootenai River population, these individuals may be able to take advantage of suitable spawning and rearing conditions only if appropriate habitat can be identified and restored. Mature fish account for a large proportion of the female population. Critical habitat loss is considered to be a major impediment to Kootenai River White Sturgeon recovery, and management activities consist of two major thrusts: hatchery supplementation of the wild population to replace lost recruitment and habitat manipulations that are being performed to restore spawning habitat and reestablish natural reproduction[24,25,44,45].

Management efforts for the conservation of Snake River White Sturgeon include providing for a sustainable no-harvest recreational fishery, facilitating data collection, integrating and defining the role of artificial propagation, and promoting public awareness and compliance through information and education. Snake River White Sturgeon populations are defined by three categories: core conservation, conservation, and sportfish. Core conservation populations (below Hells Canyon and Bliss Dams) are self-sustaining from natural recruitment, with all age and size classes present and supporting recreational sportfisheries. Conservation populations are reaches with existing White Sturgeon populations and sportfisheries but display infrequent or no natural recruitment and unbalanced age and size structures. White Sturgeon in reaches upstream of the historical Snake River distribution are designated as sportfish populations and intended primarily to increase the diversity and fishing opportunity in existing fisheries. These categories guide management strategies based on reach designation with an emphasis on protecting habitat; genetic integrity and diversity, including measures that evaluate fishing-related mortality; population demographics and temporal trends; and use of conservation aquaculture where necessary to maintain population abundance and fishing opportunity in recruitment-limited reaches[5].

The authors gratefully acknowledge the review, editing, and significant input to this chapter by Jim Chandler and Ken Lepla of Idaho Power Company.

Chapter Citations

1. Nelson, 2006. 2. Scott and Crossman, 1973. 3. Wallace and Zaroban, 2013. 4. Wydoski and Whitney, 2003. 5. Dillon and Grunder, 2008. 6. Lebreton and Beamish, 2004. 7. International Game Fish Association, 2014. 8. Perrin, Rempel and Rosenau, 2003. 9. Nechako White Sturgeon Recovery Initiative, 2004. 10. Beamesderfer and Anders, 2013. 11. Semakula, 1963. 12. Beamesderfer, Rien and Nigro, 1995. 13. Irvine, Schmidt and Hildebrand, 2007. 14. Edson, 1956. 15. Paragamian, Beamesderfer and Ireland, 2005. 16. Brannon, et al., 1986. 17. Buddington and Christofferson, 1985. 18. Idaho Power Company, 2005. 19. Bates, 2013. 20. Duke, et al., 1999. 21. United States Fish and Wildlife Service, 1999. 22. Anders, Richards and Powell, 2002. 23. Kootenai Tribe of Idaho and Montana Fish Wildlife and Parks, 2004. 24. Paragamian, et al., 2009. 25. McDonald, et al., 2010. 26. Gadomski and Parsley, 2005. 27. Jager, et al., 2001. 28. Parsley, Beckman and McCabe, 1993. 29. McCabe and Tracey, 1994. 30. McAdam, 2011. 31. Paragamian and Kruse, 2001. 32. Israel, Drauch, Gingras, 2009. 33. Golder Associates Ltd. 2010. 34. Bentz and Lepla, 2013. 35. Wang, Binkowski and Doroshow, 1985. 36. Apperson and Anders, 1990. 37. Brannon and Setter, 1992. 38. Howell and McLellan, 2013. 39. Parsley and Popoff, 2004. 40. Parsley, et al., 2008. 41. Hatten and Parsley, 2009. 42. Bajkow, 1951. 43. Haynes, Gray and Montgomery, 1978. 44. Paragamian, Wakkinen and Kruse, 2002. 45. Rust and Wakkinen, 2007.

Additional readings for this chapter are listed in the Supplemental References section.

"There are always new places to go fishing. For any fisherman, there's always a new place, always a new horizon."

Jack Nicklaus

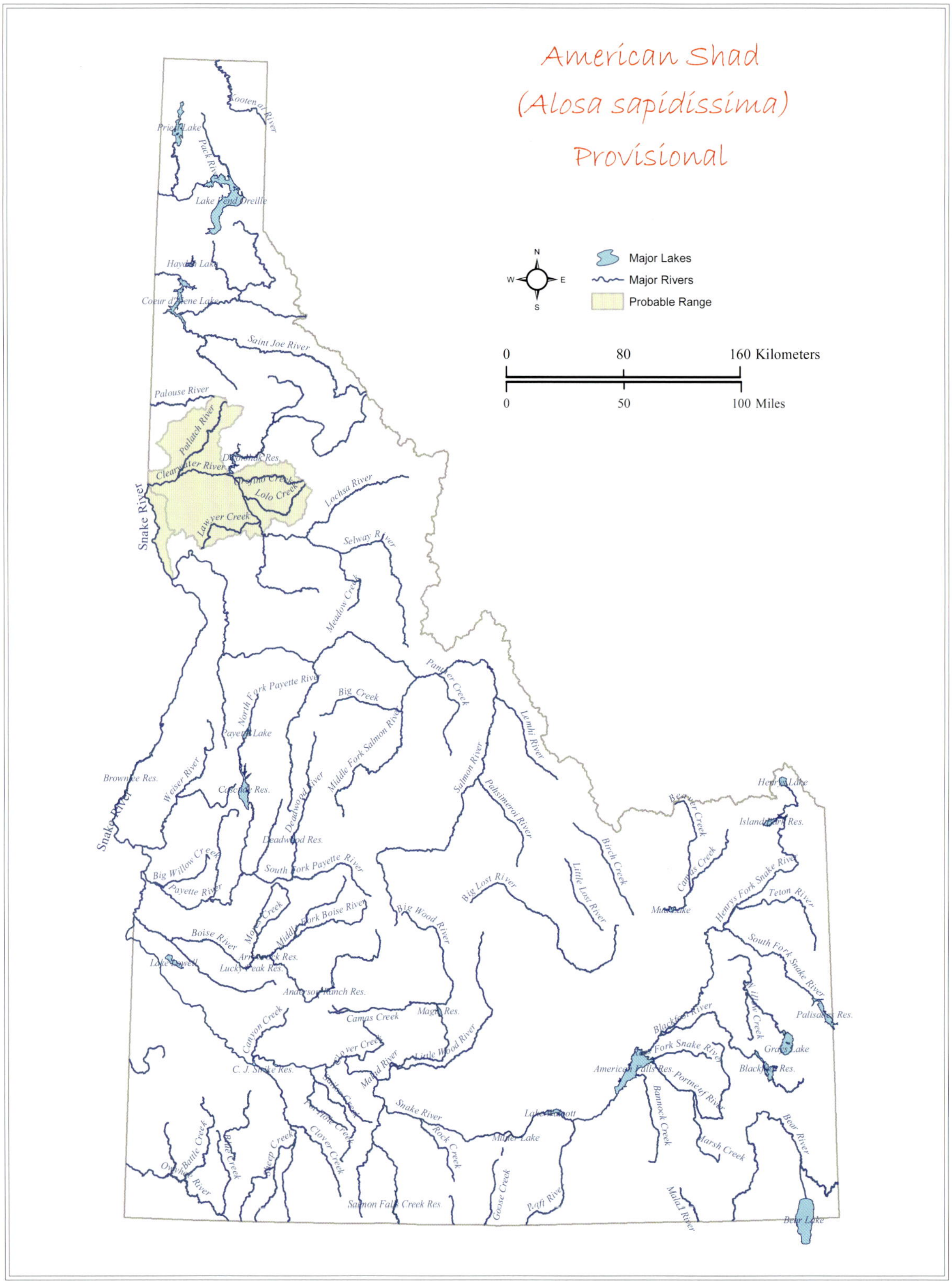
American Shad
(Alosa sapidissima)
Provisional
N
W
E
S
Major Lakes
Major Rivers
Probable Range
0
80
160 Kilometers
0
50
100 Miles
Saint Joe River
Palouse River
Potlatch River
Lolo Creek
Lawyer Creek
Lochsa River
Snake River
Selway River
Meadow Creek
Panther Creek
North Fork Payette River
Big Creek
Middle Fork Salmon River
Lemhi River
Salmon River
Weiser River
Pahsimeroi River
Deadwood River
Birch Creek
Little Lost River
Big Lost River
Big Willow Creek
Payette River
South Fork Payette River
Big Wood River
Boise River
Camas Creek
Little Wood River
Snake River
Rock Creek
Clover Creek
Goose Creek
Salmon Falls Creek Res.
Canyon Creek
C. J. Strike Res.
Anderson Ranch Res.
Lucky Peak Res.
Brownlee Res.
Palisades Res.
Teton River
Henrys Fork Snake River
South Fork Snake River
Blackfoot River
Bannock Creek
Marsh Creek
Bear River
Malad River
Bear Lake
Island Park Res.
Mud Lake
Blackfoot Res.
Portneuf River

8

CLUPEIFORMES

Herrings and Anchovies

Order Clupeiformes

Herrings and Anchovies originated in the upper Jurassic Period. These lower bony fishes have soft rays and cycloid, typically deciduous scales. This order includes Herrings and closely related species. All members of this order are silvery and have compressed keel-like bodies. The gas bladder extends into the skull and terminates in small bulbae on each of two branches. All fins are without spines, and there is always a single dorsal fin. Pectoral fins are located ventrally behind the head. Most are plankton feeders, with long and sometimes very numerous gill rakers that serve as plankton straining devices[1]. These fishes are physostomous[2–4].

Family Clupeidae—Herrings

In herrings, scales in the lateral series are about 40 to 50, five to 10 branchiostegal rays, vertebrae usually 37 to 59. A valued commercial fishery exists for clupeids in many parts of the world[1]. While a few landlocked populations exist, most are anadromous and are primarily represented by marine species. They are small to medium-size fish and are silvery and pelagic. The cycloid scales are deciduous. A lateral line is absent; teeth are small or absent. They occur throughout the world's oceans. Approximately 190 species exist worldwide[2–4]. In Idaho, one species may be present.

American Shad

Alosa sapidissima

(Wilson, 1811)

Etymology

Alosa = Saxon "allis," old name of the European Shad *Alosa alosa; sapidissima* = most delicious.

Importance

Within their native range, American Shad are of variable value in commercial fisheries. They are esteemed as game fish in many parts of the eastern United States, and there is a sportfishery for them in some stocked populations on the West Coast of the United States[2].

Description

American Shad are a rather elongate species, being highly compressed laterally with a moderate snout. Overall, American Shad are silvery with a blue or blue-green shiny luster on the back. Scales are large, and the fish can generally be distinguished from other species by the sawlike serrated edge along the midline of the belly. A longitudinal row of black spots appears at the shoulder under the scales that is apparent when the deciduous scales are lost (as in handling). The caudal fin is distinctly forked, and the lateral line is not developed; there are about 50 to 55 midlateral scale rows[2,5].

Range and Distributional Status

This species is not native to Idaho. American Shad are anadromous on the Atlantic Coast of North America from Newfoundland, Canada, to Florida. They were widely introduced into the Sacramento River and Columbia River on the Pacific Coast as early as 1871. Populations are now established from southern California to Cook Inlet, Alaska, and onto the Kamchatka Peninsula. They were reportedly stocked in the Bear River and Bear Lake, Utah-Idaho, during the 1890s and have become established in the Snake River below Lewiston, Idaho[2,5,6].

Size and Longevity

Average length of adult American Shad is reported to be 381 mm (15.0 in), and they may grow to lengths of 101 to 127 mm (4.0 to 5.0 in) in freshwater before migrating to the sea. Mature males may reach a length of 762 mm (30.0 in) and weights of 6.8 kg (15.0 lb). Female American Shad from the Columbia River range in length from 432 to 559 mm (17.0 to 22.0 in) and weights of 1.6 to 2.3 kg (3.5 to 5.1 lb). Males from this fishery are 406 to 495 mm (16.0 to 19.5 in) and weigh 1.1 to 1.8 kg (2.4 to 4.0 lb). Larvae are about 9 to 10 mm (0.35 to 0.39 in) long when hatched and reach sizes of 51 to 76 mm (2.0 to 3.0 in) when they enter the sea. Two-year-old fish average about 230 mm (9.1 in) and four-year-old individuals about 430 mm (17.0 in)[2,5].

Food and Feeding

American Shad feed on plankton. Both larvae and young American Shad eat copepods and related crustaceans and insect larvae (largely chironomids) while resident in freshwater. Copepods and mysids are the principal food when the species is in the sea. Migrating individuals eat little but resume feeding during the postspawning downstream migration[2].

Limiting Factors

Seals may be the principal predator on American Shad, but otherwise adults seem to have few predators. Young shad may fall prey to a variety of predators in freshwater. American Shad do not carry heavy parasite infestations[2,5].

Reservoirs on the Columbia River drainage have allowed increases in American Shad and access to river reaches previously inaccessible due to natural barriers. Within the native range of American Shad, dams have restricted access to some historical spawning grounds. In areas where commercial fishing is a factor, there is little evidence that the species can be overfished under current management protocols[2,5].

Breeding Habits

American Shad males in Washington first mature at age three and females at age four. In the Columbia River, optimal spawning temperatures of 15.6 to 18.3°C (60 to 65°F) occur in June and August, and peak spawning varies with location[5]. Females may produce 20,000 to 150,000 eggs. Fertilized eggs are 2.5 to 3.5 mm (0.1 to 0.14 in) in diameter, transparent, pale pink, or amber and nonadhesive. Hatching takes place in eight to 12 days at temperatures around 10°C (50°F)[2,5].

American Shad are a wide-ranging anadromous fish that exhibit variation in demographic histories and reproductive strategies and provide a unique perspective on the evolutionary processes that govern the genetic architecture of anadromous fishes. Examining the magnitude and spatial distribution of genetic variation among 33 populations across species revealed that patterns of genetic diversity and differentiation among populations varied widely and reflect the differential influences of historical demography, microevolutionary processes, and anthropogenic factors across the species' range. Sequential reductions of diversity with latitude among formerly glaciated rivers are consistent with stepwise postglacial colonization and successive population founder events. Weak differentiation among some US populations may be a consequence of human-mediated gene flow, while weak differentiation among other populations probably reflects natural gene flow. Evidence for an effect of reproductive strategy on population structure suggests an important role for environmental variation and that the factors responsible for shaping American Shad life history patterns may also influence population genetic structure[7].

Habits and Habitat

In a fashion similar to Steelhead and salmon, American Shad are anadromous, spending the first summer after hatching in the natal stream and then migrating to the sea. On the West Coast, American Shad mature after three or four years at sea and return to the natal stream to spawn[2,5].

Management and Conservation Status

As a nonnative species, American Shad may present as yet unforeseen impacts upon native species and ecosystems. In the Columbia River Basin below the confluence with the Snake River, both commercial and sportfishing efforts exist, in some instances providing significant recreational benefits. American Shad were evaluated as a potential forage fish for Smallmouth Bass in Brownlee Reservoir but were not introduced by the IDFG[8].

Chapter Citations

1. Nelson, 2006. 2. Scott and Crossman, 1973. 3. Bond, 1979. 4. Sigler and Sigler, 1996. 5. Wydoski and Whitney, 2003. 6. Simpson and Wallace, 1978. 7. Hasselman, Ricard and Bentzen, 2013. 8. Dillon and Myers, 1990.

"The most important single ingredient in the formula of success is knowing how to get along with people."

Theodore Roosevelt

9

CYPRINIFORMES

Carps, Minnows, Suckers & Loaches

Order Cypriniformes

Worldwide this order is composed of six families, 321 genera, and more than 3,000 species[1]. Cypriniform fishes are one of the largest and most important orders of fishes in the world, second only in number and faunal dominance to the Order Perciformes. Many of these species are important both ecologically and economically[2]. Worldwide, this order includes great diversity, from the freshwater flying fish of South America to the Electric Eel (*Electrophorus electricus*) that is capable of generating 650 volts in specialized organs. In North America, members of two families—Cyprinidae (carps and minnows) and Catostomidae (suckers)—constitute a significant portion of the fish species in any given ecosystem. Most species of cyprinids have pharyngeal teeth but a lesser number of jaw teeth. All members of the Cypriniformes are physostomous[2].

Together with a number of smaller families, such as the suckers (Catostomidae), the loaches (Cobitidae), and the river loaches (Balitoridae), minnows (Cyprinidae) make up the order Cypriniformes. In this order and two large related orders, Characiformes and Siluriformes, the first four or five vertebrae and associated ligaments are modified to provide a direct link between the swim bladder and the inner ear. This structure, referred to as the Weberian apparatus, apparently enhances sound detection in water[3].

In North America, earliest records of minnows are from the mid-Oligocene Epoch (28.5 mya) from the Pacific Northwest. There are almost 300 indigenous species of minnows in North America. Introduced species occur widely[3].

In Idaho, 28 species in 20 genera are reported from three families (Cyprinidae, Catostomidae, and Cobitidae), with nine non-native species: Goldfish, Grass Carp, Common Carp, Golden Shiner, Fathead Minnow, Spottail Shiner, Tui Chub, Tench, and Oriental Weatherfish.

Family Cyprinidae—Carps and Minnows

Cyprinidae is the largest family of freshwater fishes and the largest family in the Order Cypriniformes[1]. More than 2,000 species in 220 genera of this family are found worldwide, and many inhabit freshwaters of North America, Eurasia, and Africa. Fins of North American forms are all soft-rayed; introduced forms such as Common Carp and Goldfish have spines. In North America, there are 235 species, mostly occurring east of the Continental Divide, and all are found in freshwater. Family members occupy a highly diverse set of ecological niches and vary in size from less than 28 g (1 oz) to approximately 136 kg (300 lb). Across their native ranges, they are important economically and ecologically as well as for food (wild and cultivated), forage, pets, and research. Cyprinids are widely distributed throughout the geologic record and are known from the Paleocene of Europe, Eocene of Asia, and Miocene of North America[2,4]. In Idaho, there are 19 species reported in 16 currently recognized genera:

Acrocheilus alutaceus	Agassiz & Pickering, 1855	Chiselmouth
Carassius auratus	(Linnaeus, 1758)	Goldfish
Couesius plumbeus	(Agassiz, 1850)	Lake Chub
Ctenopharyngodon idella	(Valenciennes, 1844)	Grass Carp
Cyprinus carpio	Linnaeus, 1758	Common Carp
Gila atraria	(Girard, 1856)	Utah Chub
Lepidomeda copei	(Jordan & Gilbert, 1881)	Northern Leatherside Chub
Mylocheilus caurinus	(Richardson, 1836)	Peamouth
Notemigonus crysoleucas	(Mitchill, 1814)	Golden Shiner
Notropis hudsonius	(Clinton, 1824)	Spottail Shiner
Pimephales promelas	Rafinesque, 1820	Fathead Minnow
Ptychocheilus oregonensis	(Richardson, 1836)	Northern Pikeminnow
Rhinichthys cataractae	(Valenciennes, 1842)	Longnose Dace
Rhinichthys falcatus	(Eigenmann & Eigenmann, 1893)	Leopard Dace
Rhinichthys osculus	(Girard, 1856)	Speckled Dace
Rhinichthys umatilla	(Gilbert & Evermann, 1894)	Umatilla Dace
Richardsonius balteatus	(Richardson, 1836)	Redside Shiner
Siphateles bicolor	(Girard, 1856)	Tui Chub
Tinca tinca	(Linnaeus, 1758)	Tench

"To those devoid of imagination a blank place on the map is a useless waste; to others, the most valuable part."

Aldo Leopold, A Sand County Almanac.

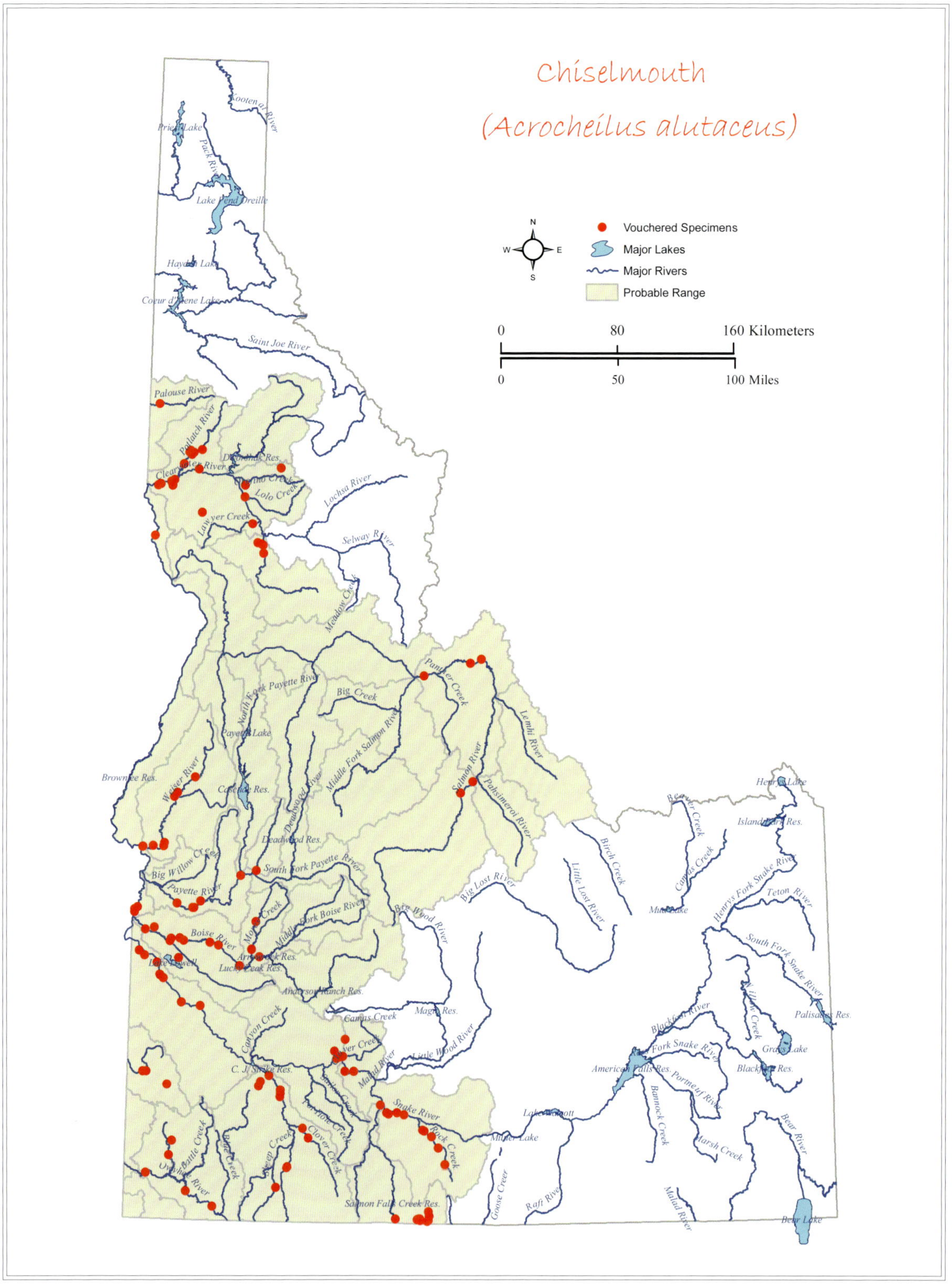
Chiselmouth
(Acrocheilus alutaceus)
N
W
E
S
Vouchered Specimens
Major Lakes
Major Rivers
Probable Range
0
80
160 Kilometers
0
50
100 Miles
Palouse River
Lochsa River
Selway River
Lolo Creek
Lawyer Creek
Big Creek
Payette Lake
Weiser River
Deadwood Res.
South Fork Payette River
Big Willow Creek
Payette River
Boise River
Lucky Peak Res.
Anderson Ranch Res.
Canyon Creek
C. J. Strike Res.
Snake River
Rock Creek
Salmon Falls Creek Res.
Lemhi River
Pahsimeroi River
Salmon River
Birch Creek
Little Lost River
Big Lost River
Big Wood River
Little Wood River
Camas Creek
Magic Res.
Malad River
Lake Walcott
Raft River
Goose Creek
Henrys Fork Snake River
Teton River
South Fork Snake River
Palisades Res.
Blackfoot River
American Falls Res.
Portneuf River
Bannock Creek
Marsh Creek
Bear River
Bear Lake
Henrys Lake
Island Park Res.
Grays Lake
Blackfoot Res.
Middle Fork Salmon River
Meadow Creek
Saint Joe River
Lake Pend Oreille
Pack River
Priest Lake
Brownlee Res.
Middle Fork Boise River
Arrowrock Res.
Mud Lake
Camas Creek
Beaver Creek

Chiselmouth

Acrocheilus alutaceus

Agassiz & Pickering, 1855

Etymology

Acrocheilus = sharp; *alutaceus* = leathery.

Importance

Taxonomically, Chiselmouth are of interest to naturalists as they are the only living species in this genus. Chiselmouth provide a vital link between primary plant producers and piscivorous fishes. A unique upper- and lower-jaw adaptation allows them to feed by scraping algae from stones, which is the source of the common and generic scientific names[2,4,5]. They may have been a food item in aboriginal diets within their native range[2,4,6].

Description

Chiselmouth bodies are elongate and slightly compressed, with a blunt head and rounded snout. Coloration is rather drab, with lighter sides and dark brown anteriorly. Young may have a poorly defined dark spot at the base of the caudal fin, and adults may have several to many small black dots that are lighter on the lower sides. Both pectoral and pelvic fins have an orange coloration in the axils, and the caudal fin is distinctly forked. The caudal peduncle is very narrow. A single dorsal fin has 10 rays, and the anal fin has nine or 10 rays. Scales are cycloid and small and number from 85 to 93 in the lateral series; the lateral line is complete. Chiselmouth have a bluntly rounded snout, and the lower jaw has a distinctive sharp, hard plate[2,4,6].

Although sexual differences in size are not obvious in Chiselmouth, adult females are generally slightly larger than adult males. The most apparent external difference between the sexes in Chiselmouth is pectoral fin length. In adult males, pectoral fins are longer than the head and extend back almost to the origin of the pelvic fins. In adult females, on the other hand, pectorals are shorter than the head and stop well in front of the pelvic fins. Both sexes develop tubercles during breeding on the pectoral and pelvic fins, the head, the operculum, and the back. Tubercles are less obvious in females than in males[3].

Range and Distributional Status

This species is native to Idaho. Chiselmouth are strictly a Pacific drainage species, occurring in the Fraser and Columbia River drainage in Nevada, Idaho, Oregon, Washington, and British Columbia. They occur in Malheur Lake, Oregon[2,4,7]. In Idaho this species occurs in the Snake River and tributaries below Shoshone Falls[5]. In Washington, Chiselmouth are reported in the Pend Oreille and Spokane River drainages[6].

Size and Longevity

Newly hatched larvae are about 8 mm (0.3 in) long, and adult Chiselmouth may reach total lengths of 305 mm (12 in) and live six years[8]. Age and growth data for this species are sketchy and have been difficult to obtain[9].

Food and Feeding

Both feeding behavior and the feeding mechanism in Chiselmouth are unique. Chiselmouth feed by scraping their chisel-like lower jaw along rocks or other hard substrate in quick, short movements of about 25.4 mm (1 in), ingesting the algae and associated zooplankton. Both the cartilaginous sheathing of the mouth and the long, convoluted intestine support this feeding-behavior adaptation[4,5,8].

Limiting Factors

Chiselmouth are preyed upon by piscivorous fishes and are affected by parasites. Hybridization may occur with Northern Pikeminnow[5,8].

Breeding Habits

Male Chiselmouth attain sexual maturity at age two or three years, whereas females become sexually mature at age three or four years. Chiselmouth are stream and reservoir spawners, and spawning occurs in late June through early July when waters warm to about 17°C (62°F). Eggs have been found on the bottom and buried among boulders. Mean egg count (fecundity) is 6,000[4,5,8].

Habits and Habitat

Chiselmouth are found primarily in warmer reaches of rivers and streams but also in lakes. Little is known about this species in Idaho as of 2016, although it occurs in abundance in some locations[4,5,8].

In the Blackwater drainage of British Columbia, Chiselmouth were typically not detected from sites with maximum temperatures below 20°C (68°F) both within a single drainage and between larger drainages. Indices of stream size were the most significant predictors of Chiselmouth presence. In contrast, bankfull channel width and basin area were poor predictors of Chiselmouth presence in main-stem habitat within larger drainage basins throughout British Columbia. Inclusion of thermal variables, particularly degree days greater than 12°C (54°F), doubled correct classification rates of Chiselmouth presence across larger drainage basins. These habitat associations suggest that water temperature is the primary constraint on the presence of Chiselmouth populations in larger drainages, while selection of different habitat types (main-stem habitat over smaller tributaries) determines distribution within any given basin[10].

Management and Conservation Status

Chiselmouth are considered common, widespread, and abundant range-wide (G5) and within Idaho (S5)[11]. They are relatively widespread across western Idaho aquatic ecosystems and presently have little or no economic importance. Preservation of habitat and protection from nonnative introductions will likely maintain their current distribution in Idaho.

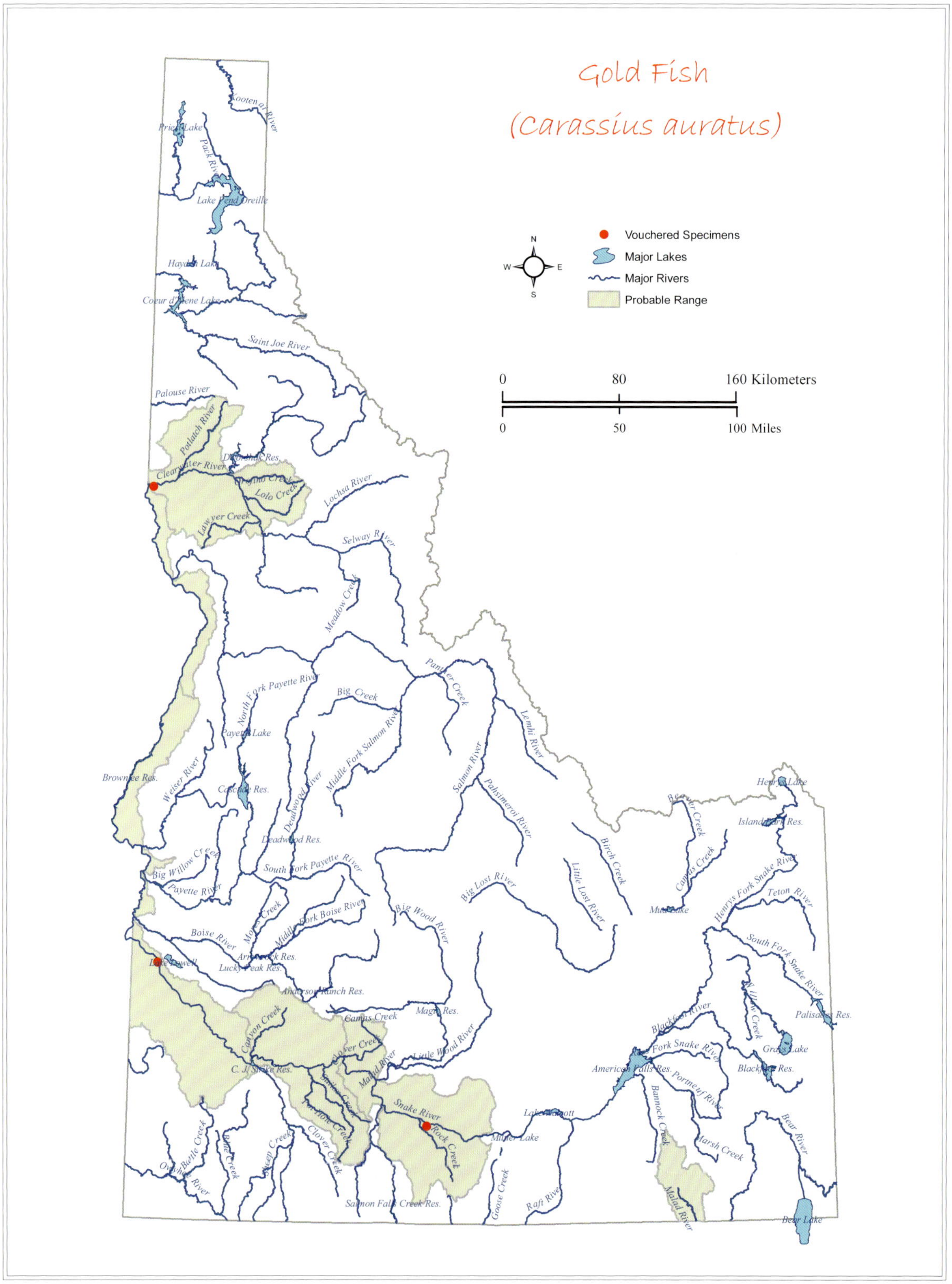
Gold Fish
(Carassius auratus)
Vouchered Specimens
Major Lakes
Major Rivers
Probable Range
0 80 160 Kilometers
0 50 100 Miles

Goldfish

Carassius auratus

(Linnaeus, 1758)

Etymology

Carassius = vernacular name for Crucian carp, C. *carassius; auratus* = gilded, referring to the distinctive color of this species.

Importance

For centuries, Goldfish have been cultured for aquarium and ornamental pools. While in Idaho Goldfish are highly regarded as both recreational aquarium and research fish, they are, as are most nonnatives, generally considered a negative factor in natural ecosystems. Although the premise is unsubstantiated by scientific data, in some ecosystems they may compete with more desirable species while providing very limited recreational fishing benefit or contributions to ecosystem sustainability. Literature on this species is extensive[2,4].

Description

Reflective of the common name, Goldfish vary in color from gold in captivity to olive green to drab green in the wild, often resembling Common Carp. Goldfish have a moderately compressed but robust body and typically have large scales. They are without jaw teeth and have a short head and oblique mouth. Goldfish have no barbels around the mouth, which helps distinguish them from Common Carp. The dorsal fin has 18 to 20 rays[2,4], and the anterior spine is serrated.

Range and Distributional Status

This species is not native to Idaho. Goldfish have been introduced into a variety of aquatic ecosystems worldwide. Native to China and eastern Asia, Goldfish were commercially raised in Maryland as early as 1889 and have been randomly introduced throughout the United States[2,4]. Goldfish have been collected from the Snake River and tributaries below Shoshone Falls.

Size and Longevity

In the wild, Goldfish typically remain small to medium size, averaging 152 to 254 mm (6.0 to 10.0 in), although individuals may reach lengths of 305 to 365 mm (12.0 to 14.4 in). Cultured individuals may even grow to larger adult sizes. Goldfish do not live past about six years of age[2,4].

Growth rates in most populations of Goldfish are variable, and stunting is common in feral populations. Typically, males mature in their second year, while females mature a year later. Individuals in wild populations probably live only four to six years. However, in aquaria, they may survive for up to 30 years[3].

Food and Feeding

Goldfish are omnivorous, surviving in most habitats on a variety of larvae and adult aquatic invertebrates and algae[2,4].

Limiting Factors

Smaller Goldfish are subject to predation, although larger individuals generally are not. They are generally infested with a variety of parasites. In the wild, they may be outcompeted by other species[2,4].

Breeding Habits

Goldfish, like Common Carp, are spring spawners at age one to three years in the wild. Spawning may last several weeks. In most ecosystems, they seek warm, weedy, shallow water in May or June when temperatures reach 17 to 20°C (63 to 68°F). Small adhesive eggs (1.2 to 1.5 mm [0.05 to 0.06 in]) in diameter are broadcast and hatch in about three days, depending on ambient temperature[2,4].

Generally, populations of Goldfish in the United States contain both males and females. However, some European populations contain only females. These females are sexual parasites on other cyprinids, that is, they reproduce by soliciting fertilization from males of other species. The sperm activates the eggs, but normal fusion of male and female nuclei never occurs. Consequently, only the female genome participates in further development. This is referred to as gynogenesis[3].

Habits and Habitat

Goldfish have successfully established wild populations in small bodies of water or low-gradient larger rivers with abundant aquatic plants[2,4].

Management and Conservation Status

Goldfish are considered common, widespread, and abundant range-wide (G5). A state conservation status rank is not applicable (SNA) because the species is not a suitable target for conservation activities[11]. In Idaho, Goldfish occur only as a direct result of illegal releases. Additional introduction or release of Goldfish into any natural Idaho ecosystem is strongly discouraged. Goldfish are considered an undesirable species in all Idaho aquatic ecosystems.

Cauldron Linn, narrowest part of the Snake River, near Murtaugh, ID. Taken in spring 2017 at approximately 18,000 cfs. Photo Courtesy John Sigler.

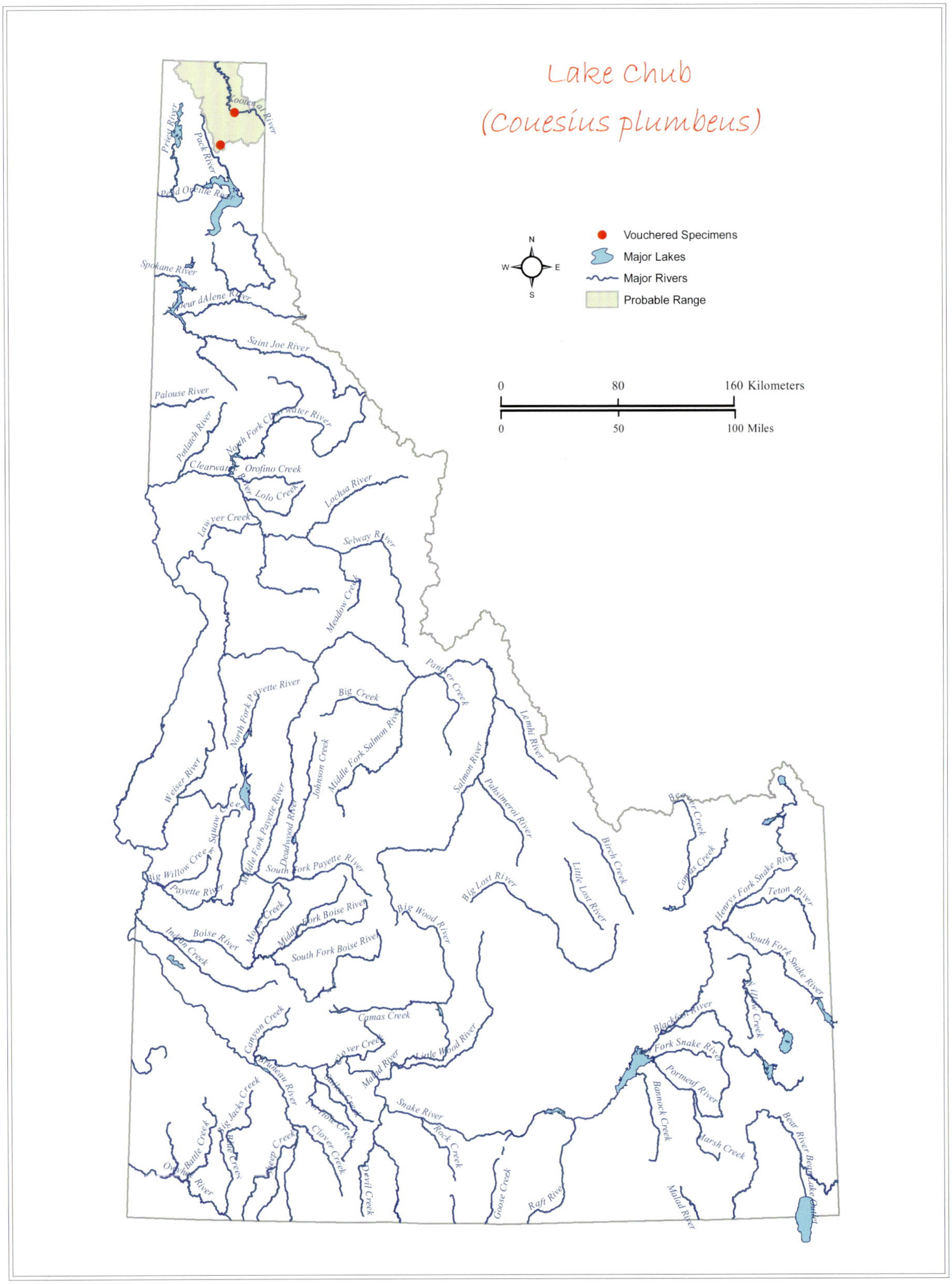
Lake Chub
(Couesius plumbeus)
Vouchered Specimens
Major Lakes
Major Rivers
Probable Range
0 80 160 Kilometers
0 50 100 Miles

Lake Chub

Couesius plumbeus
(Agassiz, 1850)

Etymology

Couesius = for Dr. Elliot Coues, collector of the holotype for this genus; *plumbeus* = lead or perhaps leaden, referring to the coloration of the sides of this species.

Importance

In some areas, Lake Chub are an important forage and bait fish. Although they are considered a nuisance species by some anglers, their sparse distribution requires close monitoring[6].

Description

Adult Lake Chub may range between 100 and 200 mm (3.9 to 7.9 in). They have an anal fin with eight rays, and the origin of the anal fin is behind the end of the dorsal fin base. A silver peritoneum is present and may be speckled with small dark spots. Pigmentation on the sides of Lake Chub includes a single dark band along the sides of most fish up to 51 mm (2.0 in) in length; the band is confined to the posterior half of the body in larger individuals. Scattered and small cycloid scales may be present on the body, and there are 58 to 70 scales in the lateral series[5,6,12].

Adult female Lake Chub are typically up to 30 percent larger than adult males. Spawning males and females both develop small tubercles on the head; however, tubercles are better developed in males than in females. The distinctive patches of rough scales on the breast of males are absent in females[3].

Range and Distributional Status

This species is native to Idaho. Lake Chub are distributed from New England west to Washington, north to Alaska, and east to Quebec. In the northern United States, Lake Chub are found irregularly as far south as New York, Iowa, South Dakota, Colorado, Wyoming, Montana, and Washington[5,6]. In Idaho they are restricted to the Kootenai River and its tributaries[5].

Morphologically, three subspecies of Lake Chub exist: *C. p. greeni* populations are in the upper Columbia River and Fraser River and adjacent Pacific slope rivers; *C. p. dissimilis* occurs east of the Continental Divide, the Great Plains in Canada, and the United States to southwestern portions of Lake Superior and Lake Michigan; and *C. p. plumbeus* occur in northeastern North American Atlantic slope rivers[13]. No subspecies designation affects this species in Idaho.

Size and Longevity

In Idaho, the average length of Lake Chub aged one to three, in mm (in), is 63.5 (2.5), 114 (4.5), and 140 (5.5). Lake Chub from most populations reach total lengths up to 102 mm (4.0 in), although the largest specimen, from Lake Mattagami, Quebec, was 226 mm (8.9 in). Females grow faster and live longer than males[6].

Food and Feeding

Previous investigations have concluded that Lake Chub are primarily sight feeders based on large optic nerves in the brain and minimal numbers of taste buds. Smaller fish feed on zooplankton such as cladocerans and algae, while larger individuals feed on larger zooplankton and terrestrial and aquatic insects[6]. Lake Chub are classified as invertivores, but larger Lake Chub may also consume small fishes[5].

Lake Chub typically have a diet consisting primarily of benthic organisms (amphipods, chironomid larvae, oligochaetes, and some plant material), but also take terrestrial insects and occasionally the fry of other fishes. In a New Brunswick, Canada, study, Lake Chub were most actively feeding at dawn and dusk. However, they may forage throughout the day and at a wide range of temperatures in other ecosystems[3].

Limiting Factors

Overfishing of some populations may occur, as this species is easy to catch. Their use as a bait fish may impact some populations locally. In Washington, studies are under way to assess their status relative to Washington's designation of endangered, threatened, or sensitive species[6].

Breeding Habits

Lake Chub are broadcast spawners, using coarse gravel to rubble substrate; they provide no parental care to the eggs. In a Saskatchewan population, an 89 mm (3.5 in) female contained 600 eggs, while a 130 mm (5.1 in) female contained 2,400 eggs. Significant post-spawning mortality has been suggested. Spawning occurs in the spring from April through June when water temperatures range from 14 to 19°C (57 to 66°F). Most Lake Chub are sexually mature at age three or four years[5,6].

Habits and Habitat

Lake Chub can be found in both lakes and streams but appear to prefer lakes. Typically, they are found near the bottom of both shallow and deep habitats. Lake populations may migrate into tributary streams to spawn; adults move into deeper waters in the summer[5,6].

Management and Conservation Status

Lake Chub are considered common, widespread, and abundant range-wide (G5). The conservation status of Lake Chub is unranked (SNR) in Idaho[11]. Little information is available on Idaho populations of Lake Chub, and no definitive studies have been conducted in Idaho to assess their population status. As noted above, individual populations may be impacted by overfishing[5].

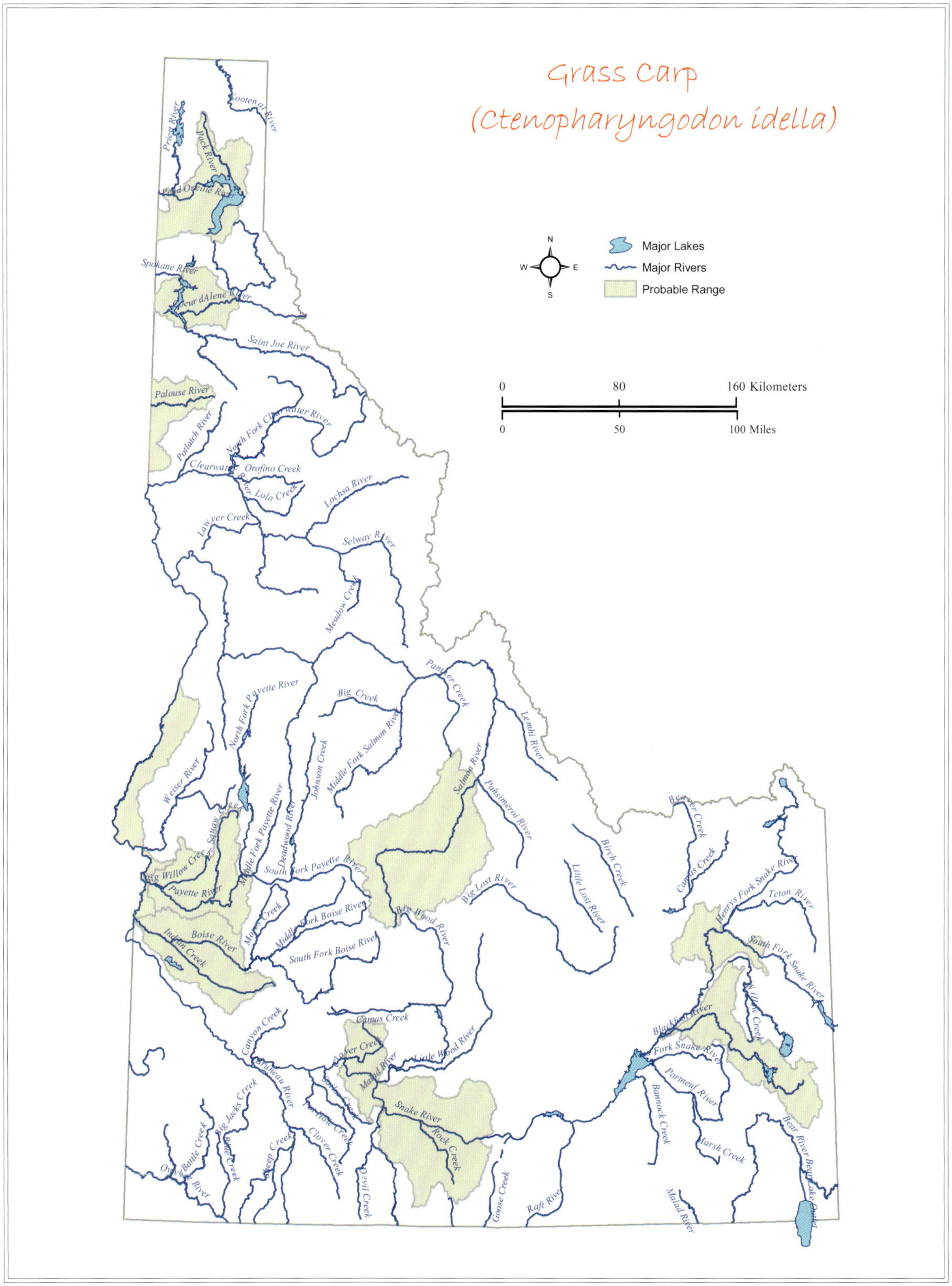
Grass Carp
(Ctenopharyngodon idella)
N
W
E
S
Major Lakes
Major Rivers
Probable Range
0
80
160 Kilometers
0
50
100 Miles

Grass Carp

Ctenopharyngodon idella
(Valenciennes, 1844)

Etymology

Ctenopharyngodon = *cteno* for comb and *pharyngodon* for teeth, referring to the comblike pharyngeal teeth: *idella* = distinctive.

Importance

Originally introduced in the United States (Arkansas) in 1963 for control of aquatic vegetation, Grass Carp are now widespread as a result of both planned and unplanned releases. Escape of diploid (reproductively capable) individuals from source ponds in the Mississippi River basin resulted in widespread distribution of adults capable of reproduction. These fishes are capable of significant impacts in low-gradient lotic and lentic ecosystems, as their food selection is not limited to aquatic vegetation. Most states now restrict releases of this species and require that only certified triploid (sterile, not reproductively capable) individuals be used in aquatic vegetation-control projects. Grass Carp may be more economically used to reduce aquatic vegetation than mechanical or chemical treatments. However, potential adverse impacts on other species, including waterfowl, require careful study and consultation with knowledgeable biologists prior to any introductions. In some locales, Grass Carp are acceptable as a food fish, with taste and texture similar to Common Carp[6].

Description

Grass Carp are olive to brassy green above and silvery white to yellow below. They have seven to eight rays in the dorsal fin but lack the distinct heavy cartilaginous ray that is present in the Common Carp. The anal fin is close to the caudal fin. Pharyngeal teeth, from which the scientific name is derived, are distinctive[6].

Range and Distributional Status

This species is not native to Idaho. Grass Carp are native to the Amur River basin in eastern Asia. Most Grass Carp present in the United States today are of Chinese origin. Additional intentional stocking of potentially reproductively active fish occurred through 1978, prior to the development of triploid (sterile) forms. Distribution in Idaho is not well known, but introductions are believed to be essentially statewide, except for the Kootenai and independent river drainages.

Size and Longevity

Grass Carp commonly live 15 years but may attain ages up to 21 years. Within their native range, this species may reach lengths of 1,219 mm (50.0 in) and weights exceeding 45.4 kg (100 lb). In US ecosystems, Grass Carp can demonstrate rapid growth: one individual reportedly grew from 122 mm (4.8 in) to 655 mm (25.8 in) in one year[6].

Food and Feeding

Adult Grass Carp are mainly herbivorous and graze primarily on aquatic vegetation, including rooted macrophytes and filamentous algae. Younger fish feed on phytoplankton but begin to eat rooted aquatic plants when they are about 25 mm (1.0 in) long. Grass Carp exhibit a ravenous appetite and can consume the equivalent of their own body weight in food in a single day. Feeding decreases when water temperatures are below about 14°C (57°F). However, when their commonly used food items, such as aquatic vegetation, are consumed, Grass Carp will readily switch to invertebrates and small fishes, making them a potential competitor with native animals, including various fishes and waterfowl[6]. This capability makes them a threat to native species.

Limiting Factors

Occasionally, Grass Carp are caught by hook-and-line anglers, mainly if aquatic vegetation has been depleted. Grass Carp are likely susceptible to the same limiting factors as Common Carp: predation on juveniles by other fishes and birds, parasites, and adverse ecological conditions, particularly temperature[6,8].

Breeding Habits

Grass Carp spawn in large rivers under high flows when temperatures are about 14 to 21°C (57 to 69°F). The semi-buoyant eggs float in the current and hatch in 16 to 60 hr at temperatures between about 17 and 30°C (63 and 86°F). Incubation is faster at higher temperatures and slower at lower temperatures. Grass Carp fecundity is high: a 760 mm (30.0 in) female can produce about 800,000 eggs. High fecundity provides the potential to rapidly spread within any ecosystem when adequate food sources are present[6].

Habits and Habitat

In their native range in Asia, Grass Carp are found in large, generally slow, warm rivers. In the United States, they have been introduced into a variety of pond systems from which many reproductively capable individuals have escaped. They now occur in ponds, lakes, streams, and rivers across the country[6]. Grass Carp have been introduced into at least 200 ponds and lakes in neighboring Washington over the past 20 years[6].

Management and Conservation Status

Grass Carp are considered common, widespread, and abundant range-wide (G5). A state conservation status rank is not applicable (SNA) because the species is not a suitable target for conservation activities[11]. Grass Carp have been extensively stocked into various aquatic ecosystems in the United States to control aquatic vegetation. As with all nonnative species, additional introductions of Grass Carp into Idaho waters are strongly discouraged. Special-need introductions for control of aquatic vegetation should be closely coordinated with regional fisheries personnel, and only certified triploid (sterile) individuals should be released. Time and study will afford fisheries managers a more informed perspective on how Grass Carp can used for vegetation control in North American and Idaho waters.

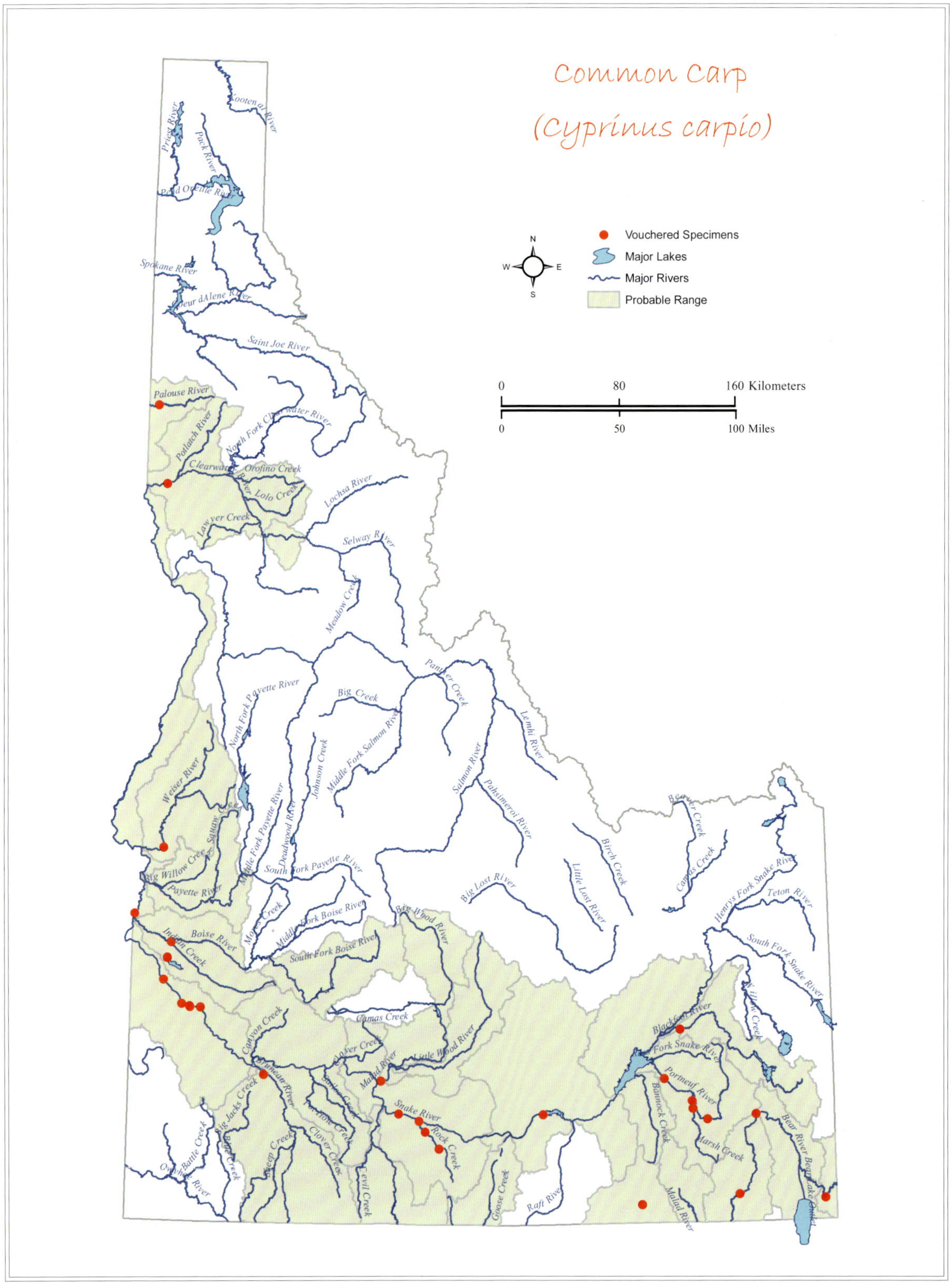
Common Carp
(Cyprinus carpio)
Vouchered Specimens
Major Lakes
Major Rivers
Probable Range
N
S
E
W
0
80
160 Kilometers
0
50
100 Miles

Common Carp

Cyprinus carpio

Linnaeus, 1758

Etymology

Cyprinus = ancient name for carp, which may be derived from Cyprus, abode of the goddess Venus, alluding to their fecundity; *carpio* = Latinization of the name carp.

Importance

Common Carp have been cultivated for human consumption in their native range for thousands of years. While the species was prized as a delicacy in the United States from about 1876 to 1910, few presently consider Common Carp an acceptable table food in western North America. Some larger individuals, for instance from Utah Lake, Utah, are captured and transported to large cities in California as food for religious ceremonies. Some are smoked, which increases their desirability throughout the United States. Since their introduction into North America, Common Carp have been used as food for poultry, furbearers, and trout. In addition to their failings as a food fish in the United States, Common Carp are destructive to habitat in ecosystems where other more desirable species exist and can compete for food with native species, destroying vegetation and increasing turbidity through their feeding and breeding behaviors. On a positive note, young carp can provide limited forage for some nonnative game species such as Northern Pike, Muskellunge, and Largemouth Bass. In various locations, Common Carp provide sportfishing opportunities with either bow and arrow or rod and reel, using a variety of locally dictated bait[2,4,6]. A variety of YouTube presentations, including some from Joe Tomelleri, show various aspects of this sport. An investigation of the tremendous biomass produced by Common Carp populations included efforts to study more efficient uses of this species as a food stuff, either for humans or for domestic animals[14].

Nonnative fish introductions disrupt ecosystem processes and can drive native species to local extinction. Common Carp are one of the most widespread introduced species in the United States. In North and South Americas, these introductions stem from aquaculture facilities as well as historical introductions for food or recreational angling. An emergent field of ecological niche modeling provides robust predictions of the geographic potential of alien species in an effort to better understand their capacity to become established at broad scales. Model results indicated that distributional potential of Common Carp in the Americas covers most temperate regions and high-mountain tropical aquatic systems. Effects on native aquatic faunas can be profound throughout temperate regions[15].

Description

Common Carp are deep-bodied, with a distinctive triangular head. Adults are olive green to gold on the back, becoming yellow to pale yellow below. Paired barbels at the margins of the mouth, one shorter than the other, are distinctive. On the anterior surface of the first ray of the dorsal fin, there is a serrated ray that is not a true spine. A similar spine-like ray occurs on the anal fin. The dorsal fin has 17 to 21 rays. The lower halves of the caudal fin and the anal fin often have a reddish hue that is stronger in large individuals. Pharyngeal teeth are molar-like and used to grind food[2,4,6]. Common Carp have a laterally compressed body that may range from fully scaled to scaleless or nearly so. A variety of local and regional names have been applied to irregularly scaled individuals, such as mirror carp (few large scales) or leather carp (scaleless)[2,4,6]. Koi are Common Carp bred for ornamental purposes.

Range and Distributional Status

This species is not native to Idaho. Common Carp are native to Asia, where they have been cultivated for human consumption for generations. Imported into Germany and other European countries hundreds of years ago, they were imported into the United States as a potential food fish from 1876 to 1910 and were stocked in waters across the North American continent. Introductions in Idaho occurred in the early 1880s. Presently, Common Carp range over most warm- and cool-water ecosystems of the 48 lower states and continue to spread throughout a variety of aquatic ecosystems. In Idaho they are found in a variety of ponds, streams, and lakes in the Snake, Bear, and Palouse drainages. They are widely distributed throughout Canada[2,4,6].

Size and Longevity

Common Carp are relatively long lived and fast growing, factors that contribute to the species' ability to rapidly and adversely affect ecosystems. Growth rates depend on food availability and water temperatures. Lengths, in mm (in), for Common Carp taken from the Ogden Bay Bird Refuge near Ogden, Utah, for individuals ages one to eight was 173 (6.8), 363 (14.3), 518 (20.4), 635 (25.0), 691 (27.2), 717 (28.2), 724 (28.5), and 722 (28.4). Common Carp in nearby Bear River Migratory Bird Refuge, at the mouth of the Bear River, near Brigham, Utah, grow slightly slower. Common Carp in Bear Lake, Idaho-Utah, grow to lengths of 678 mm (26.7 in), but it takes them 13 years. The world-record Common Carp was taken near Pretoria, South Africa, and weighed 37.4 kg (82 lb, 7 oz), while the reported Idaho record was caught in 2011 and weighed 30.7 kg (67 lb, 10.4 oz)[2,4,6].

Food and Feeding

Common Carp are voracious feeders in any ecosystem and thrive on a wide variety of food types. They are characteristically omnivorous and opportunistic, but primarily feed on aquatic invertebrates, mostly insects but also algae. They may eat their young under some circumstances. Insects, primarily midge larvae and pupae, make up as much as 87 percent of the diet in the Bear River Migratory Bird Refuge population. Carp have been observed feeding throughout the water column but typically feed on the bottom, where they uproot aquatic vegetation and disturb sediments, contributing to ecosystem disruption and habitat destruction[2,6].

Common Carp have a well-deserved reputation as bottom feeders, and when rooting about in the substrate of most ecosystems, they ingest a mixture of organic and inorganic particles. Adult Common Carp use a filtration system to concentrate small particles in the pharynx and expel inorganic particles out through the gill slits. Large food particles are retained by special palatal protrusions, and large inorganic particles are spit out through the mouth. Diet changes with age: young-of-the-year Common Carp feed primarily on plankton and algae, whereas juveniles gradually shift to the adult diet[3].

Limiting Factors

Common Carp can be fed upon by a variety of fishes and birds. When spawning, Common Carp lose their customary wariness and can be easily taken with a variety of bow-and-arrow techniques. They are also vulnerable to seining during spawning activity. Parasites may be a factor in some populations[2,4,6]. Fly-fishing for large Common Carp is becoming locally popular in many regions.

Breeding Habits

Common Carp are spring spawners and begin to exhibit spawning behavior when water temperatures reach 10 to 19°C (50 to 67°F), typically May to late August in any given ecosystem. When spawning, Common Carp move into shallow, warmer areas with weedy vegetation and spawn day or night. Spawning may occur in water so shallow that the backs of the adults are exposed, making them vulnerable to predators and humans. Common Carp build no nest and provide no care for embryos or young. Females broadcast 500 to 600 adhesive eggs at one time over vegetation, rocks, or debris and ultimately deposit up to 100,000 eggs. Spawning groups typically consist of one to three females with two to three males, but up to 15 males may be involved. Common Carp spawn for the first time at age two to four years and are repeat spawners. Embryos hatch in three to six days after fertilization, depending on water temperatures[2,4,6].

Habits and Habitat

As is evidenced by their widespread distribution across North America, Common Carp can adapt to a variety of habitats, from large cool rivers and lakes to small farm ponds and other shallow waters. Unfettered populations may achieve biomass of 450 to 900 kg (1,000 to 2,000 lb) of fish per 0.4 ha (1.0 ac) of surface in farm ponds[2,4,6].

Recent Canadian research examined the effects of adult Common Carp on shallow aquatic ecosystems and compared them to the effects of Channel Catfish. Common Carp were positively related to total phosphorus, turbidity, suspended solids, and zooplankton biomass and negatively to macrophyte and macroinvertebrate abundance. Suspended solids in the carp treatments consisted primarily of inorganic particles. Common Carp were either positively or negatively related to phytoplankton, depending on zooplankton abundance. Higher biomasses of Common Carp had greater effect on nutrients, turbidity, and suspended solids. Channel Catfish were positively related to total phosphorus concentrations and altered zooplankton composition but did not affect turbidity, suspended solids, macroinvertebrates, and macrophytes. Results suggest that Common Carp have a stronger influence on water quality and aquatic community structure than Channel Catfish[16].

Management and Conservation Status

Common Carp are considered common, widespread, and abundant range-wide (G5). A state conservation status rank is not applicable (SNA) because the species is not a suitable target for conservation activities. In Idaho there is no current need to preserve or protect Common Carp populations[11]. Efforts to reduce or eliminate Common Carp from various ecosystems have met with only limited success and are generally expensive and time consuming. In Utah Lake, Utah, extensive studies estimated the standing Common Carp population at 171 million individuals aged two or older. Mechanical removal (seining operations) may be able to successfully manage this population.

It might be argued by both professional fisheries biologists and anglers alike that Common Carp are the most significant nongame fish in North America. Common Carp were introduced into most western states in the late 1800s and early 1900s by the US Bureau of Fisheries. These introductions were made to supply food fish for expanding western populations, and most if not all landowners wanted Common Carp planted in their farm ponds. Current thinking on the presence of Common Carp in almost any water body in the western United States is highly negative.

"We have an unknown distance yet to run, an unknown river to explore. What falls there are, we know not; what rocks beset the channel, we know not; what walls ride over the river, we know not. Ah, well! we may conjecture many things."

John Wesley Powell

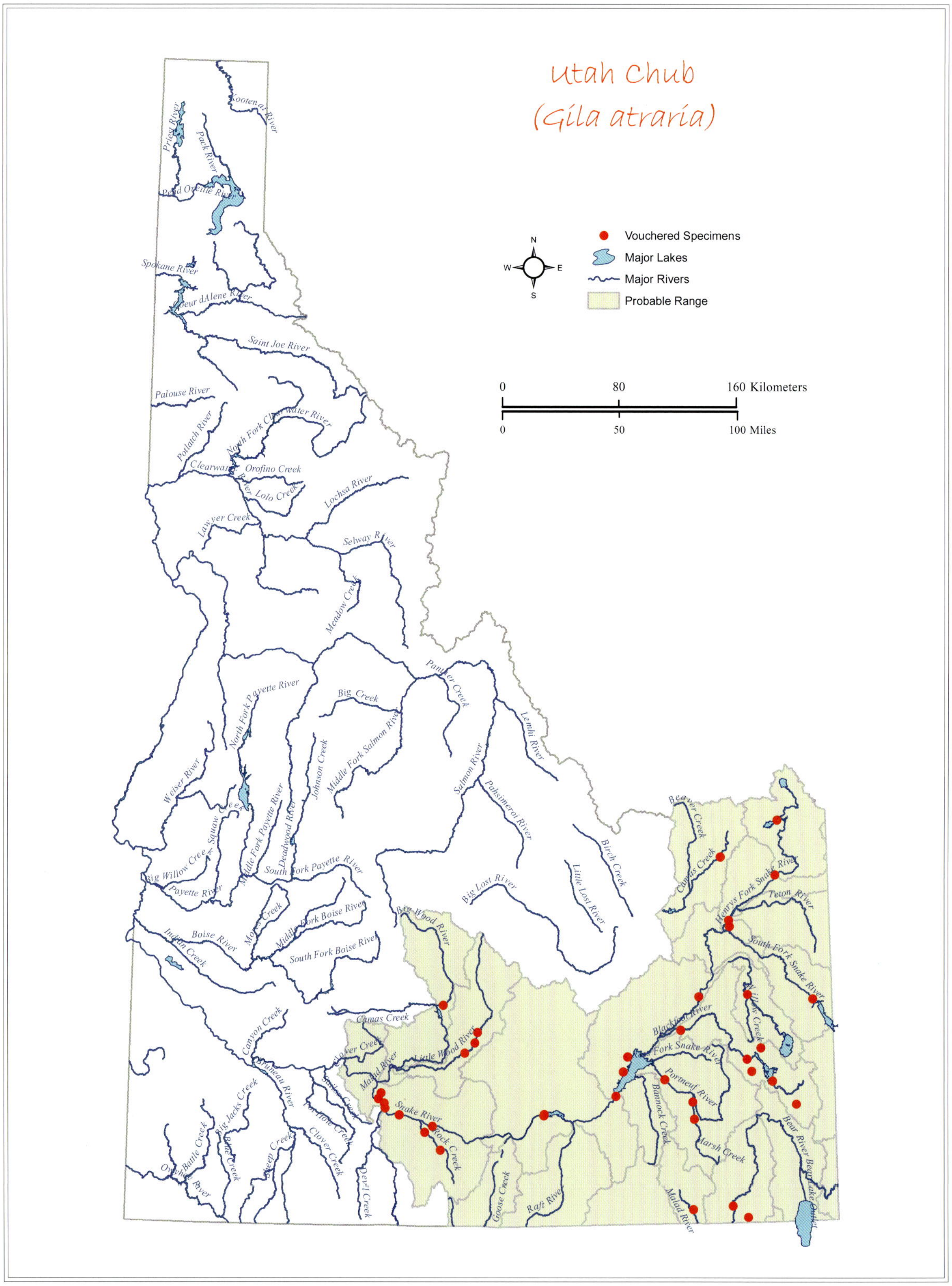

Utah Chub
(Gila atraria)
Vouchered Specimens
Major Lakes
Major Rivers
Probable Range
N
S
E
W
0
80
160 Kilometers
0
50
100 Miles
Pack River
Saint Joe River
Palouse River
Potlatch River
Orofino Creek
Lochsa River
Meadow Creek
Big Creek
Middle Fork Salmon River
Johnson Creek
Weiser River
Payette River
Boise River
South Fork Boise River
Lemhi River
Salmon River
Pahsimeroi River
Big Lost River
Little Lost River
Birch Creek
Camas Creek
Canyon Creek
Snake River
Rock Creek
Goose Creek
Raft River
Bannock Creek
Marsh Creek
Malad River
Teton River
South Fork Snake River
Blackfoot River
Portneuf River
Clover Creek
Devil Creek
Big Jacks Creek

Utah Chub

Gila atraria

(Girard, 1856)

Etymology

Gila = named for the Gila River, Arizona (probably a misnomer); *atraria* = from *ater* and *arius*, one who is black, perhaps in reference to the coloration of this species in the waters of the type locality.

Importance

In some ecosystems, Utah Chub are a serious competitor of game fishes, particularly in ecosystems where they are not native or in areas where ecosystem modification has been significant. Competitive factors include their great abundance in disturbed ecosystems and their general feeding habits. These effects are particularly true for trout populations in disturbed ecosystems[2].

Utah Chub are taken easily on hook and line and are therefore considered a nuisance for trout anglers. They can provide important forage for game fishes in some ecosystems and may contribute to record size trouts. Their abundance, distribution, and ease of capture make them an attractive species for younger anglers[2].

Description

Utah Chub exhibit highly variable coloration that changes with both age and habitat type. In general, the back is dark olive brown or sometimes bluish; the sides are silver, brassy, or golden. Coloration of the belly is whitish or silvery in most individuals but black in some. They may have a narrow golden stripe along the upper side, brighter in males than in females. Fins may be dull olive, yellow, or golden, and the dorsal fin may have a blue sheen. Breeding males may have a trace of orange or yellow in the axil of the pectoral fins, around the mouth, and along the preopercle, being more highly colored than females. Utah Chub have small and oblique mouths[2,5].

Utah Chub have rather robust bodies. The dorsal fin originates directly over the origin of the pelvic fins and has nine rays. The anal fin has eight rays. Lateral series scales number 45 to 65. Gill rakers vary from eight to 16 and are most numerous in populations that occupy lake ecosystems[2,5].

Range and Distributional Status

This species is native to Idaho. Ancient Lake Bonneville's historic drainage area defines most of the native range of Utah Chub in Utah and the portions of Idaho, Wyoming, and Nevada covered by that system[2,5].

In Idaho Utah Chub are found in the Bear River drainage and the Snake River upstream of Shoshone Falls. They are present in the lower Wood River system which was connected to the upper Snake River by canals in the early 1900s. On the Henrys Fork of the Snake River, they are restricted to the area downstream of Mesa Falls and recently have been transplanted to the Island Park Reservoir and other small reservoirs in the area by anglers. Reports of the species' presence in the Snake River near Boise have been noted but unconfirmed. Utah Chub have been inadvertently introduced into various waters in Montana, Wyoming, Colorado, and Utah[5,17].

Extensive hydrological changes in the Bonneville Basin and upper Snake River in Idaho during the late Pleistocene (0.8 mya) potentially influenced geographic distribution and evolutionary processes of numerous aquatic species, including Utah Chub. Examination of 16 populations of Utah Chub that span the natural distribution of this species across the Bonneville Basin and upper Snake River was completed to assess the effect of hydrologic changes on the distribution and evolutionary processes of this species. Two distinctive groups of Utah Chub that represent an early Pleistocene (2.4 mya) split between the upper Snake River and Bonneville Basin were noted. Within these two groups, geographic structuring was consistent with the hydrological history of late Pleistocene (0.011 mya) Lake Bonneville and the upper Snake River, suggesting that glacial-induced shifts in climate and unpredictable geological events have played a major role in shaping genetic subdivision among populations of Utah Chub. These findings suggest that historical events responsible for population fragmentation may have also contributed to expressed shifts in life histories, both indirectly by limiting gene flow among populations and directly by altering the selective environments where populations persisted[18].

Size and Longevity

Bear Lake has produced an 11-year-old Utah Chub that was 590 mm (23.2 in) long and weighed 1.36 kg (3.0 lb). Elsewhere in historic Lake Bonneville waters, Utah Chub rarely reach more than 406 mm (16.0 in) and 0.9 kg (2.0 lb)[2]. More common sizes are 127 to 203 mm (5.0 to 8.0 in)[2]. Utah Chub aged one to seven in Panguitch Lake, Utah, achieved lengths, in mm (in), of 33 (1.3), 58 (2.3), 91 (3.6), 125 (4.9), 145 (5.7), 160 (6.3), and 173 (6.8)[2,19]. Normal longevity is five to eight years or less[2,5]. The reported Idaho record for Utah Chub is a 1.3 kg (2 lb, 13 oz), 424.2 mm (16.7 in) individual taken from Lake Walcott in 2010[37].

Food and Feeding

Larger Utah Chub are omnivorous, feeding on aquatic vegetation, insects, snails, and crustaceans, while juvenile Utah Chub feed on zooplankton until they reach lengths of 150 to 175 mm (5.9 to 6.9 in) and then become omnivorous. Occasionally, larger ones will feed on small fishes and fish eggs[5].

Observations on Utah Chub in Fish Lake, Utah, documented that they eat generally the same food as small trouts[20]. As Utah Chub increase in size, they are able to feed on a greater variety of foods. In desert springs, they have been observed eating green algae and feeding on their own eggs[2].

Limiting Factors

Utah Chub are preyed upon by other fishes and some birds and mammals, but persist under adverse ecological conditions. Fluctuating water levels during the spawning season may trap (desiccate) and kill eggs and young[2].

Breeding Habits

Utah Chub generally spawn in water less than 0.6 m (2.0 ft) deep at temperatures of 15 to 30°C (59 to 86°F) in the spring and later summer[2,5]. Water temperatures during spawning in Scofield Reservoir, Utah, varied from 11 to 17°C (52 to 62°F)[2]. During spawning a female is attended by two to six males; females produce 5,000 to 25,000 eggs; the number is dependent on body size. Female Utah Chub mature at age three to four, while males mature at age two to three with some ecosystem variability[2,5]. Yellowish eggs are broadcast over a variety of substrates, and no nest is prepared, nor is care provided to embryos or young. In the laboratory, eggs hatched within nine days at 18°C (65°F) and six days at 20°C (67°F). Lake populations migrate from deep to shallow water in the late spring[2].

Habits and Habitat

Utah Chub thrive over a wide range of temperatures and occur in a wide range of ecosystem types. They are found in irrigation ditches, ponds, sloughs, creeks, and large rivers. They prosper in streams with either slow or fast currents and in ecosystems with bottom types that range from clay, mud, sand, gravel, marl, and peat to rubble[2,5,12].

Utah Chub are typically found near the surface with dense vegetation[2,5] in waters 0.46 to 1.22 m (1.5 to 4.0 ft) deep. However, in Bear Lake, they can be found in water more than 31 m (100 ft) deep[21].

NOAA Fisheries researchers compared eight populations of Utah Chub, isolated since the late Pleistocene (0.011 mya) to evaluate the relationship between predation by Cutthroat Trout and Utah Chub life history processes. Utah Chub populations that historically coexisted with Cutthroat Trout showed higher juvenile growth rates, delayed age at maturity, larger size at maturity, and lower female reproductive effort than isolated populations without predators. These results are consistent with life history models of size-selective predation and suggest that divergence among Utah Chub populations may best be explained by differences in predator-mediated mortality[22].

Management and Conservation Status

Utah Chub are considered common, widespread, and abundant range-wide (G5), and within Idaho Utah Chub are considered uncommon but not rare (S4); there is some cause for long-term concern due to declines or other factors[11]. Utah Chub vary widely in abundance throughout their native range in the Bonneville Basin. In Idaho they are apparently secure, and their range appears to have expanded, likely due to the actions of humans. However, no studies are known to have been conducted to document their status.

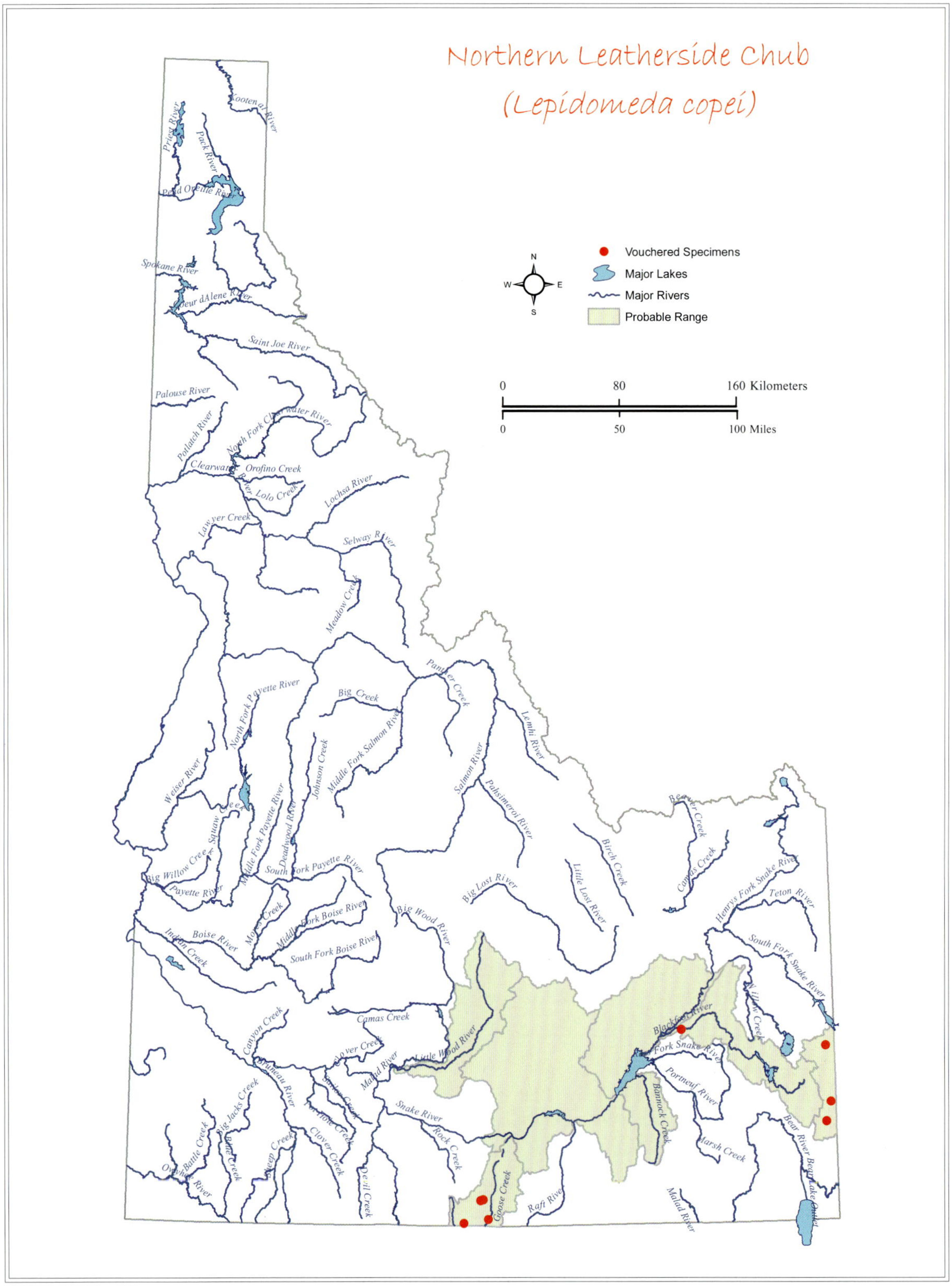
Northern Leatherside Chub
(Lepidomeda copei)
Vouchered Specimens
Major Lakes
Major Rivers
Probable Range
N
W
E
S
0
80
160 Kilometers
0
50
100 Miles
Priest River
Pack River
Spokane River
Saint Joe River
Palouse River
Potlatch River
Orofino Creek
Lolo Creek
Lochsa River
Lawyer Creek
Selway River
Meadow Creek
Big Creek
Middle Fork Salmon River
Johnson Creek
Salmon River
Lemhi River
Pahsimeroi River
Weiser River
Squaw Creek
Middle Fork Payette River
Deadwood River
Payette River
Boise River
Indian Creek
South Fork Boise River
Big Wood River
Big Lost River
Little Lost River
Birch Creek
Camas Creek
Henrys Fork Snake River
Teton River
South Fork Snake River
Willow Creek
Canyon Creek
Little Wood River
Snake River
Rock Creek
Goose Creek
Raft River
Portneuf River
Bannock Creek
Marsh Creek
Malad River
Bear River
Clover Creek
Big Jacks Creek
Sheep Creek

Northern Leatherside Chub

Lepidomeda copei
(Jordan & Gilbert, 1881)

Etymology

Northern Leatherside Chub were formerly included in the genus *Gila* = named for the Gila River, Arizona (likely a misnomer). Transferred into the genus *Lepidomeda* = Latin for scaled; *copei* = named for Edward Drinker Cope (a noted U.S. ichthyologist of the late 1800s).

THE NAME OF THE LEATHERSIDE CHUB, *SNYDERICHTHYS COPEI*, is surrounded by controversy. It was transferred from the genus Gila to the genus *Snyderichthys* by Miller[23] on the basis of external traits, but was later moved back to the genus *Gila* by Uyeno[24] in his analysis of bones of that genus. Coburn and Cavender[25] analyzed the phylogeny of North American cyprinids based on morphological and osteological characters and confirmed that *G. copei* was in *Gila*. The Spine Daces, Plagopterini, are a tribe of six species in the lower Colorado River drainage, unique among the 1500 species of minnows in the world, in the possession of strong spines in the pelvic and dorsal fins[26]. *Snyderichthys copei* has no hint of those spines, but has small scales, causing Miller and Hubbs[26] to consider a possible distant relationship with the small-scaled plagopterins. Later, DeMarais[27] found several similar protein characters and Dowling et al.[28] and Smith et al.[29], found similar mitochondrial DNA (mtDNA) in *Snyderichthys* and the plagopterine *Lepidomeda*, suggesting either retained ancestral similarities or transfer of DNA by hybridization (see *Chasmistes,* later in this chapter). Later, Simmons[30] and Johnson et al.[31] went further and considered the DNA similarity a better indicator of true relationship than the fins and bones of the fishes (but they did not examine bones). Other authors in unpublished manuscripts have noted many jaw, teeth, pharyngeal, branchial, and fin traits that suggest that the fish's true relationships are best represented by morphology, with some evidence of introgression—not DNA similarities, which are markers reflecting their own separate histories, and contribute little to construction of the fish that we see in nature. More extensive molecular work will provide better answers to this controversy. GRS.

Recent research on this species indicates that this taxon previously considered as Leatherside Chub, is actually composed of two species, the Northern Leatherside Chub (*Lepidomeda copei*) and the Southern Leatherside

Chub (*L. aliciae*). Results of both genetic and ecological research prompted this separation into two evolutionarily distinct species[28,31,32].

Importance

Northern Leatherside Chub are of great ecological interest because of their diverse occurrence in varied ecological situations. Northern Leatherside Chub likely provide forage for some game fishes[2]. Northern Leatherside Chub current population structure is representative of both historical habitat variations and recent habitat fragmentation (human caused). These fragmented populations are more susceptible to extirpation, resulting in an increased conservation value for the populations that are still intact[33].

Description

Northern Leatherside Chub are dark bluish or greenish gray dorsally and on the upper sides and silvery below. Males may have an orange to reddish hue on the lower sides, on the lower lobe of the caudal fin, and at the base of the anal fin. The dorsal fin originates posterior to the origin of the pelvic fins, and the dorsal and anal fins are rounded. The anal fin contains eight rays. Northern Leatherside Chub have small, oblique mouths, not distinctly inferior, and no barbel is present[5]. There are 75 or more scales in the lateral line series.

Range and Distributional Status

This species is native to Idaho. Historically, the Northern Leatherside Chub range encompassed the northeastern margins of the Bonneville Basin in Idaho, Utah, and Wyoming and; within the Snake River drainage, Goose Creek, Wood River, and Raft River drainages in Idaho. They are also present in the Snake River basin upstream of Shoshone Falls in Idaho and Wyoming[5,34]. Currently, extant populations of Northern Leatherside Chub are patchily distributed and largely isolated across their native range in the Great Basin and adjacent watersheds, likely a reflection of historical and contemporary processes. Both genetic and geologic evidence suggests historical hydrologic connectivity between populations in the Snake River, Columbia River Basin, Bear River, northern Bonneville Basin, and Green River watersheds. As an ancient tributary to the Snake River, the Bear River is thought to share part of its history with the Snake River until about 600,000 to 50,000 years ago when lava flows diverted the Bear River southward toward the Bonneville Basin. It is surmised that fauna transfer occurred during this early period of diversion in the Bear River. This hypothesis is further supported by similarity in genetics among other species, including Green Sucker, Utah Sucker, Cutthroat Trout, Speckled Dace, Utah Chub, and Mountain Whitefish[33].

Size and Longevity

Northern Leatherside Chub typically live less than five years and range from 76 to 165 mm (3.0 to 6.5 in) TL[5].

Food and Feeding

Food habits have not been reported for this species in Idaho. Their ecological occurrence suggests that they feed on small drift organisms, algae, and aquatic insects[2,5].

Limiting Factors

Threats to Northern Leatherside Chub populations include stream dewatering and instream barriers, which interrupt stream flow and isolate populations within stream reaches, causing population fragmentation. Predation by nonnative fishes (particularly Brown Trout) is an additional factor threatening Northern Leatherside Chub[34].

Breeding Habits

Northern Leatherside Chub spawn over small cobble substrate in the summer when water temperatures range from 15 to 20°C (59 to 68°F). Northern Leatherside Chub provide no care for eggs, embryos, or young[5].

In research on Northern Leatherside Chub, conducted at the Fisheries Experiment Station in Logan, Utah, spawning occurred between the end of April and early September. Fertilized eggs are translucent, whitish, and 1.25 to 1.5 mm (0.05 to 0.06 in) in diameter. Eggs are adhesive and can stick together on rocks in clumps of up to 90 eggs, although most eggs were found singly and unattached. Embryos began hatching four days after being collected and had all hatched by six days after collection, indicating a hatching time of four to six days post-fertilization at 18°C [72–108 degree days] (64.5°F). Northern Leatherside Chub fry were 6.0 mm (0.2 in) TL immediately after hatching and remained on the bottom of the tank unless disturbed. Swim-up occurred approximately six days post-hatch when fry were approximately 8.0 mm (0.3 in) TL. Fins were first noticeable at four days post-hatch on some fry and were noticeable on most fry by eight days post-hatch. Fry grew quickly, nearly tripling in size after 40 days at 18°C (64.5°F), which equates to an absolute growth rate of 0.3 mm/day (0.01 in/day). Significantly more eggs were found in the small cobble substrate than in the other three substrate options. The four substrates were pebble (0.5 to 1.5 cm in diameter, mean 0.9 cm/0.2 to 0.6 in, mean 0.35 in), small cobble (2.1 to 4.8 cm, mean 3.1 cm)/0.8 to 1.9 in, mean 1.2 in), large cobble (5.8 to 11.2 cm, mean 8.0 cm/2.3 to 4.4 in, mean 3.1 in), and a coarse artificial spawning mat. Temperature affected the timing of the onset of spawning but did not increase the duration of reproduction. Spawning first occurred just two days after the water temperature was increased from 19 to 24°C (66 to 75°F). Reproduction in other tanks at 19°C (66°F) occurred two weeks after the temperature was increased. Duration of spawning ranged from 93 to 118 days, lasting longer in the spawning systems with water temperature of 19°C (66°F). Temperature did not appear to affect progeny production but did affect time to first feeding and larval and juvenile growth[32].

Habits and Habitat

Northern Leatherside Chub inhabit pools and low-velocity runs of creeks and small to medium-sized rivers. They inhabit desert streams that have a broad range of widely varying physical conditions, including high variability of stream flow, annual precipitation, gradient, elevation, conductivity, and pH. Substrates in these ecosystems have lower percentages of sand-silt and gravel. Loss of habitat heterogeneity caused by erosion, removal of riparian vegetation, and channelization creates unfavorable ecological conditions for Northern Leatherside Chub[34–36].

Typically, Northern Leatherside Chub are found in cool to cold streams, riffles, or pools, although tending to prefer pools. Young seek quiet pockets or brush areas near shore. This species may also inhabit rivers. Water quality for Northern Leatherside Chub may be clear or moderately turbid and flow over an entirely gravel bottom, although there may be sand, rubble, boulders, or silt. Current velocity in Northern Leatherside Chub ecosystems is typically moderate, 13 to 23 cm/s (0.4 to 0.8 f/s) in waters ranging from 0.6 to 1.0 m (2.0 to 3.3 ft) deep or less[5].

Management and Conservation Status

Northern Leatherside Chub are considered critically imperiled to imperiled range-wide (G1G2). Northern Leatherside Chub are rare in Idaho and are considered a state species of special concern, imperiled (S2)[11]. The USFWS has determined that listing this species under the federal Endangered Species Act is not warranted[5]. No detailed conservation management plan has been developed for this species.

Fragmentation, both natural and anthropogenic, of once-connected populations makes such scattered populations more vulnerable to extirpation. Because dispersal for aquatic species is severely limited when compared with terrestrial species, isolation of occupied or suitable habitats severely restricts gene flow and recolonization. Obviously, the few populations that remain are of high conservation value, and efforts should be made to sustain habitat quantity and quality in streams where these populations exist. There is a growing body of evidence that suggests human alteration of aquatic systems within the past few hundred years has had a significant negative influence on the distribution, genetic structure, and ultimately the survival and future adaptability in many fish species[33].

Bear River leaving the Black Canyon as seen from PacifiCorp fishing access bridge at the Black Canyon Take-Out, near Grace, ID. Photo Courtesy John Sigler.

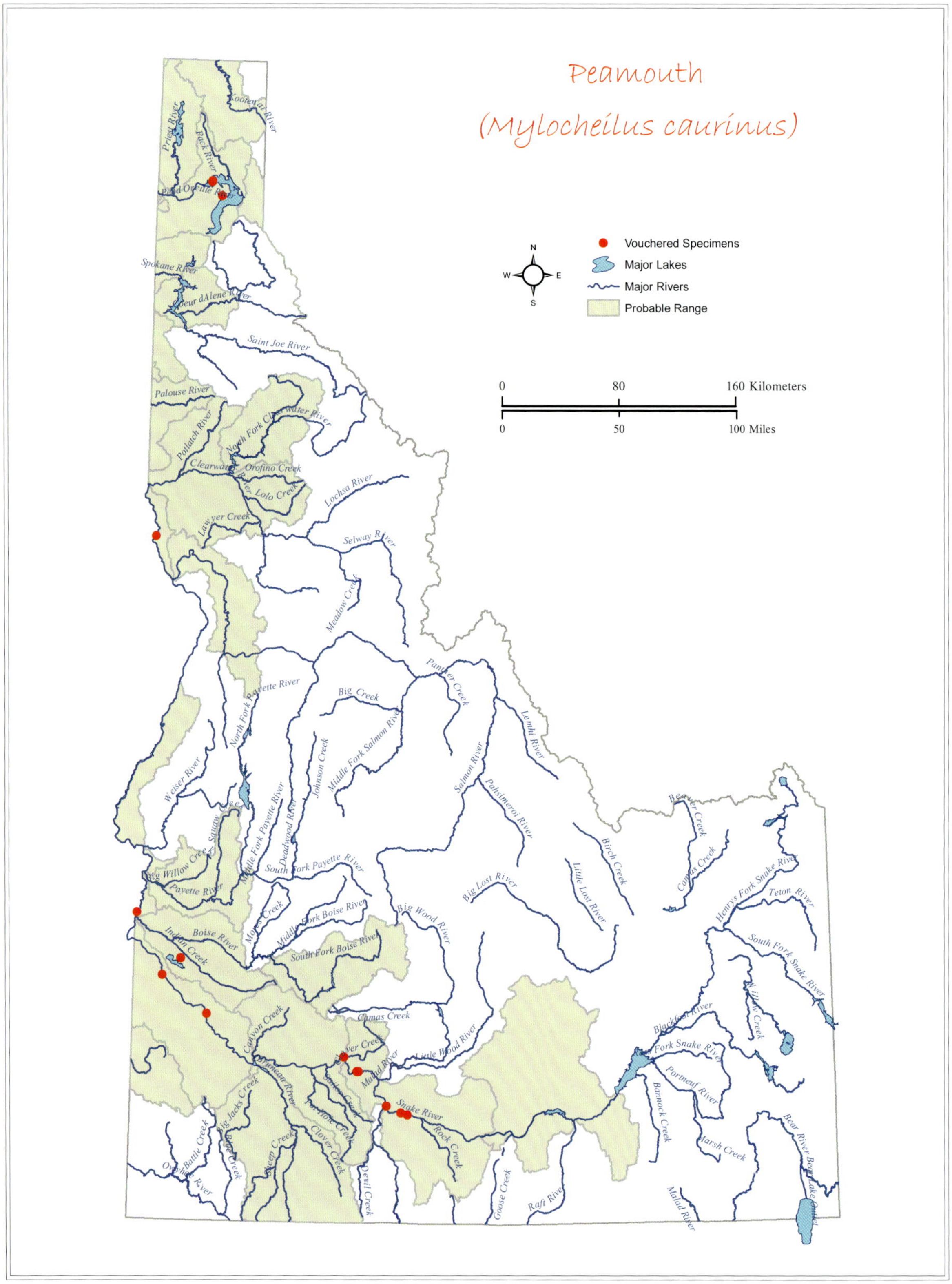
Peamouth
(Mylocheilus caurinus)
Vouchered Specimens
Major Lakes
Major Rivers
Probable Range
N
W
E
S
0
80
160 Kilometers
0
50
100 Miles
Kootenai River
Priest River
Pack River
Spokane River
Saint Joe River
Palouse River
Potlatch River
Orofino Creek
Lolo Creek
Lochsa River
Lawyer Creek
Selway River
Big Creek
Panther Creek
Lemhi River
Weiser River
Johnson Creek
Salmon River
Pahsimeroi River
Birch Creek
Little Lost River
Big Lost River
Big Wood River
Teton River
South Fork Snake River
Blackfoot River
Portneuf River
Bannock Creek
Marsh Creek
Bear River
Camas Creek
Snake River
Rock Creek
Goose Creek
Raft River
Malad River
Devil Creek
Clover Creek
Boise River
Payette River
South Fork Boise River
Middle Fork Payette River
Deadwood River
South Fork Payette River
Big Willow Creek
Indian Creek
Canyon Creek
Big Jacks Creek
Owyhee River
Battle Creek

Peamouth

Mylocheilus caurinus

(Richardson, 1836)

Etymology

Mylocheilus = grinder, lip; *caurinus* = *caurus*, the northwest wind = northwestern.

Importance

In the early 1900s, Peamouth were used for food and were served in hotels as whitefish. Some angling occurred. The species, which had many common names across its western distribution, was also believed to have fair game qualities, fighting well when hooked and biting readily[4]. Within portions of its native range, Peamouth provide a sport fishery, and they can be easily caught on artificial flies and bait. Young Peamouth provide limited forage for game species in some ecosystems[6].

Description

Peamouth have an elongate body that is not strongly compressed laterally. The small mouth is slightly inferior due to the overhanging snout, and there is a small barbel at each corner. Pharyngeal teeth are usually 1,5 - 5,1, and in adults have molariform grinding surfaces. Peamouth have a single dorsal fin, and the caudal fin is deeply forked. Pectoral fin rays are 15 to 18, but typically 16 or 17. Scales are cycloid, numbering 68 to 79 along a complete lateral line. Peamouth are brightly colored with localized areas of red at the angles of the mouth; the red coloration may extend along the cheek and opercle to the pectoral fin and along the sides. Breeding individuals of both sexes are highly colored, with red lower lips and lateral red stripes. Males have green backs and females a brown back, while immature or non-spawning fish are generally silvery, with a dark back and two distinct lateral bands[4].

Range and Distributional Status

This species is native to Idaho. Peamouth are found in the Columbia River drainages and coastal drainages of Oregon, Washington, British Columbia, and Alberta. They are found on Vancouver Island and other coastal islands off British Columbia. In Idaho this species is native to the Snake River and its tributaries below Shoshone Falls and the Spokane River, Pend Oreille River, and Kootenai River systems. They have not been recorded in other Idaho drainages[5]. They have been reported in the Snake River downstream of Lake Walcott.

Size and Longevity

Female Peamouth reportedly grow faster and larger than males. Fork lengths of individuals from Okanagan Lake, British Columbia, ages one to five, were, in mm (in), 51 (2.0), 107 (4.2), 142 (5.6), 170 (6.7), and 198 (7.8)[4]. Peamouth may live to age 13, and growth rates are markedly different across ecosystems. Idaho's angling record, is a 375 mm (14.75 in), 481 g (1 lb, 1 oz) individual from the Clearwater River, taken in 2004.

Size sexual dimorphism is expressed in Peamouth, and mature females are larger than males. This size dimorphism results from a more rapid growth rate, combined with a longer life span, in females than in males[3].

Food and Feeding

Peamouth are considered to be primarily insectivorous, feeding on a wide variety of aquatic insects (chironomids, mayflies and caddisflies), and their larvae, as well as some terrestrial insets. Midge pupae form a large part of the diet of Peamouth in Lake Washington during summer and spring. They also consume a wide variety of planktonic crustaceans, mollusks, and occasionally small fishes[4,5].

Peamouth feed in the pelagic zone as well as on the bottom. Food items of Peamouth in the Hanford Reach of the Columbia River were dominated by periphyton, snails, caddisfly larvae, and midge larvae[6].

Limiting Factors

In part due to their inshore spawning behavior, Peamouth may be preyed upon by a variety of fish-eating birds and mammals. A variety of parasites, including trematodes, nematodes, cestodes, acanthocephalans, and crustaceans, have been identified in Peamouth[4]. Predation also reduces some populations.

Breeding Habits

Peamouth spawning occurs in shallow water in late May through June at water temperatures around 12°C (54°F). Spawners aggregate in shallow water near shore, and females are crowded by two or more males into 2.5 to 5 cm (1 to 2 in) of water at the shoreline, where fertilization occurs. Eggs are gray-green when spawned, become adhesive, and sink to the bottom, where they attach to the substrate. Newly hatched young remain in schools along the shoreline and move into deeper water in late summer[4].

During daylight spawning periods in lakes and streams, Peamouth are in aggregations, but do not spawn en masse. In streams, ripe fish mill slowly about in a pool or low-velocity area below riffles until a female swims rapidly upstream into the riffle, where she is joined immediately by four to five males. The males crowd against the female and vibrate rapidly, causing her to release a few eggs. Over the course of spawning periods, females spawn many times with many different males[3].

Peamouth eggs become yellow when ripe, and embryos hatch in seven to eight days at temperatures of 12°C (54°F). Females 305 to 330 mm (12.0 to 13.0 in) long produce 11,800 to 18,900 eggs, whereas larger and older females produce higher numbers of eggs[6].

Habits and Habitat

Primarily a freshwater species that inhabits both streams and lakes, Peamouth are one of only a few minnow species that tolerate saltwater. Peamouth are a schooling species; adult fishes remain on the bottom in deep water during daylight, moving into shallows at night. At the onset of spawning, they move inshore during spring and summer[6].

Management and Conservation Status

Peamouth are considered common, widespread, and abundant range-wide (G5) and within Idaho (S5)[11]. However, limited information exists on the status of Peamouth in Idaho.

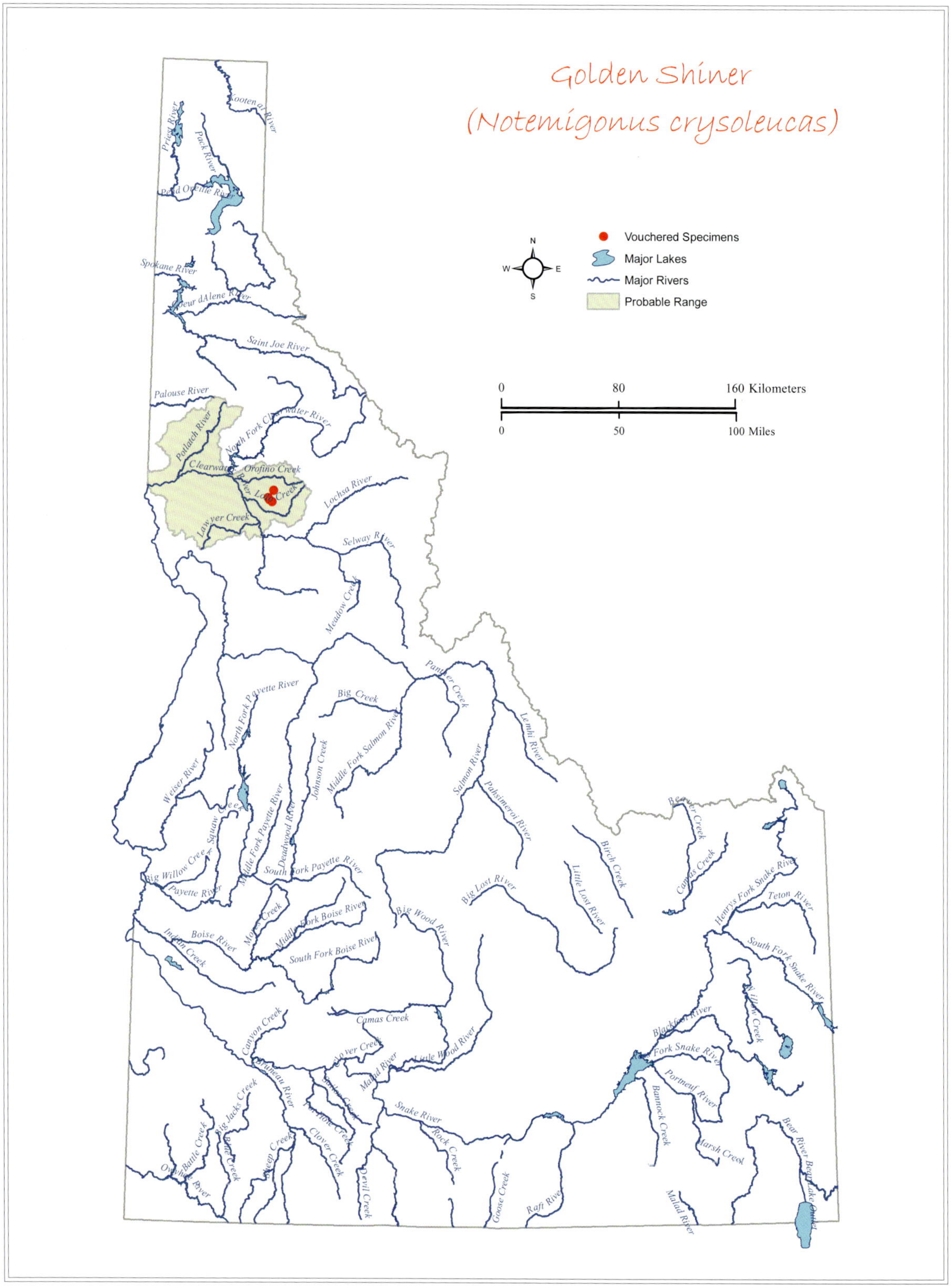
Golden Shiner
(Notemigonus crysoleucas)
N
W
E
S
Vouchered Specimens
Major Lakes
Major Rivers
Probable Range
0
80
160 Kilometers
0
50
100 Miles
Pack River
Spokane River
Saint Joe River
Palouse River
Potlatch River
Orofino Creek
Lochsa River
Selway River
Meadow Creek
Big Creek
Middle Fork Salmon River
Johnson Creek
Weiser River
Big Willow Creek
Payette River
South Fork Payette River
Deadwood River
Middle Fork Payette River
North Fork Payette River
Boise River
Indian Creek
Middle Fork Boise River
South Fork Boise River
Big Wood River
Big Lost River
Little Lost River
Birch Creek
Salmon River
Pahsimeroi River
Lemhi River
Panther Creek
Camas Creek
Canyon Creek
Snake River
Rock Creek
Clover Creek
Devil Creek
Goose Creek
Raft River
Malad River
Marsh Creek
Bannock Creek
Portneuf River
Blackfoot River
Henrys Fork Snake River
Teton River
South Fork Snake River
Bear River
Willow Creek
Little Wood River
Bruneau River
Big Jacks Creek
Battle Creek
Owyhee River
Lawyer Creek
Clearwater River

Golden Shiner

Notemigonus crysoleucas
(Mitchill, 1814)

Etymology

Notemigonus = back, half, angle—the back being almost carinated; *crysoleucas* = gold, white.

Importance

Golden Shiner have been one of the most popular of all bait fishes in North America. They have been cultivated successfully in ponds for sale as bait or by fish hatcheries as food for pond-cultured bass, particularly Largemouth Bass, for many years across the United States, from Michigan to Mississippi[4].

Under favorable environmental conditions, nearly 200,000 bait-size Golden Shiner can be produced per surface acre annually. These operations, which raise as many as 50 million Golden Shiner per year, may have an economic value exceeding two million dollars. Golden Shiner are considered the most widely distributed minnow in California because of widespread introduction from bait buckets. This is likely true in other western states as well. In the southwestern United States, introduced Golden Shiner adversely affect some native fishes. Eradication of Golden Shiner and reestablishment of native fishes have been attempted[6]. Success has not been determined.

In some ecosystems, Golden Shiner have a primary value as forage. Because of their tolerance of pollution, Golden Shiner may markedly outnumber other fishes. In a particular ecosystem, this may be a good indicator of a polluted ecosystem or one that has suffered habitat modification[2].

Description

Golden Shiner have a deep body that is strongly laterally compressed. Although variable among ecosystems, individuals average 76 to 127 mm (3.0 to 5.0 in) TL. The single dorsal fin has its origin distinctly posterior to the pelvic fin origin, and the posterior edge is straight or slightly falcate; rays generally number eight. The caudal fin is distinctly forked, with pointed lobes. Pelvic fins originate distinctly in advance of the dorsal fin origin, and rays usually number nine or rarely eight. Pectoral fins are rather short, never reaching the pelvic fin origin, and have 16 to 18 rays. Scales are cycloid and persistent. The lateral line is complete and strongly decurved. A dusky peritoneum is present, the intestine is short, and vertebrae number 37 to 39[4].

Coloration of Golden Shiner is green to olive on the back, with a faint dark stripe along the midline. Sides are golden or silver, and the belly is silvery or white. There is a fleshy, scaleless keel along the belly, from the anus to the base of the pelvic fins. The mouth is a bit upturned and small, and the eye is forward of the jaw. There are 45 to 54 scales in the lateral series. Fins have no marks[2].

Range and Distributional Status

This species is not native to Idaho. Native distribution of Golden Shiner extends from the Canadian Maritime Provinces south to the southern tip of Florida, west to eastern Texas, and northwesterly to eastern Montana and southeastern Saskatchewan, continuing eastward along the border between the United States and Canada to the Maritime Provinces. This species has been introduced into various parts of the United States generally as a bait or forage fish[6]. In Idaho, Golden Shiner have been collected in the Clearwater River drainage.

Size and Longevity

Golden Shiner are one of the larger of the small cyprinids, reaching lengths over 178 mm (7.0 in). Wisconsin Golden Shiner age one to seven had yearly lengths, in mm (in), of 28 (1.1), 58 (2.3), 84 (3.3), 127 (5.0), 155 (6.1), 173 (6.8), and 185 (7.3). Females grow larger and more rapidly than males, and at higher elevations Golden Shiner grow much slower than at lower and generally more fertile elevations. Maximum weight is generally less than 0.5 kg (1 lb, 2 oz)[2,6].

Food and Feeding

Cladocerans, particularly *Daphnia*, are important food items of Golden Shiner. They may take small fishes in some ecosystems as well as mollusks and aquatic insect larvae and small flying insects. When animal food is in short supply, Golden Shiner will switch to filamentous algae[2].

Limiting Factors

A wide variety of parasites have been identified in Golden Shiner populations in North America, including protozoans, trematodes, cestodes, nematodes, acanthocephalans, leeches, and crustaceans[4]. Golden Shiner are preyed upon by a variety of fishes and occasionally are cannibalistic. Declines of Golden Shiner populations in some ecosystems can be attributed to habitat modification, including draining of marshes and alteration of floodplains. Golden Shiner have a high tolerance to adverse water quality conditions, surviving in some badly polluted ecosystems with high turbidity or temperatures[2].

Breeding Habits

Golden Shiner spawning may occur as early as March in southern climes or as late as August in northern locations. Spawning generally occurs early in the morning when water temperatures rise to 21°C (70°F) and stops when temperatures exceed about 27°C (80°F). Males and females mature at age one, in their second year of life. In southern waters, they may spawn when only seven months old at about 64 mm (2.5 in) TL[2,6].

Golden Shiner produce adhesive eggs that are scattered in shallow water, generally over aquatic vegetation ranging from filamentous algae to rooted aquatic plants. Embryos hatch in four to five days at temperatures of 24 to 26°C (75 to 80°F), and the young form schools near the surface in shallow waters. Occasionally, they spawn in the nests of other fishes, including Largemouth Bass, where eggs have a higher survival rate because of the protection provided by male bass guarding the nest[2,6].

Habits and Habitat

The habitat commonly used by Golden Shiner is generally clean, quiet, shallow water in the littoral zone of a variety of oxbows, ponds, and lakes with heavy rooted aquatic vegetation. Golden Shiner are also found in backwaters, sloughs, and low-velocity reaches of streams, rarely being found in stream reaches with noticeable current[2,6]. Preferred water temperatures range between 17 and 24°C (63 and 75°F). They occur in small, loosely aggregated schools utilizing all depths of the water column. These schools typically occupy the littoral zone during the day, but at sunset fishes disperse into open water of the limnetic zone[6].

Management and Conservation Status

Golden Shiner are considered common, widespread, and abundant range-wide (G5), but have no official ranking in Idaho[11]. Because Golden Shiner are a nonnative species, additional introductions are strongly discouraged, and anglers are encouraged to use care with this species and all bait species to reduce unwanted introductions.

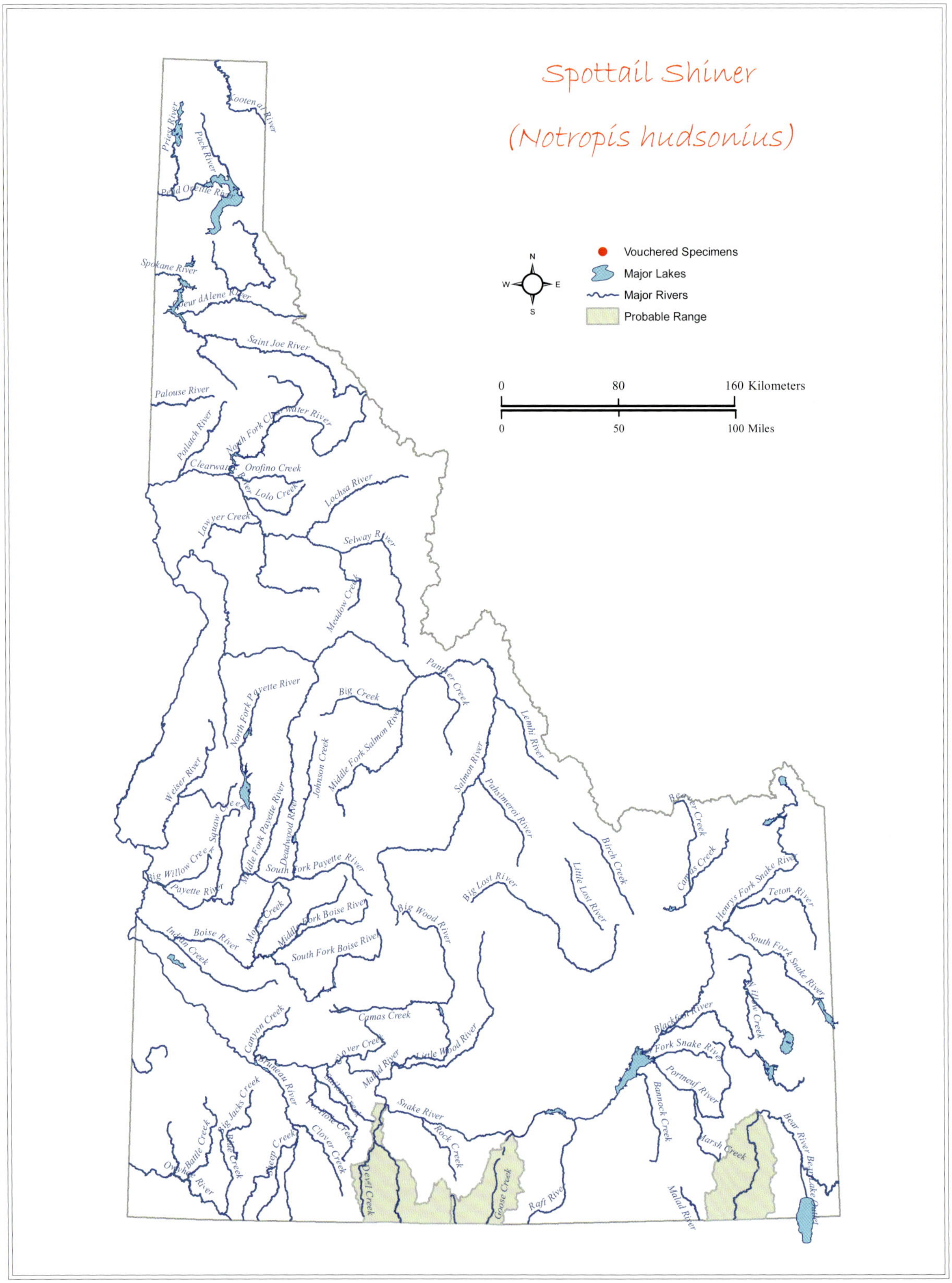
Spottail Shiner
(Notropis hudsonius)
N
W
E
S
Vouchered Specimens
Major Lakes
Major Rivers
Probable Range
0
80
160 Kilometers
0
50
100 Miles
Pack River
Saint Joe River
Palouse River
Potlatch River
Orofino Creek
Lochsa River
Lolo Creek
Selway River
Meadow Creek
Big Creek
Middle Fork Salmon River
Johnson Creek
Salmon River
Lemhi River
Pahsimeroi River
Birch Creek
Little Lost River
Big Lost River
Big Wood River
Weiser River
Middle Fork Payette River
Deadwood River
Boise River
South Fork Boise River
Camas Creek
Canyon Creek
Teton River
South Fork Snake River
Henrys Fork Snake River
Portneuf River
Bannock Creek
Marsh Creek
Malad River
Snake River
Rock Creek
Goose Creek
Raft River
Devil Creek
Clover Creek
Sheep Creek
Big Jacks Creek
Bear River

Spottail Shiner

Notropis hudsonius
(Clinton, 1824)

Etymology

Notropis = Greek for keeled back; Rafinesque apparently gave this name to the generic-type specimen that had a shriveled appearance; *hudsonius* = named for the Hudson River.

Importance

Spottail Shiner are an important forage fish for a variety of sportfishes over much of their native range in North America. They are an important bait species in many parts of their range and are the preferred bait for Walleye in many ecosystems. They are of importance in a variety of ecosystems where they are preyed upon by a variety of sportfishes[2,4].

Description

Coloration of Spottail Shiner is silvery with yellow or golden overtones, pale green to olive backs, with silvery sides and a silvery-white belly. Fins are mainly clear except for the black spot at the base of the caudal fin. Specimens larger than 127 mm (5.0 in) often show little evidence of the spot, although it is nearly always obvious after death and preservation, even on large specimens[4].

As with many members of the minnow family, female Spottail Shiner grow faster than do males and on average are larger than males of the same age. Breeding males have small tubercles on the dorsal surface of the head and upper surface of the pectoral rays; pectoral fins are longer in males than in females[3].

Spottail Shiner have a stout body that is laterally compressed and a bluntly triangular head. The single dorsal fin has its origin directly over the pelvic fin origin, and the rays are almost always eight or rarely nine; the caudal fin is distinctly forked. Pelvic fins are small, originating directly below the dorsal fin origin, and the pectoral fins are moderate in size, usually with 14 to 16 rays. Scales are cycloid and large, numbering 36 to 42 along the complete lateral line. The peritoneum is silvery, and vertebrae number 37 to 39 but typically 38[2,4].

Range and Distributional Status

This species is not native to Idaho. Spottail Shiner were introduced in Oneida Narrows Reservoir and Salmon Falls Creek Reservoir to supplement the forage base for Walleye. Spottail Shiner occur in North America from portions of Canada south in the United States to Georgia in the East and Iowa and Missouri in the West. Spottail Shiner are one of the most widely distributed native North American fishes[2,4].

Size and Longevity

Spottail Shiner rarely grow larger than 127 mm (5.0 in). Females are generally larger than males. This species is relatively fast growing but short lived. Total lengths for a Wisconsin population of Spottail Shiner aged one to five, in mm (in), were 64 (2.5), 97 (3.8), 109 (4.3), 119 (4.7), and 125 (4.9)[2].

Food and Feeding

Spottail Shiner feed on filamentous algae and plankton (*Daphnia and Bosmina)* as well as a variety of organisms, including aquatic insect larvae, crustaceans, aquatic insects, and their own eggs. Larger shiners tend to shift from small invertebrates to insects[2]. Spottail Shiner have been variously reported to feed on insect larvae and masses of filamentous algae. Algae, crustaceans, aquatic insects, and even eggs and larvae of their own species may at times be significant food items[4].

Limiting Factors

Spottail Shiner may be infected by a variety of parasitic trematodes, cestodes, nematodes, and protozoans. A variety of piscivorous fishes and birds prey upon them[2,4].

Breeding Habits

Spottail Shiner spawn in spring and early summer throughout their wide range, with specific dates depending upon latitude and seasonal weather. Yearling ripe females 70 to 90 mm (2.8 to 3.5 in) long may contain 100 to 1,400 eggs, while two-year-old females may contain 1,300 to 2,600 eggs. Eggs are 0.8 mm (.03 in) in diameter and yellow[4]. Spottail Shiner spawn after age one over gravel riffles in 0.9 to 4.6 m (3.0 to 15.1 ft) of water, generally along the sandy shoals of lakeshores. No evidence of nest preparation or care for eggs or young has been documented[2].

Habits and Habitat

Spottail Shiner prefer large lakes and rivers, but smaller lakes and other streams are also used. Generally, Spottail Shiner occur in moderately shallow water or shoals but may also occur at depths to 46 m (150 ft). In Lake Michigan, large schools occur near shorelines with sparse to moderate vegetation[2].

As adults, Spottail Shiner are typically considered an open-water species, being found in large rivers and lakes. In lakes during the day, they generally occur out beyond any emergent vegetation, but at night they move in closer to shore. In flowing water ecosystems, Spottail Shiner occupy river margins in areas with little current[3].

Management and Conservation Status

Spottail Shiner are considered common, widespread, and abundant range-wide (G5). A state conservation status rank is not applicable (SNA) because the species is not a suitable target for conservation activities[11]. Because Spottail Shiner are a nonnative species, additional introductions into Idaho aquatic ecosystems are strongly discouraged. Anglers are encouraged to use care with this species and all bait species to reduce unwanted introductions.

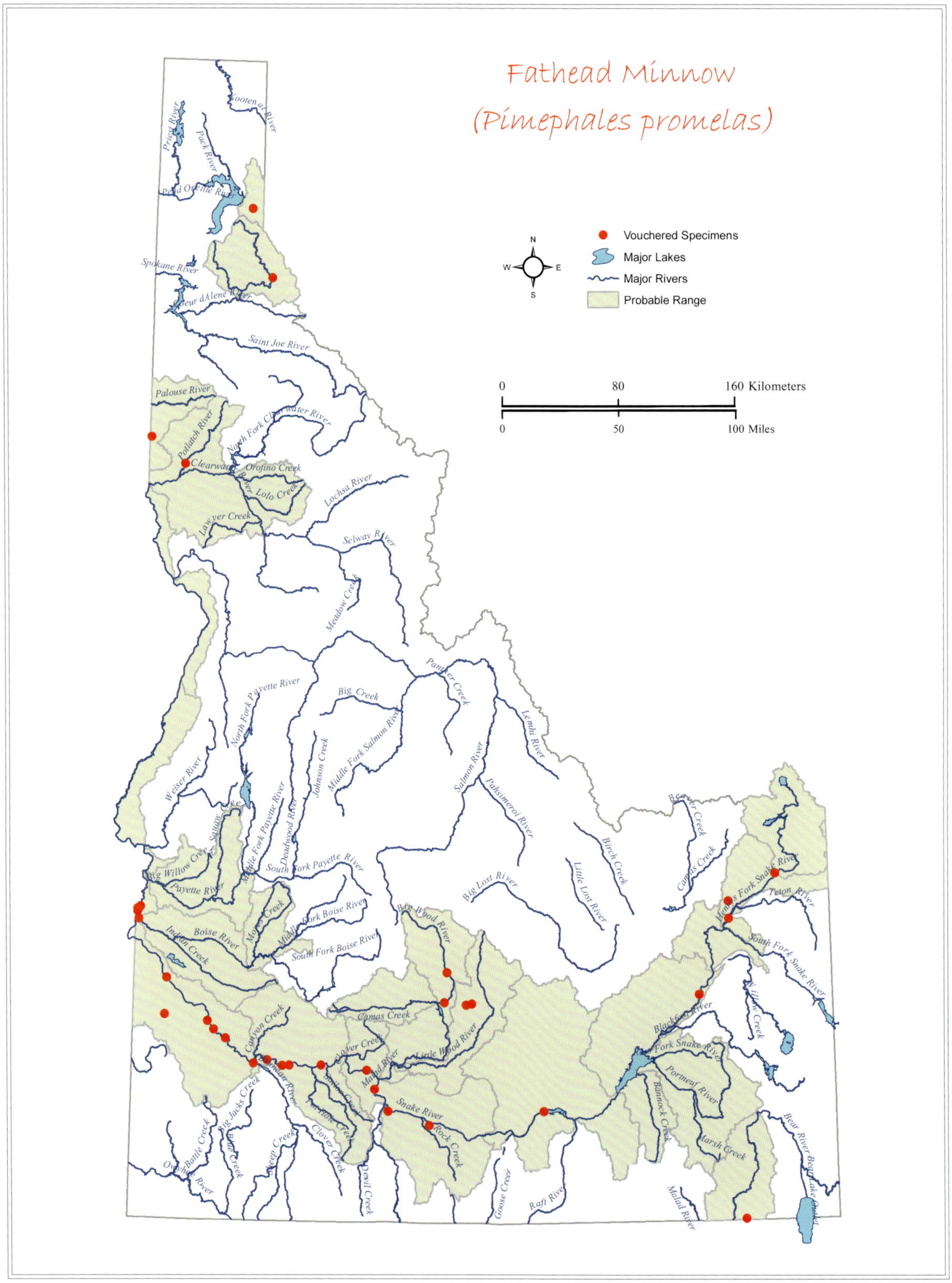
Fathead Minnow
(Pimephales promelas)
Vouchered Specimens
Major Lakes
Major Rivers
Probable Range
0 80 160 Kilometers
0 50 100 Miles

Fathead Minnow

Pimephales promelas
Rafinesque, 1820

Etymology

Pimephales = Greek for fathead; *promelas* = Greek for "before" and "black," probably referring to the blackish forespot.

Importance

Primarily, Fathead Minnow are important as a forage fish; they convert algae and organic detritus from bottom deposits and planktonic organisms into prey for other fishes. The role they play in this arena is inestimable[4]. They are used extensively as test specimens in bioassays of toxic substances and environments.

In their preferred habitat, Fathead Minnow are often the predominant fish species. Fathead Minnow are one of the most commonly used bait minnows in North America, and in their native range, they are propagated in hatchery ponds for bait and as food for various species, including Smallmouth Bass and other game fishes[2].

Description

Overall coloration of Fathead Minnow is typically dark olive green or brown above, with silvery or brassy reflections on sides and a silvery white belly. Young and nonbreeding adults may be lighter in color but exhibit a distinct lateral band that extends from the caudal peduncle to the head. Fathead Minnow have a short body that is laterally compressed, with a pronounced belly. The head is rather sharply triangular. A single dorsal fin has its origin over the pelvic fin origin, and the rays number eight, with the first ray being about half the length of the second. The caudal fin is shallowly forked, and the anal fin has seven rays, with its origin below the depressed dorsal fin. Pelvic fins are small, originating below the dorsal fin origin and usually extending to the anal fin origin in males but not in females. Fathead Minnow scales are cycloid and moderate in size. The lateral line is incomplete, usually terminating below the dorsal fin origin or before; scales in the lateral series number 41 to 54. The peritoneum is uniformly black, and the intestine is long but variable; vertebrae typically number 36 or 37, seldom 35 or 38. Nuptial tubercles appear on males only[4].

Range and Distributional Status

This species is not native to Idaho. Fathead Minnow have a native distribution range through most of central North America, from Louisiana and Chihuahua, Mexico, north to the Great Slave Lake drainage and from New Brunswick, Canada, in the East to Alberta, Canada, in the West. Fathead Minnow have been widely introduced in North America and are now likely the most widespread fish species in the United States[4,6]. In Idaho Fathead Minnow have been documented in the Clark Fork River, Coeur d'Alene River, and Clearwater River of northern Idaho and the Snake River, Big Wood River, Little Wood River, Boise River, and Bear River drainages of central and southern Idaho.

Size and Longevity

Most reports indicate that Fathead Minnow are short lived, rarely living beyond age two. Growth is rapid in warm, food-rich waters. As with many species of minnows, there is sexual dimorphism in body size. In most minnows, females are larger than males. However, male Fathead Minnow grow more rapidly and to a larger size than females, a common trait characteristic of minnows when males guard the nest and perhaps the young. This large size is probably an advantage in male-male competition for mates and spawning sites[3].

Food and Feeding

Major foods eaten by Fathead Minnow are diatoms and algae on substrates, but they also consume free-swimming zooplankton and aquatic insect larvae. One Wyoming study indicated that Fathead Minnow consume 95 percent plant matter and 5 percent animal matter, mostly insects[6].

Limiting Factors

Fathead Minnow are ideal prey for fishes and birds due to their wide distribution, small size, and highly prolific nature. A list of potential predators on Fathead Minnow would include all fish-eating birds and fishes associated with Fathead Minnow in a particular ecosystem. A wide range of parasites, including various protozoans, trematodes, cestodes, nematodes, and crustaceans, have been documented in Fathead Minnow[2,4].

Breeding Habits

Fathead Minnow begin to spawn in spring and into summer, approximately May through August, when water temperatures reach about 16°C (60°F). Spawning activity may continue for months. Fathead Minnow mature at age one or less. Spawning may cease if water temperatures drop but then resume when temperatures again become favorable. Spawning of Fathead Minnow also stops if water temperatures exceed 29°C (84°F)[2,4,6].

Fathead Minnow attach eggs to the undersides of rocks and sticks, and males rub their dorsal pad on egg deposition sites and on eggs after they are deposited. This rubbing behavior, which deposits mucous on the sites, may prepare the site for egg deposition, clean the eggs, and assist in eggs' hatching[3].

Spawning behavior in Fathead Minnow is rather involved and unique for minnows. Males select a nest site in water 0.6 to 0.9 m (2.0 to 3.0 ft) deep, typically under a log, rock, or other structure. They nudge a female into the nest. Females have an ovipositor that is used to deposit adhesive eggs in the nest. Once the adhesive eggs are deposited, they are fertilized by the male, and then he chases the female away. Following spawning, males are aggressive to the female and other fishes. Males often seek out more than one receptive female. Females may spawn with a number of males at different times during the spawning season[2,4,6].

Habits and Habitat

Fathead Minnow generally inhabit shallow lakes and ponds as well as slow-flowing stream ecosystems. They are also found in eddies in pools of smaller streams to large rivers, inhabiting quiet backwaters and sloughs. Fathead Minnow are tolerant of high water temperatures, greater than 33°C (91°F); low dissolved oxygen levels; high pH values to 9.8; and high total alkalinity to 29,000 mg/l (ppm).

Management and Conservation Status

Fathead Minnow are considered common, widespread, and abundant range-wide (G5). A state conservation status rank is not applicable (SNA) because the species is not a suitable target for conservation activities[11]. Because Fathead Minnow are a nonnative species, additional introductions are strongly discouraged. Anglers are encouraged to use care with this species and all bait species to reduce unwanted introductions.

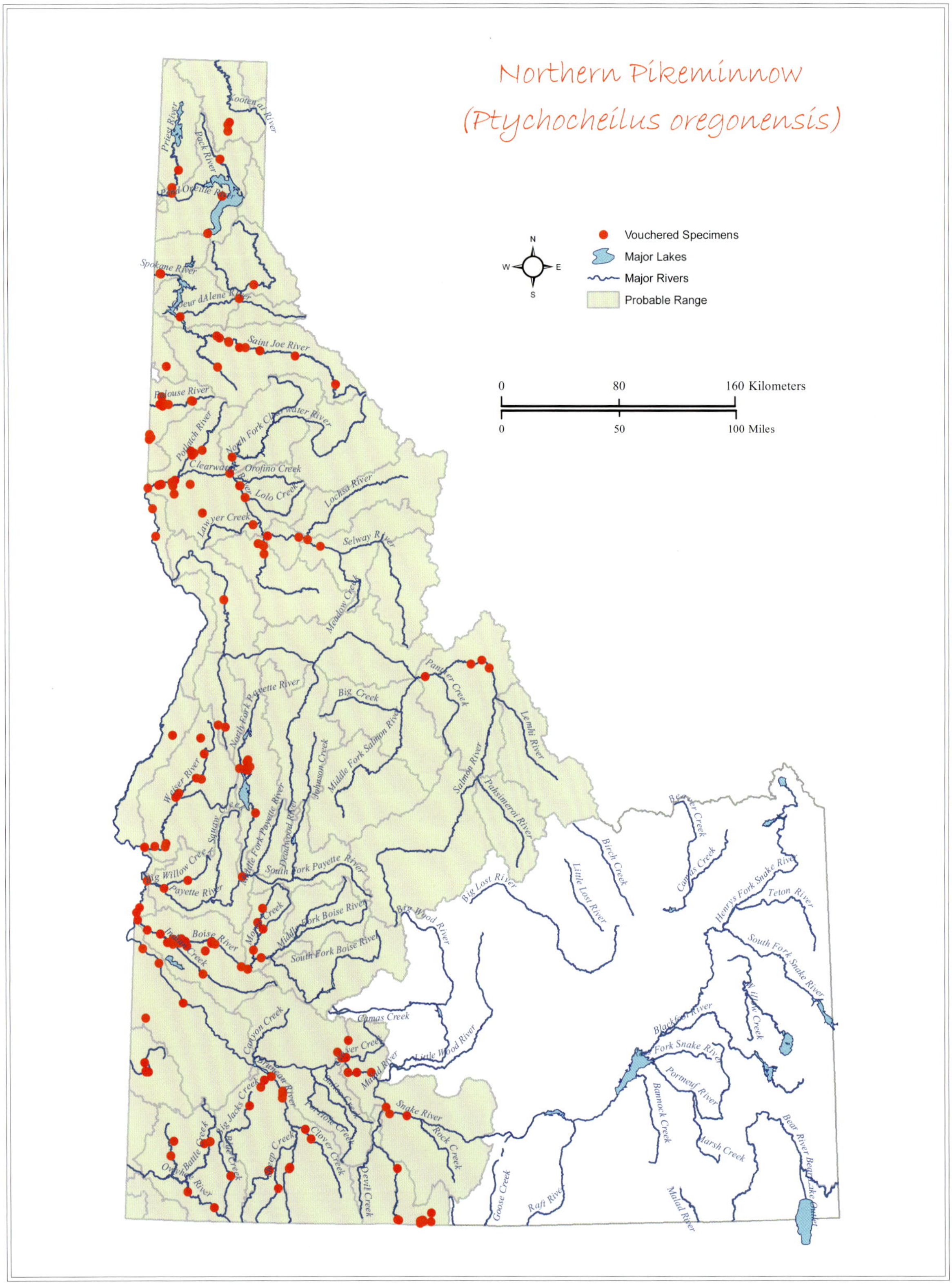
Northern Pikeminnow
(Ptychocheilus oregonensis)
N
W
E
S
Vouchered Specimens
Major Lakes
Major Rivers
Probable Range
0
80
160 Kilometers
0
50
100 Miles

Northern Pikeminnow

Ptychocheilus oregonensis
(Richardson, 1836)

Etymology

Ptychocheilus = fold, lip, the skin of the mouth behind the jaws is folded; *oregonensis* = of Oregon.

Northern Pikeminnow were previously referred to as northern squawfish. Because the former name was objectionable to Native Americans, the common name of this species was changed by the Names of Fishes Committee of the American Fisheries Society to Northern Pikeminnow in 1998. The word "squaw" was a French corruption of the Algonquin Indian term and is viewed as an obscenity[6].

Importance

As might be assumed, any North American fish species that is predatory and grows to a size in excess of 2.3 kg (5.0 lb) would be of interest to humans. Northern Pikeminnow are of interest due to either their direct or their indirect competition with sportfish, that is, as a competitor with "more desirable" game species. Northern Pikeminnow qualify, at least in a minor way, as a sportfish because of their large size. However, anglers quickly find they are not good fighters. In ecosystems where Northern Pikeminnow coexist with salmon or trout (or both), They are in conflict with the most effective and selfish predators (humans). Northern Pikeminnow do consume young salmon and trout, switching to piscivory when they are about 100 mm (3.9 in) long. Flesh of Northern Pikeminnow is soft, white, and edible; however, the abundance of bones makes it less desirable for eating[4].

With a long life span and highly predatory habits, management of Northern Pikeminnow requires consideration of the predator-prey relationship with game species, as they can potentially reduce the survival of some hatchery-released salmonids. Northern Pikeminnow over 300 mm (11.8 in) are considered the most effective predators on salmonid out-migrants in the Columbia River system. From 1991 through 2014, the primary goal of the Northern Pikeminnow Management Program was to reduce Pikeminnow populations by 10 percent using a bounty system paid to anglers and funded by the Bonneville Power Administration. Larger Northern Pikeminnow are the most effective predators on salmonid out-migrants but are also the most vulnerable to angling. For that reason, a large reduction in salmonid predation can be realized through a small reduction of Northern Pikeminnow populations of individuals more than 280 mm (11.0 in) in length[6].

Description

Northern Pikeminnow are dark green or green-brown above, lighter on the sides, and silvery white or cream colored on the belly. Males are more colorful at spawning time when their lower fins become yellow or yellow-orange. Breeding fish have a dark lateral band from the snout to the base of the caudal fin, and small nuptial tubercles form on the heads of males, giving their heads a frosty appearance. Young fish, when placed in preservative, exhibit a black spot at the base of the caudal fin and a narrow lateral band commencing before the dorsal origin and terminating at the caudal spot. Body form of Northern Pikeminnow is elongate, being somewhat laterally compressed. The head is moderately long, and the mouth is large, terminal, and horizontal. All Northern Pikeminnow fins are clear. The single dorsal fin has its origin over, or slightly behind, the origin of the pelvic fins and has nine rays, occasionally 10; the caudal is distinctly forked, and the lobes are pointed. The single anal fin has its origin below the free edge of the dorsal fin, with eight rays, occasionally nine, and the pelvics are small, having nine rays, with their origin under or slightly in advance of the dorsal fin origin. Pectoral fins are relatively short, with 15 or 16 rays. Scales are cycloid and small, numbering 65 to 77 along the complete lateral line. The peritoneum is speckled, and the intestine is short; vertebrae number 44 to 46. Nuptial tubercles are fine, occurring on the head and back and on the pectoral and pelvic fins and occasionally on the caudal fin[4].

Outside the breeding season, sexes are difficult to separate, and the only clear difference is in pelvic fin length. Pelvic fins in both males and females fall short of the vent; however, pelvic fins in mature males come within less than one eye diameter of the vent, while pelvic fins in mature females fall short of the vent by more than one eye diameter[3].

Range and Distributional Status

This species is native to Idaho. Northern Pikeminnow are native along the Pacific slope of western North America, from Oregon north to the Nass River of British Columbia. They occur east of the Continental Divide in the Peace River, Alberta. In Idaho Northern Pikeminnow are found in the Snake River below Shoshone Falls, including the Palouse drainage; and in the Spokane River, Pend Oreille River, and Kootenai River systems[5]. Recently, Northern Pikeminnow have been reported in Shoshone Falls Reservoir upstream of Shoshone Falls but have not been detected upstream of Twin Falls.

Size and Longevity

Northern Pikeminnow are long lived and reach large sizes, females growing slightly faster and living longer than males. Northern Pikeminnow have been documented to age 19 in Montana and 16 years in the Snake River, Idaho. Growth rates of Northern Pikeminnow vary among ecosystems. Growth of the Northern Pikeminnow was variable for Columbia River reservoirs, Columbia River downstream from Bonneville Dam, and for Snake River reservoirs[6].

Northern Pikeminnow have been reported to reach weights of 13.2 kg (29 lb), with a maximum total length of 700 mm (27.6 in). The record length and weight of Northern Pikeminnow in Idaho is 640 mm (25.2 in) and 3.6 kg (7 lb, 15 oz), taken from the Snake River in 2004[37].

Food and Feeding

Young Northern Pikeminnow 25 to 102 mm (1.0 to 4.0 in) TL feed heavily on insects, but as they grow larger they become heavily piscivorous. Very large Northern Pikeminnow feed almost exclusively on other fishes and at times crayfishes. Males have been observed cannibalizing embryos when spawning[4].

Small Northern Pikeminnow, 102 to 254 mm (4.0 to 10.0 in) long, feed primarily on both aquatic insects (mayflies, caddiesflies, damselflies, water boatman, and dipterans) and terrestrial insects (grasshoppers and beetles). For example, the diet of fish 102 to 152 mm (4.0 to 6.0 in) long was one percent fish, whereas the diet of fish larger than 365 mm (14.4 in) was 60 percent fish in a Washington study[6].

In a 2000 Lake Pend Oreille study, Northern Pikeminnow consumed a variety of prey items. Insects were the principal food item of individuals 100 to 150 mm (3.9 to 5.9 in) TL (87.9 percent) and 150 to 305 mm (5.9 to 12.0 in) TL (25.6 percent). Kokanee were the principal food item of Northern Pikeminnow larger than 305 mm (12.0 in). Anglers in various ecosystems have speculated that Northern Pikeminnow consume large quantities of salmonids. In Lake Pend Oreille, this assumption is specifically related to Kokanee. Previous research in Cascade Reservoir, Priest Lake, and Lower Granite Reservoir found that Northern Pikeminnow actually consume mostly crustaceans and insects and to a lesser extent salmonids, being, in general, opportunistic feeders that consume a wide range of dietary items. In the Lake Pend Oreille ecosystem, only individuals larger than 305 mm (12.0 in) regularly consumed Kokanee[38].

Limiting Factors

Northern Pikeminnow may be infested with a variety of parasites, including trematodes, cestodes, nematodes, acanthocephalans, leeches, crustaceans, and mollusks. Eggs or embryos and potentially young Northern Pikeminnow may be eaten by their own species or other fishes in a particular ecosystem[4].

In British Columbia populations, occasional hybrids are found between Northern Pikeminnow, Peamouth, Redside Shiner, and Chiselmouth. Typically, but not always, these hybrids are morphologically intermediate between the parental species. Hybrids between Northern Pikeminnow and Chiselmouth are an ecologically and behaviorally unusual hybrid combination, one parent being a periphyton scraper and the other a piscivore[3].

Breeding Habits

Northern Pikeminnow spawn from late May to July over gravelly shallows, sometimes near shore in lakes, sometimes in lakes near a tributary stream, and sometimes a short distance upstream in a tributary. Spawning fish tend to gather in large numbers; no nest is built. Females may be accompanied by a few to many males, and the small eggs and sperm are released close to the bottom. Eggs, which are adhesive and demersal, are about 1 mm (0.04 in) in diameter and settle in the gravel. Coloration has been observed to be pale orange. Eggs hatch in about one week[4,6].

Sexual maturity of Northern Pikeminnow is attained in about six years when individuals are about 305 mm (12.0 in) TL. In some populations, males mature at age three and females at age four. Water temperatures during spawning have been recorded as 14 to 18°C (57 to 64°F) in lower Columbia River locations and elsewhere. Egg production is 6,700 to 83,000 eggs per female, with larger females generally producing more eggs. Mean fecundity of 197 Northern Pikeminnow from the Columbia River reservoirs was 30,679 eggs per female. Eggs hatch in seven days at 18°C (65°F), and young become free swimming in 14 days. Yolk sac larvae, 9 to 10 mm (0.35 to 0.4 in) SL, drifted from one to three days after hatching in the Columbia River before occupying the warmer, low-velocity shoreline and backwater habitats for rearing[5,6].

Habits and Habitat

Typically, Northern Pikeminnow inhabit lakes and areas of slow to moderate currents in streams. During summer, fish are found either in the shallows or may move to the surface or the pelagic zone of lakes where water temperatures are similar to those in littoral areas. During winter they occupy benthic habitats in deep water, although some individuals may be found up in the water column. Movement is apparently slight, as tagged Northern Pikeminnow have been recaptured in close proximity to where they were originally captured. In studies in the Pend Oreille River, most Northern Pikeminnow were recaptured where they were tagged[6].

Northern Pikeminnow in British Columbia are typically found in lakes and large, slow-moving rivers. Tests of their sustained swimming ability suggest that water velocities greater than 1.0 m/s (3.3 ft/s) may exclude this species. In lake ecosystems during summer, adults often cruise the littoral zone about 1 m (3.3 ft) above the bottom and on the offshore side of weed beds. In winter this species moves into deeper water and is not as bottom oriented[3].

Management and Conservation Status

Northern Pikeminnow are considered common, widespread, and abundant range-wide (G5) and within Idaho (S5). There are no apparent threats to their continued existence in these ecosystems[11]. Numbers have declined in some waters, possibly related to Smallmouth Bass introductions.

"We abuse land because we regard it as a commodity belonging to us. When we see land as a community to which we belong, we may begin to use it with love and respect."

Aldo Leopold

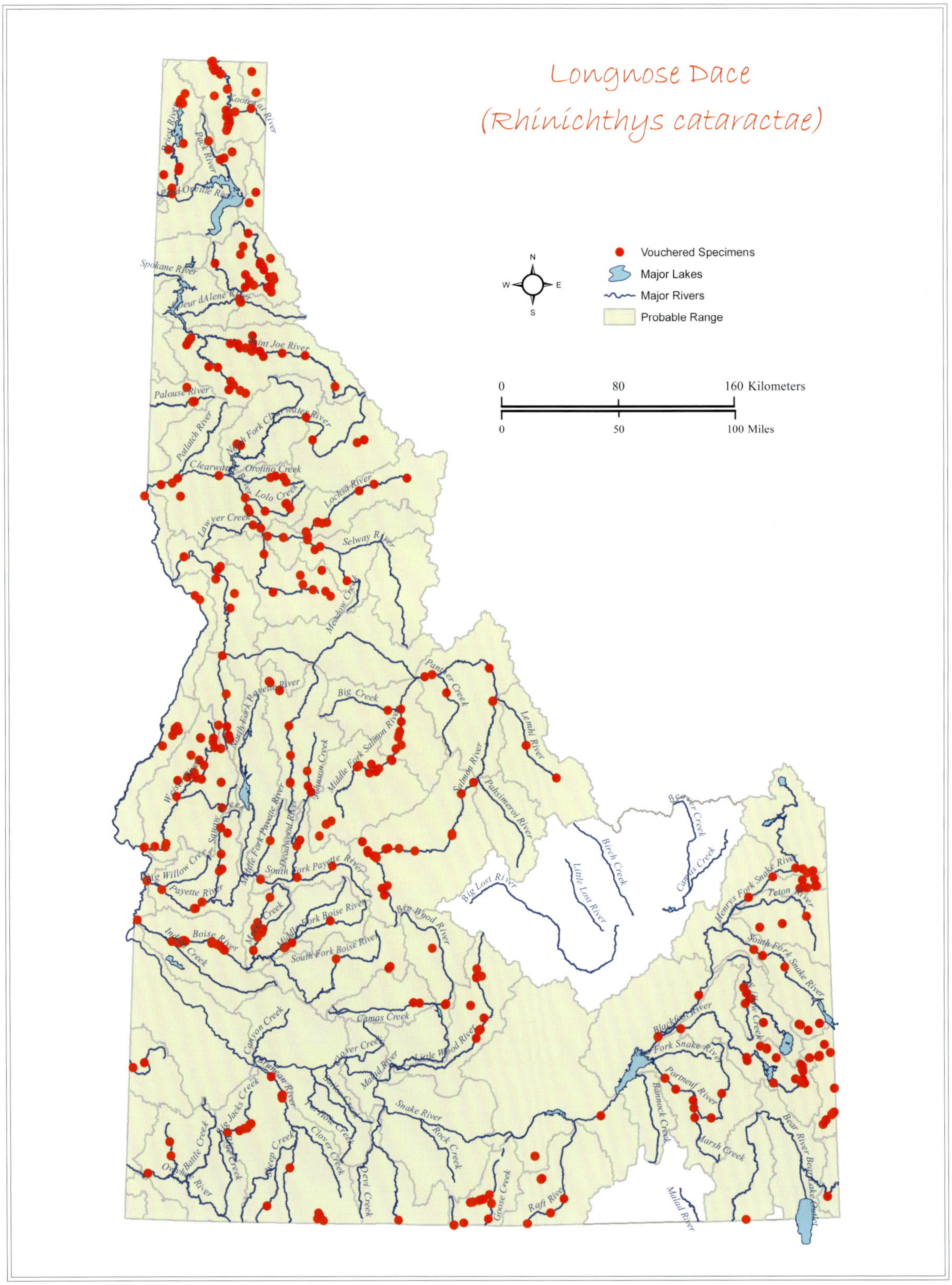
Longnose Dace
(Rhinichthys cataractae)
Vouchered Specimens
Major Lakes
Major Rivers
Probable Range
N
S
E
W
0
80
160 Kilometers
0
50
100 Miles

Longnose Dace

Rhinichthys cataractae
(Valenciennes, 1842)

Etymology

Rhinichthys = snout, fish—this species has a prominent snout; *cataractae* = of the cataract; the original type specimen was from Niagara Falls. The name "dace" originated in England as a modification of the word "darce," referring to the quick movement characteristic of minnows. This term now applies to several species[2].

Importance

Longnose Dace have a wide distribution in Canada and the United States. In Wyoming, Iowa, and Utah, they have been widely used as a bait fish, primarily for bass and catfish. As a small pelagic fish, this species may serve as food for juvenile trout and other predatory game fish[2,4,6].

Description

Longnose Dace vary in color from olive green to brown on the back and shade into cream or silvery white on the belly, with the dark dorsal color often ceasing abruptly below the midpoint of the sides. Lake populations have typically grayish backs. Coloration is variable, as some have a dark stripe ahead of the eye, a dark lateral line from the gill cover to the caudal base, but only on the posterior half of others, or absent in some populations. Bodies of Longnose Dace are stout; the head is triangular and broad. Scales are cycloid and small, numbering 61 to 72 along the straight, or nearly so, and complete lateral line. The peritoneum is silvery, speckled with brown, and the swim bladder is not typically well developed. Vertebrae number 37 to 41. The single dorsal fin has its origin posterior to the pelvic fin origin, with eight rays; the caudal is shallowly forked in the lobes; the rounded anal fin has its origin approximately below the insertion of the dorsal fin, with seven rays; and pelvic fins are small, with the origin in advance of the dorsal origin. Pectoral fins are short, rounded, and somewhat paddle-like, with 13 or sometimes 12 rays. Pharyngeal teeth are usually 2,4 - 4,2, occasionally 1,4 - 4,1, but always four in the main row[4].

Longnose Dace have a long body with a thick caudal peduncle. The snout overhangs the subterminal mouth; the upper lip is attached to the snout by the frenum where the groove between the upper lip and the snout is not complete and does not extend onto the dorsal surface. Morphological variation has been reported among populations of Longnose Dace in various ecosystems in the Pacific Northwest[6].

Breeding males have bright orange to scarlet on the lips and insertions of the lower fins. Breeding tubercles develop on the head and body of males as well as on some females[2,4,6]. Females are not as brightly colored as males. Longnose Dace are nearly round in cross-section. Dorsal and pelvic fins have eight rays, and the pectoral fins have 13 rays. There are 61 to 72 cycloid scales along the lateral line. Longnose Dace may be separated from other species of dace in Idaho by the presence of the frenum that interrupts the upper lip and a snout that extends well beyond the mouth[5].

Range and Distributional Status

This species is native to Idaho. Longnose Dace are widely distributed from the Pacific Ocean to the Atlantic Ocean, ranging as far north as Labrador and the Northwest and Yukon Territories in Canada. Their range extends southward in the Appalachian Mountains to Georgia and in the Rocky Mountains to northern Mexico. In Idaho they are common in every river system, with the exception of the independent (lost) drainages, the upper Malad River, and Curlew Valley[5].

Longnose Dace are endemic to North America, where they have the widest geographic distribution of any indigenous minnow. They occur in rivers from the Atlantic to the Pacific Coasts and from the Arctic to northern Mexico. Given the extensive geographic range of this species, it is not surprising that Longnose Dace have been divided into subspecies. However, there are unresolved taxonomic problems with the Longnose Dace. The supposed western subspecies (*Rhinichthys cataractae dulcis*) was named from a Missouri River tributary, the Sweetwater River in Wyoming. It is thus unlikely that this subspecies occurs west of the Continental Divide. Additionally, mitochondrial DNA data indicate about a 4 percent sequence difference in Cytochrome *b* between Longnose Dace on the eastern and western sides of the Continental Divide. These data argue that the two forms diverged sometime before the beginning of the Pleistocene Epoch (0.8 mya). In addition, behavioral observations indicate major differences in spawning behavior and nuptial coloration between the eastern and northwestern Longnose Dace[3].

Size and Longevity

Longnose Dace have a fairly long life span, with some individuals living to at least five years. Age and growth of Longnose Dace from different ecosystems are remarkably consistent across their range. Both sexes grow at equal rates, but the older and larger fish are generally females, an indication that females may live longer than males. One of the larger Longnose Dace collected in Idaho was a 158 mm (6.2 in) individual collected from the Henrys Fork near Ashton, Idaho[6].

Food and Feeding

Longnose Dace are benthic feeders, as indicated by the subterminal mouth. They are classified as invertivores; young dace feed on drift organisms, but adults feed mainly on immature aquatic insects from gravel, rocks, and the rubble of hard substrate[5].

Studies on Longnose Dace in the Yellowstone River, Montana, concluded that availability and abundance of food organisms were the primary factors governing diet; simulid larvae from fast-flowing water were eaten mainly by the larger dace. Other studies found midge larvae, adult midges (Chironomidae), and mayflies (Ephemeroptera) constituted nearly 90 percent of the food consumed[4].

Nocturnal foraging by Longnose Dace may be related to predation. In Alberta, Canada, Longnose Dace in streams began foraging about 30 minutes after sunset and ceased feeding about 30 minutes before sunrise. In contrast, in the laboratory this species also fed during the day. Feeding efficiency was affected by light levels[6].

Limiting Factors

Longnose Dace have been reported to be infested with trematodes, nematodes, protozoans, mollusks, and crustaceans[4]. Smaller pelagic fry of Longnose Dace probably serve as forage for juvenile trout and other predatory game fish[2,6].

Breeding Habits

Spawning of Longnose Dace generally occurs between May and July when water temperatures reach about 12°C (53°F). Water velocity in spawning areas is 0.5 to 1.0 mps (1.5 to 3.3 fps). From 200 to 1,200 transparent, adhesive eggs are broadcast, and hatching occurs in seven to 10 days at about 16°C (60°F). No nest is built, but territories are established and defended by one parent, which also guards the embryos. This guardianship of eggs and young may, however, be an incidental by-product of male territoriality. The yolk sac is absorbed in about seven days, and once the swim bladder is inflated, young rise to the surface. Young are pelagic for about four months, living primarily in the midwater column, unlike adults, and inhabit low-velocity waters near shore[4].

Habits and Habitat

Longnose Dace are inhabitants of swift water in streams, preferring summer temperatures between about 13 and 21°C (55 and 70°F). In Oregon Longnose Dace inhabit streams that were less than 10 m (33 ft) wide and in water 15°C (59°F) or warmer. Water velocity in these streams was 0.9 mps (3.0 fps) or faster[6]. In lakes Longnose Dace live near shore over rocky, gravelly bottoms and move into deeper water in hotter weather[2,6].

Management and Conservation Status

Longnose Dace are considered common, widespread, and abundant range-wide (G5) and within Idaho (S5)[11]. They require habitat protection for continued abundance throughout the state, including acceptable water quality.

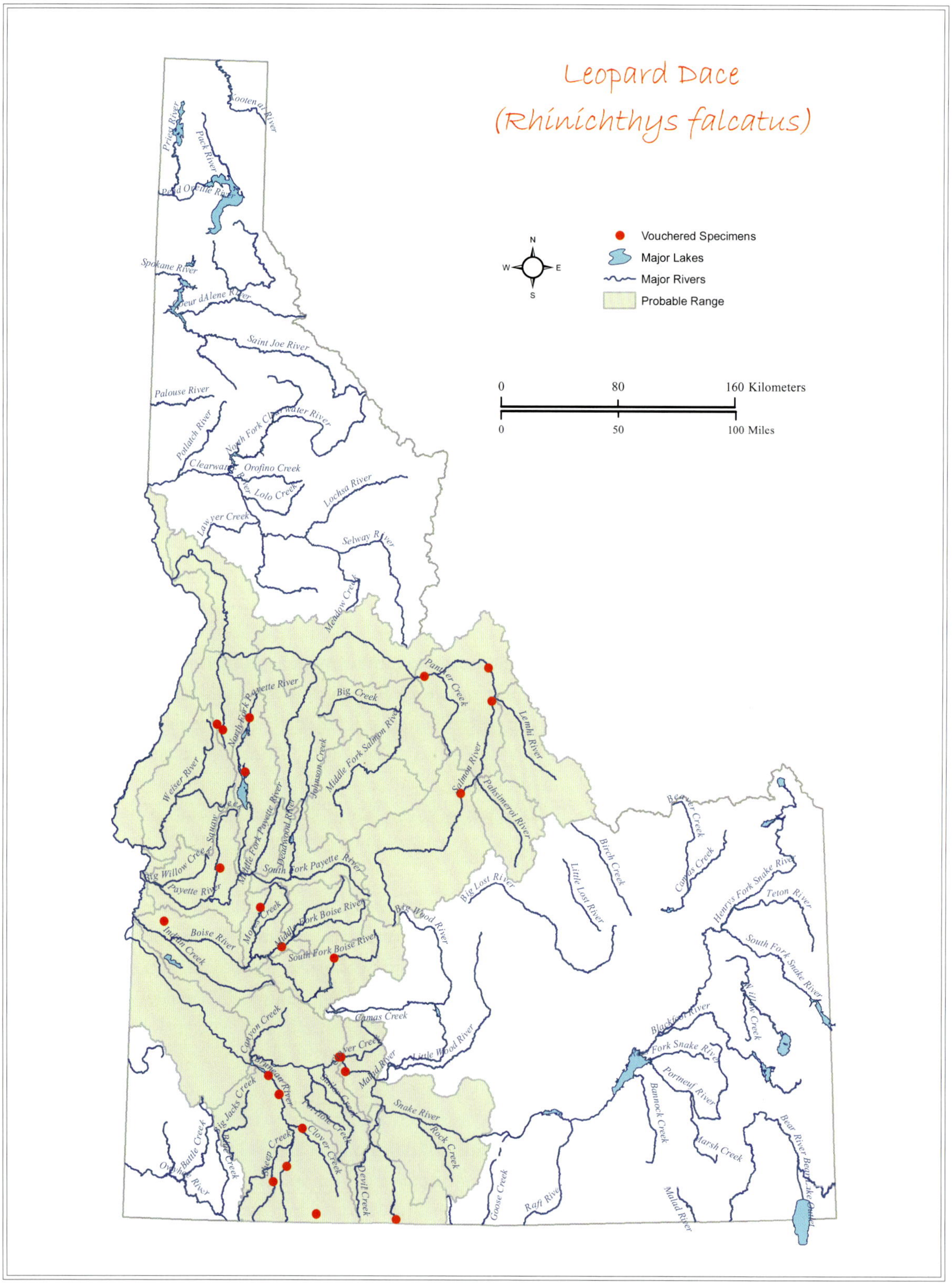
Leopard Dace
(Rhinichthys falcatus)
Vouchered Specimens
Major Lakes
Major Rivers
Probable Range
N
S
W
E
0
80
160 Kilometers
0
50
100 Miles
Kootenai River
Priest River
Pack River
Spokane River
Saint Joe River
Palouse River
Potlatch River
Orofino Creek
Lochsa River
Selway River
Big Creek
Middle Fork Salmon River
Lemhi River
Salmon River
Pahsimeroi River
Weiser River
Payette River
Boise River
Indian Creek
South Fork Boise River
Middle Fork Boise River
Big Wood River
Big Lost River
Little Lost River
Birch Creek
Camas Creek
Henrys Fork Snake River
Teton River
South Fork Snake River
Blackfoot River
Portneuf River
Bannock Creek
Marsh Creek
Bear River
Malad River
Snake River
Rock Creek
Goose Creek
Raft River
Canyon Creek
Big Jacks Creek
Clover Creek
Devil Creek
Willow Creek
Bear Lake Outlet

Leopard Dace

Rhinichthys falcatus
(Eigenmann & Eigenmann, 1893)

Etymology

Rhinichthys = snout, fish—the snout is prominent; *falcatus* = sickle shaped.

Importance

Leopard Dace, like Longnose Dace, probably serve as a link in the food web between invertebrates and piscivorous fishes[6], although little information is available regarding the role of this species in Idaho ecosystems.

Description

Coloration on Leopard Dace is highly variable. However, they are generally light colored and creamy on the sides, being somewhat darker on the back, especially on the head. They are silvery white below, and the sides are marked with large irregularly shaped blotches. Breeding males have fins with a yellow tinge; lips and the base of the pelvic fins are orange-red. The body of Leopard Dace is elongate and distinctly heaviest before the dorsal fin, while being somewhat laterally compressed. The single dorsal fin has a slightly concave trailing margin (hence the sickle-shaped name), and the origin of the dorsal is slightly behind the origins of the pelvic fins, with nine rays. The caudal fin is deeply forked and attached to a slender caudal peduncle; a single anal fin has its origin slightly in advance of the dorsal fin insertion, with seven rays. Pelvic rays are uniquely equipped, with fleshy "stays" (attachments) joining the base of the fin rays to the surface of the body, with eight rays; pectorals are rather small, more or less rounded, typically with 14 rays. Scales on Leopard Dace are cycloid and number 52 to 57 along the complete and almost straight lateral line. The swim bladder in the Leopard Dace is well developed[4].

Leopard Dace and Umatilla Dace were once considered subspecies of Speckled Dace. Morphological differences between these two species are the basis to separate them: Leopard Dace have conspicuous stays or fleshy attachments to the rays of the pelvic fins that are less developed or absent on the pelvic fins of Umatilla Dace. Leopard Dace have blotches on their sides, whereas Umatilla Dace from the type locality near Umatilla, Oregon, have some dark individual scales, giving a freckled appearance[39]. Leopard Dace have a completely inferior mouth; mouth is only slightly inferior in the Umatilla Dace[6]. Leopard Dace have distinctly falcate dorsal fins, whereas Umatilla Dace from the type locality do not have this feature[39].

Range and Distributional Status

This species is native to Idaho, the type specimen being from the Boise River at Caldwell. Leopard Dace occur in the Columbia River system east of the Cascade Mountains and the Fraser River system in British Columbia. Leopard Dace are strongly associated with the main-stem Columbia River and Fraser River and their larger tributaries[3].

Size and Longevity

Leopard Dace grow approximately 20 mm/yr (0.8 in/yr). Fork length (FL) of young-of-the-year in the Fraser River, British Columbia, were 10 to 18 mm (0.4 to 0.7 in) by mid-August, while an adult was 120 mm (4.7 in) FL. Length-frequency analysis suggests that Leopard Dace may live to five years in the wild, but potential longevity may be six to seven years. While males and females are approximately of equal sizes at the end of each growing season, females longer than 46 mm (1.8 in) are heavier than males of the same age[6].

Food and Feeding

While young-of-the-year and age-one Leopard Dace feed primarily on aquatic insects, about 85 percent of the stomach volume in older fish contains terrestrial insects. There is a marked differentiation between Longnose Dace and Leopard Dace feeding patterns in the same section of an ecosystem. Resource partitioning occurs between Longnose and Leopard Dace, as Leopard Dace feed largely on terrestrial insects, while adult Longnose Dace feed primarily on immature aquatic insects. These feeding patterns are typical of insectivore fishes[5,6].

Limiting Factors

Little specific information is available on parasitic infestations in Leopard Dace; they are likely subject to the same variety of parasites as related species. Given their size and habitat preference, they are likely preyed upon by predatory fishes and perhaps birds and mammals.

Breeding Habits

Leopard Dace spawning activity occurs between May and July, depending on ecosystem location. Breeding males have bright orange to scarlet lips and similar scarlet coloration on the insertions of the lower fins. Breeding tubercles develop on the head and body of males and some females. Spawning habitats are likely similar to those of other dace that spawn in riffles where several males spawn with one female. Adhesive eggs are broadcast onto small substrates, such as gravel and stones. Males remain on the spawning site, probably to spawn with other females, as opposed to actually guarding the site, the eggs, embryos, or young[6].

Habits and Habitat

Leopard Dace occur in stream ecosystems and sometimes in lake ecosystems. At night some individuals move into deeper water, while some adults move into shallower water. Age-0 Leopard Dace in a British Columbia stream were collected mostly from backwater pools at depths of less than 46 cm (1.5 ft), while adult Leopard Dace were found at depths of slightly more than 91 cm (3.0 ft) over hard substrates covered by fine sediment. In these stream ecosystems, habitat segregation occurred between Leopard Dace, Umatilla Dace, and Longnose Dace based on water velocities[6].

In rivers juvenile Leopard Dace occupy similar habitats to adults but at sites with lower water velocities and shallower water depths than those occupied by adults. In late summer, juveniles are more common on the slower side-channel sides of bars than are adults, and at these sites the substrate is typically coarse gravel or cobbles. During freshets juveniles move into quiet water, seeking shelter in flooded vegetation along the river edges[3].

Management and Conservation Status

Leopard Dace are considered uncommon but not rare; there is some cause for long-term concern due to declines and other factors range-wide (G4). In Idaho its conservation status is not yet assessed (SNR)[11]. Habitat requirements for Leopard Dace appear similar to the requirements for salmonids in Idaho, although their population and conservation status in Idaho are unknown. No defined studies have been conducted in the state to determine population status.

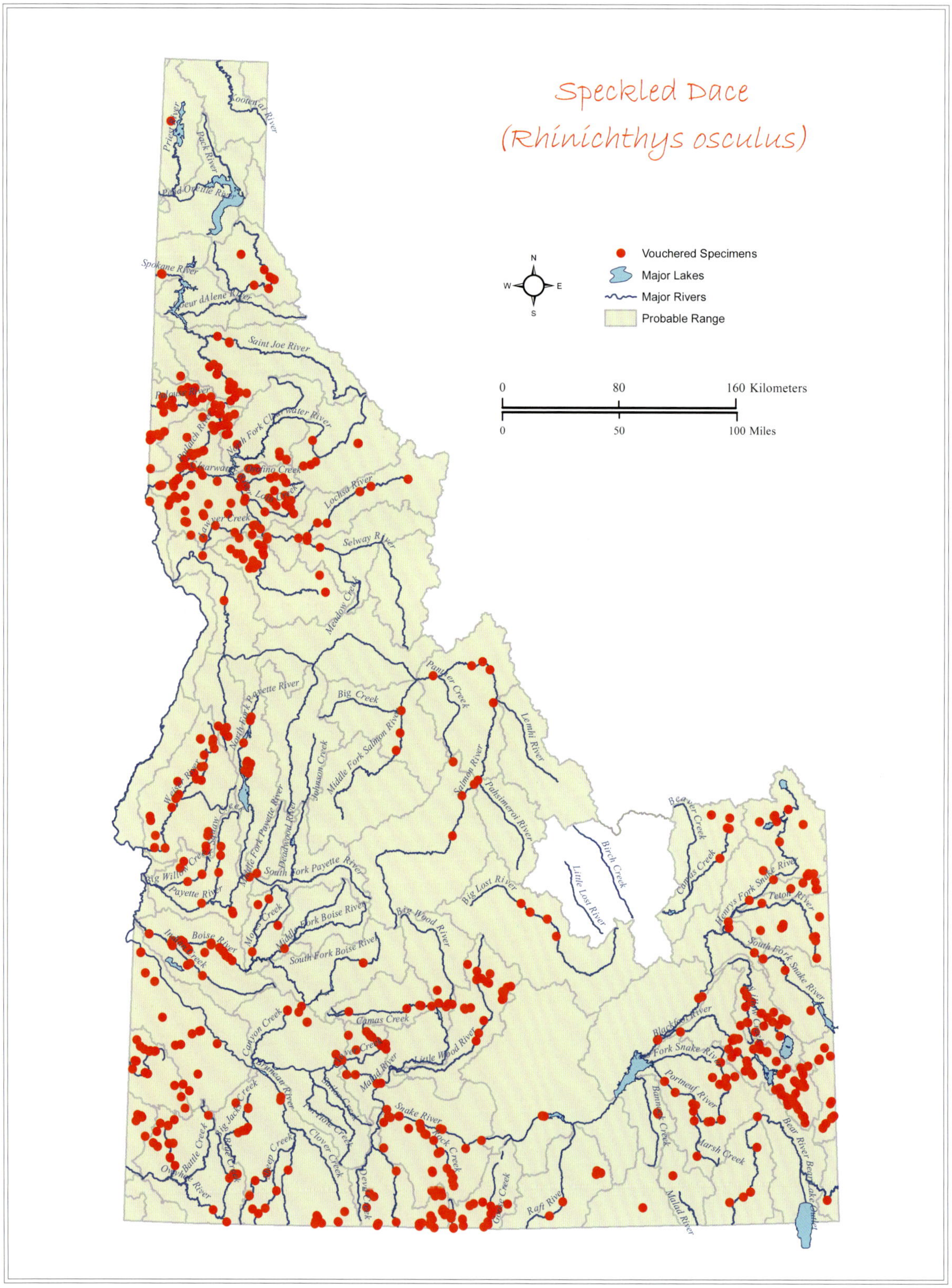
Speckled Dace
(Rhinichthys osculus)
N
W
E
S
Vouchered Specimens
Major Lakes
Major Rivers
Probable Range
0
80
160 Kilometers
0
50
100 Miles
Saint Joe River
Lochsa River
Selway River
Big Creek
Middle Fork Salmon River
Salmon River
Lemhi River
Pahsimeroi River
Birch Creek
Little Lost River
Big Lost River
Big Wood River
South Fork Boise River
South Fork Payette River
Camas Creek
Little Wood River
Snake River
Raft River
Malad River
Bear River
Teton River
South Fork Snake River
Portneuf River
Marsh Creek
Clover Creek
Owyhee River
Boise River
Payette River

Speckled Dace

Rhinichthys osculus
(Girard, 1856)

Etymology

Rhinichthys = snout, fish—the snout is prominent; *osculus* = small mouth.

Importance

Speckled Dace are small and thus provide forage for trout and other game fishes. Speckled Dace have been used as a bait fish in western states, including Wyoming and New Mexico, and in portions of the lower Colorado River[6].

Description

Speckled Dace have overall coloration of gray brown or gray with scattered and vague darker flecks, usually above the midline of the sides; the lower sides and belly are somewhat yellowish or creamy white. In some specimens, a lateral band begins weakly at the dorsal fin and extends onto the caudal peduncle and terminates with a diffuse spot on the caudal fin base; the spot and the lateral band are more conspicuous on young fish. An orange-red coloration appears around the mouth, the upper part of the gill opening, the anal fin base, and the lower caudal lobe of spawning males in Utah. This coloration variant has not been reported for other populations[4].

Body form of Speckled Dace is elongate and rather heavily bodied anteriorly, but distinctly laterally compressed behind the dorsal fin. The head is bluntly triangular, with a small but distinct hump behind the head. There is a fleshy lip on the lower jaw that is overhung by the snout and extends back to about the midnostril in some populations, often with one maxillary barbel at each corner of the mouth. The single dorsal fin has its origin distinctly behind the pelvic fin origin, with a straight posterior margin and eight or nine rays. The caudal fin is forked, and the lobes are rounded; the single anal fin has its origin under the posterior base of the dorsal fin, with six or usually seven rays. Pelvic fins are relatively small and rounded, with eight or nine rays, and the pectorals are small and rounded, with 13 or 14 rays. Scales are cycloid and number about 55 to 75 in the lateral series; the lateral line is complete in most populations but may terminate before the base of the caudal fin. The peritoneum is uniformly dark brown or with darker patches; vertebrae number 37 or 38[4].

Speckled Dace can be easily confused with Northern Leatherside Chub where these species coexist. However, Northern Leatherside Chub have eight anal fin rays and typically about 80 lateral line scales, in contrast to Speckled Dace, with typically seven anal fin rays and 55 to 67 lateral line scales. Speckled Dace have a rather robust body, short caudal peduncle, and a terminal mouth, thus differing from the more definitively inferior mouth and frenum present in the Longnose Dace. A groove completely separates the upper lip from the snout of the Speckled Dace, which contrasts to the Longnose Dace, where the groove between the upper lip and the snout does not extend onto the dorsal surface of the head[6].

Range and Distributional Status

This species is native to Idaho. Speckled Dace are present in tributaries of the Snake, Bear, Spokane, Pend Oreille, and Kootenai Rivers in Idaho as well as a significant number of independent drainages[5,6]. Speckled Dace are widely distributed across western North America, from the Columbia River system to the north, south to the Colorado River system, as well as in coastal and interior drainages[5].

Rhinichthys osculus may actually include more than one taxa[62]. In Utah this taxon may represent at least two species: a southern Sevier River population and a northern Weber River population, which may have Snake River Basin attachments[2]. Throughout the arid regions of western North America, Speckled Dace as a "species" exist in many isolated populations, many of which are morphologically distinct; therefore, numerous subspecies have been named. This has resulted in the taxonomy of Speckled Dace being in disarray over much of their geographic range. Within the Columbia River system, morphological variation in Speckled Dace can be split into three broad categories. In the lower Columbia River and its tributaries, adult Speckled Dace typically possess a distinctive dark lateral band (the *nubalis* form). In the middle and upper Columbia (including the Snake River downstream of Shoshone Falls), adults are lightly speckled and usually lack the dark lateral band; in the Spokane region, some individuals lack the typical barbels; and above Shoshone Falls and in the Wood River system, adult Speckled Dace usually have light-green sides along with large black spots (the *carringtoni* form)[3].

Size and Longevity

Speckled Dace are a short-lived species; few fish live beyond age three. Total length is about 102 mm (4.0 in). After age one, females are slightly longer than males[6].

Food and Feeding

Ecologically, Speckled Dace are associated with stream bottoms. Consequently, food for older individuals is primarily of benthic origin. Young feed primarily on zooplankton, while adults feed on aquatic insects, freshwater shrimp, plant material, and plankton. Stomach contents of Speckled Dace from a Wyoming stream had 21 percent (by volume) algae. Speckled Dace feed during daylight[6].

Limiting Factors

Speckled Dace may be infested by a variety of parasites, including protozoans, trematodes, nematodes, crustaceans, and mollusks[4]. Fishes and birds prey on Speckled Dace. Survival of the species is enhanced by their overwhelming abundance and wide distribution[2].

Breeding Habits

Spawning occurs from June through August, generally with a peak in late June. In some southern locales, two spawning periods may occur, one in spring and a second in late summer. Generally, the second spawning follows periods of seasonally heavy rainfall. Speckled Dace are mature at age two; a high mortality of adults occurs after spawning. Breeding adults are deep orange to scarlet around the mouth and at the base of the anal and pelvic fins. Initial spawning activity begins with males occupying a riffle where rocks and stones are cleaned by the current and by individuals "mouthing" stones. As a ripe female reaches a spawning site, she is immediately accompanied by several males; fertilized eggs sink and adhere to rocks and in interstitial spaces. Eggs adhering to the top of rocks are generally eaten by other fishes in the ecosystem. Reported egg production varies from 170 to 448 for a 48 mm (1.9 in) female to 514 for a 71 mm (2.8 in) female. Laboratory studies indicate that eggs will hatch in six days at water temperatures of 18 to 19°C (64 to 66°F)[6].

Habits and Habitat

Speckled Dace are considered generalists, as they inhabit a wide variety of habitat types across their native range. They typically occupy colder streams and are sometimes found in lakes. Preferences are for water temperatures that range from 0 to 20°C (32 to 68°F), water velocities from 0 to 0.9 m/s (0 to 3 ft/sec), and streams no larger than 20 m (66 ft) wide. They may be the sole species in intermittent streams and in desert springs. Large schools are generally maintained, except during spawning periods. In some populations, they are active at night more than in daylight, spending days among rocks in shallow waters[2,6].

Management and Conservation Status

Speckled Dace are considered common, widespread, and abundant range-wide (G5) and within Idaho (S5)[11]. Although no definitive studies have been conducted in Idaho to determine population status, high-quality spawning and rearing habitat is required for this species.

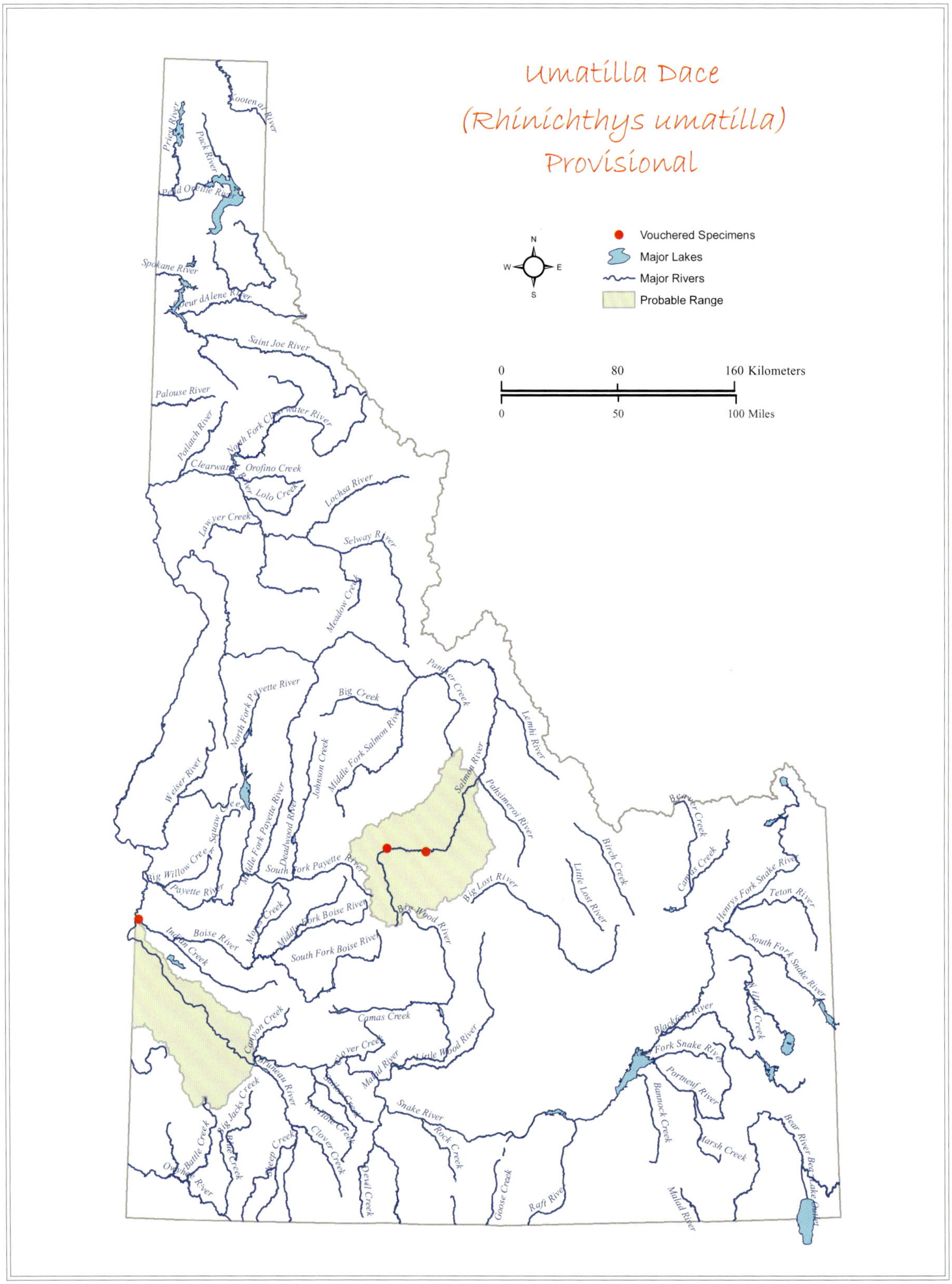
Umatilla Dace
(Rhinichthys umatilla)
Provisional
N
W
E
S
Vouchered Specimens
Major Lakes
Major Rivers
Probable Range
0
80
160 Kilometers
0
50
100 Miles
Priest River
Pack River
Saint Joe River
Palouse River
Potlatch River
Orofino Creek
Lolo Creek
Lochsa River
Selway River
Meadow Creek
Big Creek
Middle Fork Salmon River
Johnson Creek
Weiser River
Middle Fork Payette River
Deadwood River
Big Willow Creek
Payette River
Boise River
Indian Creek
South Fork Boise River
Big Wood River
Big Lost River
Little Lost River
Birch Creek
Pahsimeroi River
Lemhi River
Camas Creek
Little Wood River
Snake River
Rock Creek
Clover Creek
Devil Creek
Goose Creek
Raft River
Bannock Creek
Marsh Creek
Malad River
Teton River
South Fork Snake River
Henrys Fork Snake River
Bear River

Umatilla Dace

Rhinichthys umatilla
(Gilbert & Evermann, 1894)

Etymology

Rhinichthys = snout, fish—the snout is prominent; *umatilla* = the original description by Gilbert and Evermann was from a specimen taken near the town of Umatilla, Oregon.

Importance

Limited information is available on the Umatilla Dace. Spotty distribution of the Umatilla Dace suggests high risk in association with human development and manipulation of ecosystems within the Columbia River Basin. In Washington Umatilla Dace were listed as a "state candidate" species during 1998 because of their spotty distribution and status[6]. Umatilla Dace were recognized as a separate species in 2004 by the American Fisheries Society[5].

Umatilla Dace have a complex taxonomic history. It is noted that even the original describers (Gilbert and Evermann, 1894) expressed doubts about the specific validity of the species with allusions to its unusual morphology, opening the door to the possibility that it might be a hybrid. Umatilla Dace have been treated as a distinct species, a subspecies of Speckled Dace, or simply ignored. Most recent treatments of Umatilla Dace confer status as a distinct species[3].

While many factors are still not fully understood regarding this species and its close relatives, preservation of the species through conservation of habitat, primarily water quantity and quality, is important.

Description

Umatilla Dace have a background color that is brownish on the dorsal surface and grayish on the sides. Previously, Umatilla Dace have been considered a variant form of Leopard Dace. Both Leopard Dace and Umatilla Dace were previously considered subspecies of Speckled Dace. Umatilla Dace have a partially inferior mouth (the snout projecting slightly beyond the upper lip), a thick caudal peduncle; and rounded dorsal and anal fins[6]. Umatilla Dace will typically have more than 60 scales in the lateral series. The caudal peduncle is intermediate between Leopard Dace, which have a narrow caudal peduncle, and Speckled Dace, which have a broad caudal peduncle. Fleshy stays between the pelvic fins and the body are reduced or absent compared to the Leopard Dace[5]. Umatilla Dace from the type locality near Umatilla, Oregon, have some dark individual scales, giving a freckled appearance[39]. Other authors describe Umatilla Dace as having dark blotches nearly continuous along the lateral line. There is strong evidence that the Umatilla Dace evolved through a relatively old hybridization event involving Speckled Dace and Leopard Dace. No evidence is available of contemporary hybridization between Leopard Dace and any other species. The taxonomic status of Umatilla Dace is unresolved.

While most species of *Rhinichthys* exhibit a variety of sexual dimorphism characteristics, including spawning tubercles, spawning tubercles have not been observed and reported on Umatilla Dace. Mature males are separable from mature females by the length of the pelvic fins, with pelvic fins reaching back to the origin of the anal fin in males and falling short of the anal fin origin in females[3].

Range and Distributional Status

This taxon is believed to be native to Idaho but we list it as provisional due to the marked morphological differences between specimens from the type locality (Columbia River near Umatilla, Oregon) and those collected in Idaho. Umatilla Dace have a spotty distribution in the Pacific Northwest, being confined to the Columbia River Basin east of the Cascade Mountains. Umatilla Dace have been reported from the Boise River, Salmon River, Snake River, and Little Wood River of Idaho, but available specimens do not resemble fish from the type locality. Umatilla Dace occur in the Columbia River in Oregon and, in Washington, they have been reported from the Columbia River, Yakima River, Okanogan River, Similkameen River, Kettle River, Colville River, and Snake River[5,6].

Size and Longevity

In Washington young-of-the-year Umatilla Dace are generally under 41 mm (1.6 in) SL. Larger fish from rivers in Washington are about 99 mm (3.9 in) SL. It is estimated they live three or possibly four years in the wild; however, longevity could be between six and eight years[6].

Food and Feeding

Food and feeding habits of the Umatilla Dace are unknown and unreported. It is likely that food habits are similar to other dace that feed primarily on aquatic insect larvae[5,6].

Limiting Factors

Umatilla Dace are probably parasitized by similar organisms as other dace and are preyed upon by predatory fishes, birds, and mammals depending on ecosystem location and complexity.

Breeding Habits

In Washington Umatilla Dace apparently spawn in early to mid-July, while spawning in British Columbia probably occurs later, in mid-July and early August. Umatilla Dace and Leopard Dace are sympatric in the Columbia River Basin. However, few specimens with intermediate characteristics have been found, indicating that the two taxa remain reproductively isolated[6]. Although unreported, the breeding habits and characteristics of Umatilla Dace are likely similar to those of Leopard Dace and Speckled Dace.

Habits and Habitat

Umatilla Dace are found in relatively productive low-gradient and low-elevation streams in ecosystems where substrates have very little or no silt. Juvenile Umatilla Dace occur in shallow water with cobble and rubble substrates that may be covered with dense mats of green algae. Umatilla Dace typically occupy habitats with water velocities less than 0.5 m/s (1.6 f/s). They are often found along stream banks in shallow water less than 1.0 m (3.3 ft) deep. Larger individuals (subadults and adults) occupy deeper water habitats[6].

Management and Conservation Status

Umatilla Dace are considered uncommon but not rare; there is some cause for long-term concern due to declines and other factors range-wide (G4). In Idaho their conservation status is not yet assessed (SNR)[11] and the distribution of Umatilla Dace in Idaho is unknown.

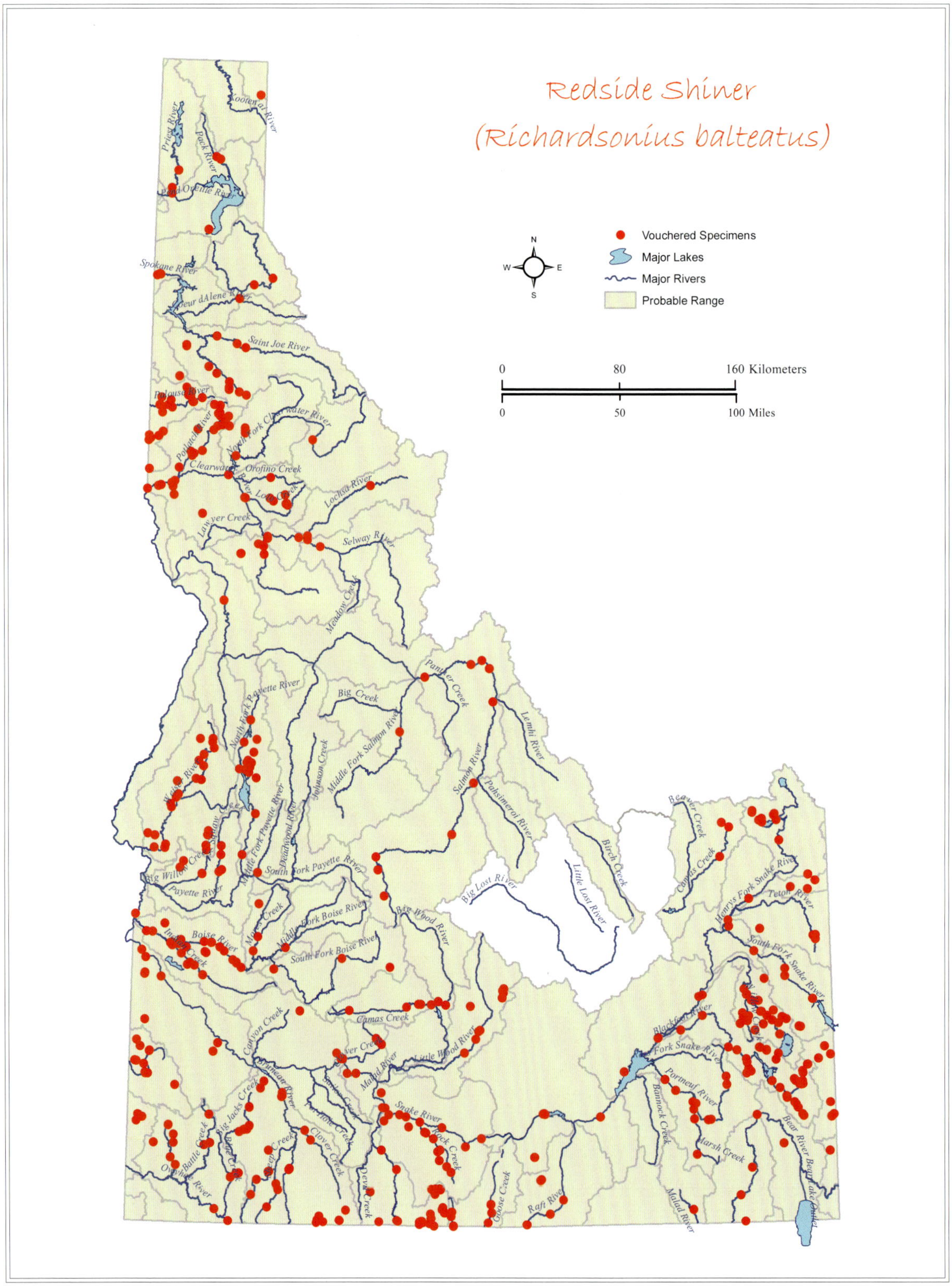
Redside Shiner
(Richardsonius balteatus)
N
S
W
E
Vouchered Specimens
Major Lakes
Major Rivers
Probable Range
0
80
160 Kilometers
0
50
100 Miles
Priest River
Pack River
Spokane River
Saint Joe River
Clearwater River
Orofino Creek
Lochsa River
Selway River
Meadow Creek
Lawyer Creek
Panther Creek
Lemhi River
Big Creek
Middle Fork Salmon River
Salmon River
Pahsimeroi River
Birch Creek
Little Lost River
Big Lost River
Big Wood River
Camas Creek
Little Wood River
South Fork Payette River
Payette River
Weiser River
Boise River
South Fork Boise River
Canyon Creek
Snake River
Rock Creek
Goose Creek
Raft River
Portneuf River
Bannock Creek
Marsh Creek
Malad River
Bear River
Teton River
South Fork Snake River
Henrys Fork Snake River
Camas Creek
Beaver Creek
Salmon Falls Creek
Clover Creek
Big Jacks Creek
Castle Creek
Owyhee River

Redside Shiner

Richardsonius balteatus
(Richardson, 1836)

Etymology

Richardsonius = named for Sir John Richardson, a surgeon-naturalist with the Franklin expeditions of the early 1800s searching for the Northwest Passage, who first described it; *balteatus* = girdled, possibly referring to the brilliant red streaks occurring on males during breeding.

Importance

Redside Shiner are of interest because of their spectacular but puzzling morphological variation and well-documented impact on sport fisheries, particularly for Rainbow Trout, in some ecosystems. Redside Shiner can seriously compete for food and space with various species of trout in a given ecosystem. Population densities as high as 2,000 to 40,000 individuals/ha (5,000 to 100,000/ac) make this species both a serious competitor and a substantial food source for Rainbow Trout or other game species occurring in the ecosystem[4].

Description

Coloration of Redside Shiner is steely blue, olive, or dark brown to black on the body and head. On flanks below the countershading of the dorsal surface is a narrow, unpigmented band running from the operculum to the dorsal fin. A dark lateral band is present on the sides, extending from the snout to the caudal peduncle, and is broadest below the dorsal fin but often inconspicuous from the head to the tip of the pectoral fins. Below the lateral line, the body is bright silver, and the ventral surface is silvery or white. Fins are clear but amber to dusky. Both sexes are more highly colored at spawning time, when the female is a golden color above and below the lateral band and on the head and the male is more strikingly colored, with a band of yellow between the dorsal countershading and the more intensified dark lateral band. A broad orange to crimson stripe is apparent below the lateral band that may run from the operculum to the anal fin. The lower part of the body is yellow, as is most of the head below the eye; the lower fins are bright yellow, especially in males[4].

The body of Redside Shiner is variable but moderately deep with a long head and strongly compressed laterally. The snout is moderately long and bluntly pointed. Pharyngeal teeth occur in several combinations but are typically 2,5 - 5,2 and 2,5 - 4,2. Gill rakers are short and stubby, numbering six to nine on the first arch. A single dorsal fin is moderately large, occurring at the midpoint of the body, and the origin is well in advance of the origin of the anal fin but behind the insertion of the pelvic fins. Dorsal fin rays number eight to 12 but typically 10. Anal fin rays number 10 or more, up to 22. The caudal fin is long and deeply forked, and the tips are pointed and markedly falcate. Caudal rays are variable, ranging from 13 to 18 but may be 10 to 24. Pelvic fins are abdominal and low, with small, rounded tips and eight or nine rays. Pectoral fins have 13 to 16 rays and are larger than the pelvics, and the tips have rounded points. Both the pelvic and the pectoral fins are longer in males than females. Redside Shiner have small cycloid scales numbering 50 to 67 along the complete and weakly decurved lateral line. The peritoneum is pale, silvery, and lightly speckled with black; vertebrae number 38 to 43. Nuptial tubercles appear only on males across the head and upper surface of the paired fins[2,4].

Redside Shiner are one of the most variable species in northwestern North America, and there are a variety of recognizable subspecies. Redside Shiner, which occur above and below Shoshone Falls on the Snake River, are separable: the form found above the falls is the subspecies *R. balteatus hydrophlox*, which is more slender bodied than the Columbia River form. *Richardsonius* are also native to a cluster of rivers on the central Oregon coast, and the Oregon coastal form was originally named as a separate species; it has not been recognized as such for many years. However, there are both morphological and mitochondrial differences between the Oregon coastal form and the Columbia Basin Redside Shiner, suggesting that the coastal form may be a valid species[3].

Range and Distributional Status

This species is native to Idaho. Redside Shiner are distributed across North America, occurring mainly west of the Rocky Mountains. Distribution extends from the Peace River in northwestern Alberta throughout much of British Columbia and south to the coastal streams of Washington and Oregon. Redside Shiner also occur in isolated drainages in eastern Oregon and in the Bonneville Basin in Idaho, Utah, and Wyoming. In Idaho, this species is common in all major river systems, except the independent drainages (known only from the Beaver Camas watershed). They have been introduced into the Colorado River[2,4,5].

Size and Longevity

Maximum total length of Redside Shiner is about 178 mm (7.0 in); however, most are less than 127 mm (5.0 in) long. Life span is rather short, with few fish living beyond age five. Specimens collected in Montana were up to 140 mm (5.5 in) long. Females are generally larger than males of the same age[6].

Food and Feeding

Redside Shiner fry feed on zooplankton and algae, while adults are active foragers, feeding primarily on aquatic and terrestrial insects and snails. In a study of 55 individuals from the Hanford Reach of the Columbia River, Redside Shiner diet consisted of adult caddis flies, caddisfly pupae, caddisfly larvae, and chironomid pupae[6]. Adult Redside Shiner are considered to be omnivorous, feeding on aquatic and terrestrial insects, snails, and zooplankton as well as fish eggs and fry, including their own when available[2].

Limiting Factors

Redside Shiner are preyed upon by predatory fishes, including Rainbow Trout, Cutthroat Trout, Smallmouth Bass, and Northern Pikeminnow. Additionally, many fish-eating waterfowl such as mergansers and loons and mammals such as mink are also predators on this species[4].

Incidence of parasitism is often high in dense schools of Redside Shiner, both in variety and in number. Redside Shiner may be infested by many varieties of trematodes, cestodes, crustaceans, nematodes, and acanthocephalans[4].

Breeding Habits

In general, Redside Shiner are sexually mature at age two or three. In the Yellowstone Lake ecosystem, this species matures between 71 and 99 mm (2.8 and 3.9 in) in length. During spawning season, males become highly colored, with crimson and gold on their sides; however, the females remain drabber. Spawning occurs in the spring and early summer, ranging from April through July, depending on location of the ecosystem. Water temperatures during spawning are generally 14 to 18°C (58 to 64°F); in cooler water, spawning occurred from mid-June until the first week of July at water temperatures between 6.8 and 10°C (44 and 50°F). Spawning occurs over gravel bottoms or in vegetation along lake shorelines. Adhesive eggs are generally broadcast at night, and egg production is related to female size. Egg production for females 88 to 104 mm (3.5 to 4.1 in) long ranged from 829 to 3,602 eggs. In laboratory tests, embryos hatched in three to seven days at water temperatures of 21 to 23°C (70 to 73°F). Redside Shiner do not exhibit territoriality on spawning sites, but a defined homing behavior was reported in British Columbia lakes. Spawning mortality is high, though some fish survive to spawn two or more successive years[6].

Habits and Habitat

Redside Shiner are nearly ubiquitous in aquatic ecosystems across their native range. This species is found in a variety of habitats where currents are slow to moderate: in large rivers, streams, spring, sloughs, irrigation ditches, ponds, and lakes where summer water temperatures are 12.8 to 20°C (55 to 68°F). They have also been reported at temperatures as cold as 6.8°C (44°F) and as high as 24°C (75°F) during the summer. Redside Shiner typically move in schools and tend to occupy habitats with aquatic vegetation when they are in shallow areas of these ecosystems. It is typical for all individuals in a particular school to be about the same size. On an annual cycle, schools move to near-shore areas in spring and remain there until July, moving to deeper water by August, when surface water temperatures become warm. They also exhibit a smaller diel movement pattern when in shallow water during the day but move into deeper water at night. In Yellowstone Lake, Wyoming, Redside Shiner spawn in shallow areas with submerged aquatic vegetation; however, during the remainder of the year, they occupy the littoral zones of lagoons and bays protected from wave action. They generally occur in water less than 3.4 m (11 ft) deep where rooted aquatic vegetation is often abundant[2,5,6].

During the summer, young-of-the-year Redside Shiner inhabit shallow waters of less than 1 m (3.3 ft) along lake and stream margins. They often occur in aquatic vegetation and are typically associated with mixed schools of under-yearling Peamouth and Northern Pikeminnow. In stream ecosystems, young Red Shiner prefer quiet waters with velocities less than 0.1 m/s (0.33 ft/s) with fine-grained substrates in water less than 0.5 m (1.6 ft) deep. They may be abundant in shallow backwaters in weedy bays and are typically associated with mixed schools of other minnows[3].

Management and Conservation Status

Redside Shiner are considered common, widespread, and abundant range-wide (G5) and within Idaho (S5)[11]. However, no defined studies have been conducted in Idaho to determine population status. Species preservation for Redside Shiner requires maintenance of good water quality and habitat in a particular ecosystem.

Kackley Preserve, Sagebrush Steppe Land Trust, Bear River, near Grace, ID. Photo Courtesy John Sigler.

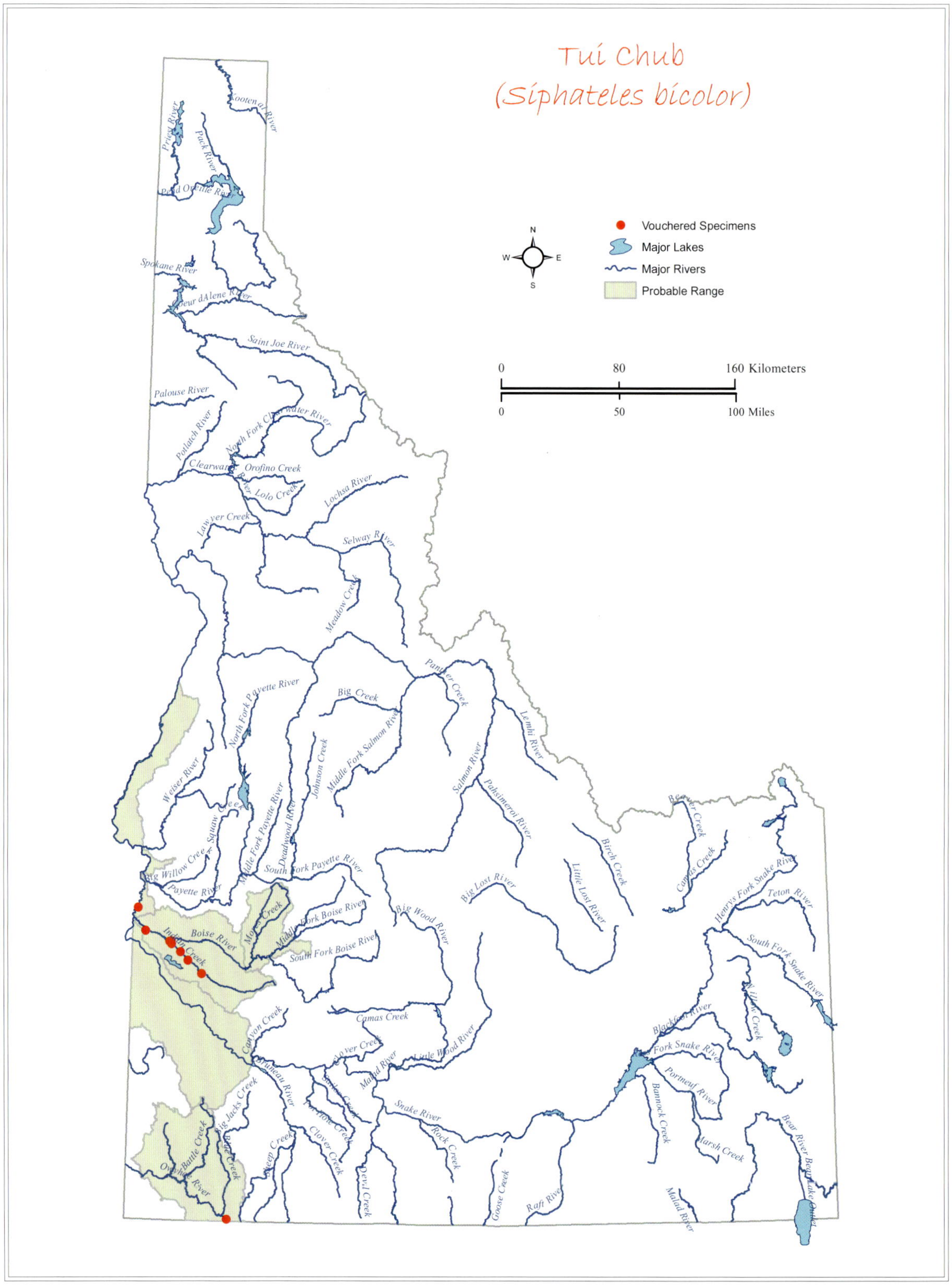
Tui Chub
(Siphateles bicolor)
N
W
E
S
Vouchered Specimens
Major Lakes
Major Rivers
Probable Range
0
80
160 Kilometers
0
50
100 Miles
Pack River
Saint Joe River
Palouse River
Potlatch River
Orofino Creek
Lochsa River
Selway River
Big Creek
Middle Fork Salmon River
Johnson Creek
Salmon River
Lemhi River
Pahsimeroi River
Birch Creek
Little Lost River
Big Lost River
Big Wood River
Weiser River
Payette River
Boise River
South Fork Boise River
Camas Creek
Teton River
South Fork Snake River
Henrys Fork Snake River
Snake River
Rock Creek
Clover Creek
Devil Creek
Goose Creek
Raft River
Bannock Creek
Portneuf River
Marsh Creek
Malad River

Tui Chub

Siphateles bicolor
(Girard, 1856)

Etymology

Previously assigned to the *Gila* genus. *Gila* for the Gila River, New Mexico, from which it was thought the first fish referred to in this genus had been obtained (type species). The specimens actually came from the Zuni River in New Mexico. *Siphateles* = incomplete siphon, meaning obscure; *bicolor* = two colored, referring to the coloration of the species. *Tui* is derived from the Native American Paiute name "tui-pagwi," where "pagwi" appears to be the Paiute word for minnow.

Importance

Tui Chub are an important forage species over most of their range; rarely are they a serious predator of game fishes. Lahontan Cutthroat Trout in Pyramid Lake, Nevada, feed on Tui Chub after reaching a length of 300 mm (11.8 in) and progressively more heavily as they become larger. Tui Chub reproduce rapidly at a young age and because of their slow growth and relatively small size are vulnerable to predation throughout their entire lives. In the Pyramid Lake ecosystem, coevolution of the Tui Chub and Lahontan Cutthroat Trout has resulted in a unique predator-prey system. Historically, large Lahontan Cutthroat Trout in Pyramid Lake grew large due to the abundance of Tui Chub in the ecosystem, resulting in a high-energy efficiency transfer and subsequently larger Cutthroat Trout[8].

Description

Tui Chub are highly successful in a wide variety of ecosystem types, and they respond to variations in ecosystem type with changes in color, shape, size, and at times morphology. Adaptability is accomplished at the subspecies level. Almost every isolated or partially isolated drainage system in California, Nevada, and Oregon supports at least one distinctive subspecies or form. This variation in morphology created considerable confusion early in the 1800s and is associated with nomenclature changes based on differences expressed in populations of Tui Chub. Coloration is deep olive dorsally, lighter laterally, and white on the belly. Upper parts and sides have a very pronounced brassy reflection, and some of the scales are darker than others; some are pinkish. Fins are olive and may be strongly tinted with red, and the belly may be yellow. Other individuals may be greener than olive, and some may have a dark lateral color extending almost to the ventral surface. Colors are more pronounced for individuals in lake ecosystems. Young of this species have a narrow dark stripe along the lateral line. The head is pointed, and the mouth is rather oblique; the eye is large, and the lateral line is slightly curved down. Peritoneum is dusky, the intestine is short, and the scale count in the lateral series ranges from 44 to 60[8].

Two forms of Tui Chub exist in Pyramid Lake, Nevada. A bottom-dwelling and coarse gill-rakered form feeds only inshore. The open-water and fine-gill-rakered form feeds more on plankton[8].

Range and Distributional Status

This species is not native to Idaho. Tui Chub are native to and widely distributed in the Lahontan Basin and central systems of the interior West and in the Owens River and Mohave River Basins of the Great Basin. This species is present in the upper Pit River and Goose Lake of the Sacramento–San Joaquin system in California and in the Klamath River system in Oregon. They are widespread in the Columbia River drainage of Washington and Oregon, to which they may have been introduced, as well as the interior basins of Oregon. They are present in drainages on both sides of the Sierra Nevada and are abundant in Pyramid Lake, Tahoe Lake, Eagle Lake, Walker Lake, Nevada reservoirs, and the Humboldt River system of Nevada[8]. In Idaho Tui Chub are known from the Boise and Owyhee River drainages.

Size and Longevity

Tui Chub are neither fast growing nor long lived. Occasionally, this species may reach a size of more than 410 mm (16.1 in), but average (mean) length is 230 to 280 mm (9.1 to 11.0 in). In Pyramid Lake, Nevada, the total length of Tui Chub aged one to seven, in mm (in), was 137 (5.4), 191 (7.5), 239 (9.4), 287 (11.3), 345 (13.6), 401 (15.8), and 417 (16.4). Females in Pyramid Lake grow faster than males and live to be seven years of age, while the oldest males are four years old[8,40].

Food and Feeding

Tui Chub eat invertebrates, vascular plants, algae, and occasionally fishes. Individuals less than 102 mm (4.0 in) long ate little plant material, but as size increases they feed more frequently on algae. In Pyramid Lake, the bottom-dwelling (coarse-rakered) form eats more plants and algae than the open-water (fine-rakered) form that feeds both inshore and offshore. Fine-rakered individuals feed more on plankton than the coarse-rakered individuals. Tui Chub of all sizes feed most heavily on zooplankton in winter and spring. Large Tui Chub occasionally feed on fish. In a Pyramid Lake study (n=551), Tui Chub ate algae, zooplankton, ostracods, midge larvae and pupae, vascular plants, fishes, terrestrial insects, and amphipods. In Lake Tahoe, inshore forms eat 89 percent invertebrates, 5 percent fish and fish eggs, 3 percent plankton, and 3 percent plants. The open-water form in Lake Tahoe feeds almost exclusively (92 percent) on plankton, mostly consisting of microcrustaceans. Young feed on crustaceans and rotifers, gradually shifting to larger organisms as their size increases. In the Pyramid Lake ecosystem, Tui Chub move to the surface to feed between eight at night and four in the morning, following the zooplankton that migrate upward at night. During summer months, they may feed in water more than 61 m (200 ft) deep and in dissolved oxygen concentrations less than two mg/l (ppm)[8,40].

Limiting Factors

Tui Chub are preyed upon heavily throughout their lives by fishes (particularly Cutthroat Trout), birds, and occasionally mammals. In some areas, water diversions seriously damage Tui Chub habitat. A dominant subspecies of Tui Chub may extirpate other subspecies in a given ecosystem[40].

Breeding Habits

Tui Chub spawn from late April to as late as early August. In Pyramid Lake, Nevada, spawning temperatures ranged from 17 to 22°C (62 to 72°F), and peak spawning activity is reached in June when water temperatures are in the mid- to high teens °C (low to mid-60s°F). Females mature at age two or three at lengths of 200 to 250 mm (7.9 to 9.8 in); occasionally, males will mature at age one, although most mature at age two at about 200 mm (7.9 in). Tui Chub females produce 6,100 to 68,900 eggs with an average of 23,300. When spawning is initiated, males move inshore ahead of females, generally in the vicinity of heavy beds of vegetation and shallow water. Females, attended by several males, deposit adhesive eggs over a wide area. Neither parent provides care of embryos or young. Young remain close to shore in the vicinity of heavy vegetation for most of the summer[40].

Habits and Habitat

Across their native range, Tui Chub are adapted to a wide range of habitats, from small streams to large lakes. They occur as a distinctive form (subspecies in many cases) in most isolated or partially isolated drainage systems in California, Nevada, and Oregon. Tui Chub have also been successful in reservoirs and are the most abundant fish in many Lahontan drainage reservoirs, such as Stampede Reservoir and Boca Reservoir on the Little Truckee River system. They prosper in cold Lake Tahoe and small streams that may attain temperatures above 20°C (70°F). During summer months in Pyramid Lake, Nevada, they form large donut-shaped schools near the surface over deep water and more than 90 m (295 ft) across. Schooling groups in shallow water move randomly or parallel to shore. Tui Chub tend to disperse into deep water in late fall; however, many remain in relatively shallow water or near the surface in deep water. Juveniles form schools and swim parallel to shore in quite shallow waters or congregate in heavy vegetation[40].

Management and Conservation Status

Tui Chub are considered uncommon but not rare; there is some cause for long-term concern due to declines or other factors range-wide (G5). A state conservation status rank is not applicable (SNA) because the species is not a suitable target for conservation activities[11]. Because of their high reproductive rate and great adaptability to varied ecosystems, Tui Chub need protection only from degradation of habitat caused by water diversions or pollution. At least 16 subspecies potentially exist. Tui Chub, being somewhat unique in their ability to adapt to a wide range of ecosystem and climatic conditions, should be preserved at the subspecies level as a representative of unique biological diversity. Overall effects of this nonnative species have not yet been assessed in Idaho waters.

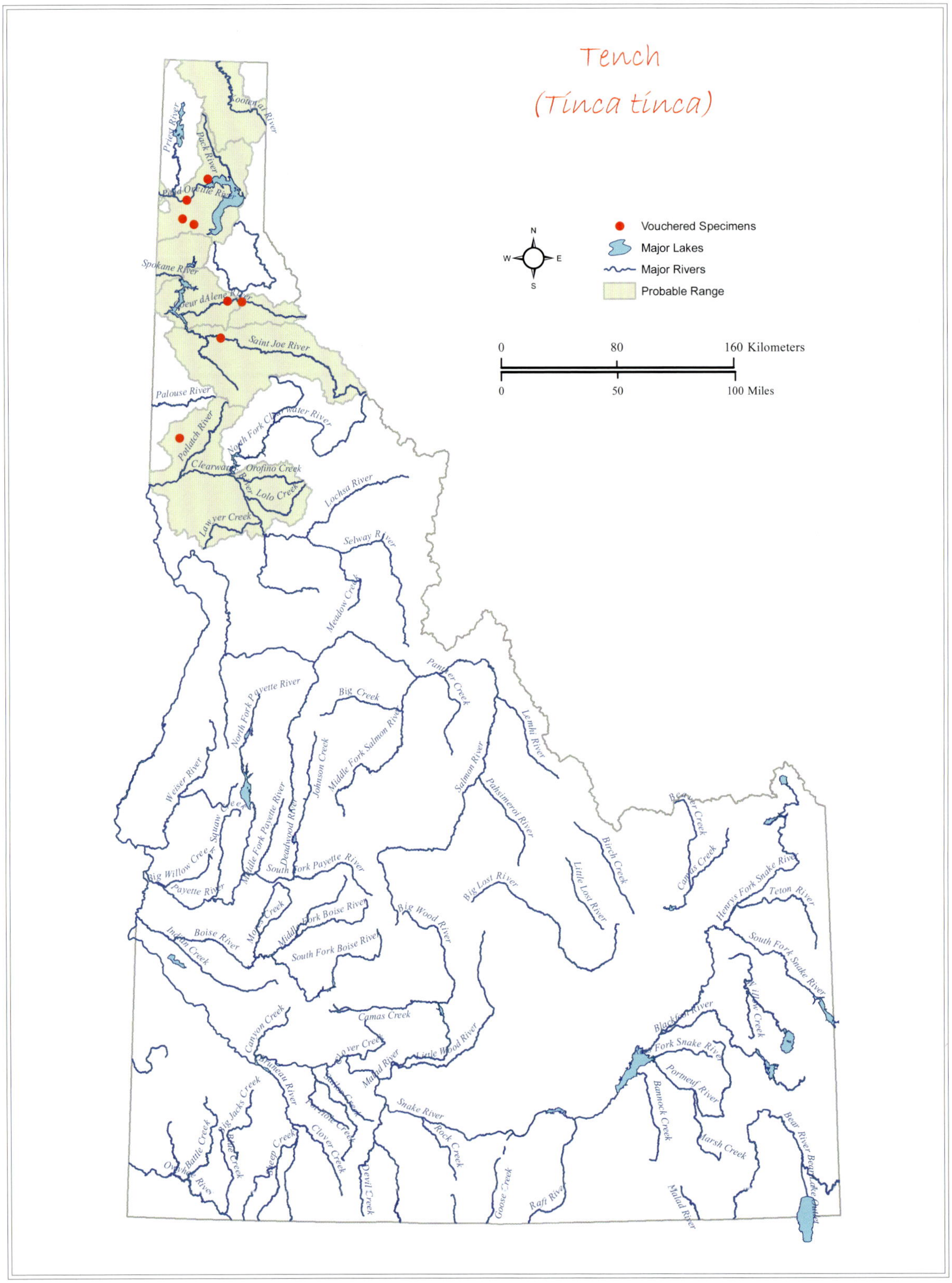
Tench
(Tinca tinca)
N
W
E
S
Vouchered Specimens
Major Lakes
Major Rivers
Probable Range
0
80
160 Kilometers
0
50
100 Miles
Priest River
Pack River
Spokane River
Saint Joe River
Palouse River
Potlatch River
Orofino Creek
Lolo Creek
Lochsa River
Meadow Creek
Big Creek
Johnson Creek
Middle Fork Salmon River
Salmon River
Lemhi River
Pahsimeroi River
Birch Creek
Little Lost River
Big Lost River
Big Wood River
Weiser River
Payette River
Boise River
South Fork Boise River
Canyon Creek
Camas Creek
Snake River
Rock Creek
Goose Creek
Devil Creek
Clover Creek
Bruneau River
Big Jacks Creek
Teton River
South Fork Snake River
Portneuf River
Bannock Creek
Marsh Creek
Malad River
Bear River

Tench

Tinca tinca
(Linnaeus, 1758)

Etymology

Tinca = Latin name for Tench.

Importance

Tench have been widely introduced in many waters in the United States. Tench are not sufficiently abundant in any US ecosystems to create any particular problems. In Canadian and most North American waters, they are ignored, whereas in Europe Tench have a following of devoted anglers and are regarded as a useful species in pond culture. Others deem it a good food fish. A golden variety is used for ornamental ponds. Young Tench may serve as forage for predaceous sportfishes[4,6].

Description

Overall coloration of Tench is olive green to golden, or at times dark green or almost black, with golden reflections laterally and darker golden reflections on the ventral surface. Fins are dark[4,6].

Tench have a thick-set, heavy, and laterally compressed body with a characteristically deep and short caudal peduncle. The head is triangular; the eye is orange-red in color and small. The mouth is terminal and small, with thick lips and a well-developed barbel at each corner. Gill rakers number about 13. All fins are dark and rounded. A single dorsal fin is rounded, with nine rays; the caudal is shallowly or not forked, and the lobes are rounded; the anal fin is rounded, with seven or eight rays. Pelvic fins are rounded; adult males have a thickened second ray not present in females or in immature fish. Pectoral fins are relatively short and rounded. Scales along the complete lateral line are small and embedded, numbering 95 to 100, and vertebrae number 38 or 39[4].

Range and Distributional Status

This species is not native to Idaho. Tench range over the flatland rivers and lakes of Europe typically northward to 61°N. In Asia they occur in the Arctic Ocean drainage, including the basins of the Ob and Yenisei, but are rare in Lake Baikal. Tench have been introduced into North America, including the Columbia River basin, lakes in the Puget Sound drainages, and elsewhere in the United States, such as the Potomac River[4]. Introduction of Tench started with their cultivation by the US Bureau of Fisheries as early as 1883. In Idaho Tench are typically found in the Spokane River and Pend Oreille River systems.

Size and Longevity

In some ecosystems, Tench grow to lengths of about 229 to 250 mm (9.0 to 9.8 in) and weights of 0.3 kg (10.6 oz) in about four years, at which time they may be mature. They have a total life expectancy of 20 to 30 years[4].

In Lake Washington and Box Canyon Reservoir in Washington, Tench about 460 mm (18.1 in) long have been collected. The largest recorded Tench weighed 5.0 kg (11.0 lb). The record Tench for Idaho is a 2.2 kg (4 lb, 14 oz) 533 mm (21 in) specimen taken from the Spokane River in 2000.

Food and Feeding

Young Tench feed primarily on small food items such as microscopic animals and algae. Larger individuals feed mainly on aquatic insects and invertebrates. Tench captured from Lake Washington contained snails and small fingernail clams as well as fish eggs, whereas those from Medical Lake and Loon Lake (Spokane County, Washington) contained cladocerans. Other food items included amphipods, clams, snails, mayflies, caddiesflies, dragonflies, damselflies, and terrestrial insects[6].

Limiting Factors

Tench may be infested with a variety of parasites, including trematodes, nematodes, crustaceans, protozoans, acanthocephalans, and leeches. Smaller Tench are likely preyed upon by predatory fish such as Northern Pike, but large individuals probably have few enemies, except humans[4].

Breeding Habits

While spawning seasons and breeding behavior in Europe are well understood, little has been written about the biology of Tench in North America. In Europe spawning typically occurs in weedy shallows during June or July, and large numbers of eggs, 123,759/kg (273,000/lb) of female body weight, are deposited. Eggs are small, about 1 mm (0.04 in) in diameter[4].

Habits and Habitat

Tench inhabit shallow areas of lakes and ponds, generally in areas with dense aquatic vegetation. Reports from Europe indicate this species is mostly inactive on the bottom but becomes quite active during the spawning season. No detailed life history studies have been made for this species in the United States[4]. Tench are reported to be rather sluggish in their movements and are not very aggressive toward other Tench or other fishes. Usually, Tench are solitary, sedentary, and nonmigratory, although they may congregate in deep holes in shady areas during warmer months, apparently seeking cooler water[41].

Management and Conservation Status

Tench are considered common, widespread, and abundant range-wide (G5). A state conservation status rank is not applicable (SNA) because the species is not a suitable target for conservation activities[11]. As a nonnative species, Tench should not be introduced into additional Idaho aquatic ecosystems. Overall effects of this nonnative species have not yet been assessed in Idaho waters.

Family Catostomidae—Suckers

Members of this family generally have 16 or more pharyngeal teeth and the lips are thick and fleshy with plicae or papillae. While the maximum lengths of members of this family may range to over 1,000 mm (39.4 in), lengths of about 600 mm (23.6 in) are the most common. Seventy-two species are represented in 13 currently recognized genera with about 45 species in the *Catostomus* and *Moxostoma* genera. Fossil genera include species from the Eocene-Oligocene[1] (55.8 to 23.0 mya).

In the Tribe Catostomini, the lateral line is present and there are more than 50 lateral line scales. Most of the species are from western North America and two are particularly widespread: *Catostomus catostomus,* the Longnose Sucker, and *Catostomus commersonii*, the White Sucker[1].

Suckers, thought to have their origin in Asia, make up a distinct element of the North American fish fauna. This group includes approximately 100 species distributed from the Arctic to the tropics. Suckers have soft fin rays, toothless jaws, and cycloid scales: the mouth is typically sucking, as the common family name implies. In most species the mouth is protrusible with papillose lips. However, the lake suckers, *Chasmistes* spp. have a short nose, terminal rather than ventral mouths, with thin lips and almost no papillae. Suckers are closely related to minnows, and juveniles of the two families are often misidentified in the field. Most suckers, run upstream in the spring to spawn, while others spawn along lake shores. Sucker larvae have terminal mouths and feed at the surface. Several species have substantial commercial value[2].

Catostomids have been historically undervalued, both economically and ecologically, and have mistakenly been assumed to be "weedy" or "trash" fish which are common and tolerant of degraded habitats. As a result, catostomids have been understudied relative to more valued families such as salmonids, but increasing concern about the conservation status of many catostomids has led to more interest in their biology and evolutionary history[42].

In this book, we have adopted a number of changes to taxa descriptions based on the recent work of Smith et al.[29] and Unmack et al.[43]. We have followed the research of Unmack et al. which supports resurrection of *Pantosteus* to full genus status and recommends changing Bluehead Sucker in the Bonneville and upper Snake River basins to Green Sucker *Pantosteus virescens*. Smith et al.[29] describe a new species, the Cascadian Sucker *Pantosteus bondi*. Therefore, we describe eight sucker species in three genera (*Catostomus*: river suckers, *Chasmistes*: lake suckers, *Pantosteus*: mountain suckers):

Catostomus ardens	Jordan & Gilbert, 1881	Utah Sucker
Catostomus catostomus	(Forster, 1773)	Longnose Sucker
Catostomus macrocheilus	Girard, 1856	Largescale Sucker
Chasmistes muriei	Miller & Smith, 1981	Snake River Sucker
Pantosteus bondi	Smith et al., 2013	Cascadian Sucker
Pantosteus columbianus	Eigenmann & Eigenmann, 1893	Bridgelip Sucker
Pantosteus platyrhynchus	(Cope, 1874)	Mountain Sucker
Pantosteus virescens	Cope, 1875	Green Sucker

"Give a man a fish and you feed him for a day. Teach a man to fish and you feed him for a lifetime."

Anne Isabella Thackeray Ritchie

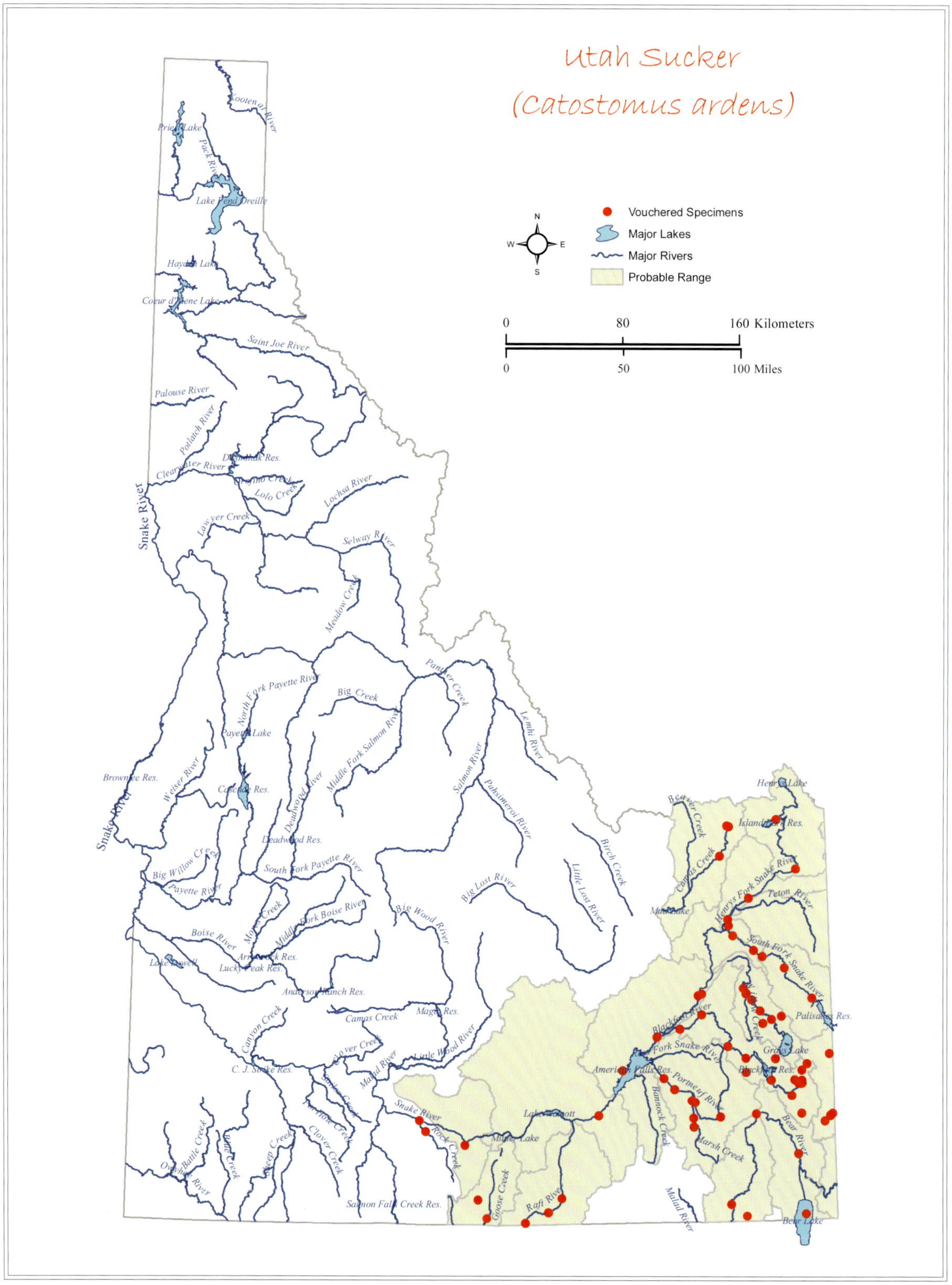
Utah Sucker
(Catostomus ardens)
Vouchered Specimens
Major Lakes
Major Rivers
Probable Range
N
W
E
S
0
80
160 Kilometers
0
50
100 Miles
Kootenai River
Pack River
Saint Joe River
Palouse River
Potlatch River
Lochsa River
Lolo Creek
Snake River
Selway River
Meadow Creek
Big Creek
North Fork Payette River
Middle Fork Salmon River
Lemhi River
Salmon River
Pahsimeroi River
Weiser River
Big Willow Creek
Payette River
Deadwood Res.
South Fork Payette River
Big Lost River
Birch Creek
Little Lost River
Big Wood River
Boise River
Lucky Peak Res.
Canyon Creek
Camas Creek
Magic Res.
Little Wood River
C. J. Strike Res.
Clover Creek
Salmon Falls Creek Res.
Owyhee River
Rock Creek
Goose Creek
Raft River
Malad River
Marsh Creek
Bear River
Bear Lake
Bannock Creek
Portneuf River
American Falls Res.
Blackfoot River
Grays Lake
Palisades Res.
South Fork Snake River
Teton River
Henrys Fork Snake River
Camas Creek
Beaver Creek
Henrys Lake
Lake Lowell
Brownlee Res.
Lake Walcott
Milner Lake
Mud Lake

Utah Sucker

Catostomus ardens
Jordan & Gilbert, 1881

Etymology

Catostomus = inferior mouth, or mouth below; *ardens* = burning, probably referring to the bright red colors of breeding males.

Importance

Historically, Utah Sucker were of considerable value as a food fish, particularly to early-day Mormon pioneers, in Utah. Settlers in Utah, and likely in Idaho, returned to harvest suckers time and again from Bear Lake. The Utah Sucker was used fresh, smoked, and brined. Today they may be used for human food, but at a much lesser level than in pioneer days. In various ecosystems, Utah Sucker are also important as forage for predaceous fishes. In the recent past, Utah Sucker accounted for the highest biomass of any species in Bear Lake and numerically ranked third in abundance. Trapper's diaries from excursions into Utah during the 1830s to 1850s tell of hundreds of Native Americans, during Native American and trapper rendezvous, dropping whatever they were doing and catching thousands of suckers on the east side of Bear Lake, likely during the spawning run when this species is highly vulnerable[2,8,21].

Description

Utah Sucker are bronze or grayish above, whitish below, and have dark fins. Males have a bright rosy lateral stripe during the breeding season. Utah Sucker have a long and somewhat compressed body; the back is broad and slightly elevated. Utah Sucker have a short, conical head that is broad, the mouth is inferior, with weak lips. There is no lateral notch between the upper and lower lips and the lower lip is typically totally cleft. The caudal fin is short, broad and weakly forked. Large cycloid scales number 60 to 70 in the lateral series and dorsal fin rays number 11 to 13, typically 12[2].

Range and Distributional Status

This species is native to Idaho. Utah Sucker are native throughout the historic Lake Bonneville Basin of Idaho, Utah, Wyoming, and Nevada. Populations of Utah Sucker occur in the Snake River drainage above

Shoshone Falls, but have been collected in Snake River tributaries below the falls. Utah Sucker have been present in the Colorado River drainage since at least the early 1980s[2].

Size and Longevity

In some ecosystems, Utah Sucker attain lengths of 648 mm (25.5 in) and weights of more than 2.3 kg (5.1 lb). Spawning adults in Bear Lake, where growth is relatively slow as a result of low water temperatures, weigh between 0.9 and 1.36 kg (1.9 and 3.0 lb). Outside of Bear Lake, they are relatively fast-growing; most live to 10 to 12 years. Yearly total lengths of Utah Sucker, age one to ten years from Bear Lake, are, in mm (in): 38 (1.5), 130 (5.1), 213 (8.4), 262 (10.3), 333 (13.1), 371 (14.6), 420 (16.5), 447 (17.6), 483 (19.0), and 559 (22). Individuals taken from Blackfoot Reservoir, Idaho reportedly attain weights of 3.63 kg (8.0 lb) and larger individuals have been taken by commercial seiners[2]. The certified Idaho record Utah Sucker is a 3.6 kg (7 lb, 14 oz) 64.8 cm (25.5 in) specimen taken from the South Fork Snake River in 2011.

Food and Feeding

Utah Sucker graze on filamentous and attached algae, and plants and animals. Both young and adults may feed on about the same items in ecosystems where food selections are minimal or limited, except for differences related to the size of the individuals. Adults feed at all depths, while larvae feed at the surface. Adults take most of their food items from the bottom in deep lakes, such as Bear Lake, feeding far and wide throughout the entire year[2].

Limiting Factors

Utah Sucker are parasitized by a variety of organisms. In Bear Lake, they are parasitized by a body-cavity tapeworm. No other Bear Lake fishes examined were as highly parasitized although follow-up studies 30 years after the initial study found no parasitized Utah Sucker. There is no explanation for this change. Utah Sucker are preyed upon by a variety of fishes, birds, mammals and humans, depending on ecosystem. Post-spawning mortality is often high and adults caught in July, are often in poor condition, indicating that spawning stress may be a factor[2].

Breeding Habits

Utah Sucker are spring spawners, moving out of lakes or large rivers into small streams when water temperatures reach about 15.6°C (60°F). When females are ready to spawn, two or more males approach them, generally one on either side and all three quiver rapidly to discharge eggs and milt. Gravel, sand, and other bottom materials are stirred by violent tail movement of the males. No nest or care of the embryos or young is provided. Sexual maturity is reached at age two or three years[2].

In St. Charles Creek, a tributary to Bear Lake, Utah Sucker initiate spawning runs in early to mid-June. Two distinct runs have been identified: one consisting of small, darker immature fish and the other large, faded and somewhat battered larger spawners[2].

Bear Lake populations of Utah Sucker, and likely other lake populations, have discrete lake-spawning segments. Groups congregate and move inshore, during both day and night hours, often just at dusk, and spawning activity becomes frenzied. In mid-June 1976, at 2:30 AM, a 4.8 km (3.0 mi) long and about 1.83 m (6 ft) wide school of Utah Sucker was observed on the east side of Bear Lake. The school was in water 15 to 61cm (6.0 to 24 in) deep. When approached, individuals in this congregation were almost without fear, continuing spawning activity even when individuals walked near them or shined a light on them[2].

Habits and Habitat

Utah Sucker are highly adaptable, prospering over a wide range of habitats and ecosystem types which vary from large, deep, cold lakes, to shallow warm lakes, to small warm streams with water temperatures above 26.7°C (80°F). Utah Sucker occur over a variety of fast and slow current conditions, turbid to clear water and bottom types which vary from mud and clay to sand and gravel. Bear Lake Utah Sucker prosper in deep water, feeding near the bottom where temperatures are often 5.6 to 7.8°C (42 to 46°F)[2].

Prior to the drought of the early 1930s Utah Sucker were present in warm, shallow Utah Lake in numbers that were absolutely astounding. Commercial catches of hundreds of thousands did not appear to depress this population for other than a short time. Presently, their numbers have been greatly reduced. It is interesting to note that, ecologically, Bear Lake and Utah Lake are extremely different but Utah Sucker prosper in both ecosystems[2].

Management and Conservation Status

Utah Sucker are considered common, widespread, and abundant range wide (G5) and within Idaho (S5)[11]. Because Utah Sucker are such an adaptable species, surviving well in a wide range of environmental conditions, they probably need no special protection in Idaho waters. However, no defined studies have been conducted in Idaho to determine population status.

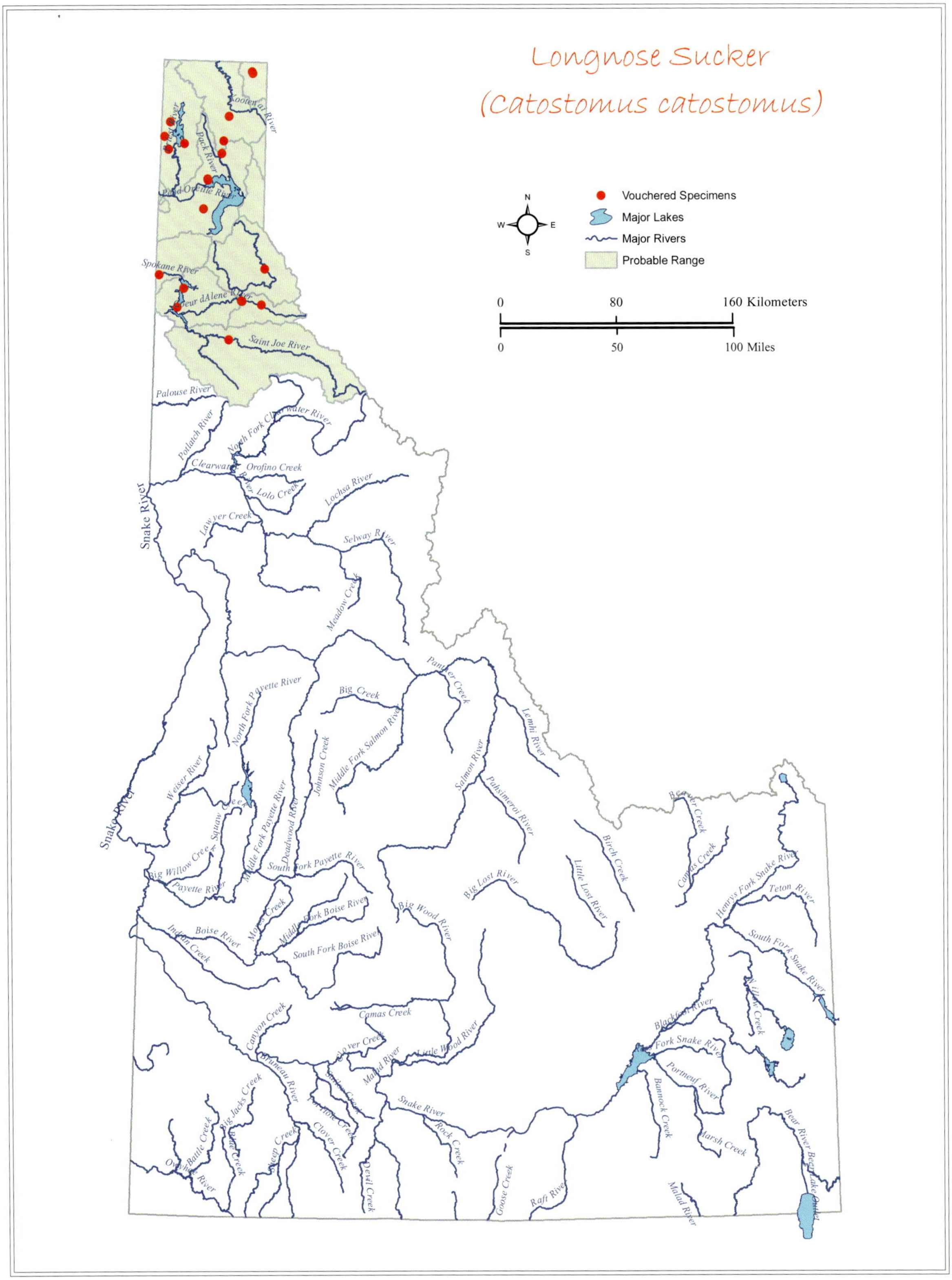

Longnose Sucker
(Catostomus catostomus)
Vouchered Specimens
Major Lakes
Major Rivers
Probable Range
0 80 160 Kilometers
0 50 100 Miles

Longnose Sucker

Catostomus catostomus

(Forster, 1773)

Etymology

Catostomus = inferior mouth, or mouth below.

Importance

In Canada, Longnose Sucker are used for food for some native peoples but more commonly as food for dogs. Flesh of Longnose Sucker is firm, white, flaky, and sweet: but has numerous bones[4]. Young are forage for other fishes and fish-eating birds[5]. Because of their feeding habits, they may compete with other species for food in some ecosystems[4].

Description

Longnose Sucker are black, dusky or dark olive above, have lighter sides and a white belly. Some larger specimens have a cream or pale yellow belly. A rosy stripe is present on the sides of breeding males which may also be reddish on the belly and lower fins. Several dark blotches occur along the back of juveniles. Longnose Sucker have no notches between the upper and lower lips, however, the lower lip is deeply incised and the snout overhangs the mouth[5].

Longnose Sucker have elongate, cylindrical, torpedo-shaped bodies, almost round in cross-section. There are no teeth in the mouth and the pharyngeal teeth are in a single row, comb-like, and ascending in length. Gill rakers are short and fleshy, numbering 23 to 30 and branchiostegal rays number three. The single dorsal fin is soft rayed and the origin is ahead of the midpoint of the body with nine to 11 principal rays. The pelvic fins are abdominal and only moderate in length while the anal fin is long with a pointed tip and seven principal rays. Scales on the Longnose Sucker are small, cycloid and crowded towards the head but larger on the caudal peduncle with 99 to 108 along the complete but inconspicuous and almost straight lateral line. Vertebrae, including Weberian ossicles, number 45 to 47. Nuptial tubercles develop on breeding males only[4,44].

Range and Distributional Status

This species is native to Idaho. Longnose Sucker have a broad natural range from Siberia and Alaska across North America to the Atlantic coast, ranging as far south as the Great Lakes, Colorado and Idaho. Populations of Longnose Sucker are found only in the northernmost drainages in Idaho, specifically, the Kootenai River, Pend Oreille River and Spokane River systems[5]. This is the only North American sucker which also occurs in Asia[4].

Size and Longevity

Adult Longnose Sucker may reach lengths of 375 to 450 mm (14.8 to 17.7 in) TL, although specimens of 675 mm (26.6 in) and three kg (6.6 lb) have been recorded. The largest specimen of this species reported in Idaho is 1.6 kg (3.5 lb)[5].

On average, mature female Longnose Sucker are larger than males. Breeding males have a distinctive red or pink lateral band that is sometimes sandwiched between an upper yellow-gold band and a lower black band. Breeding colors in females are subdued. Nuptial tubercles vary between the sexes: breeding males have large tubercles on the anal fin and the lower lobe, of the caudal fin, whereas females generally lack tubercles on these fins or have only small tubercles. The only patch of tubercles that is more strongly developed in females than in males is on the dorsal surface of the caudal peduncle immediately in front of the caudal fin. However, adults do differ in the size and shape of some fins: the anal fin and the lower lobe of the caudal fin are noticeably longer in adult males than in females, and the posterior margins of the pelvic fins are squared off in males (all rays of about equal length) and bluntly pointed in females. This difference in pelvic fin shape is reported to be a reliable indicator of sex in both immature and adult individuals outside the breeding season[3].

In British Columbia, three kinds of Longnose Sucker are present: a widely distributed "normal" form that breeds at a relatively large size of greater than 300 mm (11.8in) FL, a set of geographically scattered "dwarf" populations that breed at relatively small sizes of 150 to 250 mm (5.9 to 9.8in) FL, and a genetically distinct Salish Sucker. Whether or not these three population types represent genetically programmed life histories or are simply affected by their environment is unknown[3].

Food and Feeding

Food of Longnose Sucker is variable depending on ecosystem and varies by season and fish size. Food consists totally of bottom invertebrates: no vertebrate prey has been reported. Food items, listed in descending order of frequency of occurrence, are Amphipods, Trichoptera, chironomid larvae, Ephemeroptera, ostracods, gastropods, Coleoptera, pelecypods, copepods, cladocerans, and plants[4]. In some ecosystems, Longnose Sucker may consume fish eggs[5].

Limiting Factors

Young of Longnose Sucker probably fall prey to a wide variety of predaceous fishes and fish-eating birds. Larger Longnose Sucker may be taken by Northern Pike. Adults, during spawning season, are probably taken by bears, other mammals, and by osprey (*Pandion haliaetus)*, bald (*Haliaeetus leucocephalus*) or golden

(*Aquila chrysaetos*) eagles. As bottom feeders, Longnose Suckers compete for food with other bottom feeders. Longnose Sucker may be infected with a variety of parasites[4].

Breeding Habits

Age at sexual maturity is variable. For example, in British Columbia, first spawning occurred at five to seven years whereas in Great Slave Lake, no spawners younger than an estimated nine years were seen. Longnose Sucker may spawn more than once[4].

Longnose Sucker spawn in the spring preferably in streams or shallow areas in lakes. They enter spawning streams when water temperatures exceed 5°C (41°F), typically in mid-April to mid-May in Canada. Spawning often occurs in water 15 to 28 cm (6.0 to 11.0 in) deep, with velocities of 30 to 45 cm/sec (0.9 to 1.5 f/s) over gravel bottoms from 5.0 to 10.0 cm (1.9 to 3.9 in) in diameter. Spawning normally occurs from 6 AM to 9 PM, and involves a female moving from quiet water near shore into a group of males near the center of the stream. Up to four males crowd around one female, clasping her with or beating against her with their anal fins and thrashing about in a spawning act that lasts for three to five seconds. This act may occur from six to 40 times per hour. Following spawning, males and females separate and return to their previous stream positions. No nest is constructed and the white, demersal embryos adhere to the gravel substrate. Ripe eggs in females are yellow, 2.8 to 3.0 mm (0.11 to 0.12 in) in diameter and coated with a thin gelatinous layer. Egg numbers range from 17,000 to 60,000/female. Mortality of adults during the spawning season is estimated to be 10 to 25 percent. Experimentally, eggs hatched in eight days at 15°C (59°F) and 11 days at 10°C (50°F). Hatching likely takes two weeks at seasonally prevailing stream temperatures. Young, 10 to 20 mm (0.39 to 0.79 in) in length remain in the gravels for one to two weeks before emerging, generally in June, and migrate to the lake[4].

Habits and Habitat

Longnose Sucker are generally found in clear, cold lakes, and lake tributary streams. Young may remain in shallow, weedy areas whereas larger fish inhabit deeper water of a particular ecosystem[5].

Management and Conservation Status

Longnose Sucker are considered common, widespread, and abundant range wide (G5). Within Idaho, they are apparently secure; uncommon but not rare; some cause for long-term concern due to declines or other factors (S4)[11]. No defined studies have been conducted in Idaho to determine population status of Longnose Sucker although their abundance and wide distribution likely make additional protection unnecessary. Maintaining habitat quality is likely the most important aspect of protection for this species.

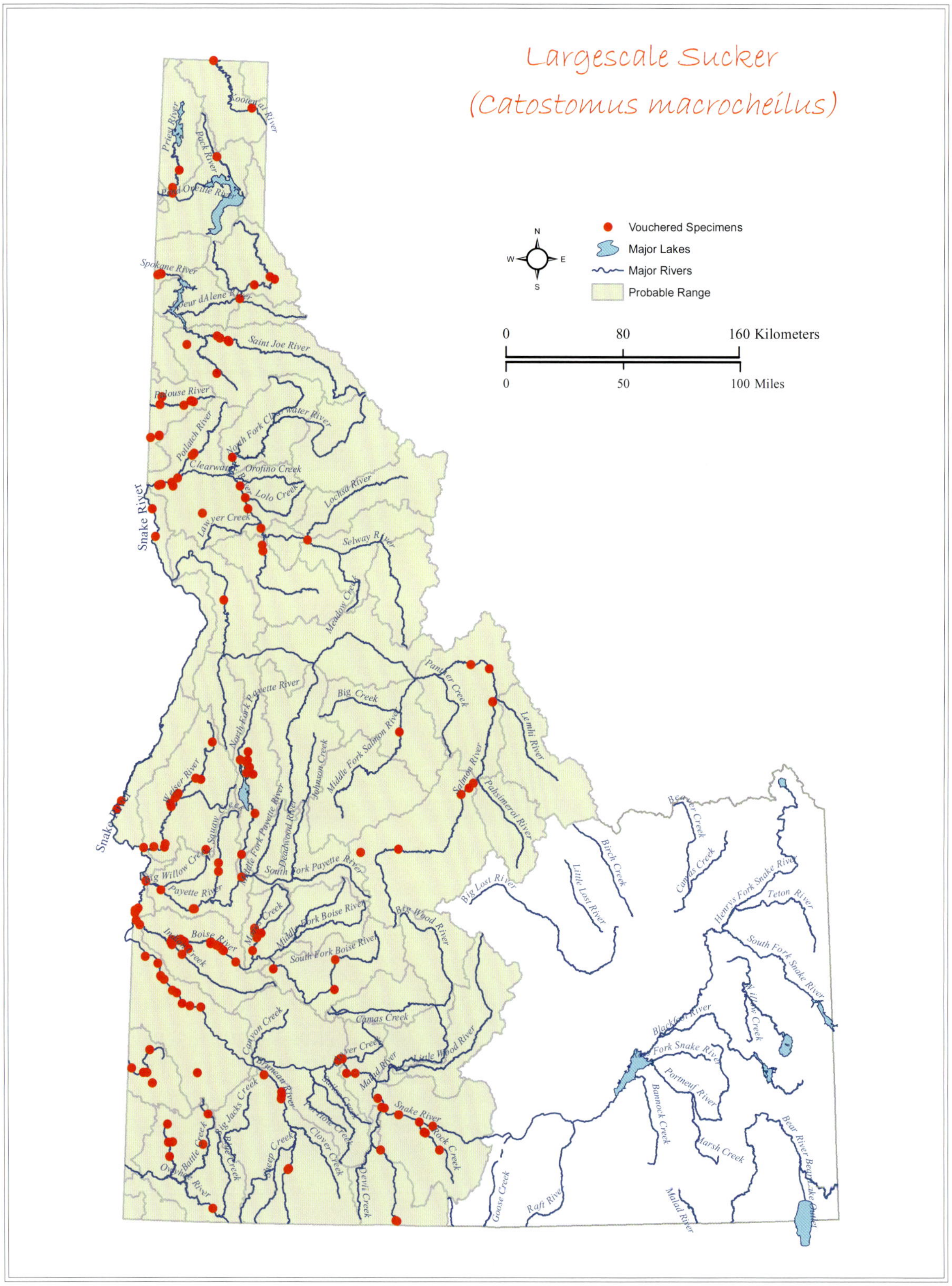
Largescale Sucker
(Catostomus macrocheilus)
Vouchered Specimens
Major Lakes
Major Rivers
Probable Range
0 80 160 Kilometers
0 50 100 Miles

Largescale Sucker

Catostomus macrocheilus

Girard, 1856

Etymology

Catostomus = inferior mouth, or mouth below; *macrocheilus* = large lip, in reference to the prominent lips.

Importance

Largescale Sucker are of little direct or indirect importance to man. While the flesh is firm, white, flaky, and edible, it is bony and not highly flavored. Undoubtedly this species historically served as food for a variety of Native American communities and their dogs. Like most other suckers, Largescale Sucker function as a competitor and may eat eggs of more highly valued sport fish. However, this claim may be exaggerated[4] and they likely have little population consequences for other fishes.

Description

Largescale Sucker adults are dark and distinctly counter shaded; the back, upper sides to just below the lateral line and the head to just below the eye are blue-gray to olive, while the lower sides and the lower part of the head and the ventral surface are creamy to whitish. Young Largescale Sucker, 152 mm (6.0 in) in length have three or four profuse dark spots on the sides. Counter shading and lateral band are darker and more pronounced in breeding males[4,5].

Largescale Sucker have moderately deep bodies and are more laterally compressed than the Longnose Sucker. The head is moderately long and the snout is bluntly rounded with a depression ahead of the nostrils; the mouth is ventral, protrusible, suctorial, but small and is overhung by the snout. There are no teeth in the mouth and pharyngeal teeth are in a single row and comb like, numbering 43 to 46+43 to 48. Gill rakers are short and fleshy, numbering 25 to 37. Branchiostegal rays number three. The single dorsal fin is soft rayed and a little ahead of the midpoint of the body with 12 to 17, typically 13 or more, principal rays. The caudal fin is moderately long with well forked tips and rounded points. The anal fin is long, pointed and has seven prominent principal rays. Pelvic fins are abdominal, low, moderately long, and somewhat pointed and broad with small axillary process and weak fleshy stays; rays number nine to 12. Pectoral fins are low, long and more or less pointed with 16 to 18 rays. Scales are cycloid and large, numbering 60 to 83, typically fewer than 80, along the complete lateral line, are somewhat crowded anteriorly and increase in size from the head to the tail. There are no pyloric caecae. Vertebrae, including the Weberian ossicles number 47 to 49. Nuptial tubercles are white, horny, and well developed on spawning males on the anal fin and lower lobe of the caudal fin while absent or weakly developed on females[4].

Mature Largescale Sucker females are, in general, larger than are mature males. In this species, sexes also differ in the extent and distribution of nuptial tubercles. Breeding males develop large tubercles on the anal and caudal fins, whereas females, usually either lack tubercles on the anal and caudal fins or have only small tubercles. As with other suckers, adults differ in the size and shape of some fins. The anal fin and lower lobe of the caudal fin are noticeably longer in mature males than in females. In addition, the posterior margins of the pelvic fins are squared off in males (all rays about equal length), and are bluntly pointed in females. This difference is apparently a reliable indicator of sex in immature individuals and non-breeding adults[3].

Range and Distributional Status

This species is native to Idaho. Largescale Sucker are limited in distribution to the Pacific Northwest, occurring from the Smoky, Skeena, Peace, Nass, Frazier, and Columbia River systems in British Columbia and Alberta, south to the Sixes River of Oregon and as far east as Idaho and western Montana. In Idaho they are common in the Snake River downstream of Shoshone Falls and also occur in the Spokane River, Pend Oreille River, and Kootenai River systems[5,6].

Size and Longevity

Annual growth is generally slow; fry growing to about 20 to 25 mm (0.8 to 1.0 in) by July. Average length of individuals, age one to 11 from a Canadian population are, in mm (in): 46 (1.8), 76 (3.0), 124 (4.9), 170 (6.7), 190 (7.5), 229 (9.0), 284 (11.2), 305 (12.0), 340 (13.4), 356 (14.0, and 376 (14.8)[4]. Largescale Sucker have been aged to 15 years in the Columbia River, and some may have been older. One 27-year-old fish was 495 mm (19.6 in) long and weighed 1.3 kg (2 lb, 14 oz). In Washington, the angling record for Largescale Sucker is a 2.3 kg (5 lb, 5 oz) individual taken from Mason Lake in 2001[6]. The current Idaho record for Largescale Sucker is a 3.8 kg (8 lb, 6 oz) 66 cm (26 in) specimen taken from Cascade Lake in 2005.

Food and Feeding

Largescale Sucker are omnivores. Young eat small zooplankton when pelagic, adults feed on aquatic insect larvae, other aquatic invertebrates, diatoms and other plant material. After hatching, fry remain in the gravel or on the surface of the sand for the first few weeks until the elongate yolk sac is absorbed; at this point the mouth is terminal and individuals become pelagic. They remain pelagic until the mouth moves to the ventral position at about 16 to 18 mm (0.6 to 0.7 in) in length[4]. Larger individuals feed on a variety of bottom organisms, such as crustaceans, aquatic insect larva, earthworms, snails, and detritus. Periphyton algae may be a major food in some ecosystems; constituting as much as 60 percent (by volume) in fish ranging in length from 41 to 150 mm (1.6 to 5.9 in) FL. Other studies have documented consumption of ostracods, copepods, amphipods, Trichoptera, Chironomidae, larva of aquatic insects, mollusks, and fish eggs, diatoms, algae, and detritus. Stream fish eat more algae, diatoms, and insects other than chironomids and fewer amphipods and mollusks than in lentic systems. In one Canadian study, a high percentage of fish eggs was present only during the fall spawning of Kokanee, and Pacific salmonids[4].

Limiting Factors

Largescale Sucker may compete for food with various sport fishes, however, fry and juveniles may also serve as forage for various fishes including White Sturgeon, Channel Catfish, Smallmouth Bass, Walleye, and Northern Pikeminnow and by birds such as mergansers (*Mergus* spp.). Adults in shallow spawning streams are likely eaten by bears and other mammals as well as osprey (*Pandion haliaetus*) and eagles (*Haliaeetus leucocephalus* or *Aquila chrysaetos*). Largescale Sucker present in large numbers in a particular ecosystem could constitute serious food competitors with young and adult salmonids. Largescale Sucker are parasitized by protozoans, trematodes, cestodes, nematodes, acanthocephalans, leeches, crustaceans. This species may hybridize with Bridgelip Sucker, diluting its genetic base[4].

Breeding Habits

Largescale Sucker spawn in the spring, usually in the deeper area of streams with sandy bottoms and may at times spawn on gravelly or sandy shoals in lakes. In the northern edge of the range in British Columbia, spawning takes place from late April to late June, but typically mid-May to late June, depending on stream temperatures. Adults enter streams when stream temperatures are 7.8 to 8.9°C (46 to 48°F) and spawn both during the day and night. Several males may accompany a female and the yellow, adhesive eggs are broadcast along the bottom[6]. Females may deposit as many as 20,000 eggs, which are demersal, yellow, and average about 2.5 mm (0.09 in) in diameter before fertilization. Eggs hatch in about two weeks and the fry remain in the gravel or on the surface of the sand for the first few weeks until the elongate yolk is absorbed[4]. In Montana ecosystems, this species does not mature until their fourth or fifth year of life. Males may at mature in three years. In British Columbia, most males probably do not mature and until five years and females a year later. Ripe males apparently stay on the spawning grounds for long periods of time, while females usually remain in pools until they are ready to spawn[5].

Habits and Habitat

Largescale Sucker inhabit both lake and stream ecosystems and at the mouths of streams entering lakes. Fry move into shallow areas during the day and back into deeper water at night. Adults have been found at depths to 10 m (33 ft) in main stream channels, moving inshore at night[6].

In a British Columbia lake ecosystem, fry were most abundant in shallow water over rock and gravel substrates in July that shifted to open sand areas in August, appearing to avoid heavily vegetated areas. Fry made diurnal onshore-offshore movements: onshore at dawn and offshore at dusk. A seasonal shift in their vertical distribution was also noted and corresponded to the diet shift from plankton to benthos. In early summer months, small schools of fry cruise close to the surface in shallow water, whereas schools were closer to the bottom in late summer[3].

Management and Conservation Status

Largescale Sucker are considered common, widespread, and abundant range wide (G5) and within Idaho (S5)[11]. No defined studies have been conducted in Idaho to determine conservation status although they are distributed over a wide range of habitats.

Introduction to *Chasmistes*

Lake suckers of the genus *Chasmistes* are mid-water planktivores with numerous dendritic gill rakers and large, terminal mouths[1,45]. This terminal position of the mouth is so exceptional among the usually ventral-mouthed suckers that it has been regarded as an extreme specialization, although certain primitive suckers (*Amyzon, Ictiobus cyprinellus)* and presumed sucker ancestors had or have terminal mouths. In early larval stages *Catostomus* and other suckers have terminal mouths[45].

In *Catostomus* and *Chasmistes* the mouth becomes ventral after individuals attain 30 mm SL (1.2 in) but the mouth of *Chasmistes* eventually becomes oblique. Bones involved in these modifications are the dentary, maxilla, and premaxilla. *Chasmistes* is the best-known genus of American Cenozoic freshwater fishes, occurring in Miocene to Pleistocene (23.0 to 0.011 mya) deposits of six western states, including a number of basins where the genus no longer survives. Divergence and extinction have occurred in the late Cenozoic lakes of western North America to which it is or was endemic. Three contemporary species once provided a large part of the subsistence fishery for Native Americans and early settlers around Utah Lake, Utah; Pyramid Lake, Nevada; and Klamath Lake, Oregon. Now, however, only one of these species, *Chasmistes cujus,* in the Pyramid Lake ecosystem, survives in pure form and it is listed as a federally endangered species. A fourth contemporary species, described below, is officially considered extinct[45]. However, fish exhibiting lake sucker attributes have recently been collected in Jackson Lake, Wyoming and Palisades Reservoir, Idaho[46], (South Fork of the Snake River). Nomenclatural history for this genus is complex. In erecting the genus, Jordan[47] diagnosed it in a footnote and designated *Catostomus fecundus* as its type. Later in the same year, Jordan gave a detailed description of *Chasmistes,* stating that it occurred only in Utah Lake and designating those collected specimens as the type species of *Chasmistes.* In an addendum to that paper, Jordan recompared *Catostomus fecundus* with the specimens on which *Chasmistes* was based and concluded that the type species of *Chasmiste*s was incorrectly identified and was undescribed, and proposed *Chasmistes liorus.* This proposal was followed by a description of *Catostomus fecundus.* Complications and protocol application resulted in assignment of this species as *Chasmistes liorus* in Utah Lake[45,47].

The living species of *Chasmistes* occurred historically in four separate drainages: the Bonneville Basin (Utah Lake) in Utah, the Lahontan Basin (Pyramid Lake) in Nevada, the Klamath River Basin (Klamath Lakes and associated drainages in Oregon and California), and the upper Snake River basin (upstream of Shoshone Falls) in Wyoming and possibly Idaho. These occurrences are associated with major remnants of pluvial lakes in the western United States[45].

Chasmistes are large robust suckers with a long, broad, and deep head; mouth terminal or sub-terminal, moderately to strongly oblique (usually about a 45° angle), and long mandible (about equal to snout length); mouth opening wide; lateral folds well developed under maxillae; lips thin, often plicate, with scattered, weakly-developed papillae that may be obsolete; lobes of lower lip widely separated (note photo on facing page); eye typically in anterior half of head; gill rakers dendritic or fimbriate, flat and paddle-like on top, 38 to 53; pharyngeal bones weaker than usual in *Catostomus*, the teeth generally smaller; dorsal and anal fins as in *Catostomus*, the dorsal with 10 to 12 rays and the anal with seven (rarely eight) rays. Jaw bones elongate, slender, the premaxillae wider than high; skull short and wide, the dermethmoid broader than long[45].

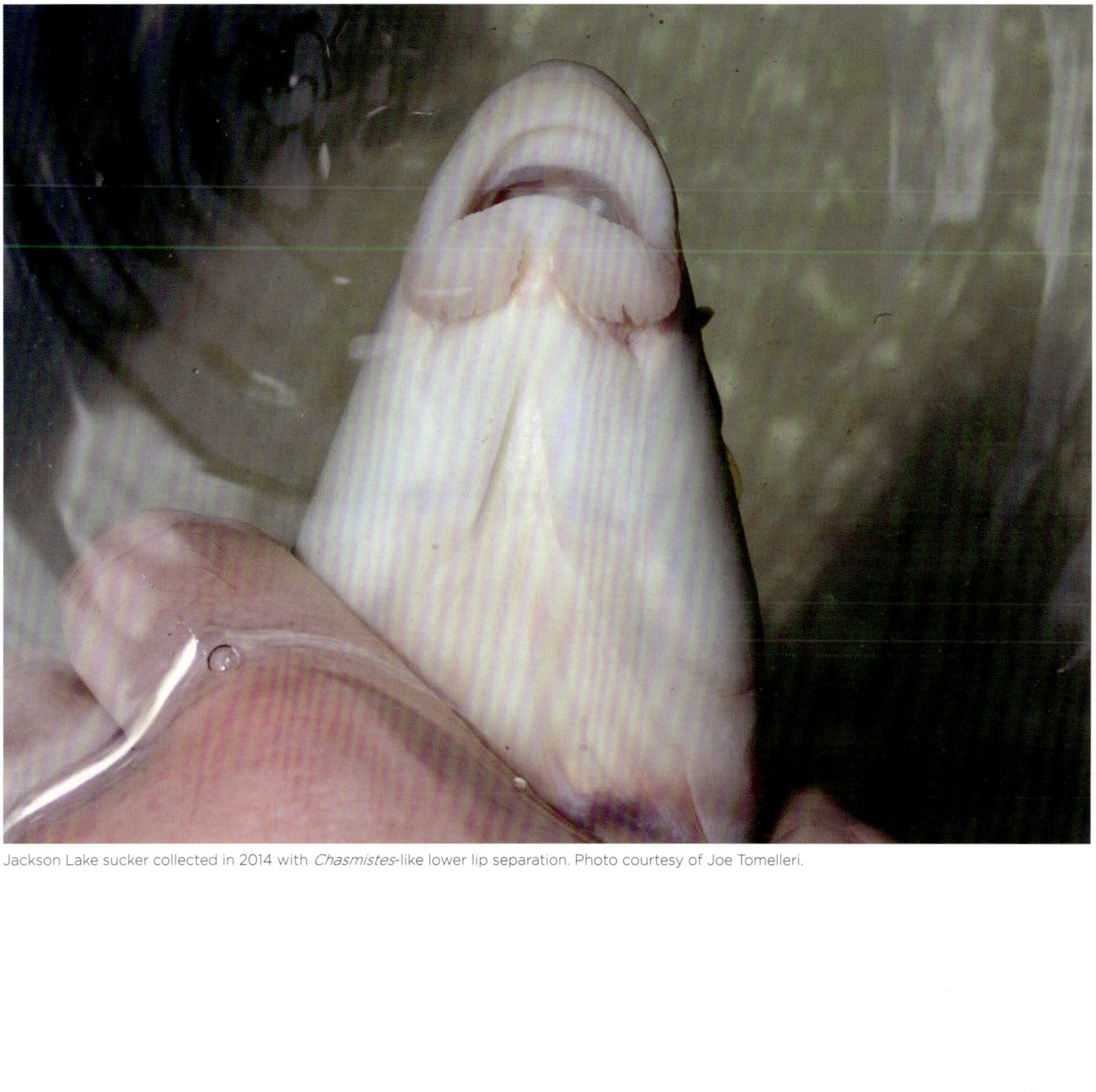

Jackson Lake sucker collected in 2014 with *Chasmistes*-like lower lip separation. Photo courtesy of Joe Tomelleri.

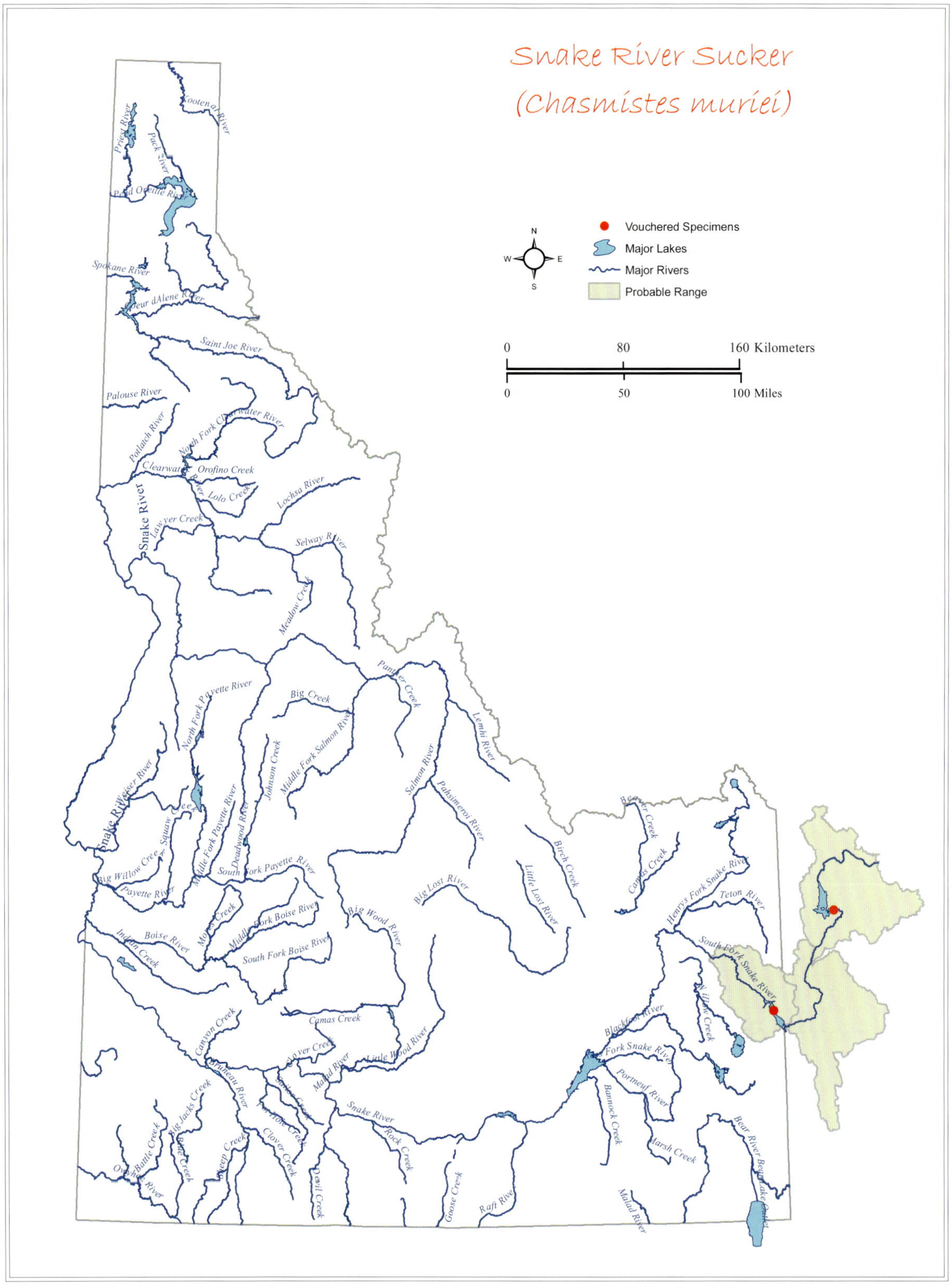
Snake River Sucker
(Chasmistes muriei)
Vouchered Specimens
Major Lakes
Major Rivers
Probable Range
N
S
E
W
0
80
160 Kilometers
0
50
100 Miles
Kootenai River
Priest River
Pack River
Pend Oreille River
Spokane River
Coeur dAlene River
Saint Joe River
Palouse River
Potlatch River
North Fork Clearwater River
Orofino Creek
Lolo Creek
Lochsa River
Snake River
Selway River
Meadow Creek
Panther Creek
Lemhi River
Big Creek
North Fork Payette River
Middle Fork Salmon River
Johnson Creek
Salmon River
Pahsimeroi River
Weiser River
Squaw Creek
Middle Fork Payette River
Deadwood River
South Fork Payette River
Big Willow Creek
Payette River
Boise River
Indian Creek
Middle Fork Boise River
South Fork Boise River
Big Wood River
Big Lost River
Little Lost River
Birch Creek
Camas Creek
Henrys Fork Snake River
Teton River
South Fork Snake River
Willow Creek
Blackfoot River
Portneuf River
Bannock Creek
Marsh Creek
Bear River
Bear Lake
Malad River
Canyon Creek
Camas Creek
Little Wood River
Snake River
Rock Creek
Goose Creek
Raft River
Clover Creek
Big Jacks Creek
Owyhee River
Sheep Creek
Devil Creek

Snake River Sucker

Chasmistes muriei
Miller & Smith, 1981

Etymology

Chasmistes = Greek, meaning one who yawns, in reference to the large, oblique mouth; muriei = named for the late Olaus Murie, biologist and wilderness advocate, who collected the only known specimen: [female, 371 mm SL (448 mm TL,), from Snake River downstream of Jackson Lake Dam, Wyoming on 13 October 1927].

Importance

As a member of the *Chasmistes*, Snake River Sucker are a natural history curiosity which can possibly illuminate origins of this genus. Study of the wide genetic and phenotypic range of suckers in the Snake River basin within and above Palisades Reservoir is recommended.

Lake sucker declines have been attributed to historical over-exploitation (commercial, recreational, and subsistence); changes in aquatic habitat (degraded water quality, flow alterations, stream channelization, loss of lake littoral zones, especially at tributary confluences); competition with and predation by non-native fishes; and hybridization with *Catostomus* (*C.* spp.). Jackson Lake and the upper Snake River basin, native range of the Snake River Sucker, represent some of the least anthropogenically-impacted lake sucker habitats[42].

Description

A potentially distinct form of *Chasmistes*[45]. A *Chasmistes* with a sub-terminal mouth and papillae on the lips, resembling *Ch. liorus* in gill-raker structure and most body proportions but more slender; (body depth about 5.0 times rather than < 4.5 times in SL), with smaller scales, 72 vs. 55 to 70 in lateral line, about 40 vs. 30 to 35 before dorsal fin, 68 vs. 52 to 61 around body, 24 vs. 19 or 20 around caudal peduncle, and having fewer gill rakers, 40 vs. 37 to 53. This species is known only from the holotype. A geographically related species was *Ch. spatulifer*, known from Lake Idaho in the ancestral upper Snake River basin. The single specimen differs from other *Chasmistes*, except hybrids (depicted in the color illustration above): in the presence of a sub-terminal mouth with papillae on the lips. It seems unlikely, on the basis of other information on this genus, that the original population of *Ch. muriei* had this combination of characters. It is possible that the specimen possesses introgressed characters from *C. ardens* (Utah Sucker)[45].

A number of characteristics are common to all described extant species of *Chasmistes*: each lake sucker population is sympatric with at least one species of *Catostomus* over all or part of their range. Where sympatry occurs, a morphological continuum exists from benthivore to planktivore, with a large proportion of individuals exhibiting intermediate morphology. Historical descriptions also describe suckers of intermediate morphology. Even the paleontological record includes specimens somewhat intermediate between limnetic and benthic forms, with some sites yielding Miocene (23 to 5.3 mya) fossils described as "primitive" *Chasmistes*. In two of the three described extant lake sucker populations, suckers of intermediate morphology are indistinguishable from the *Chasmistes* form using genetic markers. These include Upper Klamath Lake, where *Ch. brevirostris* and *C. snyderii* are sympatric, and Utah Lake, where *Ch. liorus* and *C. ardens* are sympatric. By contrast, *Ch. cujus* in Pyramid Lake shows no evidence of gene flow with its sympatric benthivore, *C. tahoensis*, and fish that appeared morphologically to be hybrids were genetically differentiable from *C. tahoensis*, but not from *Ch. cujus*[42].

Range and Distributional Status

Snake River Sucker may have been native to Idaho. They occur in Jackson Lake, Wyoming, near the headwaters of the Snake River, and *Chasmistes*-like specimens have been collected in Palisades Reservoir, Idaho[46].

Suckers morphologically resembling lake suckers have been collected from Jackson Lake and Palisades Reservoir as recently as 2016 during surveys conducted by various resource agencies. Consistent with observations of suckers in Utah Lake and Upper Klamath Lake, Oregon, those in Jackson Lake and Palisades Reservoir exhibit a continuum between benthic and limnetic forms. DNA analysis reveals that suckers spanning the benthic to limnetic spectrum sampled in the Snake River just upstream of Jackson Lake, Wyoming during the 2004 spawning season were members of the northeastern branch or group of Utah Sucker, eliminating the possibility that the *Chasmistes* forms in Jackson Lake were transplanted June Sucker, which exhibit the southwestern Utah Sucker branch form[42].

Size and Longevity

Based on the most recently collected specimens, this species reaches total lengths of at least 514 mm (20.2 in). They likely live to ages similar to other catostomids; 10 to 12 years.

Food and Feeding

Based on the unique *Chasmistes* mouth parts and location, it is presumed that this species would have had similar foraging behavior and food habits as other *Chasmistes;* June Sucker are midwater feeders that consume principally zooplankton, aquatic insects, and algae[2]. Larval, juvenile, and adult Cui-ui (*Ch. cujus*) feed primarily on algae, generally in the water column in water nine to 31m (30 to 100 ft) deep[8].

Limiting Factors

This species is presumed to be extinct. However, the presence of lake sucker forms in Jackson Lake and Palisades Reservoir suggests that Snake River Sucker, as originally described, may persist, although whether representing a genus, or even a species, distinct from *C. ardens,* remains in dispute. Numerous accounts in the literature of the rediscovery of a fish species presumed to be extinct highlight this prospect. In Utah Lake, severe drought in the early 1930s led to extremely low water levels, culminating in a devastating winterkill in 1934-35. No spawning June Sucker were observed the following spring, and they were believed extinct. Re-discovered June Sucker were subsequently classified as a different subspecies by some researchers (*Ch. liorus mictus*) from the pre-drought form due to presumed introgression from Utah Sucker, despite Jordan's (1889) much earlier description of suckers morphologically intermediate between June Sucker and Utah Sucker. The original description of the Snake River Sucker also includes reference to introgression from Utah Sucker[42].

Breeding Habits

Snake River Sucker breeding habits are presumed to be similar to June Sucker (*Ch. liorus*), extant in Utah Lake, Utah, which spawn in only one section of the Provo River, the major Utah Lake tributary. Ripe June Sucker females have been captured as early as June 22 and peak spawning activity is over a brief period of time between June 1 and June 29. Spawning activity is greatest during midday. Water depths at spawning sites range from 0.3 to 0.8 m (1 to 2.5 ft) with a mean depth of 0.52 m (1.7 ft). During spawning in the Provo River, mean daily water temperatures range from 11.7 to 12.8°C (53 to 55°F). Bottom water velocity reported for the Provo River was 0.18 m/s (0.6 f/s). June Sucker do not spawn in sand, silt, or calm backwater areas. June Sucker eggs are pale yellow and demersal with a mean diameter of 0.5 mm (0.02 in) and newly hatched larvae average 7.6 mm (0.3 in) in length. At a mean temperature of 21°C (70°F) they hatch in four days. Larvae form large schools of several hundred to several thousand. Detailed spawning and other reproductive information for Cui-ui (*Ch. cujus*) has been reported[8,40,48].

Habits and Habitat

Although three species of *Chasmistes* currently inhabit other large lakes in the western United States, habitats in the four ecosystems where *Chasmistes* is known are not all that similar, either in origin or current ecological status[42]. It is, however, assumed that Snake River Sucker have similar behaviors and use habitats similar to extant species.

Ancestors of the three other known lake suckers in Utah Lake, Upper Klamath Lake, and Pyramid Lake inhabited very large, ancient lakes that receded, eventually confining sucker populations to small remnant lakes, whereas the ancestors of the suckers in Jackson Lake invaded a newly formed lake within the last 9,000 years. Research data describing limitations to this concept are consistent with an expanding sucker population in the area following late Pleistocene (0.011 to 0.8 mya) glaciation[42].

Management and Conservation Status

Snake River Sucker are officially considered extinct and have no conservation status in Idaho. Apparent persistence in Jackson Lake and Palisades Reservoir, along with the finding of another benthic-limnetic continuum involving lake suckers and a *Catostomus* species, raises important questions about lake sucker ecology, evolution, and taxonomy.

Conservation status and management of this form should be actively addressed, and lethal capture techniques should be curtailed until these issues are more clearly understood. Study of suckers in waters above Jackson Lake is highly recommended. Further, management programs for the other lake suckers (all federally endangered) should consider the possibility that these forms may exist as a dynamic continuum with sympatric *Catostomus* species, and that these pairs may share both evolutionary histories and evolutionary futures. In all *Chasmistes-Catostomus* pairs, studies using larger numbers of genetic loci to assess divergence, as well as studies on the genetic basis of morphological and functional differences, would be informative[42]. Interested readers are referred to the several publications that address these issues[29,42,43,45,49,50].

Edson Fichter Community Fish Pond, Indian Hills area, Pocatello, ID. Photo Courtesy John Sigler.

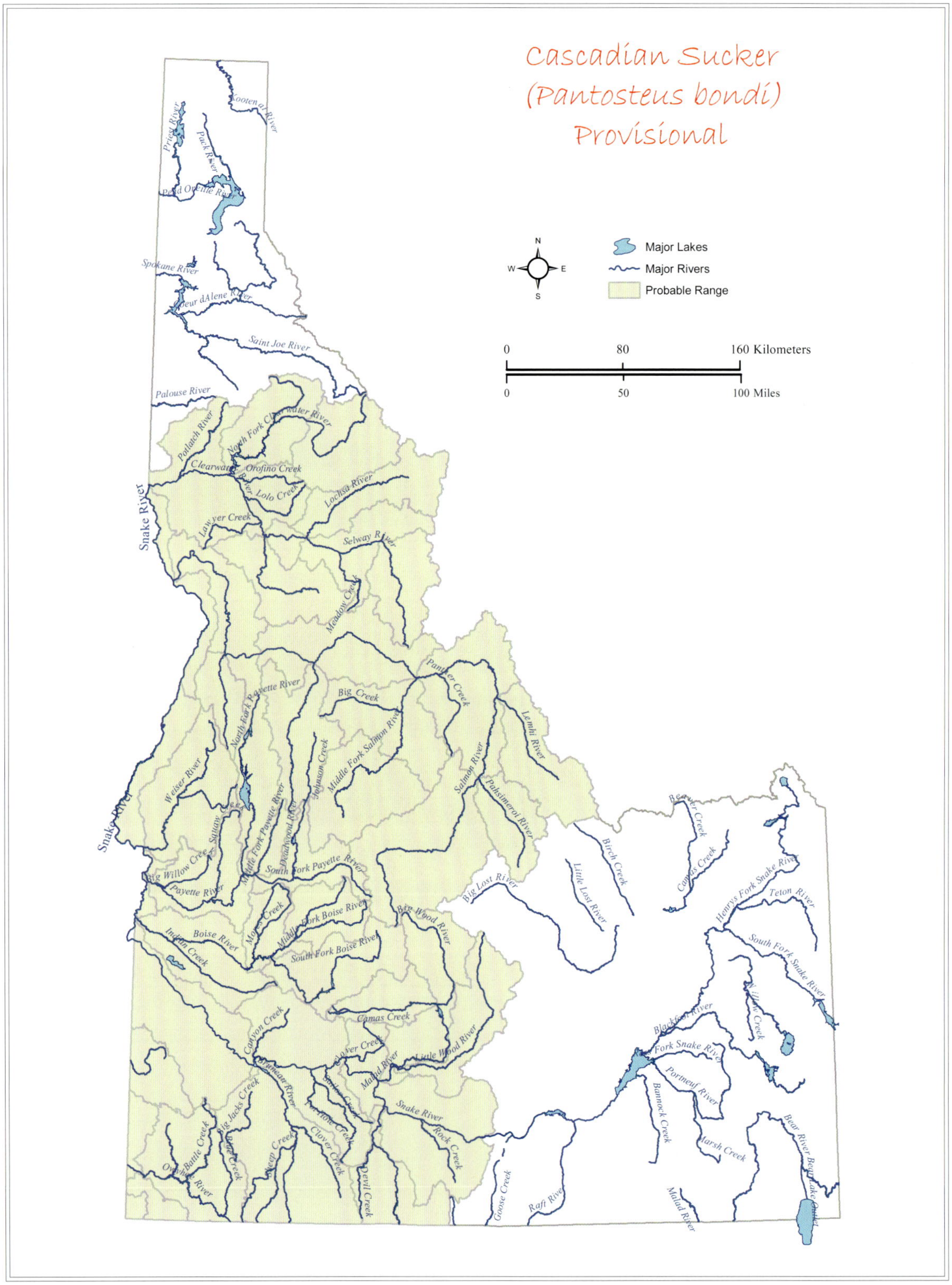
Cascadian Sucker
(Pantosteus bondi)
Provisional
N
S
W
E
Major Lakes
Major Rivers
Probable Range
0
80
160 Kilometers
0
50
100 Miles
Kootenai River
Priest River
Pack River
Saint Joe River
Palouse River
Potlatch River
Orofino Creek
Lolo Creek
Lochsa River
Lawyer Creek
Selway River
Snake River
Meadow Creek
Big Creek
Panther Creek
Lemhi River
Middle Fork Salmon River
Johnson Creek
Salmon River
Pahsimeroi River
Weiser River
Middle Fork Payette River
South Fork Payette River
Big Willow Creek
Payette River
Boise River
Indian Creek
Middle Fork Boise River
South Fork Boise River
Big Wood River
Big Lost River
Little Lost River
Birch Creek
Camas Creek
Canyon Creek
Little Wood River
Henrys Fork Snake River
Teton River
South Fork Snake River
Portneuf River
Bannock Creek
Marsh Creek
Big Jacks Creek
Clover Creek
Devil Creek
Rock Creek
Goose Creek
Raft River
Malad River
Bear River

Cascadian Sucker

Pantosteus bondi
Smith et al. 2013

Etymology

Pantosteus = all bone; *bondi* = named for the late Professor Carl Eldon Bond, co-collector of the type specimen (South Santiam River, Linn County, Oregon) in 1951.

Recent investigations provide convincing evidence that this is a new species of sucker for Idaho[29,43].

Importance

Cascadian Sucker are small and thus, readily available as forage for their entire life. When co-existing with sport fishes, they may supply an important local source of food. They may at times be used as bait. Their presumed algae diet precludes them from functioning as a competitor or predators of game fishes.

Description

Cascadian Sucker are similar in external appearance to the Mountain Sucker but generally have more gill rakers, fewer lateral line scales, and more predorsal scales (note descriptions in taxonomic key)[29]. Similar to the Mountain Sucker, Cascadian Sucker have notches at the corners of the lips, a pelvic axillary process and caudal membrane pigmentation concentrated next to the fin rays.

Range and Distributional Status

This species is native to Idaho. Cascadian Sucker are reported in the Snake River basin below Shoshone Falls including the Boise, Owyhee, Palouse, Salmon Falls Rivers and possibly the Salmon River[29]. As a newly described species, their range in Idaho is provisionally estimated. Refinement and confirmation of this estimated range awaits future study.

Size and Longevity

Size and longevity of Cascadian Sucker are presumed to be similar to that of Mountain Sucker. The accuracy of this presumption needs to be assessed in future work.

Food and Feeding

The diet and feeding behavior of Cascadian Sucker are presumed to be similar to that of Mountain Sucker. The accuracy of this presumption needs to be assessed in future work.

Limiting Factors

Biotic and abiotic factors that influence Cascadian Sucker are presumed to be similar to those that impact Mountain Sucker. The accuracy of this presumption needs to be assessed in future work.

Breeding Habits

The spawning habits and attributes of Cascadian Sucker are presumed to be similar to those of Mountain Sucker. The accuracy of this presumption needs to be assessed in future work.

Habits and Habitat

The habitat associations of Cascadian Sucker are presumed to be similar to those of Mountain Sucker. The accuracy of this presumption needs to be assessed in future work.

Management and Conservation Status

As a newly described species, the management and conservation status of Cascadian Sucker are unknown. This fish has no official status in Idaho. Their status is likely similar to that of Mountain Sucker, but the accuracy of this speculation needs to be assessed through future work.

Idaho Falls Dam, Snake River, Idaho Falls, ID. Photo Courtesy John Sigler.

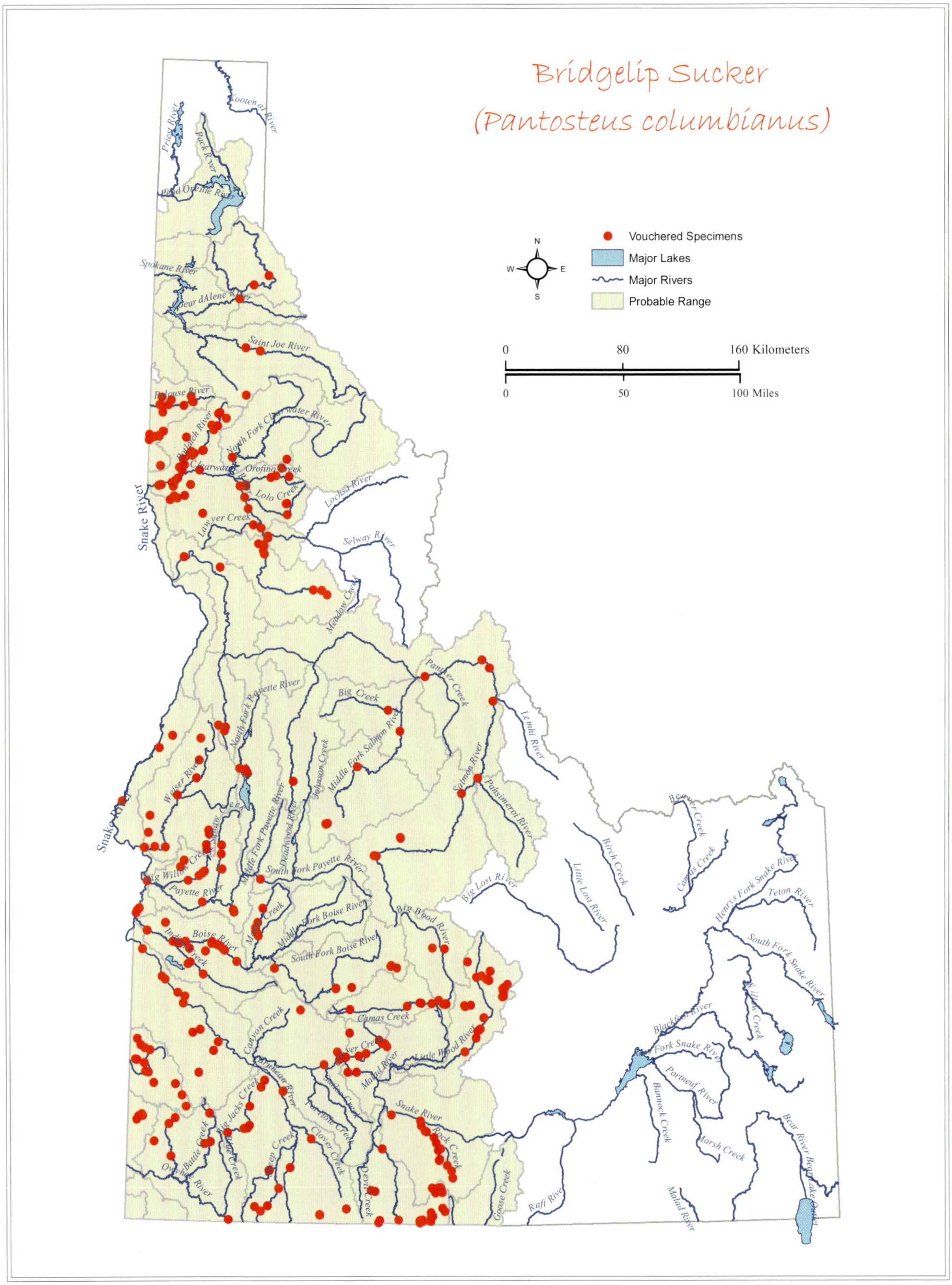
Bridgelip Sucker
(Pantosteus columbianus)
Vouchered Specimens
Major Lakes
Major Rivers
Probable Range
N
W
E
S
0
80
160 Kilometers
0
50
100 Miles

Bridgelip Sucker

Pantosteus columbianus

Eigenmann & Eigenmann, 1893

Etymology

Pantosteus = all bone; *columbianus* = of the Columbia River (type locality, Boise River, Idaho).

Importance

In the past, Bridgelip Sucker were highly prized for food by Nez Perce tribal members and were harvested from the Columbia River primarily during spawning runs in early spring by snagging or using weirs. Bridgelip Sucker fillets and eggs were eaten in the spring and sucker flesh was also canned for use during the following winter[6]. Although this species is edible, and may have constituted food for native peoples and their dogs in the past, it is not highly sought after today. Small Bridgelip Sucker may constitute forage for economically important, piscivorous sport fishes, although they may compete with sport fishes for food[4]. They are ecologically valuable in connecting algae, birds, fish and other components of the food web in a particular ecosystem[5].

Description

Bridgelip Sucker are dark brown to olive on the back and top of the head, while the upper sides are often mottled and paler brown; the lower sides, the head below the eye and the ventral surface are white or pale yellow. The lateral line is prominent near the head and paler than background; breeding males may have a prominent, orange lateral band. Sexually mature males from the Hanford Reach of the Columbia River had a bright red-orange stripe along the lateral line from the operculum to the base of the caudal fin during May through November. Young Bridgelip Sucker have a black peritoneum which is obvious in the abdominal region and often have three dark lateral blotches[4]. They are rounded, torpedo-shaped fish with a cylindrical body that is not laterally compressed. The head is moderately long. Bridgelip Sucker have a ventral mouth which is protrusible and suctorial; strongly overhung by the snout; the lips are thick and papillae are large and round; the lobes on the lower lip are rounded but not completely cleft. Gill rakers are short and fleshy. Branchiostegal rays are three. The single dorsal fin is soft rayed with a height slightly greater than the base and has nine to 12 rays, typically 12. The caudal fin is rather long, deeply forked with rounded tips. Anal fin has seven rays; the pelvic fins are abdominal and low, with the origin at the midpoint of the dorsal fin.

Pectoral fins are low, long, and pointed with 17 rays. Scales are cycloid, small, crowded anteriorly, and larger on the caudal peduncle. Lateral line scales number 87 to 111. The lateral line, at the midpoint of the body, is pale and complete but often with breaks and branches. The peritoneum is black and no pyloric caeca are present. Vertebra including the Weberian ossicles number 43 to 51. Nuptial tubercles on males often occur on the anal fin, the lower lobe of the caudal fin, and on scales of posterior half of the body[4,5].

Pantosteus columbianus is a species that arose in the Snake River below Shoshone Falls as the product of hybridization between *Catostomus macrocheilus* (Largescale Sucker) and an ancestor of *Pantosteus virescens* (Green Sucker)[29,43]. Perhaps for this reason, Bridgelip Suckers are more variable and show more geographic divergence than other species in Idaho.

The gill rakers of the Bridgelip Sucker are unique among suckers in having a cluster of spines on each raker rather than a double row of spines as in other mountain suckers. On the outer row of the first arch there are usually 30-42 gill rakers in Bridgelip Suckers over most of the state, but 25-30 in specimens from the Wood River and 33-39 in the Palouse River. On the inner row of the first arch there are usually 41-55 in most of Idaho, but 34-41 in the Wood River and 42-52 in specimens from the Palouse River. These counts apply to specimens over 70 mm (2.8 in) long, because gill raker numbers increase with early growth of the fish[51].

Bridgelip Sucker can be distinguished from Cascadian Sucker and Mountain Sucker by lip shape. Bridgelip Sucker lower lips are about as deep as they are wide and typically not or very weakly notched at the corners of the mouth, whereas lower lips of Cascadian Sucker and Mountain Sucker are wider than they are deep and are conspicuously notched at the corners of the mouth. In Bridgelip Sucker fry less than 20 mm (0.8 in) TL, the conspicuous black peritoneum clearly shows through the belly wall and distinguishes them from sympatric sucker fry[3].

Range and Distributional Status

This species is native to Idaho. Bridgelip Sucker are restricted to the freshwaters of northwestern North America from the Fraser River system in British Columbia south to central Oregon. This species occurs throughout the Columbia River system and in Idaho they are common in the Snake River and its tributaries downstream of Shoshone Falls, in the Palouse River, and Spokane River drainages[5,49]. They have been reported in the Pend Oreille system.

Size and Longevity

Bridgelip Sucker live to at least age nine and may reach a TL of 543 mm (21.4 in). In Idaho populations, juveniles sampled at the end of the first summer of life were between 64 and 79 mm (2.5 and 3.1 in) in length[6]. Young in Idaho were 7 mm (0.3 in) in July and 40 to 80 mm (1.6 to 3.2 in) by the end of the first summer after hatch. Adult maximum size has been reported from 254 to 381 mm (10 to 15 in)[4] and from 300 to 425 mm (12 to 17 in)[5]. The Washington angling record weighed 1.4 kg (3.0 lb, 1.0 oz) and was taken from Palmer Lake in 1999[6].

Food and Feeding

Bridgelip Sucker are herbivores. Mouth shape and the distinct sharp-edged lower jaw suggest a feeding process which involves scraping periphyton from hard substrates. This feeding mechanism has been verified for all sizes of Bridgelip Sucker collected from the mid-Columbia River where periphyton was the dominant food item. Invertebrates consumed by this species included chironomids (midges), zooplankton, and miscellaneous other invertebrates. In Lake Roosevelt, Washington, stomachs of Bridgelip Sucker contained cladocerans, chironomid larvae and pupae, aquatic worms, beetles, and terrestrial insects, as well as high percentages of unidentified organic remains[5,6].

Limiting Factors

Bridgelip Sucker may be parasitized by trematodes, cestodes, acanthocephalans, and crustaceans[4]. Hybridization between Bridgelip Sucker and Largescale Sucker occurs in the Hanford Reach of the Columbia River, complicating identification of some individuals[6].

Breeding Habits

Bridgelip Sucker in the mid-Columbia River spawn from mid-April until mid-June, peaking sometime during May, when water temperatures are 7.8 to 13°C (46 to 55°F). Spawning of Bridgelip Sucker in a Washington population occurred in water that was 0.15 m (0.5 ft) deep, with a current of 0.4 m/s (1.3 ft/s): substrate consisted of 43 percent pebbles, 36 percent cobble, 13 percent gravel, and 8 percent sand. Average FL of spawning fish has been reported at 404 mm (15.9 in). Spawning aggregations of Bridgelip Sucker in Washington populations consisted of one female and two to 10 males. Males in the mid-Columbia River begin to reach sexual maturity at FL of about 320 mm (12.6 in) at age five: all were mature at age six with lengths of about 360 mm (14.2 in); females mature about a year later at age six. When ripe, eggs are yellow and about 2.0 to 2.5 mm (0.08 to 0.1 in) in diameter; 366 to 460 mm (14.4 to 18.1 in) females produced between 9,955 and 21,400 eggs each. Bridgelip Sucker fry 13 mm (0.5 in) TL, emerged in mid-May, approximately 25 days after spawning. Spawning behavior included females assuming an upstream position from the males, which were positioned on either side of the female: females clean the substrate with vigorous swimming activity and rapid undulations of the caudal fin, producing spawning areas that were approximately two m long and 0.8 m (6.6 ft long and 2.6 ft) wide. Fertilized eggs settled in the interstitial spaces of the substrate and were predominantly found adhering to the underside of larger rocks[6].

Habits and Habitat

Bridgelip Sucker are seldom found in lake ecosystems in Idaho. Typically they inhabit the quiet areas in pocket waters, backwaters, or the edges of the main current of rivers and streams where there are sand, silt, or mud bottoms. Bridgelip Sucker may also be found in the colder water of fast-flowing, small rivers with gravelly to rocky bottoms[5].

Newly emerged fry of Bridgelip Sucker utilize areas inshore of the main river current. Larger young-of-the-year are found in shallow pools between 0.1 and 0.6 m (0.33 and 2.0 ft) deep, typically occurring over a mixed mud and rock bottom. Sub-adult individuals about 200 mm (7.8 in) long or more, and adults occupied the tails of pools between 1.0 and 2.0 m (3.3 and 6.6 ft) deep and deeper during daylight. At night, both sub-adults and adult are more abundant in slower, shallow water, 0.6 to 2.4 m (2.0 to 7.9 ft) deep, and near shore. Bridgelip Sucker are considered to be generalists in Oregon streams, occupying habitats with substrates of sand, boulders, and rubble, with water velocities between 0 and 1.2 m/s (0 to 4 ft/s)[6].

Management and Conservation Status

Bridgelip Sucker are considered common, widespread, and abundant range wide (G5) and within Idaho (S5)[11]. No defined studies have been conducted in Idaho ecosystems to determine their population status[41].

"Each species is a masterpiece, a creation assembled with extreme care and genius."

Edward O. Wilson

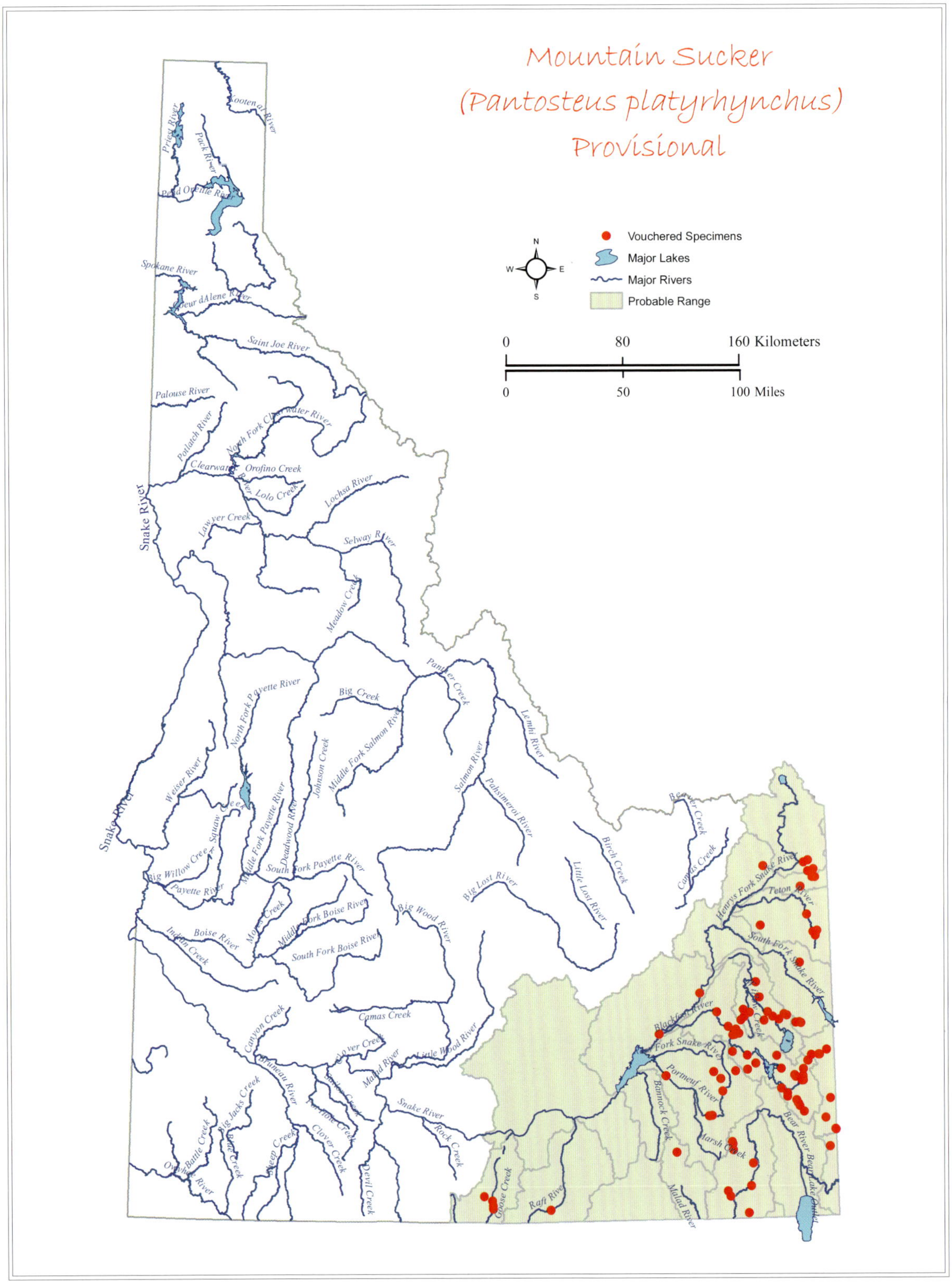
Mountain Sucker
(Pantosteus platyrhynchus)
Provisional
Vouchered Specimens
Major Lakes
Major Rivers
Probable Range
N
W
E
S
0
80
160 Kilometers
0
50
100 Miles
Priest River
Pack River
Spokane River
Saint Joe River
Palouse River
Potlatch River
Orofino Creek
Lolo Creek
Lochsa River
Snake River
Selway River
Meadow Creek
Big Creek
Middle Fork Salmon River
Johnson Creek
Deadwood River
Middle Fork Payette River
North Fork Payette River
Weiser River
Squaw Creek
Big Willow Creek
Payette River
South Fork Payette River
Boise River
Indian Creek
South Fork Boise River
Big Wood River
Big Lost River
Little Lost River
Birch Creek
Camas Creek
Lemhi River
Pahsimeroi River
Salmon River
Henrys Fork Snake River
Teton River
South Fork Snake River
Blackfoot River
Portneuf River
Bannock Creek
Marsh Creek
Malad River
Bear River
Bear Lake
Canyon Creek
Camas Creek
Little Wood River
Bruneau River
Clover Creek
Snake River
Rock Creek
Goose Creek
Raft River
Devil Creek
Big Jacks Creek
Battle Creek
Owyhee River

Mountain Sucker

Pantosteus platyrhynchus
(Cope, 1874)

Etymology

Pantosteus = all bone; *platyrhynchus* = flat snout

Importance

Mountain Sucker are small and thus, readily available as forage for their entire life. When co-existing with sport fishes, they may supply an important local source of food. They may at times be used as bait. Their algae diet precludes them from functioning as a competitor or predators of game fishes and makes them ecologically important as a link in the aquatic food chain.

Description

Mountain Sucker have backs and heads ventral to the lateral line that are green, gray, or brown and may be speckled with black. The lateral line, although present, is not prominent and the ventral surfaces of the head and body are pale yellow to white with a dark green to black lateral band or five dorsal blotches of fine black pigment. Fins are colorless or faintly red although breeding individuals are more highly colored with an orange to deep red lateral band. Young are markedly counter shaded dark green above, white below and have three dark vertical bars and a black peritoneum[4].

Mountain Sucker are a small, shallow, cylindrical fish with a generally torpedo-shaped cross-section. The body is slightly compressed laterally. The lower jaw has a cartilaginous sheath obvious even when the mouth is closed. The lower lip has a shape similar to paired wings, and the anterior edge is markedly convex and covered with large round papillae. The cleft is shallow and incomplete; there is a pronounced notch at the point of the lateral connection of upper and lower lips. Gill rakers number 23 to 37+31 to 51. Branchiostegal rays number three. A single soft-rayed dorsal fin is present and the edge is slightly emarginated with nine to 11 principal rays; occasionally eight to 13 over its entire range. The caudal fin is not long and not deeply forked and the tips are rounded points. The anal fin is long with seven prominent rays. Pelvic fins are abdominal, under the middle of the dorsal fin with rounded tips, nine or 10 rays and an axillary process is present. Pectoral fins are long with pointed tips and 15 rays. The cycloid scales are small, and crowded anteriorly. Scales number 79 to 89 or 60 to 108 along the lateral line over the range of the species, typically 80 to 85 in Idaho specimens. The lateral line is complete and straight. The peritoneum is black or dusky and there are no pyloric caecae. Vertebrae including the Weberian ossicles number 42 to 67[4].

Nuptial tubercles occur in both sexes, but are larger and more abundant in males. Tubercles are present on males on the anal fin, the lower lobe of the caudal fin, caudal peduncle, and the dorsal surface of the paired fins and over the entire surface of the body. Minute tubercles are present on females on the anal fin, the lower lobe of the caudal fin, on the caudal peduncle, and the dorsal surface of the paired fins[4].

Mountain Sucker may have dorsal and caudal fins that are greenish and the lower lobe of the caudal fin may be amber, while the paired lower fins are yellowish. Spawning males have a deep orange to reddish lateral stripe on each side above the greenish-black stripe that runs from the tip of the snout to just below the middle of the caudal fin base[2].

The upper lip is smoothed and curved upward while the lower lip has a shallow, median incision. The edges of the jaws, inside the lower lips, have cartilaginous sheaths used for scraping algae from rocks[2].

Bridgelip Sucker and Mountain Sucker can be separated by lip shape: Bridgelip Sucker lower lips are about as deep as they are wide and only slightly, if at all, notched at the corners of the mouth, whereas lower lips of Mountain Sucker are wider than they are deep and are conspicuously notched at the corners of the mouth[3].

Range and Distributional Status

This species is native to Idaho. Mountain Sucker are restricted to the freshwaters of the mountainous regions of western North America. In Idaho, Mountain Sucker are now thought to occur in the Snake River basin above Shoshone Falls and in Bear River ecosystems, separate from Cascadian Sucker[49]. This belief needs to be confirmed or refuted through future study.

Size and Longevity

Mountain Sucker are small and may reach total lengths of 230 mm (9.0 in). Females in Montana were slightly longer than males of the same age and apparently live longer, reaching approximately age nine as compared with age seven for males.

Mountain Sucker are not only small, but are slow growing. Total lengths for Montana populations, age one to five are, in mm (in): 31 (1.2), 56 (2.2), 102 (4.0), 127 (5.0), and 152 (6.0). An age eight individual was 203 mm (7.9 in)[2].

Food and Feeding

Mountain Sucker are herbivores, feeding mostly on algae scraped off rocky substrates of the ecosystem using the cartilaginous lower jaw. Small aquatic insects, rotifers, and crustaceans found in the diet of these individuals were probably living in the algae community attached to the rocks. Individuals also may feed upside-down while feeding on large boulders. Some Mountain Sucker contained animal material, principally midge larvae, and a few individuals have been found to have ingested caddisflies, mayflies, and stoneflies[6].

Limiting Factors

Water temperatures in the mid 20°C (mid to high 70°F) are harmful to Mountain Sucker. Dewatering of ecosystems may adversely affect required habitat. A variety of fishes, birds, and mammals prey on Mountain Sucker. Occasionally hybridization with Tahoe Sucker occurs where the species overlap may dilute genetic structure[2]. Mountain Sucker are preyed upon by some birds and some populations may be limited by predation, leading to extinction if suitable habitat is limited in stream ecosystems[2,6].

Breeding Habits

Sexual maturity in Mountain Sucker males occurs at two to three years when they are 127 to 152 mm (5 to 6 in) long and females at age four to five years at 127 to 178 mm (5.0 to 7.9 in). During breeding season, male Mountain Sucker develop secondary sex characteristics, differentiating them from females. Anal fin rays become enlarged and large, cone-shaped tubercles develop on the fins, but females have a small anal fin and small-to medium-sized tubercles, limited to only the posterior ray. Breeding fish develop a deep orange to red lateral band that is brighter and extends posteriorly to the caudal fin in males only. Other populations may have females which are more fecund[6], where between 900 and 4,000 eggs may be produced during spring spawning when water temperatures exceed 10°C (50°F). Spawning may terminate if temperatures reach the high 20s°C (high 60s°F). Incubation period is probably short[2].

Habits and Habitat

Mountain Sucker prosper in locales that include an abundance of periphyton and relatively cool waters. Mountain Sucker inhabit clear, cold waters of mountain streams, generally less than about 12 m (40 ft) wide with bottoms consisting of sand, gravel, or boulders. Mountain Sucker are also found in some lakes, reservoirs and large rivers. Mountain Sucker longer than about 127 mm (5.0 in) are usually found adjacent to pools where the current is about 0.5 m/s (1.5 f/s), and water is between 1.0 to 1.5 m (3.3 to 5.0 ft) deep. Adults are found in the deeper pools of streams in winter, but in spring move to areas adjacent to pools with moderate current and water depths of about 1.0 to 1.5 m (3.3 to 5.0 ft) deep. Following spawning, adults return to deeper pools with overhanging cover and aggregate in small schools. Preferred water temperatures of Mountain Sucker are between 13 and 21°C (55 and 70°F) during the summer; they can tolerate a maximum water temperature of about 28°C (82°F). Mountain Sucker that inhabit reservoirs and lakes apparently require tributary streams with suitable habitat for spawning[6].

As might be implied by their common name, Mountain Sucker prefer habitat that includes riffles and clear, cold waters, typically in creeks and rivers. Individuals seek habitats which include areas of gravel, rubble, sand, or boulders, however, they do occur in turbid streams with mud, sand, or clay bottoms and in large lakes. Although typically considered to be a high elevation species, occurring at elevations up to 2804 m (9200 ft) they also live near sea level. Preferred summer water temperatures range between 12.8 and 21.1°C (55 and 70° F) although higher temperatures are tolerated[2].

Management and Conservation Status

Mountain Sucker are considered common, widespread, and abundant range wide (G5) and within Idaho (S5)[11]. No defined studies have been conducted in Idaho to determine conservation status. Noted declines of Mountain Sucker in part of their historic range are attributed to the loss of essential spawning habitat in streams above reservoirs, land management and irrigation practices, riparian alterations, predation or competition from introduced fish species, and hybridization with sympatric suckers. Protection of stream habitat essential for natural reproduction of Mountain Sucker may be necessary to prevent extirpation.

Idaho Falls Dam, Snake River, Idaho Falls, ID. Photo Courtesy John Sigler.

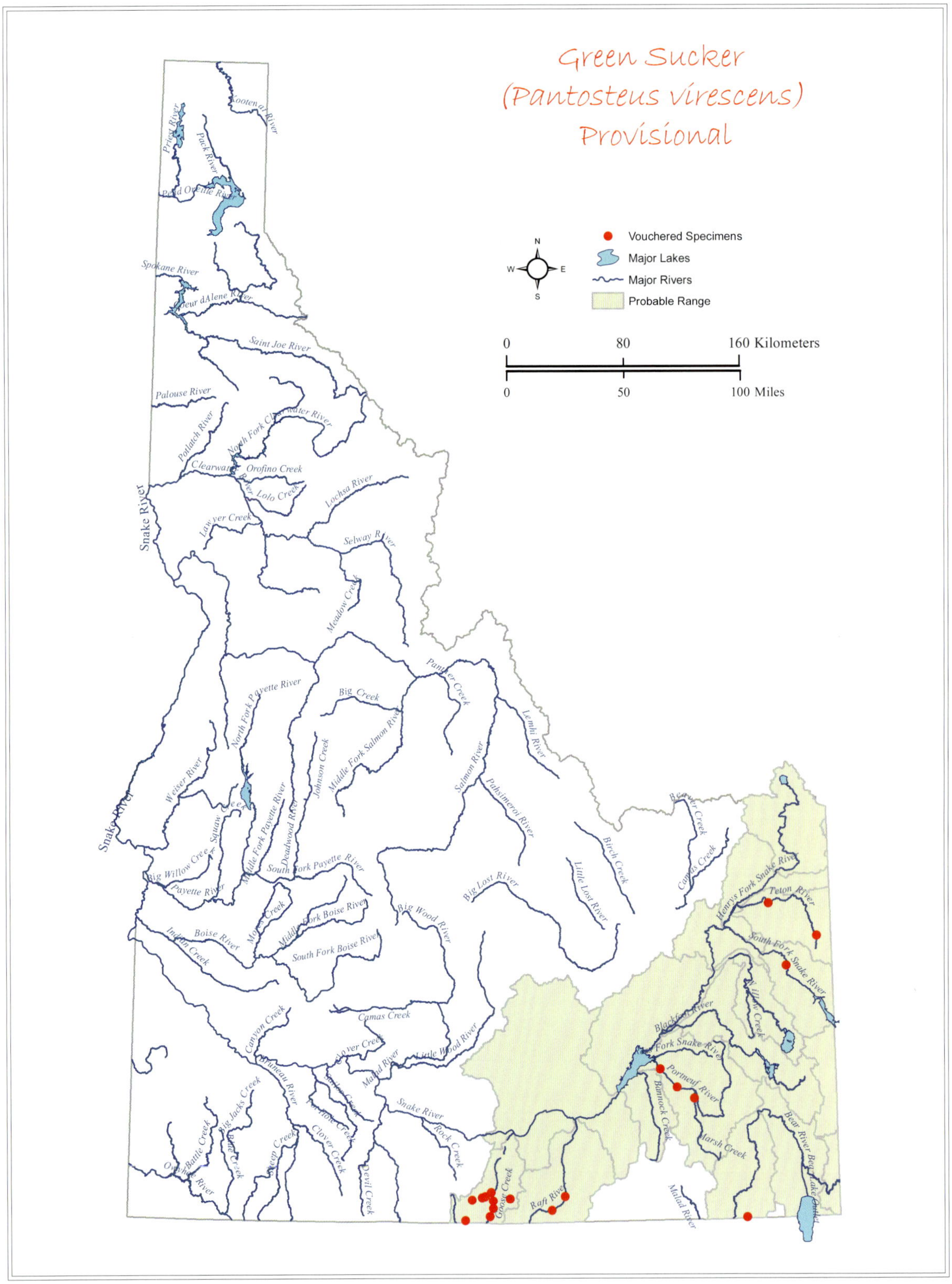
Green Sucker
(Pantosteus virescens)
Provisional
N
S
W
E
Vouchered Specimens
Major Lakes
Major Rivers
Probable Range
0
80
160 Kilometers
0
50
100 Miles
Priest River
Pack River
Spokane River
Saint Joe River
Palouse River
Potlatch River
Orofino Creek
Lolo Creek
Lochsa River
Snake River
Selway River
Meadow Creek
Big Creek
Middle Fork Salmon River
Johnson Creek
Weiser River
Payette River
Boise River
Indian Creek
South Fork Boise River
Big Wood River
Big Lost River
Salmon River
Pahsimeroi River
Lemhi River
Birch Creek
Little Lost River
Camas Creek
Teton River
South Fork Snake River
Portneuf River
Bannock Creek
Marsh Creek
Malad River
Bear River
Raft River
Goose Creek
Rock Creek
Snake River
Salmon Falls Creek
Clover Creek
Devil Creek
Canyon Creek
Big Jacks Creek
Little Wood River

Green Sucker

Pantosteus virescens
Cope, 1875

Etymology

Pantosteus = all bone; *virescens* = greenish

Importance

In some Idaho ecosystems where native, Green Sucker are a forage fish. They may compete little with sport fish populations. Where sport fish and Green Sucker habitats overlap, suckers may sustain a portion of the predation, reducing pressure on young game fishes. Green Sucker are likely valuable in converting algae into fish protein.

Description

This fish was formerly called Bluehead Sucker in Idaho. However, the type specimen described by Cope[52] in 1875 was apparently mislabeled and has not been reported from the type locality in the Arkansas River basin since. In our description of this fish, we follow the rationale of Snyder[53] and Unmack et al.[43] that Green Sucker occur in the Bonneville and upper Snake River basins whereas the Bluehead Sucker occurs in the upper Colorado River basin. Additional support for this contention is provided by Hopken et al.[53] where they report the putative Bluehead Sucker of the Bear, Weber and upper Snake River basins to be genetically distinct based on mitochondrial DNA analyses[53].

Coloration of preserved Green Sucker was originally described by Cope[52] as "*Color in spirits olivaceous; decidedly green on the head, lower surfaces and fins (narrowly on belly) yellow*". Snyder[54] describes the posterior part of Green Sucker, when in the water, as typically dense black and the anterior region is lighter. The top of the head is olive, lighter on the sides and brassy beneath the eyes, and the lips and throat are yellowish. When fresh from the water, Snyder[54] describes the Green Sucker as "bright metallic yellow on the sides, darker above and dead white below".

Snyder[54] reports Green Sucker typically have 11 (occasionally 12) dorsal fin rays, the anterior lip has four to five rows of papillae, and possess 36 to 40 gillrakers on the first arch. Green Sucker may be distinguished from the Bluehead Sucker of the Colorado River basin by a wider caudal peduncle, mean depth 8.4 to 8.8 percent SL; as opposed to 5.4 to 8.3 percent of standard length in Bluehead Sucker[29]. In specimens from the Snake River basin, vertebrae number 43 to 46, pre-dorsal scales 45 to 58, lateral line scales 78 to 97, and isthmus width is 7 to 11.5 percent of SL[55].

Range and Distributional Status

This species is native to Idaho. Green Sucker occur in the Weber and Bear River watersheds in Utah and the Snake River drainage above Shoshone Falls[43,54].

Size and Longevity

Green Sucker from the Weber River (putative Bluehead Sucker) are reported to reach at least 575 mm (22.6 in) TL. Size class structure suggests these fish are at least six years old[56].

Food and Feeding

No reports of Green Sucker diet and feeding behavior have been located, however, the relatively high number and short length of the gill rakers suggests periphyton is a major food item[55].

Limiting Factors

Green Sucker populations in the Weber River of Utah appear affected by hydrologic modification (dams and diversions), channelization, introduced species (particularly Brown Trout), siltation, habitat destruction and potentially pesticides[56–58].

Breeding Habits

Green Sucker in the Weber River spawn in the spring as evidenced by gonad development[59].

Habits and Habitat

In a study of the Weber River, Green Sucker (putative Bluehead Sucker) seemed to prefer 1 to 2 m (3.3 to 6.6 ft) pools with moderate current and swift-current runs 1 m (3.3 m) deep[57]. During another Weber River study, Green Sucker were observed to move less than 1 km (0.6 mi) up to at least 15 km (9 mi) over two sampling periods between September 2007 and March 2009[56] .

Management and Conservation Status

As a re-described taxon Green Sucker have no official status rangewide or in Idaho. Information on their status awaits official recognition of the taxon and future study.

Unmack et al.[43] report Green Sucker were formerly abundant in the Weber River, Bear River and upper Snake River basins. They recommend the Green Sucker be given high conservation significance since they represent the remaining non-introgressed populations of this declining species.

"Underwater, I experience space with my body. I'll see a school of fish gathering and moving together and I'll exclaim, 'This is architecture.'"

Antoine Predock

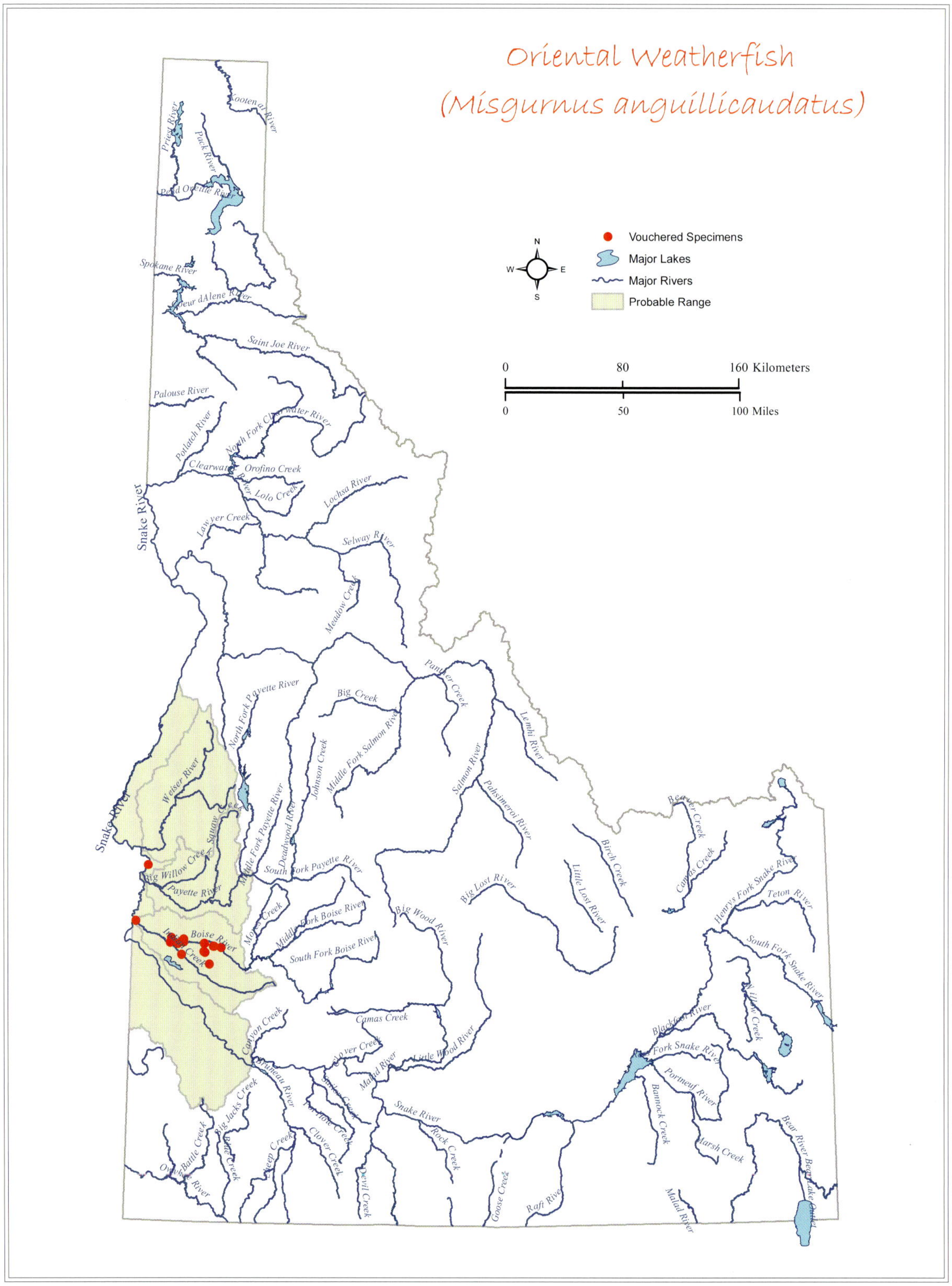
Oriental Weatherfish
(Misgurnus anguillicaudatus)
Vouchered Specimens
Major Lakes
Major Rivers
Probable Range
N
S
E
W
0
80
160 Kilometers
0
50
100 Miles
Snake River
Weiser River
Payette River
Boise River
Big Willow Creek
Salmon River
Clearwater River
Saint Joe River
Palouse River
Lochsa River
Selway River
Big Lost River
Big Wood River
Camas Creek
Bear River
Teton River
South Fork Snake River
Henrys Fork Snake River
Portneuf River
Birch Creek
Little Lost River
Lemhi River
Pahsimeroi River
Middle Fork Salmon River
Big Creek
Johnson Creek
South Fork Payette River
South Fork Boise River
Middle Fork Boise River
Malad River
Raft River
Goose Creek
Rock Creek
Bruneau River
Lawyer Creek
Orofino Creek
Lolo Creek
Palouse River
Potlatch River
Spokane River
Pack River
Priest River
Meadow Creek
Deadwood River
Middle Fork Payette River
North Fork Payette River
Canyon Creek
Big Jacks Creek
Clover Creek
Devil Creek
Little Wood River
Bannock Creek
Marsh Creek
Willow Creek
Blackfoot River
Camas Creek
Beaver Creek

Family Cobitidae—Loaches

Body is worm- or eel-like to fusiform. The mouth is subterminal. There are three to six pairs of barbels present. The erectile spine is below the eye. There is one row of pharyngeal teeth. These are popular aquarium fishes. Greatest diversity of this family occurs in Asia. Species in this family are bottom dwellers. This family includes about 26 genera and about 177 species[1].

Subfamily Cobitinae

Members of this subfamily have one pair of rostral barbels, and the caudal fin is usually rounded or slightly emarginated. There are about 19 genera, including *Misgurnus*, and about 130 species[1]. In Idaho one species is represented.

Oriental Weatherfish

Misgurnus anguillicaudatus
(Cantor, 1842)

Etymology

Etymological information on the scientific name for this species has not been found. The common name, Weatherfish, refers to the habit of this species to become excited or agitated when a change occurs in barometric pressure.

Importance

As an introduced and invasive species, Oriental Weatherfish are an undesirable species in Idaho waters. Potential impacts such as competition, indirect effects on aquatic community structure, and introduction of fish parasites are a few of the issues associated with their presence in northwestern waters[60].

Description

Oriental Weatherfish can be easily recognized by the 10 barbels around the subterminal mouth, six large and four small ones. Oriental Weatherfish have small scales that are embedded in thick skin, contrasting them with catfish, which lack scales. The elongate body of the Oriental Weatherfish is round anteriorly in cross-section but becomes compressed posteriorly. The caudal fin is rounded. Dorsal and anal fins are located posteriorly. Oriental Weatherfish swim in a rapid, undulating eel-like manner. Generally, coloration of the body is grayish brown, with small irregular spots or blotches primarily on the dorsal part of the body; individuals are pale silvery white ventrally, and females are more pale than males[6]. Males have an enlarged second pectoral fin ray[61].

Range and Distributional Status

This species is not native to Idaho. Oriental Weatherfish are native to the eastern Asia-Amur region (Siberia), China, Korea, and Japan. Oriental Weatherfish have been introduced in Australia, Mexico, Philippines, and the United States. In the United States, Oriental Weatherfish have been reported from California, Florida, Hawaii, Idaho, Illinois, Michigan, Oregon, Tennessee, and Washington. As this species is a popular aquarium fish, populations in natural waters are assumed to have originated from releases of aquarium pets[6], a practice that is highly discouraged. In Idaho they have been collected from the Boise River basin and the Snake River between the Boise and Weiser Rivers.

Size and Longevity

Adult Oriental Weatherfish are generally 100 to 203 mm (3.9 to 8.0 in) but can reach maximum sizes of 250 mm (9.8 in). In portions of the Lake Washington, Washington, drainage, Oriental Weatherfish range from 84 to 201 mm (3.3 to 7.9 in) TL. Mature female Oriental Weatherfish that averaged 170 mm (6.7 in) were larger than mature males, which averaged 142 mm (5.6 in)0[6]. In 1999 in the Lake Washington Basin, nine females averaged 171 mm (6.7 in), and eight males averaged 143 mm (5.6 in). Four immature individuals averaged 103 mm (4.1 in)[60]. Specimens up to 202 mm (8.0 in) have been collected from the Boise River drainage.

Food and Feeding

Oriental Weatherfish are omnivores, and quite opportunistic, using food that is available in their particular ecosystem. Preference appears to be for worms and larvae of aquatic insects. Oriental Weatherfish feed by "combing" sand and silt substrates that are taken into their mouths and rapidly expelling water through the gill openings, retaining larger materials, including organisms and organic matter. Oriental Weatherfish appear to use their barbels as olfactory organs to locate prey. In a portion of Portage Bay, Washington, Oriental Weatherfish consumed chironomid larvae (41 percent by number), cladocerans (46 percent), and amphipods (9 percent). In Japanese rice paddies, mosquito larvae were readily consumed, and in aquariums Oriental Weatherfish consumed live fry of mosquitofish and commercial salmon-egg fishing bait[6].

In a Washington Lake drainage study, Oriental Weatherfish consumed chironomid larvae and pupae of amphipods and cladocerans, which made up 96.3 percent of the diet (wet weight). Cladocerans (46 percent) and chironomids (40 percent) were the most frequently ingested prey. While little is known of Oriental Weatherfish diet in their natural habitat, they readily consumed mosquito larvae and have been used successfully in South Korea to control mosquitoes in rice patties. In Australian populations, where Oriental Weatherfish are also introduced, their diet consisted of algae, detritus, ostracods, and cladocerans. Additional notes included feeding on zooplankton[60].

Limiting Factors

Little is known about factors that may limit Oriental Weatherfish in northwestern aquatic ecosystems. Because Oriental Weatherfish can apparently withstand a wide range of environmental conditions and given their habitat preferences, they may be able to expand into a variety of Idaho ecosystems[60]. They would presumably be preyed upon in some portion of their life history by a variety of fish and avian predators.

Breeding Habits

Oriental Weatherfish become sexually mature at about 102 mm (4.0 in) TL. Adult males are distinguishable from adult females by the larger pectoral fins, which are longer than the head and have enlarged or thickened second rays. Oriental Weatherfish spawning occurs over a silt-mud substrate or in beds of aquatic vegetation in the spring. Gravid females collected from a Washington lake were present from April to July. Females produce small, adhesive, reddish-colored eggs. Both males and females swim around the eggs until they are covered with silt and mud but apparently do not remain at the spawning site to afford protection to the nest or young. Embryos hatch in a few weeks, hatching time being dependent on water temperature. Young Oriental Weatherfish bury in the substrate for long periods of time, likely to feed on microscopic organisms[6].

Habits and Habitat

Within the United States, habitat preferences of Oriental Weatherfish include still or slow-moving waters of lake or stream ecosystems. They are secretive and burrow in muddy bottoms, often with only the head protruding, making them difficult to observe. Oriental Weatherfish are most active at twilight. They have been found in water less than 0.9 m (3.0 ft) deep, with substrates of fine sediments and sometimes cobble substrate along main channels[6].

In their native range, Oriental Weatherfish occur in habitat where water temperatures are often cold: their preferred temperature range is 10 to 15°C (50 to 59°F). This species is able to tolerate water that is low in dissolved oxygen due to the presence of an accessory intestinal respiratory organ. Oriental Weatherfish may also rise to the surface, swallow air that is forced into the highly vascular hind gut, and release it as large bubbles through the anus[6]. In Idaho they are often found in irrigation ditches, where they bury themselves in the substrate once water is gone. Once buried, they overwinter and emerge once water returns.

Management and Conservation Status

Considered common, widespread, and abundant range-wide (G5). A state conservation status rank is not applicable (SNA) because the species is not a suitable target for conservation activities[11]. As a nonnative species, additional Oriental Weatherfish introductions are strongly discouraged. While little data are presently available on their statewide distribution and potential effects of this species on Idaho aquatic ecosystems, possible interactions with native species include the introduction of new fish parasites, competition for food or space, predation on eggs of desirable sport fisheries, and additional restructuring of the aquatic ecosystem[60]. As this species is a popular aquarium fish, populations in natural waters are assumed to have originated from releases of aquarium pets[6], a practice that is illegal and highly discouraged.

Chapter Citations

1. Nelsen, 2006. 2. Sigler and Sigler, 1996. 3. McPhail, 2007. 4. Scott and Crossman, 1973. 5. Wallace and Zaroban, 2013. 6. Wydoski and Whitney, 2003. 7. La Rivers, 1962. 8. Sigler and Sigler, 1987. 9. Moodie, 1966. 10. Rosenfeld, et al. 2001. 11. Idaho Department of Fish and Game, 2005. 12. Zaroban, et al. 1999. 13. Taylor, Darveau and Schulte, 2013. 14. Sigler, 1958a. 15. Zambrano, et al. 2006. 16. Parkos, Santucci and Wahl, 2003. 17. Baxter and Stone, 1995. 18. Johnson, 2002. 19. Neuhold, 1957. 20. Sigler, 1953. 21. McConnell, Clark and Sigler, 1957. 22. Johnson and Belk, 1999. 23. Miller, 1945. 24. Uyeno, 1960. 25. Coburn and Cavender, 1992. 26. Miller and Hubbs, 1960. 27. DeMarais, 1992. 28. Dowling, et al., 2002. 29. Smith, Stewart and Carpenter, 2013. 30. Simons and Mayden, 1997. 31. Johnson, Dowling, and Belk, 2004. 32. Billman, Wagner, and Arndt, 2008. 33. Blakney, Loxterman and Keeley, 2014. 34. Wilson, 2013. 35. Wilson and Belk, 2001. 36. Wilson and Belk, 1996. 37. Idaho Department of Fish and Game, 2015. 38. Videgar, 2000. 39. Markle, 2016. 40. Sigler and Kennedy, 1978. 41. Moyle, 2002. 42. Cole, 2014. 43. Unmack, et al., 2014. 44. Simpson and Wallace, 1978. 45. Miller and Smith, 1981. 46. Smith, et al., in preparation. 47. Jordan, 1878. 48. Sigler, Vigg and Bres, 1985. 49. Miller and Smith, 1967. 50. Cole et al., 2008. 51. Smith, 2016. 52. Cope and Yarrow, 1875. 53. Hopken, Douglas, and Douglas, 2013. 54. Snyder, 1924. 55. Smith, 1966. 56. Webber, Thompson and Budy, 2012. 57. Peters, 1974. 58. Giddings and Stephens, 1999. 59. Andreasen and Barnes, 1975. 60. Tabor, Wagner, and Hager, 2001. 61. Urquhart and Koetsier, 2011. 62. Smith, et al., 2017.

Additional readings for this chapter are listed in the Supplmental References section.

"There is nothing permanent except change."

Heraclitus

10

SILURIFORMES

Catfishes

Order Siluriformes

An adipose fin is typically present in these fishes. Spine-like rays are generally present in the dorsal and pectoral fins. Several species of catfish are known to be venomous or poisonous and can inflict severe wounds with their spines; some stings can cause death. In many areas these fishes are a valuable and popular sportfish[1]. Among the characteristics of this order is its worldwide distribution in generally warm, fresh, and salt waters. Approximately one-half of the more than 2,000 species are South American species. None of the individuals in this order have scales, but are considered naked. Highly diversified, individuals in this order include electric catfishes, and some have air breathing apparatuses that allow them to breathe out of water, a capability that often frustrates fisheries managers when the Walking Catfish (*Clarius batrachus*) stroll out of ponds from which their removal is desired[2].

This order has a Weberian apparatus; dorsal and pectoral fins are different in origin from those of spiny rayed fishes. Siluriformes have barbels around the mouth[1,2]. There are 35 families with about 446 genera and 2,867 species[1].

Family Ictaluridae—North American Catfishes

Easily recognizable by the distinctive four pairs of barbels around the mouth, these fishes are the "bullheads" of many a youthful adventure in the local pond or slow-moving stream. Although they are somewhat nocturnal, they can often be caught on simple gear. Individuals in this family are native to the warm waters east of the Rockies, except in Mexico. These are small to large fishes with a naked body and large, flattened heads. There is a spine in the dorsal fin; pectoral fins are low and armed with a serrate spine, and the pelvic fins are abdominal. The caudal fin is forked, square, or round. Seven genera are present in North America, with 46 species[1–3]. At least seven species have been introduced in Idaho.

Ameiurus melas	(Rafinesque, 1820)	Black Bullhead
Ameiurus natalis	(Lesueur, 1819)	Yellow Bullhead
Ameiurus nebulosus	(Lesueur, 1819)	Brown Bullhead
Ictalurus furcatus	(Lesueur, 1840)	Blue Catfish
Ictalurus punctatus	(Rafinesque, 1818)	Channel Catfish
Noturus gyrinus	(Mitchill, 1817)	Tadpole Madtom
Pylodictis olivaris	(Rafinesque, 1818)	Flathead Catfish

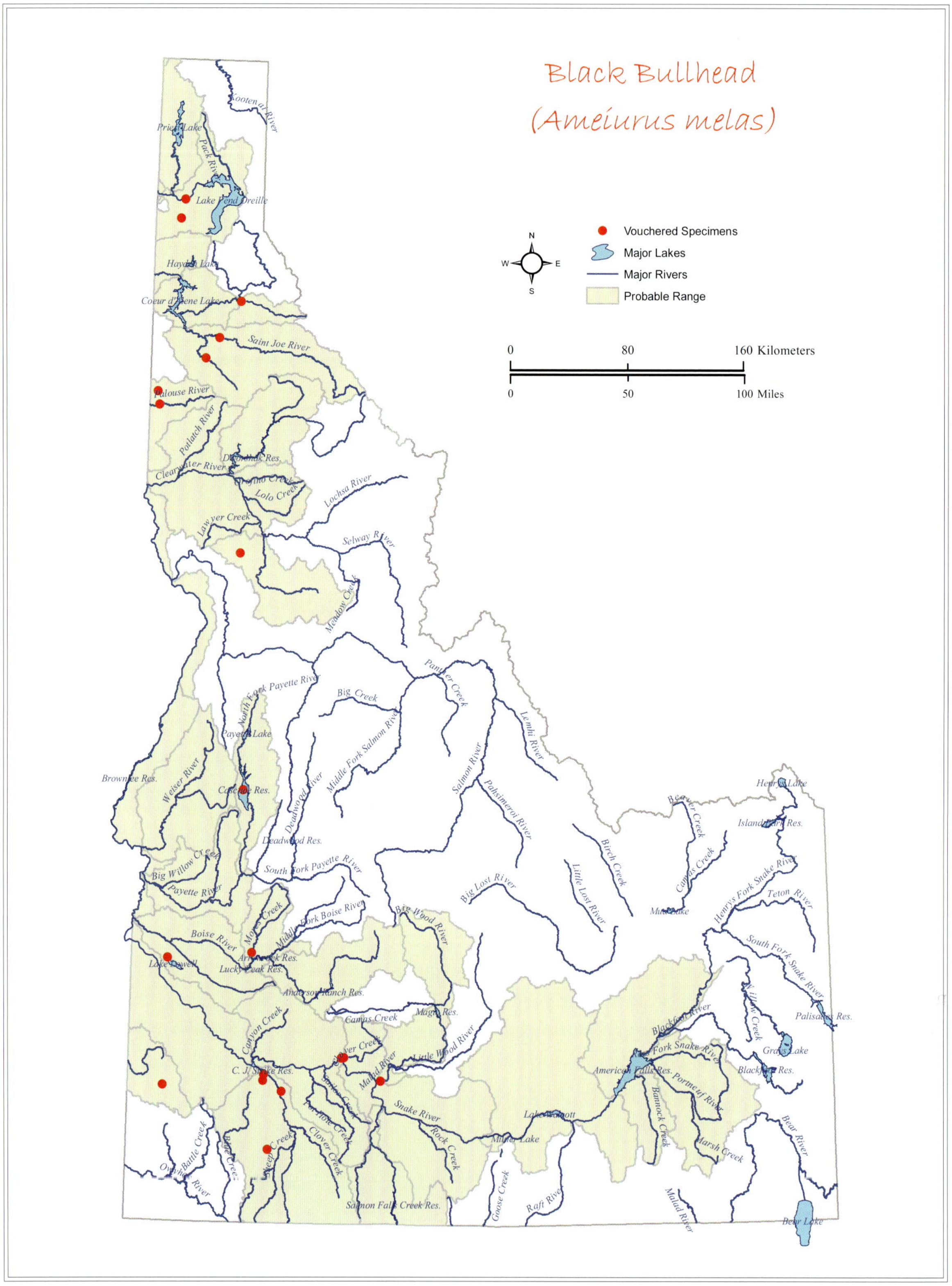
Black Bullhead
(Ameiurus melas)
Vouchered Specimens
Major Lakes
Major Rivers
Probable Range
0 80 160 Kilometers
0 50 100 Miles

Black Bullhead

Ameiurus melas
(Rafinesque, 1820)

Etymology

Ameiurus = not (a fish without) curtailed, the caudal fin is not notched; *melas* = black.

Importance

As a recreational sportfish, Black Bullhead are a perfect match for youngsters and others who are not inclined to invest a great deal of effort but anticipate catching fish regularly. When Black Bullhead strike bait, it is without subtlety but still slowly. Many hours of fishing pleasure are derived annually from this species, some in areas where few other fishing opportunities exist. In ponds where they are fed regularly, they may reach 152.4 mm (6.0 in) and 0.45 kg (15.9 oz) but often produce populations with too many small individuals to be of interest to any but the most casual angler[2,3].

Description

Adult Black Bullhead are blackish to dark olive in some ecosystems or dark brown on the back. Depending on the waters, the belly is greenish white or bright yellow. Colors are more brilliant during breeding season. Black Bullhead have a distinct pale vertical bar at the base of the caudal fin, and the barbels are darker than the chin. The fin membranes are black. One pair of the flattened maxillary barbels' length will reach the opercular opening. Chin barbels are pigmented, as are those on the Brown Bullhead, which distinguishes them from the Yellow Bullhead. Gill rakers are moderately long, pointed, and number 14 to 20, ranging from 10 to 14 on the lower limb and four to six on the upper limb. Although this fish is smaller than some of its close relatives in this family, the head is massive in proportion and very distinctive. Spines in the pectoral fins have weak or absent serrations[2–4].

Range and Distributional Status

This species is not native to Idaho. Black Bullhead are native to southern Ontario, Canada, southwest through New York to northwestern Georgia and Alabama, west to eastern New Mexico, and north to eastern Montana. Black Bullhead are now common in most of the lower 48 states in warm waters. In Idaho

Black Bullhead occur in western and northern drainages. Black Bullhead have been found in the Clark Fork, Saint Joe, and Clearwater Rivers in northern Idaho; the Payette and Boise Rivers in central Idaho; and portions of the Little Wood and Bruneau Rivers in western Idaho[2–4].

Size and Longevity

Black Bullhead are small, usually only 127 to 178 mm (5.0 to 7.0 inches), but they have a rather massive, naked (scaleless) body[3]. Some individuals may reach sizes of 356 mm (14.0 in) and about 0.9 kg (1 lb, 15.8 oz), but average weight is about half that size[2]. A 3.37 kg (7 lb, 7 oz) fish caught in 1993 from Long Island, New York, holds the all-tackle world record for this species[5]. Total average lengths for individuals at the end of years one through five are, in mm (in), about 100 (3.9), 170 (6.7), 240 (9.4), 290 (11.4), 320 (12.6), and 350 (13.8)[6].

No record is listed for Idaho, but a Utah record of 1.13 kg (2 lb, 8 oz) has been reported[2].

Food and Feeding

Black Bullhead feed opportunistically and are omnivorous scavengers. Adults tend to feed at night, while juveniles may feed mainly at twilight; this species feeds almost exclusively on or near the bottom of the ecosystem. They will feed on plant material, snails, frogs, other fishes, crayfish, and waste or detrital material. If abundant, crustaceans may make up 75 percent of Black Bullhead food in a particular ecosystem[2].

Limiting Factors

Black Bullhead are parasitized by protozoans, trematodes, cestodes, nematodes, and crustaceans[3]. Black Bullhead are preyed upon by other fishes, and predators may decimate eggs and young, even when young are guarded by the adults. Adults may be injured by fighting during the breeding season. In some ecosystems, Black Bullhead may produce dominant-year classes that result in later year classes with a paucity of young. Lethal temperatures are about 35°C (95°F)[2].

Breeding Habits

Spawning takes place in areas of moderate to heavy submerged vegetation, where in shallow water the female excavates a nest about the diameter of her length. Gravel is pushed away, and silt and debris are fanned away with her tail. Prior to spawning, pairs may butt each other and slide their barbels over one another. Egg deposition follows a head-to-tail clasping in which the caudal fin of the male is wrapped around the head of the female[3]. Spawning occurs from May to July at temperatures of 19 to 22°C (66 to 72°F). Females may produce 2,000 to 6,000 eggs that hatch in less than a week at temperatures of 22.2 to 23.9°C (72 to 75°F). Either or both of the adults may guard the nest initially, and the female may shortly depart. Young school when they leave the nest and form a jet-black compact ball that may be as much as 0.45 m (17.7 in) in diameter. The school is guarded by one or both parents until young are 25 to 51 mm (1.0 to 2.0 in). Schooling continues throughout the summer but may be permanently disrupted by repeated predator attacks[2]. Actual mating is preceded by the pair lying side by side, facing opposite directions, and with the male, mouth open, curling his caudal fin around the face of the female. Spawning may occur five times during one hour, and successive spawnings may be preceded by fewer of the lateral embraces. Females guard the nest and eggs during the first day after spawning, with the male assuming the role after the first day. Eggs hatch in five to 10 days, and free-swimming young remain in a tight school, typically guarded by the male, for about two weeks, at which time they have attained about 25 mm (1.0 in) total length. Sexual maturity is reached at about 160 mm (6.3 in)[6].

Habits and Habitat

Black Bullhead prefer shallow lakes and ponds with weedy areas[7]. Black Bullhead are warm-water fish and are found in a variety of small ponds, quarries, small reservoirs, and intermittent creeks, often occurring in muddy backwaters of various ecosystems[2]. Black Bullhead are hardy fish, able to survive in temperatures into the mid-20°C range (26.7°C/80°F) and at oxygen concentrations of less than 2 mg/l (ppm)[2].

Reporting on studies in marshes in western France, researchers provide strong evidence that Black Bullhead prefer reed-bed habitats, a habitat type that has rapidly invaded the marsh because of the decline of agricultural practices in this artificial ecosystem, suggesting that Black Bullhead may have benefited from the expansion of reed beds in view of the positive relationships found between their abundance and the presence of adjacent reed beds, the selection of peat substratum in small ditches, and the preferential use of reed beds (versus marsh meadows) during the flooding period. Some differences were found between young-of-the-year (YOY) and adult individuals as a consequence of the ontogenetic shift in habitat use from YOY (pelagic habits) to adult (benthic habits) life stages. In general, YOY and adult abundance in permanent habitats was positively correlated with trends derived from the development of reed beds. Black Bullhead are tolerant to the harsh conditions imposed in many wetlands; therefore, they are likely to benefit from such changes in the abiotic environment. It was also noted that Black Bullhead illustrate a strong affinity to reed beds as temporary flooded habitats. Furthermore, abundance increases when decreasing water depth, so the production of a large amount of litter by the reed beds would favor them by limiting water depth[8].

Management and Conservation Status

Black Bullhead are not viable competitors for food in large reservoirs, but in small pond ecosystems or small shallow lakes, they may become the single dominant species. When too many individuals are produced, the population may stunt, leaving large numbers of individuals that are of little interest to most anglers[2]. Black Bullhead are considered secure, common, widespread and abundant rangewide (G5). In Idaho, a conservation rank is not applicable (SNA) because a nonnative species is not a suitable target for conservation activities They are classified as a game fish in Idaho[13].

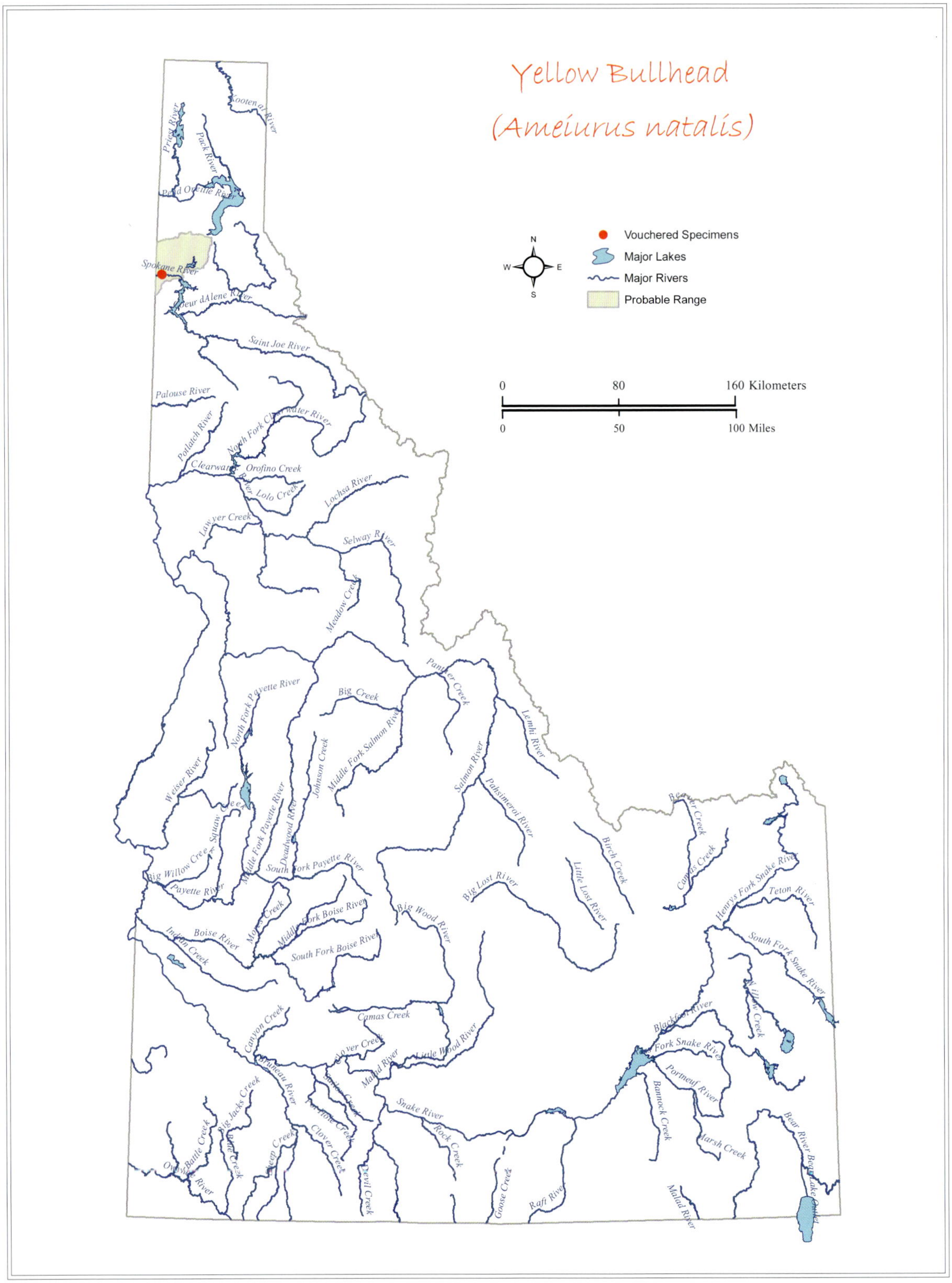
Yellow Bullhead
(Ameiurus natalis)
Vouchered Specimens
Major Lakes
Major Rivers
Probable Range
0 80 160 Kilometers
0 50 100 Miles
Kootenai River
Priest River
Pack River
Pend Oreille River
Spokane River
Coeur dAlene River
Saint Joe River
Palouse River
Potlatch River
Clearwater River
Orofino Creek
Lolo Creek
Lochsa River
Selway River
Meadow Creek
Big Creek
Johnson Creek
Middle Fork Salmon River
Lemhi River
Salmon River
Pahsimeroi River
Weiser River
Payette River
Boise River
Indian Creek
South Fork Boise River
Big Wood River
Big Lost River
Little Lost River
Birch Creek
Camas Creek
Henrys Fork Snake River
Teton River
South Fork Snake River
Blackfoot River
Portneuf River
Bannock Creek
Marsh Creek
Bear River
Snake River
Rock Creek
Goose Creek
Raft River
Malad River
Salmon Falls Creek
Bruneau River
Owyhee River
Canyon Creek
Little Wood River
Devil Creek

Yellow Bullhead

Ameiurus natalis

(Lesueur, 1819)

Etymology

Ameiurus = not (a fish without) curtailed, the caudal fin is not notched; *natalis* = having large nates or buttocks.

Importance

Few specifics are available for this species in Idaho. Within their native range, Yellow Bullhead may constitute a significant portion of the local fishery in some ecosystems. In part, this species' importance may be disguised because in some fisheries information, it is lumped with all other "bullheads."

In many instances in the western United States, Yellow Bullhead have been introduced as a sportfish. As is generally the case with introductions of nonnative species, some unanticipated consequences resulted in many ecosystems[9].

Description

Coloration of the dorsal area of the Yellow Bullhead is yellow, olive, or brown. Yellow Bullhead are distinguishable from other closely related species by the immaculate white chin barbels. They have a long, rounded caudal fin. Yellow Bullhead have 24 to 27 anal fin rays, including rudimentaries, and a vague dark stripe runs through the anal fin. The dorsal fin has six soft rays, and the pelvics have eight rays[2]. Yellow Bullhead have an upper jaw that is slightly longer than the lower jaw and thick but not prominent lips. Gill rakers are moderately long and well-spaced, numbering 14 to 16 (usually 10+4). Branchiostegal rays are usually eight or nine. The adipose fin is long, fleshy, and slightly turned up, with a free length equal to or greater than the attached base. The pelvic fins are abdominal and low, overlapping the base of the anal fin, with eight rays; the pectorals are broad[3].

Range and Distributional Status

This species is not native to Idaho. Native range of the Yellow Bullhead is from Vermont south to Florida, west to central Texas, north to eastern South Dakota, and east to New York and Vermont. They have been introduced into numerous locations in the western United States, often as a sportfish. Partly due to their preference for clear but warm-water ecosystems, they do not have an extensive distribution in Idaho, only known from the Spokane River below Coeur d'Alene Lake.

Size and Longevity

Yellow Bullhead do not generally live past five years, although some reports of individuals aged seven are noted[2–4]. Yellow Bullhead grow more rapidly than the closely related Black Bullhead and reach an acceptable size for anglers at 178 to 254 mm (7.0 to 10.0 in) in around three years[4].

A 2.89 kg (6 lb, 6 oz) Yellow Bullhead was taken in Missouri in 2006[5]. No size record information is available for Idaho. Neighboring Washington State produced a 349 mm (13.7 in), 0.74 kg (1 lb, 10 oz) individual in 1994[4]. Composite lengths for individuals from the northern United States were, for fishes ages one through four, are, in mm (in), 30 to 66 (1.2 to 2.6), 180 to 244 (7.1 to 9.6), 193 to 274 (7.6 to 10.8), and 226 to 295 (8.9 to 11.6)[3].

Food and Feeding

Yellow Bullhead feed opportunistically at night. Yellow Bullhead consume a wide range of food items on the bottom of their ecosystems, finding food items with their barbels. Food items range from crustaceans and immature aquatic insects to mollusks and potentially some fishes, including dead fishes[3,4].

Limiting Factors

Yellow Bullhead achieve a level of protection from predation by virtue of their spines and their nocturnal habits. They are preyed upon by a variety of basses, Walleye, and other catfishes. Parasites include trematodes, cestodes, nematodes, leeches, crustaceans, and cephalans[3].

Breeding Habits

Yellow Bullhead generally spawn in the spring in their native range for a period of about two weeks in late May to early June. One or both sexes build the nest, which may range from a shallow depression to a burrow more than 61 cm (2 ft) deep. Batches of creamy white glutinous eggs, usually numbering 300 to 700, are laid and fertilized during a spawning clasp. Eggs hatch in five to 10 days, and the males guard the nest and the brood of young until they are about 51 mm (2.0 in) long[3].

Habits and Habitat

Yellow Bullhead prefer slow streams with clear water and ponds or lakes with abundant vegetation. Yellow Bullhead are aggressive, but this behavior may function only as a warning, with little physical injury occurring to intruders. Aggression increases with higher water temperatures, and physical injury may occur during these interactions. Yellow Bullhead use a chemical communication system in their social interactions, which is apparently disrupted at higher water temperatures[4].

Management and Conservation Status

Yellow Bullhead are considered secure, common, widespread and abundant rangewide (G5). In Idaho, a conservation rank is not applicable (SNA) because a nonnative species is not a suitable target for conservation activities. They are classified as a game fish in Idaho[13].

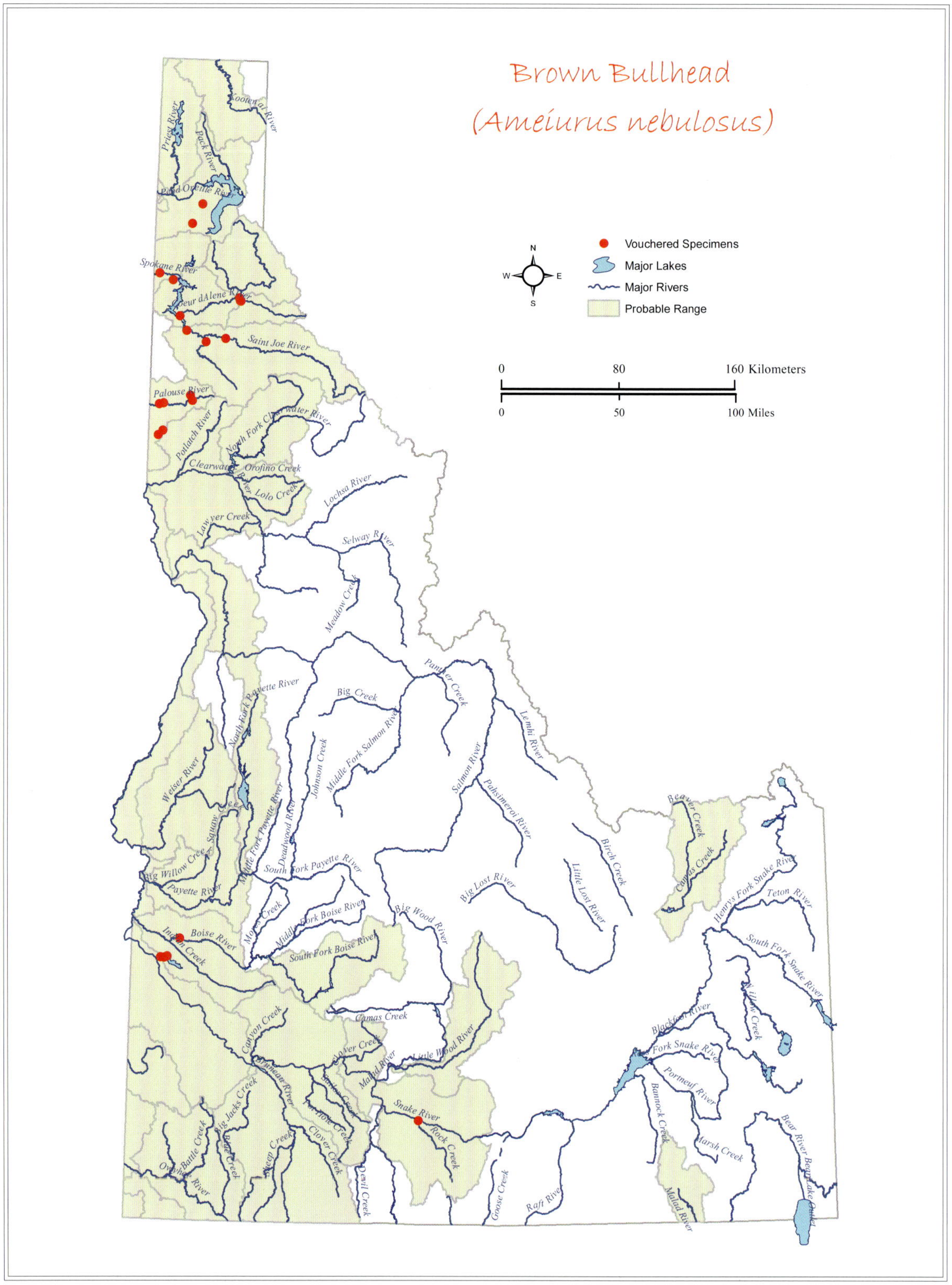
Brown Bullhead
(Ameiurus nebulosus)
N
W
E
S
Vouchered Specimens
Major Lakes
Major Rivers
Probable Range
0
80
160 Kilometers
0
50
100 Miles
Kootenai River
Priest River
Pack River
Spokane River
Saint Joe River
Palouse River
Potlatch River
Orofino Creek
Lolo Creek
Lochsa River
Selway River
Meadow Creek
Big Creek
Johnson Creek
Middle Fork Salmon River
Salmon River
Lemhi River
Pahsimeroi River
Birch Creek
Little Lost River
Big Lost River
Big Wood River
Weiser River
Deadwood River
Middle Fork Payette River
South Fork Payette River
Payette River
Boise River
South Fork Boise River
Canyon Creek
Camas Creek
Snake River
Rock Creek
Clover Creek
Big Jacks Creek
Battle Creek
Devil Creek
Goose Creek
Raft River
Teton River
South Fork Snake River
Portneuf River
Bannock Creek
Marsh Creek
Malad River

Brown Bullhead

Ameiurus nebulosus

(Lesueur, 1819)

Etymology

Ameiurus = not (a fish without) curtailed, the caudal fin is not notched; *nebulosus* = clouded, in reference to this species' mottled and gray coloring.

Importance

Brown Bullhead, were once fished commercially in lakes of the Coeur d'Alene basin, with harvests ranging up to more than 250,000 lb. Brown Bullhead may be sought in the early spring when they move into shallows to feed and spawn[7]. When Brown Bullhead exist in ecosystems near population centers, they may provide an important component for local or regional "easy access" sportfishing[4]. Brown Bullhead readily take a variety of baits and are easy to catch, potentially providing a satisfactory fishing experience for youngsters or less experienced anglers[3].

Description

Brown Bullhead are a naked, moderate-size fish, usually 203 to 356 mm (8.0 to 14.0 in). The body is massive, rounded, and heavy forward. The mouth is terminal, but the upper jaw is slightly longer than the lower jaw, and the mouth is horizontal and wide. Brown Bullhead possess one pair of long, flattened maxillary barbels that are the longest of the four pairs of barbels. Gill rakers are moderately long and pointed, numbering nine on the upper limb and four or five on the lower limb. Branchiostegal rays are usually 9,9 ; but can be 8,8; or other unequal combinations. There is a single dorsal fin that is ahead of the midpoint of the body and is soft-rayed, with the first ray modified into a stout spine. The adipose fin is long, fleshy, and somewhat curved upward. Pelvic fins are abdominal and low, with the origin behind the dorsal fin insertion. Pectoral fins are high, broad, and rounded, with a heavy anterior pectoral spine. Brown Bullhead have strong serrations on the posterior edge of the pectoral fin spines, separating them from the Black Bullhead, where the serrations are weak or absent. Pectoral fins usually have four to eight bars and eight rays. Brown Bullhead fin membranes are dark pigmented but not black.

The lateral line is complete and almost straight. The peritoneum is silvery to gray and not speckled with black. This species has no pyloric caeca. There are 34 to 39 vertebrae and Weberian ossicles[3].

Range and Distributional Status

This species is not native to Idaho. The native range for Brown Bullhead is from Nova Scotia, Canada, south to Florida and west to Louisiana, then northwest to North Dakota and Saskatchewan, Canada, and back east to Nova Scotia. Brown Bullhead have been introduced across the western United States, including Idaho, since at least 1874[3,4]. In Idaho, Brown Bullhead are known to occur in the Pend Oreille, Spokane, Bear River, and Snake River drainage below Shoshone Falls. They may be more widely distributed, since many stocking records simply list bullhead.

Size and Longevity

Brown Bullhead rarely live beyond age five, but they grow rapidly. Brown Bullhead in Lake Washington, Washington, were age seven when males were about 356 mm (14.0 in) and females were about 305 mm (12.0 in)[4]. Some Brown Bullhead may reach sizes from 178 to 381 mm (7.0 to 15.0 in) and weights up to 0.91 kg (2 lb, 0 oz), and a few may reach 406 to 457 mm (16.0 to 18.0 in) and approximately 1.36 kg (3 lb, 0 oz)[10].

A northern Wisconsin population had total lengths, at ages two through five, in mm (in), of 152 (6.0), 193 (7.6), 241 (9.5), and 267 (10.5)[3].

A 3.35 kg (7 lb, 6 oz) Brown Bullhead was taken from Mahopac Lake, New York, in 2009 and stands as the world record for this species[5]. No record is provided for Idaho[11].

Food and Feeding

Brown Bullhead are omnivorous and active scavengers. Generally, they feed on or near the bottom on plant material or almost anything that is edible, including snails, fish, and waste materials. When available, crustaceans may make up 75 percent of the total diet. Brown Bullhead tend to be nocturnal in their feeding habits but may actually feed 24 hours a day[10].

Limiting Factors

Brown Bullhead, particularly the young, are preyed upon by and suffer competition from a variety of other fish. Potential parasite infestation includes protozoans, nematodes, trematodes, leeches, mollusks, and eye flukes. They may hybridize with Black Bullhead in some ecosystems where habitat is marginal for one or both species[3].

Breeding Habits

Brown Bullhead spawn in the spring when water temperatures reach 21.1°C (70°F), generally in May and June in most ecosystems. One or both sexes clear a shallow nest that has a diameter of approximately one body length. Nests may be cleared on bottoms of mud or sand or among rooted aquatic vegetation. Water over nest sites can be as shallow as 152 mm (6.0 in). Spawning occurs in the daytime and involves caressing by both sexes with the barbels. Eventually, both sexes settle over the nest site; facing opposite directions, with bodies touching, the spawning act occurs. Eggs are pale cream colored and coated with a gelatinous material; eggs are about 3 mm (0.1 in) in diameter[3].

Females from 203 to 230 mm (8.0 to 9.1 in) contain 2,000 to 13,000 eggs. Eggs in the nests of Brown Bullhead are cared for by one or both parents, which includes fanning with the paired fins, moving or stirring with the barbels and fin spines, and picking up eggs with the mouth and then ejecting them back into the water. Eggs take six to nine days to hatch at temperatures of 20.6 to 23.3°C (69 to 74°F). Young are about 6 mm (0.2 in) when they hatch and may stay in the nest for seven days, after which they begin to swim and feed[3].

Habits and Habitat

Whereas Black Bullhead are typically found in pond and lake ecosystems[10] and Yellow Bullhead are a stream species, Brown Bullhead are present primarily in lakes[10].

Management and Conservation Status

Brown Bullhead are considered secure, common, widespread and abundant rangewide (G5). In Idaho, a conservation rank is not applicable (SNA) because a nonnative species is not a suitable target for conservation activities. They are classified as a game fish in Idaho[13].

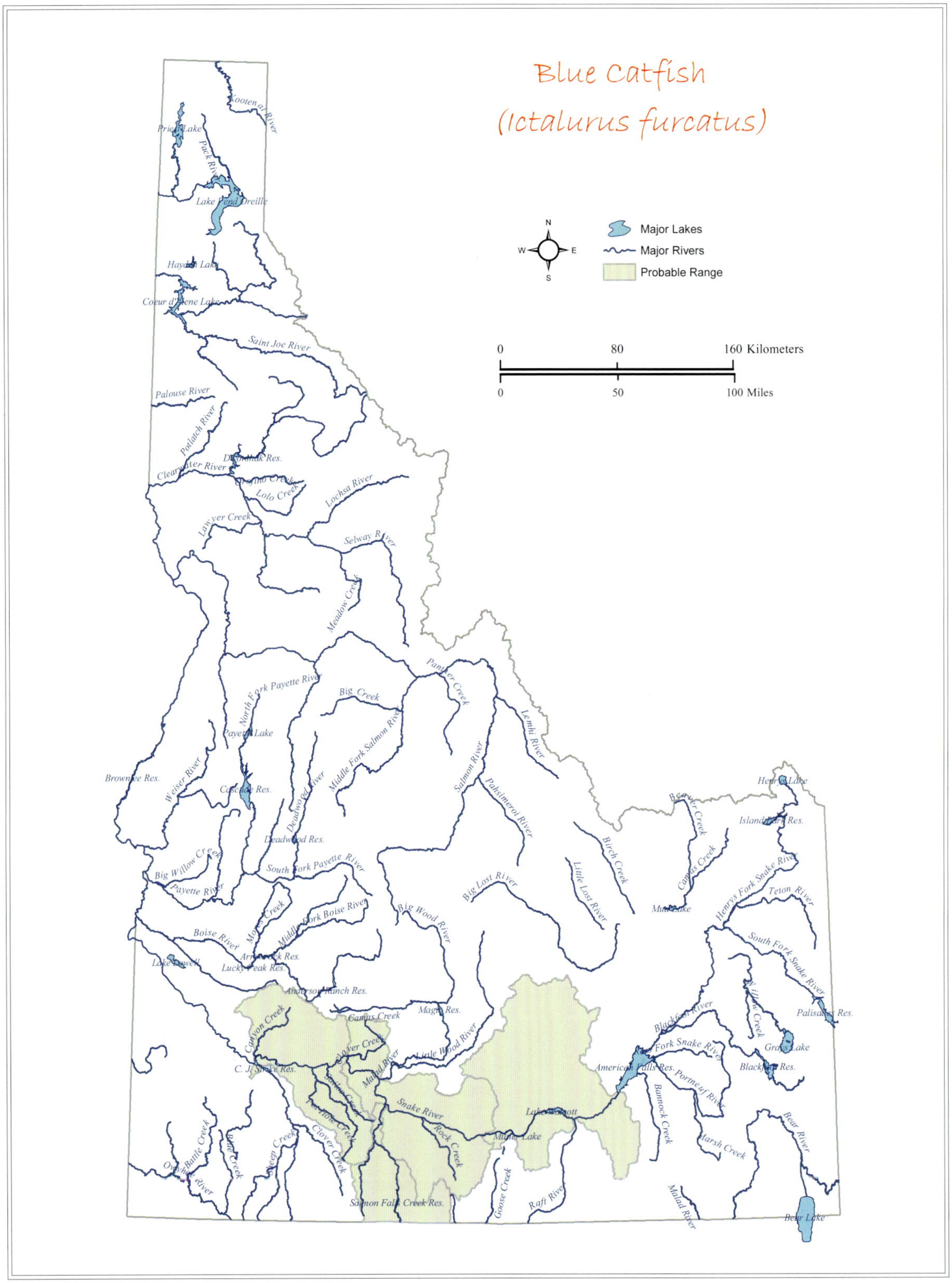
Blue Catfish
(Ictalurus furcatus)
N
W
E
S
Major Lakes
Major Rivers
Probable Range
0
80
160 Kilometers
0
50
100 Miles
Priest Lake
Pack River
Lake Pend Oreille
Saint Joe River
Palouse River
Potlatch River
Lolo Creek
Lochsa River
Selway River
Meadow Creek
Big Creek
Middle Fork Salmon River
Payette Lake
Brownlee Res.
Weiser River
Cascade Res.
Deadwood Res.
Salmon River
Pahsimeroi River
Lemhi River
Birch Creek
Little Lost River
Big Lost River
Big Wood River
Camas Creek
Teton River
South Fork Snake River
Palisades Res.
Grays Lake
Blackfoot Res.
Big Willow Creek
Payette River
Boise River
Lucky Peak Res.
Magic Res.
Little Wood River
Snake River
Rock Creek
Lake Walcott
Bannock Creek
Marsh Creek
Bear River
Bear Lake
Malad River
Goose Creek
Raft River
Salmon Falls Creek Res.
Clover Creek
C. J. Strike Res.

Blue Catfish

Ictalurus furcatus
(Lesueur, 1840)

Etymology

Ictalurus = fish cat; *furcatus* = forked, referring to the deeply forked tail fin.

Importance

Blue Catfish were apparently stocked into the Snake River, and anglers have reported catching this species in both the Snake and the Columbia Rivers. However, fishery biologists doubt that they are established in these drainages. Blue Catfish are reared in a commercial fish hatchery near Twin Falls, Idaho, and some individuals may escape into the Snake River.

Description

Blue Catfish are pale gray to dark blue dorsally and white ventrally. The fins will have dusky margins. Anal fin rays number 30 to 36. The trailing edge of the anal fin is typically straight, whereas it is curved in Channel Catfish[12]. Gill rakers number 14 to 21. The pectoral spine has well-developed posterior serrae, similar to that of Channel Catfish. The air bladder in this species is very constricted. On Blue Catfish, the back slopes steeply and in a straight line from the dorsal fin to the snout, giving the head of this species a wedge-shaped appearance[6]. Blue Catfish lack spots on the body.

Range and Distributional Status

This species is not native to Idaho. Blue Catfish are widespread in the Mississippi Basin and in the Gulf drainages from Mobile Basin west and south through Texas, Mexico, and northern Guatemala. They have been widely introduced elsewhere[6]. Simpson and Wallace[7] reported in their 1978 book that a shipment of catfish was received by the IDFG from the USFWS in 1943 that was supposed to be Blue Catfish. However, it was believed that this shipment probably contained Flathead Catfish. In 1985 more than 30,000 Blue Catfish were stocked in Idaho waters, primarily in the Snake River. None have been reported from the Idaho reaches of the Snake River hence.

Size and Longevity

Blue Catfish may reach a total length of more than 1,000 mm (39.4 in) by the end of their ninth year. Life span, while unknown, surely exceeds 20 years. Total lengths, in mm (in), for age classes one through nine are reported as 141 (5.6), 212 (8.3), 260 (10.2), 308 (12.1), 355 (14.0), 407 (16.0), 477 (18.8), 585 (23.0), and 666 (26.2). Yearly growth increments increase markedly after the second year. Maximum size is conjectural, but has been reported up to 143 kg (315 lb, 4 oz). The largest reasonably confirmed record is a 68 kg (150 lb) specimen, taken in 1879. The Tennessee record is a 59 kg (130 lb), 1,550 mm (5 ft) specimen from Fort Loudon Reservoir, taken in 1976[6]. The all-tackle record is 64.9 kg (143 lb), taken from Kerr Lake, Buggs Island, Virginia, in 2011[5].

Food and Feeding

Young Blue Catfish under 125 mm (4.9 in) feed primarily on zooplankton. Adults feed on burrowing mayflies, crayfishes, fingernail and unionid clams, and other fishes.

Limiting Factors

Ecosystem changes in the form of impoundments and siltation are having adverse effects on the Blue Catfish in their native range[6]. Parasites that afflict this species are likely similar to those noted for Channel Catfish, and small Blue Catfish are likely preyed upon by a variety of other fishes. Due to the fishes' large size, predation on adults is likely limited. Blue Catfish compete with other bottom feeders in any given ecosystem.

Breeding Habits

Spawning occurs in spring, beginning as early as April in Louisiana, and extending into June farther north[6]. No definitive information is available on the spawning habits of this species in Idaho waters.

Habits and Habitat

In their native range, Blue Catfish are inhabitants of the largest rivers, and they are successful in some reservoirs. Typically, they occur in deep waters with firm substrates and considerable current[6]. It has been noted that Blue Catfish do not adapt very well to reservoir environments, and fish that escaped from the commercial facilities would therefore be found primarily in swift runs, glides, or chutes of free-flowing portions of the Snake and Columbia Rivers[4]. In their native range, Blue Catfish are highly esteemed for their eating qualities[6].

Management and Conservation Status

Blue Catfish are considered secure, common, widespread and abundant rangewide (G5). In Idaho, a conservation rank is not applicable (SNA) because a nonnative species is not a suitable target for conservation activities. They are classified as a game fish in Idaho[13].

"All the romance of trout fishing exists in the mind of the angler and is in no way shared by the fish."

Harold F. Blaisdell, The Philosophical Fisherman, 1969

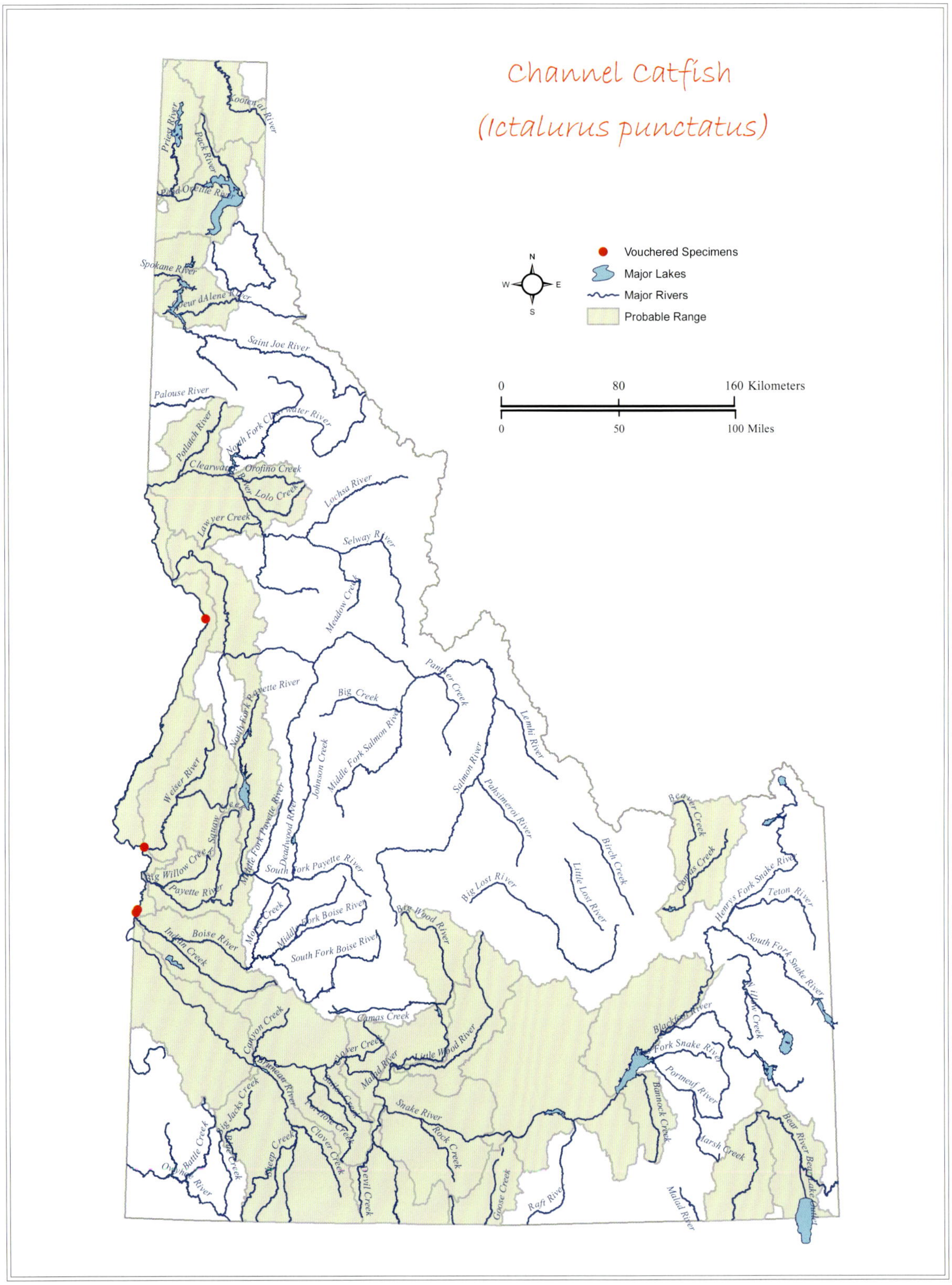
Channel Catfish
(Ictalurus punctatus)
Vouchered Specimens
Major Lakes
Major Rivers
Probable Range
N
S
E
W
0
80
160 Kilometers
0
50
100 Miles

Channel Catfish

Ictalurus punctatus
(Rafinesque, 1818)

Etymology

Ictalurus = fish cat; *punctatus* = spotted.

Importance

While Channel Catfish are of high importance to anglers in their native warm-water range, they are of limited importance to anglers in western and Pacific Northwest states. Channel Catfish often provide additional ecosystem services by removing detrital material or organic debris and other bottom materials from an ecosystem. In portions of their native range, use of fee-fishing ponds has tremendous potential[2]. Channel Catfish are raised in commercial fish farms in parts of midwestern and southern states[4]. No such industry has developed in Idaho.

Description

Channel Catfish are distinguished from other catfishes by their long anal fin and deeply forked tail[2]. As with all catfishes, they are naked, lacking scales. Channel Catfish have an inferior mouth, with an overhanging papillate upper lip. They have a pair of long, flattened, wide maxillary barbels that are the longest and largest of the four pairs of barbels on this species. There are fine cardiform teeth in several irregular rows on the dentary, on the premaxillary, and on pairs of rounded patches on the pharyngobranchial bones. Gill rakers are long, widely spaced, slender, and pointed, numbering 14 to 18 and usually nine to 12 on the lower limbs and five or six on the upper limb. Branchiostegal rays are eight or nine. The single dorsal fin is ahead of the midpoint of the body and is soft rayed, with the first soft ray modified to a stout spine. Peritoneum is speckled with black, and the intestine is well differentiated and coiled; there are no pyloric caeca[3]. The anal fin typically has 24 to 27 rays, and the trailing edge is rounded[12].

Individuals less than about 305 to 356 mm (12.0 to 14.0 in) are pale blue to pale olive with a silvery overcast. The ventral surface is silver white to milk white, and the sides have varying numbers of olive to almost black spots of varying sizes. On adults the dorsal surface of the head, back, and upper sides is steel blue to gray, with the lower sides lighter and the ventral surface of the head and body to the pelvic fins dirty white to silver white. Males at breeding time may turn brighter blue than normal[3].

Range and Distributional Status

This species is not native to Idaho. Native range of the Channel Catfish extends from the St. Lawrence River in Quebec, Canada, south on the western side of the Appalachian Mountains to southern Georgia and central Florida, west through the Gulf states to eastern Texas and northern Mexico, and then northwest through New Mexico to Montana. Channel Catfish have been widely introduced throughout the eastern and western United States[2]. They were first introduced into Idaho in 1893 when 100 fish were released in the Boise River by the US Fish Commission. In 1940 additional releases were made by the IDFG in Little Wood River, Snake River at Burley, and Snake River between Glenns Ferry and Weiser. Channel Catfish are now abundant in the Snake River, from Swan Falls to the Columbia River[7]. In Idaho Channel Catfish are present from the extreme southeast corner of the state in the Bear River drainage, throughout the lower portions of the Snake River into the Boise River drainage, and throughout the Payette, Clark Fork, Saint Joe, Spokane River, Pend Oreille, and Kootenai drainages.

Size and Longevity

This is a large species, producing individuals averaging 356 to 533 mm (14.0 to 21.0 in) and 0.91 to 1.8 kg (2.0 to 4.0 lb). A 26.3 kg (58 lb) Channel Catfish taken from Santee-Cooper Reservoir in South Carolina in 1964 is the world record for this species[5]. In Idaho a 14.1 kg (31 lb, 1 oz) fish from Mann Lake, taken in 2001, is the state record. The fish was 965 mm (38.0 in) long[11].

Growth is rapid in early years, and YOY were 51 to 102 mm (2.0 to 4.0 in) in the fall, growing to 89 to 190 mm (3.5 to 7.5 in) at age one year[3]. Age at maturity varies more than for most catfishes; average age at maturity is three to five years[2].

Food and Feeding

While young Channel Catfish feed primarily on insects, other arthropods, and other invertebrates, adult Channel Catfish are omnivorous. They may feed both in daylight and in darkness, but feeding is primarily from twilight to midnight. Bottom feeding, while characteristic of Channel Catfish, is occasionally supplemented by surface feeding. In clear water they may rely more on sight than smell, an atypical habit for a catfish. Large Channel Catfish, 152 to 305 mm (6.0 to 12.0 in), feed primarily on fish, and fish larger than 509 mm (20.0 in) are almost exclusively predatory on fishes[2].

Limiting Factors

Young Channel Catfish may fall prey to various other fishes in a given ecosystem, but the adults, primarily because of their large size, are relatively free from predation. Channel Catfish may compete for food with other bottom feeders except in ecosystems where algae and other plants form a significant portion of the diet[3].

Channel Catfish may hybridize with other catfishes such as the Flathead Catfish (*Pylodictis olivaris*), and there have been some attempts to produce hybrids artificially. Parasites noted of Channel Catfish include protozoans, trematodes, cestodes, nematodes, leeches, and crustaceans[3].

Channel Catfish are quite easy to catch, and populations may suffer if fishing pressure is heavy. Minimum acceptable oxygen content for three temperatures is as follows: at 25°C (77°F) 0.95 mg/l (ppm), at 30°C (86°F) 1.05 mg/l (ppm), and at 35°C (95°F) 1.08 mg/l (ppm)[2].

Breeding Habits

Channel Catfish spawn in the spring to early summer at temperatures in the 23.9 to 29.5°C (75 to 85°F) range. Depending on ecosystem configuration, spawners may or may not migrate into rivers or moving water to spawn. In all cases, spawning occurs in secluded, semi-dark nests that are built by the male; Channel Catfish will not spawn in clear waters. They may use ecosystem components such as logs or undercut banks or artificial components such as bottles or other debris. Females spawn over a four- to six-hour period and only once a year, while males may spawn more frequently[3].

A female Channel Catfish may lay as many as 34,500 eggs, and on average a female will produce 1,820/kg (4,000/lb). Eggs are 3 to 4 mm (0.1 to 0.2 in) in diameter before water hardening. Males guard the nest and aerate it after eggs are fertilized. Young hatch in about seven days at temperatures of 27°C (80°F)[2].

Habits and Habitat

Channel Catfish are an excellent pond-culture species, and there are extensive operations of this type in the species' native range[3]. While Channel Catfish may thrive in some lake and impoundment ecosystems, they are more adapted for life in moderately swift streams than other catfishes. This adaptation is reflected in the streamlined body ansd deeply forked tail. Channel Catfish seek clear-water lake and stream ecosystems with a variety of bottom types[2,10].

Management and Conservation Status

Channel Catfish are considered secure, common, widespread and abundant rangewide (G5). In Idaho, a conservation rank is not applicable (SNA) because a nonnative species is not a suitable target for conservation activities. They are classified as a game fish in Idaho[13].

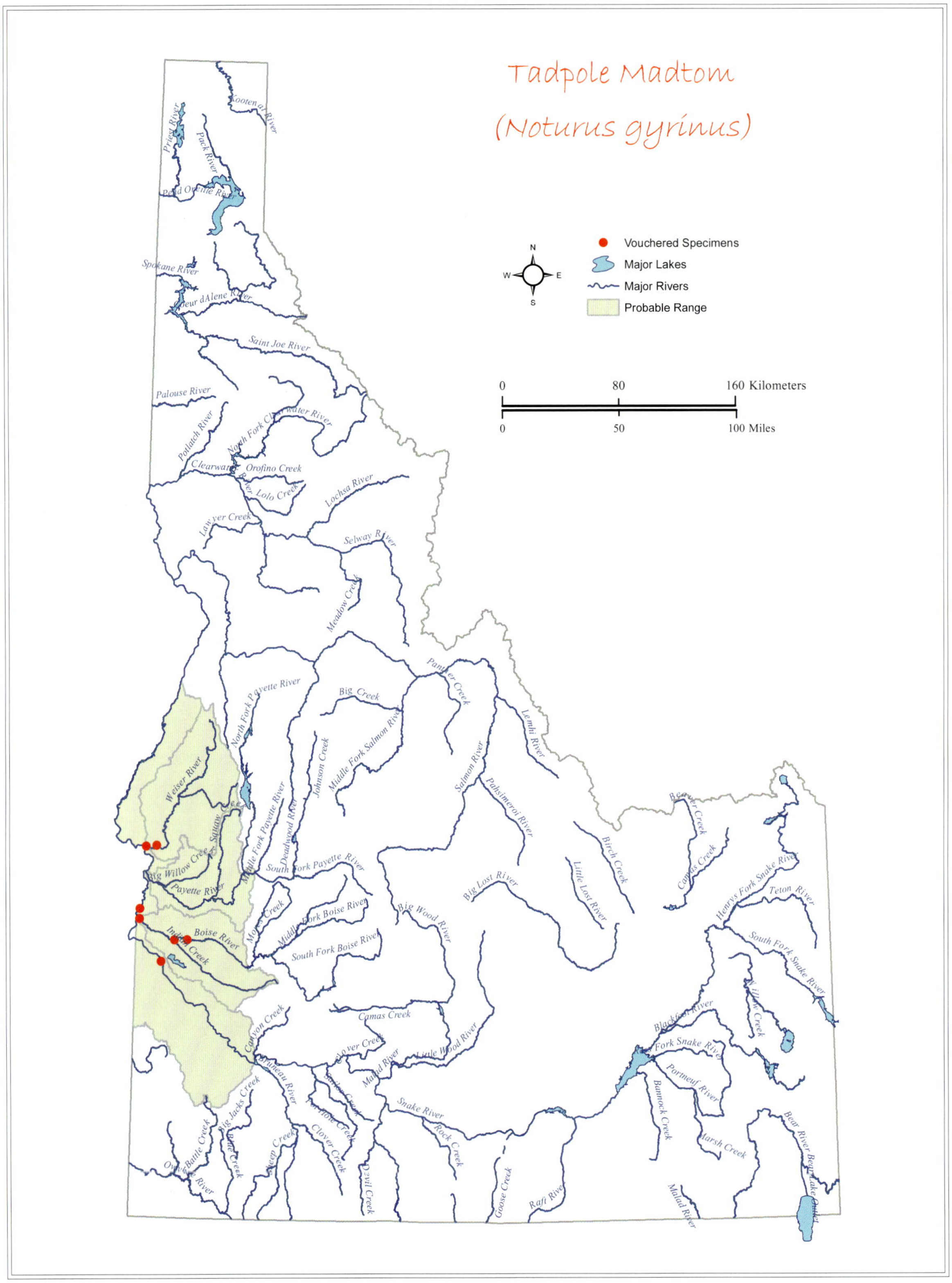
Tadpole Madtom
(Noturus gyrinus)
Vouchered Specimens
Major Lakes
Major Rivers
Probable Range
N
S
E
W
0
80
160 Kilometers
0
50
100 Miles
Priest River
Kootenai River
Pack River
Spokane River
Saint Joe River
Palouse River
Potlatch River
Orofino Creek
Lolo Creek
Lochsa River
Selway River
Meadow Creek
Big Creek
Middle Fork Salmon River
Johnson Creek
Lemhi River
Salmon River
Pahsimeroi River
Birch Creek
Little Lost River
Big Lost River
Big Wood River
Weiser River
Payette River
Deadwood River
South Fork Payette River
Middle Fork Payette River
Boise River
Middle Fork Boise River
South Fork Boise River
Camas Creek
Teton River
South Fork Snake River
Snake River
Portneuf River
Bannock Creek
Marsh Creek
Rock Creek
Clover Creek
Goose Creek
Raft River
Malad River
Bear River
Big Jacks Creek
Canyon Creek
Little Wood River

Tadpole Madtom

Noturus gyrinus
(Mitchill, 1817)

Etymology

Noturus = tail over the back, alluding to the fact that the adipose fin is joined to the caudal fin; *gyrinus* = tadpole.

Importance

Tadpole Madtom are important where they are abundant in their native range as prey and bait fish[3]. This species has venom-producing glands beneath the skin that can result in painful, though not fatal, puncture wounds from the dorsal and pectoral fins[4].

Description

This species is naked and grows to 51 to 76 mm (2.0 to 3.0 in). The body is heavy forward toward the head. There are one pair of long, flattened barbels on the snout and one pair of long, flattened maxillary barbels as well as two pairs of barbels on the underside of the head. There are cardiform teeth in several irregular rows on the lower jaw and the premaxillary. Gill rakers are six to eight, usually with six on the lower limb and one or two on the upper limb. Branchiostegal rays are usually nine. The single dorsal fin is well ahead of the midpoint of the body, with soft rays; the first of the six obvious rays is modified to a stout spine that is straight and sharply pointed but without barbs or serrae. Pelvic fins are abdominal, with the origin just behind the insertion of the dorsal fin, usually with eight or nine rays. Pectoral fins are broad and straight edged, with the tip bluntly tipped and with one spine and usually eight rays. Lateral line is incomplete, short, and shallowly curved upward above the pectoral fin. There are no pyloric caeca[3].

Range and Distributional Status

This species is not native to Idaho. The native range of Tadpole Madtom was from central North Dakota east to Quebec, south of the tip of the Florida peninsula, and west to eastern Texas. This species is absent from the Appalachian Mountains and central Pennsylvania south to central Georgia in the eastern United States. This species has been introduced into Idaho, probably with a shipment of Channel Catfish or Brown

Bullhead, and was collected from the Snake River in 1942. In Idaho the Tadpole Madtom is now found in the lower Boise River and Snake River as well as their tributaries, downstream of Swan Falls. This species' range in the Pacific Northwest is increasing slowly, and they may become established in other downstream tributaries and backwaters of the Columbia River system[4].

Size and Longevity

Adult Tadpole Madtom rarely exceed 89 mm (3.5 in) total length, and few fish live beyond three years[4].

Food and Feeding

Food of young Tadpole Madtom is primarily zooplankton such as cladocerans and ostracods, while larger fish feed primarily on larval insects, small crustaceans, and other benthic invertebrates[4]. This species forages for all food items on the bottom at night[3].

Limiting Factors

This species is parasitized by protozoans, trematodes, cestodes, nematodes, and crustaceans. It may hybridize with the Brindle Madtom (*Noturus miurus*) where the species overlap[3].

Breeding Habits

Most Tadpole Madtom probably become sexually mature at age two. Like other catfishes, Tadpole Madtom spawn in the spring (May and June). Spawning may continue as late as June or July in some locations, depending on temperatures. Eggs are generally deposited in a gelatinous mass under rocks and logs, but other objects such as boards, pieces of crockery, and cans are also used. Eggs are attended by at least one of the parents[4]. Eggs are light yellow in color and about 3.5 mm (0.1 in) in size. Eggs adhere to each other and to the substrate, and the whole egg mass is surrounded by a gelatinous envelope[3].

Habits and Habitat

This species prefers clear water, soft muddy bottoms, and extensive vegetation. This species is generally found in low-gradient areas with little current, such as quiet pools of smaller streams and sloughs and quiet backwaters of larger rivers. Tadpole Madtom tend, however, to be intolerable of high turbidity caused by siltation. This is a secretive species that takes cover under rocks, logs, and debris during the day and is active only at night[4].

Management and Conservation Status

Tadpole Madtom are considered secure, common, widespread and abundant rangewide (G5). In Idaho, a conservation rank is not applicable (SNA) because a nonnative species is not a suitable target for conservation activities. They are classified as a game fish in Idaho[13].

Palouse Rolling Hills, near Moscow, ID. Photo Courtesy John Sigler.

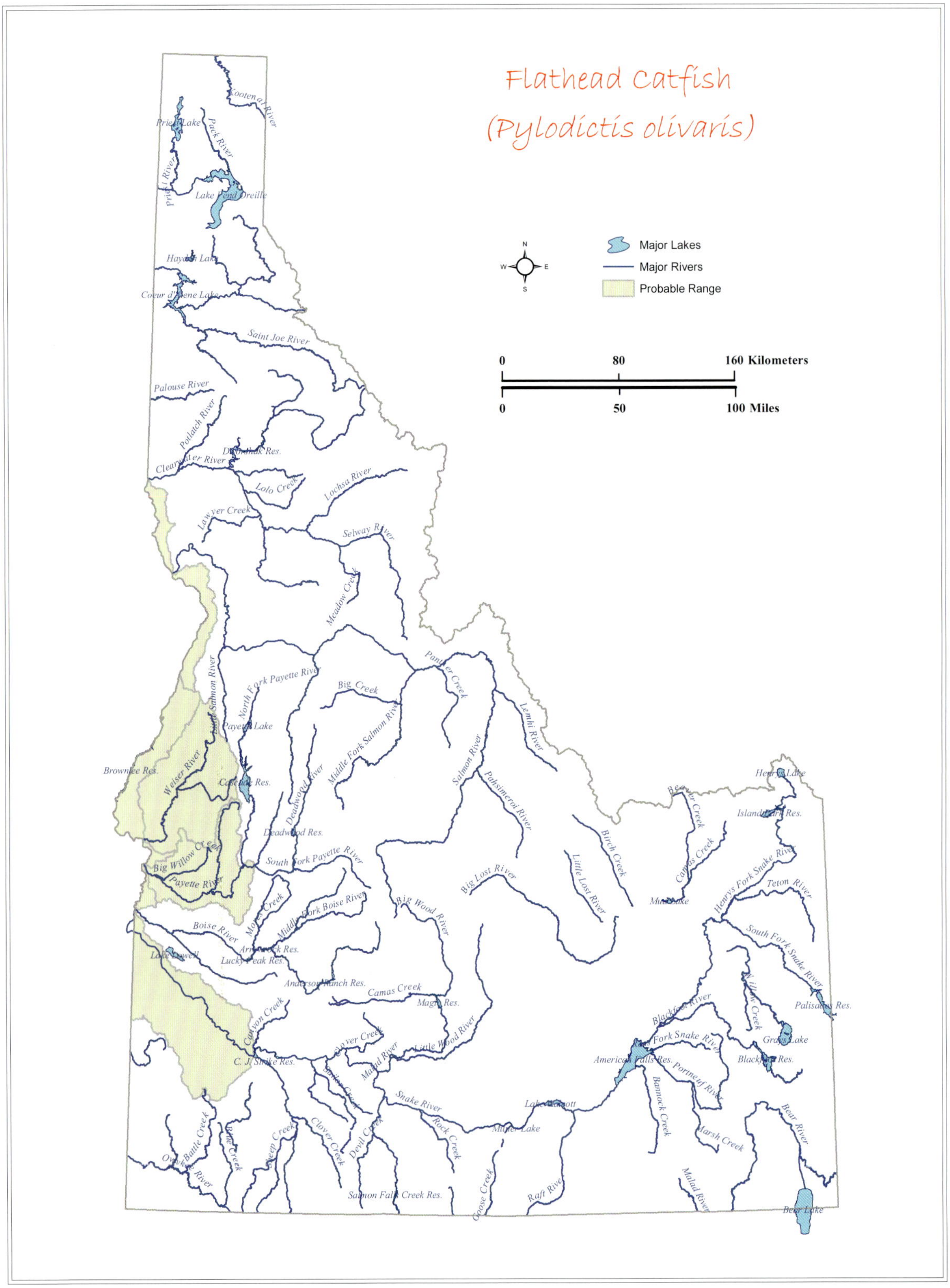
Flathead Catfish
(Pylodictis olivaris)
Major Lakes
Major Rivers
Probable Range
N
S
E
W
0
80
160 Kilometers
0
50
100 Miles
Pack River
Saint Joe River
Palouse River
Potlatch River
Lochsa River
Selway River
Meadow Creek
North Fork Payette River
Big Creek
Middle Fork Salmon River
Little Salmon River
Weiser River
Lemhi River
Salmon River
Pahsimeroi River
Birch Creek
Little Lost River
Big Lost River
Big Wood River
Camas Creek
Teton River
South Fork Snake River
Big Willow Creek
Boise River
Canyon Creek
Camas Creek
Snake River
Rock Creek
Bannock Creek
Marsh Creek
Bear River
Malad River
Goose Creek
Clover Creek
Salmon Falls Creek Res.
Lemhi River

Flathead Catfish

Pylodictis olivaris
(Rafinesque, 1818)

Etymology

Pylodictis = pylos = mud + ichthys = fish; *olivarus* = olive colored.

Importance

Flathead Catfish are an important sport- and commercial fish in large rivers of the Mississippi, Missouri, and Ohio River basins in the central United States. In the Pacific Northwest, Flathead Catfish are not abundant, so sportfishing is limited. In Idaho only a small number of anglers fish specifically for Flathead Catfish. Most fish harvested weigh about 0.45 to 2.3 kg (1 to 5.1 lb)[4]. Within their native range, Flathead Catfish are an important commercial species that are taken primarily on trail lines, set lines, and trammel nets. This species is also actively sought by anglers where it is abundant, using live or fresh fish as the preferred bait. This species is highly esteemed as a food fish throughout its native range[6].

Description

Flathead Catfish differ from the Black, Yellow, and Brown Bullhead by having a lower jaw that projects beyond the upper jaw, an upper jaw with a U-shaped pad of premaxillary teeth, and an anal fin that is shorter than the head, with 14 to 17 anal rays[4]. The caudal fin of Flathead Catfish is slightly notched. In larger Flathead Catfish, the large, very flat head is prominent. The back and sides are olive-yellow to light brown in color, with dark mottling that differs from most bullheads, which have a more uniform coloration[4]. Gill rakers are nine to 28. There is a premaxillary tooth patch with posterior extensions present[6]. Juveniles are often nearly black dorsally, and both juveniles and young adults may have a white blotch on the tip of the dorsal lobe of the caudal fin[6].

Range and Distributional Status

This species is not native to Idaho. The native distribution of Flathead Catfish was larger rivers of the Mississippi, Missouri, and Ohio River basins in the central United States and the Rio Grande in New Mexico. This species has been introduced into other parts of the United States, such as the Gila River and lower

Colorado River. Although no records of intentional introductions to waters of the northwestern United States exist, this species is found in the Snake River downstream of Swan Falls in Idaho and in the Snake and Columbia Rivers in Washington. Flathead Catfish have been present in the Columbia River since the mid-1970s. This species may have unknowingly been introduced when a shipment of what was thought to be Blue Catfish was planted in the Snake River[4].

Size and Longevity

Flathead Catfish live to at least 20 years of age and are the largest member of the catfish family[4]. Total lengths for individuals ages one through eight years are, in mm (inches), 64 to 239 (2.5 to 9.4), 127 to 353 (5.0 to 13.9), 185 to 475 (7.3 to 18.7), 259 to 579 (10.2 to 22.8), 320 to 648 (12.6 to 25.5), 376 to 780 (14.8 to 30.7), 421 to 907 (16.6 to 35.7), and 465 to 985 (18.3 to 38.8). Other research indicates Flathead Catfish up to 18 years of age and over 1 m (39.4 in) in length. Within their native range, Flathead Catfish have been reported at weights exceeding 45.4 kg (100 lb), which are quite likely valid[6]. The all-tackle world record is a 55.8 kg (123 lb) individual caught from the Elk City Reservoir, Kansas, in 1998[5]. Brownlee Reservoir on the Snake River produced the Idaho record Flathead Catfish, which weighed 26.5 kg (58 lb, 7 oz) and was 1.2 m (48 in) long, captured in 1994[11].

Food and Feeding

Juveniles feed mostly on aquatic insects in riffle areas. At lengths of 100 mm (3.9 in) and larger, their food habits shift almost exclusively to other fishes, with an occasional crayfish. Flathead Catfish feed heavily on shad where they are sympatric. It has been noted that a possible feeding strategy of adults is to lie motionless with their mouths open until another cover-seeking species swims into this apparent refuge[6].

Limiting Factors

Due to their large size, Flathead Catfish adults are not often prey for other species. Young and eggs, however, may be food for other species in a given ecosystem.

Breeding Habits

Spawning occurs in cave-like depressions in the bank or in a cavity hollowed out by either one or both parents or under a board, boulder, log, or similar object. Within the native range of Flathead Catfish in Tennessee, spawning probably takes place from June into July. Compact egg masses laid in the nest may contain 4,000 to 100,000 eggs. After spawning, the male guards the nest and the newly hatched young. Males become aggressive toward the females[6]. Flathead Catfish eggs hatch in six to eight days, depending on temperature[4].

Habits and Habitat

Flathead Catfish are a solitary and secretive species occupying medium to large rivers and reservoirs. Individuals typically spend daylight hours associated with cover such as caves, undercut banks, brush piles, or log jams. Juveniles are typically encountered in riffle areas of these ecosystems[6].

Management and Conservation Status

Natural reproduction of Flathead Catfish has been documented in portions of the Snake River in Idaho. It is now an established member of the Snake River ecosystem. Flathead Catfish are considered secure, common, widespread and abundant rangewide (G5). In Idaho, a conservation rank is not applicable (SNA) because a nonnative species is not a suitable target for conservation activities. They are classified as a game fish in Idaho[13].

Chapter Citations

1. Nelson, 2006. 2. Sigler and Sigler, 1996. 3. Scott and Crossman, 1973. 4. Wydoski and Whitney, 2003. 5. International Game Fish Association, 2015. 6. Etnier and Starnes, 1993. 7. Simpson and Wallace, 1978. 8. Cucherousset, et al., 2006. 9. Schade and Bonar, 2005. 10. Sigler and Sigler, 1987. 11. Idaho Department of Fish and Game, 2015. 12. Becker, 1983. 13. Idaho Department of Fish and Game, 2005.

Additonal readings for this chapter are listed in the Supplmental References section.

"There is certainly something in angling that tends to produce a serenity of the mind."

Washington Irving

11

OSMERIFORMES

Smelts

Order—Osmeriformes

This is an order of ray-finned fishes that includes the true or freshwater smelts. This order includes fishes that are circumpolar and may occur in fresh, marine, or brackish water environments. Some species are anadromous[1]. An adipose fin may be present or not, and radii are absent on the scales. This order has two superfamilies, three families, 22 genera and 88 species[2].

Family—Osmeridae—Smelts

Smelt occur in marine, anadromous, and coastal freshwater forms in the Northern Hemisphere in the Arctic, Atlantic, and Pacific Oceans. An adipose fin is present, and the lateral line, while present, is typically incomplete. Branchiostegal rays are five to 10, and there are teeth on the premaxilla, maxilla, dentary, and inner mouth bones. Pyloric caeca number zero to 300 and vertebrae 51 to 782. Smelt are small, generally elongated fishes that are laterally compressed and have large mouths. Coloration is generally silvery. Smelt are circumpolar but restricted to the Northern Hemisphere. Osmeridae includes six genera and 10 species.

Rainbow Smelt

Osmerus mordax
(Mitchill, 1814)

Etymology

Osmerus = osmé = smell or scent; *mordax* = biting.

Importance

In Canada smelts have supported commercial fisheries in marine waters for well over 100 years. In many parts of Canada, smelt are an important sportfish. Smelt are caught either by angling through the ice in winter or by dipnetting or seining during spawning runs in the spring. Data from 1960 indicated that 68,000 smelt fishermen captured 2.4 million kg (5.3 million lb) from Lake Erie. As a food fish, Rainbow Smelt are generally well accepted[1]. In Idaho Rainbow Smelt are not listed in the current fisheries management plan[3].

Description

Rainbow Smelt have elongated bodies and average only 177 to 203 mm (7 to 8 in) in length as adults. This species is laterally compressed. Gill rakers are 8-11 + 18-24. Rainbow Smelt have a single dorsal fin that is soft rayed and located at midbody with its origin over the origin of the pelvic fins and eight to 11 rays. An adipose fin is present and well developed. Scales are cycloid and number 62 to 72 in the lateral series; the lateral line is incomplete. Rainbow Smelt are slender, mostly silvery fish with pale green on the back and purple, blue, or pink iridescent flashes on the sides[1].

Range and Distributional Status

This species is not native to Idaho. Rainbow Smelt occur in Atlantic drainages from Lake Melville, Newfoundland, to Delaware River, Pennsylvania; Arctic and Pacific drainages from Bathurst Inlet, Northwest Territories, to Vancouver Island, British Columbia; and the Pacific drainages of Asia[4]. Rainbow Smelt are believed to extirpated in Idaho, therefore no distribution map is provided.

Size and Longevity

Rainbow Smelt do not achieve large sizes in any population. Male and female Rainbow Smelt ages one to five had lengths, in mm (in), of (male) 66 mm (2.6), 152 (6.0), 185 (7.3), 201 (7.9), and 218 (8.6) and (female) 66 (2.6), 150 (5.9), 196 (7.7), 221 (8.7), and 239 (9.4) in a Lake Superior population[1].

Food and Feeding

Rainbow Smelt are carnivorous and feed on a wide variety of smaller organisms such as *Mysis* shrimp, amphipods, ostracods, aquatic insect larvae, and aquatic worms. In the Great Lakes, fish occur only rarely in the diet, but individual sculpin, Burbot, White Bass, various whitefish, and Emerald Shiner have been reported[1]. *Mysis* are an important prey item for Rainbow Smelt in the profundal zone of southern Lake Ontario[5].

A Canadian field survey noted the documentation of the feeding rhythm of larval Rainbow Smelt in natural conditions. Smelt larvae fed during daylight hours, as do many other fish larvae; and they fed on a flooding (incoming) tide. Parasitized larvae ingested half of the food consumed by nonparasitized larvae. This result is consistent with observations of cestode-infected herring larvae in the North Sea. Parasites affect the condition of the host, and parasitized larvae may physically not be able to ingest more food because of the lack of space in the gut[6].

Equipped with large teeth, Rainbow Smelt are known to be piscivorous. Rainbow Smelt have frequently been implicated in declines of nongame and game fishes, including Walleye. However, direct evidence of predation on game fish is rare, and the only piscivory observed in Colorado's Horsetooth Reservoir was cannibalism by adult Rainbow Smelt on age-zero smelt. No instances of Rainbow Smelt predation on Walleye fry were observed[7].

Limiting Factors

Rainbow Smelt are preyed upon by a diverse set of predators. Larger Rainbow Smelt may consume members of their own species. This species is host to a variety of parasites[1,6].

Breeding Habits

Rainbow Smelt are anadromous. In the spring, this species leaves either the resident lake or the sea and ascends freshwater streams to spawn. Rainbow Smelt that are ripe (ready to spawn) move into spawning streams not long after the ice is off in the spring, in March, April, or May, at temperatures near 10°C (50°F). In the Great Lakes, Rainbow Smelt may spawn in off-shore gravel beds in lakes rather than streams. Postspawning mortality can be high[1].

Spawning occurs mainly at night and may last for up to three weeks; spawners return to the lake during the day. Spawning behavior includes two or more males in position against a female in swift water; eggs are released in clusters along with milt. Eggs become adhesive after extrusion and attach to bottom gravels. Number of eggs produced by females is size related, and 127 to 209 mm (5 to 8.2 in) females may produce 8,500 to 69,660 eggs. Number of eggs per female is variable from one watershed or population to another[1].

In their native range, anadromous Rainbow Smelt depend on discrete spawning riffles near the tidal interface to deposit demersal, adhesive eggs. In many rivers, these locations are also centers of human development where watershed alterations have degraded water and substrate quality. The degradation of spawning habitat could reduce a fitness advantage in the life history of diadromous fish that depend on the integrity of freshwater habitats[8].

Habits and Habitat

Rainbow Smelt are pelagic schooling fish that inhabit the midwaters of lakes or inshore coastal waters. This species does not inhabit flowing streams except during spawning activity[1].

Adult Rainbow Smelt can negatively affect native fishes through predation and competition for ecosystem resources. Adult Rainbow Smelt are usually distributed in cool metalimnetic and hypolimnetic areas, and interactions with other species may be related to the extent of spatial overlap between them and native fishes. Age-0 Rainbow Smelt may interact with different fishes than do the adults because the young live in warm epilimnetic waters and use different resources[9]. Laboratory experiments in Canada showed that an increase in turbidity significantly enhanced the growth rate of smelt larvae[6].

Management and Conservation Status

As a nonnative species, Rainbow Smelt are afforded no special protections in Idaho waters. Approximately 15 million eggs were planted in lakes and streams of the Sawtooth Mountains in May 1930[9]. We know of no reports of established populations from this introduction[10]. Rainbow Smelt are included in this book provisionally. Additional introductions are discouraged.

Invasive fishes are continuing to colonize aquatic ecosystems worldwide, particularly due to introductions by humans. In considering the effects of an exotic fish, emphasis should focus on its life stages, especially when the species has a strongly segregated life history. The potential instability in native fish populations confronted by intentional stocking of nonnative species should serve as a warning against such practices[9].

Chapter Citations

1. Scott and Crossman, 1973. 2. Nelson, 2006. 3. Idaho Department of Fish and Game, 2013. 4. Page and Burr, 1991. 5. Walsh, et al., 2008. 6. Sirois and Dodson, 2000. 7. Johnson and Goettl, 1999. 8. Chase, 2009. 9. Hrabik, Carey, and Webster, 2001. 10. Simpson and Wallace, 1978.

"A river cuts through rock, not because of its power, but because of its persistence."

Jim Watkins

Link Jackson
©

12

SALMONIFORMES

Whitefishes, Chars, Graylings, Trouts & Salmons

Order—Salmoniformes

These taxa are restricted to the family Salmonidae[1]. These soft-rayed, bony fishes are mostly physostomous (having a connection between the gas bladder and the esophagus), but a gas bladder may not be present in all species. An adipose fin is present in the majority of species in this order and in all members of the salmonid family. This order originated in the Cretaceous, possibly as long as 135 mya[2]. These taxa are now restricted to a single family, Salmonidae[1], where previous phylogeny included the pike family (Esocidae) which has been elevated to Order Esociformes[3].

No other order of fishes contains as many recreational sportfishes. Many of these species are popular with fishing enthusiasts and curious citizens.

Family Salmonidae—Whitefishes, Graylings, Chars, Trouts, and Salmons

Salmonids include freshwater and anadromous fishes that range widely across the waters of the Northern Hemisphere. Salmonids are medium to large terete-bodied fishes that express a variety of shapes and appearances. Scales are cycloid and confined to the body and not on the head. A lateral line is present, as are pyloric caeca that are well developed and numerous. Fin rays are all soft, there are no spiny rays, and individuals have a physostomous gas (air) bladder[2,5,6].

As a family, salmonids are economically the most important group of cold-water fishes in the United States. Salmonids are extensively reared in state, federal, and private hatcheries for stocking into recreational fishing waters and for direct consumption[2,5,6]. The Salmonid family is composed of 40 species in seven genera[3]. Twenty-two taxa (not including extirpated taxa), in six genera are known from Idaho; eight are introduced; three are endemic to Bear Lake, Idaho-Utah.

Biological diversity in this family is greater than is recognized in our nomenclature-limited current taxonomy[1]. While the American Fisheries Society (AFS) list[3] does not address subfamilies, Nelson notes three subfamilies with 11 genera and about 66 species and makes a compelling case for adopting the three subfamilies discussed below[1]. AFS's list[3] includes both ciscoes and whitefishes in the genus *Coregonus* and the genus *Prosopium*, a point that Nelson[1] discusses at length, highlighting the imperfections of the current understanding of the true phylogenetic relationships of these genera. Here we present all of the Salmonidae species in the context of three subfamilies: Coregoninae, Thymallinae, and Salmoninae.

The diversity of trouts in western North America resulted from evolutionary patterns in response to different selective processes operating in this geographically complex region. Prominent specializations are associated with anadromy, movements among fluvial and lacustrine habitats, and other ecosystem factors[7].

Subfamily Coregoninae

Whitefishes

Individuals in this subfamily are characterized by having fewer than 16 dorsal fin rays, large scales with fewer than 110 along the lateral line, no teeth on the maxilla, and vomer usually small and without teeth. Worldwide, there are three genera and about 66 species[1]. In Idaho there are representatives of two genera, *Coregonus* and *Prosopium*, with six species, three of which are endemic to Bear Lake.

Coregonus clupeaformis	(Mitchill, 1818)	LakeWhitefish
Prosopium abyssicola	(Snyder, 1919)	Bear Lake Whitefish
Prosopium coulterii	(Eigenmann & Eigenmann, 1892)	Pygmy Whitefish
Prosopium gemmifer	(Snyder, 1919)	Bonneville Cisco
Prosopium spilonotus	(Snyder, 1919)	Bonneville Whitefish
Prosopium williamsoni	(Girard, 1856)	Mountain Whitefish

Subfamily Thymallinae

Graylings

Individuals in this subfamily are characterized by having more than 17 dorsal fin rays, teeth on the maxilla, no orbitosphenoid or suprapreopercular bones. One genus (*Thymallus*) occupies freshwater in the Northern Hemisphere[1].

Thymallus arcticus	(Pallas, 1776)	Arctic Grayling

Subfamily Salmoninae

Chars, Trouts, and Salmons

The subfamily Salmoninae is usually considered to contain six genera: *Brachymystax*, *Hucho*, *Oncorhynchus*, *Salmo*, *Salmothymus*, and *Salvelinus*. While the genetic boundaries in portions of this subfamily are still unclear, genetic boundaries of the Salmoninae native to North America are relatively firm. There are three native genera in North America: the Pacific trout and salmon (*Oncorhynchus*), the Atlantic trout and salmon (*Salmo*), and the char (*Salvelinus*). These are all coldwater species that breed in freshwater; however, some of these species, and many populations, are anadromous[4]. Below we list the extant and extirpated taxa reported in Idaho.

Char

Salvelinus alpinus oquassa	(Girard, 1854)	Sunapee Trout
Salvelinus confluentus	(Suckley, 1859)	Bull Trout
Salvelinus fontinalis	(Mitchill, 1814)	Brook Trout
Salvelinus namaycush	(Walbaum, 1792)	Lake Trout

Atlantic Trout and Salmon

Salmo salar (extirpated in Idaho)	(Linnaeus, 1758)	Atlantic Salmon
Salmo trutta	(Linnaeus, 1758)	Brown Trout

Pacific Trout and Salmon

Oncorhynchus aguabonita	(Jordan, 1892)	Golden Trout
Oncorhynchus clarkii (5 subspecies in Idaho)	(Richardson, 1836)	Cutthroat Trout
Oncorhynchus clarkii bouvieri	(Bendire, 1882)	Yellowstone Cutthroat Trout
Oncorhynchus clarkii behnkei	Montgomery, 1995	Snake River Fine-Spotted Cutthroat Trout
Oncorhynchus clarkii henshawi	(Gill and Jordan, 1878)	Lahontan Cutthroat Trout
Oncorhynchus clarkii lewisi	(Girard, 1856)	Westslope Cutthroat Trout
Oncorhynchus clarkii utah	(Suckley, 1874)	Bonneville Cutthroat Trout
Oncorhynchus keta (extirpated in Idaho)	(Walbaum, 1792)	Chum Salmon
Oncorhynchus kisutch	(Walbaum, 1792)	Coho Salmon
Oncorhynchus mykiss	(Walbaum, 1792)	Redband Trout and Steelhead
Oncorhynchus nerka	(Walbaum, 1792)	Sockeye Salmon/Kokanee
Oncorhynchus tshawytscha	(Walbaum, 1792)	Chinook Salmon

Salmonid Hybrids

Oncorhynchus clarkii × *O. mykiss*	Cutthroat Trout × Rainbow Trout	Cutbow Trout
Salmo trutta × *Salvelinus fontinalis*	Brown Trout × Brook Trout	Tiger Trout
Salvelinus fontinalis × *S. namaycush*	Brook Trout × Lake Trout	Splake
	Brook Trout × Bull Trout	

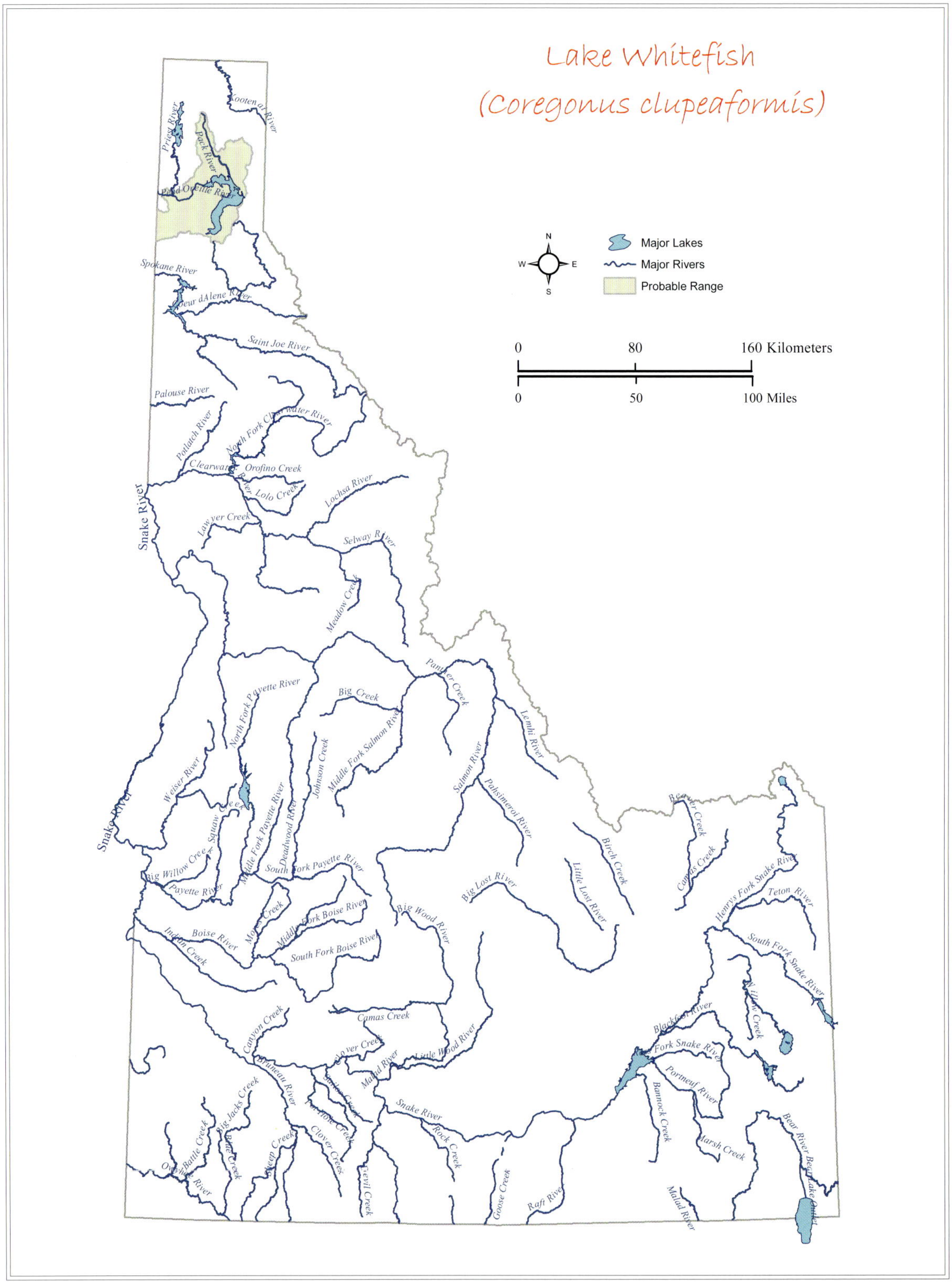
Lake Whitefish
(Coregonus clupeaformis)
Major Lakes
Major Rivers
Probable Range
0
80
160 Kilometers
0
50
100 Miles

Subfamily Coregoninae

Whitefishes and Ciscoes

Lake Whitefish

Coregonus clupeaformis
(Mitchill, 1818)

Etymology

Coregonus = angle-eye, coined by Artedi from the Greek words meaning "pupil" (of the eye) and "angle"; *clupeaformis* = herring shaped.

Importance

In Canada and the Great Lakes, this species has been the most valuable commercial freshwater fish over many decades. It is not native west of the Continental Divide[5]. Scott and Crossman[5] report on the exceptionally fine flavor of this species, noting that Richardson (1836)[8] stated, "*Although we cannot verify that one can eat it for months without tiring, we can say from personal experience that a diet of whitefish alone, with no other food, can be eaten for days, without losing its appeal.*" Whitefish eggs, when properly prepared, can be used as caviar.

Description

Lake Whitefish are elongate and generally small and slender. Larger individuals may be somewhat ovate in lateral view. Gill rakers may range from 19 to 33 but are seldom fewer than 22. Scales are cycloid and large and range from 70 to 97 in the lateral line. Pyloric caeca are 140 to 222; vertebrae are 55 to 64. This fish is generally silvery with a pale greenish-brown back or light brown and dark in some inland lakes in the native range. Fins are generally clear or only lightly pigmented in Great Lakes populations and may be black tipped in some populations[5].

Lake Whitefish are difficult to differentiate as to sex outside the breeding season; however, like many other fish species, females are larger on average than are males. Breeding Lake Whitefish develop "pearl" organs or spawning tubercles that form hard white lumps on scales along the midflanks but may also be present on the head back and fans[4].

Range and Distributional Status

This species is not native to Idaho. Lake Whitefish are widely distributed in North American freshwater from the Atlantic Coast watershed west across Canada and the northern United States, into Alaska[5].

In Idaho the U.S. Fish Commission introduced this species into Lake Pend Oreille, Coeur d'Alene Lake, and Hayden Lake in 1889. Additional introductions were made into Bear Lake. This species is established in Lake Pend Oreille, but there is no evidence to indicate it is established elsewhere[9].

A variety of gene allele frequencies governing electrophorectic phenotypes of Lake Whitefish have previously been shown to be useful tools for characterizing populations of this species. Using these multiple genetic characters[10] the genetics of which were determined previously, proved useful tools for characterizing populations. Malate dehydrogenase (MDH) described post-glacial routes of gene flow among western Canadian Lake Whitefish populations and related the biochemical data to the glacial refugia in which they are believed to have survived the Wisconsin glaciation. Logically, most Lake Whitefish populations west of the Ontario-Manitoba boundary and east of the Rocky Mountains were derived from a Mississippi refugium stock with some input from the Bering refugium. At an early stage in the mixing of the two stocks, headwaters of the Peace River, Athabasca River, and Mackenzie River and the Fraser River system were invaded. Subsequently, these areas were cut off from main-stem rivers of the plains, and an allele not found in early emigrants from the Mississippi refugium appeared and spread throughout the major river systems of central Canada. No evidence was found that any Mississippi-stock Lake Whitefish were able to invade the upper Liard River and Yukon River system, including the area of the Bering refugium. Bering-stock Lake Whitefish probably emigrated from that refugium by way of headwater transfer from the Yukon River to the Liard River and possibly, by way of the Porcupine and Peel Rivers, from the Yukon River to the lower Mackenzie River. Routes of post-glacial dispersal suggested by geological information, rather than selection, appear to provide adequate explanations for the distributions of the biochemical characters in present-day Lake Whitefish populations. This explanation of the biochemical data is also in accord with the distribution of modal gill-raker counts in Lake Whitefish populations and also with the distribution of other freshwater fishes in western Canada[10].

Size and Longevity

The reports of record-sized Lake Whitefish are contradictory. In past decades, Lake Whitefish attained sizes of 9.1 kg (20 lb), and a record was reported caught off Isle Royale, Lake Superior, at 19.1 kg (42 lb)[5]. Washington populations reportedly reach lengths up to 600 mm (24 in) and weights of 1.4 to 2.3 kg (3 to 5 lb)[11]. A world record and all-tackle record for Lake Whitefish is a 6.52 kg (14 lb, 6 oz) fish from Meaford, Ontario, Canada, caught in 1984[12]. In 2010 Lake Pend Oreille produced the Idaho record Lake Whitefish, a 2.9 kg (6 lb, 8 oz), 660 mm (26 in) individual[13].

Food and Feeding

Adult Lake Whitefish are bottom feeders, consuming a wide variety of benthic invertebrates and small fishes. Previous studies documented a correlation between gill-raker length and food type: fish with short gill rakers and high gill-raker counts ate a higher proportion of benthic foods. Diet varies with season; copepods are important for young fish, and cladocerans become significant later in spring. By early July, bottom organisms begin to enter the diet. Adult food varies regionally and may include aquatic insect larvae, mollusks, and amphipods. Small fishes and fish eggs are sometimes consumed as well[5].

Limiting Factors

Small Lake Whitefish may fall prey to other fish species such as Lake Trout, Northern Pike, and even Lake Whitefish themselves in some ecosystems. Generally, the species is subject to many parasites[5].

Breeding Habits

Lake Whitefish mature at four or five years; some males may mature at age two. Spawning is somewhat dependent on local conditions and generally occurs from October to January in water temperatures of 7.8°C (46°F)[2, 6].

Eggs hatch in April or May with water temperatures near 10°C (50°F). Males generally precede females to spawning areas. Lake Whitefish spawn in the evening or night over rocky gravel or sandy bottoms. Spawning generally occurs in shallow waters at depths less than 7.6 m (25 ft). Eggs are deposited more or less randomly. Leaping, thrashing, and jumping out of the water have been reported by some researchers[5].

When water hardened, Lake Whitefish eggs average 2.9 mm (0.1 in) in diameter. Fecundity varies within and among populations, but average counts range from 10,643 to 44,000 eggs. Approximately 22,000 eggs are produced per kg (10,000 eggs per lb) of body weight[11]. Survival is best at incubation temperatures ranging from 3 to 6°C (37 to 43°F), and hatching times vary from about 40 days at 10°C (50°F) to about 180 days at 0.5°C (33°F). Newly hatched larvae are about 10 to 12 mm (0.4 to 0.5 in) long and are positively phototactic[4].

Habits and Habitat

Lake Whitefish are a coldwater species usually regarded as schooling fish. Throughout most of the Great Lakes, these fish move from deep to shoal waters in early spring and back to deeper waters as warming occurs. In the fall they move to shallow water to spawn[5].

Adaptive behaviors with a genetic basis that are implicated in population divergence remain poorly understood. Recent studies[14] provide evidence of a phenotype-environment association for differential adaptive swimming behaviors through experiments conducted on dwarf, normal, and hybrid Lake Whitefish. Highly significant differences for depth selection, directional changes, and burst swimming were observed, implicating a genetic basis for these behaviors. Hybrid crosses revealed that depth selection is under additive genetic control, while dominance effects were suggested for directional changes and burst swimming. Estimates for the genetic basis of behavioral differentiation from an animal model were consistent with these observations. Comparative estimates of behavioral differentiation (Q_{ST}) against neutral expectations (F_{ST}) revealed pronounced departures from neutral expectations in all three behavioral phenotypes, consistent with the hypothesis that directional selection has driven the divergence of behavior in dwarf and normal Lake Whitefish ecotypes[14].

Management and Conservation Status

Lake Whitefish are considered common, widespread, and abundant range wide (G5). A state status rank is not applicable (SNA) because the species is not a suitable target for conservation activities[15]. An evaluation of the Lake Whitefish population by the IDFG in 2005 indicated the population could conservatively support a sustainable harvest of about 39,000 kg (86,000 lb) annually. In 2008 the Idaho Fish and Game Commission added Lake Whitefish as a species available for commercial fishing in Lake Pend Oreille. The intent of this action was to allow for the possibility that incidental harvest could help offset the cost of a commercial lake-trout removal effort. Subsequent evaluation of the cost of the commercial netting program and the potential value of the Lake Whitefish bycatch indicated minimal benefit, and the interest in a commercial whitefish net fishery has diminished. There is still some interest in a commercial rod-and-reel fishery. The IDFG will evaluate the potential benefits and pitfalls of a commercial rod-and-reel fishery for Lake Whitefish and recommend either implementation of a commercial fishery on a trial basis or removal of them from the commercial species list[16].

"Things fishermen know about Trout aren't facts, but articles of faith."

John Gierach

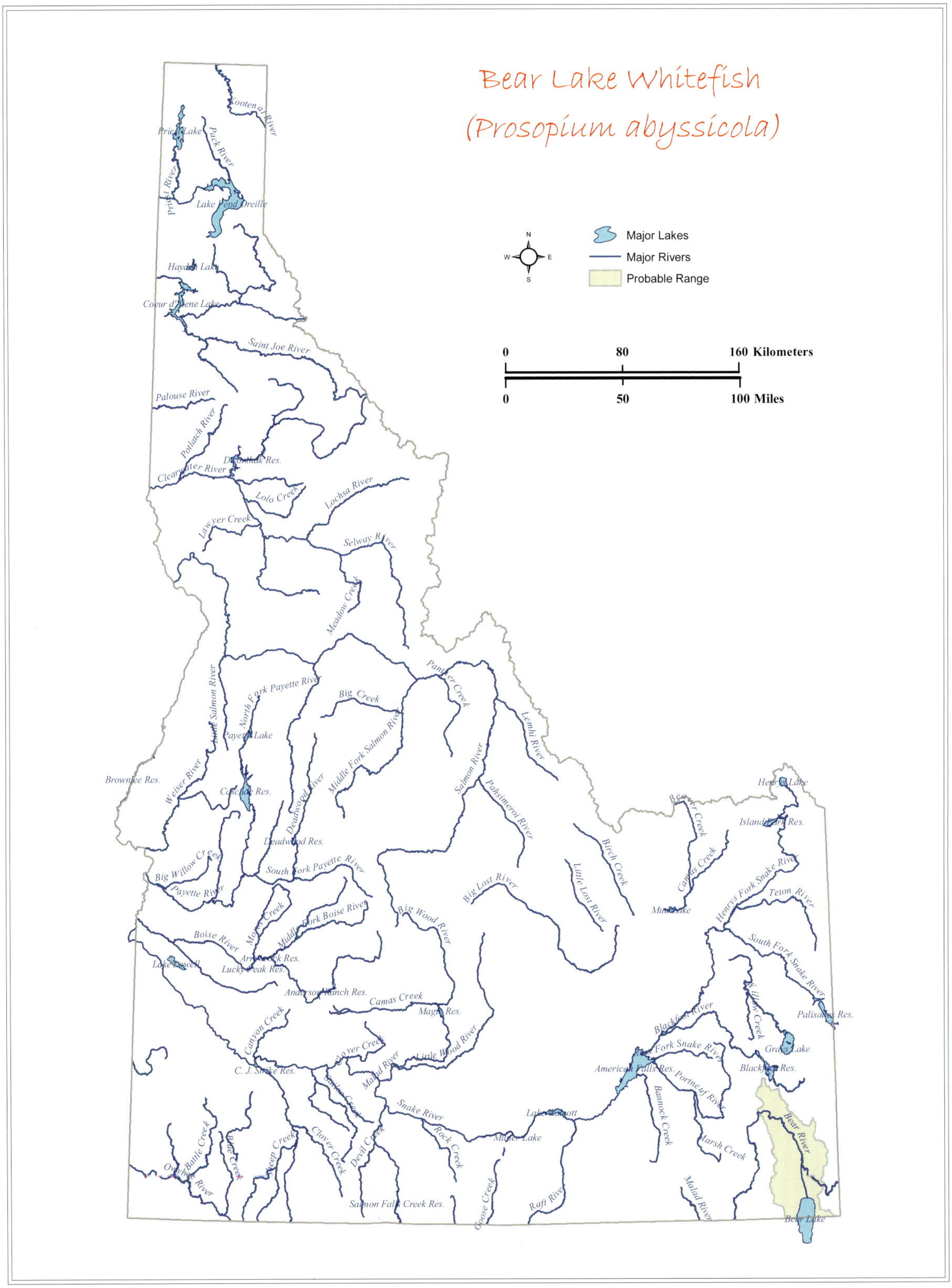
Bear Lake Whitefish
(Prosopium abyssicola)
N
W
E
S
Major Lakes
Major Rivers
Probable Range
0
80
160 Kilometers
0
50
100 Miles
Pack River
Lake Pend Oreille
Saint Joe River
Palouse River
Potlatch River
Lochsa River
Lolo Creek
Selway River
Meadow Creek
Little Salmon River
North Fork Payette River
Big Creek
Middle Fork Salmon River
Panther Creek
Lemhi River
Salmon River
Pahsimeroi River
Weiser River
Deadwood River
Deadwood Res.
South Fork Payette River
Big Willow Creek
Payette River
Birch Creek
Little Lost River
Big Lost River
Big Wood River
Camas Creek
Beaver Creek
Camas Creek
Henrys Fork Snake River
Teton River
South Fork Snake River
Palisades Res.
Boise River
Lucky Peak Res.
Canyon Creek
C. J. Strike Res.
Little Wood River
Snake River
Rock Creek
Clover Creek
Salmon Falls Creek Res.
Goose Creek
Raft River
Bannock Creek
Marsh Creek
Malad River
Bear River
Bear Lake
Blackfoot River
Portneuf River
American Falls Res.

Bear Lake Whitefish

Prosopium abyssicola

(Snyder, 1919)

Etymology

Prosopium = a small mask, from the large preorbitals; *abyssicola* = for their presence in the deep lake they inhabit (Bear Lake, Idaho-Utah).

Importance

Bear Lake is the single lake in North America that supports four endemic fishes: Bear Lake Whitefish, Bonneville Cisco, Bonneville Whitefish, and Bear Lake Sculpin. All four species originated some 10,000 to 35,000 years ago. Bear Lake Whitefish provide forage for introduced Lake Trout and native Bonneville Cutthroat Trout. They provide only rare opportunities for hook-and-line recreational fishing[2]. During the early colonization of the Bear Lake area, Bear Lake Whitefish were fished commercially, primarily as a bait fish for large Bonneville Cutthroat Trout in the lake[6]. Although the economic contributions of this species are limited, its value as a unique ecological entity in a unique ecosystem is of great intrinsic worth[17].

Description

Snyder's (1919) original description[18] includes a discussion of the possible relationship between this species and other similar species both in Bear Lake and elsewhere in western waters[17]. Bear Lake Whitefish are greenish gray on the dorsal surface, silvery on the sides, and white on the ventral surface. Scales along the lateral line number 67 to 78, and there are no more than 17 rows of scales above the lateral line, which is fairly straight. Total lengths (TL) of Bear Lake Whitefish rarely exceed 270 mm (10.6 in)[6].

Bear Lake Whitefish are not casually distinguished from their close cousins the Bonneville Whitefish. Bear Lake Whitefish scales are larger than Bonneville Whitefish scales, and there are fewer in the lateral line; thus, scale counts can serve as a useful meristic characteristic in the field if individuals of both species are in hand. One characteristic that is useful in separating the species is the "Roman nose" of the Bear Lake Whitefish. Previously, the difference in spotting patterns was a potentially useful feature for separating these two species, the characteristic being that Bonneville Whitefish have spots until they reach lengths of about 254 mm (10 in). Bear Lake Whitefish were not found to have spots[2,17]. Utah Division of Wildlife

Resources (UDWR), however, has captured Bear Lake Whitefish (positively identified since it was during the spawning season) that were spotted. Bear Lake Whitefish were raised in captivity in the aquatics laboratory at Utah State University in the late 1990s that were fully spotted. Spotting is a characteristic of juvenile whitefish (both Bonneville and Bear Lake species) found in Bear Lake[19,20].

Range and Distributional Status

This species is native to Idaho. Bear Lake Whitefish are restricted to Bear Lake[2,17]. No successful introductions into other ecosystems have been made[6].

Bear Lake Whitefish and Bonneville Whitefish are thought to have speciated from the more common Mountain Whitefish during the past 30,000 years. Mountain Whitefish are found in the Bear River and some Bear Lake tributaries, but not in Bear Lake itself[21].

Size and Longevity

Bear Lake Whitefish are considered a dwarf species, rarely reaching total lengths of more than 270 mm (10.6 in)[6]. Sigler[22] reports Bear Lake Whitefish aged one to eight had the following year-end sizes, in mm (in): 33 (1.3), 76 (3.0), 112 (4.4), 132 (5.2), 150 (5.9), 165 (6.5), 178 (7.0), and 191 (7.5). Snyder (1919), in his original description, reported a 305 mm (12 in) fish that was 13 years old[2]. Recent information suggests Bear Lake Whitefish are a slow-growing, long-lived species, possibly living as long as 37 years[6]. UDWR has aged fish in excess of 20 years using otoliths[20]. This growth strategy is consistent with the cold, oligotrophic conditions that exist in Bear Lake.

Food and Feeding

Implications of variation in diet between Bonneville Whitefish and Bear Lake Whitefish which may have ecological significance are presented here. In spring Chironomidae (midges), Ostracoda (mussel shrimp), Pisidae (spider crabs), and terrestrial insects contributed at least 5 percent each to the total diet of whitefish. Bonneville Whitefish had all of these food items in their stomachs, but Chironomidae were the dominant food group, followed by Pisidae. At a given depth, Chironomidae were present in most Bonneville Whitefish stomachs and made up between 50 and 85 percent of the biomass on average for an individual fish. Unlike Chironomidae, Pisidae were present in few stomachs, but they made up a high percentage of the biomass when present. Ostracoda and terrestrials were present in few diets and made up small percentages of the diet when present. In contrast to Bonneville Whitefish, spring diets of Bear Lake Whitefish consisted mostly of Ostracoda, with Chironomidae contributing to the weight in a couple of stomachs at shallower depths. At the depths with the highest abundance (50 m and 55 m [164 and 181 ft]) of Bear Lake Whitefish, Ostracoda averaged between 85 and 95 percent of their diet[23].

Chironomidae were present in a small percentage, and Pisidae and terrestrial groups were found in very small percentages. At depths at which Bonneville and Bear Lake Whitefish overlapped in low numbers (35 m and 40 m [115 and 131 ft]), diets were still different. However, a few Bear Lake Whitefish diets were similar to those of Bonneville Whitefish. Some Bear Lake Whitefish collected at depths dominated by Bonneville Whitefish had diets similar to those of Bonneville Whitefish, whereas some were more like the diet of the majority of Bear Lake Whitefish[23].

In summer, large diet differences exist between species, and diets within a species were similar between seasons. In most Bonneville Whitefish stomachs, Chironomidae to make up very high percentages of the biomass, and Pisidae make up a high percentage of the biomass in some stomachs. Also, Ostracoda and terrestrial insects were found at low percentages, similar to diet composition in spring. In contrast to the Bonneville Whitefish diet but again similar to the spring diet, Bear Lake Whitefish diets were dominated by Ostracoda at deeper depths. However, in this season, Ostracoda also dominated diets of the few Bear Lake Whitefish found at depths with high abundances of Bonneville Whitefish[23].

Bear Lake Whitefish are invertivores[6], feeding heavily on several groups of invertebrates such as ostracods, copepods, midge larvae, and aquatic oligochaetes. Occasionally, fishes and aquatic insects may enter the diet. Adult Bear Lake Whitefish feed heavily on ostracods and chironomid larvae[2].

Limiting Factors

Bear Lake Whitefish appear to be affected only by predation and senility[2].

Breeding Habits

UDWR has documented spawning from mid-February to mid-March[20]. Bear Lake Whitefish reach sexual maturity at age three, with some males maturing at age two[6]. Large schools are formed during the spawning process, and spawning occurs at depths of 18 to 31 m (59 to 102 ft)[6]. Female Bear Lake Whitefish of 200 mm (7.9 in) total length produce approximately 2,000 eggs. Spawning occurs at water temperatures of 2 to 4°C (35.6 to 39.2°F)[6].

Spawning periods and depth distribution of the closely related Bear Lake Whitefish and Bonneville Whitefish are distinct and contribute to reproductive isolation between the species. Spawning of Bonneville Whitefish occurs in December on rocky shoreline areas and the Rock Pile near Gus Rich Point, Utah. Bear Lake Whitefish spawn on the Rock Pile during February[21].

Habits and Habitat

Bear Lake Whitefish are endemic to Bear Lake[6]. Bear Lake is a cold, infertile (oligotrophic) lake that is approximately 12.9 km (8 mi) wide by 32.3 km (20 mi) long with a maximum surface elevation (by law) of 1,803 m (5,924 ft) above sea level. Slightly more than a third of the volume of the lake is in Utah, the remainder in Idaho. Approximately half of the lake has a depth greater than 31 m (100 ft)[24].

Based on recent research, Bonneville and Bear Lake Whitefish have very different benthic distributions in both spring and summer. In spring average Bonneville Whitefish abundance was highest at 5 m (16 ft) and decreased gradually with depth to 35 m (115 ft). Below 35 m (115 ft), in the deep profundal zone, few Bonneville Whitefish were caught. Bear Lake Whitefish abundance had the opposite pattern. Few Bear Lake Whitefish were caught until 40 m (131 ft), and abundance increased rapidly from 45 m (148 ft), with the highest abundance being at 55 m (181 ft)[23].

In summer Bonneville Whitefish abundance was low at 5 m (16 ft), then increased to a maximum at 15 m (49 ft), and then decreased to 35 m (115 ft). As in spring, few Bonneville Whitefish were caught deeper than 35 m (115 ft) in the deep profundal zone. Unlike Bonneville Whitefish, Bear Lake Whitefish abundance had a similar pattern in summer as in spring. Bear Lake Whitefish abundance was low until 40 m (131 ft) and then increased rapidly with depth to 55 m (181 ft)[23].

Recent studies support the idea that the two benthic whitefish in the Bear Lake fish community are ecologically very different. These species inhabited different depths, with Bonneville Whitefish dominating the upper strata and Bear Lake Whitefish dominating the deep profundal zone. Additionally, Bonneville Whitefish that exceed 300 mm (11.8 in) TL feed predominantly on Bear Lake Sculpin, further expanding the differences between these species[23].

Management and Conservation Status

Bear Lake Whitefish are ranked as critically imperiled range wide (G1) and statewide (S1) since they only occur in Bear Lake[15]. No reliable estimate of Bonneville Whitefish or Bear Lake Whitefish numbers has been made, but relative catch rates suggest that several million Bonneville Whitefish and close to one million Bear Lake Whitefish are present in the lake[21].

Bear Lake faces many potential threats, including nutrient loading, invasion of nonnative species, and stocking of piscivorous fishes. Distinctive ecologies of each whitefish species in Bear Lake indicate that these threats will likely have different effects on each species, ultimately requiring complex management strategies. For example, increased nutrient loading could result in deoxygenation of the profundal zone and may have severe consequences for Bear Lake Whitefish by causing death of individuals, egg mortality, or immigration into shallower areas. Boats from many areas throughout Utah and Idaho are launched daily onto Bear Lake throughout the summer. This could lead to nonnative benthic invertebrates becoming established, thereby changing the prey base on which these whitefish rely[25].

Bear Lake is stocked frequently with piscivorous native Bonneville Cutthroat Trout and nonnative Lake Trout (triploid and sterile), which could directly reduce the population size of one or both whitefish species or cause a change in habitat use. A shift in either fish's distribution from any one of these threats may cause overlapping distributions and increased risk of competition and hybridization. Conservation of these fishes requires protection of the diverse habitat that allows them to coexist. Studies examining how each of the threats might impact Bear Lake Whitefish are needed to improve conservation strategies for these endemic fishes[23].

Both Idaho and Utah classify this species as a game fish. Bear Lake Whitefish are an important forage species, but all indications are that the population is stable and abundant. Protection from changes in water quality and restriction of exotic species introductions as well as reducing change within the Bear Lake ecosystem will continue to be important factors in preserving this unique species.

In 2010 the IDFG and UDWR collaborated on a Bear Lake Fisheries Management Plan. Specifically for Bear Lake Whitefish, the agencies have planned the following management scenario:

Management Goal: Maintain a healthy, self-sustaining population of Bear Lake Whitefish.

Monitoring and Evaluation: (1) Set gill nets annually in the spring, summer, and fall in order to monitor Bear Lake Whitefish populations by season and depth; and (2) whitefish less than 250 mm (9.8 in) TL should continue to be subsampled from gill nets so the percentage composition of the total gill-net catch of Bonneville and Bear Lake Whitefish can be computed in order to track individual population trends over time[26].

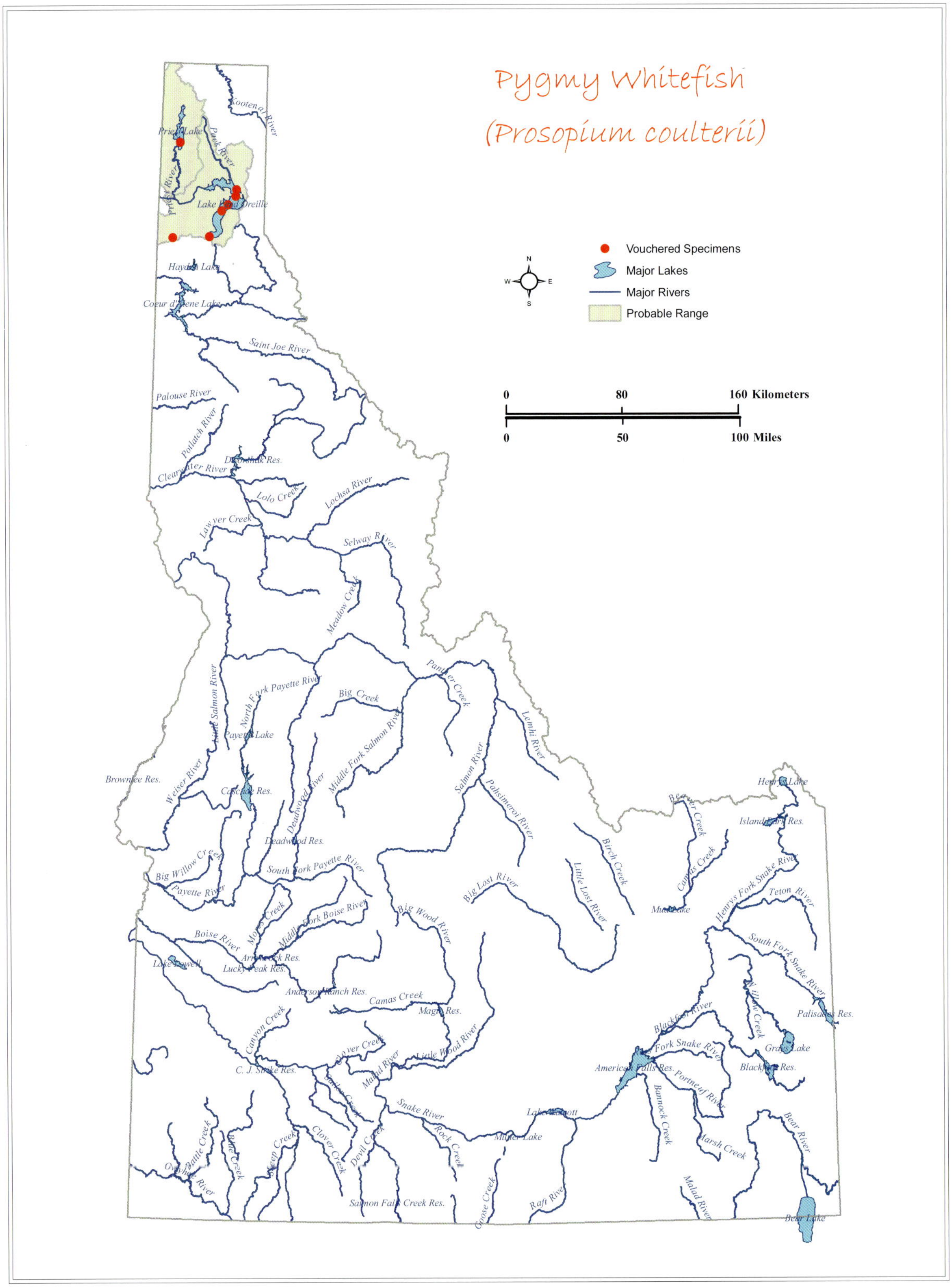
Pygmy Whitefish
(Prosopium coulterii)
Vouchered Specimens
Major Lakes
Major Rivers
Probable Range
N
S
E
W
0
80
160 Kilometers
0
50
100 Miles
Priest River
Pack River
Kootenai River
Saint Joe River
Palouse River
Potlatch River
Lolo Creek
Lochsa River
Selway River
Meadow Creek
North Fork Payette River
Big Creek
Middle Fork Salmon River
Salmon River
Lemhi River
Pahsimeroi River
Birch Creek
Little Lost River
Big Lost River
Big Wood River
Weiser River
Brownlee Res.
Payette River
Boise River
Camas Creek
Snake River
Rock Creek
Clover Creek
Canyon Creek
C. J. Strike Res.
Salmon Falls Creek Res.
Goose Creek
Raft River
Bannock Creek
Marsh Creek
Malad River
Bear River
Teton River
South Fork Snake River
Henrys Fork Snake River
Palisades Res.
Camas Creek
Little Wood River
Milner Lake

Pygmy Whitefish

Prosopium coulterii

(Eigenmann & Eigenmann, 1892)

Etymology

Prosopium = a small mask, from the large preorbitals; *coulterii* = after Dr. J. M. Coulter, a distinguished botanist.

Importance

Little scientific information is available for this species. They are of no known direct economic importance, and, although listed as a game species in Idaho, they are seldom caught by anglers. Across portions of their native range, they are consumed by Bull Trout, Arctic Char, Lake Trout, and Rainbow Trout[5,6].

Description

These whitefish are brownish on the dorsal surface, silvery on the sides, and white on the ventral surface. Dorsal, caudal, and pectoral fins are generally clear, while the anal and pelvic fins are whitish and immaculate. The presence of a faint dark spot is sometimes noted at the base of the caudal fin, and a series of 12 to 14 similar spots occurs along the midline of the back; seven to 14 dark, round to oval parr marks with diffuse borders occur along the lateral line of young and subadults that are vague or absent in adults[5]. Pygmy Whitefish and Mountain Whitefish have a single flap between the nostrils, but Lake Whitefish have two flaps[11].

As with all of the whitefishes, Pygmy Whitefish have an elongate, almost cylindrical body, which is only slightly laterally compressed. The gill rakers are short, totaling 11 to 21, with three to seven on the upper limb of the gill arch and eight to 13 on the lower limb. Teeth are small and present on the tongue. Branchiostegal rays are six to nine and six to nine with some variation across populations. Scales are large and cycloid, with 50 to 70 pored scales in the complete lateral line. Pyloric caecae are simple and low in number, 13 to 33. Vertebrae number 50 to 55 but may vary across populations. There is a small adipose fin, and the single dorsal fin has 10 to 12 rays. Pelvic fin rays are nine to 11, and pectoral rays are 13 to 18; pectoral fins are thoracic[5].

Range and Distributional Status

This species is native to Idaho. Pygmy Whitefish are discontinuously distributed across northern North America, primarily in deep lakes that are remnants of the last Pleistocene period Ice Age, about 10,000 years ago[5,11]. In Idaho, Pygmy Whitefish are native to Priest Lake, Upper Priest Lake, Lake Pend Oreille, and Spirit Lake[6].

Pygmy Whitefish are a particularly enigmatic member of the genus and are generally considered to be a glacial relict that inhabits cold, deep lakes. Pygmy Whitefish are a western North American species with populations scattered from Washington, Idaho, and Montana north to British Columbia, Alaska, and the Yukon Territory. They do occur, however, in a few scattered localities east of the Continental Divide (e.g., western Lake Superior, Lake Athabasca, Great Bear Lake, and Waterton Lake), as well as on the Chukotsk Peninsula, Siberia. Studies of the phylogeography of the Pygmy Whitefish across their North American range in 2011 demonstrated that there are two major genetic lineages. These two major lineages and their geographic association with areas that served as glacial refugia support the idea that the current range disjunctions in North America probably result from historical patterns of isolation and recolonization during the glacial cycles of the Pleistocene[27].

Size and Longevity

As can be assumed from the common name, Pygmy Whitefish are a diminutive species, rarely reaching lengths greater than 152 mm (6 in). Larger specimens have been collected in Washington State, with total lengths of 247 to 290 mm (9.7 to 11.4 in). Growth of Pygmy Whitefish is variable across ecosystems and is most likely related to water temperatures and productivity (food availability)[11].

Food and Feeding

Like many pelagic species, Pygmy Whitefish are invertivores, feeding on a variety of benthic aquatic invertebrates, insect larvae, crustaceans, and mollusks. Feeding exclusively during daylight hours, Idaho populations in Lake Pend Oreille feed primarily on zooplankton and aquatic insect larvae, primarily midges[6].

Limiting Factors

Few studies have been completed to document the factors that may adversely affect Pygmy Whitefish. Idaho lists the species as of greatest conservation concern but in a category that lacks essential information pertaining to status in Idaho populations[6]. Little information has been published on limiting factors such as diseases or parasites.

Breeding Habits

Across their native range, 50 percent of male Pygmy Whitefish are sexually mature at age two, and 50 percent of females are sexually mature at age three[11]. Spawning likely occurs in October, November, or December, varying regionally across the native range. Spawning occurs in shallow water in streams and lakes where eggs are scattered over coarse gravel. Egg development occurs over winter with emergence from gravels in early spring. Eggs of this species are large when ripe, measuring 2 mm (0.08 in) to 2.4 mm (0.09 in). Females may produce 440 to 580 eggs; egg number is dependent on female size[5,11].

Habits and Habitat

Pygmy Whitefish most often inhabit deep, primarily oligotrophic (unproductive) lakes. In lakes, these fish usually occur at depths greater than 6.1 m (20 ft)[11]. In Idaho, Pygmy Whitefish have been found only in lakes where they occupy cold-water areas at depths over 300 m (984 ft) in Lake Pend Oreille and at depths of 20 to 50 m (66 to 164 ft) in other lakes[6].

Pygmy Whitefish are a cold-water species, and in British Columbia they are usually found in water less than 10°C (50°F), occurring in both lakes and rivers. In Priest Lake, in the fall, Pygmy Whitefish demonstrate a diel movement of adults onshore in late afternoon or early evening and back into deeper water at dawn, suggesting that the onshore movements may be associated with spawning activity. In rivers Pygmy Whitefish are found in both turbid and clear waters, with moderate to swift current and usually over gravel or cobble substrate[4].

Management and Conservation Status

Pygmy Whitefish are considered common, widespread, and abundant range wide (G5). Their status in Idaho is not yet assessed (SNR)[15]. IDFG has conducted abundance surveys of Pygmy Whitefish populations in northern Idaho lakes and found them to be relatively abundant. IDFG will continue to periodically evaluate Pygmy Whitefish populations.

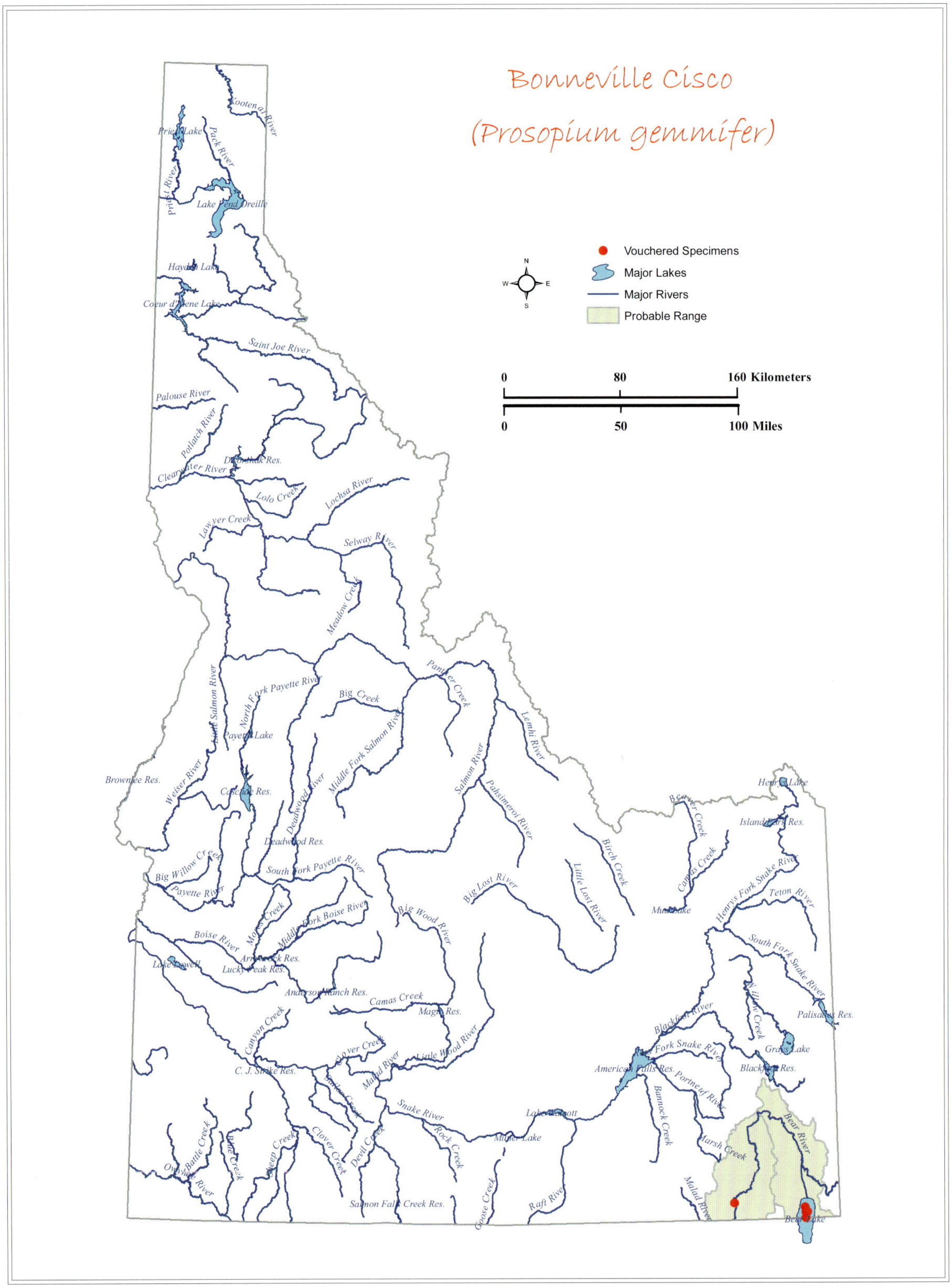
Bonneville Cisco
(Prosopium gemmifer)
N
S
W
E
Vouchered Specimens
Major Lakes
Major Rivers
Probable Range
0
80
160 Kilometers
0
50
100 Miles
Kootenai River
Pack River
Saint Joe River
Palouse River
Potlatch River
Lolo Creek
Lochsa River
Lawyer Creek
Selway River
Meadow Creek
Big Creek
Middle Fork Salmon River
Salmon River
Lemhi River
Pahsimeroi River
Birch Creek
Little Lost River
Big Lost River
Big Wood River
Camas Creek
Payette River
Weiser River
Big Willow Creek
Boise River
Canyon Creek
C. J. Strike Res.
Snake River
Rock Creek
Raft River
Goose Creek
Clover Creek
Bannock Creek
Bear River
Teton River
South Fork Snake River
Palisades Res.
Malad River

Bonneville Cisco

Prosopium gemmifer
(Snyder, 1919)

Etymology

Prosopium = a small mask, from the large preorbitals; *gemmifer* = set with gems, due, it is believed, to the highly reflective scales on the sides of the species.

Importance

Bonneville Cisco, occurring only in Bear Lake, Idaho-Utah, are a unique natural history entity. During the approximately 16-day period when they can be dipnetted along the eastern shore of Bear Lake, they provide thousands of hours of recreational fishing, food, and enjoyment for anglers. Historically, they provided a commercial fishery that has been discontinued. Bonneville Cisco are important forage fish for Bear Lake (Bonneville) Cutthroat Trout. They are almost never caught on hook and line[22,28].

Description

Bonneville Cisco are herring-like, that is, cylindrical in body shape. The nose is distinct, leading to the local name "peak nose." Body is pale moss green on the dorsal surface. Sides are silvery with a pearly iridescence, and the ventral surface is white. During spawning season a streak of gold and yellow extends backward from the pectoral fin to the anal region. The eye is large, and the maxillary is entirely in front of the eye. They have no teeth[2,6].

Range and Distributional Status

This species is native to Idaho. Bonneville Cisco are endemic to Bear Lake and occur in no other ecosystem. Attempts have been made to introduce the species into Lake Tahoe, Nevada-California; Twin Lakes, Colorado; and Flaming Gorge Lake, Utah-Wyoming. No survival has been documented in any other of these ecosystems. Fossil remains of Bonneville Cisco have been discovered in prehistoric Lake Bonneville, which was greatly larger than Bear Lake[2,29]. Connections between Bear Lake and ancient Lake Bonneville have been significantly researched, and some recent research concludes that the Bear River did flow into Bear Lake connecting it to Lake Bonneville during the Pleistocene. These intermittent fillings raised the water level of Bear Lake until it rose to the valley threshold at Nounan Narrows. Several lake-level fluctuations in the tens of meters (30+ ft) occurred[30].

Size and Longevity

Bonneville Cisco are small fish, rarely reaching total lengths of more than 216 mm (8.5 in) or weights over 7 g (2.5 oz). While not normally living past age seven, a few individuals have been aged at 11[2,31].

Food and Feeding

Zooplankton are the major food of Bonneville Cisco, with one copepod, *Epischura*, being the predominant food of importance all year. Bonneville Cisco also feed on cladocerans, principally *Bosmina*, during the winter. Bonneville Cisco feed year-round[2].

Bonneville Cisco are most generally found at depths of 15.2 to 30.4 m (50 to 100 ft) during the summer. They may leave cool water and move to warmer surface waters to feed[2,31].

Limiting Factors

Predation, from Bear Lake (Bonneville) Cutthroat Trout, and fishing pressure are the primary limiting factors for Bonneville Cisco. Although some individuals are infested with stomach nematodes, disease apparently is a minor factor in mortality[2].

Breeding Habits

When Bonneville Cisco spawn, hundreds of thousands of them may come out of the depths of the lake to shallower waters. Although the primary recreational fishing effort is along the eastern shore, cisco also spawn at other locations around the lake's edge[2,31]. Spawning occurs during a roughly two-week period in late January to mid-February when water temperatures average 2 to 4°C (35.6 to 39.2°F) but may range from 0.5 to 5.6°C (33 to 42°F)[2,6,31]. Both sexes mature at age three. Females produce between 2,000 and 3,600 eggs each. Spawning occurs in waters ranging in depth from surface waters to 12 m (3 in to 40 ft). Males move inshore ahead of females and remain near shore for the duration of the spawning run. Females move inshore in groups when they are ripe, and then return to deeper water. While spawning, large schools of Bonneville Cisco move parallel to shore at depths of 0.9 to 2.4 m (3 to 8 ft), depending on water turbulence, the presence of anglers, and the presence or absence of ice. Each female is generally attended by three to five males. When a group of individuals stops to spawn, they are easy prey for dipnetters[2,31]. Eggs are broadcast, settle to the bottom, and become attached to substrate[6].

Early life history aspects of Bonneville Cisco are reasonably well known. Fertilized eggs deposited in January are subject to predation from a variety of egg predators, and less than 5 percent of fertilized eggs may survive to hatching[21,32]. Age-0 cisco hatch in early May. These larvae swim to the surface water and begin consuming small crustacean and ciliate zooplankton. Growth is rapid in the warmer, food-rich surface waters of the lake. Predation on eggs and larvae generally exceeds 99 percent of fertilized eggs and likely determines cohort success in this species[21].

Habits and Habitat

Bonneville Cisco prefer temperatures less than 14.4°C (58°F). During summer months when the lake temperature warms, Bonneville Cisco retreat to depths and cooler water. At temperatures lower than 14.4°C (58°F), they are scattered throughout the lake. Bonneville Cisco are most generally at depths of 15.2 to 30.4 m (50 to 100 ft). During the summer, they may leave cool water and move to warmer surface waters to feed[2,6].

Estimates of Bonneville Cisco abundance for Bear Lake ranged from 1.9 million in 1991 to 9.7 million in 2001. These abundance estimates, developed using hydroacoustic methodology, indicated that cisco were most abundant in mid-water depths in the eastern and southern regions of the lake. High population variability is thought to be associated with the impact of water-level fluctuations in the lake on cisco spawning success[21].

Management and Conservation Status

Bonneville Cisco are ranked as critically imperiled range wide (G1) and statewide (S1) since they only occur in Bear Lake[15]. Bonneville Cisco are managed as a game species by both Idaho and Utah. Sampling indicates the Bonneville Cisco population is abundant and stable. The management goals for Bonneville Cisco are to (1) maintain a healthy, self-sustaining population; and (2) maintain a sport fishery. The popular dip-net fishery for this species provides a unique opportunity to capture this endemic species. Reduced water levels or changes in water quality and introduction of nonnative species are threats to Bonneville Cisco. The management agencies closely monitor the status of Bonneville Cisco in Bear Lake, and if population levels drop below a minimum population level, stocking of predatory sportfishes will be discontinued until cisco abundance rebounds[26].

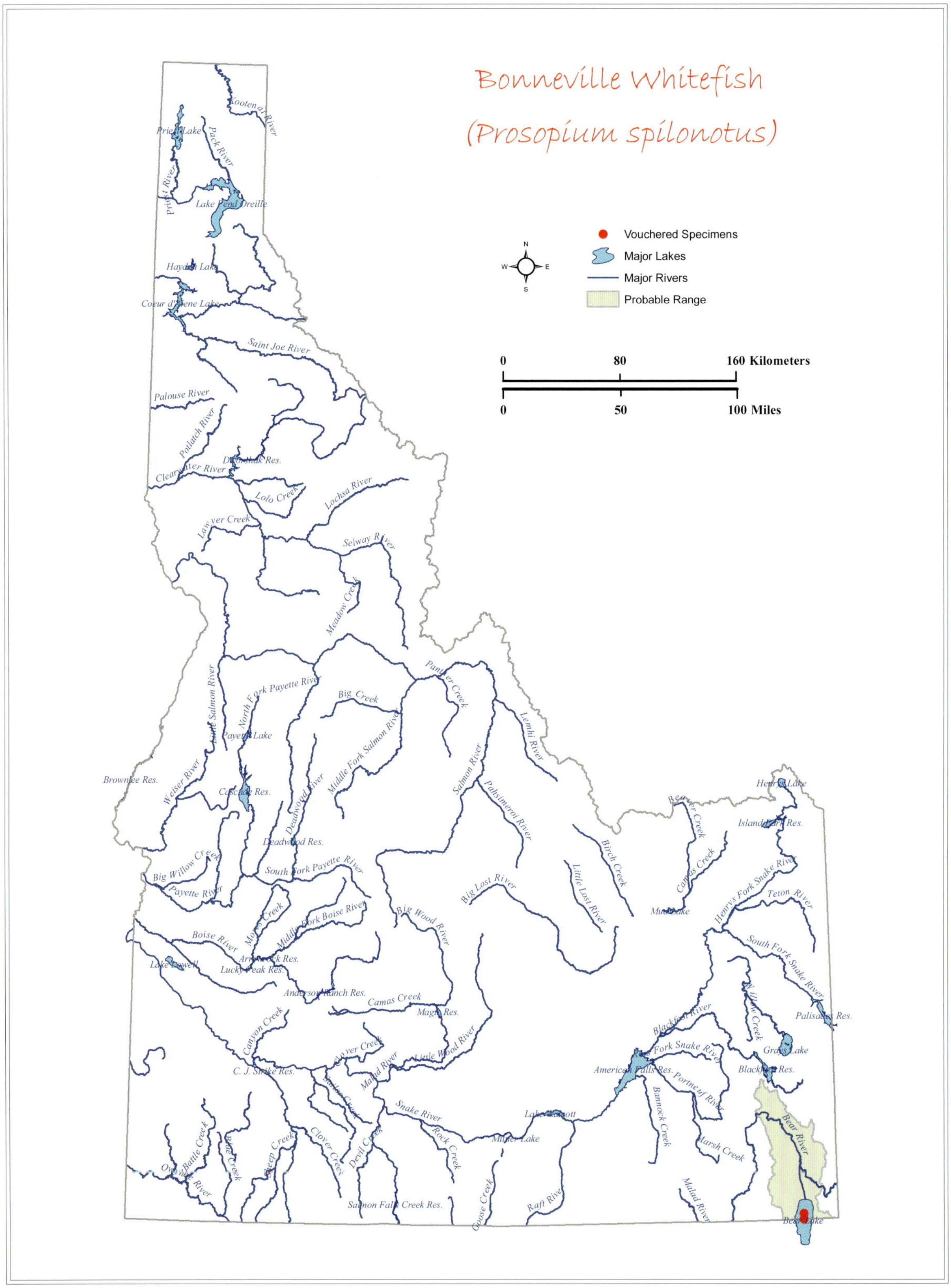
Bonneville Whitefish
(Prosopium spilonotus)
Vouchered Specimens
Major Lakes
Major Rivers
Probable Range
N
S
E
W
0
80
160 Kilometers
0
50
100 Miles
Kootenai River
Priest Lake
Pack River
Priest River
Lake Pend Oreille
Saint Joe River
Palouse River
Potlatch River
Lolo Creek
Lochsa River
Selway River
Meadow Creek
Big Creek
Middle Fork Salmon River
Salmon River
Lemhi River
Pahsimeroi River
Birch Creek
Little Lost River
Big Lost River
Big Wood River
Camas Creek
Weiser River
Big Willow Creek
Payette River
South Fork Payette River
Deadwood Res.
Boise River
Lucky Peak Res.
Canyon Creek
C. J. Strike Res.
Snake River
Rock Creek
Clover Creek
Devil Creek
Goose Creek
Raft River
Malad River
Marsh Creek
Bannock Creek
Portneuf River
Blackfoot River
Teton River
South Fork Snake River
Palisades Res.
Grays Lake
Bear River

Bonneville Whitefish

Prosopium spilonotus
(Snyder, 1919)

Etymology

Prosopium = a small mask, from the large preorbitals; *spilonotus* = spot + back.

Importance

Like the Bonneville Cisco and the Bear Lake Whitefish, Bonneville Whitefish, as one of four endemic Bear Lake fishes, are a unique natural history entity. Within the recreational fisheries arena, they are of little direct importance, as they are only infrequently caught on hook and line. This species does, however, provide forage for large Bear Lake (Bonneville) Cutthroat Trout and Lake Trout, thereby contributing indirectly to highly prized recreational fishery opportunities[2,6].

Description

As with the other Bear Lake *Prosopium* species, Bonneville Whitefish are cylindrical, grayish white on the dorsal surface, silvery on the sides, and white on the ventral surface[2,6]. Bonneville Whitefish do not have the classic "Roman nose" present in the Bear Lake Whitefish, and this trait allows rapid separation of individuals in the field[2].

Spotting is a characteristic of juvenile whitefish (both Bonneville and Bear Lake species) found in Bear Lake[19,20]. Bonneville Whitefish have a short pointed head with the snout extending slightly beyond the lower jaw. Lateral line scales typically number 76 to 90 (74 to 94). Gill rakers are 19 to 26. Dorsal fin rays are 10 to 12 and anal fin rays nine to 11[2,6].

Range and Distributional Status

This species is native to Idaho. Bonneville Whitefish are endemic to Bear Lake, occurring in no other ecosystems[2,6].

Bear Lake Whitefish and Bonneville Whitefish are thought to have speciated from the more common Mountain Whitefish during the past 30,000 years. Mountain Whitefish are found in the Bear River and some Bear Lake tributaries, but not in the lake itself[21].

Size and Longevity

This is the largest of the endemic Bear Lake whitefishes. Adults may attain total lengths of 457 to 559 mm (18 to 22 in), although most are 356 to 406 mm (14 to 16 in). Bonneville Whitefish can live to twenty years[6].

Food and Feeding

In Bear Lake, Bonneville Whitefish eat primarily chironomid larvae and pupae. A variety of aquatic and terrestrial insects and worms make up the remainder of the diet. Occasionally, Bonneville Whitefish may prey on other fish species and may prey on the eggs of other whitefishes[2,23].

Limiting Factors

No single mortality factor would account for a decrease in Bonneville Whitefish. Only very few Bonneville Whitefish are caught on hook and line, and an unknown but likely minor number are prey for other fishes and predatory birds. Advanced age is likely the most common mortality factor[2].

Breeding Habits

Bonneville Whitefish become sexually mature at ages three and four. Spawning typically occurs in late November and early December in shallow areas; eggs are deposited in rocky areas or on sandy bars[6]. Mature Bonneville Whitefish are 203 to 254 mm (8 to 10 in) in total length. When a female stops to spawn, five to six males attend her. After spawning, which lasts only five to 15 seconds, the spawners return to the school and continue traveling. Temperature preference for spawning is around 7.2°C (45°F)[2]. Females of spawning size produce 600 to 900 eggs,[6] but one 229 mm (9 in) female contained 1,200 eggs[22].

Spawning periods and depth distribution of the closely related Bear Lake Whitefish and Bonneville Whitefish are distinct and contribute to reproductive isolation between the species. Bonneville Whitefish spawning occurs in December on rocky shoreline areas and the Rock Pile near Gus Rich Point, Utah. Bear Lake Whitefish spawn on the Rock Pile during February. No reliable estimate of Bonneville Whitefish or Bear Lake Whitefish abundance has been made, but relative catch rates suggest that several million Bonneville Whitefish and close to one million Bear Lake Whitefish are present in the lake[21].

Habits and Habitat

Recent studies support the idea that the two benthically oriented whitefishes in the Bear Lake fish community are ecologically very different. First, these species inhabit different depths, with Bonneville Whitefish dominating the upper strata and Bear Lake Whitefish dominating the deep profundal zone. Second, these species have very different diets even at depths at which the species overlap, with Bonneville Whitefish eating a variety of benthic invertebrates, but mostly Chironomidae, and Bear Lake Whitefish eating mostly

Ostracoda. Additionally, Bonneville Whitefish larger than 300 mm (11.8 in) TL have been shown to eat primarily sculpin, further expanding the differences between these species[23].

Bonneville Whitefish inhabit the 12.2 to 30.5 m (40 to 100 ft) layer of water most of the time in Bear Lake, but may also be much deeper. Adults are in shallow water more often than young and inhabit shallow water both summer and winter, more often than either the Bonneville Cisco or the Bear Lake Whitefish[2,22].

Management and Conservation Status

Bonneville Whitefish are ranked as critically imperiled range wide (G1) and statewide (S1) since they occur only in Bear Lake[15]. Both Idaho and Utah manage Bonneville Whitefish as a game species. Sampling indicates that the population of Bonneville Whitefish in Bear Lake is both abundant and stable. Threats to this unique endemic species include changes in water quality, varied lake levels due to drawdowns for irrigation, and the introduction of additional nonnative species.

Bear Lake faces many potential threats, including nutrient loading, invasion of nonnative species, and stocking of piscivorous fishes. Distinctive ecologies of each whitefish species in Bear Lake indicate that these threats will likely have different effects on each species, ultimately requiring complex conservation strategies. Boats from many areas throughout Utah and Idaho are launched daily onto Bear Lake throughout the summer. This could lead to nonnative benthic invertebrates becoming established, thereby changing the prey base on which these whitefish rely. Additionally, Bear Lake is stocked frequently with piscivorous native Bonneville Cutthroat Trout and nonnative Lake Trout, which could directly reduce the population size of Bonneville Whitefish or cause a change in habitat use. A shift in distribution from any one of these threats may cause overlapping distributions and increased risk of competition and hybridization. Conservation of these fish depends on preserving the diverse habitat that allows them distribute differently than the other whitefishes. Studies examining how each of the threats might impact all of Bear Lake's whitefish are needed to improve conservation strategies for these endemic fishes[23].

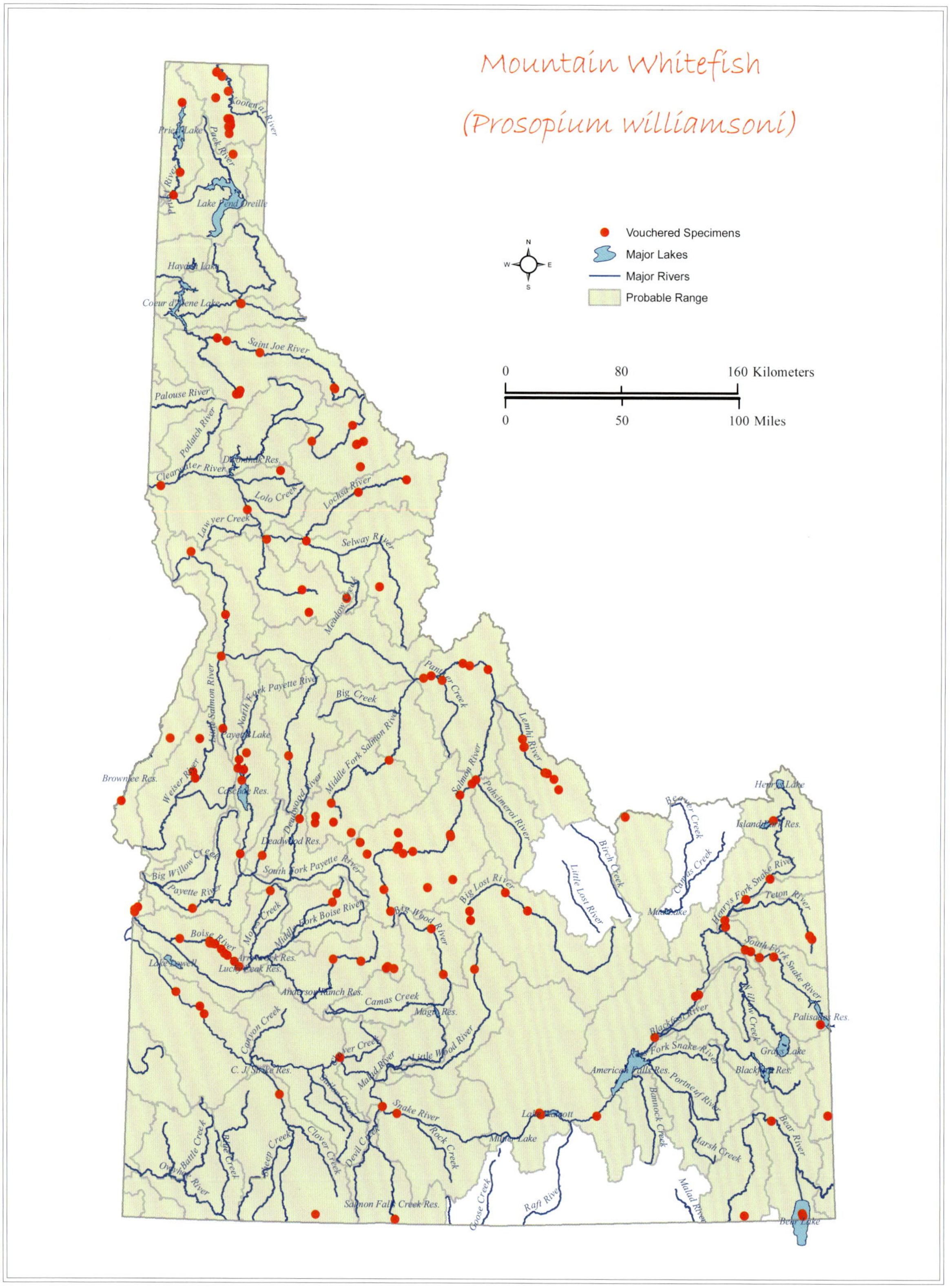
Mountain Whitefish
(Prosopium williamsoni)
Vouchered Specimens
Major Lakes
Major Rivers
Probable Range
0
80
160 Kilometers
0
50
100 Miles
Kootenai River
Priest Lake
Pack River
Priest River
Lake Pend Oreille
Hayden Lake
Coeur d'Alene Lake
Saint Joe River
Palouse River
Potlatch River
Clearwater River
Dworshak Res.
Lolo Creek
Lochsa River
Lawyer Creek
Selway River
Meadow Creek
Panther Creek
North Fork Payette River
Big Creek
Little Salmon River
Payette Lake
Lemhi River
Brownlee Res.
Weiser River
Cascade Res.
Middle Fork Salmon River
Deadwood River
Salmon River
Pahsimeroi River
Deadwood Res.
South Fork Payette River
Big Willow Creek
Payette River
Big Lost River
Little Lost River
Birch Creek
Camas Creek
Henrys Lake
Island Park Res.
Henrys Fork Snake River
Teton River
Middle Fork Boise River
Big Wood River
Boise River
Arrowrock Res.
Lucky Peak Res.
Lake Lowell
Anderson Ranch Res.
Camas Creek
Magic Res.
South Fork Snake River
Palisades Res.
Blackfoot River
Willow Creek
Grays Lake
Blackfoot Res.
American Falls Res.
Portneuf River
Little Wood River
Malad River
C. J. Strike Res.
Canyon Creek
Snake River
Lake Walcott
Bannock Creek
Bear River
Rock Creek
Marsh Creek
Clover Creek
Devil Creek
Battle Creek
Owyhee River
Salmon Falls Creek Res.
Goose Creek
Raft River
Malad River
Bear Lake

Mountain Whitefish

Prosopium williamsoni
(Girard, 1856)

Adult

Juvenile

Etymology

Prosopium = a small mask, from the large preorbitals; *williamsoni* = after Lieutenant R. S. Williamson of the United States Pacific Railroad Exploration (ca 1853-1855).

Importance

From a sportfishing perspective, Mountain Whitefish can be regarded only as a mixed bag. One group's perspective is that this species is a good food fish and a good sportfish. A second group's perspective would be just the opposite: that they are not good eating and interfere with populations of more desirable fish, particularly trout[2].

Across their native range, Mountain Whitefish have been used as food by Native peoples and European settlers for many years. In Salt Lake City, Utah, in the 1870s, thousands of Mountain Whitefish were sold in local markets[5,11].

Mountain Whitefish were commercially fished in the large lakes of northern Idaho, where they were once abundant. Acceptance of Mountain Whitefish as a sportfish is varied in Idaho, but they are popular during the winter when they feed more actively than trout. In summer they can be caught on small flies and artificial lures[6].

Life history characteristics and habitat requirements of most native salmonids in western North America have been the subject of much research. This is not so for the Mountain Whitefish, in large part because they have been considered less appealing to anglers, therefore receiving far less attention from fishery managers and researchers. In fact, from the 1950s into the 1970s, Mountain Whitefish removal programs were undertaken by government agencies in some areas of the western United States because they were perceived by anglers and biologists to limit trout production via competition for food and space, an assertion that has not been substantiated[33].

Description

Mountain Whitefish are typically brown or olive to silvery gray on the back; the sides are silvery, grading to white on the lower sides and the ventral surface. No spots are present on adults[6].

Mountain Whitefish have a slender, elongate body. Teeth are small and restricted to a small patch on the tongue and on gill rakers of adults. Gill rakers are short, numbering 19 to 26, usually eight to 11 on the lower limb of the gill arch and 11 to 15 on the upper limb. Branchiostegal rays 7 + 7 to 10 + 10. A relatively large adipose fin is present, and the dorsal fin has 11 to 15 rays. Pelvic fins are abdominal, with 10 to 12 rays, and a distinct pelvic axillary process is present. Pectoral fins have 14 to 18 rays. Scales on Mountain Whitefish are large, cycloid, and usually outlined with a dark pigment. The lateral line is complete with 74 to 90 scales. Pyloric caecae are variable, 50 to 146, and vertebrae are 53 to 61[5].

Two morphological variants have been documented recently. The rounded snout form, which is the type most anglers see, and a "long-nose" or "Pinnochio" form. The Pinnochio form is present across southern Idaho. This long-nosed form has in the past been described as a separate species but is confirmed as *Prosopium williamsoni*. This form has been observed to use its long snout to root around in the substrate when feeding on aquatic insects, whereas the "normal" form feeds on drift or at the surface[6,34].

In the field, juvenile Mountain Whitefish are easily confused with adult Pygmy Whitefish. However, dorsal fin ray numbers separate the species: 11 to 15 rays in Mountain Whitefish and nine or 10 in Pygmy Whitefish. In Mountain Whitefish, the anterior lateral-line scales are noticeably smaller than the scales immediately above and below the lateral line. By contrast, the anterior lateral-line scales of Pygmy Whitefish are about the same size as the scales immediately above and below the lateral line. Juvenile Mountain Whitefish up to about 200 mm (7.8 in) have well-developed parr marks[4].

Range and Distributional Status

This species is native to Idaho**.** The species is widely distributed across western North America, occurring both east and west of the Continental Divide and from Canada to the Lahontan and Bonneville Basins of Utah and Nevada. In Idaho they occur in most major river drainages[6].

Within the reaches of the upper Snake River basin, in 119,453 km (74,225 mi) of streams in 21 major rivers where Mountain Whitefish were present, they were rarely caught in small tributary streams, but instead were usually found in wider, lower-gradient main-stem reaches. Considering only sites with water, Mountain Whitefish were present in first- through third-order streams in only 18 of 767 (2 percent) study sites, whereas in fifth- through seventh-order streams, they were present in 52 of 72 (72 percent) study sites.

Similarly, Mountain Whitefish were present in 65 (88 percent) of 74 study sites where width was greater than 15 m (49 ft), but only 23 (3 percent) of 808 study sites where width was less than 10 m (33 ft). Gradient was also strongly correlated with distribution, as Mountain Whitefish were most likely to be present when gradient was less than 1 percent. Population estimates from this research estimated that there were approximately 4.7 ($\pm$ 1.8) million Mountain Whitefish within the reaches of the upper Snake River basin. Most of this abundance stems from fifth- through seventh-order streams, which constituted only 13 percent of the total stream kilometers but accounted for 93 percent of the total abundance. Average abundance among all 106 study sites containing Mountain Whitefish was 2.2/100 m^2 (2.2/1076 ft^2) or 78.3/100 m^2 (78.3/ 328 ft^2) of stream[23].

Size and Longevity

Most Mountain Whitefish live to be eight or nine years old, although some live to be over 10 years of age. Reports in the literature note much older individuals in some populations[2]. Average yearly length for Mountain Whitefish in the Logan River, Utah, for fish aged one to nine years is in mm (in): 117 (4.6), 206 (8.1), 259 (10.2), 295 (11.6), 325 (12.8), 358 (14.1), 391 (15.4), 417 (16.4), and 442 (17.4)[35]. Mountain Whitefish from Canada, for fish aged one through nine, reached total lengths, in mm (in), of 66 to 135 (2.6 to 5.3), 106 to 224 (4.2 to 8.8), 163 to 297 (6.4 to 11.7), 196 to 328 (7.7 to 12.9), 221 to 330 (8.7 to 13.0); 285 to 358 (11.2 to 14.1), 325 to 391 (12.8 to 15.4), 351 to 417 (13.8 to 16.4), and 376 to 442 (14.8 to 17.4). Various populations across the native range do not grow as rapidly as did the Logan River population, but slow-growing populations tend to reach the same final size as fast-growing populations[2]. This is a relatively long-lived species, with 10- to 20-year-old individuals present in many populations. Oldest reported age for Mountain Whitefish is 29 years[6].

Growth of Mountain Whitefish varies according to ecosystem productivity. In the North Fork Clearwater River, Idaho, growth of Mountain Whitefish is fairly rapid for two years, with individuals reaching lengths of 100 to 130 mm (3.9 to 5.1 in) total length in the first year and 178 to 209 mm (7.0 to 8.2 in) in the second year. Growth rate then remains fairly constant[6].

Newly emerged Mountain Whitefish are 16 to 20 mm (0.6 to 0.9 in) TL and positively phototactic. By late summer, Mountain Whitefish fry reach sizes of about 60 to 100 mm (2.4 to 4.0 in), and growth is rapid for the first four years of life. In the upper Columbia River and Fraser River systems, most individuals are reported to be sexually mature by age six; however, males mature about a year earlier than females[4].

The all-tackle record for Mountain Whitefish is a 2.5 kg (5 lb, 8 oz) fish taken from the Elbow River, Calgary, Alberta, Canada, in 1995[12], however, the Idaho record Mountain Whitefish is a 2.7 kg (5 lb, 15 oz), 572 mm (22.5 in) individual from Island Park Reservoir, caught in 1997[13].

Food and Feeding

In order of importance, by volume, food of Mountain Whitefish from the Logan River, Utah, is caddisfly (Trichoptera) larvae, true fly (Diptera) larvae, mayfly (Ephemeroptera) nymphs, and stonefly (Plecoptera) nymphs. Terrestrial insects are also consumed. Mountain Whitefish feed more actively in winter than in summer, and twilight and night are the most frequent feeding times. They may surface to feed on drift or terrestrial insects and opportunistically feed on salmon eggs and hatchery salmon fry in the Columbia River[2,6,35].

Limiting Factors

Few predators exist for the Mountain Whitefish except humans[5]. For most populations, current levels of either commercial or sportfishing have not substantially reduced them. Local fishing pressure may reduce populations in ecosystems where they are eagerly sought. Small Mountain Whitefish may be preyed upon by various trout species that exist within a particular ecosystem[2]. Where ranges overlap, Bull Trout commonly feed on Mountain Whitefish.

Breeding Habits

Variation of spawning times across Mountain Whitefish populations is dependent upon latitude and falling temperatures in autumn[2]. Spawning is typically initiated from September through November at temperatures of 0 to 11°C (32 to 51.8°F). Mountain Whitefish build no nest for eggs but rather disperse eggs into rocky gravel substrates in riffle areas. After spawning, individuals quickly migrate downstream to overwintering areas. Spawning occurs primarily at night. Eggs hatch in late winter or early spring and is dependent on water temperature[6].

Sexual maturity for Mountain Whitefish is typically reached at age three years when adults are 200 to 250 mm (7.9 to 9.8 in). Females produce an average of 5,000 eggs, with large females producing as many as 20,000 to 50,000 eggs[6].

While spawning-run distances are not great for many populations[2], Mountain Whitefish often make complex annual migrations for feeding, spawning, and overwintering. In the Lochsa River and North Fork Clearwater Rivers, Idaho, migrations up to 88 km (54.7 mi) have been documented[6].

Habits and Habitat

Mountain Whitefish inhabit a wide range of altitudes and stream sizes but are most typically found in larger streams, rivers, and large lakes. There appears to be a preference for larger cold streams. Young of this species use shallow backwater areas and very shallow riffles, while larger individuals occupy deeper mid-channel waters[6].

In stream ecosystems immediately following emergence, Mountain Whitefish fry drift downstream before moving into shallow, low-velocity areas along the river's edge, into side channels, and into backwaters. Mountain Whitefish occur in both lakes and rivers. Fluvial populations may demonstrate complex multiple migrations of over 100 km (62 mi). In rivers adult Mountain Whitefish occur in loose aggregations in runs or pools, and adults are usually close to but not on the bottom in water ranging from 1 to 2 m (3.3 to 6.6 ft) deep over coarse gravel or cobble substrate. Average water depth, velocities, and substrate size all increase with body size of individuals. In lake ecosystems, adult Mountain Whitefish generally occur at depths of less than 20 m (66 ft) but exhibit seasonal changes in adult preferred habitats that are associated with temperature changes. Adults are generally found in shallow water in the spring, moving to deeper water as summer progresses, then returning to the littoral zone in the fall. After spawning, they again move to deeper water to overwinter[4].

Management and Conservation Status

Mountain Whitefish are considered common, widespread, and abundant range wide (G5) and statewide (S5)[15]. Mountain Whitefish are a recognized game fish in Idaho although often overlooked and under-utilized by anglers. Generally, anglers prefer trout to whitefish because of the perceived ease of hooking and the more spectacular fighting abilities of trouts. However, pickled, canned, or smoked, Mountain Whitefish are excellent eating[2].

Mountain Whitefish are widely distributed in Idaho's rivers and lakes, and they require clean, cold water. As a game fish, they readily take artificial flies or bait and put up a good fight when hooked. During the winter when most fishing activity slows down in Idaho, Mountain Whitefish provide some good fishing because of their active winter feeding habits.

Mountain Whitefish populations are adversely affected by similar factors that impact trout populations, including water management, channel degradation, water pollution, disease, and nonnative species interactions. Because they are geographically and physically isolated, Mountain Whitefish in the Big Lost River basin are genetically distinct from other Snake River populations. The IDFG, USFS, BLM, and other stakeholders have implemented a number of identified conservation actions proposed in a state management plan[36].

Diversions in the Big Lost River during Mountain Whitefish spawning, when movement is suspected to increase, will entrain more fish. After absorbing the egg sac, fry and juveniles may be entrained with high flows, thus reducing recruitment to adulthood. In the fall, when Mountain Whitefish are spawning, entrainment may increase because adults are migrating to spawning locations and juveniles are migrating from rearing locations[37].

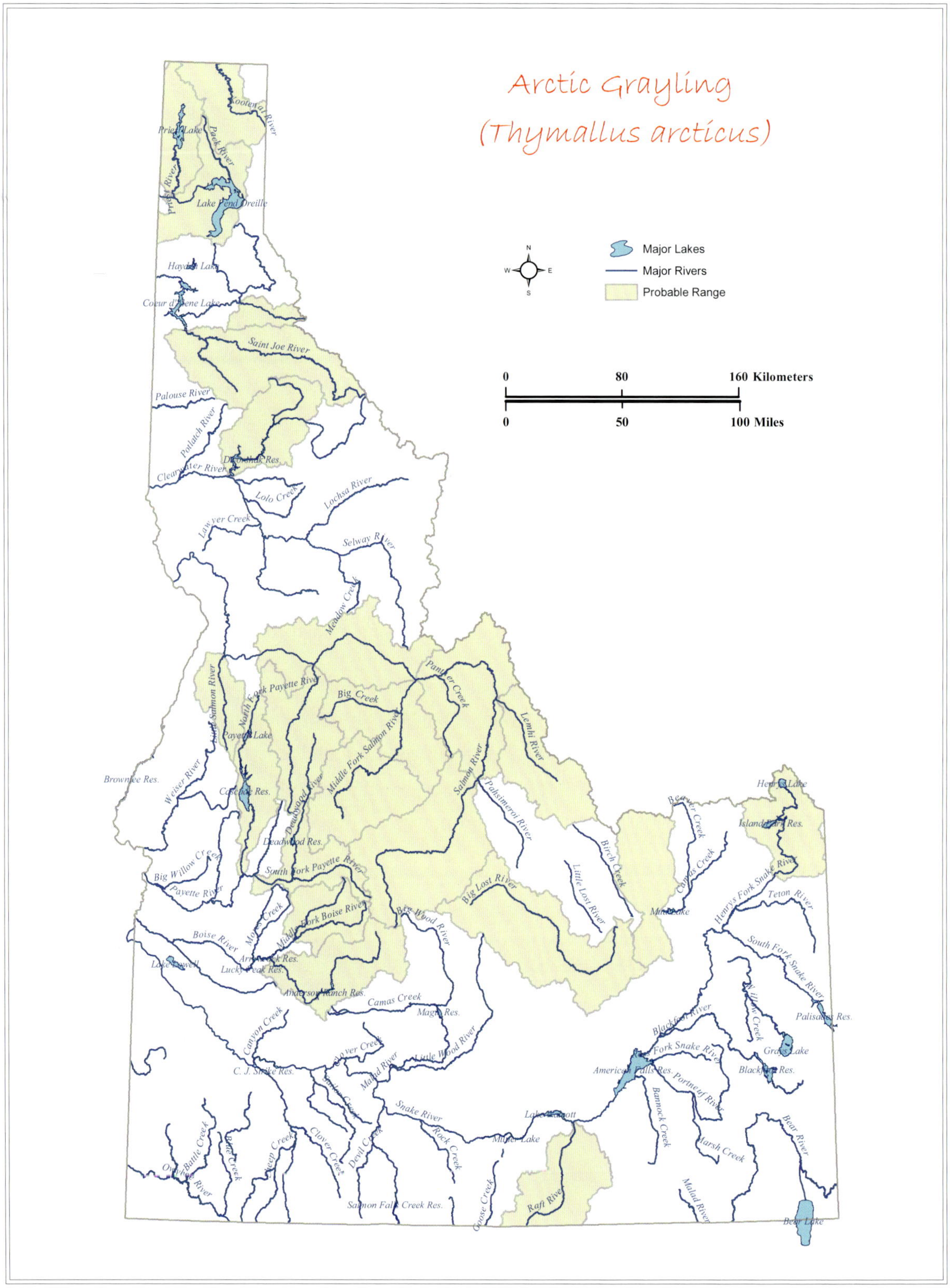
Arctic Grayling
(Thymallus arcticus)
Major Lakes
Major Rivers
Probable Range
N
W
E
S
0
80
160 Kilometers
0
50
100 Miles
Saint Joe River
Palouse River
Potlatch River
Lolo Creek
Lochsa River
Lawyer Creek
Selway River
Meadow Creek
Big Creek
Middle Fork Salmon River
Salmon River
Lemhi River
Pahsimeroi River
Birch Creek
Little Lost River
Big Lost River
Little Salmon River
Weiser River
Brownlee Res.
Cascade Res.
Deadwood Res.
South Fork Payette River
Payette River
Big Willow Creek
Boise River
Lake Lowell
Lucky Peak Res.
Anderson Ranch Res.
Camas Creek
Magic Res.
Big Wood River
Little Wood River
Canyon Creek
C. J. Strike Res.
Snake River
Rock Creek
Clover Creek
Salmon Falls Creek Res.
Goose Creek
Raft River
Blackfoot River
American Falls Res.
Portneuf River
Bannock Creek
Marsh Creek
Bear River
Malad River
Bear Lake
Teton River
South Fork Snake River
Henrys Fork Snake River
Grays Lake
Blackfoot Res.
Camas Creek
Beaver Creek

Subfamily Thymallinae

Graylings

Members of the Grayling subfamily are characterized by having more than 17 dorsal fin rays, teeth on the maxilla, orbitosphenoid absent, and suprapreopercular absent. One genus, *Thymallus*, inhabits freshwater in the Northern Hemisphere[1]. In Idaho one species represents this subfamily.

Arctic Grayling

Thymallus arcticus

(Pallas, 1776)

Etymology

Thymallus—thyme-like, referring to the reported odor of thyme associated with this species; *arcticus* = of the Arctic.

Importance

Arctic Grayling are valued by anglers for their remarkable beauty and the prestige associated with creeling them. However, most people have never seen an Arctic Grayling due to their distribution in remote areas[2]. Arctic Grayling have been taken by North American Native Americans for food for themselves and their dogs. Their habit of jumping, with the sail dorsal fin extended, makes them an attractive sportfish[5].

Description

Arctic Grayling are silvery to light purple on the sides and bluish white on the belly. They are a relatively slender fish and are most easily distinguished from trout, salmon, and char by the distinctive sail-like dorsal fin. The upper margin of this fin has gray and rose bands and is green with pink or red spots. Pectoral fins are blue and intermixed with pink at the tip. A black slash, similar to that of the Cutthroat Trout, runs along their chin[2].

Arctic Grayling have a lower jaw that is sometimes longer than the upper jaw. Teeth are rather small and occur on both jaws (premaxillary, maxillary, dentary), tongue, head of the vomer, and palatines. There are no basibranchial teeth. Gill rakers are of moderate size, usually 16 to 23, with five to seven on the upper limb and 11 to 15 on the lower limb. Branchiostegal rays are seven to nine. A dorsal adipose fin is present.

The high-sail dorsal fin is unique among salmonids and has 17 to 25 total rays. The caudal fin is broad and deeply forked with pointed tips. The anal fin is small with a square edge and 11 to 15 rays. Pelvic fins are abdominal with 10 or 11 rays, occurring under the center of the dorsal fin, the axillary process is small. Pectoral fins are low, long, and pointed, with 14 to 16 rays. Scales are cycloid and moderately large, and the lateral line is complete with 77 to 98 pored scales. Pyloric caecae are 14 to 21, and vertebrae are 58 to 62[5].

Range and Distributional Status

This species is not native to Idaho. Arctic Grayling have a Holarctic distribution and occur in northern freshwater ecosystems from Hudson Bay west to include all of Alaska, St. Lawrence Island, and the Bering Sea to the Kara River and Ob River of northern Eurasia. Grayling occur in the Missouri River headwaters above Great Falls, Montana. They are widely distributed across western Canada[5]. In Idaho, Arctic Grayling have been introduced into primarily alpine lakes in the Priest, Pend Oreille, Saint Joe, Snake, Salmon, and independent (Lost) drainages.

Continuing uncertainty is associated with the precise number and locations of Arctic refugia and their influence on extant patterns of phylogeographical variation in Arctic Grayling and other species. Recent investigations have helped to clarify some issues with respect to Arctic Grayling. Fossils, available genetic information, and historical distributions all point to Arctic Grayling surviving glaciation in a complex series of glacial refugia. Late Pleistocene fossils prove survival in the Yukon, the Great Lakes region, and southern Alberta, Canada. Genetic data suggest separate northern and southern Beringian refugia as well as Nahanni and upper Missouri refugia. Presence of several now extirpated populations in the Great Lakes region implies survival of Arctic Grayling in an eastern refugium, and isolated populations in the upper Missouri system in Montana indicate the existence of a Great Plains refugium[4]. For example, three Arctic Grayling distributions are related to specific geographic locations that served as refugia: the North Beringia lineage in the Brooks Range Refuge, the South Beringia lineage in the Yukon River Valley Refuge, and the Nahanni lineage in the Nahanni River Refuge[38].

The decline of genetic diversity with distance from the lower Yukon River suggests that post-glacial dispersal of Arctic Grayling from South Beringia into southern and eastern regions of North America occurred as the glaciers retreated during the Holocene and is consistent with theoretical expectations[38].

Size and Longevity

Arctic Grayling are slow growing and moderately short lived: most live less than six years, but a few may survive to age ten. Yearly length of grayling from Yellowstone National Park in mm (in) for fish aged one to six is 89 (3.5), 203 (8), 279 (11), 330 (13), 368 (14.5), and 394 (15.5). Average size is about 0.45 kg (1 lb)[2]. Nez Perce Lake produced the Idaho record Arctic Grayling in 1992 that weighed 1.1 kg (2 lb, 7 oz) and was 461 mm (18.1 in) long[13]. The all-tackle hook-and-line record for Arctic Grayling is 2.69 kg (5 lb, 15 oz) with a length of 759 mm (29.9 in) from the Katseyedie River, Northwest Territories, Canada, in 1967[12].

Food and Feeding

Arctic Grayling feed primarily on surface-drifting insects in streams. In addition to aquatic insect larvae, pupae, and adults, they feed on cladocerans and terrestrial insects. Arctic Grayling feed primarily at the surface to mid depths, rarely feeding on the bottom except when drift is reduced in the fall. They begin feeding earlier than many salmonids, having rapidly absorbed the yolk sac[2,39].

Limiting Factors

Arctic Grayling are particularly sensitive to habitat degradation. Arctic Grayling are easy to catch and may therefore be easily overfished[2,39].

Breeding Habits

Arctic Grayling spawn in the spring from April to June, based on water temperature. Spawning is initiated when water temperatures are in the 7.5 to 10°C (45 to 50°F) range. Similar to Lake Trout, Arctic Grayling do not build a nest or redd; rather, they spawn over gravel and rocky areas, and the eggs drop into crevices in the substrate. Mature eggs of Arctic Grayling are small, 2.5 mm (0.1 in) in diameter. Arctic Grayling have a higher relative fecundity than trout and salmon; about 11,000 to 13,000 eggs per kg (5,000 to 6,000 eggs per lb) of female body weight. Arctic Grayling embryos hatch rapidly, after about 15 days at water temperatures of 10°C (50°F), about half the time of Rainbow Trout and Cutthroat Trout. Newly hatched Arctic Grayling are only about 8 mm (0.3 in) in length, less than half that of trout and salmon[7].

Habits and Habitat

Arctic Grayling prefer water temperatures of 7.8 to 10°C (46 to 50°F) for spawning but live in warmer waters during parts of the year. While Arctic Grayling may live in lakes, they more commonly inhabit streams. Arctic Grayling stay in the open areas of streams, often in sight, even when approached. In the fall, Arctic Grayling leave tributary streams and overwinter in lakes or deep pools of streams[2].

Arctic Grayling display three separate life-history patterns: most common is a riverine form; however, both adfluvial (populations that migrate between lakes and streams) and lacustrine populations occur throughout most of the species' geographic range[4].

Management and Conservation Status

Arctic Grayling are considered common, widespread, and abundant range wide (G5). A state status rank is not applicable (SNA) because the species is not a suitable target for conservation activities[15]. Arctic Grayling can be easily overfished. Stocking should be in waters where there is a minimum or no competition from other species. In some Idaho aquatic ecosystems, this species provides a unique fishing opportunity in alpine lakes and is highly valued by anglers.

The upper Missouri River population has been proposed for protection as threatened or endangered under the Endangered Species Act (ESA)[40].

Subfamily Salmoninae

Chars, Trouts, and Salmon

Members of this subfamily are characterized by having fewer than 17 dorsal fin rays, with small scales, more than 110 along the lateral line; teeth on the maxilla; orbitosphenoid present; and suprapreopercular present. Worldwide, there are seven genera and about 30 species[1]. For broad-ranging presentations on the North American species, see Behnke[7,34].

For Idaho we report on 18 taxa in three genera and acknowledge that not all fishery professionals will agree with the subspecific treatments offered here for either Cutthroat Trout or Rainbow Trout.

Chars

Salvelinus alpinus oquassa	(Girard, 1854)	Sunapee Trout
Salvelinus confluentus	(Suckley, 1859)	Bull Trout
Salvelinus fontinalis	(Mitchill, 1814)	Brook Trout
Salvelinus namaycush	(Walbaum, 1792)	Lake Trout

Atlantic Trouts and Salmons

Salmo salar (extirpated in Idaho)	Linnaeus, 1758	Atlantic Salmon
Salmo trutta	Linnaeus, 1758	Brown Trout

Pacific Trout and Salmons

Oncorhynchus aguabonita	(Jordan, 1892)	Golden Trout
Oncorhynchus clarkii	(Richardson, 1836)	Cutthroat Trout
Oncorhynchus clarkii bouvieri	(Bendire, 1882)	Yellowstone Cutthroat Trout
Oncorhynchus clarkii behnkei	Montgomery, 1995	Snake River Fine-Spotted Cutthroat Trout
Oncorhynchus clarkii henshawi	(Gill and Jordan, 1878)	Lahontan Cutthroat Trout
Oncorhynchus clarkii lewisi	(Girard, 1856)	Westslope Cutthroat Trout
Oncorhynchus clarkii utah	(Suckley, 1874)	Bonneville Cutthroat Trout
Oncorhynchus keta (extirpated in Idaho)	(Walbaum, 1792)	Chum Salmon
Oncorhynchus kisutch	(Walbaum, 1792)	Coho Salmon
Oncorhynchus mykiss	(Walbaum, 1792)	Redband Trout / Steelhead
Oncorhynchus nerka	(Walbaum, 1792)	Sockeye Salmon / Kokanee
Oncorhynchus tshawytscha	(Walbaum, 1792)	Chinook Salmon

Salmonid Hybrids

Oncorhynchus clarkii × *O. mykiss*	Cutthroat Trout × Rainbow Trout	Cutbow Trout
Salmo trutta × *Salvelinus fontinalis*	Brown Trout × Brook Trout	Tiger Trout
Salvelinus fontinalis × *S. namaycush*	Brook Trout × Lake Trout	Splake
	Brook Trout × Bull Trout	

Chars—Introduction to the Genus *Salvelinus*

This interesting and diverse group of fishes includes both freshwater residents and anadromous populations in the Northern Hemisphere. Worldwide, there are nominally eight species, including Arctic Char, Dolly Varden, and Bull Trout[1]. In Idaho there are nominally four taxa in the genus *Salvelinus*. Some disagreement exists about the phylogenetic relationships between the observed populations.

In their landmark work "Fishes of North and Middle America," Jordan and Evermann spelled the common name as "charr. "Charre" is of Celtic origin from the sixteenth century and describes a blood-red color[34]. Simpson and Wallace include the Arctic Char (as *Salvelinus alpinus*) in their 1978 volume, but based on that account it is likely that these individuals were Sunapee Trout from New Hampshire, stocked between 1925 and 1928[41].

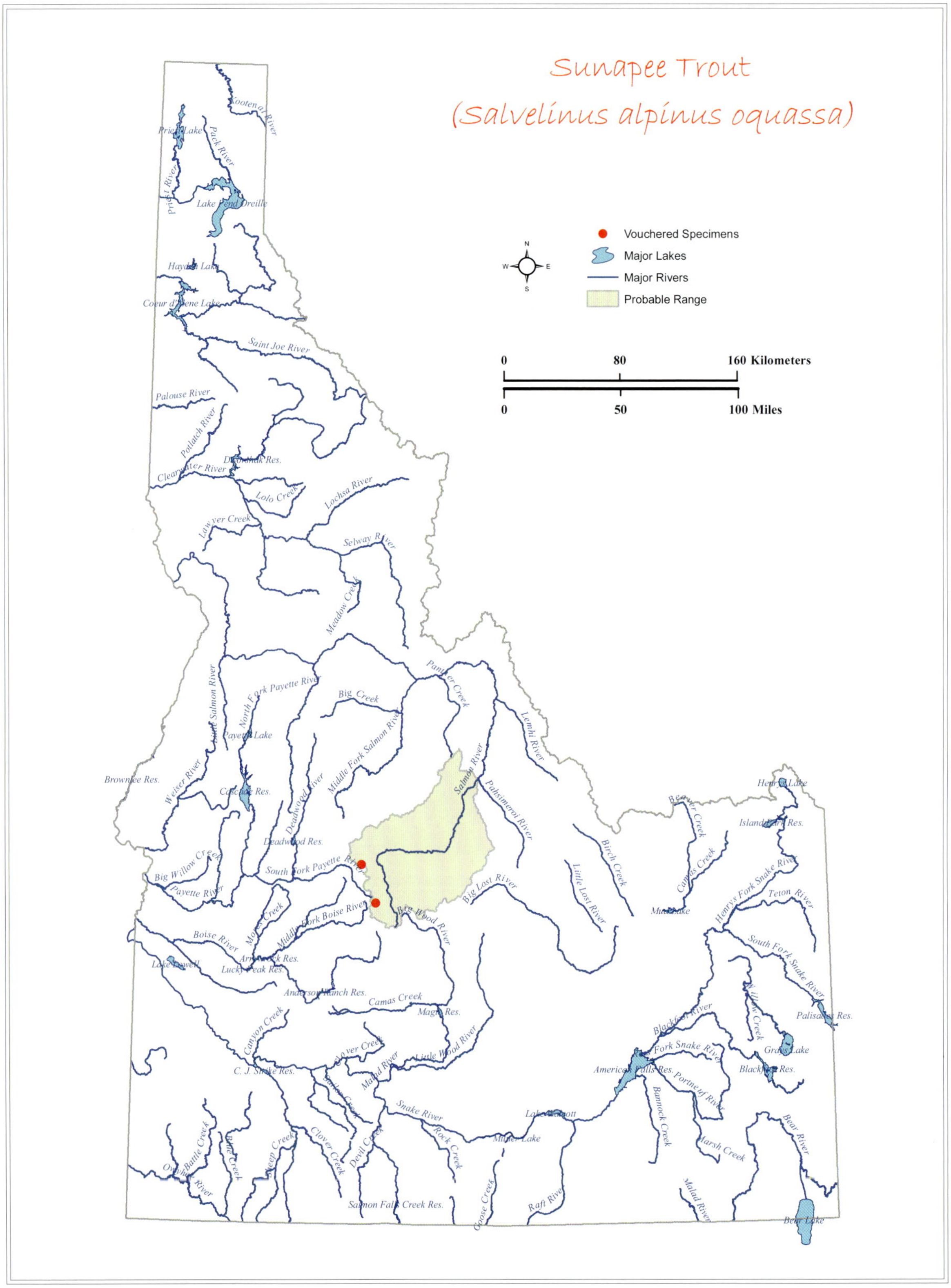
Sunapee Trout
(Salvelinus alpinus oquassa)
Vouchered Specimens
Major Lakes
Major Rivers
Probable Range
N
S
W
E
0
80
160 Kilometers
0
50
100 Miles
Pack River
Lake Pend Oreille
Saint Joe River
Palouse River
Potlatch River
Lolo Creek
Lochsa River
Selway River
Meadow Creek
North Fork Payette River
Big Creek
Payette Lake
Little Salmon River
Weiser River
Middle Fork Salmon River
Deadwood River
Lemhi River
Pahsimeroi River
Salmon River
Big Lost River
Little Lost River
Birch Creek
Camas Creek
Teton River
Henry's Fork Snake River
South Fork Snake River
Big Willow Creek
Payette River
Boise River
Big Wood River
Little Wood River
Camas Creek
Canyon Creek
Snake River
Rock Creek
Clover Creek
Bannock Creek
Marsh Creek
Bear River
Malad River
Raft River
Goose Creek
Salmon Falls Creek Res.
Blackfoot River
Portneuf River
Milner Lake
Brownlee Res.
Anderson Ranch Res.

Sunapee Trout

Salvelinus alpinus oquassa
(Girard, 1854)

Etymology

Salvelinus = an old name for char; *alpinus* = alpine (likely for its northern circumpolar distribution) (Linnaeus [1758] was credited for the original description of this taxon at the species level [i.e., *Salvelinus alpinus*]); *oquassa* = likely for Lake Oquassa, Maine.

Use and description of the name *oquassa* was published in 1854 for the Blueback Trout of Lake Oquassa, one of the Rangeley Lakes of Maine. The native char of the Rangeley Lakes was extinct by the early 1900s. The "golden trout" of Sunapee Lake, New Hampshire, was named *aureolus* in 1888 and still persists in Flood Pond, Maine[34].

In recent times, various ichthyologists have revived studies of relict Arctic Char. Agreement was reached that there are no consistent differences in taxonomic characters between Sunapee Trout, Blueback Trout, and Québec Red Trout. All should be grouped together and classified as one subspecies of Arctic Char: *Salvelinus alpinus*. Three names were originally proposed: *oquassa* in 1853, for Blueback Trout; *aureolus* in 1887, for Sunapee Trout; and *marstoni* in 1893, for Québec Red Trout. Following existing rules of taxonomy, the Sunapee Golden Trout, Blueback Trout, and Québec Red Trout are all classified as *Salvelinus alpinus oquassa*. This classification has been supported by quantitative genetic studies comparing protein systems that indicate virtually no differences among the relict Arctic Char[42]. A full if somewhat disjointed discussion of the original populations of Sunapee Trout is provided by Kendell[43], and a summation/review of the species is provided in Behnke's articles and books [7,34,42].

Original use of the subspecies *oquassa* was associated with a fish described from Lake Oquassa, Maine[34]. According to the USGS website for this species:[44]

(http://nas.er.usgs.gov/queries/FactSheet.aspx?speciesID=936) for more information.

> There has been considerable confusion and disagreement among taxonomists concerning the status of the Sunapee and blueback. For instance, some authors have treated each as distinct species, the blueback as *Salvelinus oquassa* and the Sunapee as *S. aureolus*. Quadri[45] suggested that the Sunapee and

blueback, along with the Quebec red (sometimes listed as *S. marstoni*), are conspecific and should be synonymized as a subspecies of the Arctic char (i.e., *S. alpinus oquassa*). According to Behnke both the Sunapee and blueback trout should be recognized as *S. aureolus oquassa*. Behnke later decided on a different name for this species. In his 2002 book, *The Trout and Salmon of North America*, Behnke[34] listed Sunapee Golden Trout as *Salvelinus alpinus oquassa*, stating that it likely arrived in New England by crossing the North Atlantic during preglacial times and is probably more closely related to European subspecies of *S. alpinus* than it is to other North American subspecies. Kendall[43] provided color plates showing both the Sunapee and Blueback Trout and gave a table to distinguish between the two forms. He also provided detailed descriptions and morphometrics. Scarola[46] provided a key for the Sunapee Trout. Everhart[47] gave a photo of Blueback Trout and a key to both Sunapee Trout and Blueback Trout.

Importance

This species is a representative of the group of alpine or northern chars, which at the time of its first description was known from no other place in the country than the Rangeley Lakes, Maine. It was later recognized as a close relative of certain Arctic forms. Even in the period when Kendall[43] was writing (1914), it was suggested that all the nominal species of these groups of fishes were really specifically identical but subject to considerable local variation in form, color, and habits[43]. In Kendall's documentation, he notes that the settlers prepared members of this species from the spawning beds in the fall for their use as food in the following winter and summer. Some were cured by salting, others by drying, and still others by smoking. In addition, Kendall reports that Girard, reporting in 1853, noted that the flesh of this fish is highly flavored and more delicate than that of the Brook Trout in Europe and America. Other individuals are quoted as saying that they were a nice pan fish, juicy, tender, and delicate[43].

The Idaho portion of the Sunapee Trout story actually begins in 1925 and involves three Idaho personalities. William Keil, the fish commissioner for the state of Idaho in the early 1900s, obtained Sunapee Trout eggs from the New Hampshire Fish and Game Department in exchange for fish eggs from Idaho. These fish were raised in Idaho hatcheries and stocked in four Idaho lakes. In 1977 Kent Ball, an IDFG biologist, while surveying mountain lakes for the state of Idaho discovered thriving populations of Sunapee Trout in two lakes, a fact uncovered or rediscovered by Dick Wallace[9] and reported to Behnke in conjunction with the Sunapee Trout article in the magazine *Trout Unlimited* in 1984[42].

Description

Coloration of *Salvelinus alpinus oquassa* is variable from lake to lake at different stages of sexual maturity. Yellow and orange spots on the sides and the ventral region are common, but in many ecosystems, particularly lake ecosystems, red is more characteristic. Background body coloration is silvery gray, tinted with blue. Toward the dorsal region, coloration darkens. This is an example of where local color variations have given rise to common names such as Golden, Silver, Blueback, and Red Trout. Spots on the body of this species range from pale white to bright red. The subspecies averages about 20 to 21 gill rakers, the lowest number found in any of the Arctic Char morphotypes. Vertebra average about 64, also at the lower end of the range for Arctic Char. Typical pyloric caeca accounts are about 35 to 40[34].

> *This is one the most rare and beautiful charrs in the world. It is indigenous to only four lakes and all in the New England states. It is supposed to have evolved from the Greenland charr and to have been carried*

down from the Arctic region during the last great glacial disturbances. I will tell you in beauty of form and delicate coloring it surpasses all other trout and charrs. No cut or even color plates can do justice to the dainty pastel shades of coloring with which this fish is clothed. In the adult form, specimens fresh from the water show the following coloration: back a dark olive, blending into cream colored sides with bellies of pale lemon. Each scale on the sides seems flecked with pure gold. The lower fins are yellow with faint indications of pink in margined with white bands. The immature examples appear not unlike a yearling eastern brook with pronounced finger marked blotches across the sides. The adult fish has indistinct markings on the back with numerous orange yellow spots on the sides. Many attempts been made from time to time to establish this charr in other water thought suitable for its development and reproduction, but no apparent results have ever followed such introductions with the exception of those planted by the Idaho Fish and Game Department into several vacant lakes of the Sawtooth country[41].

Range and Distributional Status

This species is not native to Idaho. Sunapee Trout were native to Sunapee Lake, New Hampshire; Averill Pond, Vermont; Big Dan Pond, New Hampshire; and Floods Pond, Maine[34]. Genetic studies on this and related species may enlighten both past and future stocking efforts in Idaho and elsewhere[48].

In the 1925, eggs of the Sunapee Lake char were shipped to Idaho. The resulting fry were stocked in four lakes in the Sawtooth Mountains. Brook Trout were later stocked, and only one population is known to have persisted and now includes Brook Trout × Sunapee Trout hybrids. In this cold mountain lake in Idaho, unlike the conditions in most New England lakes, the Sunapee Char reaches sizes of 457 to 483 mm (18 to 19 in), a greater length than the Brook Trout[34].

Through an exchange arranged by the writer with the Fish Commissioner of New Hampshire, 5000 eggs of this species were received from Lake Sunapee, New Hampshire, in March, 1925. They were hatched at the Hayspur State Hatchery and, as three-inch fingerlings, were planted in four lakes previously selected. In August 1928, these fish were in evidence in three of lakes planted, and several specimens taken. Fishing for them has been discouraged as much as possible until they have had an opportunity to spawn, which should be during the fall of 1928. This is an interesting experiment and probably would've been followed by failure had these lakes been inhabited by fish of any kind. Little will be known of what coloration or form they may assume in their new environment until another season has passed and it reached physical maturity. This charr spawns on rocky reefs in the lake itself and does not require running streams or gravel beds for successful reproduction[41].

This species of char (now known to be Sunapee Trout, *Salvelinus alpinus oquassa*) was introduced into the Snake River drainage below Shoshone Falls, Idaho[49]. This species was introduced into several lakes at the headwaters of the Salmon River in Idaho, including Alice, Vernon, Big Redfish, and Sawtooth Lakes[44,48,50,51]. Regional Fisheries Management Investigation Reports from Idaho Fish and Game for the period July 1, 1987, to June 30, 1988, note that eight lakes in the Redfish Lake Creek drainage were surveyed for Sunapee Trout (*Salvelinus alpinus oquassa*), but the species was documented only in alpine lakes. Sunapee Trout had also been documented in Alice Lake in previous surveys[52]. Documentation also included input from Robert Behnke, to whom specimens were shipped, confirming that these were Sunapee Trout.

Size and Longevity

Sunapee Trout typically reach lengths of 254 to 381 mm (10 to 15 in) and weights of 198 to 454 g (7 to 16 oz). They may live to age seven or eight. In some ecosystems, with enhanced prey bases, Sunapee Trout have been reported at 4 kg (8 lb, 13 oz)[34]. Kendall provides a quotation from Jordan and Evermann's *American Food and Game Fishes* (1904) that this species formerly attained a length of only about 150 to 250 mm (6 to 10 in) and four to six to the pound; they are now caught weighing as much as two pounds and are known only from the Rangeley Lakes[43].

No record individuals for Sunapee Trout have been officially cataloged, either in Idaho or in the IGFA online system. Reports of a 910 mm (36 in) and 9 kg (20 lb) fish have been noted[34].

Food and Feeding

Kendall notes that the only mention of food of this species is the statement that their teeth were very fine and plentiful and that they evidently live on ground feed, which is plentiful in the lakes and is explained as *"an important element with all fish, being composed of insectivorous varieties and largely of viscous matter, which settles profusely"*. The stomachs of specimens examined in the 178 to 229 mm (7 to 9 in) range were full of insect larvae of various kinds and a large number of Entomostraca[43].

Limiting Factors

Limiting factors for this species are unknown but likely similar to other small members of this group. Many individuals are infested with a variety of parasites, as are most trouts, but given that they are present in Idaho in only small, isolated lakes at high altitudes, other mortality factors are likely limiting.

Breeding Habits

Girard's accounts of the habit of this species, as reported by Kendall, state that this species remains in the deep water of the lake from near the middle of November until the middle of October when individuals ascend brooks in countless numbers to spawn, the run lasting about three weeks. Apparently, this species ascends the spawning tributaries in pairs, male and female together, using spawning beds that have previously been cleared and vacated by Brook Trout. These runs occur at night, but in the height of the season many fish remain up throughout the day, hidden beneath rocks and stumps[43].

Habits and Habitat

In their native range, Sunapee Trout were described from a variety of small lake or pond ecosystems. In Idaho they have been planted in a variety of small high mountain lakes. Little is known about the behavior or habits of this species. From the reports listed in Kendall[43], it may be surmised that this species was fairly secretive in its native ecosystems in Maine, being observed and caught only during very limited times of the year, primarily during the spawning season.

Management and Conservation Status

Sunapee Trout are ranked as imperiled globally (G5/T2) due to restricted range. In Idaho a conservation ranking is not applicable since these fish are introduced[15]. Investigations into the high mountain lake ecosystems in which Sunapee Trout exist should be continued. Documentation of both spawning habits and age and growth information would allow appropriate management for this unique species in its Idaho setting[48].

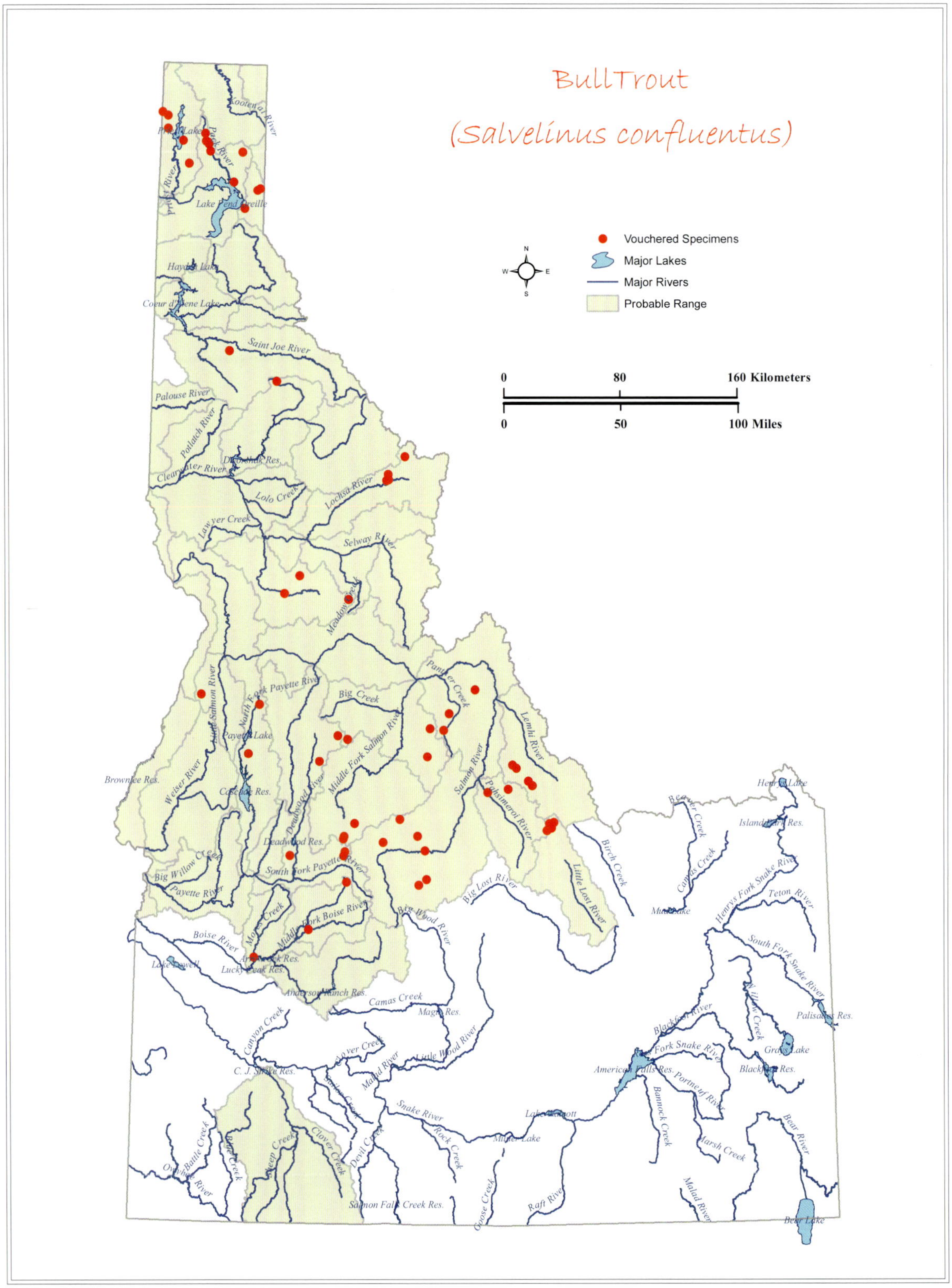
BullTrout
(Salvelinus confluentus)
Vouchered Specimens
Major Lakes
Major Rivers
Probable Range
0
80
160 Kilometers
0
50
100 Miles

Bull Trout

Salvelinus confluentus

(Suckley, 1859)

Adfluvial

Resident

Etymology

Salvelinus = an old name for char; *confluentus* = flowing together.

For many years Bull Trout and Dolly Varden were confused with one another and often considered as one species. They are very similar in external appearance and in the number of gill rakers, vertebrae, and pyloric caecae. Cavender's 1978[53] paper provided adequate evidence for separation of the two species. Subsequent genetic analysis has verified that Bull Trout and Dolly Varden are distinct species in the genus *Salvelinus*[11]. Page et al.[3] list both species separately in the 2013 publication. As McPhail[4] points out, the relationships between Bull Trout and other North American char are complex, and the relationships have been confounded by periods of range fragmentation through glaciation and accompanying refugia isolation.

The relationship between Bull Trout and other North American chars (especially Dolly Varden and Arctic Char, *Salvelinus alpinus*) is complex. Detailed accounts have been presented by various authors[53–55].

Description

Bull Trout coloration is highly variable. A Bull Trout's body is mostly gray infused with green, but lake populations may be silvery gray. Bull Trout are covered with relatively small white or pale yellowish spots that are intermingled with pink or red spots. Ventrally, coloration can change from white to orange. Lower fins may develop orange and red colors. Bull Trout typically have 15 to 19 gill rakers, 63 to 66 vertebrae, and 22 to 35 pyloric caecae. Dorsal and caudal fins of Bull Trout as well as Bull Trout × Brook Trout hybrids have coloration that is not as pronounced as those in Brook Trout. Bull Trout's lower fins have only white anterior borders[34].

Bull Trout are char with pale pink, lilac, or red spots along the flanks. Often confused with Dolly Varden where the two species overlap, the two species differ in head shape, upper jaw length, and other features. Where they coexist, the two species have dramatically different life histories. Dolly Varden do not occur in Idaho[4].

In most populations of Bull Trout, breeding males develop a hooked lower jaw (kype) that may, in some extreme cases, fit into an indentation on the tip of the snout. Breeding males are usually more vividly colored than females, especially the markings on the paired fins. Outside of the breeding season, sexes are difficult to distinguish, although males generally have larger adipose fins than females of the same size[4].

Bull Trout, similar to other char, are distinguished from other salmonids by having light spots on a dark background and no teeth on the roof of the mouth. Bull Trout in Idaho are distinguished from other char in Idaho by having no spotting on their dorsal or caudal fins. Young Bull Trout have eight to 12 dark-colored parr marks on the sides[6].

Range and Distributional Status

This species is the only char native to Idaho. Historical native range and overall distribution of Bull Trout are somewhat obscured due to the taxonomic confusion with Dolly Varden. Bull Trout are native to western North America and are found in the headwaters of the Mackenzie River and Yukon River in the north as well as other coastal rivers in British Columbia, that drain to the Pacific Ocean, Puget Sound, the Columbia River basin, and west to Montana. They occur in the Klamath River basin of Oregon and the Jarbidge River, Idaho, and in north-central Nevada[6]. Although there is no evidence to indicate that they survived to modern times, there is a possibility, based on fossil fragments, that Bull Trout occurred in the Bonneville Basin of Idaho, Utah, and Wyoming[34].

In Idaho, Bull Trout occur in drainages of the Snake River below Shoshone Falls including the Boise River, Payette River, Weiser River, Jarbidge River, Salmon River and Clearwater River, and drainages of northern Idaho including the Kootenai River, Pend Oreille River, and Coeur d'Alene River. A population of Bull Trout is found in the Little Lost River basin above Shoshone Falls. It was originally thought that the Bull Trout in the Little Lost River basin were native and that they had been naturally established through a headwater stream capture event from the Salmon River basin. However, more recent research indicates Bull Trout are not native to the Little Lost River basin, or any of the other Lost Streams[56]. This work suggests Bull Trout may have been introduced into the Little Lost River basin in the late 1800s or early 1900s either through stocking by settlers or through a canal system that originated in the Pahsimeroi River basin[56].

We have removed the Bull Trout occurrence in the Beaver-Camas drainage indicated in the *Fishes of Idaho*[9]and *Native Fishes of Idaho*[6]. Since the publication of these texts, an August 23, 1977 letter from James Simpson tio Richard Wallace has been found that changes the Beaver Creek occurrence from the

Beaver-Camas to the Middle Fork Salmon River drainage. Why this correction was not made prior to the publication of *Fishes of Idaho*[9] in 1978 and in the 1982 second edition is unknown.

Size and Longevity

Adult Bull Trout may achieve sizes exceeding 9 kg (20 lb), providing adequate prey species are available. While longevity is variable, Bull Trout tend to live to ages that exceed most other salmonids. Resident individuals may be eight to 10 years old and only 250 mm (9.8 in) long, while adfluvial populations may exceed 12 years[6].

A distinct size differential exists between the resident form of Bull Trout found in small streams versus those migratory forms inhabiting lakes (adfluvial) and large rivers (fluvial). In small streams they may be 203 to 254 mm (8 to 10 in) long and reach weights of 91 to 170 g (4 to 6 oz). In larger streams and lakes they may reach sizes of 305 to 457 mm (12 to 18 in) and weights up to 1.1 kg (2 lb, 7 oz). Maximum sizes are 711 to 914 mm (28 to 36 in) and weights of 2.3 to 9 kg (5 to 20 lb)[34].

In Idaho the record Bull Trout caught by rod and reel was taken from Lake Pend Oreille in 1949. This fish stands as the world record for this species at 14.51 kg (32 lb)[12].

Food and Feeding

Bull Trout are opportunistic feeders with voracious appetites. Young Bull Trout feed on aquatic and terrestrial invertebrates in streams but will convert to fish as they get larger. Large Bull Trout will feed on other species of trout, char, salmon smolts, whitefish, minnows, suckers, and sculpins[6].

In a 2011 study of Bull Trout and Westslope Cutthroat Trout diets, analysis of Bull Trout stomach samples revealed that 74 percent of the prey items consumed by Bull Trout were fish; there was no significant correlation between prey length and predator body size. An additional 20 percent of the Bull Trout diet was composed of aquatic invertebrates, mainly Ephemeroptera (13 percent). In contrast, stomach samples from Bull Trout in tributaries revealed a diet composition that was considerably different. Stomach samples from Bull Trout in tributaries were dominated by aquatic invertebrates (87 percent of prey items)—mainly Trichoptera (46 percent), Ephemeroptera (18 percent), and Diptera (17 percent)—but also showed no significant correlation between prey length and predator body size. Terrestrial insects accounted for an additional 11 percent of the prey items consumed by Bull Trout sampled in tributaries, whereas fish prey contributed only 2 percent[57].

Limiting Factors

The ongoing fragmentation and isolation of Bull Trout habitat has led to reductions in population size and presumably in effective population size for many populations. A resulting loss of genetic variation can influence the dynamics and persistence of populations through at least three mechanisms: inbreeding depression, loss of phenotypic variation and plasticity, and loss of evolutionary potential. Loss of genetic variation in populations with small effective population size may experience reduced fitness through the so-called inbreeding effect of small populations and lead to an accelerating decline toward extinction in a process termed "extinction vortex." Because of the importance of effective population size, the so-called

50/500 rule has emerged as general guidance in conservation management. The generally accepted view is that an effective population size of less than about 50 is vulnerable to the immediate effects of inbreeding depression and a minimum of 500 individuals is needed to reduce genetic drift" after depression[58].

Some believe that Brook Trout (and other introduced species) are one of the primary causes of decline in Bull Trout throughout its range. Anecdotal evidence of a widespread loss of native species precipitated by Brook Trout seems strong. Brook Trout now occur in most of the US Geological Survey 5th Code (~15,000 ha [37,066 ac]) watersheds of the interior Columbia River basin. Managers throughout the region have experimented with Brook Trout removal and installation of in-stream barriers in hopes of stemming Brook Trout invasions[59].

Empirical evidence is less clear. Some studies have demonstrated a progressive shift in numerical dominance from Bull Trout to Brook Trout at sampling sites in the zone of sympatry, but a population-level effect could not be discerned. Others have found evidence of direct interaction between Brook Trout and native species, but the results are variable and some invasions appeared to be stalled. It appears that co-occurrence of Brook Trout and native salmonids within individual streams may be common. Although co-occurrence does not imply indefinite coexistence, it does suggest that elimination of native forms is not inevitable throughout the range of streams accessible to Brook Trout[59]. Bull Trout can be adversely impacted by a variety of anthropogenic factors, among them nonnative species stocking (particularly Brook Trout), increased sediment loads, and increased water temperatures[34].

Breeding Habits

Age at first spawning for Bull Trout is four to seven years. Spawning occurs primarily in headwater streams from September through October when water temperatures fall below 9°C (48°F). Females produce 1,300 to 8,000 eggs, depending on size. Both fluvial and adfluvial Bull Trout populations depend on tributaries for spawning and juvenile rearing habitat. Seasonal movements of Bull Trout to spawning areas may exceed 160 km (99 mi)[6].

In a 2011 Salmon River study, home ranges of Bull Trout and Westslope Cutthroat Trout were of similar size and often included similar habitats (for instance, the use of common spawning tributaries and wintering locations). Timing of movements throughout these home ranges was dissimilar. Bull Trout movement was limited during the winter, and some individuals remained within the same pool for long periods. Renewed movement generally began in April and May as water temperature and flow began to increase. Bull Trout began entering tributaries on the descending limb of the hydrograph and were often in proximity to spawning areas by late July. Movement in August was significantly less than that in all other summer months besides April. During this time, Bull Trout remained near spawning locations. This behavior likely reflects an absence of the appropriate environmental cues to initiate spawning activity. Spawning began in late August and was generally completed by mid-September. Bull Trout remained in tributaries for two to three months before spawning. After spawning activity ceased, Bull Trout made rapid migrations downstream to the Salmon River[57].

Generally, Bull Trout spawning is thought to occur later in the fall (October–November), but by returning to the Salmon River by mid-September Bull Trout may be able to capitalize on two seasonally abundant food resources. In this time frame spawning activity of Chinook Salmon peaks in the upper Salmon River, and a secondary peak in the seaward migration activity of juvenile Chinook Salmon occurs. The presence of spawning Chinook Salmon (and subsequently their eggs and decaying carcasses) and the out-migration of Chinook Salmon smolts may influence Bull Trout migration patterns and timing of spawning activity. After returning to the Salmon River, Bull Trout moved to over-wintering locations during September to November. Average monthly movement of radio-tagged Bull Trout decreased during December–February; this decrease may be related to lower metabolic rates associated with decreased water temperature[57].

In a British Columbia study, redds were found fairly close to cover; spawning occurred at temperatures below 10°C (50°F); average nose velocity at the upstream edge of the redds was 0.39 to 0.45 m/s (1.3 to 1.5 ft/s). Large females, in the 350 to 650 mm (14 to 26 in) range, produce up to 9,000 eggs that are about 4.0 to 6.3 mm (0.16 to 0.25 in) in diameter. Smaller females in the < 250 mm (9.8 in) range produce fewer than 1,000 eggs. Eggs hatch in 119 to 126 days at temperatures of 2°C (36°F), and survival to hatching drops significantly at temperatures above 8°C (46°F). Wild newly emerged fry are about 25 mm (1 in) total length[4].

Similar to all other char, Bull Trout spawn in the fall, and, except for stream-resident populations that spawn locally, a migration typically proceeds spawning activity. British Columbia fluvial populations usually begin migrating from larger rivers into smaller streams in late August when temperatures are high and water levels are low. Round-trip distances of these spawning migrations may exceed 400 km (250 mi). In late summer or early fall, adfluvial populations migrate from lakes or reservoirs into spawning streams, and some of these migrations may be very short, perhaps less than 10 km (6 mi), while others may be substantial, ranging up to 100 km (62 mi)[4].

Habits and Habitat

Bull Trout populations occur in both large and small streams and in lakes with clean, cold water in the 10 to 12°C (50 to 53.6°F) range. This preference for cold-water habitats generally limits Bull Trout to headwater sections of large rivers and their tributaries and deep cold lakes and reservoirs. Large submerged woody debris and undercut banks are additional important habitat elements[6].

Bull Trout likely have important effects on the structure of communities and the flow of energy and nutrients in the ecosystems they inhabit. Through their roles as predators on invertebrates and other fishes, these char have the capacity to indirectly regulate organisms at lower trophic levels. For instance, two Japanese studies showed that when terrestrial invertebrate prey were not available, Dolly Varden intensified their foraging on benthic invertebrates, which triggered an increase in the growth of algae but also reduced the emergence of adult aquatic insects and the abundance of spiders in the riparian forest. Studies like these have not been conducted for Bull Trout, but similar indirect effects on algae have been described for Brook Trout in a Canadian study. If predation by these char can regulate prey fish populations in lakes, they could indirectly control phytoplankton dynamics, as has been described for many other piscivorous fishes, including Lake Trout and Arctic Char. In addition, these char, through their migratory life histories, can play roles yet to be described to link the food webs of multiple habitats, and they may also transport energy and nutrients, as has been found for other migratory fishes. These processes indicate that char play important ecosystem roles, and local extirpations or declines in these species may have much wider impacts than is commonly recognized[60].

Bull Trout are cold-water species, rarely encountered in ecosystems where temperatures exceed 15°C (59°F) for prolonged periods of time. Habitat use by Bull Trout fry, juveniles, and adults is strongly influenced by life-history type and the presence or absence of other species. As with most fish, Bull Trout shift habitats as they increase in size[4].

Regardless of life-history type, most Bull Trout rear in streams for periods of two to four years, with seasonal and diel shifts in the microhabitats used. Juvenile Bull Trout have a strong association with pools and deep side channels. At night Bull Trout emerge from cover and shift into shallow waters away from cover. Juveniles in adfluvial populations migrate to a lake in their third or fourth year. In lakes juvenile Bull Trout are rarely found in the littoral zone[4].

Management and Conservation Status

Bull Trout are ranked as vulnerable to extinction or extirpation range wide (G3) and statewide (S3)[15]. Bull Trout were listed as a threatened species under the federal Endangered Species Act in 1998. Bull Trout populations have declined across most of the native range due to habitat loss and degradation, overharvest, and loss of stream connectivity. The presence of nonnative species has decimated some populations through competition, predation, and hybridization. There is evidence in Idaho that some populations are expanding in size and range (e.g., Salmon River basin), probably due to a combination of the no-harvest angling rules and proactive conservation management occurring in many parts of the basin[61].

An important point for consideration in management of Bull Trout populations is that maintaining the natural connections and potential for gene flow among populations can be critical. Fishery managers should seek to conserve the full expression of life history and processes influencing natural dispersal and gene flow among populations. Removal of barriers to migration, restoration of habitat conditions in migratory corridors, and the maintenance of at least some populations large enough to act as sources for dispersal is important[58].

Many case studies suggest that even large populations can become highly vulnerable if present conditions change. For example, Bull Trout in Flathead Lake and the Flathead River system in northwestern Montana were once very abundant, and it was thought the population was secure. These populations quickly crashed in the early 1990s, due to major ecosystem changes and cascading food-web interactions as a result of the introduction of a single nonnative invertebrate species, the Opossum Shrimp *(Mysis relicta)*. This introduction disrupted trophic relationships between native (Bull Trout and Westslope Cutthroat Trout) and nonnative fishes (Lake Trout, Kokanee, and Lake Whitefish) that had been relatively stable for nearly a half century prior to the *Mysis* introduction[60].

Idaho has not allowed the harvest of Bull Trout since 1996. The lack of black spots on the dorsal fin of Bull Trout has led to the fishery managers' advice to anglers: "No black, put it back."

"It has always been my private conviction that any man who pits his intelligence against a fish and loses has it coming."

John Steinbeck

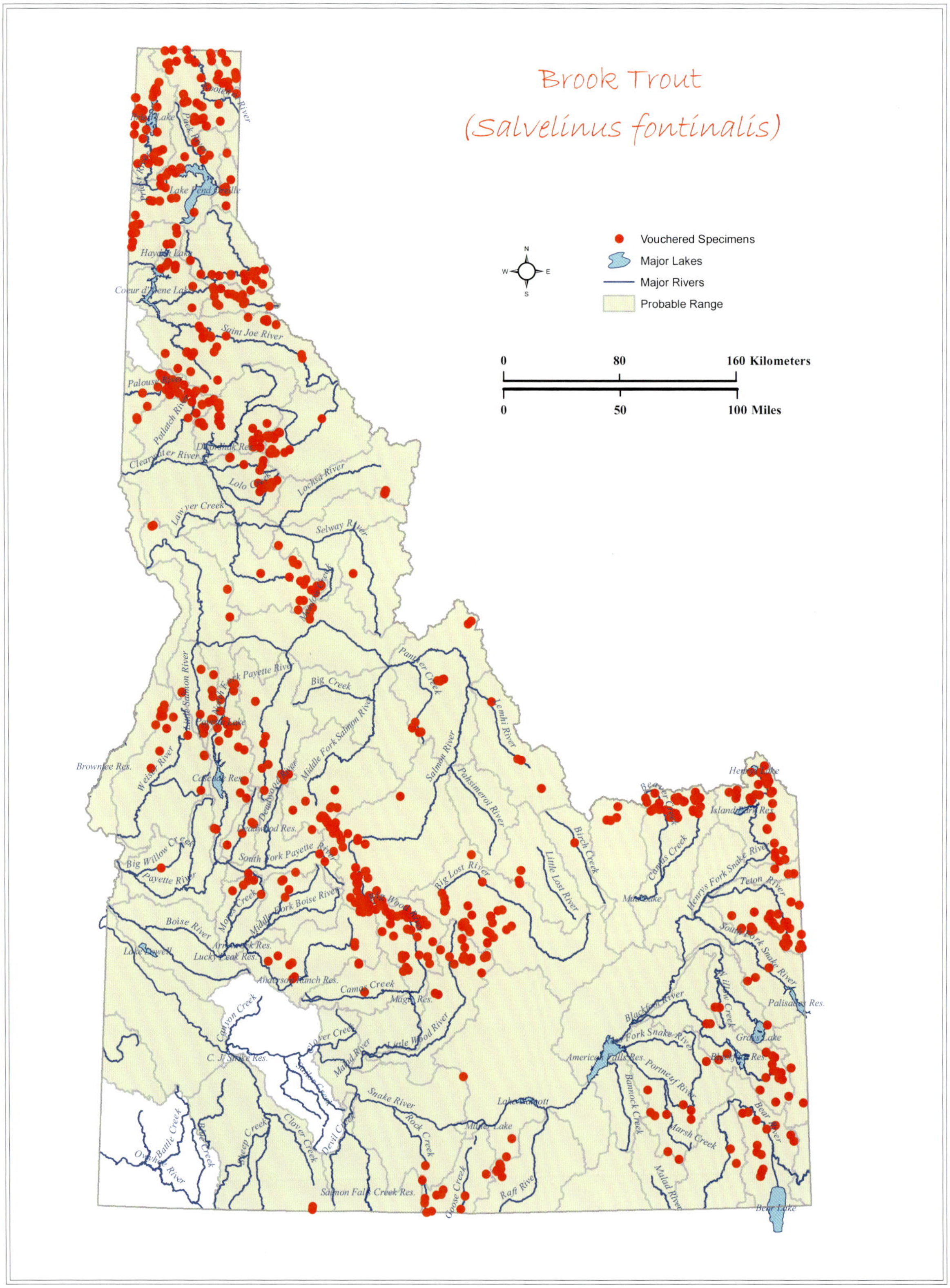
Brook Trout
(Salvelinus fontinalis)
Vouchered Specimens
Major Lakes
Major Rivers
Probable Range
0
80
160 Kilometers
0
50
100 Miles
Saint Joe River
Palouse River
Potlatch River
Clearwater River
Lochsa River
Selway River
Lawyer Creek
Panther Creek
Big Creek
Lemhi River
Middle Fork Salmon River
Salmon River
Pahsimeroi River
Birch Creek
Little Lost River
Big Lost River
Brownlee Res.
Weiser River
Cascade Res.
Deadwood Res.
South Fork Payette River
Big Willow Creek
Payette River
Boise River
Middle Fork Boise River
Arrowrock Res.
Lucky Peak Res.
Lake Lowell
Anderson Ranch Res.
Camas Creek
Magic Res.
Little Wood River
C. J. Strike Res.
Snake River
Rock Creek
Goose Creek
Raft River
Salmon Falls Creek Res.
Lake Walcott
Milner Lake
American Falls Res.
Portneuf River
Bannock Creek
Marsh Creek
Malad River
Bear River
Bear Lake
Grays Lake
Blackfoot River
Palisades Res.
South Fork Snake River
Teton River
Henrys Fork Snake River
Island Park Res.
Henrys Lake
Camas Creek
Mud Lake
Beaver Creek
Hayden Lake
Coeur d'Alene Lake
Lake Pend Oreille
Priest Lake
Dworshak Res.
Lolo Creek

Brook Trout

Salvelinus fontinalis

(Mitchill, 1814)

Male

Female

Etymology

Salvelinus = an old name for char; *fontinalis* = living in springs.

Importance

Mitchill's original description includes glowing reviews of how great a sportfish Brook Trout were in New York in the 1800s, even though the size of the fish creeled was small[2]. Brook Trout are one of the most highly esteemed sportfishes in North America and have been pursued with all types of fishing gear, from artificial flies to spinners and live bait[5]. While Brook Trout may not be an important sportfishing component in any given ecosystem, they are popular because they often occur where no other trout species are present, particularly in high mountain lakes and streams. Brook Trout are one of the few salmonids that can provide self-sustaining populations in lakes with no tributary streams, making them a valuable tool in sport fisheries management[2] and increasing opportunities for fishing enthusiasts.

Although Brook Trout have a long history of being stocked as a sportfish in Idaho, they have been identified as a species that has contributed to the decline of many native Cutthroat Trout populations in the state. Much recent effort has been undertaken to develop methods to remove Brook Trout from areas where they have caused the decline of Cutthroat Trout populations through competition and predation[62].

Description

Brook Trout are highly variable in coloration. The back is usually olive green to dark brown or even black, while the sides are lighter with a white ventral surface. Distinctively, Brook Trout have vermiculations or wavy lines that are cream colored to light green color on the sides. There are small discrete red spots surrounded by bluish halos on the sides. Anal, pelvic, and pectoral fins have an immaculate white leading edge followed by a black stripe, a characteristic that helps distinguish them from trout. Brook Trout have heads that are large but not long, and the body is about five times as long as it is deep[2,5].

Mature Brook Trout males have longer paired fins and a longer upper jaw than females. Breeding males also develop a small kype or hooked lower jaw. During spawning both sexes develop bright red sides, and the paired fins and anal fin are red with a conspicuous white leading edge, followed by a thin band of black pigment. All of these colors are more vivid in males than in females[4].

Brook Trout typically have elongated bodies that are only moderately laterally compressed. There are well-developed teeth on the upper and low jaws (premaxillary, maxillary, and dentary), on the head of the vomer but not on the shaft, on the palatines, and on the tongue in two rows. Basibranchial teeth are usually lacking. Gill rakers are 14 to 22 (four to seven on the upper arch and 10 to 15 on the lower arch). Branchiostegal rays are nine to 13, usually fewer on the right side: 10 to 13 on the left side, nine to 12 on the right side. A dorsal adipose fin is present. Dorsal fin rays number 10 to 14, and the caudal fin has a shallow fork. Major anal rays are nine to 13, and pelvic rays are eight to 10, with a distinct pelvic axillary process. Pectoral fin rays are 11 to 14. Brook Trout scales are small and cycloid, in about 230 rows at the lateral line, which is straight with 110 to 130 pore scales. Pyloric caecae average 23 to 55 but typically fewer than 50. Vertebrae are 58 to 62[5].

Range and Distributional Status

This species is not native to Idaho. Brook Trout are native to northeastern North America, from the North Atlantic seaboard south to Cape Cod, in the Appalachian Mountains south to Georgia, west of the upper Mississippi and Great Lakes to Minnesota, and north to Hudson Bay. Brook Trout have been stocked in South America, New Zealand, Asia, and Europe. They were stocked extensively in the western United States since at least 1875[2] and in Idaho since at least 1908[63]. Brook Trout are ubiquitous in Idaho, with populations established in most waters of the state except in the far southwestern corner.

Size and Longevity

Brook Trout growth is more variable than that of other char. Abundances of prey, coupled with moderate temperatures, result in rapid growth in most ecosystems. In cold high mountain ecosystems, Brook Trout may reach a size no greater than 203 mm (8 in). Optimum temperatures for growth are 12.8 to 15.6°C (55 to 60°F)[2].

While there is some variability among ecosystems and populations, few Brook Trout live past age four; exceptions up to 15 years of age have been reported. Where adequate food is not available (infertile waters), Brook Trout populations will stunt and overpopulate the ecosystem, reducing the value for sportfishing[11].

Brook Trout over 4.5 kg (10 lb) are rare but occasionally reported. In a Canadian freshwater population, the following standard lengths in mm (in) were reported for fish aged 0+ to 5+: 39 (1.5), 109 (4.3), 156 (6.2), 178 (7.0), 224 (8.8), and 274 (10.8)[5].

Two ecologically distinct forms of Brook Trout have been identified: a smaller short-lived form that lives three to four years and reaches sizes of 203 to 254 mm (8 to 10 in) and resides in small cold streams and lakes; and a larger form that lives eight to 10 years, reaches weights of 4.1 to 6.4 kg (9 to 14 lb), and is associated with large lake, stream, and estuary ecosystems. Most Brook Trout are not fast growing, nor are they long lived. Yearly total lengths of a stream population, in mm (in), were 61 mm (2.4), 127 mm (5.0), 180 mm (7.1), 234 mm (9.2), 274 mm (10.9), and 328 mm (12.9)[2].

Idaho's record Brook Trout was taken from Henrys Lake in 1988 and weighed 3.2 kg (7 lb, 1 oz) and was 597 mm (23.5 in) long[13]. Nipigon River, Ontario, Canada, holds the all-tackle world record of 6.57 kg (14 lb, 8 oz), caught in 1916[12].

Food and Feeding

Brook Trout have the reputation for eating anything; they are carnivorous, voracious feeders. Brook Trout eat insects throughout their lives when other prey is not available. Young in streams feed primarily on plankton. A 25.4 mm (1 in) Brook Trout may live on entomostracans, while 50.8 mm (2 in) individuals appear to prefer midges and 102 mm (4 in) individuals seek mayfly nymphs and caddisfly larvae. For larger individuals the dominant food is aquatic insects, including caddisfly (Trichoptera) larvae, pupae, and adults; stonefly Plecoptera nymphs; midge larvae; and mayfly Ephemeroptera nymphs. Terrestrial worms, mollusks, crustaceans, and fishes are also consumed. On some rare occasions, Brook Trout may prey on frogs, salamanders, snakes, and small mammals such as field mice and shrews. It has been reported that not even large Brook Trout are piscivorous[2]; however, in some areas, Brook Trout larger than 152 mm (6 in) consumed moderate numbers of fish (up to 33 percent by volume), including sculpins, minnows, mud-minnows, and sunfishes[11].

Limiting Factors

Among the more serious predators on Brook Trout are fish-eating birds, including kingfishers and mergansers. Brook Trout are infested with a variety of parasites, including trematodes, tapeworms, and nematodes. At times there is cannibalism on eggs during spawning and on fry during emergence[5]. In some high mountain ecosystems, they may live only to three years because spawning so depletes energy levels that they cannot survive the winter. Lethal high temperatures are around 26.7°C (80°F). From embryo to the first fall season, there may be only a 2 percent survival rate[2].

Breeding Habits

Initiation of spawning may vary from late summer to autumn, depending on latitude and temperatures of the local ecosystem. In general, spawning occurs over gravel beds in shallow water of headwater streams. Gravelly shallows of headwater streams with strong spring water flow are particularly favored. Mature individuals, when ready to spawn, may travel many miles upstream to the spawning grounds, and males generally arrive first. Individual males display some aggressiveness when alone, but the level of aggression increases when they are joined by a female. Females clear away debris from the redd area with a series of rapid fanning movements of the caudal fin while lying on their side. Actual spawning events consist of one male and one female, but both males and females may spawn with more than one individual[5].

Eggs are large, measuring 3.5 to 5.0 mm (0.12 to 0.19 in), and adhesive for a short time after extrusion. Number of eggs per female is size dependent but ranges from 100 for a 144 mm (5.7 in) female to 5,000 for a 565 mm (22.2 in) FL individual. Eggs incubate in the gravel and hatch based on time and temperature. When hatched the sac fry remain in the gravel until yolk absorption is complete and emerge at about 38 mm (1.5 in)[5].

Optimal substrate size ranges from 3 to 51 mm (0.13 to 2 in)[64]. Eggs of fall and winter spawning populations overwinter at low temperatures. Development takes 35 days at 12.8°C (55°F), but takes 68 days at 7.2°C (45°F). Optimum egg incubation temperatures are in the 4.4 to 11.7°C (40 to 53°F) range[2].

Habits and Habitat

Brook Trout are a cold-water species preferring 3.3 to 16.7°C (38 to 62°F) and seeking temperatures of 10 to 14.4°C (50 to 58°F). Brook Trout reach greatest abundance in clean, cool, clear headwater streams and spring-fed streams or lakes. Warmer temperatures cause stress or even mortality. In streams, both young and adult Brook Trout are territorial and defend their territories against all fishes, even other Brook Trout. In lake ecosystems, they are less territorial but may school when frightened or near prime habitat such as springs[2].

In streams and rivers, adult Brook Trout appear to prefer low-velocity habitats of about 0.1 m/s (0.3 ft/s). Dominant fish, however, are found in water velocities as high as 0.34 m/s (1.1 ft/s). Cover plays an important role for Brook Trout and includes pools, undercut banks, or areas with overhanging stream-bank cover. Brook Trout habitat use is also strongly influenced by the presence of other salmonids. During winter at water temperatures of 1 to 9°C (34 to 48°F), stream-dwelling Brook Trout are active at night and stay concealed in cobble-boulder substrate or under woody debris during the day[4].

Management and Conservation Status

Brook Trout are considered common, widespread, and abundant range wide (G5). A state status rank is not applicable (SNA) because the species is not a suitable target for conservation activities[15]. Brook Trout, as with all nonnative species, should not be stocked on top of native salmonid species, particularly Bull Trout. Unfortunately, recent range-wide status assessments done for native species such as Cutthroat Trout have documented that Brook Trout presence is often ubiquitous, and remedies to address the threat posed by them to native fishes are not often effective. The general bag limit for Brook Trout in Idaho is 25 fish per day, as the state is seeking to encourage harvest of this nonnative fish. At Henrys Lake, the IDFG stocks sterile Brook Trout that grow to large sizes in this productive lake. Brook Trout remain a popular sportfish with anglers, and Idaho will continue to manage Brook Trout populations as appropriate. In other states the stocking of Brook Trout in high-elevation lakes has led to the demise of native amphibians and large-bodied aquatic insects.

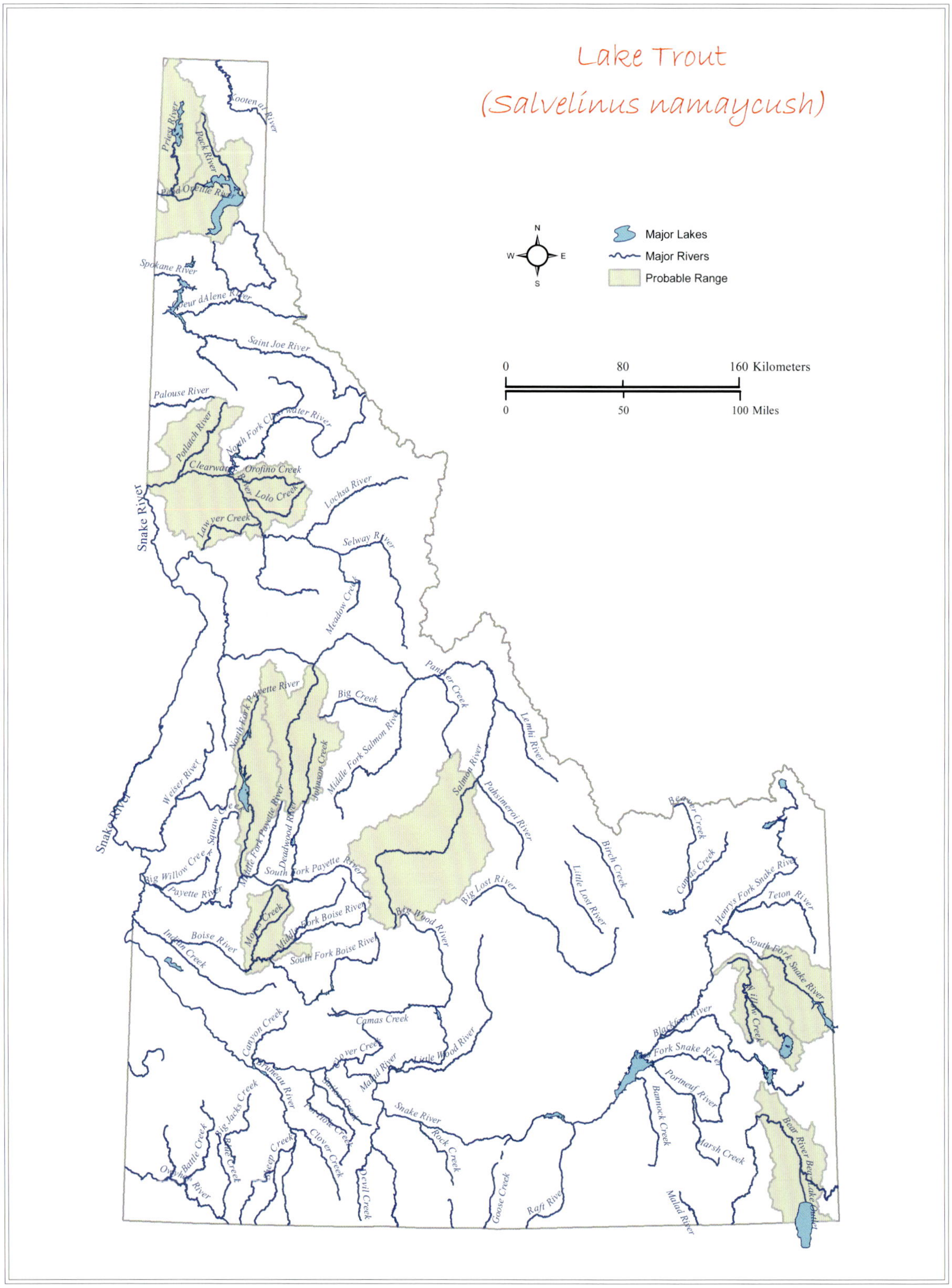
Lake Trout
(Salvelinus namaycush)
N
W
E
S
Major Lakes
Major Rivers
Probable Range
0
80
160 Kilometers
0
50
100 Miles
Kootenai River
Priest River
Pack River
Spokane River
Saint Joe River
Palouse River
Potlatch River
Orofino Creek
Lolo Creek
Lochsa River
Snake River
Selway River
Meadow Creek
Big Creek
Lemhi River
Middle Fork Salmon River
Johnson Creek
Salmon River
Pahsimeroi River
Weiser River
Big Willow Creek
Payette River
Middle Fork Payette River
Deadwood River
South Fork Payette River
Big Lost River
Little Lost River
Birch Creek
Henrys Fork Snake River
Teton River
South Fork Snake River
Boise River
Indian Creek
South Fork Boise River
Big Wood River
Camas Creek
Canyon Creek
Little Wood River
Malad River
Bruneau River
Clover Creek
Snake River
Rock Creek
Portneuf River
Bannock Creek
Marsh Creek
Devil Creek
Goose Creek
Raft River
Malad River
Bear River
Owyhee River
Big Jacks Creek
Battle Creek

Lake Trout

Salvelinus namaycush
(Walbaum, 1792)

Etymology

Salvelinus = an old name for char; *namaycush* = a Native People's name.

Importance

Lake Trout are highly valued as a sportfish because of their large size and associated trophy value. While catch per unit of effort may be small for this char, their fighting ability once hooked is considered worth the time. In their native range they are commercially important as well. In Bear Lake, Idaho-Utah, they produce a world-class trophy fishery. Large individuals may be too oily to be eaten fresh, but small ones are delectable either fresh or smoked[2]. Like Brook Trout, Lake Trout, although popular with many sportsmen, have created significant ecological problems in ecosystems where native Cutthroat Trout dominated prior to Lake Trout stocking. Efforts to remove Lake Trout from several of the large lakes in northern Idaho are ongoing or being planned. Many of the lakes where Lake Trout are causing problems were not actually stocked; Lake Trout either drifted downstream from upriver or were the result of illegal introductions.

Description

The Lake Trout body form is elongate and typically trout-like. Teeth are developed on upper and lower jaws (premaxillary, maxillary, and dentary) on the head of the vomer only and on the basibranchial (hyoid). Gill rakers total 16 to 26. Branchiostegal rays total 10 to 14 (11–14 + 10–13), usually with fewer on the right side. A dorsal adipose fin is present. The dorsal fin has eight to 10 major rays. The caudal is distinctly and deeply forked. Major anal rays are eight to 10, and pelvic rays are eight to 11, with a small pelvic axillary process. Pectoral rays are 12 to 27. Scales are cycloid and small, and the pore scales along the lateral line number 116 to 138; the lateral line is slightly decurved anteriorly, then straight. Pyloric caecae are 93 to 208, apparently increasing with size. Vertebrae are 61 to 69[5].

Sexual dimorphism, even in spawning individuals, is not as pronounced as it is in other char, though the largest fish in most populations are female. In Lake Trout, there are no bright spawning colors, but the pectoral, pelvic, and anal fins usually develop white leading edges, followed by a dusky pigment line. Absence of dark colors may be associated with the nocturnal spawning habitats of this species[4].

Range and Distributional Status

This species is not native to Idaho. Lake Trout occur naturally over a significant portion of North America, in the North from Alaska to Labrador and in the South from the Frazier and Columbia Rivers and across Canada, including all of the Great Lakes to the northern tier of the New England states. Their distribution in Canada includes Nova Scotia, New Brunswick, Quebec, Ontario, portions of Alberta and northern British Columbia, and northern Manitoba and Saskatchewan as well as many Arctic islands and the Yukon and Northwest Territories[2]. Lake Trout were likely introduced into Bear Lake in 1911.

In Idaho they are present in Bear Lake; portions of the Bear River and Snake River; and the South Fork and main-stem Salmon Rivers. They are also present in Lake Pend Oreille, Priest Lake, Payette Lake, and Palisades Reservoir.

Size and Longevity

Lake Trout are large fish. Individuals weighing in excess of 22.7 kg (50 lb) are present in many Canadian lakes, and one individual exceeded 45.4 kg (100 lb). In most inland lakes, individuals taken by anglers average less than 4.5 kg (10 lb)[5].

For Lake Trout aged one to six years from Montana, total yearly length, in mm (in), was 76 (3), 152 (6), 229 (9), 356 (14), 508 (20), 635 (25). Lake Trout are slow-growing and long-lived fish. Large individuals may dominate their year class in a particular ecosystem and turn to piscivory early in life, allowing them to reach larger sizes. Larger individuals may never have spawned, encouraging the increase in size[2].

Priest Lake yielded the Idaho record Lake Trout, a 26.1 kg (57 lb, 9 oz), 1,245 mm (49 in) fish caught in 1971[13]. Great Bear Lake in the Northwest Territories, Canada, produced the all-tackle world record Lake Trout, a 32.65 kg (72 lb), 1,499 mm (59 in) fish[12].

Food and Feeding

Lake Trout eat small insects, crustaceans, and other zooplankton when young. Lake Trout larger than about 0.9 kg (2 lb) may feed almost entirely on fish. Fish eaten by Lake Trout in Bear Lake are sculpin, the three whitefishes, Utah Suckers, and some Utah Chubs. In most lakes, Lake Trout have a tendency to move inshore in the spring and feed until the water warms above 10°C (50°F)[2].

In the Bear Lake population of Lake Trout, diets of individuals over 400 mm (15.8 in) consisted entirely of fish, including Bear Lake Sculpin, Bonneville Cisco, Bonneville Cutthroat Trout, and indistinguishable whitefishes[65].

Limiting Factors

Lake Trout are a true cold-water fish, preferring 3.3 to 16.7°C (38 to 62°F), and do not tolerate temperatures over 21.7°C (71°F). Lake Trout have a low salinity tolerance for a salmonid: 1,000 to 13,000 mg/l (ppm). In some ecosystems, small Lake Trout may be preyed upon by both game and nongame species. They can be overfished. In ecosystems with limited spawning habitat, Lake Trout may suffer low or zero reproduction in some years[2].

Breeding Habits

Spawning activity is initiated by Lake Trout when temperatures decline in the fall, approaching 12.8 to 10°C (55 to 50°F). Sexually mature individuals move to inshore locations at dusk and initiate spawning in shallow water, 3.1 m (10 ft), to relatively deep water, 12.2 m (40 ft). Spawning depths to 45.7 to 54.8 m (150 and 180 ft) have been reported. Lake Trout are generally sexually mature at ages of five to eight years old, but a few mature younger and a few at older ages[2].

Lake Trout eggs are large, ranging from 4 to 5 mm (0.15 to 0.2 in), but fecundity is low. Fecundity, which is dependent on female body size, varies from 500 to 20,000 eggs, and females in many populations do not spawn every year. Development rate of fertilized eggs is a function of incubation temperature, though development rates differ among populations, and hatching times range from about 50 days at 10°C (50°F) to about 156 days at 2°C (36°F). Alevins are about 21 mm (0.8 in) FL and negatively phototactic, remaining in the substrate as the yolk sac is absorbed and emerging from the substrate at night[4].

One or two males may spawn with one female in an area where the rocks have been cleaned by brushing with the body or tail fin or rubbing them with the snout. Males and females spawn together, extruding eggs and sperm over the rocky bottom. Fertilized eggs fall into crevices between the rocks. There is some evidence that Lake Trout home to a natal area, returning to the same spawning beds year after year. After spawning, Lake Trout disperse throughout the lake and may move distances up to 161 km (100 mi)[5].

Eggs remain in the rocks for many weeks; eggs spawned in October hatch from February to March. Young seek deeper water within a month of hatching, after the yolk sac is absorbed[5].

Habits and Habitat

In general, Lake Trout live only in deep, cold, stratified lakes. Populations in lakes spend most of their time near the bottom of the hypolimnion. In Lake Tahoe, Nevada-California, Lake Trout have been reported at depths of 427 m (1,400 ft). They are mostly solitary except during the spawning season[2].

Management and Conservation Status

Lake Trout are considered common, widespread, and abundant range wide (G5). A state status rank is not applicable (SNA) because the species is not a suitable target for conservation activities[15]. They provide an exciting sport fishery for those with the patience to learn their habits.

Lake Trout often lead to the demise of other trout species where introduced. For instance, introductions into Lake Tahoe led to the elimination of the native Lahontan Cutthroat (*Oncorhynchus clarkii henshawi*).

Lahontan Cutthroat Trout were abundant in the lake in 1907, with only an occasional Lake Trout reported. By 1938 the cutthroat had virtually disappeared, and by 1966 Lake Trout constituted 70 percent of angler catch in the lake. Stocked Lake Trout have replaced native cutthroats in deep Rocky Mountain lakes. This trout has also virtually eliminated Westslope Cutthroat Trout and Bull Trout in Flathead Lake, Montana, and Lake Pend Oreille, Idaho. Yellowstone Lake native *O. c. bouvieri* is similarly threatened. Introduction of Lake Trout led to the extirpation of three of the four populations of Sunapee Trout (Sunapee Lake and Dan's Hole Pond in New Hampshire and Averill Pond in Vermont; the sole surviving population is presumed to be in Floods Pond, Maine). Lake Trout introduced into Flaming Gorge Reservoir, Utah, were found to prey heavily on the Utah Chub (*Gila atraria*). Predation by Lake Trout was shown to be a major factor in the decline of Kokanee (*O. nerka*) in Lake Chelan, Washington, and Lake Pend Oreille, Idaho.

Atlantic Salmon

Salmo salar

Linnaeus, 1758

Etymology

Salmo = Latin name for the salmon of the Atlantic; *salar* = old name from "salio," to leap.

Importance

Atlantic Salmon are extirpated in Idaho. Abbreviated information on this species is included here for historical perspective. Atlantic Salmon have a special place in both fish lore and the marketplace. Atlantic Salmon are a world-renowned species, both for game fishing and as commercial catch. This species has been revered and used by cultures since the Gauls and the Romans. It has been exploited to extinction in many locations but has been stocked in Pacific drainages as well as Great Lakes locations[5,11].

Description

Atlantic Salmon have a typical trout-like body, elongate and somewhat laterally compressed. In the sea, Atlantic Salmon are silvery on the sides and silvery white below, while the back may be brown, green, or blue. After entering freshwater to spawn, adults lose the silvery appearance and become darker. When the time to spawn is near, they become bronze and dark brown, some with reddish spots on the head and body.

Young have eight to 11 narrow pigmented parr marks on the sides with a single red spot between each parr mark along the lateral line[5].

Range and Distributional Status

This species was not native to Idaho and is considered to be extirpated in the state. Native range of Atlantic Salmon is the basin of the North Atlantic Ocean from the Arctic Circle to Portugal in the eastern Atlantic and from Iceland and southern Greenland to the Connecticut River in the United States. Landlocked populations occur across Newfoundland, Labrador, and Quebec, Canada[5].

Atlantic Salmon were unintentionally introduced into the Pacific Northwest via escapement from commercial net pens in Puget Sound in July 1996. Atlantic Salmon were stocked into Deadwood Reservoir within the South Fork Payette River drainage in Idaho in 1990[11]. Atlantic Salmon have not been stocked in Idaho recently.

Size and Longevity

In 1928 the world record Atlantic Salmon was caught in the Tana River, Norway, and weighed 35.9 kg (79 lb, 2 oz)[12]. In 1995 the Idaho record Atlantic Salmon was caught in Deadwood Reservoir and weighed 6.0 kg (13 lb, 3 oz) and was 756 mm (29.75 in) long[13].

Habits and Habitat

Anadromous Atlantic Salmon spend portions of their life in both fresh and marine environments. Young, after emerging from the gravel, remain in freshwater for two or three years and descend natal streams to the ocean as smolts. While at sea they may range distances up to 2,414 km (1,500 mi). When temperatures fall in autumn, spawn-ready individuals return to their natal streams to start the cycle over[5].

Management and Conservation Status

Atlantic Salmon are no longer believed to be present in Idaho. Information on this species is included here for historical perspective only.

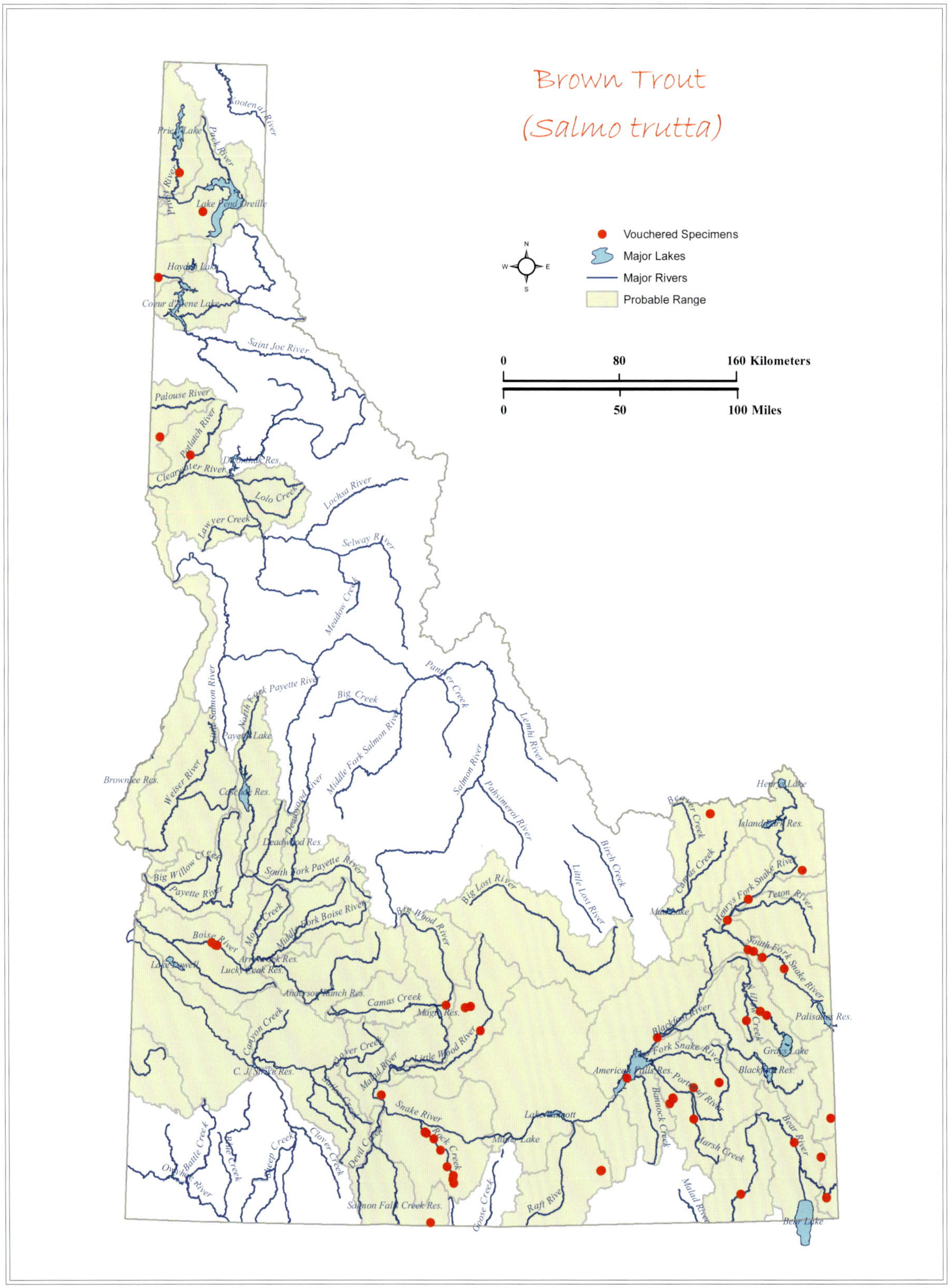
Brown Trout
(Salmo trutta)
Vouchered Specimens
Major Lakes
Major Rivers
Probable Range
N
S
E
W
0
80
160 Kilometers
0
50
100 Miles
Kootenai River
Priest Lake
Priest River
Pack River
Lake Pend Oreille
Saint Joe River
Palouse River
Potlatch River
Clearwater River
Lolo Creek
Lochsa River
Selway River
Meadow Creek
Big Creek
Lemhi River
Salmon River
Pahsimeroi River
Birch Creek
Little Lost River
Big Lost River
Big Wood River
Camas Creek
Little Wood River
Weiser River
Brownlee Res.
Payette Lake
Cascade Res.
Deadwood Res.
Big Willow Creek
Payette River
Boise River
Middle Fork Salmon River
Mores Creek
Lucky Peak Res.
Anderson Ranch Res.
Canyon Creek
C. J. Strike Res.
Malad River
Snake River
Rock Creek
Salmon Falls Creek Res.
Clover Creek
Goose Creek
Raft River
Lake Walcott
American Falls Res.
Blackfoot River
Bannock Creek
Portneuf River
Marsh Creek
Malad River
Bear River
Bear Lake
Grays Lake
Blackfoot Res.
Palisades Res.
South Fork Snake River
Teton River
Henrys Fork Snake River
Island Park Res.
Henrys Lake
Beaver Creek
Camas Creek
Mud Lake
Willow Creek
Lake Lowell
Owyhee River
Battle Creek

Brown Trout

Salmo trutta

Linnaeus, 1758

Etymology

Salmo = Latin name for salmon of the Atlantic Ocean; *trutta* = Latin name for trout.

Importance

Brown Trout have had a mixed reputation since introductions were made in North America in the 1800s. Sportsmen often complained that this species was more difficult to catch than the native species, particularly Brook Trout. Fishery managers were concerned that the more aggressive Brown Trout would displace native Brook Trout, through either competition or direct predation[2]. More recently, attitudes have changed, and some of the finest wild trout fishing in North America is now provided by self-sustaining populations of Brown Trout. Brown Trout are more able to adapt to degraded ecosystem conditions, and although they are more difficult to catch, they provide excellent fishing for the accomplished fisherman and a challenge for the less experienced[2].

Description

Brown Trout have a typical trout-like body, with a laterally compressed, elongate form. Mature males develop a pronounced kype (hook) on the lower jaw. Teeth are present on both the upper and the lower jaw, and there are many teeth on the head and shaft of the vomer, on the palatines, and on the tongue in two rows. Gill rakers total 14 to 17. Branchiostegal rays are usually 10. There is a dorsal adipose fin, and the single dorsal fin has 12 to 14 major rays. Pelvic rays are nine or 10, with a distinct pelvic axillary process. Pectoral rays are 13 or 14. Scales on Brown Trout are cycloid and small, numbering 120 to 130 in the lateral line[5] (115 to 135)[2], which is decurved anteriorly. Pyloric caecae are 30 to 60, and vertebrae number 56 to 61[5].

In streams, Brown Trout have a light-brown or tawny overall coloration with brown on the back and silvery on the sides. Pronounced black spots occur on the back, sides, and head, tending to spread below the lateral line anteriorly but less so posteriorly. These distinctive spots are often surrounded by a lighter ring or halo. Fins of adults have a smoky, opaque appearance with sometimes yellowish coloration. Young Brown Trout have nine to 14 short, narrow parr marks and a few reddish spots along the lateral line[5]. Leading edges of the pelvic fins and the anal fin are often outlined with a yellow margin. In juveniles the adipose fin is orange and has no dark spots or margin outlines[2].

Adult Brown Trout differ in appearance from Atlantic Salmon in tail shape and upper jaw length. Tails of Atlantic Salmon are forked but almost square in Brown Trout. The jaw extends back to about the center of the eye in Atlantic Salmon and past the eye in Brown Trout[4].

Range and Distributional Status

This species is not native to Idaho. Brown Trout are native to only European and western Asian waters. Brown Trout were introduced into North American waters as early as 1883 (New York and Michigan)[5]. Subsequent introductions were made over much of Canada and the United States, and they have been introduced into South America, Africa, India, Australia, and New Zealand[2].

Brown Trout have been introduced into at least twenty-four countries outside Europe and now have a worldwide distribution. The earliest introductions occurred in 1852 in eastern Russia; New Zealand in 1867–85; the United States and Canada in 1883 and 1887, respectively; Australia in 1888; South Africa in 1890; Japan in 1892; and South America between 1904 and 1938. In less than 90 years (1852–1938), Brown Trout changed from a European to a global species[66].

The reasons for the wide geographical distribution of this species are its ecological variability, excellent ability to spread and colonize new water courses, and reputation as fine food and good sport. The species varies in size, growth rate, feeding niche, and habitat use within and among water courses. The phenotypic appearance varies largely with habitat and to some degree also with its ecology. Given suitable spawning substrate, proper temperature conditions, and sufficiently good water quality, Brown Trout can occupy habitats from very small brooks to the largest rivers and coastal waters within its range. The species appears to have evolved through five different evolutionary lineages in geographical isolation during the Pleistocene and remained largely allopatric since then[66]. Variability among this species is so great that about 50 species have been described for varieties of Brown Trout, all grouped together in one polymorphic species[67].

In Idaho they are distributed across the southern third of the state in the Snake River and its tributaries, including the Boise River. In north Idaho, Brown Trout occur in the Spokane River, Pend Oreille, and portions of the Priest Lake drainage.

Size and Longevity

Brown Trout are fast growing and moderately long lived (> 13 years), growing best at temperatures of 18.3 to 22.2°C (65 to 72°F). Generally, both sexes grow at the same rate, but populations where one sex grows more rapidly than the other have been reported. In a Utah stream study, Brown Trout aged one to six years had yearly lengths, in mm (in), of 89 (3.5), 152 (6), 254 (10), 356 (14), 406 (16), and 457 (18). In the Logan River, Utah, at the end of each year one through eight, Brown Trout had total lengths, in mm (in), of 102 (4), 175 (6.9), 246 (9.7), 307 (12.1), 396 (15.6), 465 (18.3), 648 (25.5), and 704 mm and (27.7in)[68].

The Brown Trout record for Idaho is a 12.4 kg (27 lb, 5 oz), 940 mm (37 in) individual caught in Ashton Reservoir in 2007[13]. Washington State produced a record Brown Trout of 10 kg (22 lb) from Sullivan Lake in 1965[11]. The all-tackle world record for Brown Trout is a 19.1 kg (42 lb, 2 oz) individual from Ohau Canal, New Zealand, caught in 2013[12].

Food and Feeding

Brown Trout begin feeding on fish at an earlier age than do other trout[2]. Large Brown Trout feed more extensively on fish than most other trout. Forage fishes include sculpins, minnows, suckers, darters, and trouts[11]. Brown Trout are carnivorous, feeding on aquatic and terrestrial insects as well as mollusks, frogs, and rodents. Both lake and stream populations of young Brown Trout feed on zooplankton and insects[2].

Newly emerged fry and fingerling Brown Trout feed on minute bottom organisms such as blackflies, mayflies, and stoneflies. In winter fingerlings feed on freshwater shrimp and isopods[11].

In lakes zoobenthos in shallow waters are the major food, but the importance of pelagic food such as chironomid pupae and zooplankton as well as surface insects increases with increasing size and age. Apparently, the particle size eaten by similar-size fish is smaller in lakes than in streams. In lake populations, Brown Trout sometimes feed almost exclusively on zooplankton. In freshwater Brown Trout may turn to piscivory at approximately 150 mm (5.9 in)[66].

Limiting Factors

Mergansers may prey on Brown Trout in some ecosystems. Large Brown Trout may feed upon smaller ones. A variety of parasites infect Brown Trout, depending on ecosystem conditions, and Brown Trout may be infected with furunculosis (*Bacterium salmonicida*)[5]. Low water temperatures restrict distribution of Brown Trout within specific ecosystems, and they do not occupy cold high-altitude reaches of mountain streams. Brown Trout do not occur in waters where temperatures are too high: 25 to 26°C (77 to 79°F)[2].

Excessive mortality on Brown Trout has been related to heavy fishing pressure and stream ecosystem alterations such as channelization[2]. Disturbance of spawning gravel decreases survival of eggs[69].

Breeding Habits

Brown Trout spawn in the fall from October to December at temperatures of 10 to 7.2°C (50 to 40°F). Age of maturity is two to three years, and males often mature one year ahead of females. Spawning occurs during daylight hours[2].

When spawning is initiated, Brown Trout move out of deep pools or lakes upstream to the nearest acceptable riffle areas. Females select redd sites and dig the redds with their tails. Female activity on the redd attracts one or more males, and the dominant male defends the female and redd from other males. When the redd is complete, both adults move over the nest and eggs and sperm are deposited. After spawning, females cover the redd by moving upstream and fanning gravel into the redd. Water depths at redd sites are from 28 to 61 cm (11 to 24 in). Larger females breed first, and a female may lay 200 to more than 6,000 eggs, which hatch in 41 days at constant temperatures of 10°C (50°F)[2]. Eggs are generally amber colored, with a diameter of 4 to 5 mm (0.16 to 0.19 in). Egg size may vary from ecosystem to ecosytem[5].

Brown Trout redds range in width from less than 31 cm (12 in) to more than 102 cm (40 in). Size range of preferred gravel is from 0.8 to 7.6 cm (0.3 to 3 in), but larger gravels may be used[64].

As adult male Brown Trout near spawning condition, their heads broaden and flatten, and the lower jaw develops a strong upturned hook (kype), absent or not obvious in females, and small mature male parr. Brown Trout spawn in the fall or early winter, when water temperatures drop to below 10°C (50°F). Females select the nest site. Typically, nests are constructed at the tail-out of the pool just before the water shallows and becomes turbulent as it enters the next riffle. Mean water velocities at spawning sites are reported to vary from about 3.0 to 76 cm/s (0.1 to 2.5 ft/s)[4,5] and 4.6 to 76 cm/sec (18 to 30 in/sec), and substrate is usually coarse gravel. Fecundity is a function of female size and varies from a few hundred eggs in stream population to several thousand in large fish. Egg size also varies among populations and increases with female size within populations. Typically, eggs are about 4 to 5 mm (0.16 to 0.2 in) in diameter. Incubation time is temperature dependent and varies from 148 days at 2°C (36°F) to 38 days at 10.6°C (51°F). Also dependent on temperature is alevin emergence from the gravel to begin exogenous feeding, which occurs at about three to six weeks after hatching. Newly hatched larvae are about 20 mm (0.8 in) TL and average about 26 mm (1.0 in) at swim up[4].

Habits and Habitat

Brown Trout tend more toward big waters than most trouts. They do not seek small streams and do not do well in small ponds. Adult Brown Trout in streams inhabit glide or rubble areas or the bottom of deep pools, while the young are in shallow pools, close to shore in either shallow water or riffles. There may be an altitudinal limitation within a drainage[2]. Several researchers have found that movement distances increase with fish length for Brown Trout[70].

Management and Conservation Status

Brown Trout are considered common, widespread, and abundant range wide (G5). A state status rank is not applicable (SNA) because the species is not a suitable target for conservation activities[15]. Brown Trout are more difficult to catch with rod and reel than most other trouts. Brown Trout should not be stocked where native species occur, as they may outcompete native species. They are considered to be a significant predator on native nongame fishes such as Northern Leatherside Chub[71].

"Be patient and calm - for no one can catch fish in anger."

Herbert Hoover

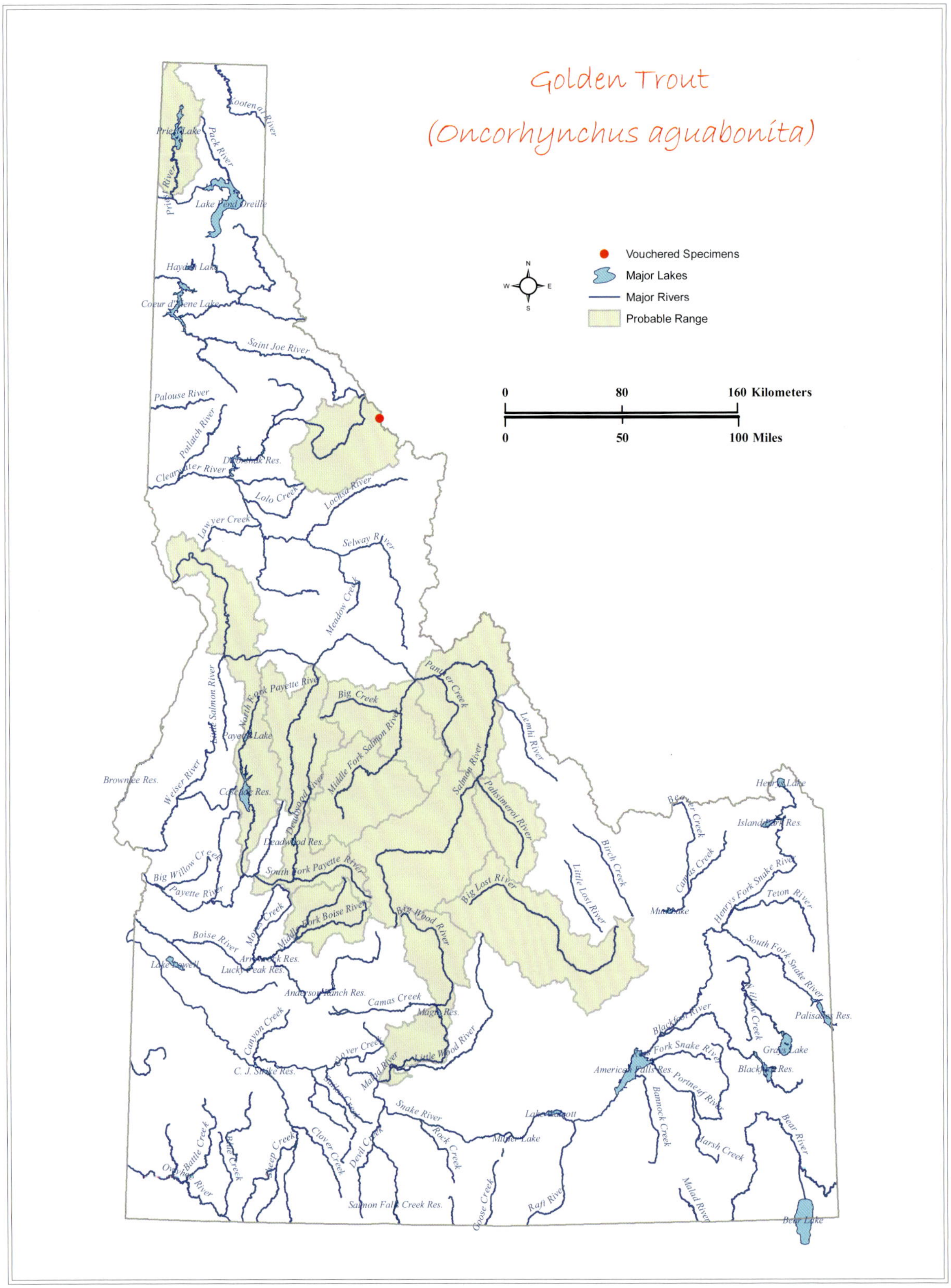
Golden Trout
(Oncorhynchus aguabonita)
Vouchered Specimens
Major Lakes
Major Rivers
Probable Range
N
S
E
W
0
80
160 Kilometers
0
50
100 Miles

Pacific Trouts and Salmons

Golden Trout

Oncorhynchus aquabonita
(Jordan, 1892)

Etymology

Oncorhynchus = hooked snout, referring to the condition of the migrating adults in Pacific salmon runs; *aquabonita* = Latin for "beautiful water, the name of a cascade on Volcano Creek, near which this trout abounds."

This attribution to Volcano Creek is an error of origin as noted by Jordan, Evermann, and Clark (1930) in their report: in his 1896 description, Jordan attributed the type specimen to Volcano or Whitney Creek, California, but the 1930 report clarified that the type specimen came from Cottonwood Lakes on the east side of Mount Whitney, into which it had been introduced from the Kern River. The species does not [*sic*] occur in Volcano Creek or Whitney Creek[72,73].

Importance

Relative scarcity and brilliant coloration are primary attractants for anglers for Golden Trout. While few people catch or even see this species, it provides an added element to the sport fishery in some ecosystems. This species has been extensively stocked in high mountain lakes of many western states[1,8].

Description

Golden Trout have a red-orange midlateral band, while the back is deep olive green. Generally, 10 parr marks are present in both young and adults. Lack of the classic cutthroat mark and of basibranchial teeth distinguish them from their close relative the Cutthroat Trout[2].

Jordan and Evermann (1904) call it the most beautiful of all the many western trouts[74]. Intensity of color in this species is influenced by age, sex, and diet[34].

Dorsal, anal, and pelvic fins have white tips offset by a black bar. Lateral-series scales number from 173 to 210, scales above the lateral line 34 to 45, pelvic rays 8 to 10, gill rakers 17 to 21, and pyloric caeca 25 to 40. Coloration of the cheeks and belly of the Golden Trout is bright red to orange; the lower sides are bright gold. In most fisheries, this is the most brightly colored species. Dark spots are chiefly posterior, large and well-marked, some as large as the pupil on tail and posterior part of body, smaller and well-marked on the dorsal[2,72].

Range and Distributional Status

This species is not native to Idaho. Native to the upper Kern River of California, Golden Trout have been stocked extensively throughout the western United States. In Idaho they were stocked as early as 1929–30 in lakes north of Imogene Lake and in lakes on the South Fork of the Payette River. This species is currently present in high alpine lakes in Idaho[2,9,20]. They are present in the Panther Creek drainage of the Salmon River.

Size and Longevity

Golden Trout aged one to five have average lengths, in mm (in), of 38 (1.5), 140 (5.5), 213 (8.4), 251 (9.9), and 264 (10.4). While Golden Trout may average 254 to 305 mm (10 to 12 in), fish as large as 610 mm (24 in) are also present in many fisheries. Golden Trout are short lived and slow growing[2]. A 4.98 kg (11 lb), 711 mm (28 in) Golden Trout was caught in Cooke Lake, Wyoming, and is the all-tackle hook-and-line world record[12]. The record Golden Trout from Idaho is a 2.3 kg (5.13 lb) individual taken from White Sands Lake in 1958.

Food and Feeding

Diet may include aquatic and terrestrial invertebrates. In lakes, Golden Trout may feed primarily on copepod and cladocerans zooplankton, apparently preferring caddisfly larvae and midges when available[2,34].

Limiting Factors

Golden Trout are protected from overfishing by the generally remote ecosystems they occupy; otherwise, they are among the easiest trouts to catch. When interbreeding with Rainbow Trout, genetic identity is lost; introduced lake populations that migrate into tributary streams are generally poor competitors[2,7].

Behnke reports that he knows of no example where introduced California Golden Trout were able to coexist with Brown Trout or Brook Trout[34].

Breeding Habits

When water temperatures reach 7.2 to 10°C (45 to 50°F) in late May to August, Golden Trout initiate spawning. These fish mature in the third or fourth year. Females produce 300 to 2,300 eggs. Gravel riffles are a requirement for successful stream spawning, and although spawning has been observed in lakes, it is unclear if actual progeny were produced[2].

Habits and Habitat

In their native range, Golden Trout are restricted to small-stream habitats. This factor limits growth. Golden Trout, a true wilderness species, does best in swift high-altitude streams and in lakes ranging from 1,829 to 3,505 m (600 to 11,500 ft) altitude[2].

Management and Conservation Status

Since being redescribed as a species rather than a subspecies of Rainbow Trout, the global conservation status of Golden Trout has not been evaluated. As an introduced fish, an Idaho status rank is not applicable (SNA) because the species is not a suitable target for conservation[15]. They are a highly sought-after game fish in Idaho's alpine lakes, but the limited and inconsistent availability of eggs for culture and stocking purposes limits the program's scope.

Cutthroat Trout

Oncorhynchus clarkii

(Richardson, 1836)

Species Introduction

In the 2013 edition of the American Fisheries Society Special Publication 34, one species of Cutthroat Trout is listed, and authorship is attributed to Richardson[3]. Here we present a discussion of Cutthroat Trout currently found in Idaho and provide information on four subspecies and one morphological variant. Yellowstone Cutthroat Trout, Westslope Cutthroat Trout, and Bonneville Cutthroat Trout are widely recognized subspecies native to Idaho. Snake River Fine-Spotted Cutthroat Trout are recognized as a morphotype of the Yellowstone Cutthroat Trout. Lahontan Cutthroat Trout are not native to Idaho, and they occur in limited locations in the state. While there remains some vagary associated with the taxonomic and phylogenetic status of these taxa, we believe a presentation of both what is known and what is postulated for the taxa serves the greater good and should be presented here. Selected references are provided within each of the presented subspecies life history accounts should the reader wish to review original research findings regarding these taxa and their status. As the Idaho state fish, Cutthroat Trout hold special significance to both resource managers and anglers alike.

Past taxonomic evaluations of Cutthroat Trout subspecies have listed up to 14 different subspecies of Cutthroat Trout in the western United States. Cutthroat Trout, native to western North America, are widely distributed along a north-south axis from Alaska to New Mexico. Populations of Cutthroat Trout are found in very diverse habitats ranging from coastal temperate rain-forest watersheds to desert lakes and streams. Not surprisingly, Cutthroat Trout exhibit dramatic morphological and life history diversity across these widely divergent and contrasting ecological conditions, resulting in early ichthyologists dividing them into as many as 14 different subspecies in an attempt to account for the observed morphological diversity[75].

Using advanced genetic procedures, recent studies documented differences and similarities in samples from eight major subspecies populations that are commonly recognized: Bonneville Cutthroat Trout (*Oncorhynchus clarkii utah*), Coastal Cutthroat Trout (*O. c. clarkii*), Colorado River Cutthroat Trout (*O. c. pleuriticus*), Greenback Cutthroat Trout (*O. c. stomias*), Lahontan Cutthroat Trout (*O. c. henshawi*), Rio Grande Cutthroat Trout (*O. c. virginalis*), Westslope Cutthroat Trout (*O. c. lewisi*), and Yellowstone Cutthroat Trout (*O. c. bouvieri*). Samples of Cutthroat Trout from populations considered by some to be separate subspecies, including Willow/Whitehorse (*O. c. ssp.*), Paiute (*O. c. seleneris*), Humboldt (*O. c. humboldtensis*), and Snake River Fine-Spotted Cutthroat Trout (*O. c. behnkei*), were included. Two other subspecies, the Alvord Cutthroat Trout (*O. c. alvordensis*) and Yellowfin Cutthroat Trout (*O. c. macdonaldi*), are thought to be extinct. Populations were sampled from locations that are believed to be representative of native populations[75].

Of the eight main groups, four, including Coastal, Westslope, Lahontan, and Colorado River Cutthroat Trout, occurred as monophyletic (single-stock) lineages and corresponded to subspecies designations previously used to describe intraspecific variation in Cutthroat Trout. Two groups (Rio Grande Cutthroat Trout and Greenback Cutthroat Trout) formed separate but not definitive groups. Bonneville and Yellowstone Cutthroat Trout did not form monophyletic lineages. The Bonneville-Yellowstone grouping included the majority of the sampling locations for these subspecies; however, a distinct and divergent group is also present within the Bonneville-Yellowstone distribution (the Great Basin lineage) and is more closely related to populations from the Colorado River watershed despite its geographic proximity to the upper Snake River and Bear River. The Paiute and Whitehorse Cutthroat Trout were nested within the Lahontan Cutthroat Trout grouping of the phylogeny, and fish representing Snake River Fine-Spotted Cutthroat Trout had the same haplotype (gene grouping) as other Yellowstone Cutthroat Trout[75].

Data from this research facilitated estimation of ancestral relationships between Rainbow Trout and Cutthroat Trout and the divergence of these two species. Based on the assumption of six million years between Rainbow Trout and Cutthroat Trout divergence, it is estimated that all Cutthroat Trout shared a common ancestor approximately 2 mya, when there was a primary divergence between the lineage that includes the Bonneville-Yellowstone, Lahontan, Westslope, and Coastal Cutthroat Trout groups versus the lineage including the Greenback, Rio Grande, Great Basin, and Colorado River Cutthroat Trout groups. Within each of these major Cutthroat Trout groups, the Bonneville-Yellowstone lineage was estimated to have diverged from other members of its group at 1.9 mya, followed by the coastal group at 1.85 mya, and then between the Westslope and Lahontan groups about 1.43 mya. Colorado River Cutthroat Trout are estimated to have diverged within its subgroup at 1.41 mya, followed by the Great Basin lineage at 1.38 mya, and finally between the Greenback and Rio Grande lineages at 0.94 mya[75].

Trotter[76] reports that the name "cutthroat" was first used by Charles Hallock in an article he wrote for the October 4, 1884, issue of the *American Angler*. This common name appearance occurred some 79 years after Meriwether Lewis first recorded the twin slashes of reddish color below the lower jaws of this species. Hallock was reporting on a fishing trip he had taken up Rosebud Creek from the Crow Indian Agency at Stillwater, Montana[34,76].

Agonistic behavior among individual cutthroat (characteristic nipping, chasing, posturing, and direct fighting) may include prominent display of the characteristic slash marks during male interactions, including territory defense.

Etymology

Oncorhynchus = hooked snout, referring to the condition of the migrating adults in Pacific salmon runs; *clarkii* = one of the two leaders of the Lewis and Clark expedition to western America in 1804–6.

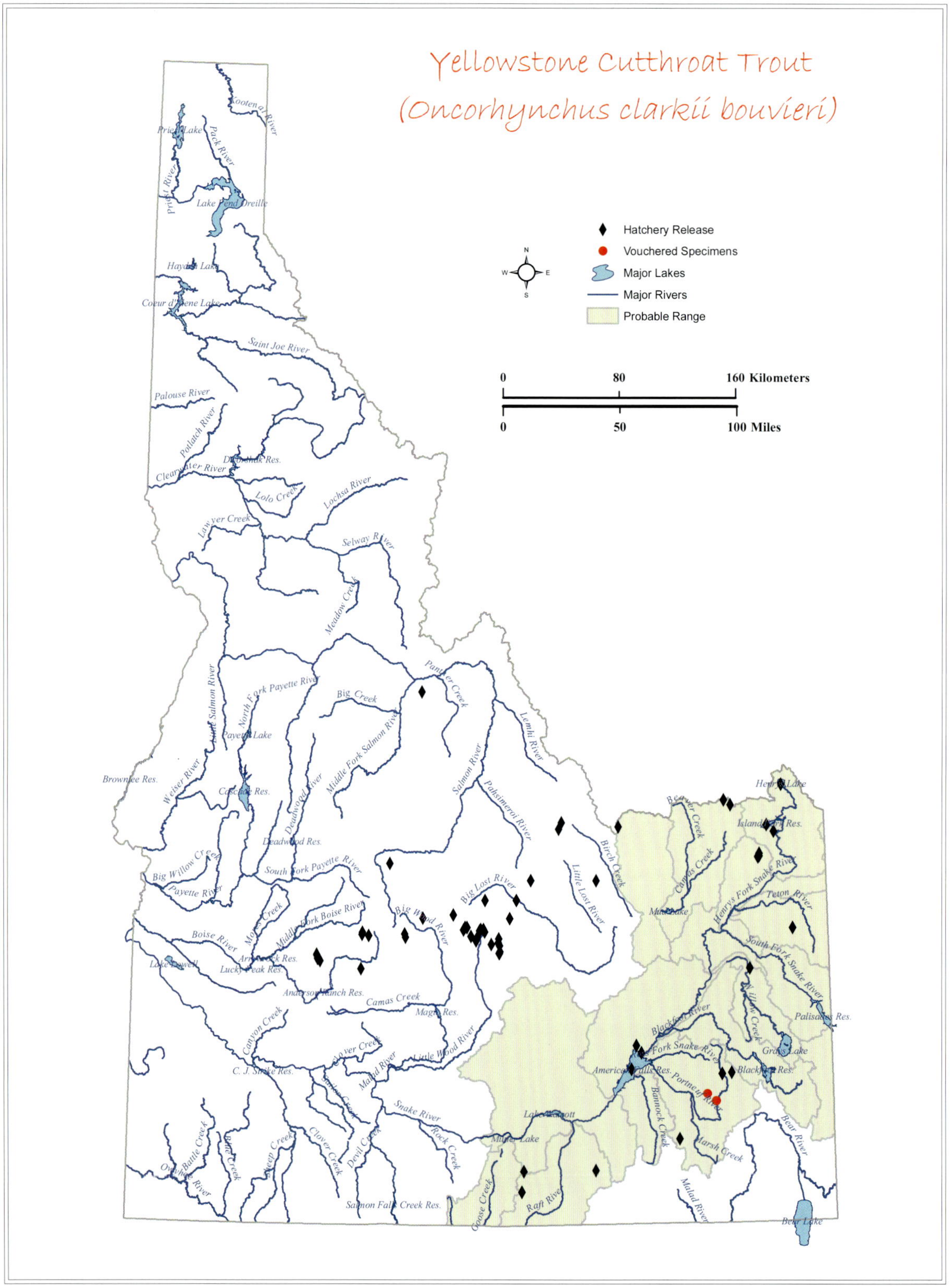
Yellowstone Cutthroat Trout
(Oncorhynchus clarkii bouvieri)
Hatchery Release
Vouchered Specimens
Major Lakes
Major Rivers
Probable Range
N
S
W
E
0
80
160 Kilometers
0
50
100 Miles
Kootenai River
Pack River
Priest River
Saint Joe River
Palouse River
Potlatch River
Lolo Creek
Lochsa River
Selway River
Lemhi River
Little Lost River
Big Lost River
Camas Creek
Snake River
Rock Creek
Clover Creek
Raft River
Goose Creek
Bear River
Bear Lake
Malad River
Teton River
Palisades Res.
Grays Lake
Bannock Creek
Big Willow Creek
Payette River
Boise River
Weiser River
Brownlee Res.
Little Salmon River
Salmon River
Pahsimeroi River
Birch Creek
Henrys Fork Snake River
South Fork Snake River
Canyon Creek
Salmon Falls Creek Res.
C. J. Strike Res.
Lucky Peak Res.
Anderson Ranch Res.
Magic Res.
Big Creek
Middle Fork Salmon River

Yellowstone Cutthroat Trout

Oncorhynchus clarkii bouvieri
(Bendire, 1882)

Etymology

Oncorhynchus = hooked snout, referring to the condition of the migrating adults in Pacific salmon runs; *clarkii* = one of the two leaders of the Lewis and Clark expedition to western America in 1804–6; *bouvieri* = Captain Bendire explored the area of Waha Lake in 1869 and apparently applied the name of an army colleague, Captain Bouvier, to this subspecies[76].

Importance

While few other western fish species elicit as much discussion and passion as the "native," it is the Yellowstone Cutthroat Trout that is the best-known and most widely recognized subspecies. Average anglers and elitists alike pursue Cutthroat Trout throughout the state of Idaho. Contributions to the economies of localities where Yellowstone Cutthroat Trout are abundant cannot be overstated. Thousands of anglers pursue this species across its current range, often traveling thousands of miles for the opportunity to land a specimen, even in a "catch-and-release" fishery. Early in the history of Yellowstone National Park, fish culturists recognized the abundance of Yellowstone Cutthroat Trout in Yellowstone Lake as a source of highly desirable planting stock. Between 1889 and 1955, some 818 million Yellowstone Cutthroat Trout eggs were taken, fertilized, and shipped to most US states, Canada, and several foreign countries. Abundance was in part measured by the sheer numbers of this species in Yellowstone Lake. In the 1950s and 1960s, before the more restrictive fishing regulations, more than 300,000 trout were taken from the lake annually. A roughly equal amount was taken by fish-eating birds and bears[34].

Recreational potential (economic value) of Yellowstone Cutthroat Trout is extensive. Based on surveys conducted in the 1980s, Varley and Gresswell report that Yellowstone Cutthroat Trout were rated higher than any of the other seven sportfish species found in Yellowstone National Park[77]. Economic value of Yellowstone Cutthroat Trout recreational fisheries far exceeded the fiscal capability of any public natural resources agency to replace this species in the wild[77].

Description

Coloration of the Yellowstone Cutthroat Trout is one of the features of this species that make it a premier sportfish throughout its range. Generally, this species is yellowish brown, silvery, or bronze. Some populations exhibit a more yellow-gold coloration. Coloration is typically less intense than its subspecies cousin the Westslope Cutthroat Trout[34]. Color and spotting are variable between watersheds and even local populations as well as among subspecies. Splashes of orange may be found on the lower fins, and rose tints may occur along the lateral line. Gill covers and sides of the head become intense red in breeding males. Prevalent black spots are small to large, pronounced and rounded, concentrated on the posterior part of the body. Few, if any, spots appear on top of the head or snout[6,34].

Classically, a red or orange slash mark, present in adults but sometimes lacking in young, runs along both sides of the lower jaw and is the iconic mark giving rise to the common name for this species[2]. Young have nine or 10 oval parr marks along the lateral line[6]. Although elongate and typically trout-like, the body is rounded and slightly compressed; scales in the lateral series are cycloid and small to medium, numbering 125 to 190 in the lateral series. Mouth is rather large and terminal. Jaw, vomer, and tongue have sharp teeth. A median basibranchial plate is present, typically with teeth, between the lower ends of the gill arches. Typically, there are 16 to 28 gill rakers, with fewer on the upper than the lower limb[2]. Yellowstone Lake individuals typically have more gill rakers on the anterior portion of the first gill arch (20 to 22 versus 17 to 20 for other Yellowstone Cutthroat Trout and other subspecies) and five to 15 gill rakers on the posterior portion of the first gill arch (versus zero to three)[34]. They also typically have more basibranchial teeth, averaging 20 or more, versus five to 10 found in other Cutthroat Trout subspecies[34].

Due to hybridization and the meristic complexity of Cutthroat Trout subspecies and their near relatives, the rainbow or redband species, it is often difficult or impossible to separate these species or subspecies in the field. As a general guideline, Cutthroat Trout usually have nine or fewer pelvic fin rays, whereas Rainbow Trout usually have 10 or more. A variety of techniques have been developed to facilitate separation in the laboratory. One of the traits that is useful is the number of chromosomes; in Cutthroat Trout, chromosome numbers vary from 64 to 68. In Rainbow Trout, the most common numbers are 58 to 60, with the exception of a few coastal populations[34].

Range and Distributional Status

This subspecies is native to Idaho. Historically, the range of the Yellowstone Cutthroat Trout included the Yellowstone River drainage in Montana and Wyoming and portions of the Snake River drainage in Wyoming, Idaho, Nevada, Utah, and perhaps Washington. Historical distribution included both sides of the Continental Divide[6,34,77]. This subspecies is native to the Snake River drainage above Shoshone Falls and to the Yellowstone River system downstream to the Tongue River, Wyoming. Historical occurrence of the Yellowstone Cutthroat Trout includes a small area of northeastern Nevada drained by Goose Creek and a small section of the Raft River drainage in northwestern Utah[34].

The type locality for the subspecies *O. c. bouvieri* is Waha Lake, an isolated lake in the hills above the lower Snake River, not far from Lewiston, Idaho, but miles downstream from Shoshone Falls. This population is most certainly extinct[76].

The Yellowstone Cutthroat Trout Coordination Group provided a 2007 status report that included the best available information on Yellowstone Cutthroat Trout in currently occupied stream segments and resulted in a determination that 7,527 miles of habitat were occupied by Yellowstone Cutthroat Trout (43 percent of historically occupied stream habitat). Yellowstone Cutthroat Trout currently occupy about 4,048 miles in Wyoming (23 percent of historical stream habitat in Wyoming), 2,033 miles in Idaho (31 percent of historical stream habitat in Idaho), 1,339 miles in Montana (31 percent of historical stream habitat in Montana), about 58 miles in Nevada (52 percent of historical stream habitat in Nevada), and 49 miles in Utah (38 percent of historical stream habitat in Utah)[78].

It is postulated that many of the processes controlling the present-day patterns of genetic diversity and gene flow in Yellowstone Cutthroat Trout across their range in Idaho are contemporary (recent droughts, habitat alteration, water diversions). Examining the distribution, diversity, and divergence of genetic lineages clearly demonstrates that the patterns of genetic variation across Yellowstone Cutthroat Trout known range have also been strongly shaped by long-term historical processes. Throughout much of the range of Yellowstone Cutthroat Trout in the Snake River Basin in Idaho and Bonneville Cutthroat Trout in the Bonneville Basin in Idaho and Utah, the dominant process has been basaltic volcanism that influenced the shared hydrological history of the Bear, Portneuf, and Blackfoot River drainages. The geological-hydrological history shared by these three drainages is evident in the distribution and divergence patterns of the dominant genetic patterns found in fish from these basins[79].

Specific genetic patterns cluster together and are predominantly found within the Portneuf, Blackfoot, and Bear River drainages. Other genetic patterns are significantly diverged from all other "Yellowstone" genetic patterns and are found throughout the Bear River Basin. Based on existing data, it is possible that these various genetic patterns represent lineages that were isolated as part of the initial diversion of the Bear River from the Snake River sometime during the past 600,000 years[79].

Absence of these genetic patterns in Portneuf and Blackfoot River individuals and their relatively large divergence from all other genetic patterns suggest that they originated in Lake Thatcher and that these rivers had to have been isolated from the Bear River Basin for a very long time[79]. Catastrophic effects associated with the Bonneville flood provided the most recent, but short-term, opportunity for faunal exchange between the Bonneville Basin and the Snake River basin and is likely the reason for the presence of at least two of the Yellowstone Cutthroat Trout genetic patterns in the Portneuf River drainage. These particular genetic patterns are widely divergent from other "Yellowstone" genetic patterns and are clearly associated with a "southern" Bonneville Cutthroat Trout genetic pattern found fixed within Bonneville Cutthroat Trout from populations in west-central and southern Utah[79].

The lack of distribution of specific genetic patterns outside the Portneuf River drainage in the Snake River basin suggests that major upstream dispersal barriers have existed since their introduction into the Portneuf River. This is likely given the large number of current and suspected historical waterfalls present in the main Snake River basin in Idaho[79].

Yellowstone Cutthroat Trout evolved in the upper Snake, Yellowstone, and Bear River basins of Idaho, Wyoming, and Utah, possibly diverging from an ancestor in the main Bonneville Basin during the early Pleistocene epoch (2.5 mya). The climatic and geological events in these basins throughout the Pleistocene were particularly complicated and chaotic. In the southern part of the Yellowstone Cutthroat Trout's range, massive basaltic eruptions and lava flows blocked and redirected entire river drainages, as exemplified by the history of the Bear River, which currently flows into the Great Salt Lake in the Bonneville Basin. However, it is believed that historically it flowed north to the Snake River, most likely through the Portneuf River canyon or possibly through the Blackfoot River drainage. About 500,000 years ago, basaltic volcanism in both the Gem and the Blackfoot River valleys produced large lava fields. One theory is that these lava flows repeatedly blocked the northward drainage of the Bear River. These events and other factors probably contributed to the formation of Lake Thatcher. Sedimentological evidence suggests that over the next 500,000 years, the Bear River reversed course multiple times as a result of basalt flows, alternately flowing south into the Lake Thatcher area or north toward the Snake River basin via surface or subsurface flows. Lake Thatcher's connection with the Bonneville Basin is not clear during much of this time period. However, it is known that by about 20,000 years ago, Lake Thatcher filled and eventually overflowed into the Bonneville Basin after the evacuation of Oneida Narrows, dramatically increasing the volume of water entering the basin and historic Lake Bonneville. Since that time the Bear River has flowed into the Bonneville Basin except for a very brief period between 14,000 and 11,000 years ago, when a catastrophic flood drained Lake Bonneville into the Snake River basin[79]. See the Biogeography Chapter of this volume for more detailed information.

Native populations of Yellowstone Cutthroat Trout are present in Idaho waters, including the Henrys Fork, Teton, Snake, Willow, Blackfoot, Portneuf, Raft, and Goose Creek drainages. Historically, this subspecies has been widely stocked throughout Idaho. More recently, widespread introductions have been reduced as the uniqueness of all of the Cutthroat Trout subspecies has been more widely recognized[6].

Size and Longevity

Growth of Yellowstone Cutthroat Trout, as with all of the subspecies, is variable and dependent on both population characteristics and environmental conditions[6]. Yellowstone Cutthroat Trout typically grow faster and larger in lower-elevation ecosystems; migratory populations attain size more rapidly than resident populations. In Henrys Lake, Idaho, Yellowstone Cutthroat Trout attain total lengths of 457 to 508 mm (18 to 20 in), while populations from tributaries above Shoshone Falls may attain total lengths of 600 mm (23.6 in) or greater[6]. Cutthroat Trout in high cold-water areas usually attain a length of 356 to 406 mm (14 to 16 in) and weights of 0.45 kg (1 lb) within four to five years[17].

In Yellowstone Lake, Yellowstone Cutthroat Trout can live to eight or nine years of age and grow to total lengths of about 530 mm (21 in). Yellowstone Cutthroat Trout stocked in Utah's Strawberry Reservoir have been documented to reach 12.3 kg (27 lb)[34]. Back-calculated total lengths for Yellowstone Cutthroat Trout in Yellowstone Lake, for fish aged one through six, are, in mm (in), 99 (3.9), 180 (7.1), 241 (9.5), 311 (12.2), 371 (14.6), and 409 (16.1)[77].

Growth rate of Cutthroat Trout in a 2014 study on the upper Snake River was significantly different between large main-stem rivers and smaller tributary streams. Larger main-stem rivers supported larger Cutthroat Trout over all ages, and smaller tributary streams contained smaller fish at a given age. This suggests that the ability to migrate from smaller tributary streams to larger main-stem rivers may permit Cutthroat Trout to maintain or increase their growth rate over their life span. As in other studies, populations of salmonids from the upper Snake River watershed that were resident in small streams had a smaller body size and slower growth rate at a given age than fish found in larger main-stem rivers. In addition, the lack of fish older than three years of age in isolated tributaries suggests that such streams are unable to support older and larger salmonids[80].

Food and Feeding

Yellowstone Cutthroat Trout eat primarily aquatic and terrestrial invertebrates and become piscivorous as they grow larger. Shifting to piscivory is a highly variable process across populations based on ecosystem differences. Like most Cutthroat Trout, Yellowstone Cutthroat Trout are opportunistic feeders, consuming food items that are readily available. Young feed mainly on aquatic insect larvae[6,34].

In a 2014 study of Cutthroat Trout in the upper Snake River, Idaho-Wyoming, diet was dominated by three orders of insects: Trichoptera, Diptera, and Ephemeroptera. These three dominant orders made up 84 percent of the invertebrates collected in Cutthroat Trout stomach samples. No other order contributed more than 8 percent to the total number of prey items, and terrestrial input was less than 0.1 percent. Analysis of the stomach samples indicated that mean, maximum, and minimum prey length expanded with increasing fish size; however, mean and minimum prey length tended to level off or decrease for the largest fish sampled. Maximum prey length increased with fish size but did so as a linear function of fish size. There was a similar increase in mean, maximum, and minimum prey width with increasing fish size. Both mean and minimum prey width increased as a nonlinear function of fish size, whereas maximum prey width increased as a linear function of fish size[80].

Limiting Factors

Yellowstone Cutthroat Trout are limited within their native range by a variety of anthropomorphic and ecosystem factors. Stream dewatering, habitat degradation, overfishing, competition, and hybridization with nonnative salmonids are all factors influencing Yellowstone Cutthroat Trout populations. In prime Yellowstone Cutthroat Trout habitat, they may outcompete introduced species[6,7,34]. Avian predation on certain populations of Yellowstone Cutthroat Trout is a growing concern in Idaho.

Breeding Habits

Yellowstone Cutthroat Trout spawning is variable and dependent upon such factors as latitude, altitude, stream flow, and water temperature and generally occurs from March to August. Mature Yellowstone Cutthroat Trout, whether lake or stream residents, spawn in fluvial environments. In spring older and larger adults migrate first, with more males present in most populations. These larger individuals tend to migrate farther up natal tributaries than smaller individuals[6,77].

Male Yellowstone Cutthroat Trout typically move into the spawning area when water temperatures are greater than 4.4°C (40°F). Female Yellowstone Cutthroat Trout fan out a redd with their tails in gravel and deposit eggs. Number of eggs per female is highly variable based on individual size and other ecological factors and may vary from a few to more than 8,000. Immediately after egg deposition, a male fertilizes the eggs and typically guards the redd to prevent other males from fertilizing the eggs as well. Following fertilization, the female moves upstream and uses her tail to cover the eggs with 13 to 18 cm (5 to 7 in) of gravel, in some cases depositing additional eggs in the newly created depression. Redds are typically 0.76 m (30 in) long by 0.46 m (18 in) wide and in 18 to 60 cm (7 to 24 in) of water. Optimal incubation temperature for embryos is approximately 10°C (50°F), and the upper safe temperature range for gravid females and embryos is 13.9°C (57°F)[6,11,17]. Egg hatching is usually within 28 to 40 days but is variable based on actual temperatures[77].

Habits and Habitat

Dramatic examples of habitat shifts by salmonids occur in anadromous populations. Similar shifts can also occur in strictly freshwater populations. Salmonids typically migrate from main-stem rivers or lakes to spawn in headwater tributaries where juveniles then feed and grow until they themselves shift downstream into larger streams or lakes. These movements or migrations are often correlated with enhanced body size in salmonids and may therefore occur as a means of maintaining or increasing growth. In contrast, individuals from populations isolated above movement barriers on headwater streams exhibit reduced maximum size and age at maturity. Although some populations of salmonids are able to switch to piscivory as a means of achieving larger size, most stream populations feed predominantly on drifting aquatic invertebrates and are able to attain a larger size in main-stem habitats. Differences in maximum body size among populations of salmonids occupying streams of increasing size and past observations of factors influencing growth in fishes suggest that habitat conditions related to prey size and abundance as well as temperature-related differences may influence the variation in growth among salmonid populations[80].

For most stream-dwelling Yellowstone Cutthroat Trout, ecosystem characteristics include clear (low-turbidity), cold (range 4.5 to 15.5°C [40 to 60°F]) water with good cover and riffle pool ratios of approximately 1:1. Altitudinally, Yellowstone Cutthroat Trout flourish in highly oligotrophic high mountain lakes and streams and in relatively shallow and highly productive waters such as Henrys Lake and the main-stem Snake River above Shoshone Falls[6,34].

In large lakes, Yellowstone Cutthroat Trout prefer areas near cover such as rubble and deep water. Yellowstone Cutthroat Trout are rarely present in areas where temperatures exceed 22°C (72°F), although they have been found in thermal areas in Yellowstone National Park where temperatures reach 27°C (81°F). Yellowstone Cutthroat Trout can survive and thrive in streams, rivers, lakes, and ponds, and fry do best in stream areas with about 40 to 50 percent pool areas[6,34].

Management and Conservation Status

Yellowstone Cutthroat Trout are ranked as apparently secure range wide (G4) and imperiled (S2) in Idaho due to restricted range or other factors[15]. Yellowstone Cutthroat Trout in Idaho are threatened by a variety of factors, many of them related to human activities. For example, mining operations have generated conservation concerns for Yellowstone Cutthroat Trout in the Blackfoot River drainage. Although habitat degradation and angler overexploitation have contributed to decline of Yellowstone Cutthroat Trout in their native range, a far greater threat is hybridization with nonnative Rainbow Trout and competition from other nonnative trout species. This decline has been particularly severe for stream-dwelling populations, and it is estimated that only about 10 percent of the Yellowstone Cutthroat Trout's original range is presently occupied[81]. Brook Trout have displaced Yellowstone Cutthroat Trout in many portions of the Yellowstone Cutthroat Trout native range.

In the South Fork of the Snake River, a three-pronged management approach is being used to help maintain a viable population of Yellowstone Cutthroat Trout in the presence of Rainbow Trout. Traps are set on the major Yellowstone Cutthroat Trout spawning tributaries (Palisades Creek, Rainey Creek, Pine Creek, Burns Creek) to allow for removal of Rainbow Trout; freshets are released from Palisades Reservoir in an attempt to duplicate pre-dam hydrology and potentially limit spawning success of Rainbow Trout, which spawn early in the spring, prior to the freshet; and no harvest limit and a reward-tag program to encourage anglers to remove all Rainbow Trout caught. At present, these practices have resulted in a balance among Yellowstone Cutthroat Trout (40 percent of population), Rainbow Trout (40 percent), and Brown Trout (20 percent) that appears to be sustainable under current conditions[82].

On the Henrys Fork of the Snake River, native Yellowstone Cutthroat Trout were replaced decades ago by Rainbow Trout and other nonnative species. Ecosystem considerations, primarily groundwater-dominated hydrology, allow high Rainbow Trout spawning success and have made the Henrys Fork a world-class Rainbow Trout fishery. From both an ecological and a sociological standpoint, the Henrys Fork will remain a blue-ribbon Rainbow Trout fishery, while nearby Henrys Lake provides an opportunity to pursue Yellowstone Cutthroat Trout–Rainbow Trout hybrids and native Yellowstone Cutthroat Trout.

Within the various aquatic ecosystems inhabited by Yellowstone Cutthroat Trout in Idaho, they should be protected by fishing and water-management protocols. These protocols should promote sustainable populations throughout the native range of the subspecies. Idaho has these protective management protocols in place.

For those Yellowstone Cutthroat Trout populations capable of sustainable angling opportunities, IDFG is exploring options for enhancing its involvement in habitat-restoration activities for native fish. However, this will require additional personnel and funding to capitalize on available federal programs to assist in habitat-restoration efforts.

IDFG will also track and respond to any effort to list the Yellowstone Cutthroat Trout under the Endangered Species Act. At this time, IDFG does not believe listing the Yellowstone Cutthroat Trout is warranted. The USFWS, after a thorough review of all available scientific and commercial information, found that listing of the Yellowstone Cutthroat Trout as either threatened or endangered was not warranted[83].

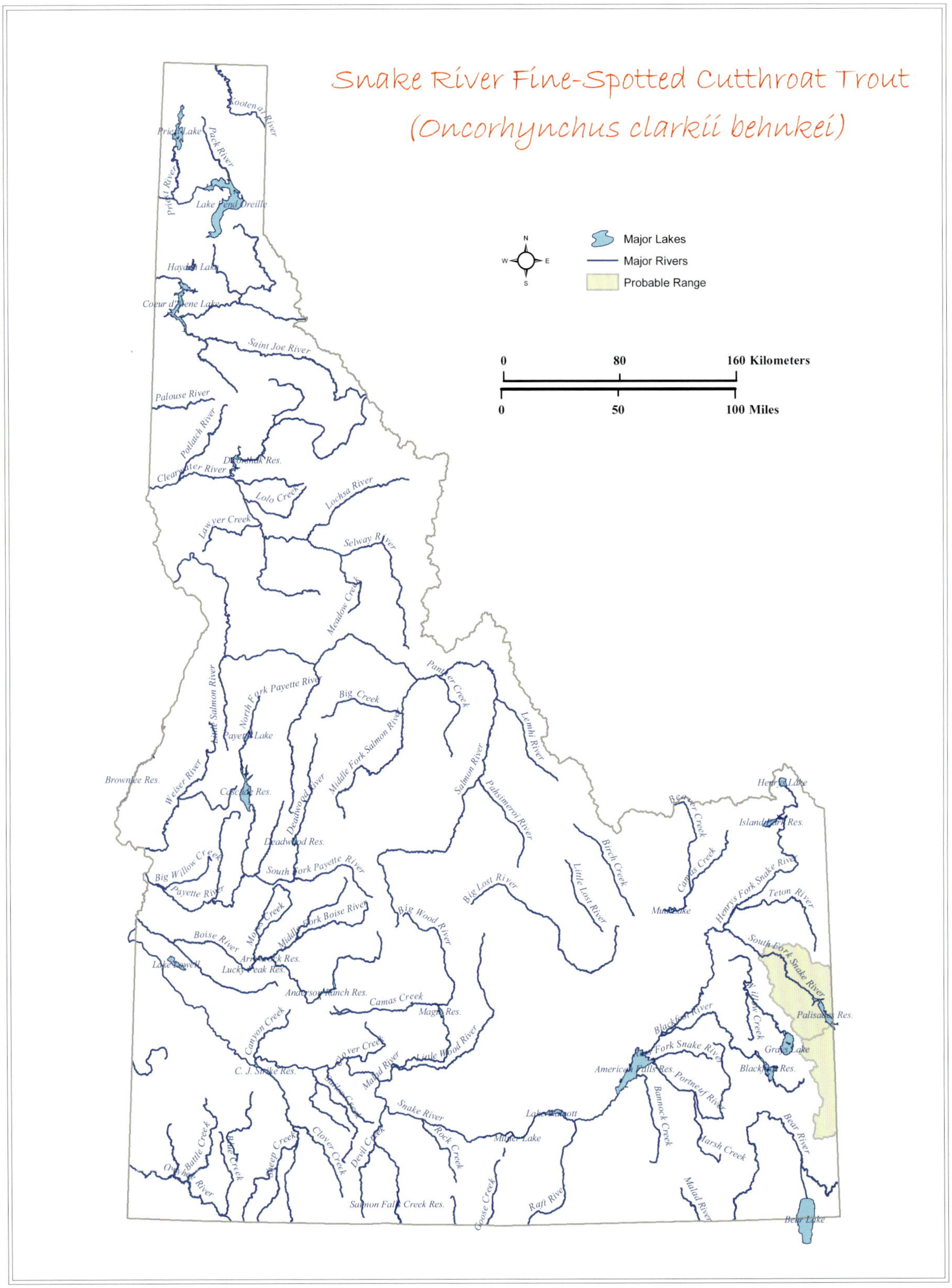
Snake River Fine-Spotted Cutthroat Trout
(Oncorhynchus clarkii behnkei)
N
S
W
E
Major Lakes
Major Rivers
Probable Range
0
80
160 Kilometers
0
50
100 Miles
Pack River
Saint Joe River
Palouse River
Potlatch River
Lolo Creek
Lochsa River
Selway River
Big Creek
Lemhi River
Salmon River
Pahsimeroi River
Birch Creek
Little Lost River
Big Lost River
Big Wood River
Camas Creek
Snake River
Rock Creek
Raft River
Bear River
Teton River
Bannock Creek
Marsh Creek
Malad River
Boise River
Canyon Creek
Clover Creek
Bruneau River
Weiser River
Brownlee Res.
Payette River
Middle Fork Salmon River
North Fork Payette River
Little Salmon River

Snake River Fine-Spotted Cutthroat Trout Morphotype

Oncorhynchus clarkii behnkei

Montgomery, 1995

Status of this form is controversial among fishery researchers. While there are some distinctive differences that proponents of designating this as a separate subspecies point to as a rationale for separation, a paucity of sound scientific data makes others skeptical of the need for and reality of subspecies status. Here we report what has been previously presented in various literature sources and leave the final status determination to a later date when additional valid scientific information may be developed.

Etymology

Oncorhynchus = hooked snout, referring to the condition of the migrating adults in Pacific salmon runs; *clarkii* = one of the two leaders of the Lewis and Clark expedition to western America in 1804–6; *behnkei* = provisional subspecies named for longtime Colorado State University Cutthroat Trout researcher Robert Behnke. Individuals interested in a more detailed presentation regarding the trinomial for this taxon should see Montgomery's *Many Rivers to Cross*[84].

Importance

Genetic origins of the Snake River Fine-Spotted Cutthroat Trout likely occurred in the later stages of the glacial period ending about 20,000 years ago. Were it not for the distinctiveness of the spot sizes between the fine-spotted and large-spotted individuals in the Snake River system, these two potential subspecies would be lumped together and classified as one. While the expressed differences in these forms are associated with the classic spotting patterns, the consistent and stable pattern of non-overlapping distributions, at a fine scale, of the two forms makes the differentiation of the morphotype important. This paradigm is confounded by HUC4 data noting overlaps of the two morphotypes at some locations. It must be accepted here that there is a hereditary (genetic) foundation for the differences between these two forms that is reflected in subtle, but observable, differences in life history, behavior, and habitat preferences that maintain the separate distributions of the forms[34]; these variations are worth preserving on a philosophical basis, and so here we provide a distinction. Management of these two closely allied forms as one is the current practice in Idaho[6]. Colorado scientists have documented that the Snake River Fine-Spotted Cutthroat Trout has the highest resistance to the debilitating whirling disease, for reasons unknown[34].

Description

Snake River Fine-Spotted Cutthroat Trout are distinctive among Cutthroat Trout with respect to spotting size and pattern. Snake River Fine-Spotted Cutthroat Trout have the smallest spots of any of the subspecies, and the spots are profuse, having been described as "a heavy sprinkling of ground pepper on the sides of the fish"[34]. Sides of the Snake River Fine-Spotted Cutthroat Trout are brownish yellow, underlaid with silver or green to bronze coloration; a purple tint may be present[6]. Anal fins of this form have intense red or orange coloration. Sides of the head and gill covers become intense red in breeding males. Lower fins are typically bright orange or red[6]. Lateral series scales are 136 to 188 with mean values of 153 to 176; pyloric caeca range from 32 to 51 with means of 39 to 46. Vertebrae counts range from 60 to 65[76].

Range and Distributional Status

This form is native to Idaho. Present distribution of Snake River Fine-Spotted Cutthroat Trout is from Palisades Reservoir on the South Fork of the Snake River, Idaho, to Jackson Lake in Grand Teton National Park, Wyoming, and in tributary streams from the Gros Ventre River, Wyoming, downstream to Palisades Reservoir and in the Salt River, Wyoming[6,34]. Snake River Fine-Spotted Cutthroat Trout range in Idaho and Wyoming overlaps with Yellowstone Cutthroat Trout distributions. Morphological differences between this form and Yellowstone Cutthroat Trout suggest they are reproductively isolated[6].

Size and Longevity

Maximum size for Idaho Snake River Fine-Spotted Cutthroat Trout is 483 to 508 mm (19 to 20 in) and 1.4 kg (3 lb). They may reach 686 to 711 mm (27 to 28 in) and 4.1 to 4.5 kg (9 to 10 lb) under ideal ecological conditions[34].

Food and Feeding

Snake River Fine-Spotted Cutthroat Trout, like all Cutthroat Trout, are opportunistic feeders, taking invertebrates and occasionally other fish. They may consume both aquatic and terrestrial invertebrates, fishes, amphibians, and crustaceans[6,34].

Limiting Factors

Effects of hybridization, principally with nonnative Rainbow Trout, may decimate Snake River Fine-Spotted Cutthroat Trout populations. Overfishing could occur if not for angling regulations that are protective of this form[6,34].

Breeding Habits

The most distinctive aspect of Snake River Fine-Spotted Cutthroat Trout breeding habits is their ability to maintain separation from the Yellowstone Cutthroat Trout. No explanation for this feature has been accepted by the scientific community, but Behnke offers an explanation: the two forms are sympatric (occurring in the continuous environment of the upper Snake River without physical separation), but they are not syntopic (they do not occur at the same place at the same time). Aside from this novel aspect of reproduction, Snake River Fine-Spotted Cutthroat Trout breeding habits are very similar to those of other Cutthroat Trout[34].

Aside from the subspecies/form overlap feature, Snake River Fine-Spotted Cutthroat Trout spawn in a fashion similar to other Cutthroat Trout. Males typically move into the spawning area when water temperatures are greater than 4.4°C (40°F). Females fan out a redd with their tails in gravel and deposit eggs. The number of eggs per female is highly variable based on individual size and other ecological factors and may vary from a few to more than 8,000. Immediately after egg deposition, a male fertilizes the eggs and typically guards the redd to prevent other males from fertilizing the eggs as well. Following fertilization, the female moves upstream and uses her tail to cover the eggs with 12.7 to 17.8 cm (5 to 7 in) of gravel, in some cases depositing additional eggs in the newly created depression. Redds are typically 76 cm (30 in) long by 46 cm (18 in) wide and in 17.8 to 61 cm (7 to 24 in) of water. Optimal incubation temperature for embryos is approximately 10°C (50°F), and the upper safe temperature range for gravid females and embryos is 13.9°C (57°F). Egg hatching is usually within 28 to 40 days but is variable based on actual temperatures[2,6].

Habits and Habitat

Snake River Fine-Spotted Cutthroat Trout habitat are limited to the main-stem and tributary streams of the upper Snake River basin in Wyoming and Idaho[6] similar to Yellowstone Cutthroat Trout.

Management and Conservation Status

Currently, Idaho does not differentiate Snake River Fine-Spotted Cutthroat Trout from Yellowstone Cutthroat Trout for management purposes. Restrictive sportfishing regulations protect this form, and stocking in systems where Snake River Fine-Spotted Cutthroat Trout occur is done with sterile triploid Rainbow Trout hybrids where feasible.

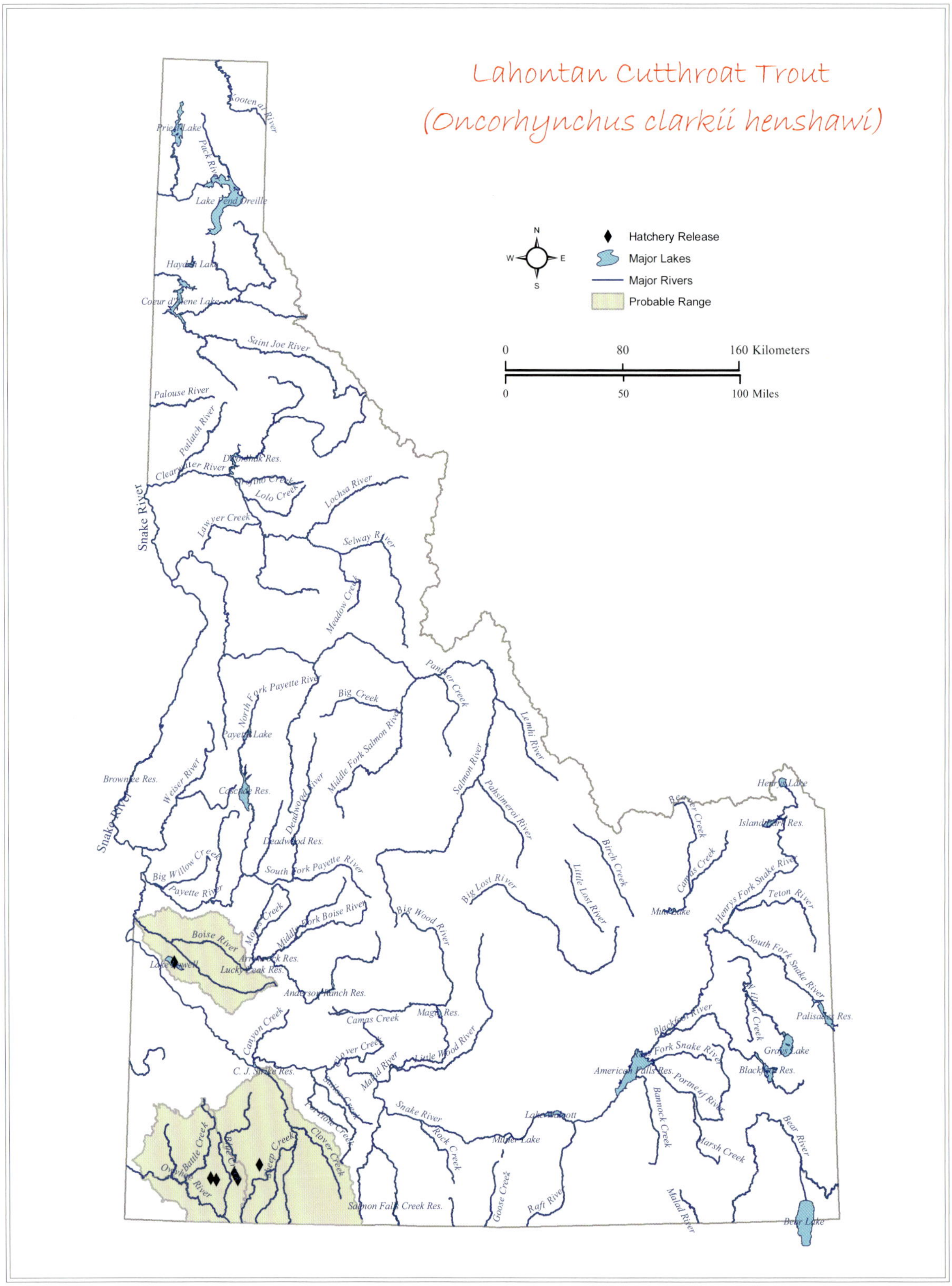
Lahontan Cutthroat Trout
(Oncorhynchus clarkii henshawi)
N
S
W
E
Hatchery Release
Major Lakes
Major Rivers
Probable Range
0
80
160 Kilometers
0
50
100 Miles
Lake Pend Oreille
Saint Joe River
Palouse River
Potlatch River
Lochsa River
Lolo Creek
Snake River
Lawyer Creek
Selway River
Meadow Creek
North Fork Payette River
Big Creek
Middle Fork Salmon River
Panther Creek
Lemhi River
Salmon River
Pahsimeroi River
Weiser River
Deadwood River
Big Willow Creek
Payette River
South Fork Payette River
Birch Creek
Little Lost River
Big Lost River
Camas Creek
Henrys Fork Snake River
Teton River
South Fork Snake River
Big Wood River
Boise River
Mores Creek
Anderson Ranch Res.
Camas Creek
Canyon Creek
C. J. Strike Res.
Little Wood River
Malad River
Blackfoot River
Portneuf River
Bannock Creek
Marsh Creek
Bear River
Snake River
Rock Creek
Goose Creek
Raft River
Malad River
Salmon Falls Creek Res.
Clover Creek
Battle Creek
Owyhee River
Island Park Res.
Grays Lake
Bear Lake
Magic Res.
Lucky Peak Res.

Lahontan Cutthroat Trout

Oncorhynchus clarkii henshawi
(Gill and Jordan, 1878)

Etymology

Oncorhynchus = hooked snout, referring to the condition of the migrating adults in Pacific salmon runs; *clarkii* = one of the two leaders of the Lewis and Clark expedition to western America in 1804–6; *henshawi* = named by Gill and Jordan for naturalist H. W. Henshaw from the Wheeler Survey of the 100th meridian[85]. Lahontan = Baron Louis Armand La Hontan (1666–1713) was an early French explorer in America. He did not enter the Great Basin, but his name was applied to the area of Pleistocene Lake Lahontan.

Importance

Perhaps the most striking contribution of the Lahontan Cutthroat Trout subspecies is related to its evolution in the highly saline Pyramid Lake ecosystem that produced the largest Cutthroat Trout in the world, making it a unique natural history entity[17,34]. Lahontan Cutthroat Trout in Pyramid Lake have always been of substantial importance to the Pyramid Lake Paiute Indian tribe and to sportsmen from both near and far. Prior to the population crash, Lahontan Cutthroat Trout were caught commercially and shipped to markets as far away as San Francisco, California[86,87].

Both the high total dissolved-solids tolerance and the evolutionary pressures that led to the large size of this subspecies were lost when the Newlands Project significantly reduced flows of the Truckee River into Pyramid Lake in 1905. Estimates that the Pyramid Lake Lahontan Cutthroat Trout fishery was capable of sustaining annual removals of 454,000 kg (1,000,898 lb) were made for the nineteenth century[11].

Description

Lahontan Cutthroat Trout have greenish bronze coloration over the dorsal portion of the body, and the sides are light yellowish copper. Fair rose colors occur around the lateral line. Genetic markers distinguish Lahontan Cutthroat Trout from other subspecies, but so do their morphological and meristic characteristics: relatively large but rounded spots occur on the sides of the body in even distribution, gill-raker numbers are higher in the Lahontan Cutthroat Trout than in any other subspecies of cutthroat or rainbow (21 to 28), and numbers of pyloric caeca are higher in this subspecies than others, typically 50 to 60 (40 to 75) rather than 25 to 40. Large, round rather dullish red spots that appear on the head and the caudal peduncle, and occasionally ventrally, are often a good field distinguishing characteristic[11,34,86,87]. Snyder's original description of the Lahontan basin Cutthroat Trout, as presented in La Rivers, is a worthy read for any aficionados of the more poetic descriptions of fish used in the 1900s[85].

Range and Distributional Status

This subspecies is not native to Idaho. Ancestral Cutthroat Trout probably moved into ancient Lake Lahontan from the Columbia River basin and eventually became what is now known as Lahontan Cutthroat Trout. Major lake populations, one of the two evolved populations, were present in Pyramid Lake, Walker Lake, Independence Lake, Lake Tahoe, and Summit Lake[86,87]. While preying upon the abundant forage fishes in these lakes, these populations remained obligate stream spawners. Spawning runs occurred in the Truckee River to Lake Tahoe and in the Walker River. Stream populations occurred in the Truckee, Carson, Walker, and Quinn Rivers. There are a number of stream-dwelling Lahontan Cutthroat Trout populations in the western United States, distributed over the past 100-plus years from various hatchery operations[17,34,87]. In Idaho, Lahontan Cutthroat Trout have been stocked in a number of water bodies since the late 1980s in hopes of providing a recreational fishery. Currently, they are stocked only in several small reservoirs in southwest Idaho.

Size and Longevity

Original stocks of Lahontan Cutthroat Trout from Pyramid Lake, Nevada, were the largest Cutthroat Trout in the world, including the official angling record of 18.6 kg (41 lb). This fish was recorded as 991 mm (39.1 in). More recent record catches range around 12.7 kg (28 lb)[17]. The largest post-1943 Lahontan Cutthroat Trout was 12.7 kg (28 lb), caught in April 1976. A Lahontan Cutthroat Trout from Lake Tahoe caught in 1911 weighed 14.3 kg (31 lb, 8 oz). Pyramid Lake Lahontan Cutthroat Trout are fast growing and short lived. A total of 562 Lahontan Cutthroat Trout ranged in size from 191 mm (7.5 in) and 82 g (2.9 oz) to 787 mm (31 in) and 6.1 kg (13 lb, 8 oz). Yearly total lengths for Lahontan Cutthroat Trout aged one to seven, in mm (in), were 231 (9.1), 312 (12.3), 386 (15.2), 462 (18.2), 523 (20.6), 610 (23.7), and 648 (25.5)[86,87]. Stream Lahontan Cutthroat Trout typically grow slower than the lake forms, but overall size as well as growth rate are dependent on both the type and the abundance of food.

Food and Feeding

All adult Cutthroat Trout are omnivorous, feeding readily on invertebrates but preferring fish. Small Lahontan Cutthroat Trout in Pyramid Lake, Nevada, feed on insects and zooplankton. Feeding on fish begins at lengths of 305 to 330 mm (12 to 13 in), and adults eat an ever-increasing proportion of fishes, primarily Tui Chub (*Siphateles bicolor*), as they increase in size. An abundant supply of Tui Chub allowed historic Lahontan Cutthroat Trout to achieve the enormous (18.6 kg [41 lb]) sizes reported pre-1944[17,86,87].

Limiting Factors

Hooking may be a significant factor in mortality of Lahontan Cutthroat Trout in Pyramid Lake. Spawning-related mortality and diseases are additional factors. Hybridization by introduced species, particularly

Rainbow Trout, can reduce Lahontan Cutthroat Trout viability. Construction of the Derby Dam by the BOR in 1905 contributed significantly to the demise of the original strain of Lahontan Cutthroat Trout[17,34,87].

Breeding Habits

Cutthroat Trout, including Lahontan Cutthroat Trout, are obligate stream spawners and are thus subject to the vagaries of both water quantity and quality, particularly temperature fluctuations. Limited evidence for Lahontan Cutthroat Trout in Pyramid Lake indicates that females do not spawn more than one time[87]. Pyramid Paiute Indians noted early in the 1900s that two distinct spawning runs occurred from Pyramid Lake. Large fish spawned in November, December, and January; a second spawning run, of smaller individuals, occurred in the spring[17,86].

Egg production by Lahontan Cutthroat Trout females is significantly related to age, fork length, and weight. Fecundities of Lahontan Cutthroat Trout from Pyramid Lake varied from 1,241 to 7,963 and averaged 3,815 eggs per female. Other populations have lower average fecundities. Female Lahontan Cutthroat Trout in Pyramid Lake mature at three or four years of age at lengths of 352 to 484 mm (13.9 to 19.1 in) FL, while males mature at ages two or three at fork lengths of 299 to 445 mm (11.8 to 17.6 in). Mature eggs of Pyramid Lake Lahontan Cutthroat Trout ranged in diameter from 2.76 to 5.09 mm (0.11 to 0.2 in)[17,86,87].

Habits and Habitat

Deep and cold, the waters of Pyramid Lake, and the associated areas of Walker Lake, Independence Lake, Lake Tahoe, and Summit Lake as well as the Truckee, Carson, Walker, and Quinn Rivers, were the original habitat of Lahontan Cutthroat Trout. Perhaps the most remarkable characteristic of Pyramid Lake is the very high level of total dissolved solids (TDS) present. Early data indicate a TDS in the early 1900s of nearly 4,000 mg/l (parts per million)[87].

Pyramid Lake, at a surface elevation of 1,157 m (3,796 ft) above sea level, has a volume of 26.4 km^3 (6.3 cubic mi or 21,288,960 acre ft). Pyramid Lake is 40 km long (25 mi) with a mean depth of 59 m (194 ft) and a maximum depth of 103 m (338 ft)[17,87].

Chemical composition of terminal lakes is determined by many factors: precipitation, temperature, evaporation, basin sediments, and influent composition, to name a few. Saline terminal lakes are typically well buffered with stable hydrogen ion concentrations[86].

Management and Conservation Status

Lahontan Cutthroat Trout are classified as a vulnerable subspecies range wide (G4T3). A state conservation status ranking is not applicable (SNA) since these fish are introduced[14]. In Idaho, Lahontan Cutthroat Trout provide a limited but unique addition to the recreational sport fisheries.

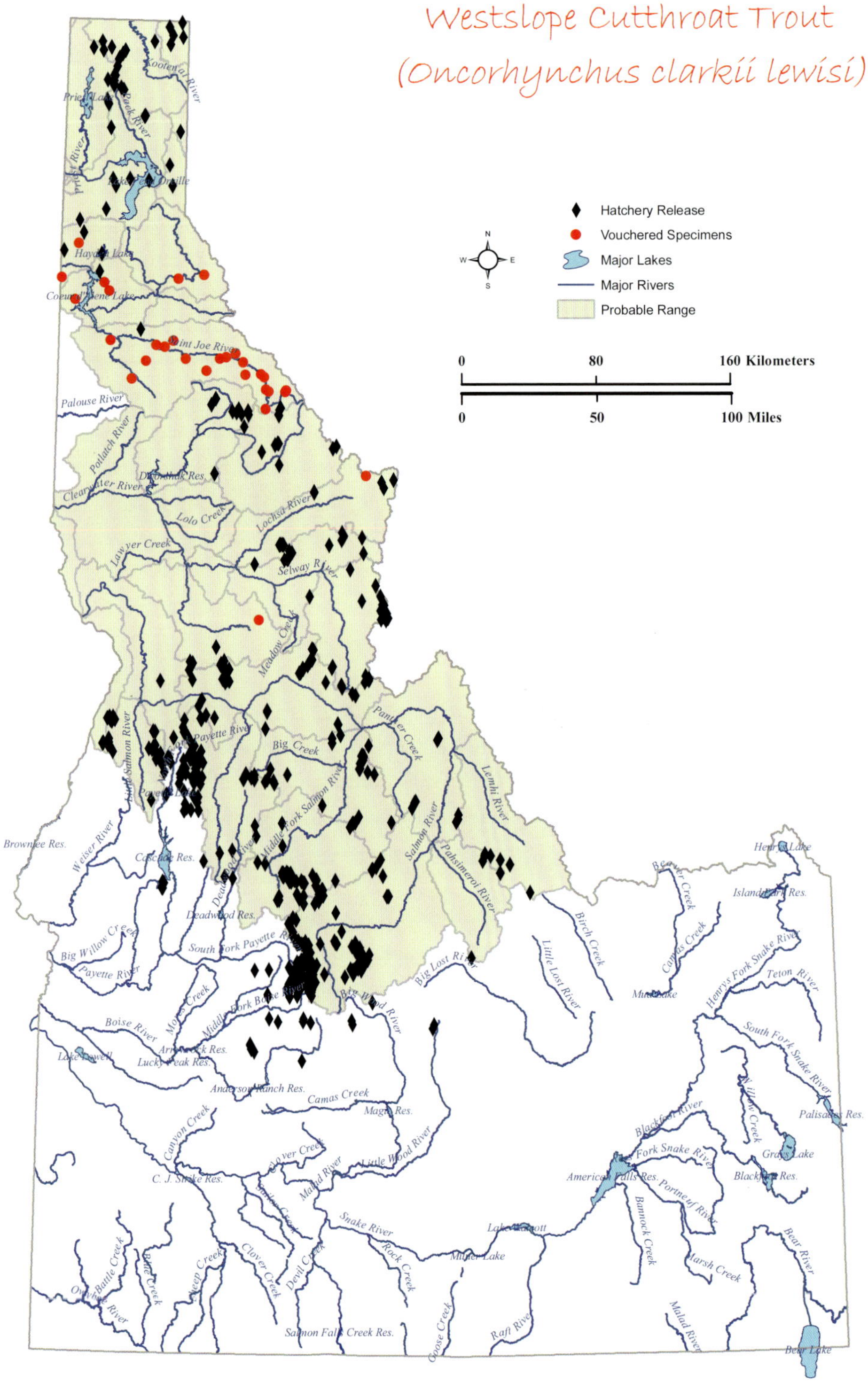
Westslope Cutthroat Trout
(Oncorhynchus clarkii lewisi)
Hatchery Release
Vouchered Specimens
Major Lakes
Major Rivers
Probable Range
N
W
E
S
0
80
160 Kilometers
0
50
100 Miles
Kootenai River
Palouse River
Potlatch River
Clearwater River
Lolo Creek
Lochsa River
Lawyer Creek
Selway River
Meadow Creek
Big Creek
Little Salmon River
Weiser River
Brownlee Res.
Cascade Res.
Deadwood Res.
South Fork Payette River
Big Willow Creek
Payette River
Boise River
Lucky Peak Res.
Anderson Ranch Res.
Camas Creek
Magic Res.
Canyon Creek
C. J. Strike Res.
Little Wood River
Snake River
Rock Creek
Clover Creek
Devil Creek
Salmon Falls Creek Res.
Goose Creek
Raft River
Milner Lake
Lake Walcott
American Falls Res.
Bannock Creek
Portneuf River
Marsh Creek
Malad River
Bear River
Bear Lake
Blackfoot Res.
Grays Lake
Palisades Res.
South Fork Snake River
Teton River
Henrys Fork Snake River
Island Park Res.
Henrys Lake
Mud Lake
Camas Creek
Beaver Creek
Birch Creek
Little Lost River
Big Lost River
Big Wood River
Pahsimeroi River
Salmon River
Lemhi River
Panther Creek
Middle Fork Salmon River
Payette River
Lake Lowell

Westslope Cutthroat Trout

Oncorhynchus clarkii lewisi
(Girard, 1856)

Adult

Juvenile

Etymology

Oncorhynchus = hooked snout, referring to the condition of the migrating adults in Pacific salmon runs; *clarkii* = one of the two leaders of the Lewis and Clark expedition to western America in 1804–6; *lewisii* = one of the two leaders of the Lewis and Clark expedition to western America in 1804–6. Westslope = George Suckley, a surgeon-naturalist working for the US Army in the 1850s, captured individuals of this species below the Great Falls of the Missouri in 1853 and noted the similarity of their appearance with fish he had captured in the lower Columbia River drainage[34].

Importance

Westslope Cutthroat Trout, despite some confusion associated with their common name, contribute some unique traits to the overall Cutthroat Trout mystique. Westslope Cutthroat Trout tend to have fewer meristic elements than Yellowstone Cutthroat Trout and uniquely have 66 chromosomes, a trait not shared with any other subspecies of Cutthroat Trout or Rainbow Trout[34]. Preservation of this unique subspecies is important, as strong populations still exist in Idaho in about 50 percent of their historical range[6].

Description

Although there are no meristic characteristics such as fin-ray counts, vertebrae, or number of scales that separate Westslope Cutthroat Trout from Yellowstone Cutthroat Trout, Westslope Cutthroat Trout tend to have fewer of these characters in general[34]. Westslope Cutthroat Trout are greenish blue to steel gray on back and sides; some individuals may be more silvery. Yellowish tints may be present or bright yellow; orange and red may be present. This coloration is more apparent than on Yellowstone Cutthroat Trout. Young have nine to 10 oval parr marks along the sides of the body, and the number of gill rakers is typically 18 to 19[6]. Vertebrae are 59 to 63, typically 60 to 61, and lateral line series scales are 150 to 200 or more, with mean values from 165 to 180[11,34]. Gill rakers are 17 to 21, pyloric caeca number 25 to 50, and basibranchial teeth are about five to 10, as opposed to five to 15 in Yellowstone Cutthroat Trout. With regard to the spotting pattern and efforts to separate Yellowstone Cutthroat Trout and Westslope Cutthroat Trout in the field: Westslope Cutthroat Trout have an arc area without spots that runs from the anterior base of the anal fin forward to the pectoral fin[34]. Westslope Cutthroat Trout can be distinguished from Yellowstone Cutthroat Trout by a combination of these traits. Westslope Cutthroat Trout have irregular spots, primarily above the lateral line and on the caudal peduncle but absent from the anal fin. These spots are reportedly smaller on this subspecies than any other. Chromosome count in Westslope Cutthroat Trout is unique at 66, which does not occur in any other subspecies of Cutthroat Trout or Rainbow Trout[6,34,77]. Westslope Cutthroat Trout form a unique monophyletic lineage from other Cutthroat Trout subspecies based on DNA sequence data[88].

Range and Distributional Status

This subspecies is native to Idaho. Naturally, Westslope Cutthroat Trout were more widely distributed than all other subspecies of Cutthroat Trout, occurring over portions of British Columbia, Alberta, Washington, Oregon, Idaho, Montana, and Wyoming, including a portion of Yellowstone National Park[11,34]. In the 2009 status update of the Westslope Cutthroat Trout Conservation Team, Westslope Cutthroat Trout were estimated to currently occupy about 33,608 miles (58 percent) of the nearly 58,030 miles of historically occupied habitats[89].

In Idaho, Westslope Cutthroat Trout are native to the Kootenai River, Pend Oreille River, Spokane River, Clearwater River, and Salmon River drainages, including Coeur d'Alene, Pend Oreille, Priest, and other lowland lakes throughout the Idaho Panhandle region[6].

Size and Longevity

Westslope Cutthroat Trout typically attain a smaller maximum size than other subspecies, which may be an evolutional aspect of past competition and coevolution with Bull Trout and Northern Pikeminnow in the upper Columbia River basin[34]. Westslope Cutthroat Trout in small streams may reach sizes of only 150 to 230 mm (6 to 9 in) in length and weights of only 28 to 142 g (1 to 5 oz), whereas individuals in larger rivers and lakes may attain sizes of 410 to 460 mm (16 to 18 in) and 0.45 to 1.4 kg (1 to 3 lb)[34]. Westslope Cutthroat Trout exhibit three distinctive life history forms: resident (nonmigratory) in small streams, a fluvial form that migrates between small streams for spawning and larger streams for the remainder of its life, and an adfluvial form that lives in lakes. Westslope Cutthroat Trout live to age six to eight; shorter life spans are exhibited by individuals in small streams[34].

Food and Feeding

Westslope Cutthroat Trout eat primarily invertebrates, with insects and crustaceans making up significant portions of the diet in lakes. Westslope Cutthroat Trout are rarely piscivorous. Fingerlings feed on aquatic insect larvae and terrestrial insects. Aquatic flies (Diptera) and mayflies (Ephemeroptera) are primary food for small fingerlings, while larger individuals feed on caddisfly (Trichoptera) larvae[6,11,34].

In a 2011 Salmon River study, diet of Westslope Cutthroat Trout was dominated by aquatic invertebrates (89 percent of prey items), mainly Ephemeroptera (49 percent), Trichoptera (26 percent), and Plecoptera (9 percent). Terrestrial insects accounted for 10 percent of prey items in the stomach contents of Westslope Cutthroat Trout, while fish prey (less than 1 percent) were rarely encountered in their stomach contents. The length of prey consumed by Westslope Cutthroat Trout showed no significant correlation with predator body size[57].

Limiting Factors

Westslope Cutthroat Trout are highly susceptible to replacement by nonnative species, particularly Brook Trout in small streams and Brown Trout in larger streams and rivers. While Westslope Cutthroat Trout coevolved with native Rainbow-Redband Trout in some ecosystems, nonnative Rainbow Trout stocked over native Westslope Cutthroat Trout always pose a threat through hybridization and dilution of the Westslope Cutthroat Trout gene pool[34]. Small size and slow growth rates make Westslope Cutthroat Trout vulnerable to Bull Trout predation in some headwater streams[11,34]. Habitat degradation associated with dam operation or reduced water quality is a decimating factor. As with all Cutthroat Trout, Westslope Cutthroat Trout are vulnerable to over exploitation in unregulated sportfishing[6].

Reporting on Westslope Cutthroat Trout restoration efforts in a small stream within the Missouri River basin, Montana, researchers documented that the population of Westslope Cutthroat Trout increased dramatically, at least sevenfold, following removal of Brook Trout, with the most pronounced response seen for age-zero Westslope Cutthroat Trout. The portion of the study stream where habitat was restored supported similar abundances as found in natural stream reaches above and below this restored portion. Following Brook Trout removal, standing crops of Westslope Cutthroat Trout in allopatry were similar to combined standing crops of Brook Trout and Westslope Cutthroat Trout in sympatry prior to initiation of Brook Trout removal[90].

Introductions of nonnative fishes often result in unintended consequences for freshwater biota and ecosystems, including hybridization among native and nonnative fishes and the potential loss of locally adapted gene complexes in native populations. Hybridization is perhaps the primary threat facing Westslope Cutthroat Trout. A Montana study examined spatial and temporal patterns of hybridization between *O. c. lewisi* and nonnative Rainbow Trout (*Oncorhynchus mykiss*) within a reported stronghold for the natives in the Flathead River system[90].

Although all invasive taxa are subject to environmental constraints during establishment and expansion, hybridization presents the additional requirement of genetic compatibility. However, if this genetic criterion is satisfied, the invasive genome may benefit by its integration into the locally adapted native species. In this study, the initial establishment of *O. mykiss* may be limited by the suitability of introgression and may not be subject to such restrictions once source populations are established[90].

Additional study findings indicate that (1) *O. mykiss* introgression is spreading rapidly in the study area, (2) native *O. c. lewisi* populations are declining rapidly in the study area, (3) hybridization is spreading rapidly in the study area and in an upstream direction from the Flathead River main stem, and (4) environmental factors should not be expected to prevent the additional spread of hybridization in the study area[91].

In a 2014 study of mechanisms relating introgression between Westslope Cutthroat Trout and Rainbow Trout, researchers found that the occurrence and level of hybridization in Westslope Cutthroat Trout were not related to whether the study site was above a movement barrier. Levels of introgression were related to the distance to the nearest Rainbow Trout stocking location, the number of Rainbow Trout stocking events within the nearest watershed, and the total number of Rainbow Trout stocked in that watershed. Data indicate that Westslope Cutthroat Trout located farther from Rainbow Trout stocking have a lower risk of introgression, but those isolated above movement barriers should not be considered free from introgression until a thorough genetic assessment of the location has been made[92].

In North America, many populations of native Cutthroat Trout are threatened by introgressive hybridization with nonnative salmonids, yet little is known about the reproductive factors influencing the spread of hybridization in the natural environment. Recent data illustrate that contact between previously allopatric species results in hybridization and introgression due to incompletely evolved reproductive isolation mechanisms. In a studied population, the timing and location of spawning varied spatially and temporally, producing zones of contact (e.g., hybrids zones) between parental species, which results in loss of integrity in, for instance, the native Westslope Cutthroat Trout gene pool. Spawning distributions and movement patterns indicate that hybridization is promulgated by long-distance movements of individuals with a high proportion of Rainbow Trout genetic material. Additionally, the spatial distribution of later-generation backcrosses suggests that hybridization is also spreading by dispersal between neighboring populations[93].

Increased straying rates in the invasive taxon can contribute to extinction of native species by hybridization. Native Westslope Cutthroat Trout populations exhibit a high degree of genetic divergence over short geographical distances, indicating low straying rates among populations. Conversely, the rapid spread of hybridization indicates a high rate of dispersal and gene flow from hybrid and Rainbow Trout sources. Data from the Flathead River watershed indicate that hybridization is spreading upriver, in part by long-distance movements of individuals with high amounts of Rainbow Trout admixture that likely strayed into streams containing nonhybridized Westslope Cutthroat Trout[93].

In Idaho, distribution (and abundance) of Westslope Cutthroat Trout and Redband (Steelhead/Rainbow) Trout are naturally co-occurring in the Clearwater and Salmon River basins. There are strong Westslope Cutthroat Trout populations that live in harmony with Redband Trout and Steelhead populations, so there must be some form of spatial/temporal spawning segregation occurring.

Breeding Habits

Typical Westslope Cutthroat Trout mature at age three with their first spawning occurring at age four or five. Spawning occurs in smaller headwater streams beginning in April and ranging into June, depending on local water temperatures and flow. Spawning occurs at temperatures from 6 to 9°C (43 to 48°F)[6].

Habits and Habitat

Westslope Cutthroat Trout are the most widespread of all of the Cutthroat Trout subspecies, resulting in a broad diversity of habitat across both their original and their current distribution. Habitat for Westslope Cutthroat Trout includes lakes, rivers, and streams with clear, cool water of less than 15°C (59°F)[34]. Complex habitat elements that mix pools, riffles, runs, pools, and pocket waters are important, as are deeper pools or pools with extensive woody debris. Interstitial spaces in the gravel of streambeds are important habitat for smaller Westslope Cutthroat Trout as temperatures drop in the fall. Low fine-grained sediment content is therefore important for both juvenile overwintering and survival of eggs and fry after spawning[6].

Management and Conservation Status

Westslope Cutthroat Trout are ranked as apparently secure but vulnerable (G4T3) range wide and vulnerable (S3) in Idaho[15]. The IDFG top priority for Westslope Cutthroat Trout conservation is habitat protection and enhancement, particularly for core conservation and conservation populations. IDFG fishery management staff work with other agencies and stakeholders to identify, plan, fund, and implement habitat-enhancement measures across the range of Westslope Cutthroat Trout in Idaho.

As with all Cutthroat Trout subspecies, genetically pure Westslope Cutthroat Trout populations have been adversely impacted by habitat loss and degradation and by competition and predation by nonnative species. In Idaho, Westslope Cutthroat Trout exist in an indeterminate percentage of their original range, largely in undeveloped areas within the vast network of protected federal lands in the state. IDFG estimates they currently occupy 27,784 km (17, 265 mi) of the historical 34,661 km (21, 637 mi), or 80 percent. However, they appear to be doing quite well[94]. Idaho classifies Westslope Cutthroat Trout as a game species, and management protocols to protect this subspecies have been implemented.

Effects of high-density stocking of juvenile salmon and Steelhead annually in the Salmon River and Clearwater River on native populations of Westslope Cutthroat Trout are unknown, and studies documenting such interactions are lacking. However, researchers have documented declines in resident salmonid stocks (numbers and biomass), probably associated with the cumulative impacts from multiple stockings of Chinook Salmon smolts[95].

In a 2003 study in the Clearwater River basin, Cutthroat Trout × Rainbow Trout hybrids were distributed in a broad hybrid zone in the study area, and introgression was detected at 64 percent of the sites sampled. The distinct distribution pattern of introgression in Westslope Cutthroat Trout populations in the Clearwater River basin has several implications for management. Data suggest that biological factors, rather than stocking practices, may be responsible for the observed distribution pattern. Biological barriers to introgression, such as habitat selection, assortative mating, and losses in fitness, have been found in several hybridized species of native interior trout. Further understanding and identification of these barriers limiting gene exchange are essential to the management and conservation of native trout populations and would allow managers to perform risk analyses of the potential for invasion of nonnative trout or introgression in different stream habitats. These tools would be useful when determining the need for various management actions, such as artificial barriers often used to protect Cutthroat Trout populations from invasion by nonnative trout[91].

Brook Trout are present in a high proportion of streams across the range of Westslope Cutthroat Trout and have replaced Cutthroat Trout in many streams across their historical range in Idaho[96]. Brook Trout occupy an estimated 9,000 km (almost 60,000 mi) of streams in Idaho within the range of Westslope Cutthroat Trout. Nonnative Rainbow Trout pose a more significant threat to Westslope Cutthroat Trout than do Brook Trout. Other nonnative species present in Idaho that potentially compete with or prey upon Westslope Cutthroat Trout include Lake Trout (*Salvelinus namaycush*), Walleye (*Sander vitreus*), Northern Pike (*Esox lucius*), and Smallmouth Bass (*Micropterus dolomieu*).

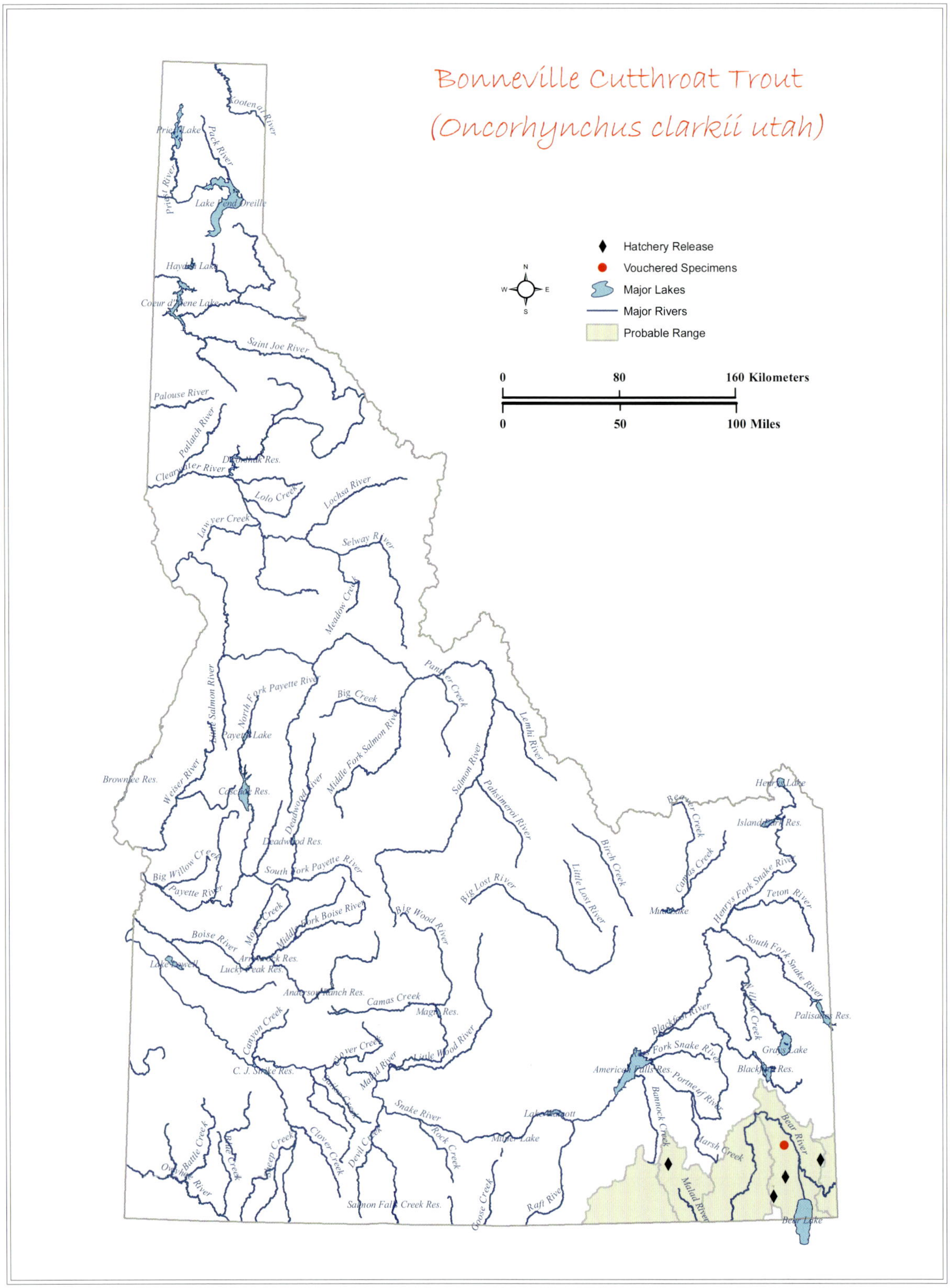
Bonneville Cutthroat Trout
(Oncorhynchus clarkii utah)
Hatchery Release
Vouchered Specimens
Major Lakes
Major Rivers
Probable Range
N
W
E
S
0
80
160 Kilometers
0
50
100 Miles
Pack River
Saint Joe River
Palouse River
Potlatch River
Lolo Creek
Lochsa River
Selway River
Meadow Creek
Big Creek
Middle Fork Salmon River
Lemhi River
Salmon River
Pahsimeroi River
Little Salmon River
Weiser River
Birch Creek
Little Lost River
Big Lost River
Big Wood River
Camas Creek
Teton River
Henrys Fork Snake River
South Fork Snake River
Payette River
Boise River
Canyon Creek
Snake River
Rock Creek
Clover Creek
Raft River
Goose Creek
Bannock Creek
Marsh Creek
Malad River
Bear River
Bear Lake
Palisades Res.
Deadwood Res.
Brownlee Res.

Bonneville Cutthroat Trout

Oncorhynchus clarkii utah
(Suckley, 1874)

Etymology

Oncorhynchus = hooked snout, referring to the condition of the migrating adults in Pacific salmon runs; *clarkii* = one of the two leaders of the Lewis and Clark expedition to western America in 1804–6; *utah* = name provided by George Suckley in his *On the North American Species of Salmon and Trout*, published in 1874. Suckley intended the subspecies designation "*utah*" to apply only to the Cutthroat Trout found in Utah Lake, Utah, because they were "less spotted" than other Cutthroat Trout from the Bonneville Basin. See Behnke for a more complete explanation of Cutthroat Trout subspecies nomenclature[34].

Importance

As the dominant native aquatic predator in the massive Bonneville Basin, the Bonneville Cutthroat Trout holds a special place among western fishes. Likely an offshoot of the Yellowstone Cutthroat Trout, Bonneville Cutthroat Trout evolved in the vast (51,800 km^2 [20,000 mi^2]) Lake Bonneville that covered more than half of the present state of Utah as well as portions of Idaho, Nevada, and Wyoming until approximately 15,000 years ago. Desiccation of Lake Bonneville left the remnant Utah Lake and the smaller Sevier Lake, which desiccates except in periods of high water[34]. Evolution of the Bonneville Cutthroat Trout in the Bonneville Basin leaves four distinct morphotypes that persist in Idaho and Utah: Bear Lake Cutthroat Trout, those populations in the main Bonneville Basin, Snake Valley populations along the Utah-Nevada border, and those individuals in the Bear River drainage, Idaho, Utah, and Wyoming. Of specific importance are the hereditary distinctions of the Bear Lake morphotype that facilitate their survival in Bear Lake, Idaho-Utah: lacustrine life history, older age at maturity, longer life span, and highly predatory (piscivorous) feeding habits primarily on Bear Lake Sculpin, Bonneville Cisco, and Bonneville Whitefish. Life history adaptations of the Bear River (harsh, highly fluctuating environment) and the Bear Lake (piscivorous and ability to survive under massive stocking of nonnative species) of this subspecies make additional study of this subspecies essential[2,6,34].

Wild populations of genetically pure lake strains of Bonneville Cutthroat are not believed to be found anywhere within their historical range except Bear Lake. Spawners are taken at the Swan Creek Trap in Idaho each spring. Fertilized eggs are taken to hatcheries for rearing. Most of the young are returned to Bear

Lake, but others are used to replenish other natal populations and are stocked out of their natural range as sportfish[21].

Description

Generally, Bonneville Cutthroat Trout and other Cutthroat Trout subspecies are similar except that Bonneville Cutthroat Trout have a tendency to appear silvery gray versus the yellowish brown of other subspecies. Spots on the Bonneville Cutthroat Trout are generally evenly distributed and large; however, great variation exists among the separate populations noted above[34]. Gill rakers are typically 18 to 19, except the Snake Valley populations typically have 20 to 23[7]. While most Bonneville Cutthroat Trout have basibranchial teeth averaging 20 to 28, most other Cutthroat Trout populations average five to 10[6]. Bonneville Cutthroat Trout do not develop brilliant crimson colors as do the other subspecies of Cutthroat Trout. Bear Lake Cutthroat Trout were given the name "bluenose" due to the azure blue color of the entire dorsal surface of this morphotype. The caudal fin is moderately forked. Compared to other Cutthroat Trout, Bonneville Cutthroat Trout native to the Bear River drainage have more scales along the sides: 165 to 180 versus 150 to 165 for other populations of the subspecies. Pyloric caeca for Bear River drainage populations are more numerous (40 to 45 versus 30 to 40), except Bear Lake Bonneville Cutthroat Trout have 50 or more[34].

Bonneville Cutthroat Trout generally have large, evenly distributed spots, but there can be a high degree of intra-basin variation. Bonneville Cutthroat Trout tend to develop large pronounced spots that are more evenly distributed on the sides of the body rather than concentrated posterior as in the Yellowstone subspecies. Coloration in Bonneville Cutthroat Trout is generally dull compared to other Cutthroat Trout subspecies with no bright colors except for the classic orange cutthroat mark. Mature males may develop a rosy tint along the body but coloration can vary depending on environmental conditions and local genetic composition. Vertebrae typically number 62-63, slightly higher than in other subspecies. Scales in lateral series average 150 to 170. Pyloric caeca number 25 to 55 with a mean of 35, except in the Bear River drainage population, which typically averages more than 40 caeca. Bonneville Cutthroat Trout average 16 to 21 gill rakers, with a mean of 17 to 19, except the Snake Valley type which have 18 to 24 (mean, 20 to 22). Another important characteristic of all cutthroat subspecies is the presence of basibranchial teeth which are absent in Rainbow Trout. Numbers of basibranchial teeth provide information about subspecies derivation and relatedness. Yellowstone Cutthroat Trout have profuse basibranchial teeth, averaging 20 to 28, while most Bonneville Cutthroat Trout average 5 to 10[7,97].

Range and Distributional Status

This species is native to Idaho. Bonneville Cutthroat Trout are distributed unevenly across the area covered by Pleistocene Lake Bonneville 15,000 years ago, including the western half of what is now the state of Utah and portions of Idaho, Nevada, and Wyoming. Presently, Bonneville Cutthroat Trout occur in the Bear River basin in Idaho, Wyoming, and Utah; the Bonneville Basin in Utah; the Snake River Valley in Nevada; and in Idaho in the Bear and Malad River drainages and in Bear Lake[6,34]. Recent genetic studies of southern and western populations of Bonneville Cutthroat Trout indicate that they may belong to an entirely different lineage of Cutthroat Trout than the current classification recognizes[75].

Lake-dwelling populations of Bonneville Cutthroat Trout are found in Wyoming in Lake Alice. Lake Alice is located on Poker Creek, tributary to Hobble Creek, a tributary to the Smiths Fork River, about 35 miles northeast of Cokeville, Wyoming. Lake Alice is about 4.8 km (3 mi) long and is natural lake that was formed by a landslide[98].

Size and Longevity

Much like their close relatives in Pyramid Lake, Nevada, Bear Lake Cutthroat Trout, one of the three forms of Bonneville Cutthroat Trout, grew to extraordinary sizes prior to ecosystem disruption and stocking of nonnative species. Reports of Bear Lake Cutthroat Trout range from 10.4 to 11.8 kg (23 to 26 lb) in previous centuries. Reductions in size to approximately 610 mm (24 in) and 2.3 kg (5 lb, 1 oz) for the current population are likely a result of competition with the introduced Lake Trout[34]. IDFG data from past studies on Bonneville Cutthroat Trout in Bear Lake reflect that Bonneville Cutthroat Trout, aged two to 14, had lengths, in mm (in), of 298 (11.7), 371 (14.6), 426 (16.7), 468 (18.4), 524 (20.6), 542 (21.3), 556 (21.9), 567 (22.3), 575 (22.6), 581 (22.9), 585 (23.0), and 589 (23.2)[99]. Idaho lists an 8.6 kg (18 lb, 15 oz) 1970 catch from Bear Lake as the record for this subspecies.

Food and Feeding

Bonneville Cutthroat Trout in Bear Lake are the only trout known to increase weight during winter months, similar to Arctic Char, when temperatures are below 4°C (39°F). Other forms of cutthroat as well as many other species of trout have been experimentally determined to stop growing at this temperature. A clear reason for the Bear Lake Cutthroat Trout's ability to gain weight during winter has not been presented. Behnke has speculated it may be the result of evolutionary programming associated with the Bear Lake Cutthroat Trout's ability to feed on spawning aggregations of Bonneville Cisco during the winter[34].

Bear Lake is an oligotrophic (unproductive) ecosystem and thus provides a scarcity of aquatic vertebrates for young Bonneville Cutthroat Trout to feed on. Juvenile Bonneville Cutthroat Trout in Bear Lake rely on terrestrial insects, which may constitute 50 percent of the trout's total food supply. As Bonneville Cutthroat Trout sizes increase to about 225 mm (8.9 in), Bear Lake Sculpin enter the diet. At lengths of 406 mm (16 in) or more, Bear Lake Cutthroat Trout feed heavily on fishes, primarily Bear Lake Sculpin and Bonneville Cisco, which may make up 95 percent of their diet[34].

Interestingly, Bonneville Cutthroat Trout in Bear Lake are more able to compete with the introduced Lake Trout than Cutthroat Trout in other ecosystems. Lake Trout were introduced into Bear Lake in 1911 and significant numbers have been stocked routinely since then. Lake Trout have not produced large numbers of surviving adults in Bear Lake, perhaps due to predation on Lake Trout eggs and young by Bear Lake Sculpin and the endemic Bonneville Cisco and Bear Lake whitefishes[6,34].

Limiting Factors

As with other Cutthroat Trout subspecies, competition with stocked nonnative species may reduce the ability of Bonneville Cutthroat Trout to survive in any of the existing populations. Interestingly, in Bear Lake, competition with the Lake Trout is related to early life history stage competition for food. Unlike the situation with Yellowstone Cutthroat Trout in Yellowstone Lake, Lake Trout consume less than 10 percent Cutthroat Trout in Bear Lake[34]. Previously, the lack of suitable in-lake spawning sites for Lake Trout may have reduced potential recruitment. Currently, only triploid/sterile Lake Trout are stocked in Bear Lake.

In other populations, water diversions, poor water quality, habitat fragmentation, hybridization, and competition from nonnative species are major factors adversely affecting Bonneville Cutthroat Trout. An estimated 31,000 Bonneville Cutthroat Trout were present in the Bear Lake in 1995. Populations of Bonneville Cutthroat Trout were in decline due to dewatering of spawning habitat and increased human exploitation.

Human changes to aquatic ecosystems have long concerned fishery researchers due to their potential to affect various aspects of fish life history. Temperature changes are among the greatest concerns in streams where lethal temperatures can develop during any part of the species life history. In concert with flow regulation, temperature can limit available habitat at critical times. In a 2014 study of Bonneville Cutthroat Trout in regulated and unregulated segments of the Bear River, Idaho and Wyoming, during the warmest period of the summer (July 1–August 15), Bonneville Cutthroat Trout used habitats with temperatures that were cooler or similar to most available habitat depending upon the study segment they were in. Each of the three methods used to measure temperatures in available habitats identified similar trends in the thermal profile of the Bear River during the summer. In the regulated segment, Bonneville Cutthroat Trout used habitats with average temperatures cooler than average temperatures in available habitats during the last two weeks of July and first week of August. In the unregulated segment, Cutthroat Trout used habitats with average temperatures around 21.5 to 22.0°C (70 to 72°F) during the first four weeks of the study period. The third week in July was the warmest week in the unregulated segment, and during this period Bonneville Cutthroat Trout used average temperatures in habitats that showed the largest difference between average temperatures measured in available habitats[100].

To compare the frequency of temperatures used by Bonneville Cutthroat Trout with their corresponding frequency of occurrence in the Bear River, researchers compiled the distribution of temperatures using thermal imagery data and the distribution of temperatures used by Bonneville Cutthroat Trout during the warmest period of the summer (July 1–August 15) recorded from radiotelemetry data. In the regulated segment, thermal imagery data recorded a range of temperatures from 18.0 to 28.1°C (64 to 83°F). Only 0.02 percent of the available habitat had temperatures less than the stressful limit of 22°C (72°F), and 0.12 percent had temperatures below the lethal limit of 24.2°C (76°F). In contrast, Bonneville Cutthroat Trout used habitats that had temperatures ranging between 11.8 and 24.3°C (53 to 76°F). Sixty-three percent of the habitat used by Bonneville Cutthroat Trout had temperatures below 22°C (72°F), and 98 percent had temperatures below 24.2°C (76°F). In the unregulated segment, thermal imagery data ranged between 17.5 and 28.5°C (64 to 83°F). Twelve percent of the available habitat had a temperature below 22°C (72°F), and 99 percent was below 24.2°C (76°F). Temperature in habitat used by Bonneville Cutthroat Trout in the unregulated segment ranged from 18.5 to 24.8°C (65 to 77°F). Researchers noted that 59 percent of the habitat used by Bonneville Cutthroat Trout had a temperature below the stressful limit of 22°C (72°F), and 97 percent was below the incipient lethal temperature of 24.2°C (76°F). The distribution of Bonneville Cutthroat Trout during peak summer temperatures was correlated with tributary confluences in the regulated segment, whereas in the unregulated segment Bonneville Cutthroat Trout distribution was more dispersed[100].

Breeding Habits

Both age of reproductive potential and time of spawning are highly variable across all Bonneville Cutthroat Trout populations. Generally, Bonneville Cutthroat Trout spawn from April through June, when water temperatures are consistent at 4 to 10°C (39.2 to 50°F). Populations at lower elevations spawn earlier. Large adfluvial Bonneville Cutthroat Trout spawn in large gravels or cobbles, and large numbers travel up Swan Creek and St. Charles Creek from Bear Lake, to spawn. Smaller stream-resident Bonneville Cutthroat Trout spawn over small gravels or coarse sand in some ecosystems in streams with gradients of 1.5 to 3 percent[6].

While most populations of Bonneville Cutthroat Trout reach sexual maturity at age two to three, the Bear Lake population matures later. Typically, Bear Lake Bonneville Cutthroat Trout reach sexual maturity at six to nine years of age, when they are 559 to 610 mm (22 to 24 in) long, weighing 1.4 to 2.3 kg (3 to 5 lb). Fecundity of Bonneville Cutthroat Trout averages 1,800 to 2,000 eggs per kg (300 to 900 per lb) of body weight[6].

Typical of all Cutthroat Trout, Bonneville Cutthroat Trout fry emerge from the gravel in mid- or late summer, depending on temperatures and time of spawning. Adfluvial Bonneville Cutthroat Trout in Bear Lake may spend one or two years in the natal stream before migrating to Bear Lake proper. Progeny of fluvial spawners may stay in tributary streams for up to three years before migrating to larger main-stem streams or river reaches. While individuals in some populations may display seasonal movements of as much as 128 km (80 mi), most resident Bonneville Cutthroat Trout move less than 1 km (0.62 mi)[6,34].

Both the age at maturity and the annual timing of spawning vary geographically with elevation, temperature, and life history strategy. Spawning by Bonneville Cutthroat Trout usually occurs during the spring and early summer at higher elevations at temperatures ranging from 4-10°C. Bonneville Cutthroat Trout in Bear Lake began spawning in late April and completed spawning in June[7,97].

Post-spawning movement distances for Bonneville Cutthroat Trout in a Bear River tributary ranged from less than 0.5 km to 82.0 km (0.31 to 51 mi) downstream from spawning locations. No fish moved upstream. Post-spawning movement distances formed a continuum, with two fish even leaving the Thomas Fork and entering the Bear River system. There was a significant positive linear relationship between post-spawning distance moved and fish length. Despite the range of movement of up to 82 km (51 mi) observed after spawning, this research notes that the same fish showed little movement during the summer at a scale larger than 0.5 km (0.31 mi). None of 11 fish monitored daily had a displacement distance of more than 0.3 km (0.19 mi) throughout the summer of observations[70].

Habits and Habitat

Bonneville Cutthroat Trout in Idaho exhibit three life history strategies: stream resident, fluvial, and adfluvial. Only the Bear Lake population is adfluvial[6,34]. Ages zero to one Bonneville Cutthroat Trout prefer pools and stream runs where root wads and woody debris occur in abundance. In Bear Lake, Bonneville Cutthroat Trout are typically found in the littoral (shallow area along the shore) and pelagic (deep water) sections of the lake most of year, with separation of juveniles and adults apparent[34]. This distribution of adults and juveniles among habitat sectors appears to be based on temperature. Fluvial Bonneville Cutthroat Trout appear to tolerate higher water temperatures than other subspecies of Cutthroat Trout. Fluvial individuals have been observed in streams where the mean water temperature during summer is 17 to 18°C (63 to 64°C). Presence of preferred habitat in both streams and lakes may be a factor in fish density[6].

Schrank and Rahel[70], reporting on post-spawning and summer movement patterns of Bonneville Cutthroat Trout, note that the extent of migratory behavior within a population appears to depend on the degree of habitat complementation that exists in a drainage. If all the habitat fish need to complete their life history is in close proximity, then movement will be minimal. However, if spawning, feeding, and overwintering habitats are located far apart, then movement distances will be high. In the Thomas Fork of the Bear River, no Bonneville Cutthroat Trout spawning activity was observed in the main stem; however, spawners were frequently observed in the smaller tributaries. Other researchers have found many young-of-the-year Bonneville Cutthroat Trout in the tributaries but few in the main stem of the Thomas Fork. Data showed that of 50 fish radio-tagged on the spawning grounds, 75 percent moved out of the tributaries within 30 days of spawning. It is postulated that fish in the Thomas Fork need the combination of main-stem habitat for feeding and overwintering and tributary habitat for spawning. Furthermore, for some fish these habitats are located a long distance apart, requiring a high degree of movement.

With regard to habitat consistency, studies on the Logan River, Utah, note it remains largely connected across 43 river km (27 river mi), including the three primary tributaries. Accordingly, much higher densities of 100 mm (4 in) and larger Bonneville Cutthroat Trout were observed at certain sites in this system than have been observed in other such populations and other subspecies of Cutthroat Trout. Densities of Bonneville Cutthroat Trout in the Logan River averaged 694 fish/km (430 fish/mi) across survey sites and years, while densities of this subspecies in northern (129 fish/km or 80 fish/mi) and southern (228 fish/km or 137 fish/mi) Utah as well as Wyoming (128 fish/km or 79 fish/mi) are much lower. Logan River Bonneville Cutthroat Trout densities are as much as 25 times the densities of Colorado River Cutthroat Trout (*O. clarkii pleuriticus*) in similar, albeit slightly smaller-size, streams. The Logan River drainage provides a diversity of habitat types within close proximity that supply adequate food and cover and are suitable for different life stages[101].

Bonneville Cutthroat Trout have been found to survive in what is considered marginal salmonid habitat conditions (e.g. turbid water, fine sediments, warmer temperatures, poor habitat structure). This may be because Bonneville Cutthroat Trout have evolved in a desert environment where climate can cause fluctuations in water, sediment regimes, and environmental condition[7]. It has been reported that Bonneville Cutthroat Trout did not emigrate from warm stream reaches or experience mortality despite maximum daily water temperatures as high as 27°C (80°F). Observed fish did not appear to be dependent upon localized cool water refugia[7,97].

Management and Conservation Status

Bonneville Cutthroat Trout are ranked as apparently secure range wide (G4T4) but vulnerable (S3) in Idaho[15]. Unique adaptations of Bonneville Cutthroat Trout in both Bear River and Bear Lake, populations lend significant credence to the need for aggressive conservation programs for these subpopulations, as well as Bonneville Cutthroat Trout in general. Once believed extinct as a result of land-management practices, water diversions, hybridization, and competition, Bonneville Cutthroat Trout have rebounded in multiple ecosystems across their original distribution. In part, this recovery is due to the efforts of state, federal, and tribal agencies participating in the range-wide Conservation Agreement and Strategy[102]. Actions identified in the strategy have been successful in reducing past impacts. In Idaho, Bonneville Cutthroat Trout are classified as a game fish and are afforded protection through angling regulations and other management practices. IDFG's top priority for Bonneville Cutthroat Trout conservation is habitat protection and enhancement, particularly for core conservation and conservation populations. IDFG fishery management staff work with other agencies and stakeholders to identify, plan, fund and implement habitat-enhancement measures across the range of Bonneville Cutthroat Trout in Idaho.

Bonneville Cutthroat Trout are classified as a game fish in Idaho and are protected with restrictive angling regulations, tributary closures, and stocking of sterile hybrids of other game species[6]. Management and conservation efforts by the UDWR, IDFG, and USFWS have reinvigorated the species, and populations have been on the rise since 2001. Regulations are in place to protect native fish and allow harvest of only marked hatchery-raised fish. A limit of two total Bonneville Cutthroat Trout has been instrumental in increasing population levels[21].

Extensive movements such as those noted for Bonneville Cutthroat Trout in a Bear River tributary have important implications or both management and conservation of fish. From a management perspective, vulnerability to harvest may vary among habitats. For example, large trout can be highly vulnerable to angling during the period they spend in shallow tributary streams, and thus seasonal closures, increased size limits, or reduced creel limits may be needed to reduce harvest. Management activities that increase habitat complementation, such as restoration of pool habitat in shallow tributary streams or creation of spawning areas lower in the drainage, can improve local population size by reducing the need for fish to emigrate to upstream or downstream reaches. From a conservation perspective, maintaining connectivity between complementary habitats will ensure that fish have access to all habitats necessary to sustain the population. Furthermore, it is important to note that movement pathways can be disrupted not only by blocking upstream spawning movements but also by connecting to sink habitats, such as irrigation canals, as fish return downstream[70].

Within a given segment of stream habitat, water temperature influences movement patterns. In some cases, salmonids move to cold-water refuge areas when remaining areas of a stream become too warm. However, in a study of Bonneville Cutthroat Trout in the Thomas Fork drainage of the Bear River, monitored individuals remain in tributaries during the summer even when daytime water temperatures reach 27 °C (81°F). The short duration of these high water temperatures, which last only a few hours during each diel cycle, may allow fish to tolerate them until thermal conditions cool. The lack of movement by larger fish in this watershed suggests that an effective management action would be a restoration of habitat to increase pools. Historically in these tributaries, beaver ponds were common and likely provided pool habitat for large fish; however, with increases in grazing and an absence of willows, beaver are less common. Previous examination of Bonneville Cutthroat Trout populations before and after habitat improvement on a Bear River tributary found that trout in the greater than 150 mm (5.9 in) size class increased from 22 trout/km (12 trout/mi) before habitat improvement to 106 trout/km (65 trout/mi) 11 years after habitat improvement. This suggests that habitat limits the abundance of large fish in the tributaries, which would be especially true in dry years[103].

The continuous and connected nature of the aquatic habitats is of primary importance for the success of this subspecies. Reports for the Logan River, Utah, note the diverse habitat and large area of the remaining habitat area as vital for sustaining populations. In addition to high population densities, relatively high annual growth and survival rates overall, especially at upper-elevation sites, were observed. At upper-elevation main-stem sites, fish grew as much as 0.5 g/d, and annual survival rates of adult fish were as high as 75 percent. In contrast, observed survival rates at sites where exotic Brown Trout were present were nearly half those observed at sites where Brown Trout were absent or occurred in extremely low abundance. This pattern is similar to other reported studies where survival rates of juvenile and age-one Colorado River Cutthroat Trout increased from 23 percent to 42 percent when exotic, sympatric Brook Trout were reduced in density. Higher survival rates for older age classes and an increase in individual survival with condition were noted as well. The relation between survival and condition makes biological sense in that individuals that are heavier per unit length must have either a higher intrinsic growth potential or greater access to food resources; both factors typically lead to increased survival and often to increased fitness[101].

Chum Salmon

Oncorhynchus keta

(Walbaum, 1792)

Etymology

Oncorhynchus = hooked snout, referring to the condition of the migrating adults in Pacific salmon runs *keta* = is the original North Asian name applied by Walbaum.

Range and Distributional Status

Chum Salmon were reportedly introduced into Bannock County in 1939 but did not persist and are extirpated in Idaho[50]. Abbreviated information on this species is included here for historical perspective. This species occurred in eastern Oregon and western Idaho in the Miocene. Based on information presented below, we presume Chum Salmon were present in Idaho about 6-8 mya.

Miocene salmon were more specialized than modern salmon in their feeding apparatus. Research on two late Miocene Pacific salmon reveal unusually long, closely-spaced, and numerous gill rakers (plankton straining structures in the pharynx). Middle Miocene *Oncorhynchus rastrosus* also has more numerous and more finely spaced gill rakers than living species. Fossil specimens described here[104] appear to be related to modern Sockeye Salmon and Chum Salmon, differing most in the numbers and morphology of gill rakers. These structural adaptations to plankton feeding in a group of fish otherwise adapted to fish predation are concordant with oceanographic evidence for remarkably high plankton productivity in the north Pacific Ocean during the middle Miocene.

The five species of Pacific salmon (*Oncorhynchus*) in North America are fundamental parts of marine, freshwater, and forest ecosystems but little is known about their evolutionary history. Two well-preserved forms of late Miocene (6–8 mya) Pacific salmon have been recovered from freshwater sediments in the Chalk Hills Formation of the western Snake River plain, Oregon and Idaho. These specimens show unexpectedly long gill rakers and high gill raker numbers, features exhibited in extreme form by the Miocene Pacific tusk-tooth salmon, *Oncorhynchus* (*Smilodonichthys*) *rastrosus*. Gill rakers are highly specialized, comb-like, anterior projections of bony gill arches in the pharynx, used to strain food and direct it back through the pharynx for ingestion. Because gill rakers play an important role in the feeding ecology of salmon, they may be indicators of evolutionary history and paleoecology. Numerous long, closely-spaced gill rakers are efficient at straining small plankton from the water column, whereas short, coarse gill rakers work to process larger food particles such as fish. Long, fine gill rakers are associated with small teeth (e.g., *Oncorhynchus nerka*, Sockeye Salmon), whereas short, coarse gill rakers (e.g., *Oncorhynchus keta*, Chum Salmon) are usually associated with large teeth used to catch larger prey. The two fossil forms from the Chalk Hills Formation were proposed by Smith[105] (1992) to be relatives of modern Sockeye Salmon and Chum Salmon on the basis of small teeth and fine gill rakers in one form and large teeth and apparently coarse gill rakers in the other. The specimens were presumed to indicate diversification of Sockeye and Chum Salmons prior to 6 mya. They are related to *O. rastrosus*, a large, anadromous (sea-run) salmon from the Miocene of Oregon, California, and Washington, that had more than 100 fine gill rakers on each arch. Middle Miocene (ca 11–12 mya) specimens of *O. rastrosus* from the Clarendonian Monterey Formation of California are the oldest known *Oncorhynchus*). A dwarfed and land-locked form of this species, also presumed to have had planktivorous gill rakers, is found in the Miocene Chalk Hills lake beds of southwest Idaho). Phylogenetic analysis indicate that *O. rastrosus* is a relative of *O. nerka* (Sockeye Salmon), *O. gorbuscha* (Pink Salmon), and *O. keta* (Chum Salmon)[104].

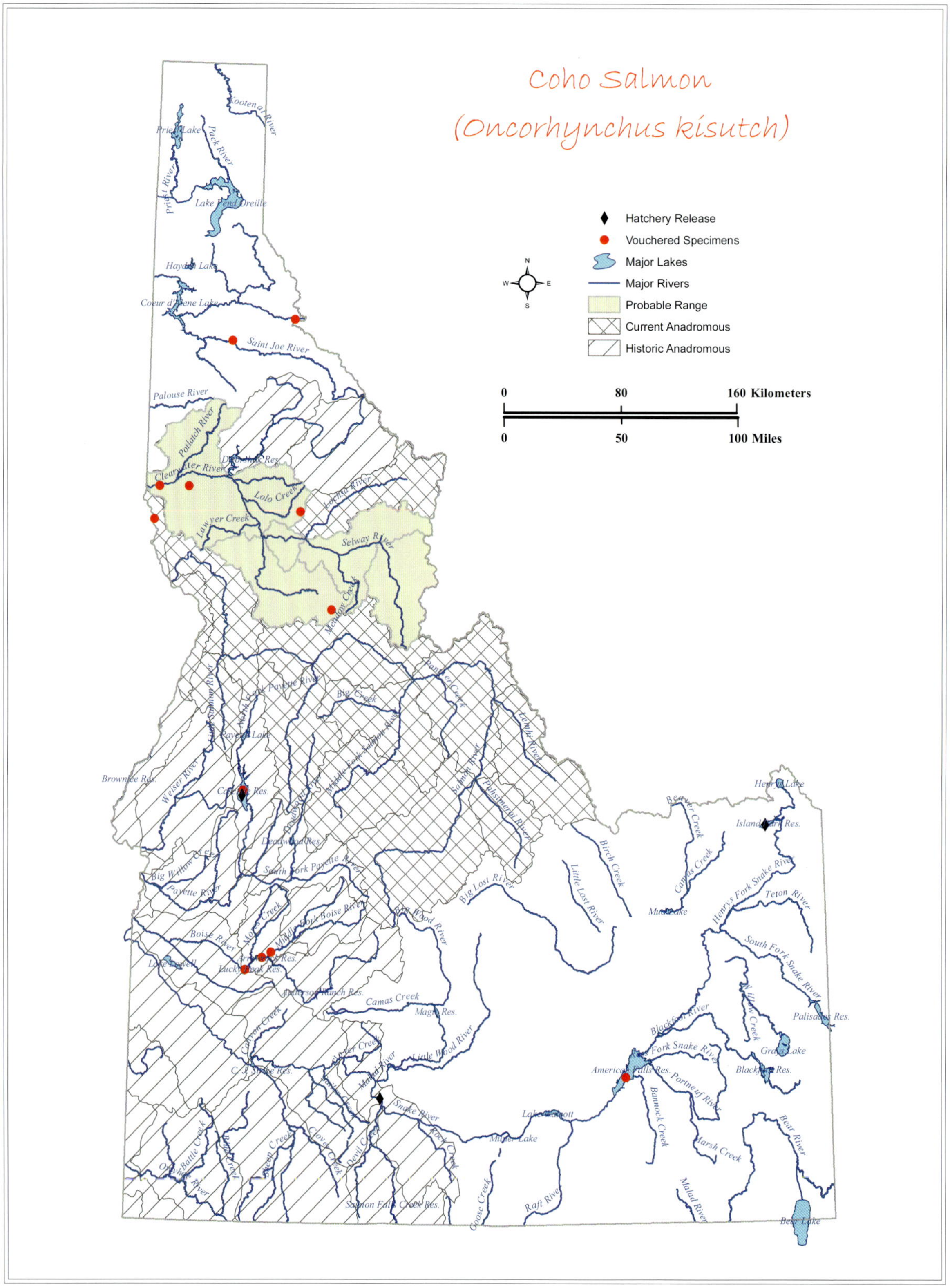
Coho Salmon
(Oncorhynchus kisutch)
Hatchery Release
Vouchered Specimens
Major Lakes
Major Rivers
Probable Range
Current Anadromous
Historic Anadromous
N
S
E
W
0
80
160 Kilometers
0
50
100 Miles
Kootenai River
Priest Lake
Pack River
Priest River
Lake Pend Oreille
Coeur d'Alene Lake
Saint Joe River
Palouse River
Potlatch River
Clearwater River
Lolo Creek
Lochsa River
Lawyer Creek
Selway River
Big Creek
Weiser River
Payette River
Boise River
Camas Creek
Magic Res.
Little Wood River
C.J. Strike Res.
Snake River
Rock Creek
Milner Lake
Goose Creek
Raft River
Lake Walcott
Bannock Creek
Marsh Creek
Malad River
Bear River
Bear Lake
Birch Creek
Little Lost River
Big Lost River
Camas Creek
Mud Lake
Henrys Lake
Island Park Res.
Henrys Fork Snake River
Teton River
South Fork Snake River
Palisades Res.
Grays Lake
Blackfoot River
Blackfoot Res.
Portneuf River
American Falls Res.
Salmon Falls Creek Res.
Clover Creek
Devil Creek
Salmon River
Pahsimeroi River
Lemhi River
Brownlee Res.
Lake Lowell
Lucky Peak Res.
Anderson Ranch Res.
Deadwood Res.
South Fork Payette River
Middle Fork Boise River
Big Wood River
Beaver Creek
Bruneau River
Owyhee River
Battle Creek

Coho Salmon

Oncorhynchus kisutch
(Walbaum, 1792)

Ocean Phase

Spawning Male

Etymology

Oncorhynchus = hooked snout, referring to the condition of the migrating adults in Pacific salmon runs; *kisutch* = vernacular name for this species in Kamchatka.

Importance

Coho Salmon have been of great commercial and sport import for many years in British Columbia, Canada, and Washington State[5,6,11]. Native runs were lost in Idaho in the mid-1980s. Several restoration programs have attempted to restore Coho Salmon to Idaho. From 1962 to 1968, IDFG planted eyed-eggs in the Clearwater River. An extensive hatchery program operated by the Nez Perce Tribe using eggs from Columbia River Coho stocks resulted in Idaho opening its first sport fishery targeting Coho Salmon in 2014. In Washington State, Coho Salmon are the fourth most abundant Pacific salmon and provide a significant portion of the recreational fishing harvest[6,11]. Commercial harvest of Coho Salmon in the Columbia River in Washington State has varied greatly since the early 1900s[11]. Coho Salmon are the most popular species of Pacific salmon due to the willingness to take flies and lures near the surface and their wide distribution; the bright pink or orange-red flavorful flesh of the Coho Salmon increases its popularity with consumers[11,34].

Description

Coho Salmon are bright silver in ocean waters, but coloration darkens in males during spawning runs. Numerous black spots may occur on the back and dorsal lobe of the caudal fin. As spawning time approaches, dull greenish brown or maroon coloration becomes more predominant. Male Coho Salmon develop a kype (hooked jaw) and a red stripe that may spread over most of the body. Coho Salmon have white gums on the lower jaw, a trait that distinguishes them from Chinook Salmon, whose gums are black[11,34]. This species has an elongate body that is streamlined and somewhat laterally compressed. Juvenile Coho Salmon have eight to 12 long, narrow parr marks along the sides[34]; Coho Salmon parr marks typically are more narrow than the spaces between them, while Chinook Salmon parr marks typically are wider than the spaces between them. Coloration is blue-green on the back with silvery sides. Pectoral, pelvic, and anal fins are tinged with orange. An adipose fin is present. The single dorsal fin is soft rayed, with nine to 12 rays; the caudal fin is broad and usually emarginated; the adipose fin is small; the caudal fin is only slightly forked; and the anal fin has 13 to 16 rays. Anterior rays of the anal fin are typically much longer than the other rays[5,6,11,34].

Gill rakers are rough and widely spaced, 18 to 25, with 12 to 16 on the lower limb and six to nine on the upper limb. Branchiostegal rays are 11 to 15. Scales are small and cycloid; the lateral line is complete and almost straight; there are 121 to 148 lateral line scales. Pyloric caeca are 45 to 114, and vertebrae are 61 to 69[5,6].

Range and Distributional Status

This species is native to Idaho. Native range of the Coho Salmon is the Pacific Ocean and its tributaries. Coho Salmon spawning runs originally extended from Monterey Bay, California, north to Point Hope, Alaska. In Asia the distribution of Coho Salmon is from Hokkaido in northern Japan to the Bering Sea basin. Coho Salmon have been introduced to the Great Lakes. Coho Salmon were at one time abundant in the Columbia and Snake River drainages[5,6,11,34].

Historical information on anadromous Coho Salmon distribution in Idaho is rare. It is most likely that Coho Salmon spawned in lower tributaries to the Clearwater River and possibly the lower Snake River near Lewiston, Idaho. Coho Salmon in Idaho were extirpated by the construction of the Washington Water Power Dam (a.k.a. Lewiston Dam) in 1927, four miles upstream from the mouth of the Clearwater River. Although a fish ladder was present, it was felt to be inadequate for most adult salmon passage. The current distribution of anadromous Coho Salmon in Idaho includes the lower Clearwater River with spawning in Clear, Lapwai, and Lawyer Creeks. The current presence of Coho Salmon in Idaho is a result of a reintroduction effort conducted by the Nez Perce Tribe's Fisheries Division[6].

Size and Longevity

Most Coho Salmon migrate to the ocean as smolts after one or two years' residence in the natal stream; smolts are 100 to 150 mm (4 to 6 in) long at migration. After spending one to two years in the ocean[6,11], Coho Salmon may reach sizes of 965 mm (38 in) and weights up to 14 kg (31 lb). Most Coho Salmon reach sexual maturity at sizes of 3.5 to 5.5 kg (7.7 to 12.1 lb) before returning to the natal stream to spawn. Total life span is two to four years, with a maximum of six[5,6,11,34].

A 2.72 kg (6 lb), 597 mm (23.5 in) Coho Salmon from Cascade Reservoir was the previous Idaho record. In 2014 a new anadromous Coho Salmon was entered into the record books. Caught in the Clearwater River, the fish was 762 mm (30 in) long and weighed 4.1 kg (9 lb)[13]. A 15.08 kg (33 lb, 4 oz) Coho Salmon from the Salmon River, Pulaski, New York, is the current world record[12].

Food and Feeding

Coho Salmon are voracious feeders. While in freshwater streams, juveniles feed primarily on insects, including true flies (Diptera), mayflies (Ephemeroptera), and stoneflies (Plecoptera); however, if fish fry are present, they will prey upon them as well[6]. In the ocean life phase, Coho Salmon are primarily piscivorous, feeding on herring, anchovy, rockfish, and planktonic invertebrates such as crab larvae, shrimp, and squid. Coho Salmon introduced to inland lakes and reservoirs feed primarily on insects and zooplankton. Spawning fish do not feed[5–7,11].

Limiting Factors

As young fry and smolts, Coho Salmon, competing with a variety of other species for food, may be prey to other species in a given ecosystem. Birds such as kingfishers and mergansers may prey upon some populations. A variety of parasites may infect Coho Salmon[5]. Hybridization with other species, most commonly Chinook Salmon, may result in genetic dilution in very localized areas[11]. Due to their dependency on small streams and tributaries, Coho Salmon are vulnerable to impacts from human endeavors such as dam diversions and urbanization[34].

Impairment of smoltification of Coho Salmon is reached at a temperature threshold of 15°C (59°F). Reasonable limits to prevent impairment may be 12 to 15.5°C (53.6 to 59.9°F). Adults migrate between 7.2 to 15.6°C (45 to 60°F), and adults migrating through waters frequently warmer than 20°C (68°F) had reduced quality and rapid deterioration of eggs. Reported temperatures for spawning range from 4.4 to 9.4°C (39.9 to 48.9°F). A typical acceptable temperature range for spawning may be 4.4 to 13.3°C (39.9 to 55.9°F)[106].

Breeding Habits

Spawning fish enter rivers in September through October or until January[5]. Coho Salmon do not generally migrate more than 200 km (124 mi) upstream to spawn[107]. Coho Salmon returning to Idaho streams obviously migrated distances exceeding 200 km (400 mi) just to reach Lewiston, Idaho, before moving to natal streams to spawn. Normal life history attributes of Coho Salmon include emergence of fry from gravels in the natal stream two to three weeks after hatching, which requires six to eight weeks depending on water temperature[5]. Coho Salmon generally seek streams where temperatures seldom exceed 21°C (70°F)[107]. Migration to the ocean as smolts follows after one to two years, with many populations averaging 18 months. Some fry may migrate immediately after emergence from the gravel[5,11].

Egg survival for Coho Salmon is consistently greater at constant temperatures of 2.5 to 6.5°C (36.5 to 43.7°F). Egg survival may still be acceptable for many stocks at temperatures of 1.3 to 10.9°C (34.3 to 51.6°F). Alevin and fry survival and health may be best at constant temperatures of 4 to 8°C (39 to 46°F), but survival may remain acceptable up to 10.9°C (52°F). A constant 12°C (54°F) may form the upper threshold for optimal development of Coho Salmon eggs and alevin. Increased mortality occurs at 11°C (52°F). Egg mortality has been reported at 14°C (57°F)[5,11,106].

During spawning, both sexes are aggressive[6,34]. During the spawning act, females build the nest (redd) by lying on their sides and fanning the gravel with their tails. Several males may attend each female, but one is generally dominant. Hovering above the redd, females and males simultaneously release the gametes, and the fertilized eggs fall into the redd. Females move upstream and cover the redd by displacing gravel with their tails. Egg production by female Coho Salmon varies with individual size and has been reported to vary from 1,440 to 5,700 in some populations and from 2,100 to 2,800 in other populations. Eggs are large and range in size from 4.5 to 6.1 mm (0.18 to 0.24 in) and are demersal. They are red-orange in color. Females guard nests as long as possible but may spawn in as many as four nests. Both males and females die following spawning[5].

Habits and Habitat

In native populations, the only freshwater habitat of Coho Salmon is the natal stream in which fry rear and to which adults return to spawn[5]. Natal streams are generally small streams tributary to large rivers with gravelly substrate and where the current is fairly swift[5,6]. Fry and later parr are found in pool edges, and size dictates location within the available habitat[6,11]. Temperature may be the dominant factor in habitat selection between Coho Salmon and other species such as Chinook Salmon[106].

Management and Conservation Status

Coho salmon are ranked as apparently secure range wide (G4). A state status rank is not applicable (SNA) because the species is not a suitable target for conservation activities. Coho Salmon are managed as a gamefish in Idaho[15]. Construction of the Washington Water Power Company Dam (a.k.a. Lewiston Dam) in 1927 allowed some enumeration of Chinook Salmon abundance, but timing and method prevented documentation or enumeration of Coho Salmon. Reliable counts started in 1950, and Coho Salmon were not documented until 1965. The Coho Salmon documented in the late 1960s were considered to be the product of the eyed-egg plants. The initial eyed-eggs for the reintroduction effort were collected from Eagle Creek Hatchery in Oregon. The 1968 eyed-eggs were supplied by the US Bureau of Fisheries from eggs collected at Little White Salmon and Washougal National Fish hatcheries. The eggs were planted in controlled-flow hatch channels in Crooked River (1966, 1967, 1968) and Red River (1963, 1964, 1965)[108] and fry were outplanted to local tributaries in the South Fork Clearwater River after hatching[109]. From 1962 to 1968, the IDFG conducted a Coho Salmon restoration effort using eyed-egg plants. In 1968 the Coho Salmon run peaked at 325 adults counted at Lewiston Dam. The program was terminated in the early 1970s due to low adult returns at Lewiston Dam[108,109].

In 1985 Coho Salmon were declared extirpated (functionally extinct) from Idaho. In 1994 the Nez Perce Tribe initiated a Coho Salmon restoration program using eggs from lower Columbia River stocks and funds from the Mitchell Act and the Pacific Coast Salmon Recovery Fund. Over time the program began to use the returning adults from the program to supplement the lower Columbia eggs. Adults are trapped at Dworshak and Kooskia National Fish hatcheries and at a trap site on Lapwai Creek. Smolts are reared at Dworshak National Hatchery. The current restoration program releases more than one million smolts annually. In 2014 the run of Coho Salmon exceeded 14,500 adults counted at Lower Granite Dam. Idaho anglers shared the bountiful returns when the state opened its first sportfishing season on anadromous Coho Salmon in the Clearwater River in 2014.

Coho Salmon are dependent on small stream and tributary habitat for spawning and access to those streams from larger rivers. Culverts, forest management practices, diversion dams, and other obstructions are all potentially detrimental to Coho Salmon.

Rainbow Trout Species Introduction

While the early descriptive work of Jordan, Evermann, and Clark[73] and the more recent descriptions of Rainbow Trout, Redband Trout, and Steelhead of Behnke[34] were exhaustive, Rainbow Trout retain a level of taxonomic and evolutionary perplexity. There is no question that this species complex has provided millions of hours of recreation throughout the West. "Rainbow Trout" found across the West have been classified either as one species to as many as six major subspecies since the 1930s[7,34]. In the 2013 AFS-ASIH fish list[3], only one species is listed, Rainbow Trout (*Oncorhynchus mykiss*), with attribution to Walbaum in 1792.

Rainbow Trout (*Oncorhynchus mykiss)* are one of the most widely distributed and diverse groups of native salmonids in western North America. Although the taxonomic nomenclature for this species remains unresolved, patterns of genetic diversity show major genetic differences between coastal and inland groups, representing major phylogenetic divisions within the species and among inland groups. Native Rainbow Trout occurring west of the Cascade Range and the Sierra Nevada are currently classified as coastal Rainbow Trout (*O. m. irideus*), whereas Rainbow Trout occurring east of these mountain ranges are classified as Redband Trout. Previous researchers[7], identified three subspecies of Redband Trout: (1) Columbia River Redband Trout (*O. m. gairdneri*), which occur in the Columbia and Fraser Rivers; (2) northern Great Basin and Upper Klamath Lake Redband Trout (*O. m. newberrii*); and (3) Sacramento Redband Trout (*O. m. stonei*), which occur in the Pit and McCloud Rivers. Similarly, other researchers found the greatest evolutionary divergence among Redband Trout groups in three major river basins: the upper Sacramento, Klamath, and Columbia Rivers, suggesting that Redband Trout should be segregated into at least four groups: (1) Columbia River populations, *O. m. gairdneri*; (2) populations from Goose Lake, the Warner Lakes, and the Chewaucan Basin, which are evolutionarily closest to *O. m. stonei*; (3) populations from Upper Klamath Lake and the upper Klamath River (the type location of *O. m. newberrii*); and (4) an unnamed subspecies in the headwaters of the upper Klamath basin and coastal Klamath Mountain. Populations from other isolated pluvial lake basins in Oregon appear to be evolutionarily closest to *O. m. gairdneri* or could not be unambiguously assigned to one of the above groups. Genetic and morphological data suggest that these major evolutionary groups warrant subspecies recognition[110].

Based on the realities of fish in Idaho streams, two definitive subspecies can be identified: interior Redband Trout (*O. m. gairdneri)* and coastal Rainbow Trout (*O. m. irideus*), likely resulting from stocked rainbow previous to the current documentation and understanding of genetic diversity. Columbia River Redband Trout (*O. m. gairdneri*) is a major assemblage of Rainbow Trout found primarily in the Fraser and Columbia Rivers east of the Cascade Mountains, whereas coastal Rainbow Trout (*O. m. irideus*) are found to the west[111].

Rainbow Trout distribution is a complex mix of species evolution, glaciations and flood events, and, more recently, human efforts to plant Rainbow Trout in all waters of the United States and around the world. Readers interested in more detail on the evolutionary and distributional history of Rainbow Trout, Redband Trout, and Steelhead are referred to Behnke[7,34], Halverson[112], Wydoski and Whitney[11], and Muhlfeld[110].

A recent summary of existing knowledge of the potential historical range and the current distribution and status of nonanadromous interior Redband Trout in the United States portion of the interior Columbia River basin and portions of the Klamath River and Great Basins estimates that the potential historical range included 5,458 subwatersheds and represented about 45 percent of the species' North American range[113]. Here we provide only a brief statement of past events that have affected the status of Redband/

Rainbow Trout. As noted, the American Fisheries Society[3] recognizes only one species of Rainbow Trout; other researchers advocate the divisions presented here, although the argument about the scientific data to support these divisions is not resolved.

Toward the end of the Pliocene epoch, about two mya, a common ancestor to Cutthroat Trout and Rainbow Trout diverged. Between then and now, some highly convoluted evolutionary paths developed, resulting in the modern rainbow. Current taxonomic differences are related to those multiple divergences[34]. Here we discuss: the coastal Rainbow Trout (*Oncorhynchus mykiss irideus)* and Redband Trout—Steelhead (*O. m. gairdneri*), which has no formal standing with the American Fisheries Society. All native populations in Idaho, regardless of life history attributes, are considered to be Redband Trout.

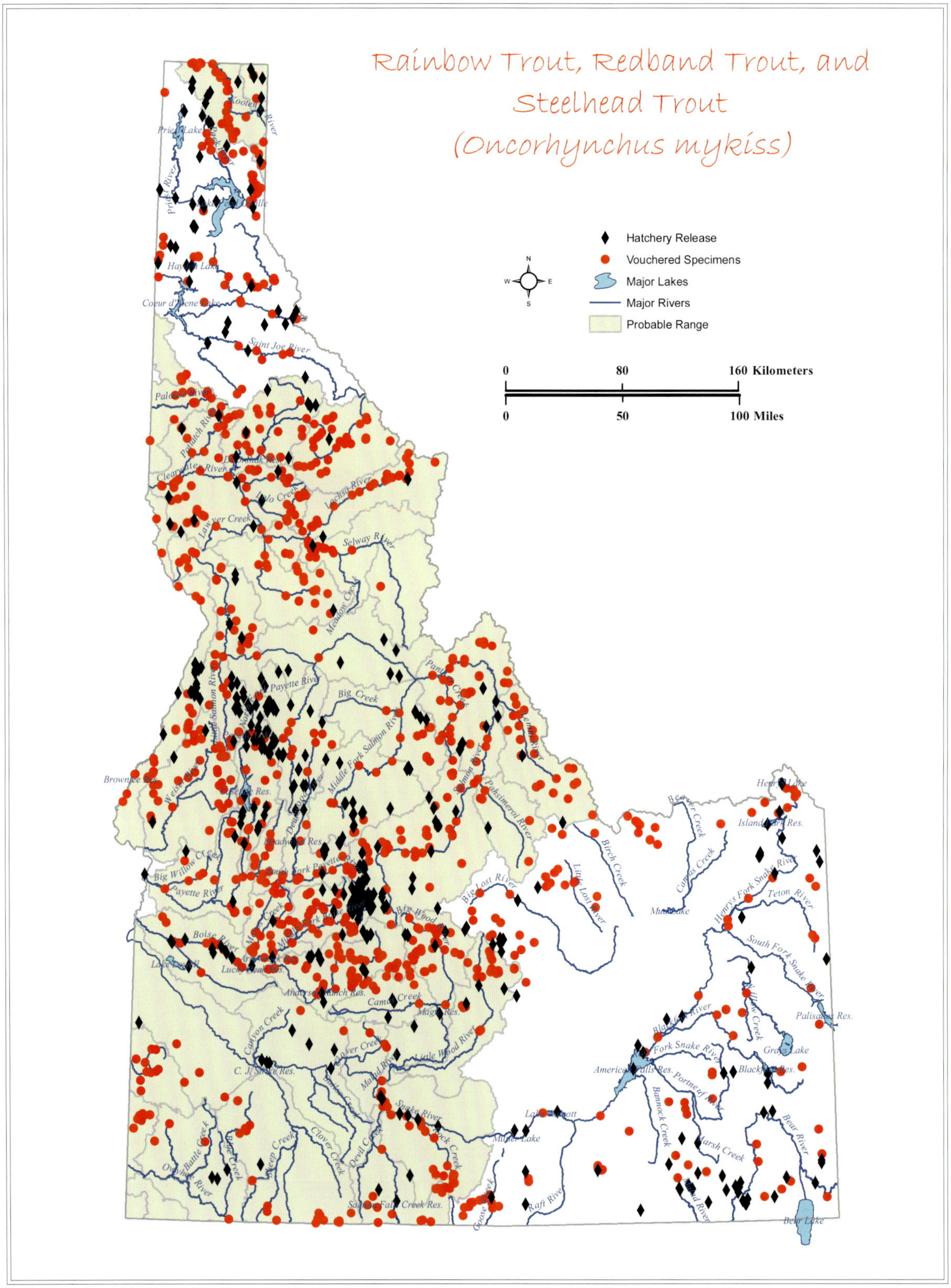
Rainbow Trout, Redband Trout, and Steelhead Trout (Oncorhynchus mykiss)
Hatchery Release
Vouchered Specimens
Major Lakes
Major Rivers
Probable Range
0 80 160 Kilometers
0 50 100 Miles
N
S
E
W
Kootenai River
Priest Lake
Priest River
Hayden Lake
Coeur d'Alene Lake
Saint Joe River
Palouse River
Potlatch River
Dworshak Res.
Clearwater River
Lochsa River
Selway River
Meadow Creek
Payette River
Big Creek
Middle Fork Salmon River
Salmon River
Pahsimeroi River
Lemhi River
Brownlee Res.
Weiser River
Big Willow Creek
Payette River
Boise River
Lucky Peak Res.
Lake Lowell
Anderson Ranch Res.
Camas Creek
Magic Res.
Canyon Creek
C. J. Strike Res.
Little Wood River
Big Wood River
Big Lost River
Little Lost River
Birch Creek
Beaver Creek
Camas Creek
Mud Lake
Henrys Lake
Island Park Res.
Henrys Fork Snake River
Teton River
South Fork Snake River
Palisades Res.
Willow Creek
Grays Lake
Blackfoot River
Blackfoot Res.
American Falls Res.
Portneuf River
Bannock Creek
Marsh Creek
Bear River
Bear Lake
Lake Walcott
Milner Lake
Snake River
Rock Creek
Raft River
Goose Creek
Salmon Falls Creek Res.
Devil Creek
Clover Creek
Deep Creek
Battle Creek
Owyhee River
Jordan Creek

Rainbow Trout

Oncorhynchus mykiss

(Walbaum, 1792)

Columbia River Redband

Duncan Creek Redband

South Fork Salmon River Steelhead Male

Etymology

Oncorhynchus = hooked snout, referring to the condition of the migrating adults in Pacific salmon runs; *mykiss* = vernacular name of the Asian form of this species, previously recognized as a separate species; and previously *gairdneri* = after Dr. Meredith Gairdner, a naturalist in the employ of the Hudson Bay Company, who compiled an early description of this taxon. *Oncorhynchus mykiss gairdneri* (Jordan & Evermann 1896), Redband-Steelhead Trout. Redband Trout (*Oncorhynchus mykiss gairdneri*). Coastal Rainbow Trout, *O. m. irideus* = from the Latin word for rainbow and the Greek goddess of the rainbow (Iris) and refers to the bright red-pink band along the lateral line and gill covers that is especially prominent in spawning males.

Importance

In the western United States, when anglers and many natural resources professionals talk about "fish" and fishing, or "trout," they are typically referring to Rainbow Trout. Based only on their contribution to sport-fishing in the western United States, Rainbow Trout can be characterized as the most important trout in the United States[2] and arguably one of the three or four most important species of game fish in the state of Idaho. Rainbow Trout are renowned and recognized around the world, and indeed they have been introduced in all cool- and cold-water habitats suitable for their survival around the world. Rainbow Trout are by a large measure the most significant species of trout raised commercially for both fish planting and commercial restaurant offerings[112]. Fighting ability when hooked and the vivid colors of most populations add credibility to the various legends associated with this species[2].

Stocking of nonnative Rainbow Trout has been implicated in the decline of many native trout populations. In Idaho hybridization often occurs with native Cutthroat Trout whenever Rainbow Trout have been introduced to areas where they are not native or in areas where they probably only infrequently encountered Cutthroat Trout.

Some Idaho anglers also specifically target native trouts for their uniqueness and perceived fighting abilities when hooked. Pursuit of native trouts by anglers adds great value to Idaho's economy[16].

Description

As is obvious to even casual observers, one of the notable characteristics of Rainbow Trout is the reddish band on the lateral surface of the body. Across the whole of the Rainbow Trout complex, there is such a great range of color and spot patterns that the only true mechanism for separation of some regional populations is genetic markers. This species' coloration is highly variable with habitat[34].

All members of these taxa, as with all salmonids, have an adipose fin. The dorsal fin occurs at the midpoint of the body and is neither high nor long; the dorsal is soft rayed with 10 to 12 principal rays. The caudal fin is broad but not long and only moderately forked. Pelvic fins are abdominal and rather small, typically with 10 or more rays. The lateral line is complete and slightly curved, with 100 to 150 pored lateral line scales. Pyloric caeca number 27 to 80, and vertebra number 60 to 66[5].

Redband Trout that are permanent stream residents have a distinctive spotting pattern that consists of a profusion of large spots on the body combined with this species' classic coloration. A highly distinctive brick-red lateral band, often with tints of yellow or orange, is predominant. Some of these individuals may display a yellow or orange "cutthroat" mark. Redband Trout have an elongate but moderately compressed and deep body. The scales are cycloid and number 120 to 160 in the lateral series. There are small sharp teeth on the jaws and tongue, but teeth are poorly developed on the vomer and absent from the basibranchial plate[6]. The head is short, and the snout is rounded. Young Redband Trout have silvery sides and blue ranging to olive on the back; the belly is generally white, and the back has five to 10 irregular marks. From five to 10 oval parr marks are spaced along the sides, straddling the lateral line[3]. Redband Trout have 11 or 12 rays in the dorsal fin and 10 pelvic fin rays; there are fewer than 150 scales in the lateral series. There are typically narrow white borders on the pelvic and anal fins[6,11].

Steelhead, in contrast, have a distinctive silvery sheen, particularly when fresh from marine waters. This coloration characteristic occurs on the Kamloops trout as well and often completely obscures the spotting pattern[34]. Steelhead have a less rounded snout, a hooked jaw, and following return to freshwater develop more vibrant colors as they get closer to spawning.

Richardson's original 1836 description of what is now classified *Oncorhynchus mykiss gairdneri* was a Steelhead (anadromous). That of David Starr Jordan's 1892 description was an inland lake form[5].

Range and Distributional Status

This species is native to Idaho, but its distribution is confusing to many because it has been widely stocked outside of its native range, even within Idaho (i.e. above Shoshone Falls). As with all nonnative stocking efforts, especially those that were completed prior to a fuller understanding of genetic diversity, problems with introgression and hybridization are common.

Redband Trout have persisted for millennia in a dynamic landscape that has undergone tremendous geological and hydrological changes from volcanism, continental glaciation, periodic major flooding associated with glacial Lake Missoula, and the formation and desiccation of large pluvial lakes. Their historical distribution spans six states in the United States (Nevada, California, Oregon, Washington, Idaho, and Montana) and encompasses a wide array of habitats, ranging from the Finlay River drainage in northern British Columbia south to desert streams in arid regions of northeastern California and northern Nevada and pluvial lake basins in Oregon and east to montane streams in the headwaters of the Columbia River drainage in Montana and Idaho[110].

Native range of all varieties of these taxa was initially the west coast of North America and freshwaters west of the Rocky Mountains and west to the Pacific Coast. Rainbow Trout have been stocked in greater quantities in more locations than any other species in the western United States[2].

Reports dating to mid-1875 indicate that 500 or so Rainbow Trout eggs were shipped to New York from California, constituting the first record of shipment out of their native range. Remarks associated with this initial shipment indicate that the easterners thought the California trout lacked beauty but were equal to the local char (Brook Trout) in flavor and taste and were much hardier in hatchery rearing operations. These California trout were also rated highly with respect to fighting capability after taking a fly[112].

By the year 1887, Livingston Stone and the McCloud River Hatchery, California, had cultured more than three million local Rainbow Trout eggs. Stone was successful in not only propagating the local Rainbow

Trout but also shipping them to diverse locations. In a dozen years, the US Fish Commission had sent McCloud River Rainbow Trout to 33 of the 38 states in the Union, and the commission also sent these fish to England, France, Germany, Switzerland, Canada, and Mexico. Stone was so capable at breeding and distributing Rainbow Trout that there is a fair chance that any Rainbow Trout you might catch today can trace at least some portion of its ancestry to the McCloud River project[112]. It is likely that both resident and anadromous forms of the fish in the McCloud River were used in the propagation process by Livingston Stone[34].

Rainbow Trout are found in the interior Columbia River Basin upstream to the geologic barriers of Shoshone Falls on the Snake River and Kootenai Falls on the Kootenai River. Rainbow Trout are present in both the Salmon River and the Clearwater River drainages, both of which are occupied by Steelhead as well. Resident Rainbow Trout are present above Hells Canyon Dam on the Snake River and above Dworshak Dam on the Clearwater River[6]. Redband Trout are found in the interior Columbia River basin from east of the Cascades upstream to geologic barriers such as Shoshone Falls on the Snake River and Kootenai Falls on the Kootenai River and in the upper Fraser River. However, they are not present in the Clark Fork and Coeur d'Alene River drainages. Redband Trout are present in the Salmon and Clearwater drainages along with anadromous Steelhead.

Size and Longevity

Rainbow Trout in small streams reach typical sizes of 150 to 250 mm (6 to 10 in) and weights of less than 0.23 kg (8 oz). Residents in larger streams and lakes may reach lengths of 360 to 460 mm (14 to 18 in) and weights of 0.45 to 1.4 kg (1 to 3 lb)[34].

Steelhead are notably larger than resident Rainbow Trout. Steelhead range in size from 640 to 910 mm (25 to 36 in) and 2.7 to 8.2 kg (6 to 18 lb). Steelhead from Idaho waters range thousands of miles in the ocean and migrate distances exceeding 1,450 km (900 mi) to return to natal streams to spawn[34]. It is the availability of virtually unlimited food supplies in the ocean that leads to the large size of Idaho Steelhead.

If the Kamloops variant of these taxa is included in size data, a remarkable increase in both length and weight is apparent. Kamloops, a Shuswap Native American word meaning "meeting of the waters," were originally found in Kamloops Lake in British Columbia, Canada. A 24 kg (52 lb, 15 oz) Kamloops of approximately eight years was caught in 1932. These Gerrard Kamloops trout were stocked in Lake Pend Oreille, in 1942 and with the availability of Kokanee as forage grew rapidly. In 1946 a four-year-old Kamloops was caught in Lake Pend Oreille that weighed 14.5 kg (32 lb), and in 1947 a five-year-old fish weighing 17 kg (37 lb, 8 oz) and measuring 1,029 mm (40.5 in) was taken from Lake Pend Oreille. Smaller Kamloops are now present in Lake Pend Oreille[34].

In Idaho, the Rainbow Trout record fish was taken from the Snake River in 2009. This individual weighed 9.1 kg (20 lb, 1 oz) and was 870 mm (34.3 in) long. The Idaho state record for Steelhead is from the Clearwater River, taken in 1973, that weighed 13.7 kg (30 lb, 3 oz) and was 1,118 mm (44 in) long[13]. A 16.8 kg (37 lb) Kamloops (likely a Gerrard) was taken from Lake Pend Oreille in 1947[13]. A 21.8 kg (48 lb) Rainbow Trout was taken from Lake Diefenbaker, Canada, in 2009[12].

Food and Feeding

Rainbow Trout, Redband Trout, and Steelhead juveniles feed on food items drifting on the surface, in the water column, or along the bottom of the rearing stream. Aquatic insects such as mayflies, stoneflies (Plecoptera), and beetle larvae, as well as amphipods, aquatic worms, and fish eggs, are elements of the diet. Opportunistically, juveniles of these taxa may consume smaller fishes[11].

Visual predators, these fish feed upon plankton and insects in lakes, and where they occur with Kokanee they may become piscivorous at an early age, facilitating development of very large individuals. As size increases, Rainbow Trout will feed at the surface on emerging or egg-laying insects, a familiar sight to fly-fishing enthusiasts[6]. Availability of other fishes as food is considered necessary for the species to attain large size in resident populations.

Limiting Factors

As with most salmonid species, all forms of Rainbow Trout are subject to predation by char, other trouts, and salmon. Diving birds and a number of mammals may also take small numbers in some ecosystems[5]. Altered ecosystem water temperatures may reduce growth and spawning productivity. Emerging fry can be adversely affected by both the quantity and the quality of rearing stream water and altered habitat. Impoundments on the Snake River and Columbia River reduce Steelhead smolt migration success[11].

Rainbow Trout are subject to a great diversity of parasites that may affect local populations. Pacific Lamprey prey on Steelhead. Stocking of nonnative (hatchery) forms of these taxa on native populations adversely affects them[5,6].

Habitat degradation, habitat fragmentation, nonnative species introductions, hybridization, and climate change were identified as primary threats to existing populations of Redband Trout. Land- and water-use practices have degraded aquatic habitats and severed the connections between aquatic habitats that are necessary for the long-term persistence of Redband Trout[110].

Breeding Habits

Genetically, all wild members of these taxa are predisposed to spring spawning, when water temperatures in natal streams are increasing. Spawning time, which is dependent on water temperature, may occur as early as January or February in Pacific Coast populations and as late as May or June in more severe climates. Temperatures associated with the initiation of spawning are 6 to 7°C (42 to 44°F). Unlike Pacific salmon, Rainbow Trout do not necessarily die after spawning[34].

Other than the accommodations associated with size and the ability of larger individuals to move larger gravel on the spawning grounds, spawning behavior between resident Rainbow Trout and other forms is remarkably similar[11]. Males are aggressive on the spawning ground, and dominant males will defend territories against other males during the breeding process. Females dig redds by lying on their side and fanning gravel to create a slight depression in the gravel. Males enter the redd area and court the digging female by sliding along and across the female's body and rubbing his snout across the female's caudal peduncle. Each female is generally attended by more than one male, but typically one is larger and dominant. At the moment of spawning, the female rests on the bottom of the redd, and the dominant male presses his body against the female; both fish gape, arch their bodies, and vibrate, releasing eggs and milt. Eggs fall into spaces between the gravel and are covered by the female digging at the upper end of the redd[5]. Steelhead returning to Idaho streams from the ocean show a skewed sex ratio, with females outnumbering males. Steelhead redds have been reported to average 2.3 m (7.6 ft) long and 1.1 m (3.6 ft) wide, larger than redds produced by resident Rainbow Trout. Steelhead redds are most commonly found in substrate (gravels) averaging 1.27 to 9.9 cm (0.5 to 3.9 in) in diameter. Mean depth of these redds has been reported to be 0.2 to 0.41 m (0.7 to 1.34 ft), with water velocities of 0.55 to 0.7 m/s (1.8 to 2.3 ft/s)[11].

Females may deposit as many as 800 to 1,000 eggs in each of several redds during the spawning event, with the same or different males. Number of eggs per female is size related and varies from as low as 200 to as high as 12,000 or more. Eggs are 3 to 5 mm (0.12 to 0.2 in) in diameter, demersal, and pink to orange in color[5].

It takes approximately four to seven weeks for eggs to hatch, and an additional three to seven days is required for sac fry to absorb the yolk before becoming free swimming. There is a great deal of variability in the hatching time due to differences in region, habitat, and temperature histories of the eggs. Fry begin to eat approximately 15 days after hatching[5].

Unlike Pacific salmon, Rainbow Trout are not genetically programmed to die after the initial spawning. Repeat spawners in Steelhead are well documented. However, due to the difficulties associated with out-migration from Idaho streams and a second return across four main-stem dams on the Columbia River and four main-stem dams on the Snake River, repeat Steelhead spawner numbers are significantly reduced from preimpoundment[6].

Habits and Habitat

As is notable from their wide and varied distribution within their native range, and the nearly worldwide distribution through stocking, Rainbow Trout are highly adaptable. Optimum temperature for growth has been reported at 15.5°C (60°F), and higher temperatures can be tolerated if oxygen is plentiful. Rainbow Trout are tolerant of a wide range of salinities[2,11]. Resident Redband Trout are found in a range of stream habitats in Idaho, from desert streams in the South to forested mountain streams in the North[6].

Stream-dwelling Rainbow Trout exhibit more aggressive territorial behavior than do lake-dwelling forms. Feeding territories in streams are established and defended[2]. Total area inhabited by stream-dwelling Rainbow Trout may be only a few hundred yards of the natal stream and often is restricted to a few pools and the riffles above them[17]. In contrast, anadromous Steelhead may travel thousands of miles from its native stream to the ocean, range over vast areas of the marine environment, and return many hundreds of miles upstream to spawn[5].

The "B"-run Steelhead are defined as individuals greater than or equal to 790 mm (31 in) in length crossing Bonneville Dam between July 1 and October 31. To the Idaho angler, "B"-run Steelhead are found primarily in the upper Clearwater River basin, the South Fork Salmon, and Middle Fork Salmon drainages, while the smaller "A"-run Steelhead are distributed in the Hells Canyon, lower Clearwater River, and upper Salmon River drainages[114].

Management and Conservation Status

Redband Trout are ranked as apparently secure (G5T4) range wide and apparently secure (S4) in Idaho, whereas Steelhead are ranked as vulnerable to imperiled (G5T2T3) range wide and vulnerable (S3) in Idaho[15]. Habitat degradation, habitat fragmentation, nonnative species introductions, and climate change were identified as primary threats to existing populations of Redband Trout. Land- and water-use practices have degraded aquatic habitats and severed the connections between aquatic habitats that are necessary for the long-term persistence of Redband Trout. Extensive translocations of nonnative species have displaced and caused extensive hybridization with native Redband Trout populations in many watersheds. Compounding the effects of these stressors is climate change, which is expected to increase water temperatures, modify hydrologic regimes, and increase disturbance events throughout the western United States, likely putting many populations at further risk of decline and extirpation[110].

Populations of Redband Trout above Hells Canyon and Dworshak Dams are locally abundant, as is the case in the Boise River, Weiser River, Payette River, Bruneau River, Owyhee River, and Wood/Malad River drainages.

IDFG recommends the following management actions for Redband Trout: (1) continue statewide population and trend monitoring, (2) continue stocking sterile fish in areas where Redband Trout and introduced hatchery fish overlap, (3) maintain or reestablish connectivity of current Redband Trout metapopulations, (4) publish a state status assessment for Redband Trout, and (5) complete a management plan. IDFG continues to introduce Rainbow Trout above Shoshone Falls and actively manages for removal of Rainbow Trout below Palisades Reservoir in the Snake River.

The decline in abundance of Steelhead in Idaho led to their listing as "threatened" under the Endangered Species Act in 1997. About 60 percent of the historical Steelhead habitat in Idaho is still available, primarily in the Salmon River and Clearwater River drainages. About 30 percent of Idaho's existing Steelhead habitat is included within designated wilderness or wild and scenic river corridors. There is a mix of natural and hatchery Steelhead production strategies in Idaho, ranging from wild refugia to large-scale hatchery programs, to provide harvest opportunities. Areas managed for wild Steelhead include the Lochsa River and the Selway River drainages of the Clearwater River, the Middle Fork and South Fork drainages of the Salmon River, Rapid River, tributaries of the Salmon River downstream of the Middle Fork Salmon River, and tributaries of the Clearwater River downstream of the South Fork Clearwater[115].

Since the 1960s, the composition of the Steelhead run entering Idaho has changed. The proportion of hatchery origin Steelhead has steadily increased due to declining returns of natural fish and development of hatcheries. During the 1960s, the Snake River Steelhead run was essentially 100 percent wild fish. From 1975 to 1979, the Steelhead run at Lower Granite Dam averaged 59 percent naturally produced fish, and from 1985 to 1989 the run averaged 24 percent naturally produced fish. From 1990 to 2010, the run averaged 14 percent naturally produced Steelhead. The run has averaged 23 percent naturally produced fish in the recent four years (2001–2 to 2013–14). This was the first time the wild run exceeded 20 percent of the total return for four consecutive years since spawn years 1981–84.

The numerous hatchery mitigation programs operating in Idaho were designed to replace Steelhead and fishing opportunities lost as a result of construction of the Hells Canyon Dam complex, Dworshak Dam, and the dams of the Federal Columbia River Power System located on the lower Snake River in Washington.

Idaho Steelhead stocks are regionally important because Idaho Steelhead must migrate through the Columbia River to return to their natal spawning grounds in Idaho[115]. Recent research documented the significant contribution of Idaho Steelhead stocks to sportfishing and tribal Steelhead harvests in the Columbia and Snake Rivers in Oregon and Washington. In the Columbia River sportfishing harvest, more than 75 percent of the harvest in the Columbia River was attributed to Steelhead originating in Idaho. For the Tribal fisheries originating upstream of Bonneville Dam, 95 percent of the adipose-clipped hatchery Steelhead originated in Idaho[115].

Steelhead fisheries in the Columbia River are managed under the *US v. Oregon* management agreement that establishes sliding-scale harvests based on abundance of Steelhead at the mouth of the Columbia River. The interjurisdictional nature of fishery management led to the development of the "A"- and "B"-run Steelhead nomenclature where fish greater than or equal to 780 mm (31 in) in length arriving at Bonneville Dam between July 1 and October 31 were considered "B" run, with smaller earlier arriving fish considered "A" run.

"Angling may be said to be so like the mathematics that it can never be fully learnt."

Izaak Walton

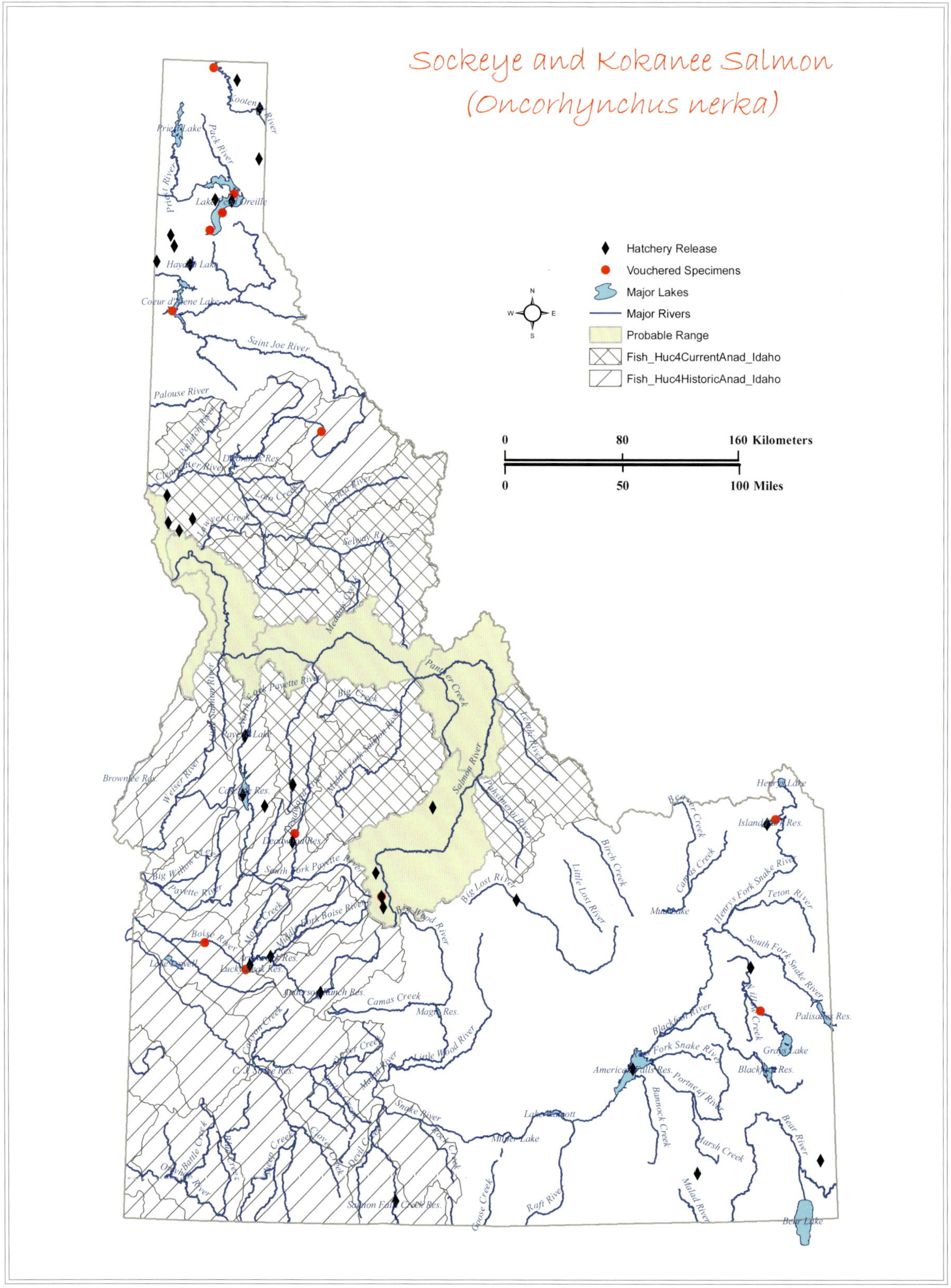
Sockeye and Kokanee Salmon
(Oncorhynchus nerka)
Hatchery Release
Vouchered Specimens
Major Lakes
Major Rivers
Probable Range
Fish_Huc4CurrentAnad_Idaho
Fish_Huc4HistoricAnad_Idaho
0 80 160 Kilometers
0 50 100 Miles

Sockeye Salmon

Oncorhynchus nerka

(Walbaum, 1792)

Sockeye Male

Idaho Kokanee

Species Introduction

Page et al.[3] list *Oncorhynchus nerka* (Walbaum, 1792) as Sockeye Salmon, and there is no mention of the Kokanee form of this species. Sockeye Salmon occur typically as a sea-run form or, alternatively, as a permanent lake resident known as Kokanee, though a nonmigratory or "residual" form of Sockeye Salmon is a known life history variant associated with some sea-run population, including Redfish Lake, Idaho[116]. Patterns of genetic affinity have been reported[117] noting that there are two major groups of North Pacific *O. nerka:* a northwestern group consisting of fish from Kamchatka, western Alaska, and northwestern British Columbia and a southern group consisting of Sockeye Salmon and Kokanee populations from the Fraser and Columbia River systems. The two major genetic groups identified probably stem from historical isolation and dispersal of *O. nerka* from two major Wisconsinan glacial refugia in the North Pacific. Genetic affinity patterns suggest that Kokanee have arisen from sea-run Sockeye Salmon several times independently in the North Pacific. The researchers concluded that the present geographic distribution of the two ecotypes results from parallel evolutionary origins of Kokanee from Sockeye Salmon throughout the North Pacific[117].

Here we discuss the taxa as one species and acknowledge variations in morphotype and behavior.

Kokanee Male

Kokanee Female

Etymology

Oncorhynchus = hooked snout, referring to the condition of the migrating adults in Pacific salmon runs; *nerka* = Russian name for the anadromous form: *kennerlyi* = by Suckley for Dr. C. R. B. Kennerly, a naturalist on the Northwest Boundary Commission under Archibald Campbell[85]. *Oncorhynchus nerka kennerlyi*[118], Kokanee.

Importance

Sockeye Salmon have long been the mainstay of Pacific Coast salmon fisheries[5,11]. Idaho's Sockeye Salmon are a unique entity, migrating from the ocean to Redfish Lake, more than 1,448 km (900 mi), to an elevation of 1,996 m (6,549 ft). This is the longest and highest elevation migration of any Sockeye Salmon population in the world. Some Chinook Salmon and Steelhead in Idaho migrate similar distances to spawn. The severe decline in abundance coupled with this unique migration led to the Endangered Species Act listing of the Snake River Sockeye Salmon as endangered in 1991[6,119].

Kokanee, the landlocked form of this species, are a sportfish of great interest across their native range. This lake-resident form is stocked in great numbers in western North America. They were commercially fished in Lake Pend Oreille as long ago as the 1940s[5,34].

The Kokanee fishery in Lake Pend Oreille was at one time unique among such fisheries. During the 1950s and 1960s, a hand-line fishery was catching fish at the rate of four fish per hour with maximum catches as high as 20 fish per hour[9]. In some ecosystems Kokanee may compete with trout species, to the detriment of the trout[2].

Description

There are no spots on immature Kokanee and Sockeye Salmon. The body is streamlined and laterally compressed and usually 203 to 229 mm (8 to 9 in) long, but when mature, Sockeye Salmon may reach lengths of 610 mm (24 in) when returning from the ocean to spawn[5,6]. Kokanee populations in southern Idaho routinely approach median lengths of 406 mm (16 in), with 457 mm (18 in) specimens not uncommon[114].

Gill rakers on this species are long, slender, serrated, and closely spaced, usually 31 to 44, a number higher than other salmon species. The adipose fin is small; and the single dorsal is at the center of the body, soft rayed with a slightly squared to emarginated edge with 11 to 13 principal rays (13 to 18 in Sockeye Salmon). The caudal fin is broad and moderately forked with pointed tips; there are 11 to 16 rays (13 to 18 in Sockeye Salmon). Scales are small and cycloid; lateral line scales number 121 to 140 in the complete lateral line. Pyloric caeca number 50 to 87 (45 to 115 in Sockeye Salmon), and vertebrae number 62 to 67 (56 to 67 in Sockeye Salmon)[5,6].

As both Sockeye Salmon and Kokanee mature, significant bodily form and color changes occur. Males of both forms develop a hooked jaw (kype) and a humped back. Bodies of both males and females turn reddish except for the tail and head, which remain greenish. Early spawners turn bright red, leading to the common name "red fish." Irregular spots become apparent near the tail and dorsal surface in mature fish[6,34].

Range and Distributional Status

This species is native to Idaho. Kokanee are likely present from the Klamath River, California, to Point Hope, Alaska, and in Asia from Kokkaido, Japan, to the Anadyr River, Russia. Kokanee occur naturally in Japan, Russia, Alaska, Yukon Territory, British Columbia, Canada, and Washington, Oregon, and Idaho in the United States. Kokanee have been widely introduced across North America, including the East Coast, and western states such as Utah, Nevada, and Wyoming[5,34].

In Idaho there were native Sockeye Salmon in Payette Lake and Warm Lake in the South Fork Salmon River basin. Current Snake River Sockeye Salmon recovery efforts have focused on Redfish, Pettit, and Alturas Lakes in the Stanley Basin. This population of Sockeye Salmon migrates farther and to a higher elevation than any other known Sockeye Salmon population[34].

Native Kokanee populations are present in Payette Lake and Stanley Basin lakes. Kokanee also occur in the Kootenai River drainage during migrations from Kootenay Lake, British Columbia, Canada, for spawning. Populations of Kokanee have been established in the Pend Oreille, Spokane, Clearwater, and Snake River (including above Shoshone Falls) drainages[34]. These Kokanee are closely related to their sympatric Sockeye Salmon populations[120].

Size and Longevity

Mature Sockeye Salmon may reach lengths of 530 to 660 mm (21 to 26 in) and weights of 1.8 to 3.2 kg (4 to 7 lb)[34]. Growth is rapid but highly variable based on ecosystem characteristics. Fish from Lake Pend Oreille, Idaho, had the following lengths, in mm (in), at ages one through four: 51 to 76 (2 to 3), 178 to 203 (7 to 8), 203 to 229 (8 to 9), and 254 to 305 (10 to 12)[5].

Stunted populations may mature at 190 to 230 mm (7.5 to 9.1 in). In Idaho, Kokanee in the 300 to 360 mm (11.8 to 14.2 in) size range are common, and fish over 500 mm (19.7 in) are considered to be trophy class[6].

In 1975, Priest Lake produced a 3 kg (6 lb, 10 oz), 622 mm (24.5 in) record[13]. A 6.9 kg (15 lb, 3 oz) Sockeye Salmon was taken from the Kenai River, Alaska, in 1987. This fish was 717 mm (28.3 in) long[12].

Food and Feeding

Kokanee are primarily pelagic plankton feeders, feeding on crustaceans and zooplankton. Bottom organisms may form a significant portion of the diet[5]. Sockeye Salmon feed largely on the same types of food items during the freshwater phase of their development[11]. Sockeye Salmon may consume squid, fishes, and a variety of amphipods and euphausiids in the ocean[34]. Gill-raker structure in Kokanee and Sockeye Salmon is uniquely efficient at capturing small crustaceans. Only this species has more than 30 gill rakers to assist filter feeding, a trait that distinguishes it from all other Pacific salmon[6,34].

Limiting Factors

Both Kokanee and Sockeye Salmon may be infected with a variety of parasites[5]. Much of the former population of Sockeye Salmon in the Columbia River drainage has been lost as a result of dams that either greatly reduce migration success or present complete barriers to migrants[7].

Breeding Habits

Sockeye and Kokanee are fall spawners. Age of maturity varies between two to five years but occurs primarily at three to four years. Populations in unproductive waters with slower growth rates may not be sexually mature until five years of age[6]. Sockeye Salmon reach sexual maturity at three to five years of age, having spent one or two years in freshwater and an additional one to three years in the ocean prior to returning to the natal stream to spawn. In Idaho, Sockeye Salmon runs are primarily four-year-old fish. Both Sockeye Salmon and Kokanee spawn only once, with both sexes dying within a few days of spawning[6].

On the spawning grounds, both males and females may be aggressive to other spawning males. Females prepare a redd in the stream or littoral gravel of a lakeshore, generally in pea-sized gravel, by lying on her side and fanning the gravel with her tail, forming a depression. The resulting depression is usually larger than the female and some 51 to 102 mm (2 to 4 in) deep. Both the male and the female swim into the depression where their mouths gape, they vibrate, and the eggs and sperm are deposited. The nest and eggs are covered by the female by digging at the upper end of the nest. Females may spawn in different redds with different males[5].

Egg number is highly variable over the range of the species and is related to the size of the females. Sockeye Salmon eggs are 4.5 to 5 mm (0.18 to 0.2 in) in diameter. Average number of eggs produced in Kokanee ranges from 350 to over 1,700[5,6]. Average fecundity for anadromous Sockeye Salmon in Idaho has ranged from 2,500 to more than 3,200 eggs.

Hatching time is related to water temperature, and fry emergence from the gravel is in March to May. Newly hatched fry move into a lake associated with the spawning stream[5,11].

For Sockeye Salmon, smoltification occurs after spending one or two years in a lake associated with the spawning stream. Smoltification begins in the spring, prior to individuals entering the ocean[34].

Habits and Habitat

Both Kokanee and Sockeye Salmon are pelagic inhabitants, spending the majority of their lives in open waters of oceans or lakes. Time spent in streams is strictly related to either downstream migrations of fry and smolts or upstream migrations of adults returning to the natal stream to spawn[6].

Management and Conservation Status

Sockeye Salmon are ranked as secure (G5) range wide, but the Snake River populations are ranked as critically imperiled (T1) and in Idaho they are ranked as critically imperiled (S1)[15]. In 1991 Snake River Sockeye Salmon were listed as endangered under the ESA. Just prior to this listing, a captive broodstock program was implemented that began to collect out-migrating juveniles, residual, and all anadromous returning adults to be spawned and held until maturity in captivity. Unlike a traditional salmon hatchery that captured adults returning from the ocean, the precarious nature of the small population led to complete captivity in two separate hatcheries to reduce the probability of catastrophic loss. The near-term goal of the captive broodstock program was to avoid extinction of this unique salmon and to maintain genetic diversity. The long-term goal is to rebuild the populations to facilitate delisting and increase abundance[119].

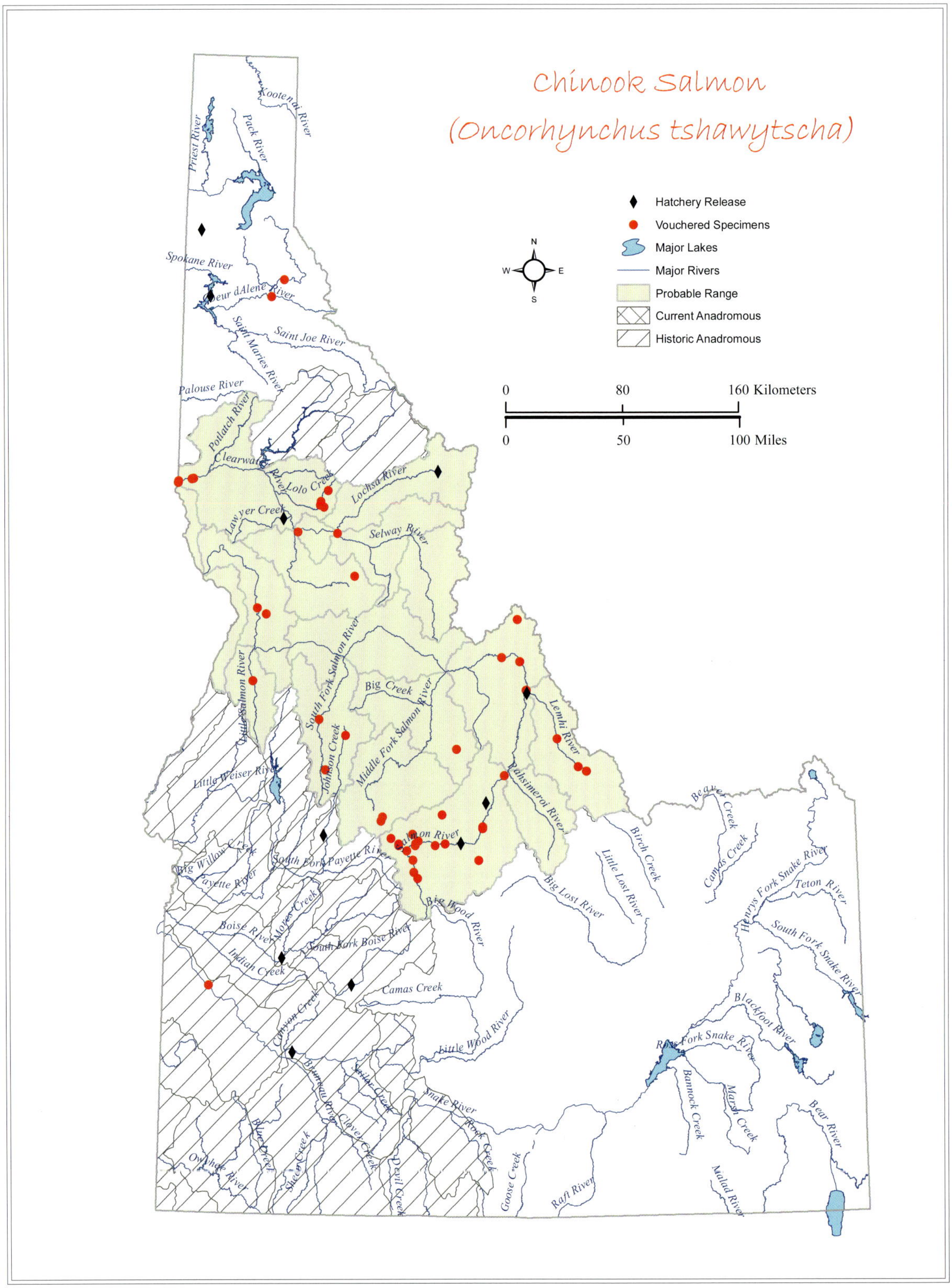
Chinook Salmon
(Oncorhynchus tshawytscha)
Hatchery Release
Vouchered Specimens
Major Lakes
Major Rivers
Probable Range
Current Anadromous
Historic Anadromous
N
W
E
S
0
80
160 Kilometers
0
50
100 Miles
Kootenai River
Priest River
Pack River
Spokane River
Coeur dAlene River
Saint Joe River
Saint Maries River
Palouse River
Potlatch River
Clearwater River
Lolo Creek
Lochsa River
Lawyer Creek
Selway River
South Fork Salmon River
Little Salmon River
Big Creek
Middle Fork Salmon River
Lemhi River
Pahsimeroi River
Johnson Creek
Little Weiser River
Salmon River
Big Willow Creek
Payette River
South Fork Payette River
Boise River
Mores Creek
South Fork Boise River
Indian Creek
Big Wood River
Big Lost River
Little Lost River
Birch Creek
Beaver Creek
Camas Creek
Henrys Fork Snake River
Teton River
South Fork Snake River
Blackfoot River
Little Wood River
Canyon Creek
Bruneau River
Salmon Falls Creek
Snake River
Rock Creek
Owyhee River
Jordan Creek
Sheep Creek
Clover Creek
Devil Creek
Goose Creek
Raft River
Bannock Creek
Marsh Creek
Malad River
Bear River
Portneuf River

Chinook Salmon

Oncorhynchus tshawytscha

(Walbaum, 1792)

Female

Male

Etymology

Oncorhynchus = hooked snout, referring to the condition of the migrating adults in Pacific salmon runs; *tshawytscha* = vernacular name of this species in Kamchatka.

Importance

For generations, Chinook Salmon have been of great importance to the Native Peoples food fishery and in the commercial and sport fisheries of the Pacific Coast of North America. Since the mid-1800s, Chinook Salmon have been commercially fished in California[5]. Designation as "king salmon" for this species is a reflection of its relationship to man and particularly to Native Peoples. Chinook Salmon are named for the Chinook people of the Columbia River basin[34]. Chinook Salmon populations in Idaho declined precipitously resulting in the listing of both the fall run and the spring-summer runs of Chinook Salmon as threatened under the Endangered Species Act in 1992. Numbers of Chinook Salmon in both wild and hatchery populations have rebounded in recent years, but wild stocks are still below the levels necessary for removal from Endangered Species Act protection.

Chinook Salmon are a high-quality food fish and are highly prized for the flavorful and attractive pink to red flesh. Artificial rearing of Chinook Salmon for the restaurant trade often depresses the price of wild-caught Chinook Salmon[34].

Description

In the ocean, as adults, Chinook Salmon have iridescent green to blue-green backs, tops of heads, and upper sides, all with gold flecking or visible sheen. On the sides below the lateral line, they are silvery, and the belly is silvery to white; some black spots may appear on all fins, including the caudal, and the back and top of head[5]. Lower gums on Chinook Salmon are black, distinguishing it from those of the Coho Salmon, which are white[6].

Bodies of Chinook Salmon are streamlined and fusiform and moderately laterally compressed. In breeding males the snout becomes greatly extended, narrowed, and turned down at the tip[6]. Gill rakers on Chinook Salmon are rough and widely spaced, usually 16 to 26, with 10 to 16 on the lower limb and six to 10 on the upper limb. Branchiostegal rays are 13 to 19. Teeth are sharp and well developed on both jaws, the head, and the shaft of the vomer palatines and tongue. Juvenile Chinook Salmon have a more deeply forked caudal fin than do adults[5,6].

A dorsal adipose fin is present, and the single dorsal fin is soft rayed with a square edge and usually 10 to 14 principal rays; the pectoral fins are abdominal, and the pelvic fins are low and rather long, with a somewhat pointed tip. The anal fin has 14 to 19 rays. Scales on Chinook Salmon are cycloid and deeply embedded; scales in the complete, almost straight lateral line are 130 to 165. Pyloric caeca number 90 to 240, and vertebrae number 67 to 75[5,6].

Range and Distributional Status

This species is native to Idaho. As marine adults, Chinook Salmon range far and wide in the Pacific Ocean and to some extent the Arctic Ocean and the Sea of Japan. Both young and spawning adults can be found in rivers flowing into these oceans from the Ventura River, California, north to Point Hope, Alaska, and in northeastern Asia from northern Hokkaido Island to the Anadyr River[5]. Through stocking programs, populations of Chinook Salmon have been established in the Great Lakes, and they are now artificially propagated and reared in pens to meet consumer demand[34].

In Idaho, prior to construction of the main-stem Snake River and Columbia River dams, Chinook Salmon ranged up the Snake River to Shoshone Falls and throughout the major tributaries. After construction of Hells Canyon Dam, access was restricted to below that point. Hatchery programs have helped reestablish Chinook Salmon in the Clearwater River that were extirpated by the construction and operation of Lewiston Dam[6]. Chinook Salmon have been stocked in both Lake Coeur d'Alene and Deadwood Reservoir to provide trophy sportfishing opportunities.

Size and Longevity

Chinook Salmon may live to be three to six years old, and the known maximum age is eight years[34]. Adult size is variable, but 839 to 915 mm (33 to 36 in) is typical. In the past the Great Lakes populations have produced specimens weighing 13.6 kg (30 lb) and measuring almost 900 mm (35 in). The large size achieved by this species, as well as their flavorful flesh, has resulted in the designation "king salmon"[5,11].

A record 19.1 kg (42 lb) freshwater Chinook Salmon was taken from Lake Coeur d'Alene in Idaho in 1987, and a 24.5 kg (54 lb) fish was taken from the Salmon River in 1956[13]. The all-tackle record for Chinook Salmon is 44.11 kg (97 lb, 4 oz), taken from the Kenai River, Alaska, in 1985[12].

Food and Feeding

As young in the natal streams, Chinook Salmon feed on small invertebrates composed primarily of aquatic insect larvae and terrestrial insects. In saltwater juveniles initially feed on small crustaceans and other invertebrates but rather quickly turn piscivorous. Samples of Chinook Salmon from the Oregon and Washington coasts were found to have between 75 and 95 percent fish in their stomachs by weight[5,6,11]. Chinook Salmon stop feeding during their spawning migration when the fish enter freshwater[34].

Limiting Factors

Young Chinook Salmon are preyed upon in freshwater by Rainbow Trout, Cutthroat Trout, Bull Trout, Coho Salmon, Northern Pikeminnow, sculpins, and a variety of birds. Adults in the oceans may be preyed upon by Pacific Lamprey and large mammals. Spawning-run individuals may be preyed upon by bears and a variety of larger birds[5].

Across their entire native range, Chinook Salmon have declined greatly, primarily due to a variety of human activities[34]. Major negative impacts to Chinook Salmon spawning grounds in Idaho have resulted from logging, mining, grazing, road building, and agricultural practices, primarily in the form of increased fine-grained sediment, which reduces spawning success, and loss of cover with accompanying increases in stream temperatures. Chinook Salmon returning to Idaho must pass eight main-stem dams in the Columbia and Snake Rivers to reach natal streams. Adult passage has been improved in recent years as the biological requirements of the species became better understood. Downstream migration of smolts is a continuing problem as a result of direct mortalities at the dams and increased time for out-migration due to the static non-flowing reservoir waters[6]. Smoltification has been reported to be impaired in spring Chinook Salmon at temperatures of 12°C (53.6°F) and at 17°C (62.6°F) or above in fall Chinook Salmon[106].

Breeding Habits

Adult Chinook Salmon return to natal streams and rivers to spawn in gravel and cobble riffles; substrate size is positively related to fish size, larger individuals using larger substrate for spawning[6].

Chinook Salmon tend to spawn in larger rivers with deeper water and larger gravel than other salmon. Males and females are both aggressive on the spawning grounds. Females prepare redds by lying on their side and thrashing their tail up and down to move gravel downstream. Redds of Chinook Salmon can be up to 3.55 m (12 ft) long and 305 mm (1 ft) deep. Generally, each female is attended by a single larger dominant male and one to several smaller males. To commence spawning, both the male and the female swim into the prepared redd and, with bodies in close proximity, gape their mouths, quiver, and expel eggs and sperm. Smaller males may enter the redd and release sperm also. Following spawning, the female covers the redd by digging at the head of the redd. Females may dig more than one redd and spawn with more than one male[5]. Some Chinook Salmon populations in Idaho spawn in streams no more than 1 m (3.3 ft) deep[121].

Egg production by female Chinook Salmon ranges between 2,250 to 7,750, averaging 5,000, and fecundity is related to female size, with more eggs produced by larger females. Dependent on water temperatures, eggs hatch in about two months, and alevins may remain in the gravel for another two to three weeks after hatching[11].

Chinook Salmon are commonly designated as spring, summer, fall, and winter runs, depending on when they enter freshwater on their spawning runs. Spring, summer, and fall runs spawn during the summer to fall period, and spring runs spawn earliest; fall runs spawn in October to November. Spawning is usually initiated two to three weeks after the fish reach the spawning grounds. Generally, spring Chinook Salmon select the upper reaches of tributary streams, while summer Chinook Salmon use the mouths of tributaries and fall Chinook Salmon use the main-stem rivers. Eggs are large, 6 to 7 mm (0.24 to 0.28 in) in diameter, demersal, and orange red in color[5,6,11].

Summary studies reported fall Chinook Salmon incubation at 10 to 16.7°C (50 to 62.1°F) or 10 to 12°C (50 to 53.6°F). Constant temperatures above 12°C (53.6°F) reduced alevin survival substantially. Constant temperatures above 9 to 10°C (48.2 to 50°F) may reduce embryo and alevin survival, with reliable survival at lower percentages when temperatures increase to 11 to 12°C (51.8 to 53.6°F). Incubation temperatures of 9.9 to 16.7°C (49.8 to 62.1°F) produced significant mortality, with complete mortality occurring at temperatures of 13.9 to 19.4°C (57 to 67°F). A maximum spawning temperature requirement below 14.5°C (58°F) has been reported for Chinook Salmon[106].

Habits and Habitat

As an anadromous species, Chinook Salmon spend a few months to two years in freshwater after emerging from the gravels of the natal stream. Migration to the ocean is followed by up to four years in the ocean environment before individuals return to their natal stream to spawn[6].

Optimal rearing temperatures for Chinook Salmon at natural feeding regimes are in the range of 12.2 to 14.8°C (54 to 58.6°F). Optimal juvenile Chinook Salmon rearing temperature is 15°C (59°F), with a range of 12 to 17°C (53.6 to 62.6°F)[106].

Management and Conservation Status

Chinook Salmon are ranked as secure (G5) range wide, but the Snake River populations are ranked as critically imperiled (T1) and in Idaho they are ranked as critically imperiled (S1)[15]. While most wild stocks of Chinook Salmon in Idaho are protected under the ESA, Idaho has had consistent sport fisheries for anadromous Chinook Salmon since the late 1990s. These fisheries are supported primarily by hatchery programs designed to replace lost adult salmon and Steelhead as a consequence of construction and operation of hydroelectric dams on the Snake River in Idaho and Washington.

SALMONID HYBRIDS

Many fish species hybridize in nature, and others have been purposely crossed in the laboratory or fish hatcheries for a variety of purposes. Although not recognized as species, hybrids are often part of both local and regional fisheries and are generally referred to by common names.

Oncorhynchus clarkii × *Oncorhynchus mykiss*
Cutthroat Trout × Rainbow Trout

Cutbow Trout

Salmo trutta × *Salvelinus fontinalis*
Brown Trout × Brook Trout

Tiger Trout

Salvelinus fontinalis × *Salvelinus namaycush*
Brook Trout × Lake Trout

Splake

Chapter Citations

1. Nelson, 2006. 2. Sigler and Sigler, 1996. 3. Page, et al., 2013. 4. McPhail, 2007. 5. Scott and Crossman, 1973. 6. Wallace and Zaroban, 2013. 7. Behnke, 1992. 8. Richardson, 1836. 9. Simpson and Wallace, 1978. 10. Franzin and Clayton, 1977. 11. Wydoski and Whitney, 2003. 12. IGFA on line, 2015. 13. Idaho Department of Fish and Game, 2015. 14. Rogers, Gagnon and Bernatchez, 2002. 15. Idaho Department of Fish and Game, 2005. 16. Idaho Department of Fish and Game, 2013. 17. Sigler and Sigler, 1987. 18. Snyder, 1919. 19. Ward, 2001. 20. Sigler and Sigler, 2014. 21. Palacios, Luecke and Robinson, 2007. 22. Sigler, 1958b. 23. Kennedy, Thompson and Luecke, 2006. 24. Sigler, 1972. 25. Higgins and Vander Zanden, 2010. 26. Utah Division of Wildlife and Idaho Department of Fish and Game, 2010. 27. Taylor, et al., 2011. 28. Sigler, 1962. 29. Colman, et al., 2006. 30. Reheis, Laabs and Kaufman, 2009. 31. Sigler and Workman, 1978. 32. Bouwes and Luecke, 1997. 33. Meyer, Ellie and Lamansky, 2009. 34. Behnke, 2002. 35. Sigler, 1951. 36. Idaho Department of Fish and Game, 2007. 37. Kennedy, 2009. 38. Stamford and Taylor, 2004. 39. Northcote, 1995. 40. Fuller, Cannister and Neilson, 2014. 41. Keil, 1929. 42. Behnke, 2007. 43. Kendall, 1914. 44. Fuller, Cannister and Nelson, 2012. 45. Quadri, 1974. 46. Scarola, 1973. 47. Everhart and Waters, 1965. 48. Kircheis, Kornfield and Seyoum, 1995. 49. Idaho Department of Fish and Game, 1990. 50. Linder, 1963. 51. Deacon, et al., 1979. 52. Curet, 2014. 53. Cavender, 1978. 54. Brunner, et al., 2001. 55. Elz, 2003. 56. Gamett, 2015. 57. Schoby and Keeley, 2011. 58. Rieman and Allendorf, 2001. 59. Rieman, Peterson and Myers, 2006. 60. Dunham, et al., 2008. 61. Meyer, Garton and Schill, 2014. 62. Dunham, et al., 2002. 63. MacCrimmon and Campbell, 1969. 64. Reiser and Wesche, 1977. 65. Ruzycki, Wurtzbaugh and Luecke, 2001. 66. Klemetsen, et al., 2003. 67. Behnke, 1986. 68. Sigler, 1952. 69. Wood and Budy, 2009. 70. Schrank and Rahel, 2004. 71. Grunder, 2014a. 72. Jordan, 1893. 73. Jordan, Evermann and Clark, 1930. 74. Jordan and Evermann, 1904. 75. Loxterman and Keeley, 2012. 76. Trotter, 2008. 77. Varley and Gresswell, 1988. 78. May and Albeke, 2007. 79. Campbell, et al., 2011. 80. Leeseberg and Keeley, 2014. 81. Gunnell, et al., 2008. 82. Van Kirk, 2014. 83. Federal Register, 2006. 84. Montgomery, 1996. 85. La Rivers, 1962. 86. Sigler and Kennedy, 1978. 87. Sigler, et al., 1983. 88. Keeley, 2015. 89. May, 2009. 90. Shepard, Spoon and Nelson, 2003. 91. Weigel, Peterson and Spruell, 2003. 92. Loxterman, Keeley and Njoroge, 2014. 93. Muhfeld, et al., 2009. 94. Grunder, Personal Communication, Westslope Cutthroat Trout, 2014. 95. Pearsons and Temple, 2010. 96. Rieman and Apperson, 1989. 97. Teuscher and Capurso, 2007. 98. Sexauer, 2015. 99. Teuscher, 2015. 100. Hillyard and Keeley, 2012. 101. Budy, Thiede, and McHugh. 2007. 102. PacifiCorp et al., 2002. 103. Schrank and Rahel, 2006. 104. Eiting and Smith, 2007. 105. Smith 1992. 106. Richter and Kolmes, 2005. 107. Moyle, 2002. 108. Lindland and Bowler, 1988. 109. Idaho Department of Fish and Game, 1969. 110. Muhlfeld, et al., 2015. 111. Kozfkay, et al., 2011. 112. Halverson, 2010. 113. Thurow, et al., 2007. 114. Hebdon, 2015a. 115. Bryne, 2014. 116. Burgner, 1991. 117. Taylor, Foote and Wood, 1996. 118. Suckley, 1862. 119. Kline and Flagg, 2014. 120. Waples, et al., 2001. 121. Hebdon, 2014.

Additonal readings for this chapter are listed in the Supplmental References section.

Pebble Creek, upper Portneuf River tributary, near Lava Hot Springs, ID. Photo Courtesy John Sigler.

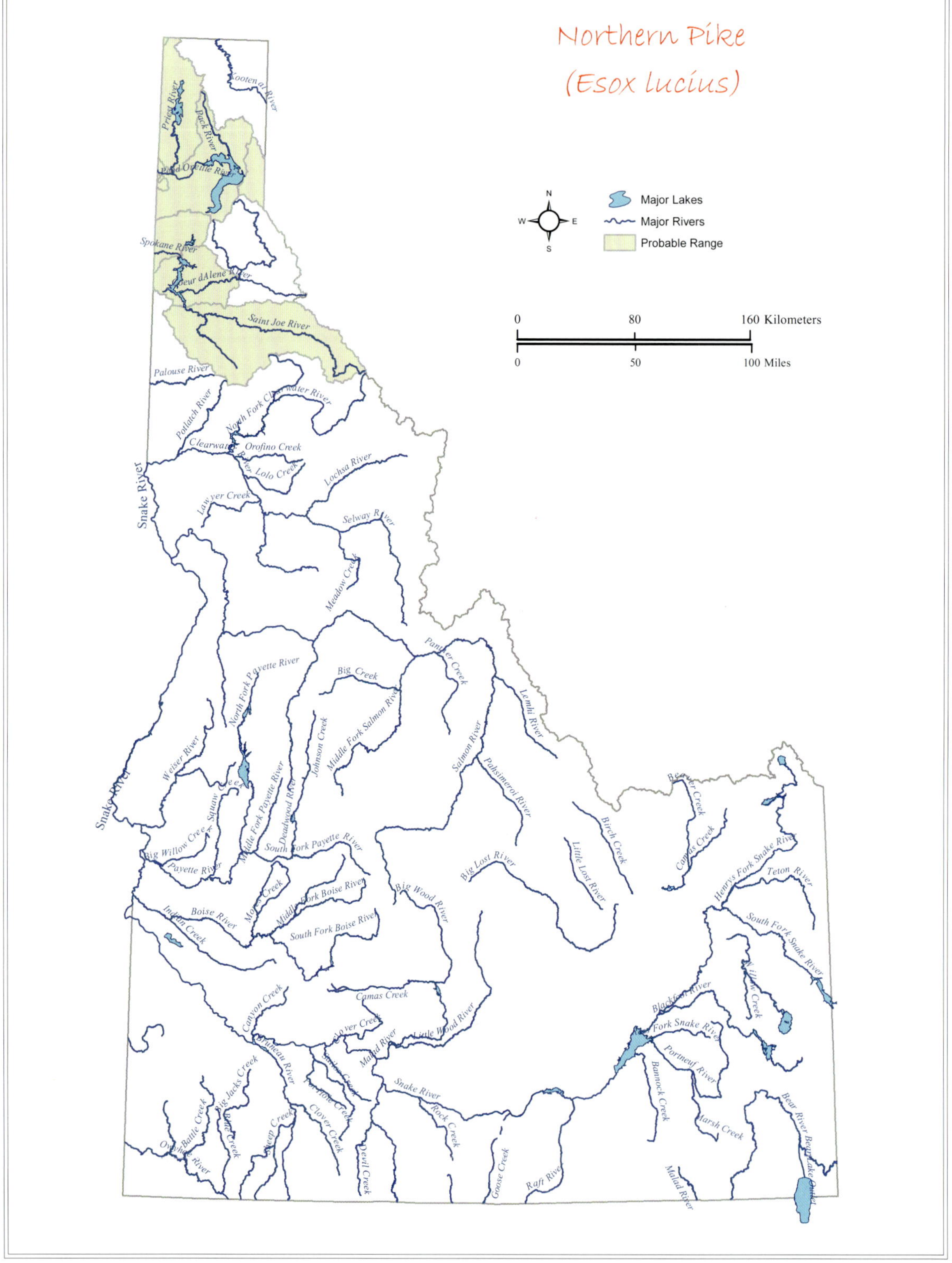

Northern Pike
(Esox lucius)
N
S
E
W
Major Lakes
Major Rivers
Probable Range
0
80
160 Kilometers
0
50
100 Miles
Priest River
Pack River
Spokane River
Saint Joe River
Palouse River
Potlatch River
Orofino Creek
Lochsa River
Lolo Creek
Snake River
Selway River
Meadow Creek
Big Creek
Middle Fork Salmon River
Johnson Creek
Salmon River
Lemhi River
Pahsimeroi River
Birch Creek
Little Lost River
Big Lost River
Big Wood River
Weiser River
Payette River
Boise River
Indian Creek
South Fork Boise River
Canyon Creek
Camas Creek
Little Wood River
Snake River
Rock Creek
Goose Creek
Raft River
Malad River
Devil Creek
Clover Creek
Teton River
South Fork Snake River
Portneuf River
Bannock Creek
Marsh Creek
Bear River

13

ESOCIFORMES

Pikes

Order Esociformes

The maxilla is toothless, there is no adipose fin, and there are no pyloric caeca. Four genera and at least 10 species are extant[1].

Previously, this order was part of the Salmoniformes and separated as suborder Esocoidei[2]. Esociformes is a relatively small order of ray-finned fishes. Some researchers[1] list two families: Esocidae (Pikes) and Umbridae (Mudminnows), while AFS[2] notes one, the Family Esocidae, which occurs in Idaho with one species.

Family Esocidae—Pikes

Pikes' distinctive duckbill-like snout and the long, sharp teeth make Esocidae species very easy to recognize. This family has members with short dorsal fins that are set far back, and the tail is forked. Only one genus, *Esox*, exists in this Holarctic Family. Pectoral fins are low, and pelvic fins are abdominal[3,4].

The lateral line is complete. The oldest members of this family are fossil species of the Late Cretaceous, and the oldest known fossil *Esox* species is from Paleocene formations of about 62 million years ago[1].

Although the native range of pikes includes much of the Northern Hemisphere, only one species occurs in Idaho.

Northern Pike

Esox lucius

Linnaeus, 1758

Etymology

Esox = an old name for pike in Europe; *lucius* = the supposed Latin name for this species which means light.

Importance

In Idaho, Northern Pike have not yet become very important as a recreational game species. Because of the large size of mature adults and the vigorous fighting ability of the species, they may be local favorites. They are highly piscivorous and have been used as management tools in some aquatic ecosystems to reduce populations of undesirable nongame species. In the original description by Linnaeus, a statement was included that translates as follows: "A most vicious fish, it can empty the fishpond; it preys upon ducks[4]." Their presence in the Columbia River presents a potential of loss of valuable salmonid stocks throughout the slow-moving reservoirs.

While some anglers regard Northern Pike as desirable sportfish because of their propensity to take live bait as well as some lures, handling and cleaning of Northern Pike for eating provides some unique challenges due to the presence of copious amounts of slime on the body and intermuscular bones that present some filleting challenges. In some ecosystems in Canada, they are looked upon as a nuisance and a decimator of more desirable species. However, they are delicious as food, and where abundant they are often fished commercially[3–5].

Description

Northern Pike have very long bodies that are somewhat laterally compressed. Cheeks are typically fully scaled, while the operculum is scaled only on the top half and the head is naked on top. Color arrangement of Northern Pike can be described as a pattern of light spots on a dark background. This species has dark brilliant green through olive green to almost brown dorsal surfaces, as do the upper sides and top and upper parts of the head. Northern Pike over 381 mm (15.0 in) have long, conspicuous irregular longitudinal rows (seven to nine) of yellow to whitish spots. In many individuals, the body has the appearance of a golden sheen that results from the presence of tiny gold spots on the tip of most body scales. The eyes are bright yellow. One soft-rayed dorsal fin is present and set far back. The caudal fin is long and moderately forked. Pelvic fins are abdominal at the middle of the body, and the pectorals are low, arising under the edge of the operculum. The lateral line is complete, and scales are cycloid and small. No pyloric caeca are present, and the species is physostomous[3,4].

Range and Distributional Status

This species is not native to Idaho. Northern Pike are circumpolar in distribution in the Northern Hemisphere. Primarily found in freshwater, there are some populations in brackish water in the Baltic Sea. In North America, they occur from Alaska south to Missouri and Nebraska east of the Rocky Mountains and west of the Appalachian Mountains[3,4]. In Idaho, Northern Pike occur in the St. Joe River drainage and north through the Coeur d'Alene, Spokane, Clark Fork, and Priest River drainages.

Size and Longevity

Northern Pike are generally considered to grow larger in Europe than in North America. Nevertheless, the angler record for Northern Pike in North America is 20.9 kg (46 lb, 2 oz) and 1,333 mm (52.5 in). Northern Pike grow fast and live to 12 or more years in the southern portion of their range. In the northern portion of their range in the Arctic, they may live to 26 years of age. Although they may be only 8.3 mm (0.33 in) when hatched, they grow rapidly and reach 152 to 305 mm (6.0 to 12.0 in) by the end of their first year. When food supplies are abundant, Northern Pike may grow even more rapidly. In Montana, the average yearly length of Northern Pike age one to six years is, in mm (in), 254 (10.0), 381 (15.0), 457 (18.0), 533 (21.0), 609 (24.0), and 762 (30.0). Ten-year-old individuals are 1,016 mm (40.0 in)[3,4]. The world record all-tackle Northern Pike is 24.9 kg (54.9 lb)[6]. In Idaho, the largest Northern Pike recorded is 18.2 kg (40 lb, 2 oz) from lower Twin Lakes in 2010[7].

Food and Feeding

Northern Pike are considered to be piscivorous throughout most of their circumpolar range. Indeed, the morphology and behavior of Northern Pike are specialized for ambushing fish prey from the cover of vegetation. Despite the piscivorous nature of Northern Pike and the advantage of piscivory for growth, predation on invertebrates (invertivory) has been documented, particularly in naturally productive systems in northwestern North America. Periodic bouts of invertivory among otherwise piscivorous Northern Pike up to 600 mm (23.6 in) have been attributed to differences in the seasonal availability of vertebrate and invertebrate prey. In other systems, predation on macroinvertebrates appears more consistent[8].

Large zooplankton and immature insects are the primary food of young Northern Pike up to 51 mm (2.0 in). Feeding begins as soon as individuals have absorbed the yolk sac. Piscivory begins soon after the juvenile stage, and by the time individuals have reached lengths of 64 mm (2.5 in), they prefer fish. As visual predators, Northern Pike are primarily active during daylight hours. Adult Northern Pike are voracious predators and will consume almost anything they can catch and swallow, a factor noted in the original description by Linnaeus. Northern Pike may remain motionless until prey is within striking distance and then move very rapidly to ingest the prey. Mice, muskrats, and ducklings may be taken if the opportunity presents itself[3–5].

Limiting Factors

Eggs of Northern Pike are subject to predation by a variety of fishes, including Northern Pike and some aquatic mammals. Stranding of young after hatching may occur if water levels drop rapidly; this condition may trap adults as well. When spawning, individuals lose some of their natural wariness and may thus be captured by bears, dogs, eagles, ospreys, or humans. Northern Pike are subject to a variety of parasites. Where they coexist with other *Esox* species, they may hybridize[3,4].

Breeding Habits

Northern Pike are spring spawners. Spawning occurs at temperatures of 4.4 to 11°C (40 to 52°F), generally immediately after ice melts. Northern Pike move into vegetated, shallow-water areas that are often no more than 18 cm (7 in) deep to commence spawning. Adults spawn repeatedly during a three- to five-day period, and then leave the spawning area, providing no care for eggs or young. Spawning activity is generally between one female and one or two of the smaller males. At random times, males and females roll, align the vents, and discharge eggs and milt with rapid vibration of the body. Five to 60 eggs may be released during each of these acts. Spawning behavior includes a thrust of the tail at the end of the spawning act, which serves to disperse the eggs. Female Northern Pike are highly prolific and may produce 4,550 to 8,200 per kg (10,000 to 18,000 eggs/lb). Mortality is high, and Northern Pike may experience a 99.8 percent mortality rate between egg and young[3].

Northern Pike eggs are 2.5 to 3 mm (0.1 to 0.12 in) in diameter when released, and the clear amber eggs hatch in 12 to 14 days at typical temperatures but may hatch more rapidly (four to five days) at temperatures of 17.8 to 20°C (64 to 68°F). Northern Pike eggs are very adhesive and demersal. When they hatch, young are 6 to 8 mm (0.2 to 0.3 in) in length[3–5].

Habits and Habitat

Northern Pike are a mesothermal, coldwater species that flourish in shallow, productive environments. Populations typically thrive in lakes with adequate flooded vegetation for spawning and recruitment, coldwater habitats for thermal refuge, and the presence of large prey. When some or all of the conditions are not met, growth may slow and size structure may be reduced. In addition, high population density may contribute to the slow growth of Northern Pike. Optimal water temperatures for adult pike growth are 8 to 18°C (46 to 64°F) and for juveniles 19 to 25°C (66 to 77°F)[9].

Northern Pike occur in a wide range of habitat types, from large, relatively cold, deep lakes to smaller, warm farm ponds. Their preference is coldwater environs, at 9 to 22°C (48 to 72°F). In rivers, slow, heavily vegetated habitats are preferred above fast, shallow waters. Northern Pike occur in waters with pH as high as 9.5 and alkalinities as high as 1,000 mg/l (ppm)[4].

As a predatory fish at the top of the food web, Northern Pike populations are small in comparison to other fishes further down the food chain. Northern Pike are extremely plastic in life history traits, however, and they are widely distributed in a variety of fluvial and lacustrine freshwater ecosystems. A study comparing growth and mortality of Northern Pike between river and lake populations at the same latitude, however, did not show any differences[10].

Management and Conservation Status

Northern Pike are ranked as common, wide spread, and abundant range wide (G5). A state status rank is not applicable (SNA) because the species is not a suitable target for conservation activities. Northern Pike are managed as a game fish in Idaho[12]. Management of Northern Pike in North America has increased over the past 50 years as angler pressure and harvest have increased. Although agencies may manage Northern Pike for different reasons (e.g., predator diversity or more angling opportunities), management should be evaluated based on the ecological and biological characteristics (e.g., lake physical characteristics and population dynamics) of the targeted water body. In addition, regulations need to be evaluated to determine which work best under different environmental and sociological conditions[11]. Management considerations for Northern Pike should include an appraisal by fisheries managers of the competitive and predatory nature of this species over often more desirable and locally preferred species.

Chapter Citations

1. Nelson, 2006. 2. Page, et al., 2013. 3. Scott and Crossman, 1973. 4. Sigler and Sigler, 1996. 5. Wydoski and Whitney, 2003. 6. International Game Fish Association, 2015. 7. The Idaho Department of Fish and Game, 2015. 8. Venturelli and Tonn, 2006. 9. Paukert and Willis, 2003. 10. Craig, 2008. 11. Paukert, et al., 2001. 12. Idaho Department of Fish and Game, 2005.

Additonal readings for this chapter are listed in the Supplmental References section.

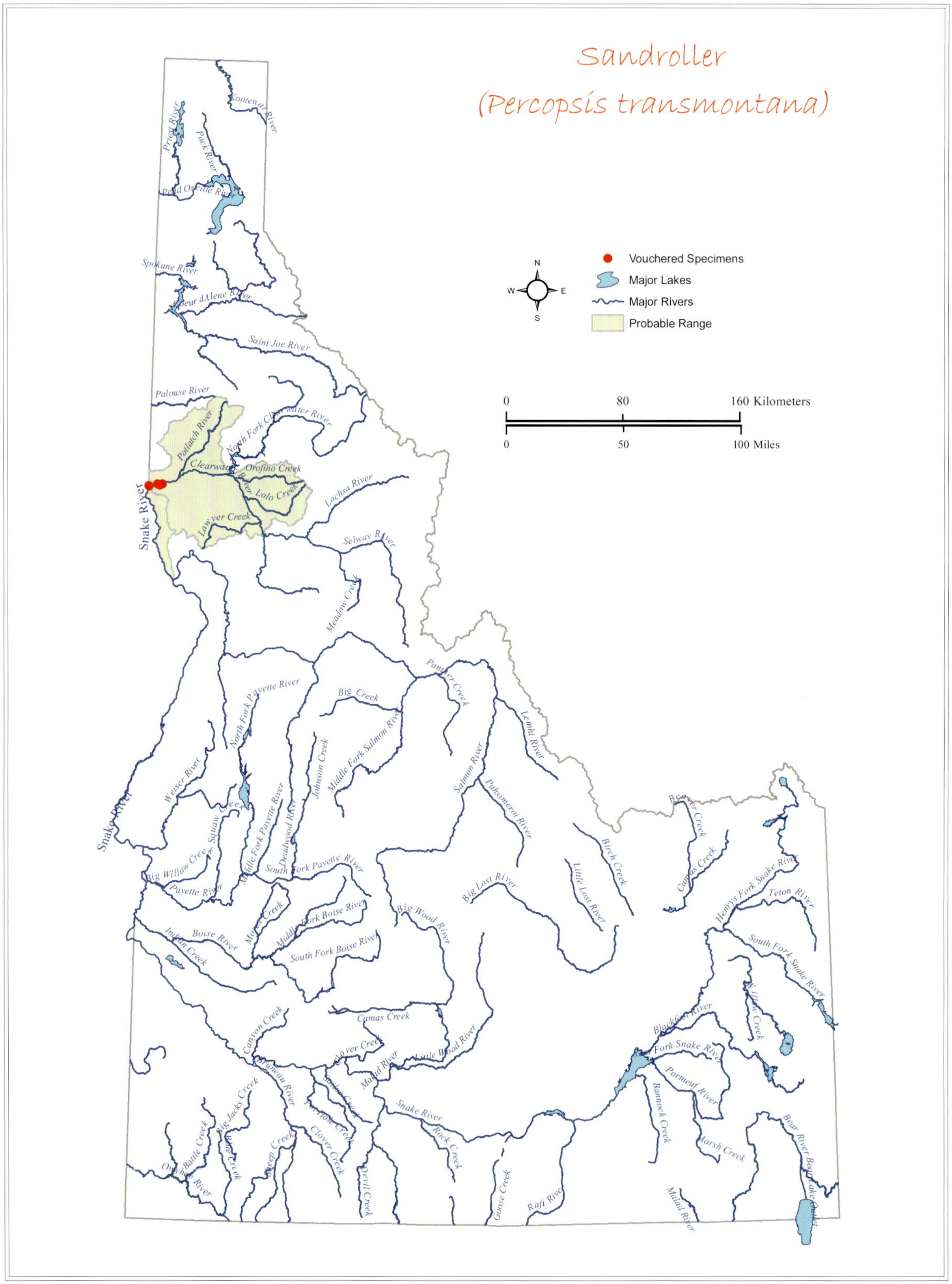
Sandroller
(Percopsis transmontana)
Vouchered Specimens
Major Lakes
Major Rivers
Probable Range
N
S
E
W
0
80
160 Kilometers
0
50
100 Miles
Kootenai River
Priest River
Pack River
Pend Oreille River
Spokane River
Coeur dAlene River
Saint Joe River
Palouse River
North Fork Clearwater River
Potlatch River
Clearwater
Orofino Creek
Lolo Creek
Lochsa River
Lawyer Creek
Snake River
Selway River
Meadow Creek
Panther Creek
Lemhi River
North Fork Payette River
Big Creek
Middle Fork Salmon River
Johnson Creek
Salmon River
Pahsimeroi River
Weiser River
Middle Fork Payette River
Deadwood River
South Fork Payette River
Birch Creek
Little Lost River
Big Lost River
Camas Creek
Henrys Fork Snake River
Teton River
Big Willow Creek
Squaw Creek
Payette River
Middle Fork Boise River
South Fork Boise River
Big Wood River
Boise River
Indian Creek
South Fork Snake River
Canyon Creek
Little Wood River
Blackfoot River
Portneuf River
Bannock Creek
Bruneau River
Malad River
Snake River
Rock Creek
Marsh Creek
Bear River
Big Jacks Creek
Owyhee River
Battle Creek
Sheep Creek
Clover Creek
Devil Creek
Goose Creek
Raft River
Malad River

14

PERCOPSIFORMES

Trout-perches

Order Percopsiformes

In this order, the premaxilla is nonprotractile; pelvic fins, if present, are behind the pectorals and with three to eight soft rays; spines are typically present in the dorsal fin. There are three families worldwide with seven genera and nine species, all confined to freshwater[1]. Percopsiformes are robust-bodied but small fishes with proportionally large heads. Pelvic fins may or may not have small spines, have seven or eight soft rays, and are abdominal to subabdominal. The pectoral fins are lateral[2].

Family Percopsidae—Trout-perches

Both ctenoid and cycloid scales are present in this family; the head is naked, and the lateral line is complete or nearly so; the vertebrae are 33 to 36[1]. Trout-perches are so named because of the interesting mix of characteristics from both trout and perch. These fishes have large heads, and both dorsal and anal fins are preceded by one or two spines. An adipose fin is present. These small, relatively dark-bodied fishes are represented in North America by only one genus with two species[2]. In Idaho, one native species is present.

Sand Roller

Percopsis transmontana
(Eigenmann & Eigenmann, 1892)

Etymology

Percopsis = Greek, *perke* = perch + Greek, *opsis* = appearance; *transmontana* = obscure, as the original collection location was from the Umatilla River and the Wallula River of the Columbia River basin.

Importance

Because of their small size and restricted distribution, the ecological significance of Sand Roller in Idaho is unknown. Abundance appears to be increasing in some Snake River locations, and evidence indicates that Sand Rollers are a "preferred" prey item for Smallmouth Bass in Lower Granite Reservoir[3].

Description

Sand Roller are a small perch-like fish with spiny rays and speckles on both the sides and the fins. This is a small species, rarely reaching 100 mm (3.9 in). Sand Rollers have blue-green coloration and ctenoid scales, the lateral line is incomplete, and an adipose fin is present. These fish are stockier than their close relative the Trout-perch. Males tend to be smaller and more slender than the females. They have spines in both the dorsal and anal fins[4–6]

Range and Distributional Status

This species is native to Idaho. In the United States, this species is restricted to the Columbia River drainage, including tributaries. This species is common in the lower Columbia River downstream of John Day Reservoir. Sand Rollers are endemic to the Columbia River drainage. They have been collected from western Idaho, southern Washington, and northern and western Oregon. In Oregon Sand Rollers have been collected from the Columbia River, the Willamette River, and several of its tributaries[4–6]. In Idaho they have been reported in the lower portions of the Clearwater River and have been collected from the Snake River near its confluence with the Clearwater River[6]. In 2015 reports, Sand Rollers have been observed at increased frequency in Lower Granite Reservoir on the Snake River[3].

Size and Longevity

These are small fish, usually less than 127 mm (5.0 in). Average fork lengths for Sand Roller ages one through six were, in mm (in), 45.7 (1.8), 66 (2.6), 76 (3.0), 84 (3.3), 94 (3.7), and 94 (3.7). One Sand Roller from the Columbia River was estimated to be six years old[4].

Food and Feeding

Little is known regarding feeding habits in the wild. In laboratory studies, individual Sand Roller consumed tubifex worms. They feed principally on aquatic insect larvae (Diptera and Trichoptera) and crustacean zooplankton[4–6].

Limiting Factors

Based on their small size, Sand Roller may be preyed on by a variety of larger species, especially outside their native range. Recent evidence indicates they are the preferred prey for Smallmouth Bass in Lower Granite Reservoir[3].

Breeding Habits

Sand Roller spawning commences in the spring and continues through midsummer. Sand Roller have small-diameter (0.14 to 0.17 mm [0.006 to 0.007 in]) adhesive eggs and are thought to spawn on gravel or other rocky substrates. Egg numbers range between approximately 1,100 and 3,400 for females 76 to 91 mm (3.0 to 3.6 in). Columbia River Sand Roller mature beginning at age two; all were mature at age three[5,6].

Habits and Habitat

Slow-water habitats in low-gradient streams are the preferred habitat of Sand Roller. In Oregon field studies, Sand Roller were most frequently found in association with roots, other large woody debris, and undercut banks with sand or gravel substrates. Sand Roller are nocturnal. During daylight hours, they have been found in pools about 4 m (13 ft) deep. In large, deep rivers, Sand Roller may exhibit diel periodicity, moving into deep waters for cover during daylight and shallow waters during darkness[5,6].

Management and Conservation Status

In Idaho Sand Roller are currently classified as historical (SH)[7], possibly extinct or extirpated[7]. However, specimens collected and verified[8] in 2008 and 2010 indicate this status needs to be revised. The distribution and ecology of this fish should be investigated and documented.

Chapter Citations

1. Nelson, 2006. 2. Scott and Crossman, 1973. 3. Hebdon, 2015b. 4. Wydoski and Whitney, 2003. 5. Scheerer, et al., 2005. 6. Scheerer and Jacobs, 2006. 7. Idaho Department of Fish and Game, 2005. 8. Orma J. Smith Museum of Natural History, 2015.

An additonal reading for this chapter is listed in the Supplemental Readings section.

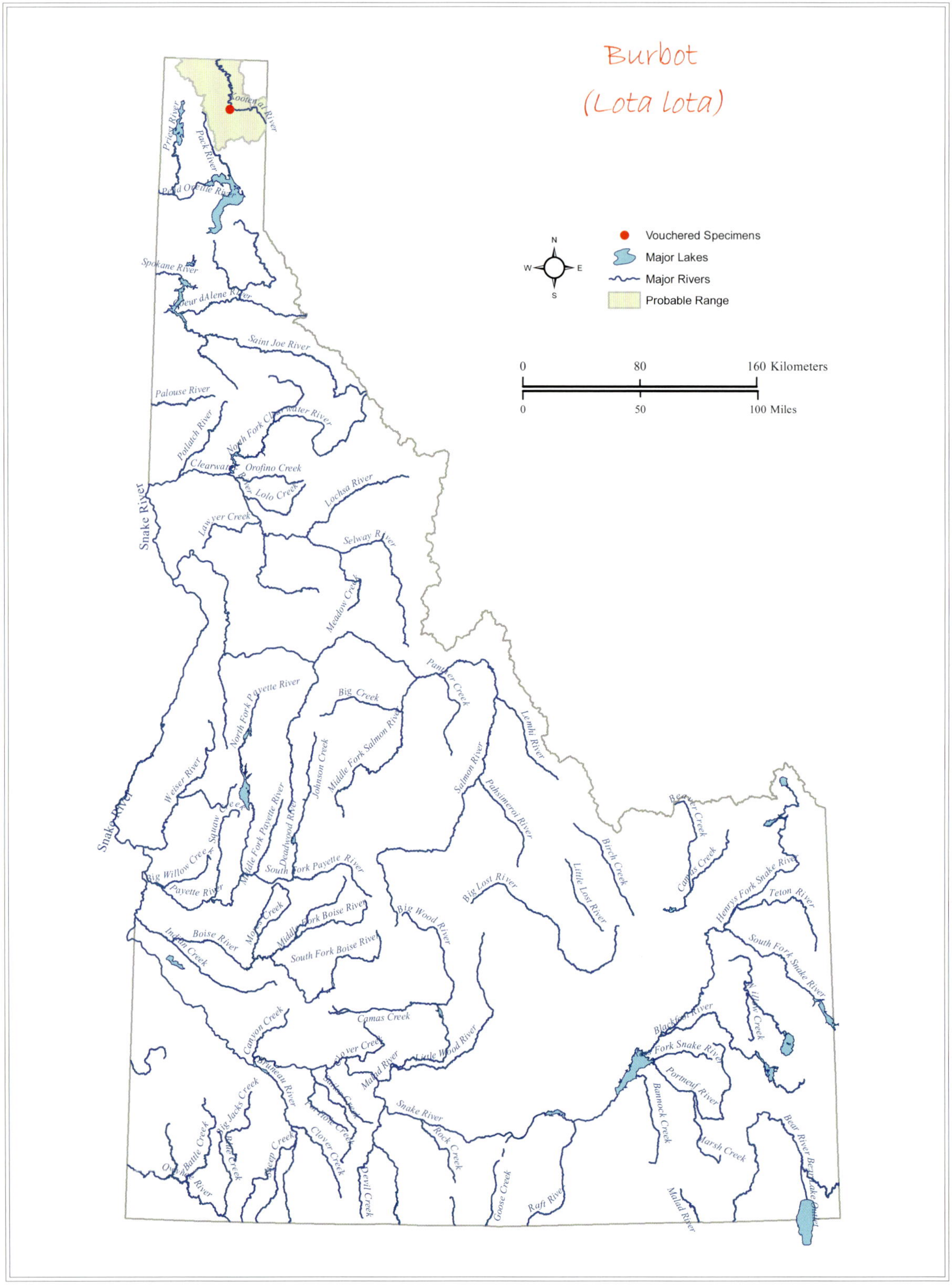
Burbot
(Lota lota)
Vouchered Specimens
Major Lakes
Major Rivers
Probable Range
N
S
E
W
0
80
160 Kilometers
0
50
100 Miles
Snake River
Saint Joe River
Palouse River
Potlatch River
Orofino Creek
Lolo Creek
Lochsa River
Selway River
Meadow Creek
Big Creek
Middle Fork Salmon River
Johnson Creek
Salmon River
Lemhi River
Pahsimeroi River
Birch Creek
Little Lost River
Big Lost River
Big Wood River
Weiser River
Payette River
Boise River
South Fork Boise River
Middle Fork Boise River
South Fork Payette River
Middle Fork Payette River
Deadwood River
North Fork Payette River
Camas Creek
Little Wood River
Teton River
South Fork Snake River
Portneuf River
Bannock Creek
Snake River
Rock Creek
Goose Creek
Raft River
Malad River
Bear River
Devil Creek
Clover Creek
Canyon Creek
Bruneau River
Big Jack Creek
Lawyer Creek
Pack River
Priest River

15

GADIFORMES

Cods

Order Gadiformes

This order consists of many of the commercially important species, constituting more than 25 percent of the world's marine fish catch. There are nine families, 75 genera, and about 555 species. Only one species is confined to freshwater. Pelvic fins, when present, are inserted below or in front of the pectoral fins. There are no true spines in the fins. Most species in this order have long dorsal and anal fins, and the scales are usually cycloid, rarely ctenoid[1].

These stout-bodied, laterally compressed, and elongate fishes are widely distributed in cool waters. Fins are all soft rayed with one, two, or three dorsal fins and one or two anal fins; pelvic fins are thoracic.

Family Gadidae—Cods

In this family, the first dorsal fin is posterior to the head. Individuals reach a maximum length of about 1.8 m (about 6 ft). There are 16 genera and about 31 species[1].

Cods, represented by only one species in Idaho, have proportionally large heads, small cycloid scales, and a single slender barbel located at the tip of the lower jaw. Burbot are the only truly freshwater species in the family[2].

Burbot

Lota lota

(Linnaeus, 1758)

Etymology

lota = the ancient name used by Rondelet. Rondelet is perhaps in reference to the sixteenth-century French physician.

Importance

Burbot have a number of unique characteristics that make them of scientific interest. Because most populations of Burbot spawn in winter, and generally under the ice, they potentially provide a unique recreational fishery opportunity. Burbot are voracious predators, and in some ecosystems this characteristic may create competition and predation problems with other locally more preferred sportfish species such as Lake Trout (*Salvelinus namaycush*). Flaky, white flesh, a characteristic of the Burbot, has not been found sufficiently palatable to gain regional favor over other species[2,3].

Description

Presence of the single slender barbel at the center of the lower jaw is perhaps the most distinguishing characteristic of Burbot. Burbot are streamlined, elongate fish with a wide flattened head that gives them an appearance similar to North American catfishes. Two dorsal fins are present, and the pelvic fins are thoracic. The first dorsal fin is short, with about 13 rays, and the second dorsal fin is long, with 55 or more rays. Gill rakers are seven to 12, and branchiostegal rays are seven, rarely eight; scales are small, cycloid, and embedded, with 27 to 29 between the second dorsal fin and the lateral line, which is incomplete. Coloration of adults is yellow light brown or tan, with the background color being overlain with a dark brown or black lace-like pattern[2,3].

Range and Distributional Status

This species is native to Idaho. Burbot are present in fresh- and brackish water in Asia, Europe, and North America. North American populations occur from Alaska to Oregon and east into Wyoming and Connecticut[3]. They occur across all of the Canadian provinces[2]. They are present in the Columbia River system. In Idaho they occur in the Kootenai River.

The natural range of Burbot has been continuously decreasing since the beginning of the twentieth century. This has been accompanied by a considerable fall in fish numbers, mainly due to years of overfishing and the lack of stocking programs[4].

Fossil evidence suggests that the genus already inhabited freshwater in the Early Pliocene, indicating that it is at least five million years old. The fossil evidence and the Holarctic genetic profile of Burbot imply that it originated in Europe and colonized North America at the beginning of the Pleistocene. Subsequently, *Lota lota* differentiated into two evolutionary lineages whose present geographical distribution corresponds to the subspecies that have previously been described as *L. lota maculosa* and *L. lota lota*, respectively, in North America below the Great Slave Lake and in Eurasia and the remainder of the Nearctic region[5].

Size and Longevity

Maximum size of Burbot reported in Canada is 937 mm (38.3 in) fork length and 8.4 kg (18 lb, 5 oz). In other populations across the native range, individuals in excess of 1,200 mm (46 in) and weights of 34 kg (74 lb, 15 oz) have been reported[2].

Growth of Burbot in ultraoligotrophic Kilpisjarvi, a lake in northern Finland, is slow. Mean lengths of age five and age nine Burbot were 198 and 217 mm (7.8 and 9.5 in), respectively[6].

Burbot from Palmer Lake, Washington, lived to be 14 years of age and were 719 mm (28.3 in) total length at age 11, 719 mm (28.3 in) at age 12, 739 mm (29.1 in) at age 13, and 861 mm (33.9 in) at age 14[3]. Young may reach lengths of 76 to 210 mm (3.0 to 8.25 in) in the first year of growth, and growth for the first four years is relatively rapid[2].

Lake Diefenbaker, Saskatchewan, Canada, produced the all-tackle world record of 11.4 kg (25 lb, 2 oz) in 2010[7]. The largest Burbot taken from Idaho waters was 6.5 kg (14 lb, 5 oz)[8].

Food and Feeding

After hatching in early to late spring, Burbot fry feed on available aquatic organisms, including insects, amphipods, snails, crayfish, zooplankton (including cladocerans and copepods), and smaller fish. Small Burbot continue to feed upon larger invertebrates, and fish become more and more important in the diet of large adults[3]. Burbot have been recorded as feeding on cisco, sculpin, whitefish, stickleback, Trout-perch, Yellow Perch, Kokanee, and smaller members of the species. Piscivory is likely a function of available prey species rather than selection by larger Burbot[2].

In Lake Kilpisjarvi, Finland, diet of Burbot varied both seasonally and with length of the fish. More Burbot ingested detritus during the open-water period than during the ice-covered period. When the lake was ice covered, more Burbot ingested mollusks, insect larvae, and benthic crustaceans[6].

Limiting Factors

Ecosystem alterations have adversely affected Burbot populations, particularly in Idaho, following the closure of Libby Dam on the Kootenai River, Montana, which transformed the Kootenai River into an impoundment[8]. Burbot harbor a wide variety of parasites[2].

Breeding Habits

Spawning under the ice during midwinter and spawning only at night are two of the interesting aspects of Burbot reproductive behavior. Burbot spawn in lentic and lotic habitats[9]. From November to May, at temperatures of 0.6 to 1.7°C (33 to 35°F), Burbot move onto the spawning grounds. Males generally arrive first and are followed by the females in three to four days. All spawning activity takes place in the dark, and documented spawning grounds are deserted during daylight hours. Actual spawning activity consists of a ball of 10 to 12 individuals intertwined and constantly moving. No care is provided for either eggs or young[2].

Habits and Habitat

Burbot are a cold-water species and prefer lakes, large rivers, and reservoirs. In Washington State, Burbot are found in a variety of lakes and reservoirs[3].

In large lakes of the Northern Hemisphere, larval Burbot hatch between March and May in the profundal zone and within a few days emerge to the surface to fill their swim bladder. After swim-bladder inflation, larval Burbot inhabit the pelagic zone, feeding on plankton. In Lake Constance, Europe, pelagic Burbot larvae have been observed to perform diel vertical migrations. From the profundal zone, Burbot presumably migrate along the slope of the lake bottom toward the littoral zone, where they arrive as juveniles. The earliest juveniles come to the littoral zone in late June, but the majority of the young of the year arrive in August or September. Burbot diel vertical migrations seem to be regulated by the light intensity above the surface, because the fish commence their ascent after sunset and descend to their daytime depth with sunrise[10].

Management and Conservation Status

At present, Burbot are protected in Idaho, and Burbot fishing has been closed since 1992. Research is ongoing to develop an understanding of the Kootenai River population dynamics. Between 1996 and 2004, mean abundance and recruitment declines were estimated to be 14 percent and 21 percent, respectively, in the Kootenai River[11]. Burbot are classified as common, wide spread, and abundant range wide (G5). Burbot are classified as S1, critically imperiled in Idaho[12].

Chapter Citations

1. Nelson, 2006. 2. Scott and Crossman, 1973. 3. Wydoski and Whitney, 2003. 4. Wolnicki, Myszkowski and Kaminski, 2001. 5. Van Houdt, Hellemans, and Volckaert, 2003. 6. Tolonen, Kjellman and Lapalainen, 1999. 7. International Game Fish Association, 2015. 8. Wallace and Zaroban, 2013. 9. McPhail and Paragamian, 2000. 10. Probst and Eckmann, 2009. 11. Paragamian, et al., 2008. 12. Idaho Department of Fish and Game, 2005.

Additonal readings for this chapter are listed in the Supplmental References section.

16

CYPRINODONTIFORMES

Top Minnows & Livebearers

Order Cyprinodontiformes

In the 2013 American Fisheries Society list of common and scientific names of fishes[1], Page et al. note changes in the sequence of families and composition of families in this order, following Nelson[2]. Killifishes have a lateral line canal and pores, chiefly on the head, and the lateral line is represented on the body only by pitted scales; the narial opening is paired; branchiostegal rays are three to seven; vertebrae are 24 to 54. There is marked sexual dimorphism, with the males often brightly colored[2]. Previously, these two families (Fundulidae and Poeciliidae) were placed in the Order Atheriniformes. That order is now composed of families and genera that do not occur naturally in North America. These two families are now placed in Cyprinodontiformes with families of fishes that are largely not found naturally in North America, with the exception of Goodeidae, which includes the Nevada springfish and poolfish and a large number of species found in Mexico[1,2].

Family Fundulidae

Topminnows occur in freshwater or brackish ecosystem and some coastal marine habitats, including the lowlands of North America from southeastern Canada to the Yucatán, including the Mississippi River drainage, Bermuda, and Cuba[2].

Fundulids are a North American family representative of Cyprinodontiformes, a large order that contains many varieties of mostly small fishes inhabiting fresh and brackish waters of much of the world, except the Australian region. Previous classifications considered several groups noted as cyprinodontiform fishes to constitute a single family, the Cyprinodontidae, which, in turn, was included among a broadly inclusive order of related fishes, the Atheriniformes. Recently, fisheries scientists noted that grouping oviparous (egg-laying) fishes previously regarded as Cyprinodontidae was unnatural unless several live-bearing families, such as the Poeciliidae, were included and thus recognized these families collectively as the Order Cyprinodontiformes. Commensurate with this action, several subgroups of the former Cyprinodontidae were elevated to familial status, including Fundulidae. This family dates to the mid-Miocene in the fossil record[3]. In Idaho, one species is known to occur.

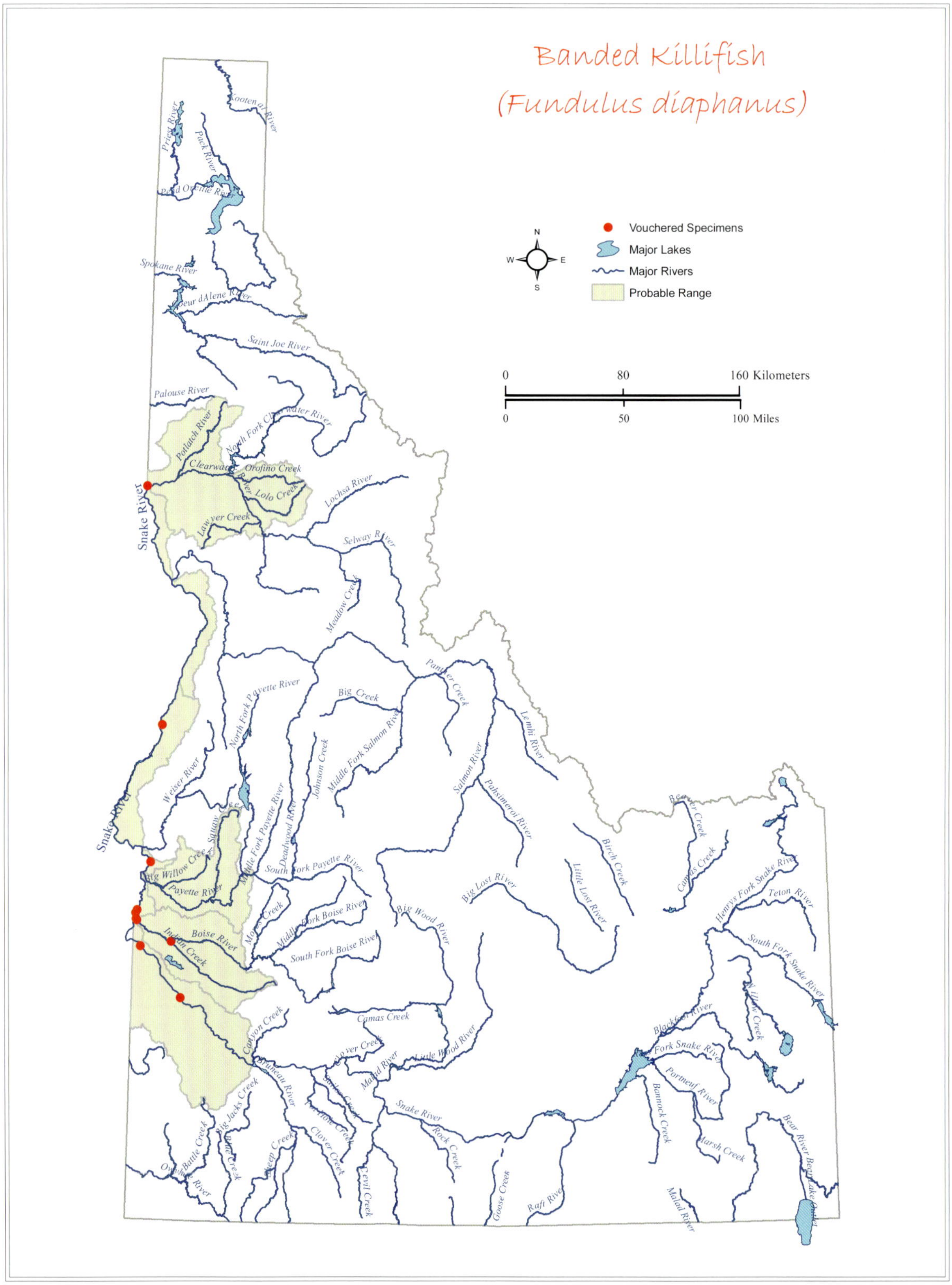
Banded Killifish
(Fundulus diaphanus)
Vouchered Specimens
Major Lakes
Major Rivers
Probable Range
N
S
E
W
0
80
160 Kilometers
0
50
100 Miles
Priest River
Pack River
Spokane River
Saint Joe River
Palouse River
Potlatch River
Orofino Creek
Lolo Creek
Lochsa River
Snake River
Selway River
Meadow Creek
Big Creek
Weiser River
Middle Fork Salmon River
Johnson Creek
Salmon River
Lemhi River
Pahsimeroi River
Birch Creek
Little Lost River
Big Lost River
Big Wood River
Payette River
Boise River
Indian Creek
South Fork Boise River
Camas Creek
Teton River
South Fork Snake River
Portneuf River
Bannock Creek
Marsh Creek
Malad River
Bear River
Goose Creek
Raft River
Rock Creek
Big Jacks Creek
Bruneau River
Clover Creek
Devil Creek

Banded Killifish

Fundulus diaphanus

(Lesueur, 1871)

Etymology

Fundulus = bottom with respect to their habitat; *diaphanus* = transparent, sheer, thin, or flimsy.

Importance

Banded Killifish may be used as bait or perhaps as forage in some ecosystems. They are apparently a hardy species, making them easy to keep in bait wells or buckets[4,5].

Description

Overall coloration of the Banded Killifish is brown or dark brown to olive green on the back, silvery or yellow silvery on the sides, with white or yellowish white on the ventral surface. Banded Killifish generally have 12 to 20 vertical bars on the sides, with fewer and more widely spaced ones on females, but these may be rather inconspicuous on living specimens. Overall coloration, including the vertical bars and the dorsal and anal fins, is enhanced on males during spawning. In particular, the dorsal fin may exhibit green-gold iridescence and a suggestion of a band or two of black pigment[4].

Banded Killifish have an elongated body that is slender and somewhat flattened at the back of the head and the nape region. The head is bluntly triangular and broad. The mouth opening is small and directed upward, with the lower jaw protruding when the mouth is closed. There are teeth on the upper and lower jaws only. Gill rakers are small, 4-7 usually 5 or 6. Branchiostegal rays are usually 6,6; but sometimes 5,6, 6,7 or 7,7 1. The single dorsal fin has its origin distinctly in advance of the anal fin origin and usually has 13 or sometimes 12 rays; the caudal fin is rounded, and the rays usually number 16 but sometimes 14, 15, or 17; the single anal fin has its origin distinctly posterior to the dorsal fin origin and with 11, 12, or 13 rays that are larger on males than females. Pelvic fins are abdominal, small and rounded, with six rays. Pectoral fins are rounded, paddle-like, with variable ray counts but usually 16, though sometimes 14, 15, 18, or 19. Scales are cycloid in 40 to 51 rows. There are lateral line pores on the head only. The peritoneum is silvery and lightly speckled, and the intestine has a single loop. Vertebrae usually number 34 but sometimes 33, 35, or 36[4].

Range and Distributional Status

This species is not native to Idaho. Banded Killifish distribution is from the coast of South Carolina north through Virginia, Maryland, Delaware, New Jersey, New York, and the New England states; to Nova Scotia and Newfoundland, Canada; west through southern Quebec and Ontario, Canada, and in the Great Lakes region to the Yellowstone River in eastern Montana; then eastward through the United States in Minnesota, southern Iowa, Wisconsin, Michigan, northern Ohio, and Pennsylvania[5].

In the Pacific Northwest, Banded Killifish were first discovered in several small creeks tributary to the Willamette River, Oregon. It is likely that Banded Killifish became established following aquarium releases. Banded Killifish are common in quiet backwaters of the lower Columbia River downstream to Vancouver, Washington, and are found downstream as far as Astoria, Oregon[5].

In Idaho Banded Killifish have been documented in the Snake River near the confluence of the Bruneau River, near the confluence of the Boise River, the Boise River; and throughout that entire section of the Snake River.

Size and Longevity

Banded Killifish are very short-lived. Adults may reach 51 to 81 mm (2.0 to 3.2 in) (TL). Banded Killifish are relatively fast growing and may reach sizes of 114 mm (4.5 in) TL[5].

From a length of 6 to 7 mm (0.2 to 0.3 in) at hatching, Banded Killifish may reach total lengths of 20 to 58 mm (0.8 to 2.3 inches) by the fall of their first year in appropriate ecosystems. Adult sizes range from 64 to 76 mm (2.5 to 3.0 inches) total length in inland waters but may be larger in various other ecosystems[4].

Food and Feeding

Newly hatched Banded Killifish, as they swim up from the spawning area, probably feed on zooplankton, since juveniles and adult individuals feed primarily on animal food organisms. Banded Killifish ranging from 31 to 41 mm (1.2 to 1.6 in) TL consumed mainly chironomid larvae, but they also fed on zooplankton, ostracods, amphipods, and even territorial insects. Adult individuals that were longer than 64 mm (2.5 in) fed on essentially the same food organisms as juveniles, but they also consumed larger aquatic insects such as newly emerged mayflies, damselflies, and dragonflies as well as small mollusks. Banded Killifish are very versatile feeders and opportunistic in their feeding habits[5].

Limiting Factors

Banded Killifish may serve as an important food of game fishes, if they occur in sufficient numbers in particular ecosystems. They are also preyed upon by kingfishers and American mergansers. Parasites include trematodes, nematodes, protozoans, larval or immature forms of cestodes, and acanthocephalans[4].

Breeding Habits

Banded Killifish become sexually mature at about 64 mm (2.5 in) total length. Spawning seasons may extend from April to September, depending on ecosystem conditions. In laboratory experiments, Banded

Killifish exhibited prespawning behavior at water temperatures of about 21°C (70°F) and spawned in the last week of May at water temperatures of about 23°C (73°F).

In natural ecosystems, males select spawning areas in quiet water with aquatic vegetation. Banded Killifish neither build a nest nor care for young, though males vigorously defend spawning areas using body postures; apparently, they do not inflict bodily harm on opponents. Eggs are deposited by females in clusters of five to 10, with individual females laying a total of about 50 eggs. Eggs attach to filamentous algae or aquatic vegetation. Banded Killifish larvae are about 6 mm (0.2 in) at hatching[5].

Banded Killifish males assume their most intensive bluish green coloration during spawning, but females remain pale. In laboratory experiments, males and females eventually paired off according to size, with each male pursuing an individual female. Female genital papillae enlarged and protruded obviously during the pursuit. After several minutes, females extruded a single egg that remained suspended by a fine, translucent thread about an inch below the papillae. Males then pursued females even more vigorously, driving females into clusters of weeds and pressing their bodies closely against theirs, holding them in this position by using both dorsal and anal fins. Females then extruded additional eggs, in clusters of five to 10. Males' bodies quivered rapidly and assumed a bow shape; at this time, milt was released and fertilization occurred. Upon fertilization, eggs usually separated from the cluster and became detached from the thread, falling into the weeds, where they became entangled by their individual threads. The entire mating performance required about 15 to 30 seconds and was repeated several times over a period of about five minutes, resulting in about 50 eggs deposited by each of the pairs. Egg diameters have been reported as 2 mm (0.08 in) and hatch in 11 to 12 days at water temperatures of 22.2 to 26.7°C (72 to 80°F). Hatching yielded fry 6 to 7 mm (0.2 to 0.3 in) long[4].

Habits and Habitat

Banded Killifish inhabit the quiet areas of ponds and lakes as well as slow-moving backwaters and sloughs of streams. Banded Killifish are usually found in areas with sand and gravel substrates and with patchy submergent aquatic vegetation. Small schools may be observed in shallow sand-bottomed habitats, and large schools may be seen in shoal areas of lake ecosystems. The species is found in brackish waters along the coast of the central and northeastern United States at salinities up to 20,000 mg/l (ppm)[5].

Management and Conservation Status

Branded Killifish are not listed in the Idaho Comprehensive Wildlife Conservation Strategy (IDFG 2005). Banded Killifish may consume mosquitoes and other small prey in some ecosystems and may therefore compete with game species in some ecosystems. In general, Banded Killifish likely have a nearly neutral effect on Idaho aquatic ecosystems.

Family Poeciliidae

Live-bearers occur in fresh and brackish water at lower elevations in the eastern United States to South America. Pectoral fins are placed high on the side of the body. Maximum length is about 200 mm (7.9 in), but most species are much smaller. There are 37 genera with about 304 species[2]. With the exception of the Western Mosquitofish, which is discussed in some detail below, only abbreviated life history information is provided for these fishes. In Idaho five species are represented in three genera:

Gambusia affinis	(Baird & Girard, 1853)	Western Mosquitofish
Poecilia mexicana	(Steindachner, 1863)	Shortfin Molly
Poecilia reticulata	(Peters, 1859)	Guppy
Xiphophorus hellerii	(Heckel, 1848)	Green Swordtail
Xiphophorus spp.		Platy(s)

"Our task must be to free ourselves, by widening our circle of compassion to embrace all living creatures, and the whole of nature...and its beauty."

Albert Einstein

Snake River rapids near Lake Walcott, Rupert, ID. Photo Courtesy John Sigler.

Western Mosquitofish

Gambusia affinis

(Baird & Girard, 1853)

Female

Male

Etymology

Gambusia = from *gambusina*, a Cuban word relating to the fact that to fish for Gambusia is to fish for nothing; *affinis* = showy or brilliant.

Importance

Across both their native and transplanted ranges, Western Mosquitofish may be used to some extent as both a bait and a forage fish. This species has been stocked throughout the world for mosquito control, particularly in areas where malaria occurs. This species has been used extensively in a variety of organized mosquito-abatement districts as well as in ornamental pools or small ponds. Their effectiveness in mosquito control where freshwater marshes are extensive is questionable[6].

Description

Coloration of Western Mosquitofish is olive or dull silvery, being darkest on the head and back and lightest on the belly. Scales are outlined by dark pigment, and there are usually several rows of black spots across the caudal fin that are more prominent in females than males. There may be a wedge-shaped bar below the eye and a very narrow dark line running along the midside from the head to the base of the caudal fin. Pregnant females exhibit a potbelly and have a black spot just above the anus. Western Mosquitofish are characterized by a rounded caudal fin, large scales on the head and body, a superior mouth, a posteriorly located dorsal fin originating behind the origin of the anal fin, and modifications in the male anal fin into a spike-like reproductive organ. Dorsal fins usually have six rays or sometimes seven, and there are about 30 scales in the lateral series. A true lateral line is not present[6].

Range and Distributional Status

This species is not native to Idaho. Western Mosquitofish naturally ranged from Florida to eastern Texas in the South and from southeastern Kansas to Illinois in the North. They have been introduced into other states as well as around the world for biological control of mosquitoes, yielding their common name. In Washington State, Western Mosquitofish were found in some spring-fed ponds near the junction of the Snake and Columbia Rivers as recently as the early 1970s[5].

In Idaho Western Mosquitofish are suspected to occur throughout portions of the lower Clearwater River Basin, and there are confirmed populations in the lower Boise River and in the Snake River upstream of its confluence with the Boise River.

Size and Longevity

Female Western Mosquitofish are larger than males due to the fact that they continue to grow throughout their lives, while males do not grow after reaching maturity, usually at four to six weeks of age. Males are generally not more than 45 mm (1.8 in) long, but large females may be up to 64 mm (2.5 in) or more. Young, when born, are likely little more than 8 mm (0.3 in) in length and can fend for themselves at birth. Most Western Mosquitofish die the same year they reach sexual maturity; a few may live 15 months or more[6].

Food and Feeding

Western Mosquitofish feed on a variety of animals such as zooplankton and small insect larvae. Their diet also includes some tiny aquatic plants such as diatoms and algae. Western Mosquitofish feed at the water surface and may extend feeding to the shoreline in water that is only 2.5 cm (1.0 in) deep. In one laboratory test, a single 58 mm (2.3 in) female ate 225 large mosquito larvae and a 32 mm (1.3 in) male ate 28 larvae in one hour. Even at birth, newly hatched Western Mosquitofish feed on threadlike mosquito larvae. Field experiments have indicated that Western Mosquitofish are effective in reducing mosquito populations by 80 to 90 percent in some ecosystems[5].

Limiting Factors

Western Mosquitofish do not tolerate sustained water temperatures much below 4.4°C (40°F). Howeve in some northern Utah ecosystems, they live in ponds that frequently form a thin layer of ice in winter. Western Mosquitofish are heavily preyed upon by both fishes and birds. Both sexual maturation and final size can be affected by stress[6].

Breeding Habits

In Western Mosquitofish, both sexes may mature as early as four to six weeks after birth, with males usually maturing when younger and smaller than females. Western Mosquitofish reproduce for a period of four to 10 weeks, during which time females may have one to five broods. Males fertilize females internally with the gonopodium, and in three to four weeks fertilized eggs hatch inside the body of the female (the term is *ovoviviparous*, which means the young are born alive and develop using the yolk sac contained in the eggs). Young are about 13 mm (0.5 in) long at birth. The average number of young produced by a female is about 50, but some may produce 100 to 300. Females can produce several broods of young from a single mating event[5].

Because Western Mosquitofish reach maturity at lengths of 25 mm (1.0 in) and ages of six weeks, with gestation periods of three to four weeks, three or four generations a year are possible in warm-water ecosystems. This characteristic provides the potential for a rapid increase in a Western Mosquitofish population. Large females may give birth to as many as 315 young, with numbers usually increasing with female size. Numbers may decrease in older females[6].

Courtship and copulation are constant among Western Mosquitofish for most of the summer. The male swims in front of the female, orienting his body at a 90-degree angle to hers, partially folding his dorsal and anal fins over her and bending his body into an *S* shape. The male then quickly swims around behind the female and attempts to insert his gonopodium into the female's genital opening. Receptive females are quickly surrounded by males. Females that have previously been bred tend to be nonreceptive but will breed again as soon as 30 minutes after giving birth. Many attempts at copulation fail in the wild because females are nonreceptive[6].

Habits and Habitat

Western Mosquitofish inhabit the quiet, shallow, densely vegetated areas of lakes, ponds, rivers, creeks, ditches, and springs. Western Mosquitofish survive, grow, and reproduce best in warm water, but they tolerate temperatures ranging from 4 to 38°C (40 to 100°F). They do not survive in water below 4°C (40°F) for long. Western Mosquitofish do well in waters with high mineral content and an abundant supply of plant and food organisms[5].

Management and Conservation Status

Western Mosquitofish are ranked as common, wide spread, and abundant range wide (G5). A state status rank is not applicable (SNA) because the species is not a suitable target for conservation activities (IDFG 2005). Western Mosquitofish need little protection in heavily vegetated, calm warm-water ecosystems with an abundance of food. They cannot survive heavy predation in open water. There is no appreciable benefit to protecting this species in Idaho aquatic ecosystems, particularly in ecosystems where there are native species.

Trail of the Coeur d'Alenes, Hayden State Park, Coeur d'Alene Lake, ID. Photo Courtesy John Sigler.

Shortfin Molly

Poecilia mexicana

(Steindachner, 1863)

Shortfin Molly are, as a group, not as popular in the aquarium trade as are Sailfin Molly, due in part to their lack of color. This genus is part of a species complex, the taxonomy of which is uncertain. In general, Shortfin Molly have negative impacts on Idaho ecosystems[7].

Shortfin Molly reach adult sizes from 31 to 51 mm (1.2 to 2.0 in), with a maximum size of 102 mm (4.0 in), and live to only two years or less in the wild[7].

Shortfin Molly feed primarily on detritus, film in the algae, diatoms, desmids, and bits of vascular plants. In their native habitat, based on observations of intestine length, Shortfin Molly are presumed to be herbivorous[7,8].

Shortfin Molly occur in coastal lagoons, estuaries, lowland ponds and rivers, and upland streams. They occur in shallow pools and on riffles or in standing water. They seem to prefer rocky bottoms covered with a rich film of filamentous algae, diatoms, protozoans, and decaying vegetation but are also common over mud, silt, sand, and rubble substrates. Water quality ranges from clear to very turbid or muddy[8].

Brood size varies from 13 to 35, but large individuals may produce as many as 105 young. Captures of young indicate an extended breeding season that probably yields offspring every month of the year[8].

Shortfin Molly produce live young. Breeding habits and courtship activity are probably similar to the Western Mosquitofish[7].

Additional introductions of Shortfin Molly in Idaho ecosystems are strongly discouraged.

Upper Falls, Cauldron Linn, Snake River, @ > 18,000 cfs, spring, 2017. Photo Courtesy John Sigler.

Guppy

Poecilia reticulata

Peters, 1859

Guppies are of questionable value in Idaho ecosystems even though they feed on mosquito larvae. They are, however, much admired as aquarium fish. In some western ecosystems, Guppy constitute a hazard to native minnows. Native range of Guppy is Venezuela, Barbados, Trinidad, northern Brazil, and Guyana. Guppy have been inadvertently stocked and established in the wild in many parts of the western United States[7].

Female Guppy are green to greenish brown with some rows of scales shining bluish in reflected light. Male coloration is highly variable, especially in aquarium specimens, but generally includes red, green, and blue, with blue dominating. In both sexes, the head is small, the mouth is dorsal, and the upturned lower jaw projects beyond the upper jaw. Mosquitofish may be confused with Guppy, though many attributes, such as lateral-series scale counts, color, and spotting characteristics of the males, can be used to distinguish these species[7].

Guppies grow rapidly, though females seldom exceed 25.4 mm (1.0 in) in length, and males are no more than 12.7 to 19.1 mm (0.5 to 0.75 in). Life span is one year or less[7].

Due to their competitive nature, and the fact that they are nonnative, additional introductions of this species into Idaho ecosystems are strongly discouraged[7].

Green Swordtail

Xiphophorus hellerii

Heckel, 1848

Green Swordtail occur on the Atlantic slope, from Rio Nautla, Mexico; southward into the Rio Usumacinta Basin in Guatemala; and into the Rio Sarstuń, Belize[8]. Green Swordtail are native to the Atlantic slope from Rio Nautla, Veracruz, Mexico, to northwestern Honduras[7].

Green Swordtail are one of the most popular aquarium species. They have been widely introduced and established in the western United States, generally to the detriment of native fishes[7].

Habitats used by Green Swordtail are diverse, including ponds, springs, shaded and rocky arroyos, ditches, open lagoons, and rivers with a variety of substrates. Water may be clear to murky, muddy, or opaque and sometimes badly polluted[8].

In flowing waters, larger adults prefer current, whereas young and juveniles remain in shallow, quiet areas near shore. Field collections suggest a long reproductive season[8].

Green Swordtail have flashes of red and yellow on the sides along the lateral line. Caudal fins of males may or may not be colored red; the female's is not. Males may also show color on the dorsal and may have a black margin on the caudal fin. The dorsal fin has 12 to 14 rays, the anal eight to 10, pectoral fins 12 or 13, and the pelvic fins six. There are 26 to 30 scales in the lateral series[7].

Green Swordtail reach lengths of about 127 mm (5.0 in) and probably are relatively short-lived, two to three years. Breeding habits of Green Swordtail are probably similar to those of Molly. Young are born alive and are immediately able to sustain themselves. In warm-water ecosystems Green Swordtail may breed more than once a year. Green Swordtail do not do well in brackish waters; freshwater is required. Optimum temperatures are 20 to 27°C (68 to 80°F)[7].

Other species of *Xiphophorus* are thought to be present in various Idaho ecosystems. Identification to species level of these fish has not yet been completed.

Chapter Citations

1. Page, et al., 2013. 2. Nelson, 2006. 3. Etnier and Starnes, 1993. 4. Scott and Crossman, 1973. 5. Wydoski and Whitney, 2003. 6. Sigler and Sigler, 1996. 7. Sigler and Sigler, 1987. 8. Miller, Minckley, and Norris, 2005.

Winter, Down River, Lower End, Black Canyon, Bear River, near Grace, ID. Photo Courtesy John Sigler.

17

SCORPAENIFORMES

Sculpins

Order Scorpaeniformes—Scorpionfishes

Classification of the Scorpaeniformes is complex and controversial. There are 26 families with 279 genera and about 1,477 species. More than 60 of these species are confined to freshwater [1,2].

Family Cottidae

There are about 70 genera[1] of sculpins found in both marine and freshwater environments in the Northern Hemisphere and eastern Australia, New Guinea and New Zealand. The bodies of sculpins often appear to be naked, but they commonly have scales or prickles. They are never completely encased in heavy, bony armor. Eyes are usually large and placed high on the head. A lateral line is present. There is one spine in the pelvic fins that has two to five soft rays; there is no spine in the anal fin. Adults do not have a swim bladder. Maximum total length is at least 225 mm (8.9 in) in Prickly Sculpin (*Cottus asper*)[3]. Sculpins are distinguished by having large flattened heads, fanlike pectoral fins, and no true scales. The cteni, or hooks, remnants of ctenoid scales, remain in some species as patches of prickles. Sculpins have two dorsal fins, the first composed of feeble dorsal spines. The skull is low (dorso-ventrally flattened), generally broad, and composed of thin bones. The preoperculum bone is often armed with spines; the eyes are in a dorsal position, high on the head, and prominent. Jaws are strong and have well-developed teeth. There are five to seven branchiostegal rays[4].

Sculpins of the genus *Cottus* are nearly exclusively freshwater residents, and there are many species, especially in western North America. Freshwater sculpins are small, generally 178 mm (7.0 in) or less. They have existed since the Oligocene epoch (23 to 34 mya)[5]. There are 126 species in 29 genera listed for North America and Mexico[6].

We provisionally describe 10 taxa of sculpins in Idaho. This count is virtually certain to increase in the near future as genetic, morphological, and meristic evidence is accumulated to further diagnose the Mottled Sculpin (*C. bairdii*) complex and additional new taxa are described. Taxonomic and systematic confusion surrounding sculpins, the poor delineation of species distributions, the difficulty of identifying individuals to species in the field, and low sampling efficiency have presumably contributed to the lack of ecological research on sculpins. Despite recent developments, the systematics of freshwater sculpins remain confusing, and new species are regularly identified. Recent work has greatly clarified phylogenetic relationships and revealed that the genus *Cottus* is not monophyletic[2,7,8].

Reports of recent mitochondrial DNA analyses suggest the Mottled Sculpin contains multiple clades and does not occur in the western United States[3,7,8]. We will continue to describe the Mottled Sculpin complex in the Snake River drainage above Shoshone Falls until the taxonomy and systematics of these sculpins are better understood and formal descriptions are published. A brief description of sculpin taxa previously proposed in western North America that are or may be relevant in Idaho follows.

The American Fisheries Society and American Society of Ichthyologists and Herpetologists once again[6] recognize the Columbia Sculpin (*C. hubbsi)*[9] and the Malheur Sculpin (*C. bendirei)*[10] as species distinct from the Mottled Sculpin. We describe the Columbia Sculpin based on its occurrence in Alturas Lake Creek. Although the Malheur Sculpin may occur in Idaho, we do not provide a description, as no confirmed reports or vouchered specimens are known.

The Bonneville Sculpin (*C. semiscaber)*[11] was described from three specimens collected at Fort Hall, Idaho. This sculpin has been included as a subspecies of the Mottled Sculpin complex[9,12] but is currently not recognized as a distinct species[6], and therefore we do not describe it in this text.

The Rocky Mountain Sculpin (*Cottus* sp.)[13] has been described as a subspecies in the Mottled Sculpin complex[14,15], and the scientific name *C. bondi* has been recently proposed[8] and used[16]. This sculpin may occur in Idaho[16], but we do not describe it in this text because this taxon is not formally described and documented occurrences or voucher specimens from Idaho are currently unknown.

The Snake River Sculpin (*C. tubulatus*)[17] was described from *"one tributary of the North Fork Clearwater River"*[17], likely the Potlatch River, near Bovill, Idaho. The common name Snake River Sculpin is not given in the original description but appears in Eddy (1957)[18]. This sculpin was synonymized with Paiute Sculpin (*C. beldingii*, Eigenmann and Eigenmann 1891) by Bailey and Bond[19]. This sculpin is not currently recognized as a distinct species, and therefore we do not describe it in this text.

Sculpins are an important component of freshwater ecosystems of the Pacific Northwest. In many habitats, they are often the most abundant fish. Because sculpins can be difficult to identify to species and are not considered commercially important, they are rarely investigated and often only in relation to salmonids. Freshwater sculpins primarily consume benthic macroinvertebrates but in some situations can be predators of salmonid eggs or juvenile salmonids[20].

Presence of sculpins in some stream ecosystems has been shown to cause salmonid fry to emerge earlier from redds and increase their downstream movement. Alternatively, cottids can be an important forage fish for stream-dwelling salmonids. Sculpins may also compete with salmonids as well as other benthic fishes. Sculpins also have other important effects on aquatic ecosystems, primarily through direct predation or indirectly by reducing the forage base of invertebrates[20].

Freshwater sculpins (*Cottus*) are a diverse and ecologically important component of cool- and cold-water ecosystems throughout the Northern Hemisphere. Despite the frequently high biomass of sculpins and their numerous ecosystem functions, the traditional fisheries management emphasis on sportfishes has led to a general neglect of small-bodied nongame fishes, such as sculpins, in both research and management[21]. In some ecosystems, sculpins may be important in transport of nutrients and energy within drainage systems or between aquatic and terrestrial ecosystems. Sculpins may host a large array of parasites and thus presumably play an integral role in ecology and distribution of parasites and diseases. As parasite hosts, sculpins contribute to the persistence of some aquatic communities[21].

Sculpins are well suited for use as bioindicators, particularly for sediment issues. Often salmonids or other sportfishes are used as "umbrella species" for management of cold-water ecosystems; however, for many management goals, focusing on or including sculpins would provide a much larger umbrella for several reasons. First, sculpins are poor swimmers relative to most large-bodied fishes, and therefore stream obstructions that do not hinder salmonid movements can prevent passage by sculpins. Fish-passage facilities that are engineered for small-bodied fishes would presumably allow most other aquatic organisms to pass as well. Second, some sculpin species are extremely sensitive to pollutants. For example, sculpins are more sensitive than salmonids to metal enrichment in the field[22] and are so sensitive to zinc that they are not adequately protected by existing US federal water quality standards for zinc[23]. Nest-site characteristics of Slimy Sculpin (*C. cognatus*) suggest that nests may be vulnerable to alterations in flow regime and increases in sediment. Finally, sculpins may prove to be effective target organisms for biological assessments of anthropogenic impacts over small areas because they are often abundant, tend to be sensitive to a variety of perturbations, and often occupy relatively small areas throughout their lives[21,24].

The fossil record for sculpins is incomplete. However, identification of preopercles from sediments from the Deer Butte Formation of Idaho points to a Miocene origin for *Cottus*. *Cottus* occurred with several other species in Lake Idaho[8,25].

Males are polygynous, building nests or using existing cavities and enticing females inside with a series of vocalizations and displays. Sculpins of the genus *Cottus* are among the most abundant vertebrates in many freshwater ecosystems. They are generally sedentary, but there are exceptions. Eggs are laid in discrete clusters on the roof of the cavity and guarded by the male[8].

The 10 *Cottus* taxa we provisionally describe in Idaho are:

Cottus bairdii complex	Girard, 1850	Mottled Sculpin complex
Cottus beldingii	Eigenmann & Eigenmann, 1891	Paiute Sculpin
Cottus cognatus	Richardson, 1836	Slimy Sculpin
Cottus confusus	Bailey & Bond, 1963	Shorthead Sculpin
Cottus extensus	Bailey & Bond, 1963	Bear Lake Sculpin
Cottus greenei	(Gilbert & Culver, 1898)	Shoshone Sculpin
Cottus hubbsi	Bailey & Dimick, 1949	Columbia Sculpin
Cottus leiopomus	Gilbert & Evermann, 1894	Wood River Sculpin
Cottus rhotheus	(Smith, 1882)	Torrent Sculpin
Cottus schitsuumsh	LeMoine et al., 2014	Cedar Sculpin

"Conservation is the foresighted utilization, preservation and/or renewal of forests, waters, lands and minerals, for the greatest good of the greatest number for the longest time."

Gifford Pinchot

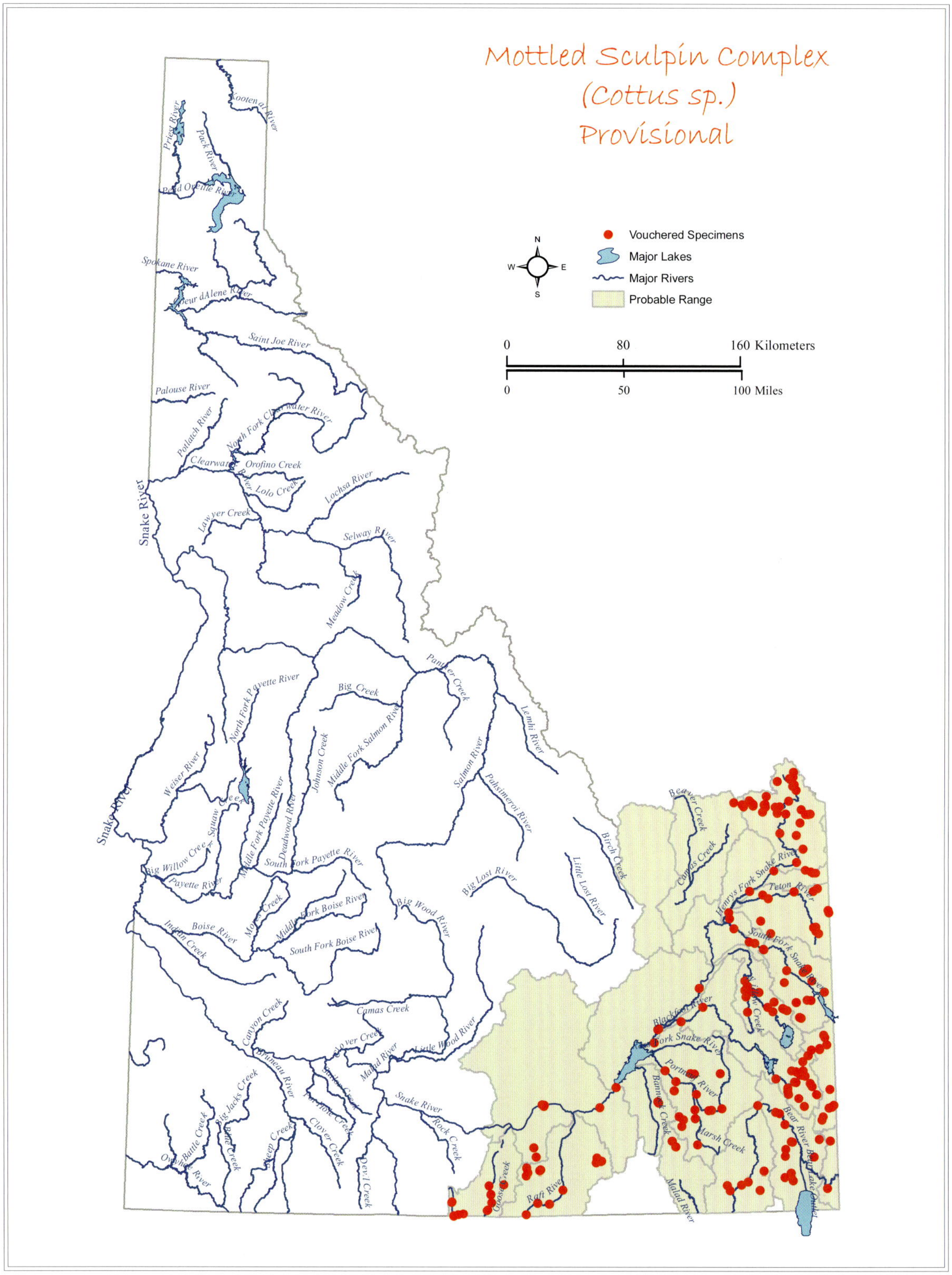
Mottled Sculpin Complex
(Cottus sp.)
Provisional
N
W
E
S
Vouchered Specimens
Major Lakes
Major Rivers
Probable Range
0
80
160 Kilometers
0
50
100 Miles
Snake River
Salmon River
Big Lost River
Little Lost River
Birch Creek
Lemhi River
Pahsimeroi River
Big Wood River
Boise River
Payette River
Weiser River
Clearwater River
Lochsa River
Selway River
Saint Joe River
Spokane River
Palouse River
Potlatch River
Kootenai River
Priest River
Pack River
Big Creek
Middle Fork Salmon River
Johnson Creek
Henrys Fork Snake River
Teton River
South Fork Snake River
Blackfoot River
Portneuf River
Bear River
Marsh Creek
Malad River
Raft River
Goose Creek
Rock Creek
Camas Creek
Beaver Creek
Owyhee River
Bruneau River
Big Jacks Creek
Canyon Creek
Indian Creek
Little Wood River
Meadow Creek
Panther Creek

Mottled Sculpin Complex

Cottus sp.

Eigenmann & Eigenmann, 1891

Etymology

Cottus = an old European name for sculpin; *bairdii* = named for the first US Fish Commissioner Spencer Fullerton Baird.

Importance

Members of the Mottled Sculpin complex provide food for trouts and other cold-water fishes. Generally, their presence is considered an indicator of quality trout waters[5].

Description

These fish have a broad, flattened head, typically 31 to 35 percent of SL. The expansive pectoral fins typically have 15 to 16 rays. The pelvic fins have four rays and a single soft spine combined with the distal ray. The body lacks scales, but may have patches of prickles behind the pectoral fins, occasionally onto the back. The anterior dorsal fin has slender spines, is joined at the base to the second dorsal fin, and is often bordered with red, brown, or cream color in adults, particularly breeding males. Females rarely show much color, and none exists in young or half-grown fish. The second dorsal fin typically has 17 or 18 rays. There are typically three preopercular spines, the strongest of which is often exposed. These fish possess large patches of palatine teeth[5,26].

Often there are three distinct "saddles" of pigment on the back and extending onto the sides of the body beneath the second dorsal fin. These fish are olive, slate, barred, and spotted. Their lower sides may have some green or gold color. The caudal fin is more or less randomly speckled, the anal fin is darkly speckled, and the pectoral fins are banded[5,26].

Range and Distributional Status

These fish are native to Idaho. The native range of Mottled Sculpin is discontinuous over a wide area throughout North America, from the Tennessee River system of Georgia and Alabama to Labrador to the north and west to the Great Lakes basin[5]. In western North America, they have historically been described in the Columbia River Basin, upper Missouri, and upper Colorado River drainages.

Reports of recent mitochondrial DNA analyses suggest the Mottled Sculpin complex contains multiple clades and does not occur in the western United States[3,7,8]. We will continue to describe sculpins collected from the Snake River drainage above Shoshone Falls and from the Bear River system that have prickles and palatine teeth as Mottled Sculpin complex until the taxonomy and systematics of these sculpins are better understood and formal descriptions are published. Below Shoshone Falls, fish historically called Mottled Sculpin will now be provisionally called Columbia Sculpin (*C. hubbsi*).

Size and Longevity

Total lengths for members of the Mottled Sculpin complex reach 152 mm (6.0 in)[5]. They may live to an age of five years. In Idaho they occasionally reach 150 mm (5.9 in) in total length, but most are in the 75 to 100 mm (3.0 to 3.9 in) range[27].

Food and Feeding

Members of the Mottled Sculpin complex are invertivores, and their diet consists mostly of aquatic insects. Other aquatic invertebrates, such as freshwater shrimp and snails, are occasionally eaten as well. Small fish, typically sculpin fry, may also be taken when available.[7] Previous research that examined stomach contents of Mottled Sculpin from the Mad River, Ontario, Canada, found chironomid larvae and mayfly nymphs to be most important in the diet of small sculpins under 51 mm (2.0 in) long. Mottled Sculpin more than 51 mm (2.0 in) long ate fewer chironomids and more mayfly nymphs, especially nymphs of the larger species[4]. Larval sculpins use the lateral line to sense food and may also use it to avoid predation[5].

Limiting Factors

Members of the Mottled Sculpin complex require clear, well-oxygenated water. They are preyed upon by other fishes and are parasitized by protozoans, acanthocephalans, and crustaceans[5].

Breeding Habits

Spawning can occur in either rapid- or slow-moving water during late winter and early spring, generally February to May, depending on the location of the ecosystem. As spawning time approaches, the male's head and body become dark colored. Males select a spawning site under a rock or other covered area. After suitable courtship, the female enters the nest, depositing salmon-colored adhesive eggs in a mass on the ceiling of the nest while upside down. After depositing her eggs, the female departs or is driven off by the male. More than one female may deposit eggs in the nest. Males fertilize the eggs and guard against predation. Males fan the eggs with their pectoral fins, presumably to keep eggs supplied with oxygen. At 11.1 to 12.8°C (52 to 55°F), eggs hatch in nine days. Adult males are larger than the females and ripen earlier.

Approximately 600 eggs per female have been noted for Utah populations of the Mottled Sculpin complex. Sexual maturity is reached at age two or three for both sexes[5].

Habits and Habitat

In the West, members of the Mottled Sculpin complex typically seek clear, cool mountain streams with moderate to rapid current. Bottom types are variable and typically include coarse gravel, small loose rocks, or rubble. However, some individuals live over sand, clay, or mud substrates and around the edges of lakes[5].

Preferred summer temperatures for these sculpins vary from 12.8 to 18.3°C (55 to 65°F). They can tolerate maximum weekly mean water temperatures up to 24°C (76°F)[28]. They live near vegetation, under stones, and part of the time in moderately swift riffles[5]. These fish frequently dominate stream-fish assemblages in both abundance and biomass. They are sit-and-wait predators of benthic invertebrates. Small individuals occupy microhabitats dominated by silt, sand, and gravel. As they reach sizes greater than 50 mm (2.0 in), these sculpins move to deeper areas dominated by cobble and boulder substrates[29].

Macroinvertebrate prey distributions are typically patchy, and their abundance within a patch is a good predictor of sculpin patch use (i.e., microhabitat). Strong evidence exists to indicate that adult density influences a variety of sculpin population and behavioral characteristics, including microhabitat use and population dynamics[29].

Management and Conservation Status

Given the taxonomic and systematic uncertainty of the Mottled Sculpin complex in Idaho, their conservation status is unknown. They are considered common, widespread, and abundant (G5/S5) and classified as unprotected wildlife by the IDFG[30]. They need clear, cold, well-oxygenated water and an abundance of benthic cover, particularly coarse substrates that provide living space. Management practices that conserve cool- and cold-water temperatures and minimize human-induced additions of fine-grained sediment are particularly important.

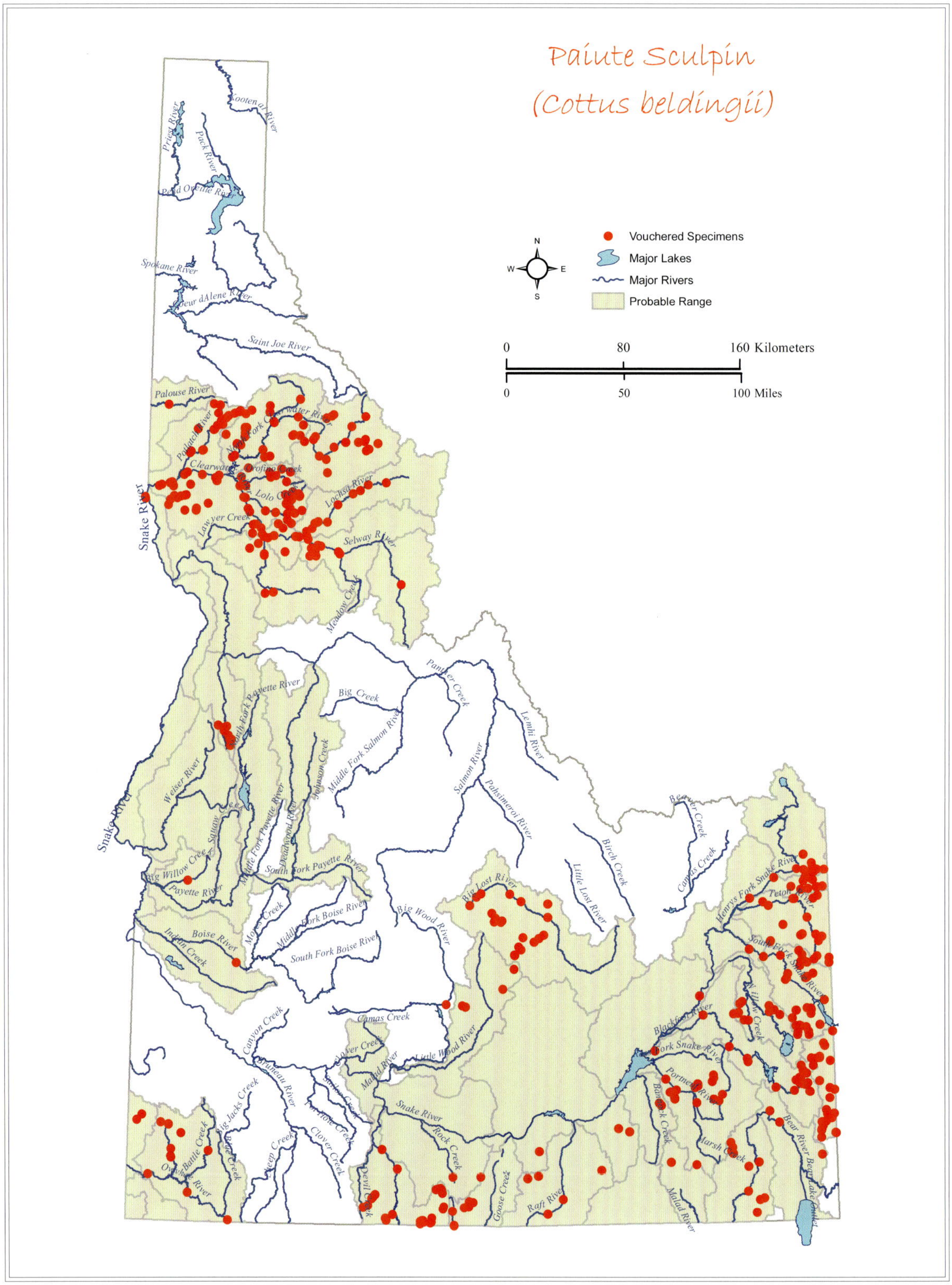
Paiute Sculpin
(Cottus beldingii)
N
S
E
W
Vouchered Specimens
Major Lakes
Major Rivers
Probable Range
0
80
160 Kilometers
0
50
100 Miles
Palouse River
Saint Joe River
Spokane River
Snake River
Selway River
Big Creek
Lemhi River
Salmon River
Pahsimeroi River
Birch Creek
Little Lost River
Big Lost River
Big Wood River
Boise River
Indian Creek
South Fork Boise River
Middle Fork Boise River
South Fork Payette River
Weiser River
Big Willow Creek
Payette River
Canyon Creek
Camas Creek
Snake River
Rock Creek
Goose Creek
Raft River
Malad River
Teton River
Blackfoot River
Portneuf River
Bannock Creek
Marsh Creek
Bear River
Clover Creek
Bruneau River
Middle Fork Salmon River
Johnson Creek

Paiute Sculpin

Cottus beldingii

Eigenmann & Eigenmann, 1891

Etymology

Cottus = an old European name for sculpin; *beldingii* = named for L. Belding, a fisheries practitioner in the early 20th century and collector of the type specimens from Lake Tahoe, California[31].

Importance

This species of sculpin is important in the food chain because of its varied diet and abundance and as prey for some species of trout, char, and other predators[27].

Description

Paiute Sculpin have a typical sculpin head that is moderately robust. They typically have a single preopercular spine, but occasionally a second blunt nub is present. Palatine teeth are typically absent, but occasionally patches of fewer than five teeth occur. The lateral line is incomplete, ending beneath the second dorsal fin. There are no prickles on the body. Fins are flecked with black, the caudal fin is rounded, the dorsal fin has 15 to 18 rays, and the anal fin has 11 to 13 rays. The first dorsal fin often has two blotches of dark pigment. The caudal peduncle is moderately thick. Paiute Sculpin have a heavily mottled body with five to seven crossbars on the back and may also have lateral banding. Overall color is brown to nearly black on the back and upper sides but somewhat paler below[5].

Range and Distributional Status

This species is native to Idaho. Paiute Sculpin occur in the Columbia River system, the Lahontan Basin of California and Nevada, and the Bonneville Basin of the western United States. They also occur in the upper Colorado River drainage. In Idaho Paiute Sculpin occur in the Snake River system, the Bear River system, and the Big Lost River drainage. Their occurrence in Idaho lakes is unknown[27]. Paiute Sculpin are found in 15 percent of sites surveyed in the upper Snake River basin of Idaho[32].

Size and Longevity

Paiute Sculpin have a short life and a slow growth rate. This species has been reported to reach lengths of 127 mm (5.0 in), but individuals of this size are rare and are generally male[5]. In Idaho adults average 75 to 100 mm (3.0 to 3.9 in) TL[27].

Food and Feeding

As with most fish species, the diet of Paiute Sculpin varies with availability of food items. Paiute Sculpin feed year-round but reduce feeding activity in winter water temperatures. There is little or no evidence of Paiute Sculpin preying on trout eggs[5].

Ambushing bottom dwellers is a principal feeding mechanism of Paiute Sculpin. When a prey organism moves within range of a largely concealed sculpin, the sculpin lunges and engulfs the prey. Paiute Sculpin forage at night. In Sagehen Creek, California, 63 percent of Paiute Sculpin diet by volume is aquatic insect larvae, especially mayflies, stoneflies, and caddis flies[33]. The remainder of the diet consists of miscellaneous bottom organisms, such as snails, water mites, aquatic beetles, and algae[5]. Paiute Sculpins are invertivores. In streams Paiute Sculpin feed primarily on aquatic insect larvae, but also take snails, water beetles, and algae[27].

Limiting Factors

Degraded habitat that adversely affects reproduction limits Paiute Sculpin populations. Some populations may suffer from excessive predation in certain ecosystems[5].

Breeding Habits

Paiute Sculpin become sexually mature at age two to three years and spawn in the spring, typically in May and June. Females produce 20 to 500 eggs, laying them in clusters on the underside of rocks. Males attend the nest and defend against other males[27]. Some spawning in deep water may take place in lakes. In streams most spawning sites are located in riffles. The number of eggs in each nest is usually 100 to 200, indicating that multiple spawning is not common. After reaching a size of 12.7 mm (0.5 in), fry are hatched and drop into the gravel and remain there for one or two weeks until the yolk sac is absorbed[5].

Habits and Habitat

Ecosystems with bottoms of rubble and gravel are the preferred habitats of Paiute Sculpin, although it is not unusual to find them living on other substrates. A typical stream habitat is rocky riffles with clear, cold water. Paiute Sculpin are generally associated with trout and trout streams. Paiute Sculpin spend daylight hours in hiding and comparative inactivity. Young Paiute Sculpin demonstrate two pronounced times of drift: one immediately following yolk-sac absorption and the other in slightly larger individuals about two weeks later[5].

Management and Conservation Status

Paiute Sculpin are common, widespread, and abundant (G5/S5) rangewide and in Idaho. They are classified as unprotected wildlife in Idaho[30]. Paiute Sculpin need a stable, unpolluted habitat, including spawning areas, in any ecosystem. No defined studies have been conducted in Idaho to determine the population status of Paiute Sculpin.

Paiute Sculpin of the Clearwater River are highly divergent (genetically) from Paiute Sculpin in the Snake River, which in turn differ greatly from those in Nevada[34]. The Clearwater population was described as a distinct species (*C. tubulatus*) by Hubbs and Schultz in 1932[17]. *C. tubulatus* were synonymized with *C. beldingii* by Bailey and Bond in 1963[19] but without supporting rationale. *C. tubulatus* is currently not recognized as a species[6]. However, given their genetic divergence from other Paiute Sculpin, *C. tubulatus* may be resurrected as a distinct species.

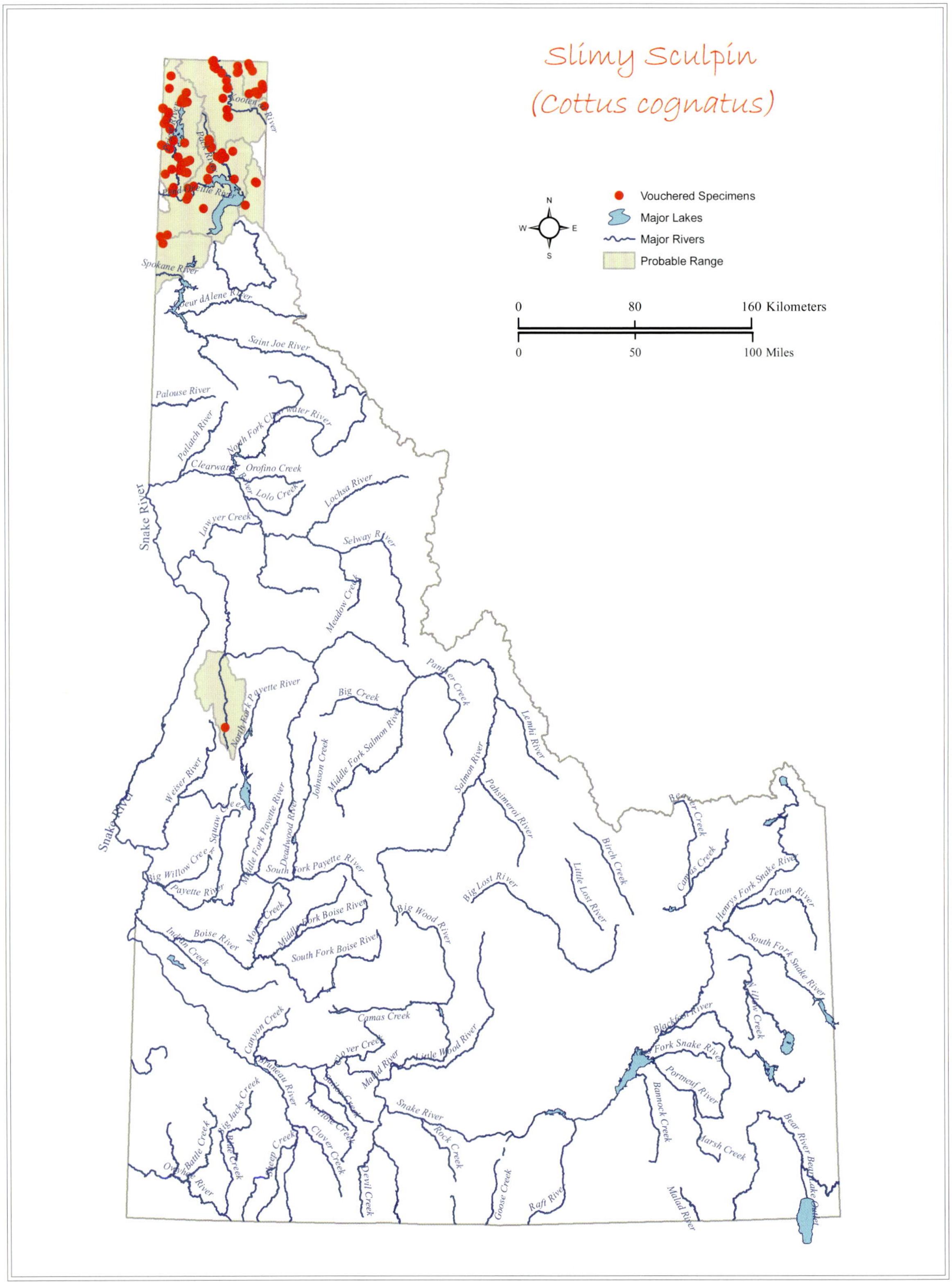
Slimy Sculpin
(Cottus cognatus)
N
S
E
W
Vouchered Specimens
Major Lakes
Major Rivers
Probable Range
0
80
160 Kilometers
0
50
100 Miles
Spokane River
Saint Joe River
Palouse River
Potlatch River
Orofino Creek
Lolo Creek
Lochsa River
Snake River
Selway River
Meadow Creek
Big Creek
Middle Fork Salmon River
Lemhi River
Salmon River
Johnson Creek
Weiser River
Pahsimeroi River
Birch Creek
Little Lost River
Big Lost River
Teton River
South Fork Payette River
Big Wood River
Boise River
Middle Fork Boise River
South Fork Boise River
South Fork Snake River
Camas Creek
Little Wood River
Portneuf River
Bannock Creek
Snake River
Rock Creek
Goose Creek
Raft River
Malad River
Devil Creek
Clover Creek
Jarbidge River
Big Jacks Creek
Bear River

Slimy Sculpin

Cottus cognatus
Richardson, 1836

Etymology

Cottus = an old European name for sculpin; *cognatus* = related (as to the European species *C. gobio*).

Importance

Most Slimy Sculpin are small and known to be eaten by trout, Northern Pike, and Burbot. They are important forage for trouts and are used as bait. Some trout anglers use flies ("Muddler Minnow" pattern) that resemble sculpins[35].

Description

Slimy Sculpin have dark-brown to black backs and sides, with various dark modeling or dark crossbars with whitish bellies. A dorsal fin that is darkly pigmented basally and almost clear marginally is typical of this species. Pectoral, caudal, and second dorsal fins typically have crossbars, and pelvic fins typically have three rays and occasionally a rudimentary fourth. Spawning males have an orange band on the margin of the first dorsal fin. This species may be confused with Shorthead Sculpin, but distinguishing features include three pelvic fin rays, typically two or three preopercular spines, and absence of palatine teeth[27].

Branchiostegal rays are 6, 6 (57), 6, 5 (1). The caudal fin is slightly rounded; the anal fin has a long base and is soft rayed. The pectorals are large and fanlike. Scales are typically absent, but small prickles may be present behind the pectoral fins. The lateral line is incomplete, terminating below the center of the second dorsal fin. Vertebrae number 31 (9), 32 (8), 33 (3), 34, or 35[4]. Caudal vertebrae number 20 to 24. The chin is uniformly pigmented, but not noticeably mottled[4].

Range and Distributional Status

This species is native to Idaho. Slimy Sculpin are widely distributed across northern North America from the East Coast of the United States and Canada to Alaska and British Columbia, Canada. In the Pacific Northwest, they are restricted to the upper Columbia River system. Slimy Sculpin exhibit a high degree of morphological variation across their range and occur in both rivers and lakes. In Idaho the species is known to occur in the Kootenai, Pend Oreille, and Little Salmon River drainages. This disjunct distribution is not currently understood[27].

Size and Longevity

Adult Slimy Sculpin are typically 50 to 75 mm (2.0 to 3.0 in) total length, but individuals over 100 mm (3.9 in) have been recorded[27]. In Montana the largest specimen was 117 mm (4.6 in) total length. Longevity to age six has been reported[35].

Food and Feeding

Slimy Sculpin are invertivores. They are opportunistic feeders, eating the most abundant food items available in a particular ecosystem. Adults may eat aquatic insects, mollusks, crustaceans, and small fishes. There are reports in the literature of this species feeding on various eggs, including trout and salmon[27]. In an Alaskan study, Slimy Sculpin from streams had a more diverse diet than did individuals taken from lake ecosystems. In Lake Ontario, Canada, smaller specimens of Slimy Sculpin fed primarily on amphipods (from 80 to 87 percent frequency of occurrence) at water depths between 35.1 to 75 m (115 and 246 ft), while larger fish fed on amphipods and mysid shrimp (74 and 26 percent of their diet, respectively) at deeper depths[35].

A New York State study showed that aquatic insects, crustaceans, small fishes, and some plant materials were consumed, but aquatic insect larvae and nymphs made up more than 50 percent of the diet and often 85 percent or more. The more important insect groups were mayflies, caddis flies, dipterous larvae, stoneflies, and dragonflies. Large fish tended to eat the larvae of larger species[4].

Limiting Factors

Most Slimy Sculpin are small and are therefore prey for trout, char, Northern Pike, and Burbot if they occur within the ecosystem[27]. A variety of parasites, including cestodes, trematodes, and nematodes, have been reported for this species[4].

Breeding Habits

Spawning behavior for Slimy Sculpin is reportedly similar to other members of this genus. Average water temperatures of 5°C (41°F) have been reported for the onset of spawning in lake ecosystems, and in spring streams spawning is initiated at 10°C (50°F) in New York[4]. Eggs are deposited under rocks and are guarded by the male, as with other sculpins. Male coloration darkens during the spawning season, and the spiny (first) dorsal fin has an orange margin. Females about 102 mm (4.0 in) in length produce approximately 1,400 eggs. Water-hardened eggs are 2.3 to 2.6 mm (0.09 to 0.1 in) in diameter and hatch in about four weeks at 7.8°C (46°F)[35].

Habits and Habitat

Slimy Sculpin are generally found in riffles among the rocks of cold and clear streams. They are abundant in shallow riffles only a few inches deep in some streams of the northeastern United States. In Lake Ontario, Canada, Slimy Sculpin have been found offshore at depths of about 9.1 to 152 m (30 to 500 ft) but were most commonly at depths of about 99 m (325 ft) or less. Slimy Sculpin have been captured with minnow traps down to 305 m (1,000 ft) in Lake Pend Oreille[35].

Management and Conservation Status

Range-wide, Slimy Sculpin are classified as secure (G5), and in Idaho they are classified as uncommon but not rare (S4) unprotected wildlife[30]. No definitive studies have been conducted in Idaho to determine conservation status of Slimy Sculpin.

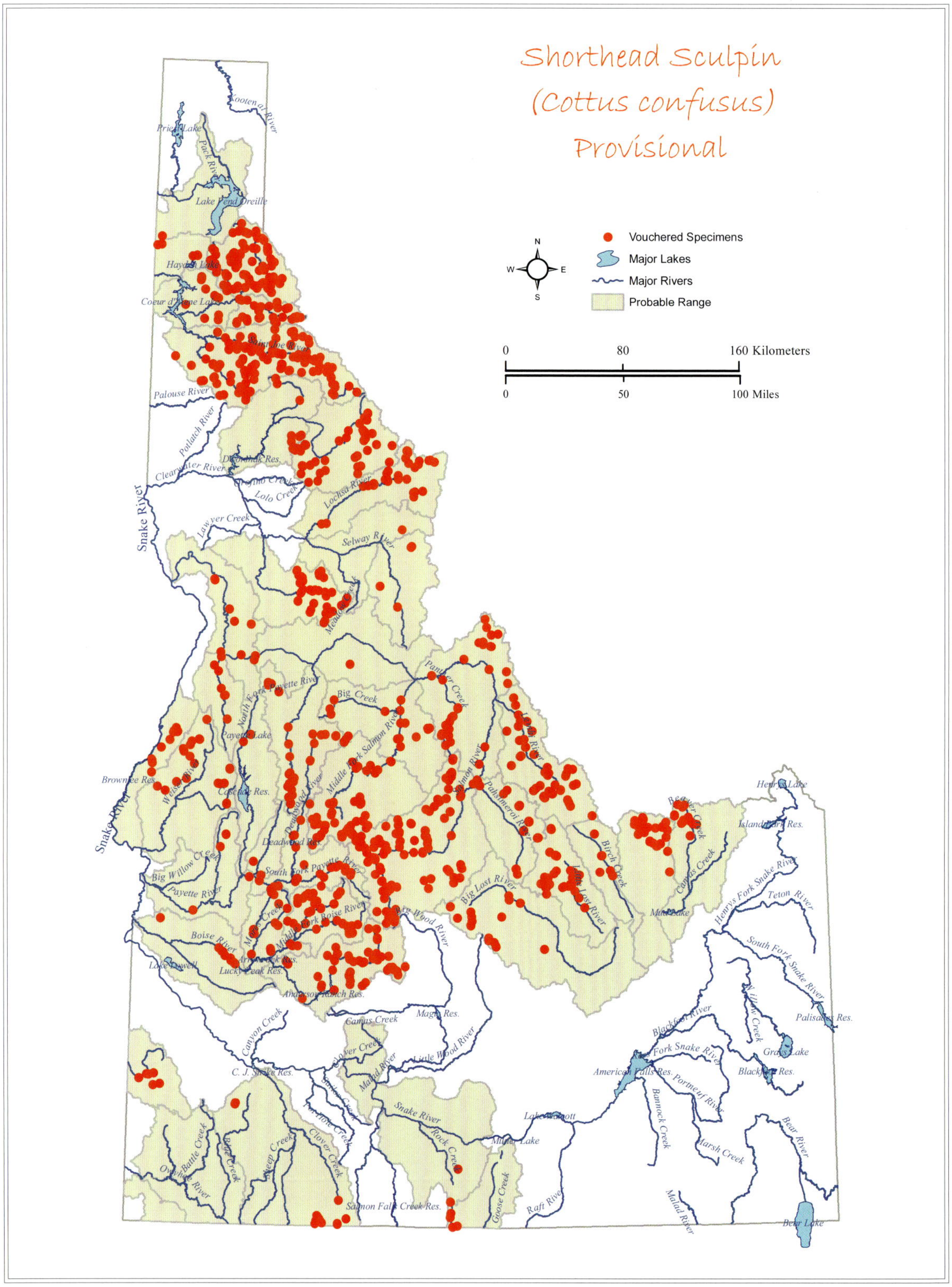
Shorthead Sculpin
(Cottus confusus)
Provisional
N
W
E
S
Vouchered Specimens
Major Lakes
Major Rivers
Probable Range
0
80
160 Kilometers
0
50
100 Miles
Kootenai River
Priest Lake
Pack River
Lake Pend Oreille
Coeur d'Alene Lake
Palouse River
Potlatch River
Clearwater River
Lolo Creek
Lochsa River
Lawyer Creek
Snake River
Selway River
Big Creek
Payette Lake
Cascade Res.
Weiser River
Brownlee Res.
Big Willow Creek
Payette River
Boise River
Lucky Peak Res.
Anderson Ranch Res.
Big Wood River
Big Lost River
Birch Creek
Pahsimeroi River
Camas Creek
Mud Lake
Henrys Lake
Island Park Res.
Henrys Fork Snake River
Teton River
South Fork Snake River
Palisades Res.
Blackfoot River
Grays Lake
Blackfoot Res.
American Falls Res.
Portneuf River
Bannock Creek
Marsh Creek
Bear River
Bear Lake
Malad River
Raft River
Lake Walcott
Milner Lake
Snake River
Rock Creek
Goose Creek
Salmon Falls Creek Res.
Clover Creek
Sheep Creek
Battle Creek
Owyhee River
C. J. Strike Res.
Canyon Creek
Camas Creek
Magic Res.
Little Wood River
Lake Lowell

Shorthead Sculpin

Cottus confusus
Bailey & Bond, 1963

Etymology

Cottus = an old European name for sculpin; *confusus* = clouded, referring to the typically irregular and indistinct body pigmentation.

Importance

No distinctive level of importance can be assigned to Shorthead Sculpin other than the fact that it is a part of the natural fauna in various ecosystems throughout the state of Idaho. They likely serve as food for a variety of predatory fishes[35].

Description

Shorthead Sculpin have a body form that is typically cottid, that is, it is heaviest forward, decreasing in size posteriorly. Shorthead Sculpin have the typical two dorsal fin arrangement; the first has a short base and is spiny with seven to 10 spines, and the second dorsal fin is longer and higher than the first and soft rayed, with 15 to 19 but typically 16 to 18 rays. The caudal fin is slightly rounded; the anal fin has a moderately long base and is soft rayed, with 12 to 14 rays. Pelvic fins are small, usually with one spine and four or occasionally five soft rays. Pectoral fins are large and fanlike, with 13 to 15 rays. Prickles are typically restricted to an area behind the pectoral fin insertion. However, individuals from the independent (Lost) drainages do not have prickles, suggesting they may be an undescribed sculpin taxa. The lateral line is never complete and terminates below ray 10 to 16 of the second dorsal fin and then is interrupted or underdeveloped[27]. Vertebrae number 33 to 35 and caudal vertebrae 22, 23, or 24[4].

Shorthead Sculpin have back and upper sides that are brownish or yellowish gray, often mottled, with cross bands that are poorly defined, alluding to the specific name. The belly is light with fine speckles. Of the two dorsal fins, the first is dark both anteriorly and posteriorly. Shorthead Sculpin have a preopercular bone typically with two spines, but occasionally with a much-reduced third. Palatine teeth are present in weak patches. Pelvic fins have four rays[27]. As the common name implies, Shorthead Sculpin have a small head, usually less than 30 percent of the standard length (range is 25 to 32 percent)[35].

Range and Distributional Status

This species is native to Idaho, the type locality being the Salmon River, Blaine County[19]. Shorthead Sculpin are a northwestern species found in the Columbia River system of Oregon, Washington, Idaho, and British Columbia, Canada. They also occur on the West Coast in streams draining into Puget Sound, Washington. In Idaho Shorthead Sculpin have been collected from the headwaters of most of the Snake River tributaries below Shoshone Falls and the independent drainages of the upper Snake River Basin[27].

Shorthead Sculpin have traditionally been reported in the Spokane River drainage of Idaho[26,27]. In 2014 a new species, Cedar Sculpin (*C. schitsuumsh*), was described from the Spokane drainage in Idaho and the Clark Fork River drainage in Montana based on genetic and morphological differences. In particular, Cedar Sculpin have one preopercular spine[36]. A review of 176 specimens from across the Spokane River drainage in Idaho reveals the majority of individuals have two or three preopercular spines (DWZ unpublished data 2016), suggesting Shorthead Sculpin or other undescribed taxa occur in addition to *C. schitsuumsh*. We will continue to provisionally call these multiple-spined fish Shorthead Sculpin until these taxonomic and systematic issues are resolved.

Size and Longevity

Mean SL at the end of each year of life for Shorthead Sculpin in the Big Lost River, Idaho, for fish aged one through six, in mm (in), is 38 (1.5), 46.7 (1.8), 58.9 (2.3), 70.7 (2.8), 82.1 (3.2), and 91 (3.6)[37].

Food and Feeding

Shorthead Sculpin, as with other cottids, typically occur in the benthic portion of a particular ecosystem, where they feed on many types of body organisms. In that light, the food of all sculpins is similar[35]. Shorthead Sculpin are invertivores, feeding primarily on benthic aquatic invertebrates, although fish eggs and fishes are occasionally taken. In the Big Lost River, Idaho, Shorthead Sculpin diet was primarily caddis flies, true flies, and mayflies[38,39].

Limiting Factors

In Idaho Shorthead Sculpin occupy the same ecosystems as some trouts and chars and serve as food for these species, especially Bull Trout[27]. Although no definitive information is available for Shorthead Sculpin populations in Idaho, it is likely that they are parasitized by the same organisms that affect other cottids.

Breeding Habits

Shorthead Sculpin spawn in the spring among the rocks and cobbles of riffles. In the Big Lost River, Idaho, spawning occurred within a two-week period in April[37]. Females collected from the Big Lost River contained 184 to 511 eggs, which is higher than fecundity reported for populations in neighboring Washington[27]. Shorthead Sculpin reach sexual maturity at age two or three. Breeding males are dark in color except for an orange margin on the first dorsal fin, while females remain mottled brown in color during the entire spawning season[35]. Observations on Shorthead Sculpin fecundity from different ecosystems indicates it is highly variable, directly related to female age and size. Shorthead Sculpin deposit adhesive light-yellow to orange eggs in clusters underneath rocks. Laboratory test rearing of eggs at 10°C (50°F) resulted in eggs hatching in 29 days. This experiment produced fry that averaged 7.6 mm (0.3 in) in length. Wild larvae are pelagic but become benthic dwellers several weeks after hatching, a life history pattern that they maintain throughout the rest of their lives[35].

Habits and Habitat

Shorthead Sculpin are typically found in riffle areas of high-elevation cold, clear streams with gravel or rubble bottoms. Older Shorthead Sculpin in the Big Lost River drainage seem to prefer pool areas with rubble substrate[38]. Shorthead Sculpin are typically found farther upstream in a given ecosystem at higher elevations than most other sculpins and are occasionally found in backwaters and along shorelines in large rivers. In Idaho's Clearwater River system, Shorthead Sculpin are common in headwaters of each major drainage above 975 m (3,200 ft) elevation, but went undetected from the drainage downstream from these areas[40]. Shorthead Sculpin seem to prefer waters with average summer temperatures less than 15°C (59°F)[27].

Management and Conservation Status

Shorthead Sculpin are considered secure range-wide (G5), and in Idaho they are classified as vulnerable (S3) unprotected wildlife[30]. Shorthead Sculpin are part of the natural ecosystem and ecology of Idaho streams. No defined studies have been conducted in most ranges in Idaho to determine population or distributional status.

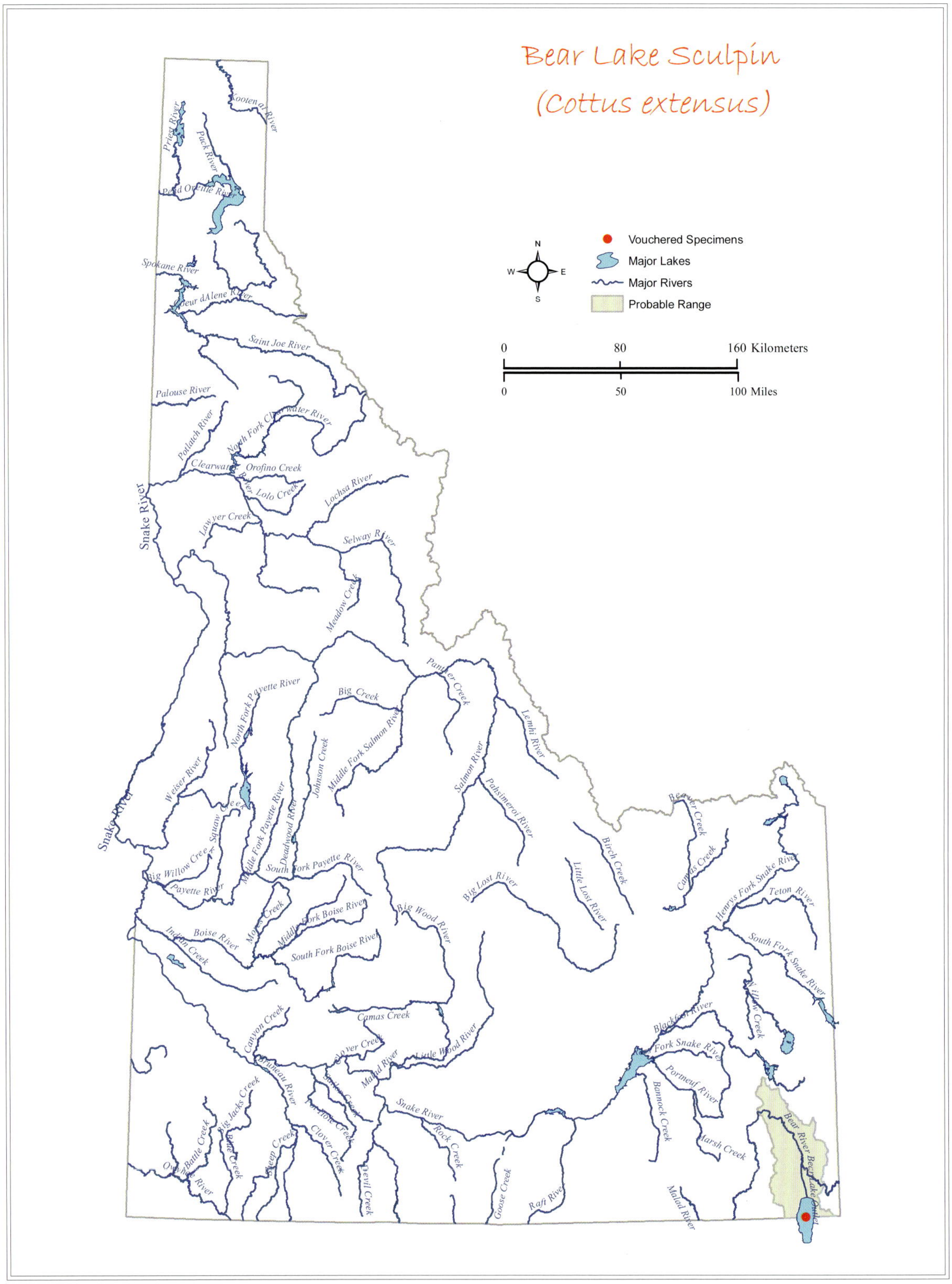
Bear Lake Sculpin
(Cottus extensus)
N
W
E
S
Vouchered Specimens
Major Lakes
Major Rivers
Probable Range
0
80
160 Kilometers
0
50
100 Miles
Kootenai River
Priest River
Pack River
Spokane River
Saint Joe River
Palouse River
Potlatch River
Orofino Creek
Lolo Creek
Lochsa River
Snake River
Selway River
Meadow Creek
Big Creek
Middle Fork Salmon River
Johnson Creek
Salmon River
Lemhi River
Pahsimeroi River
Weiser River
Middle Fork Payette River
Deadwood River
Payette River
Boise River
Indian Creek
Mores Creek
South Fork Boise River
Big Wood River
Big Lost River
Little Lost River
Birch Creek
Camas Creek
Canyon Creek
Bruneau River
Clover Creek
Rock Creek
Devil Creek
Goose Creek
Raft River
Malad River
Bannock Creek
Portneuf River
Marsh Creek
Blackfoot River
Willow Creek
Henrys Fork Snake River
Teton River
South Fork Snake River
Bear River

Bear Lake Sculpin

Cottus extensus

Bailey & Bond, 1963

Etymology

Cottus = an old European name for sculpin; *extensus* = referring to the slender form of this species.

Importance

Bear Lake Sculpin are an important food resource for Bonneville Cutthroat Trout and Lake Trout in Bear Lake. Greater than 70 percent of the diet of intermediate-age Lake Trout is composed of Bear Lake Sculpin. The proportion of sculpins in diets of older Lake Trout declines to 20 percent[41]. Intermediate-age Bear Lake (Bonneville) Cutthroat Trout, 254 to 356 mm (10.0 to 14.0 in) total length, used Bear Lake Sculpin as their major prey source throughout winter and spring months[42]. Utah State University researchers found that predators could consume a substantive portion of the annual Bear Lake Sculpin production[43,44].

Description

Bear Lake Sculpin are tan or brownish above, becoming lighter below with a few large blotches sometimes visible on the sides. Bear Lake Sculpin have short heads, normally less than one-third of the SL Pectoral fins have 15 to 16 soft rays (most often 16), and the lateral line is incomplete, with 21 to 26 pores, terminating below the second dorsal fin[27]. The preopercular bone has three spines, the principal one directed backward and slightly upward. The middle spine is directed downward and backward. The dorsal fin is in two parts, the first short and spiny, the second with 16 to 19 rays. The anal fin has 13 rays, the pectorals have 16, and the caudal fin is rounded[5]. Palatine teeth are present, and prickles are well developed on the back and sides[7].

Range and Distributional Status

This species is native to Idaho and endemic to Bear Lake. Previously, Bear Lake Sculpin were planted in Flaming Gorge Reservoir on the Green River, Utah. Their fate there is unknown[5]. Fossils of Bear Lake Sculpin have been found in Lake Bonneville deposits[45]. Connections between Bear Lake and ancient Lake Bonneville have been studied extensively, and some recent research concludes that the Bear River did flow into Bear Lake and connect it to Lake Bonneville during the Pleistocene (0.011 to 2.4 mya). These intermittent fillings raised the level of Bear Lake until the water rose to the valley threshold at Nounan Narrows. Several water-level fluctuations in the 10s of meters (30+ ft) have occurred[46].

Size and Longevity

Average size of specimens of Bear Lake Sculpin taken in gill nets is about 75 mm (3.0 in), although females of 137 mm (5.4 in) and males of 173 mm (6.8 in) have been observed[27]. Bear Lake Sculpin grow at approximately the same rate as Paiute Sculpin. Paiute Sculpin have yearly lengths for fish aged one to four years, in mm (in), of 53.3 (2.1 in), 69 (2.7 in), 84 (3.3 in), and 94 (3.7 in)[5].

These are small fish that rarely attain 102 mm (4.0 in) in total body length. They are relatively long-lived for a small fish, with some aged at eight years[47]. The population size of Bear Lake Sculpin has been estimated at one to two million fish[43,44].

Food and Feeding

Bear Lake Sculpin are invertivores, feeding primarily on the dominant bottom-dwelling zooplankton and midge larvae in Bear Lake. In a previous study, 330 sculpin stomachs contained 59 percent ostracods by weight and 39 percent adult cyclopoid copepods by weight. Although juvenile fish demonstrated a pronounced daily vertical migration, moving into the water column at dusk and returning to the bottom at dawn, they did not feed while in the water column. Bear Lake Sculpin were observed to feed only when they were on the bottom, and the stomach fullness peaked just prior to the ascent. Movement into the warmer mid-column water is evidently an adaptation to speed digestion[5].

Limiting Factors

Any water-quality degradation in Bear Lake or any drop in water levels so that the best spawning and daytime cover habitats are exposed will adversely impact Bear Lake Sculpin. Bear Lake Sculpin are preyed upon heavily by Cutthroat Trout and Lake Trout in Bear Lake[5].

Fluctuating water levels likely affect Bear Lake Sculpin populations. Spawning habitat is limited to the ring of rock habitats associated with high water elevations. When water levels are lowered, there is increasingly less habitat available for Bear Lake Sculpin spawning[48]. Water levels in the spring of 2005 were down 7.3 m (24 ft) from full pool, reducing available spawning habitat to less than 5 percent of total potential[48]. In low-water years, high predation pressure and limited spawning habitat can reduce the abundance of Bear Lake Sculpin. A computer simulation model indicated that increased Lake Trout stocking posed a greater potential threat to Bear Lake Sculpin than any other fish species in the lake during these periods of low water[48]. UDWR has reduced stocking of Lake Trout in response to concerns about prey fish abundance[43].

Breeding Habits

Bear Lake Sculpin reproduce in late spring, with adults moving into shallow, rocky areas. Males compete for nesting sites, which are the undersides of fist-size and larger rocks. When males have established territories, females circulate through the area, looking for a suitable mate. Larger males tend to successfully defend their breeding sites and thus have the most access to females. Females lay adhesive eggs on the roof of the nest, which the males then fertilize. After mating, the females leave the nest site, and males care for the eggs and young. This parental care includes protection of fertilized eggs from predators and cleaning of eggs to reduce fungus and bacteria[49]. During the four- to six-week incubation, males fan the eggs with their pectoral fins to remove waste and provide oxygen. During this time, they are not believed to leave the nest, even to feed. The fertilized eggs of Bear Lake Sculpin hatch in late May and early June, and larval sculpins move to the surface waters, where water currents can disperse them. Soon after dispersal, juvenile Bear Lake Sculpin begin a diel pattern of vertical migration. These age-zero individuals leave the bottom of the lake at night and ascend up to 40 m (130 ft) into the water column. This behavior allows these juvenile fish to occupy warmer water, thereby increasing growth rates[43,50].

Habits and Habitat

Bear Lake Sculpin are the only sculpins in the western United States that strictly occupy a deep-water lake habitat. Bear Lake Sculpin occupy the benthic regions of Bear Lake, where they feed on aquatic macroinvertebrates and zooplankton[43].

Management and Conservation Status

Bear Lake Sculpin are native and endemic to Bear Lake. Because of this restricted range, they are classified as critically imperiled, a high-risk species (G1/S1); vulnerable to range-wide extinction or extirpation; and a protected nongame species in Idaho[30]. Threats to this species include the loss of spawning and rearing shoreline habitat when Bear Lake is pumped too low to provide irrigation water and the potential threat of a change in water quality with increased development around the lake. Bear Lake Sculpin can survive in the deep-water marl bottom of Bear Lake, but without some type of inshore cover, it is doubtful they can spawn with any degree of success.

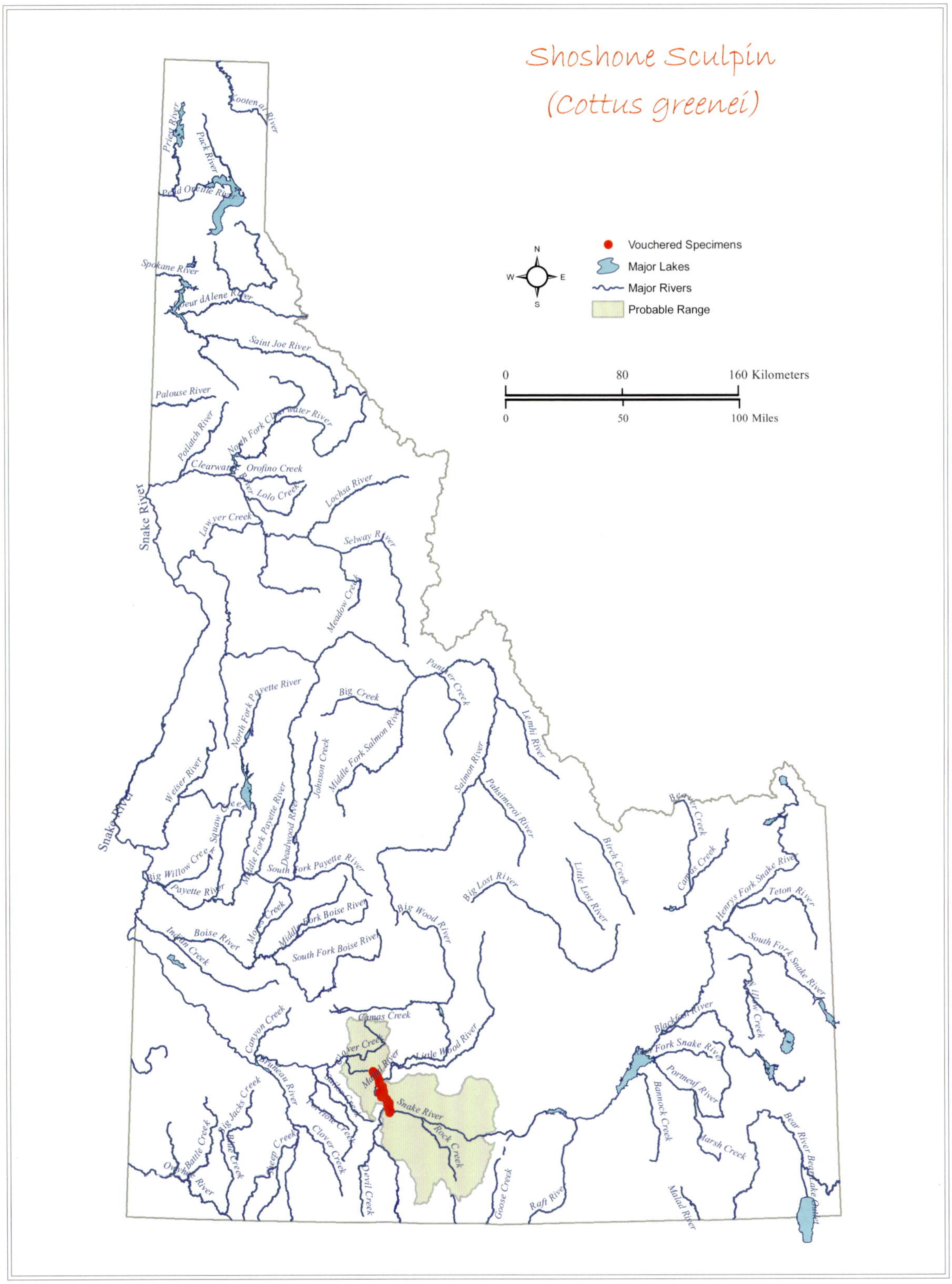
Shoshone Sculpin
(Cottus greenei)
Vouchered Specimens
Major Lakes
Major Rivers
Probable Range
N
W
E
S
0
80
160 Kilometers
0
50
100 Miles
Pack River
Saint Joe River
Palouse River
Pollatch River
Orofino Creek
Lochsa River
Snake River
Lolo Creek
Meadow Creek
Big Creek
Middle Fork Salmon River
Johnson Creek
Salmon River
Lemhi River
Pahsimeroi River
Weiser River
Middle Fork Payette River
Deadwood River
Birch Creek
Little Lost River
Big Lost River
Big Wood River
Teton River
South Fork Boise River
Boise River
Canyon Creek
Camas Creek
Rock Creek
Snake River
Clover Creek
Devil Creek
Goose Creek
Bannock Creek
Portneuf River
Marsh Creek
Malad River
Owyhee River

Shoshone Sculpin

Cottus greenei

(Gilbert & Culver, 1898)

Etymology

Cottus = an old European name for sculpin; *greenei* = named for Professor Charles Wilson Greene of Stanford University, of whom Gilbert and Culver stated, "to whose energy was due much of the success of the expedition"[51].

Importance

Shoshone Sculpin have an extremely limited native range and are therefore important as a natural resource entity within their limited ecosystem.

Description

Shoshone Sculpin are typically brownish to brownish gray. Blackish bars appear on the back and sides, and both dorsal fins and the anal fin are speckled with dark pigment. Dark bands occur on the caudal and pectoral fins. The body is short and stubby, with a very thick caudal peduncle. Shoshone Sculpin typically have one preopercular spine, weak palatine teeth, and three visible pelvic fin rays[27].

Range and Distributional Status

This species is native and restricted to the Hagerman area in south-central Idaho[52]. The type specimen was collected from Thousand Springs, Snake River, Idaho, on August 9, 1894[51]. Shoshone Sculpin are endemic to the springs and spring creeks within the Thousand Springs formation on the north bank of the Snake River in Gooding County, Idaho, between Kanaka Rapids and Bliss. They have been transplanted, and subsequently established, to an unnamed spring near Snake River at river km 965.7 (river mi 600) below Crystal Springs in Gooding County[53].

Size and Longevity

Shoshone Sculpin may live up to four years; most populations have few individuals older than age two. Shoshone Sculpin are short-lived, relatively fast-growing cottids that attain total lengths of 50 to 75 mm (2.0 to 3.0 in). The largest specimen reported in one study was 93 mm (3.7 in)[54].

Food and Feeding

Shoshone Sculpin are invertivores, feeding mostly on benthic aquatic insects and small crustaceans, such as amphipods and copepods[55]. Other food items may include ostracods, snails, and small clams. Occasionally during winter, large males have been observed taking other Shoshone Sculpin[27].

Limiting Factors

Shoshone Sculpin are most likely affected by predation in some ecosystems and may be parasitized by a variety of species. Poor water quality or lack of water may also cause mortality in some portions of this species ecosystem. The major threats to the survival of Shoshone Sculpin are probably habitat destruction and water loss related to irrigation, aquaculture, and hydroelectric projects[27].

Breeding Habits

Shoshone Sculpin become reproductively mature at age one, with females maturing at a length of 40 mm (1.6 in) and males at a length of 49 mm (1.9 in). Spawning occurs in the spring, but in at least one spring system with relatively cool water temperatures, spawning season may last from late winter to midsummer. Females produce between 20 and 120 eggs, and hatching occurs within the fourth week after being laid[27].

Habits and Habitat

Shoshone Sculpin occur in the headwaters of fairly cool 13 to 14°C (55.4 to 57.2°F), clear springs and spring-fed streams. Typical habitat for Shoshone Sculpin is characterized as highly productive, with a relatively constant temperature and abundant food[27].

In a 2011 report, Campbell[56] notes that the Shoshone Sculpin population is highly structured, with substantial genetic differentiation observed between most populations. Genetic population structure of the species refers to the amount and distribution of genetic variation within and between populations. This structuring is likely a product of a number of different influences. Freshwater sculpins are generally sedentary, with low rates of dispersal and relatively small home ranges. Evidence of isolation by districts across the range of Shoshone Sculpin is a pattern compatible with limited gene flow and random genetic drift within populations. Shoshone Sculpin are also habitat specialists, endemic to the springs and spring-creek habitats of the Thousand Springs formation along the Hagerman Valley area of the Snake River[52]. These springs are naturally fragmented and have been extensively developed for hydroelectric facilities, irrigation, and fish-culture operations. These localized anthropogenic influences, along with decreases in spring discharges (both naturally and anthropogenic), have likely further fragmented populations and reduced available habitat[56].

Management and Conservation Status

This species, an Idaho endemic, is considered imperiled (G2/S2), vulnerable to range-wide extinction or extirpation due to naturally restricted range, few populations, and other factors. They are classified as a protected nongame species by the state of Idaho[30]. Some populations of Shoshone Sculpin found in the late 1970s were not found when the surveys were repeated in the late 1980s. Loss of these populations was attributed to increased diversion of water, resulting in shallower habitat, poor quality of water, higher turbidity, and higher water temperatures.

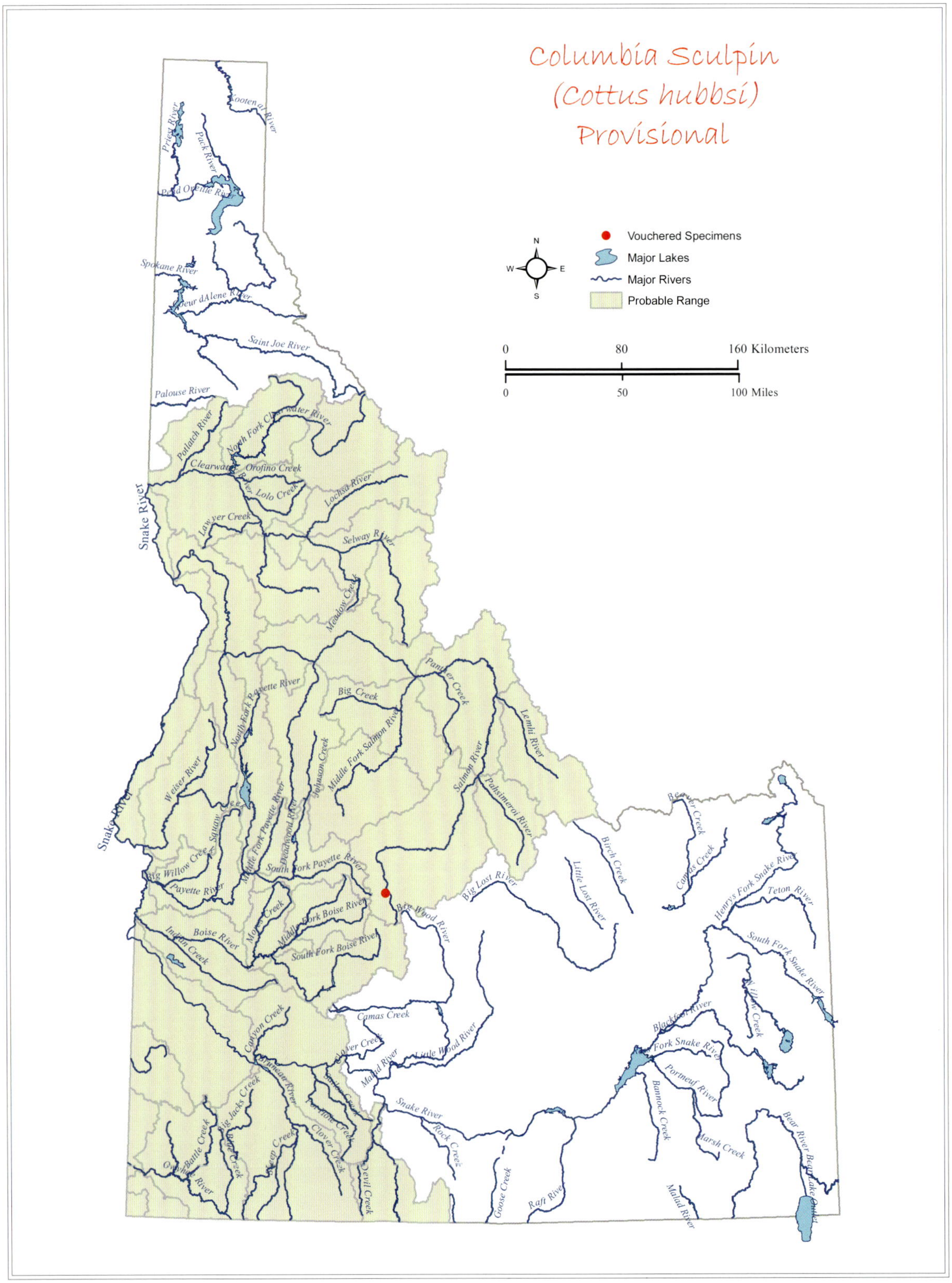
Columbia Sculpin
(Cottus hubbsi)
Provisional
N
S
E
W
Vouchered Specimens
Major Lakes
Major Rivers
Probable Range
0
80
160 Kilometers
0
50
100 Miles
Priest River
Pack River
Spokane River
Saint Joe River
Palouse River
Potlatch River
Orofino Creek
Lochsa River
Lawyer Creek
Snake River
Selway River
Meadow Creek
Big Creek
Weiser River
North Fork Payette River
Middle Fork Payette River
Deadwood River
Johnson Creek
Middle Fork Salmon River
Salmon River
Lemhi River
Pahsimeroi River
South Fork Payette River
Payette River
Big Willow Creek
Boise River
Indian Creek
Middle Fork Boise River
South Fork Boise River
Big Lost River
Little Lost River
Birch Creek
Camas Creek
Henrys Fork Snake River
Teton River
South Fork Snake River
Canyon Creek
Little Wood River
Snake River
Rock Creek
Goose Creek
Portneuf River
Bannock Creek
Marsh Creek
Bear River
Malad River
Big Jacks Creek
Clover Creek
Devil Creek

Columbia Sculpin

Cottus hubbsi

Bailey & Dimick, 1949

Etymology

Cottus = an old European name for sculpin; *hubbsi* = named for the preeminent ichthyologist Carl L. Hubbs, whose contributions to North American ichthyology are legendary.

Bailey and Dimick[9] described Columbia Sculpin from tributaries of the Columbia River in Washington and Idaho, diagnosing them on the basis of a complete lateral line and pigmentation. Bailey and Bond (1963) synonymized Columbia Sculpin and Malheur Sculpin (*C. bendirei*) with Mottled Sculpin (*C. bairdii*) but provided neither data nor a rationale for the change[8]. Consequently, the American Fisheries Society and American Society of Ichthyologists and Herpetologists no longer recognized Columbia Sculpin or Malheur Sculpin as distinct species[57–59] until both species were resurrected[60] based on the work of Markle and Hill[61] and Neely[8]. Both species are now considered separate from the Mottled Sculpin complex[3,6,60].

Importance

Columbia Sculpin are a small Columbia River endemic whose primary importance may be related to their unique natural history status. Typical of all freshwater sculpins, they serve as important forage for other piscivorous taxa, particularly native salmonids. As such, they contribute to the circulation of energy and nutrients in Idaho aquatic ecosystems.

Description

Columbia Sculpin are small, bottom-dwelling fish with a body shape that tapers from a relatively large head and pectoral fins to a narrow caudal fin. Columbia Sculpin are distinguished from other closely related cottids by a relatively long head; a complete or nearly complete lateral line; striking broad, dark bars on the caudal fin; and oblique dark bars on the anal fin[62]. They typically have three preopercular spines[9]. Columbia Sculpin have strong patches of palatine teeth and significant numbers of prickles axial to the pectoral fins and occasionally extending onto the anterior portion of the back[3].

Range and Distributional Status

This species is native to Idaho. Columbia Sculpin occur in the Columbia River and its tributaries in Washington, Oregon, and Idaho. Its range includes about 1,600 linear km (994 mi) of the Columbia and Snake Rivers from below Shoshone Falls[3] downstream to the confluence of the Umatilla and Columbia Rivers. Below this point, the species' distribution is highly fragmented[62].

In Idaho Columbia Sculpin have been collected from Alturas Lake Creek in the upper Salmon River Basin south of Stanley, Idaho[9]. Other than this occurrence, its distribution in Idaho is undocumented but provisionally believed to be all drainages of the Snake River below Shoshone Falls formerly described as containing Mottled Sculpin[26].

Size and Longevity

No size or longevity data are available for Columbia Sculpin in Idaho. Until the taxonomy and systematics of the Mottled Sculpin complex are better understood and clearly described, age and growth data are confounded by uncertain taxa identities. In British Columbia, Columbia Sculpin typically live less than five years and may reach lengths of 110 mm (4.3 in)[62].

Food and Feeding

Prey and feeding habits of Columbia Sculpin are similar to those of other cottids. Juveniles and mature fish typically feed on aquatic invertebrates, with large adults also taking small fish[3].

Limiting Factors

Drought or changes in water quality in the ecosystems occupied by Columbia Sculpin, as well as the effects of urbanization, mining, and industrial development, may adversely affect this species. It is likely that it is parasitized by organisms similar to those affecting other cottids.

Breeding Habits

In British Columbia, Columbia Sculpin females reach sexual maturity at two to three years, and males mature at two years[3] They spawn in the spring and early summer. The eggs are large and take about three to four weeks to hatch[62].

Habits and Habitat

During daylight, Columbia Sculpin shelter in riffles and runs with moderate to fast surface velocities and large rock and boulder substrates. They are most active at night. Although this species is typically associated with large rivers and their major tributaries, in places it also occurs in small streams. Columbia Sculpin are sedentary, and adults rarely move more than 50 m (164 ft)[62].

Management and Conservation Status

As a recently resurrected species, Columbia Sculpin have not been assigned a conservation status in Idaho. Globally, their conservation status is classified by the International Union for Conservation of Nature as least concern with a stable population trend. In Canada natural barriers appear to control the geographic limits of this species; there are no known populations of this species above waterfalls or in large lakes. The major threats to BC populations are drought, urbanization, industrial development, and mining activities[62]. It is likely the same suite of potentially adverse factors affects this species in Idaho ecosystems.

Unresolved taxonomic problems and the coexistence of up to five sculpin species in small streams are of scientific interest. Columbia Sculpin's somewhat limited distribution (upper Columbia River drainage) makes it a unique part of both the United States' and Canada's faunal heritage[62].

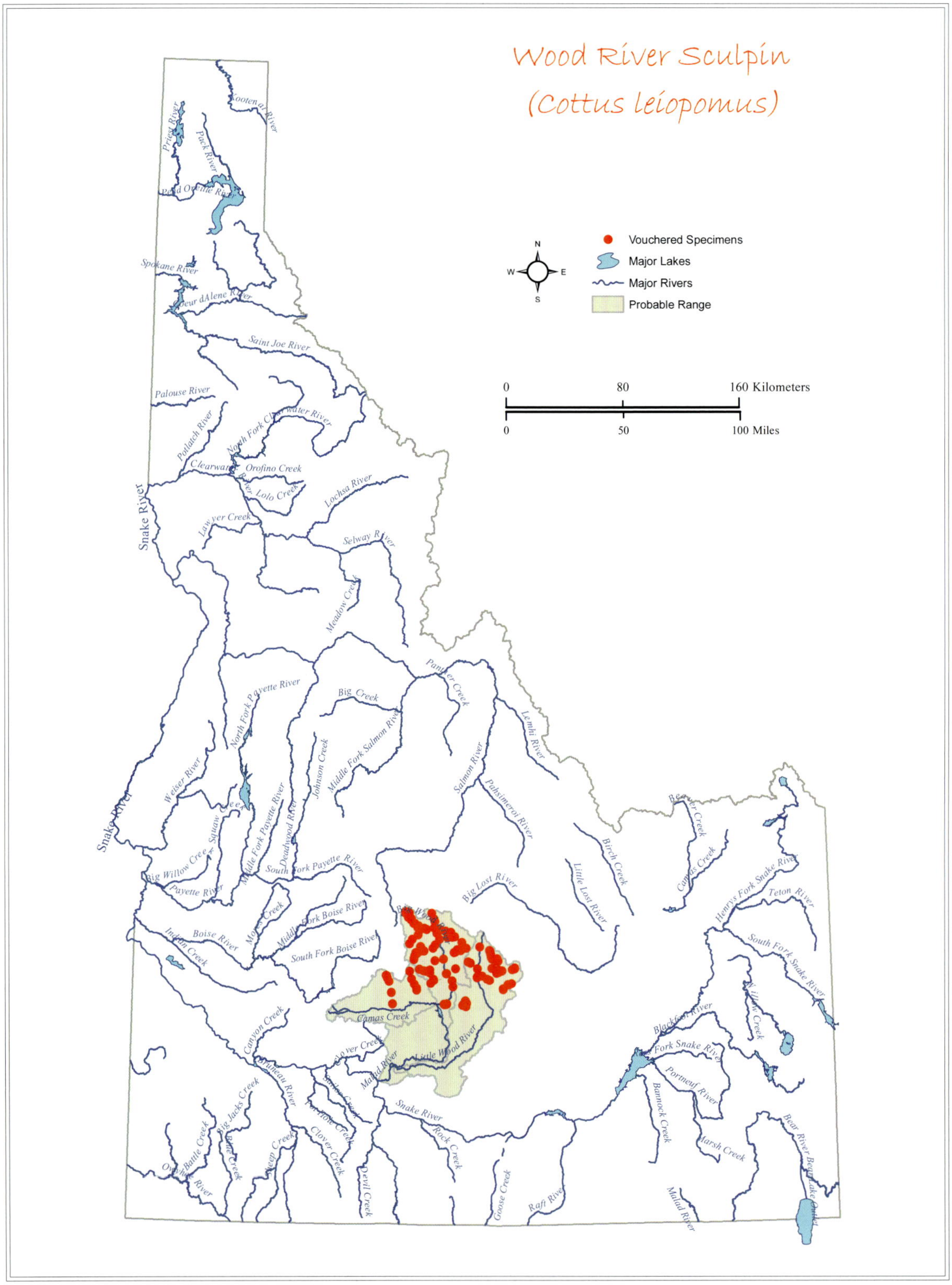
Wood River Sculpin
(Cottus leiopomus)
N
W
E
S
Vouchered Specimens
Major Lakes
Major Rivers
Probable Range
0
80
160 Kilometers
0
50
100 Miles
Saint Joe River
Palouse River
Potlatch River
Orofino Creek
Lolo Creek
Lochsa River
Snake River
Selway River
Meadow Creek
Big Creek
Middle Fork Salmon River
Johnson Creek
Salmon River
Pahsimeroi River
Lemhi River
Birch Creek
Little Lost River
Big Lost River
Weiser River
Big Willow Creek
Payette River
Boise River
South Fork Boise River
Camas Creek
Little Wood River
Canyon Creek
Snake River
Rock Creek
Goose Creek
Raft River
Blackfoot River
Portneuf River
Bannock Creek
Marsh Creek
Malad River
Teton River
South Fork Snake River
Devil Creek
Clover Creek
Deep Creek
Big Jacks Creek

Wood River Sculpin

Cottus leiopomus

Gilbert & Evermann, 1894

Etymology

Cottus = an old European name for sculpin; *leiopomus* = smooth cover = opercle.

Importance

Wood River Sculpin are endemic to the Wood River Basin of south-central Idaho. This characteristic makes them an important natural history entity[27]. They serve as important forage for other piscivorous taxa, particularly native salmonids, and contribute to the circulation of energy and nutrients in Idaho aquatic ecosystems.

Description

Wood River Sculpin are grayish olive on the dorsal side with mottling of dark brown or black. The belly ranges from white to blue. Ventral coloration is typically light cream colored with a slate-blue hue. The body, typically cottid, is deep and compressed. Preopercular spines are typically absent, though a few individuals may have a blunt nub on the preoperculum. No prickles or palatine teeth are present. The lateral line is incomplete or interrupted under the posterior end of the second dorsal fin. Wood River Sculpin have moderately large heads that are broad and oval[27].

Range and Distributional Status

This species is native to Idaho. The type specimens were collected from the Little Wood River, Shoshone, Idaho, in September 1893 by H. H. Kinsey[63]. Wood River Sculpin are endemic to the Wood River Basin of south-central Idaho, occurring in the Little Wood River drainage above the Little Wood Reservoir and in the Silver Creek watershed. In the Big Wood River and Camas Creek drainages, they currently occur above Magic Reservoir[27].

Size and Longevity

Adult Wood River Sculpin may reach total lengths of 141 mm (5.6 in). Wood River Sculpin typically live to be five years of age, but some individuals have been reported to reach age eight[27].

Food and Feeding

Wood River Sculpin are invertivores. No studies of food habits have yet been completed on this species, but examination of mouth contents of preserved specimens indicates feeding habits based on benthic aquatic invertebrates[27].

Limiting Factors

Wood River Sculpin are threatened by land development that adversely affects stream water quantity, or habitat quality[27]. Introduced predators, primarily large trout, may also be a factor.

Breeding Habits

Wood River Sculpin are reproductively mature at age three when individuals have a total length of approximately 60 mm (2.4 in). Females produce approximately 38 to 314 eggs. Spawning occurs in the spring[27].

Habits and Habitat

Wood River Sculpin are most abundant in riffles and runs with cobble or rubble substrate. This species may also be found in reduced numbers in pools, beneath undercut stream banks, and in those areas of ecosystems where submerged vegetation occurs when cobbles and rubbles are absent. Wood River Sculpin are typically found in cool and cold waters at temperatures of approximately 15°C (59°F) or less if the current is moderate. Wood River Sculpin may seasonally occupy ephemeral side channels in pools with suitable cobble and boulder substrate and are rarely found in heavily vegetated areas with fine substrates, lower water velocities, and water depths less than 10 cm (3.9 in). This species has been found at elevations above 2,374 meters (7,789 ft)[27].

Recently, the abundance of Wood River Sculpin was estimated to be at least 2.95 million fish[64]. The presence of Wood River Sculpin was positively associated with stream width to depth ratio and the percentage of cobble and boulder substrate and negatively associated with stream gradient. Results suggest that Wood River Sculpin remain relatively widespread and abundant within their endemic range, despite obvious changes in historical stream connectivity caused by irrigation diversions and other chronic habitat alterations[64,65].

Management and Conservation Status

Wood River Sculpin are considered an imperiled species (G2/S2), at risk because of restricted range and other factors, and are classified as a protected nongame species by the state of Idaho[30]. Potential threats to Wood River Sculpin include human-accelerated rates of streambed sedimentation, stream channelization, flow alteration, introductions of piscivorous fishes, and degraded water quality.

Salmon River and iconic Sawtooth Mountains, near Stanley, ID. Photo Courtesy John Sigler.

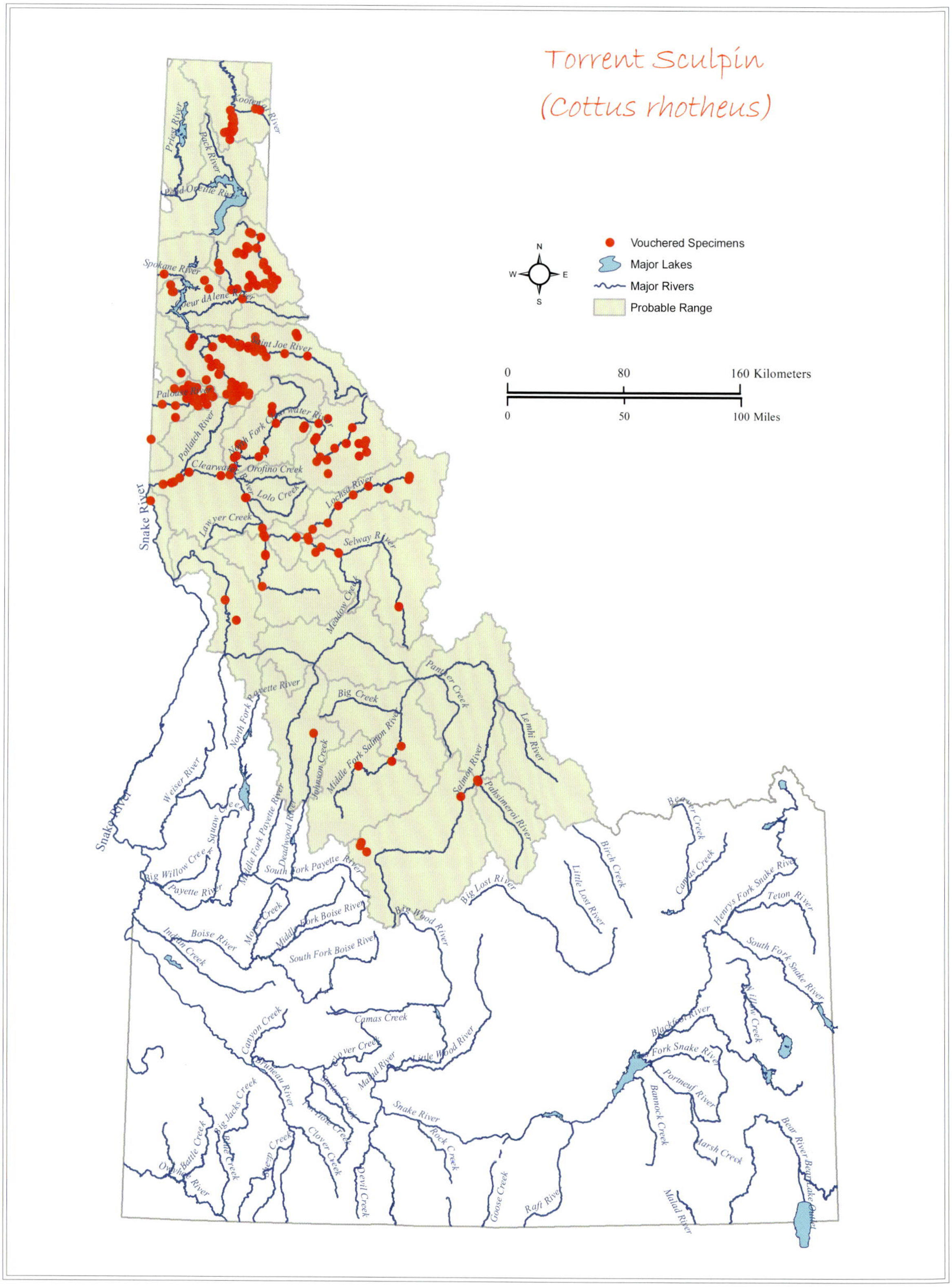
Torrent Sculpin
(Cottus rhotheus)
N
W
E
S
Vouchered Specimens
Major Lakes
Major Rivers
Probable Range
0
80
160 Kilometers
0
50
100 Miles
Priest River
Pack River
Spokane River
Palouse River
Potlatch River
Orofino Creek
Lolo Creek
Lawyer Creek
Selway River
Snake River
Big Creek
Johnson Creek
Middle Fork Salmon River
Pahsimeroi River
Lemhi River
Weiser River
Payette River
Boise River
South Fork Boise River
Big Lost River
Little Lost River
Birch Creek
Camas Creek
Teton River
Henrys Fork Snake River
South Fork Snake River
Camas Creek
Canyon Creek
Portneuf River
Bannock Creek
Big Jacks Creek
Clover Creek
Rock Creek
Goose Creek
Raft River
Malad River
Devil Creek
Snake River

Torrent Sculpin

Cottus rhotheus

(Smith, 1882)

Etymology

Cottus = an old European name for sculpin; *rhotheus* = rushing of the torrent, referring to the habitat in which this species is found.

Importance

Torrent Sculpin are little studied in Idaho, with Quintela being a recent exception[66]. They likely serve as important forage for other piscivorous taxa and contribute to the circulation of energy and nutrients in aquatic ecosystems.

Description

Torrent Sculpin are brown or grayish brown on the back and sides, with dark speckling. Lower sides are lighter, and the belly is almost white. The chin is usually moderately to darkly mottled, typically with two sensory pores. The body of this species is typically cottid, being heaviest forward and decreasing in size posteriorly to a very narrow caudal peduncle. The body is noticeably compressed laterally on the posterior portion, and the head is compressed dorso-ventrally. Dorsal, caudal, and sometimes anal and pectoral fins have darkly pigmented bars. The outer edge of the first dorsal fin becomes thickened and orange colored in spawning males. Of the two dorsal fins, the first has a short base and is spiny, with seven or eight spines or occasionally nine. The second dorsal is longer and higher than the first; soft rayed, typically with 15 rays but occasionally with 16 or 17 rays; and typically underlain with two dark, distinctive forward-slanting saddles that extend below the lateral line. The caudal fin is rounded, and the anal fin has a long base and is soft rayed, with usually 11 or 12 but occasionally with 13 soft rays. The pelvic fins are small, reduced, with one spine and four rays. Pectoral fins are large and fanlike, with 15 to 18 rays[4,27,35].

Typically, scales are absent, but with well-developed prickles scales are present, usually on the back, sides, and sometimes the caudal peduncle, but may be restricted to a patch behind the pectoral fin. The lateral line is typically complete, sometimes extending onto the caudal fin. Torrent Sculpin have large patches of palatine teeth, nearly touching the vomerine tooth patch. They typically have three preopercular spines, but occasionally a fourth is present. Geographic variation has been observed in the bodily appearance of this species[4,27,35].

Range and Distributional Status

This species is native to Idaho. Native range of Torrent Sculpin is the Columbia River drainage and rivers tributary to the Pacific Ocean from the Nehalem River, Oregon, north to the Fraser River, British Columbia, Canada. In Idaho this species occurs in the Kootenai, Pend Oreille, Spokane, Palouse, and Clearwater River Basins. It is also found in Warm Lake Creek, the main Salmon River, and the Snake River below Shoshone Falls[27].

Size and Longevity

Torrent Sculpin are a rather large and long-lived cottid species, living as long as seven years and attaining total lengths of more than 150 mm (5.9 in). Most adults are 75 to 115 mm (3.0 to 4.5 in)[27].

Food and Feeding

Young and partially grown Torrent Sculpin feed on planktonic crustaceans in very early life stages, but adults feed mainly on aquatic insect larvae, particularly midges and mayflies. At sizes larger than 55 mm (2.2 in), fish become increasingly important in the diet until fish are almost the exclusive food of sculpins 70 mm (2.8 in) or larger[4]. Large Torrent Sculpin are reported to feed on other smaller sculpins. In certain ecosystems, Torrent Sculpin may feed extensively on fry of various salmon species[35].

Limiting Factors

Torrent Sculpin may be infected with a variety of parasites, including protozoans and cestodes[4]. In some ecosystems, Torrent Sculpin may be a significant portion of the prey base for a variety of salmonids[35]. Torrent Sculpin populations will be negatively influenced by disturbances that significantly increase fine-grained sediments in substrates, reduce flows, and increase water temperatures[67].

Breeding Habits

Torrent Sculpin become sexually mature at two years of age. Torrent Sculpin spawn in late spring under stones in swift-water ecosystems. Some reports indicate that Torrent Sculpin move upstream on spawning migrations from late January to late April, moving downstream after spawning is completed in May and June. Females may produce a variety of number of eggs depending on the locality and female size. Average number of eggs per female ranges from 256 for age-two females to 226 for age-three females, 221 for age-four females, and 374 for age-five females. Egg number is highly variable[35].

Habits and Habitat

Torrent Sculpin are typically found in the middle reaches of larger rivers and in swift cool-water ecosystems with a stable gravel substrate. Substrate may also be rubble or boulders. In the Clearwater River, Idaho, Torrent Sculpin are generally found in slower water or areas bordering fast water[68]. Torrent Sculpin appear to tolerate warmer water temperatures than other cottid species found in the same drainages, having been sampled in temperatures ranging from 8 to 20°C (46 to 68°F)[27].

Management and Conservation Status

Torrent Sculpin are considered secure, an abundant species (G5/S5), and are classified as unprotected wildlife by the state of Idaho[30]. Management status for Torrent Sculpin is presently undefined, and no studies have been conducted to determine statewide population status.

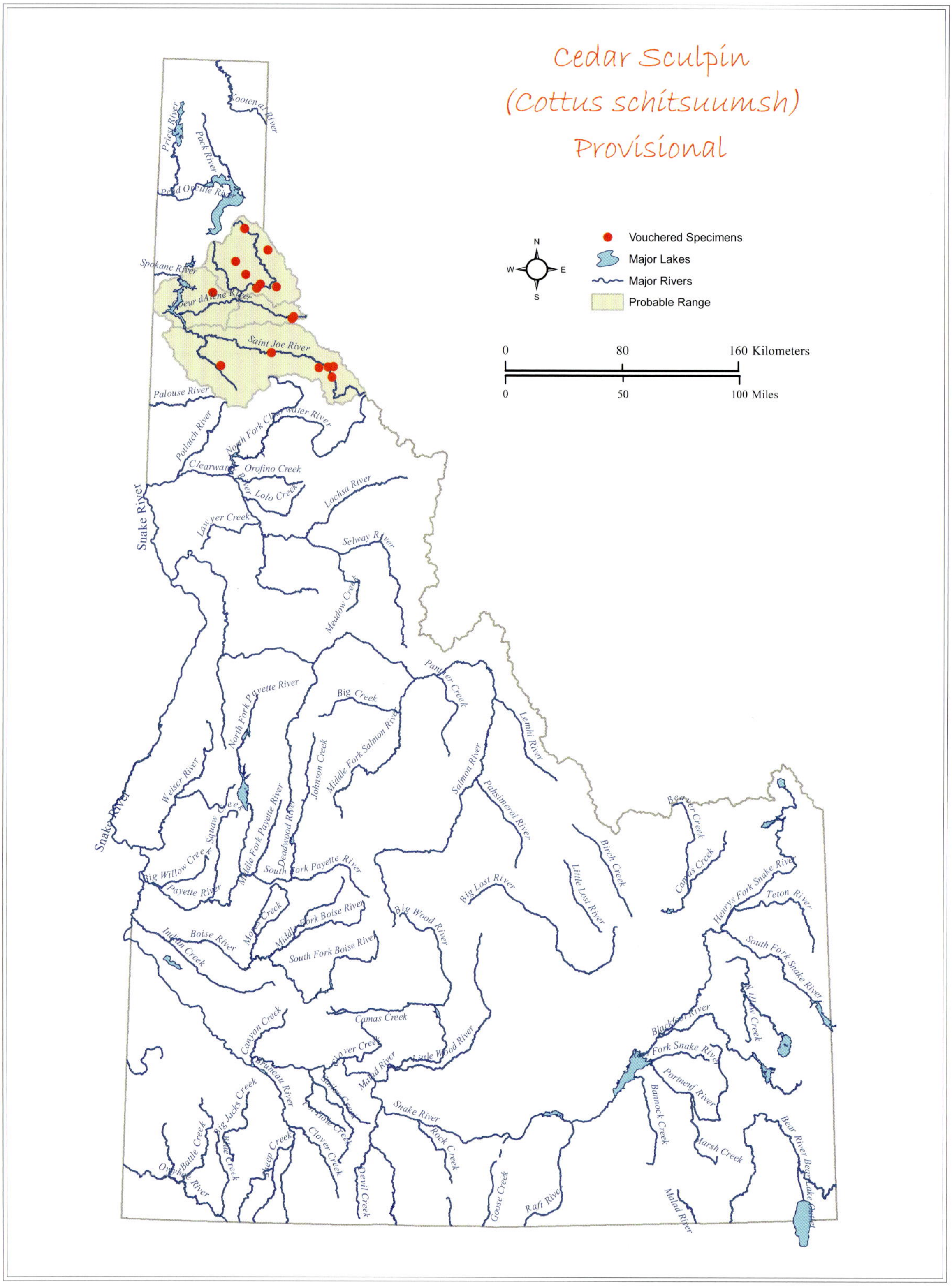
Cedar Sculpin
(Cottus schitsuumsh)
Provisional
Vouchered Specimens
Major Lakes
Major Rivers
Probable Range
N
S
E
W
0
80
160 Kilometers
0
50
100 Miles
Priest River
Pack River
Kootenai River
Spokane River
Saint Joe River
Palouse River
Potlatch River
North Fork Clearwater River
Orofino Creek
Lolo Creek
Lochsa River
Lawyer Creek
Snake River
Selway River
Meadow Creek
Big Creek
North Fork Payette River
Middle Fork Salmon River
Johnson Creek
Salmon River
Lemhi River
Pahsimeroi River
Weiser River
Squaw Creek
Middle Fork Payette River
Deadwood River
South Fork Payette River
Big Willow Creek
Payette River
Boise River
Indian Creek
Mores Creek
Middle Fork Boise River
South Fork Boise River
Big Wood River
Big Lost River
Little Lost River
Birch Creek
Camas Creek
Teton River
South Fork Snake River
Willow Creek
Blackfoot River
Portneuf River
Bannock Creek
Marsh Creek
Bear River
Malad River
Canyon Creek
Bruneau River
Big Jacks Creek
Battle Creek
Owyhee River
Jump Creek
Sheep Creek
Clover Creek
Little Wood River
Snake River
Rock Creek
Devil Creek
Goose Creek
Raft River
Camas Creek

Cedar Sculpin

Cottus schitsuumsh

LeMoine et al., 2014

Etymology

Cottus = an old European name for sculpin; *schitsuumsh* = a phonetic notation of the spoken word used by the first peoples who inhabited portions of northern Idaho and western Montana, now recognized as the Coeur d'Alene Tribe. In contemporary usage, it means "those who were found here," in reference to the Tribe (pronounced s-CHEET-sue-umsh)[36].

Importance

Fishes of the genus *Cottus* have long been a taxonomic challenge because of morphological similarities among species and their tendency to hybridize. A number of undescribed species likely remain in this genus. In 2014 Lemoine et al. used a combination of genetic and morphological methods to describe Cedar Sculpin (*C. schitsuumsh*). This is a new species from the upper Columbia River basin in Idaho and Montana. In addition to its newly discovered status, Cedar Sculpin are likely important within their native ecosystems as both a prey species and a niche species in their habitats[36].

Description

Cedar Sculpin are a moderate-size species of cottid and rarely exceed 90 mm (3.5 in) standard length. Anal fin rays are 14 (rarely 13 or 15), and the final posterior anal fin ray is commonly branched. Dorsal fin spines are eight (rarely seven or nine), and dorsal fin rays are 17 to 20 (usually 18 to 19), with the posterior dorsal fin ray branched. Pectoral fin rays are 13 to 14 (usually 14), and pelvic fin rays are four (rarely three) and of similar length. Two medial-chin pores are present. There are 10 to 12 preopercular mandibular pores and two postmaxillary pores (rarely zero). Dorsal fins of Cedar Sculpin are generally not connected but occasionally touch at the base. Prickling is consistent and restricted to a trapezoidal area, which is widest in the axillary region and narrowest along the lateral line under the beginning of the second dorsal fin. Palatine teeth are present in two or fewer rows and sometimes on only one side. Palatine tooth patches are smaller than and separated from the vomerine tooth patch and generally equal if present on both sides. The lateral line is incomplete, ending below the latter half of the second dorsal fin. One preoperculum spine was observed among all specimens. Preopercular spines are greatly reduced and skin covered. The mouth width–head fraction ranges from one-half to two-thirds[36].

Mottling of Cedar Sculpin is variable and consists of at least two forms. One form has black-brown mottling on a brown-olive background mainly above the lateral line that stops under the soft dorsal fin, and weak saddling is occasionally present. Pectoral fins on Cedar Sculpin have faint black-brown contours running vertically. The other form lacks mottling or spotting on the body, and the background color can be light brown, rust orange, or yellow. Both types have a dark blotchy marking near the base of the caudal rays, and the first dorsal fin is generally dorsally darkened. Both forms are sympatric in the upper Spokane River Basin, Idaho, but only the mottled form has been observed in Montana[36].

Range and Distributional Status

Cedar Sculpin occur in all streams sampled in the upper Coeur d'Alene River and Saint Joe River drainages of the Spokane River Basin above Post Falls, near Post Falls, Idaho, and sporadically in the middle Clark Fork River Basin between Fish Creek and Prospect Creek, approximately between Tarkio and Thompson Falls, Montana. Cedar Sculpin are common to abundant in cool to cold tributaries with cobble and gravel bottoms. They are sympatric with Westslope Cutthroat Trout and Bull Trout throughout their distribution[36].

Cedar Sculpin distribution in Idaho is consistent with that of other terrestrial and aquatic fauna that persisted immediately south of the extent of the Cordilleran glaciation in the northern Rocky Mountain refugium. The disjunct distribution in the Clark Fork River basin in western Montana, however, is puzzling. Because sculpins were a common legal bait fish in Montana until this century, it is plausible that populations in western Montana resulted from human-assisted translocations[36].

Size and Longevity

The holotype for Cedar Sculpin, from the East Fork of Steamboat Creek, a tributary of the Coeur d'Alene River at West Fork Road (Shoshone County, Idaho), was 91 mm (3.6 in). Paratypes were 57 to 73 mm (2.2 to 2.9 in), and the holotype group was 72 to 99 mm (2.8 to 3.9 in) SL. Other specimens examined during the reported research included SL ranging from 66 to 81 mm (2.6 to 3.2 in), 68 to 80 mm (2.7 to 3.1 in), 64 to 88 mm (2.5 to 3.5 in), and 69 to 96 mm (2.7 to 3.8 in)[36].

Food and Feeding

Food and feeding habits of this species are undefined. It is assumed that both food items and feeding behaviors for this species are similar to other *Cottus* species.

Limiting Factors

Limiting factors for this species have not been defined in Idaho. It is assumed that this species is limited by the same type and extent of factors that affect other members of the genus.

Breeding Habits

Breeding habits for Cedar Sculpin have not been described. It is assumed that breeding habits, timing, and behavior are similar to other members of the genus.

Habits and Habitat

Habits and habitat use have not been described for this species. Some aspects of their habitat preferences can be assumed based on their occurrence as described in specific ecosystems above.

Management and Conservation Status

Because Cedar Sculpin were described very recently, management status is presently undefined. They should be afforded every protection in Idaho waters, and further investigations into the ecology of this species should be conducted.

Chapter Citations

1. Nelson, 2006. 2. Goto, Yokoyama, and Sideleva, 2015. 3. McPhail, 2007. 4. Scott and Crossman, 1973. 5. Sigler and Sigler 1996. 6. Page, et al., 2013. 7. Kinziger, Wood, and Neely, 2005. 8. Neely, 2003. 9. Bailey and Dimick, 1949. 10. Bean, 1881. 11. Cope, 1872a. 12. La Rivers, 1962. 13. Gill, 1862. 14. Bailey, 1952. 15. Bond, 1963. 16. Montana Natural Heritage Program, Montana Fish, Wildlife, and Parks, 2015a. 17. Hubbs and Schultz, 1932. 18. Eddy, 1957. 19. Bailey and Bond, 1963. 20. Tabor, et al., 2007. 21. Adams and Schmetterling, 2007. 22. Maret and MacCoy, 2002. 23. Brinkman and Woodling, 2005. 24. Smith, 1975. 25. Kimmel, 1975. 26. Simpson and Wallace, 1978. 27. Wallace and Zaroban, 2013. 28. Eaton and Scheller, 1996. 29. Petty and Grossman, 2007. 30. Idaho Department of Fish and Game, 2005. 31. Eigenmann and Eigenmann, 1891. 32. Meyer, et al., 2012. 33. Miller, 1951. 34. Young, 2016. 35. Wydoski and Whitney, 2003. 36. LeMoine, et al., 2014. 37. Gasser, Cannamela and Johnson, 1981. 38. Cannamela, 1979. 39. Johnson, Cannamela and Gasser, 1983. 40. Maughan, 1972. 41. Ruzycki, Wurtsbaugh and Luecke, 2001. 42. Wurtsbaugh and Hawkins, 1990. 43. Palacios, Luecke and Robinson, 2007. 44. Wurtsbaugh and Luecke, 1998. 45. Smith, Stokes, and Horn, 1968. 46. Reheis, Laabs and Kaufman, 2009. 47. Ruzycki and Wurtsbaugh, 1995. 48. Albrecht, et al., 2004. 49. Ruzycki, Wurtsbaugh and Lay, 1998. 50. Nevermann and Wurtsbaugh 1994. 51. Gilbert and Culver, 1898. 52. Wallace, et al., 1984. 53. Kuda and Griffith, 1993. 54. Wallace, et al., 1982. 55. Connolly, 1983. 56. Campbell, 2011. 57. Bailey, et al., 1970. 58. Robbins et al., 1980. 59. Robbins, et al., 1991. 60. Nelson, et al., 2004. 61. Markle and Hill, 2000. 62. Committee on the Status of Endangered Wildlife in Canada, 2010. 63. Gilbert and Evermann, 1895. 64. Zaroban, 2011a. 65. Meyer, Cassinelli, and Elle, 2008. 66. Quintela, 2004. 67. Scarpella and Dudley, 1995. 68. Maughan, 1976.

Additional readings for this chapter are listed in the Supplmental References section.

18

PERCIFORMES

Sunfishes, Perches & Cichlids

Order Perciformes

Perciformes is the most diversified of all fish orders and is the largest order of vertebrates. Perciformes contains 20 suborders, 160 families, about 1,540 genera, and more than 10,000 species[1]. In Idaho there are 13 species in two families, Centrarchidae and Percidae, and several other species in the Family Cichlidae that are mostly South American and products of the aquarium trade released into Idaho waters.

Members of the Perciformes are typically spiny-rayed, bony fishes that have an elongate skull that is heavily ossified; they have strong jaws and well-developed teeth. Members of this order typically have two dorsal fins: the first is spiny, and the second is soft. Caudal fins of species in this order typically have no more than 17 rays. Pectoral fins are high; the soft-rayed anal fin is preceded by two or three spines; pelvics are thoracic or jugular. Swim bladders in members of this order are physoclistic, and they typically have ctenoid (rough) scales. While species in this order are typically marine, there are representatives in freshwaters of all continents except Australia. Members of this order date to the upper Cretaceous period[2].

Family Centrarchidae

Sunfishes occur in freshwater in North America. Suborbital bones are present in addition to the lachrymal; the lateral line is present but sometimes incomplete. Anal fin spines are usually three or fewer. Most sunfishes are nest builders. Males hollow out a small depression with their tail and guard the eggs. Centrarchids are an important recreational group. There are eight genera with 31 species[1].

This family includes some of the most highly colored and attractive North American warm-water species, and several are important sportfish. Originally restricted to eastern North America, various species, particularly the basses, have been widely introduced. Sunfishes inhabit shallows of warm, rocky, and vegetated lakes, ponds, and slow-moving streams[3].

As in other Perciform fishes, sunfishes have a spinous dorsal fin (with six to 13 spines), followed by a soft dorsal fin; the anal fin has three or more spines at its origin. The pelvic fins are beneath the pectorals and have one spine and five soft rays, and there are 17 principal caudal-fin rays. The lateral line is complete in all except Bantam Sunfish (*Lepomis symmetricus*). Scales are ctenoid except in *Acantharchus* and occur all over the body, including the breast, cheeks, and opercles[4].

Among fish species with paternal defense of offspring, the level of protection provided by a male is influenced by both the quantity and the developmental stage of his brood. Because at any time the number of currently surviving offspring determines the maximum potential fitness for each male for that particular spawning bout, males should be willing to invest more energy in defending larger broods than smaller ones. In addition, because the reproductive value of offspring increases with the age of the brood, the amount of parental care given by an attending male should increase as his brood grows older. As progeny approach independence, however, the care given by a male should decrease; further investment may not significantly increase the probability of offspring surviving to adulthood and may serve only to decrease future reproductive opportunities for that male. Such a decrease in future reproductive opportunities could be realized either in that same year through the prevention of renesting at times when females are still capable of spawning or in future years through a reduced condition of the male that might prevent nesting efforts (or even result in premature death). As a result, any event that decreases brood size or a male's energy reserves could potentially influence parental care decisions of the male.

Centrarchidae is a relatively small family of perch-like fishes. However, this family belongs to the largest and most diverse order of living vertebrates, the Perciformes. Most Perciformes fishes are marine, but centrarchids live in freshwater. They are anatomically advanced fishes with spines in their dorsal, anal, and pelvic fins; they have two united dorsal fins and lack an adipose fin; pelvic fins are placed well forward on their bodies, and they have ctenoid scales. Males of most species in this family dig nest pits in the substrate and guard eggs and fry[5].

Centrarchids were initially restricted to North America, but several species of sunfishes and basses have been planted in temperate waters throughout the world. Preglacial fossil Centrarchidae are known from Alaska, Idaho, Oregon, and Washington. There is a living remnant of the ancient western North American centrarchid Sacramento Perch (*Archoplites interruptus*) native to the Sacramento–San Joaquin drainage system of California. Species assignable to this same genus occur in Miocene and Pliocene deposits associated with the Columbia River and Snake River[5]. There are 10 species in four genera reported in Idaho:

Ambloplites rupestris	(Rafinesque, 1817)	Rock Bass
Lepomis cyanellus	Rafinesque, 1819	Green Sunfish
Lepomis gibbosus	(Linnaeus, 1758)	Pumpkinseed
Lepomis gulosus	(Cuvier, 1829)	Warmouth
Lepomis humilis	(Girard, 1858)	Orangespotted Sunfish
Lepomis macrochirus	Rafinesque, 1819	Bluegill
Micropterus dolomieu	Lacepède, 1802	Smallmouth Bass
Micropterus salmoides	(Lacepède, 1802)	Largemouth Bass
Pomoxis annularis	Rafinesque, 1818	White Crappie
Pomoxis nigromaculatus	(Lesueur, 1829)	Black Crappie

"Try to leave the Earth a better place than when you arrived."

Sidney Sheldon

Winter, Up River, Lower End, Black Canyon, Bear River, near Grace, ID. Photo Courtesy John Sigler.

Rock Bass

Ambloplites rupestris
(Rafinesque, 1817)

Etymology

Ambloplites = blunt armature; *rupestris* = among the rocks.

Importance

Rock Bass may have ecological significance in an indirect way due to their association with other species more valued by man, such as the basses. While not often sought directly by many anglers (except children), particularly in Canada, if caught incidentally to other fishes, they do have firm, white, flaky, delicious flesh. Rock Bass strike live or artificial bait hard, play hard, and provide good sport on light tackle[3].

Due in part to total lengths of 254 mm (10.0 in), Rock Bass are seldom actively sought as a sportfish. In reality, the species is a good game fish, since they are voracious feeders and will strike a lure nearly as large as the individual itself. Flesh is very tasty, though a number of individuals are generally required for a meal. Rock Bass are eaten by predatory fishes such as Walleye, Smallmouth Bass, and Northern Pike[6].

Description

General coloration of the dorsal surface of the head, body, and upper sides of Rock Bass is golden brown to olive; the lower head, the body below the pectoral fins, and the ventral surfaces are silver to white; several vague saddle marks, bars, mottling, or blotches appear over the dorsal surface to the lateral line. Eyes are usually bright red to orange (a reference to one of the common names, redeye), and the opercle flap has a vague black spot not reaching the edge[3].

The body of Rock Bass is deep and somewhat laterally compressed, but less so than most other sunfishes (*Lepomis* sp., *Pomoxis* sp.). The head is rather large, narrow, deep, and mostly scaled. Rock Bass have terminal mouths that are slightly oblique; the lower jaw is slightly longer. There are fine brush-like teeth on the palatines, vomer, tongue, and both jaws. Pharyngeal teeth are numerous, short, fine, and pointed, and the lower pharyngeal pad is long and narrow. Gill rakers number about 16, and branchiostegal rays number six or rarely five. There are two dorsal fins that are broadly joined and appear as one. The caudal fin is only moderately broad and long and little to moderately forked, with rounded tips; the anal fin has a long base and rather short spines. Pelvic fins are thoracic with one spine and five rays and a long tip with a blunt point. Pectorals are moderately high, broad, and rounded, typically with 13 rays but sometimes 12 to 14. Scales are ctenoid and moderately large, dorsally somewhat crowded forward; the lateral line is complete and high on the body, shallowly arched; lateral line scales number 37 to 51. Rock Bass have a colorless peritoneum and nine long, slim pyloric caeca. Vertebrae number 29 or 30. There are no nuptial tubercles, but color intensifies on individuals during spawning[3].

Descriptive characteristics from populations within the native range of the Rock Bass include the following: lateral-line scales typically 39 to 46; dorsal fin with 10 to 13 spines and l0 to 12 soft rays; anal fin with five to seven spines and 9 to 11 soft rays; pectoral fin rays 13 to 15; gill rakers long and slender, with seven to 10 developed; branchiostegal rays six; vertebrae 29 to 32; cheek scales in seven to eight rows; color in life olive to brassy green, with dark mottling that can change drastically with surroundings; uniformly mottled (less "blotchy") on body and fins[4].

Divergent morphological strategies have been documented in a number of centrarchids, including Pumpkinseed and Bluegill as well as species in other families and orders. Pumpkinseed and Rock Bass stream populations of both sexes were significantly different in external morphology from lake populations in a central Ontario, Canada, watershed. Predictions that stream fishes would be more slender bodied and have a more anterior placement of lateral fins than lake fishes were generally supported. The prediction that stream fishes would have a more robust caudal peduncle was partially supported. The prediction that fin size would be larger in stream fishes was not supported, as Rock Bass in lake ecosystems generally had longer and wider fins than those from stream sites. These findings suggest that in some species, smaller fins may be favored in stream-dwelling individuals because the reduction of drag during swimming more than compensates for their reduced power and propulsion efficiency in a current. Smaller fin size in stream-dwelling centrarchids may be related to their body shape or to their low usage of fast-moving water within the streams they inhabit[7].

Range and Distributional Status

This species is not native to and is likely extirpated from Idaho. Native range of Rock Bass was from southeastern Manitoba, Canada, east to southern Québec, Canada, and south through Vermont and New York to the western panhandle of Florida, west to eastern Texas, and north through eastern North Dakota. Rock Bass were introduced into the Columbia River watershed in 1893 when a group of fish described as "various sunfish" was stocked into the Willamette River, Oregon[6]. In Idaho this species was reported to have been introduced in Lattig's Pond (100 fish), near Payette, and Waller's Pond (250 fish), near Troy[8].

Size and Longevity

Rock Bass are typically rather slow growing but reach a size acceptable to anglers at age three or four. They may reach lengths of 356 mm (14 in). In general, Rock Bass of 127 to 203 mm (5 to 8 in) are considered to be of good size. Few individuals live beyond six years of age[6].

Two 1.36 kg (3 lb, 0 oz) Rock Bass hold the world all tackle record: one caught in York River, Ontario, Canada, in 1974 and one caught in Lake Erie, Pennsylvania, in 1998[9]. No Idaho record is posted. The Washington State record is a 0.6 kg (1 lb, 5 oz) individual taken from Steilacoom Lake in 1981[6].

Food and Feeding

Rock Bass are opportunistic feeders and will eat a variety of food items. Insect larvae, mollusks, crustaceans such as crayfish, and small fishes are prey of this species. Aquatic insect larvae eaten by Rock Bass include dragonflies, damselflies, mayflies, and caddis flies; in some ecosystems, crayfish may form a large part of the summer diet. Larger Rock Bass are highly piscivorous and feed on various species of small fish, especially minnows[6].

Rock Bass may eat small fishes, especially minnows and at times their own young. In a Canadian system, food taken by individuals up to 70 mm (2.8 in) was as follows: Ephemeroptera (35), Odonata (30), Cladocera (40), Amphipoda (30), Isopoda (15), and surface insects (35)[3].

Limiting Factors

Rock Bass may be parasitized by protozoans, trematodes, cestodes, nematodes, acanthocephalans, leeches, and crustaceans[3]. It is likely that this species is preyed upon by a variety of other fishes and avian predators.

Breeding Habits

Rock Bass are usually sexually mature at age two or three. Breeding males become blackish in color. Spawning generally occurs during May or June when water temperatures reach about 21°C (70°F) and continues until temperatures reach about 26°C (79°F). Adult females congregate in pools of streams until they become ripe in the period immediately preceding and during spawning. Males make the nest by digging a depression in gravel or sometimes in sand, where currents are slow. In lake ecosystems, Rock Bass make a nest in shallow water along the shoreline. Females may deposit 3,000 to 5,000 eggs on the surface of the nest, though exceptionally large females have been known to contain up to 11,000 eggs. Males attend the nest, keeping it clean and the eggs aerated by fanning the nest area with their pectoral fins. Eggs of Rock Bass are adhesive and hatch in three to four days at 20.5 to 21°C (69 to 70°F). In common with other sunfishes, male Rock Bass defend the eggs and fry vigorously. However, they provide no parental care to the young after they leave the nest[6].

Habits and Habitat

As indicated by the common name, Rock Bass generally inhabit rocky areas in shallow water in lake ecosystems and the lower warm reaches of stream ecosystems. Adults generally are found in aggregations and most often in association with other sunfishes, such as Smallmouth Bass and Pumpkinseed. Young are littoral to limnetic in various lake ecosystems[3].

Rock Bass are found in stream and lake ecosystems with clear water, rocky bottoms, and warm water temperatures (26 to 31°C [78 to 87°F]). In stream ecosystems, Rock Bass generally congregate in quiet pools with little current. They are generally found in association with cover composed of rocks, ledges, or logs with overhanging branches of trees or bushes. Rock Bass do not migrate, staying in the same general area through all seasons. In winter, however, they often rest quietly on the bottom under cover. In some stream ecosystems, Rock Bass are often found in association with Smallmouth Bass, occupying areas of lower current[6].

Management and Conservation Status

No specific management protocols have been developed for Rock Bass in Idaho. Rock Bass are believed to be extirpated in Idaho.

Sawtooth Mountains, off the Salmon River, near Stanley, ID. Photo Courtesy John Sigler.

"The charm of fishing is that it is the pursuit of what is elusive but attainable, a perpetual series of occasions of hope."

John Buchan

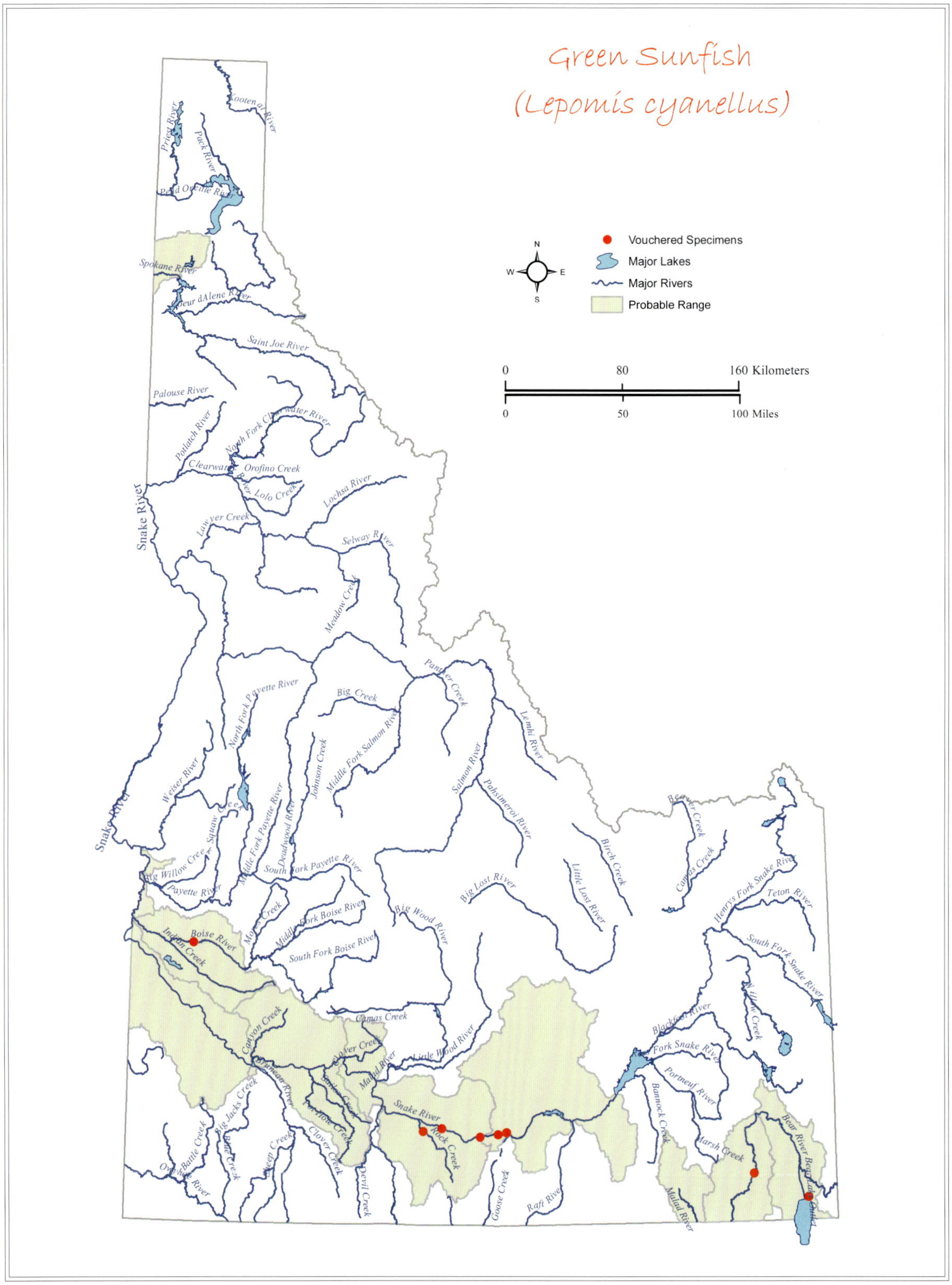
Green Sunfish
(Lepomis cyanellus)
N
W
E
S
Vouchered Specimens
Major Lakes
Major Rivers
Probable Range
0
80
160 Kilometers
0
50
100 Miles
Priest River
Pack River
Spokane River
Saint Joe River
Palouse River
Potlatch River
Orofino Creek
Lochsa River
Snake River
Selway River
Meadow Creek
Big Creek
Johnson Creek
Middle Fork Salmon River
Salmon River
Lemhi River
Pahsimeroi River
Birch Creek
Little Lost River
Big Lost River
Big Wood River
Weiser River
Payette River
Boise River
Indian Creek
South Fork Boise River
Canyon Creek
Camas Creek
Clover Creek
Malad River
Rock Creek
Goose Creek
Raft River
Big Jacks Creek
Battle Creek
Devil Creek
Bannock Creek
Portneuf River
Marsh Creek
Malad River
Bear River
Teton River
South Fork Snake River
Willow Creek

Green Sunfish

Lepomis cyanellus
Rafinesque, 1819

Etymology

Lepomis = scaled operculum; *cyanellus* = blue.

Importance

Green Sunfish are rarely important as a game species. In many states, they are regarded as a nuisance or worse because of their abundance and their ability to outcompete young of more desirable game species. Where found, however, they strike savagely at both natural and artificial baits and can provide great entertainment for youngsters. In California they sometimes contribute significantly to the fishery in new reservoirs where growth rates are better, but in general they are considered a nuisance rather than an asset. Green Sunfish, however, can tolerate extreme conditions and easily become established where introduced. Because they can spread quickly to connected waters, this species should not be introduced into any watershed. Reclamation and removal efforts in a number of ecosystems in the West have been unsuccessful. Although many Green Sunfish were killed in the process, surviving individuals replaced their numbers within a few years. Because Green Sunfish rarely get large in most waters, they do provide forage for other warm-water piscivorous species[2,6].

Description

Green Sunfish have a body that is generally brown to olive with an emerald sheen, being darker on the dorsal surface and upper sides; sides are light yellow-green; upper sides have seven to 12 dark but vague vertical bars; the ventral surface is yellow to white; the breast is not conspicuously colored. The head may have emerald spots and at times wavy, radiating emerald lines. The opercle flap has a squarish black center that is entirely edged with pale red, pink, or yellow and is brightest at the tip and below. Spawning males have dorsal, caudal, and anal fins that are dusky to olive; membranes are darker; fins are edged with white, yellow, or orange borders[3].

Green Sunfish are a deep-bodied, laterally compressed species. There are fine teeth on the palatines, vomer, and both jaws; teeth are absent on the tongue; pharyngeal teeth are larger and less numerous and conical;

the lower pad is much longer than wide. Gill rakers in the center of the arch are long and thin, numbering about 14 on the lower limb and four or five on the upper limb. Branchiostegal rays usually number six but occasionally seven[3].

There are two dorsal fins, but they are broadly joined and appear as one. The first dorsal fin is long and spiny, with 9 to 11 but typically 10 spines, and the edges are only slightly rounded; the second dorsal fin is soft rayed and higher than the first, with 10 to 12 but typically 10 or 11 rays; the edges are rounded to slightly pointed in breeding males. The caudal fin is rather broad, shallowly forked, with rounded tips. The anal fin has three sharp spines and nine or 10 soft rays; the soft-rayed part is not generally higher than the third spine. The pelvic fins are thoracic, and the origin is under the first dorsal spine; they are long, with one spine and five rays. Pectoral fins are moderately high but not long, with a broad and distinctively rounded tip and typically 13 but sometimes 12 rays. Scales are rather small and ctenoid, being somewhat crowded both in the anterior (dorsally) and the posterior; the lateral line is complete and high, with a shallow arch over the pectoral fin. Lateral line scales number 40 to 50, and there are eight to 10 scale rows above the lateral line. The peritoneum is white to silvery; the intestine is long and well differentiated; there are six to eight pyloric caeca. Vertebrae number 28 or 29. Gill rakers number 11 to 14, and the length of the longest gill rakers is about six times the basal width of the rakers (but less slender in very large adults). Color of the dorsolateral area is blue-green to dusky, usually with small, randomly scattered black flecks. The belly is white to yellowish. Females occasionally have dusky bars on the sides. A dusky basal spot is usually present in the posterior soft dorsal fin and occasionally at the posterior base of the anal fin. Soft dorsal, anal, pelvic, and caudal fins of breeding males have an orange wash or bright-orange submarginal areas and white margins. No nuptial tubercles occur, but color and pattern intensify during spawning events[3,4].

Range and Distributional Status

This species is not native to Idaho. Native distribution of Green Sunfish is from western New York, south to Louisiana, west to about central New Mexico, and northeast to Minnesota. Green Sunfish were accidentally introduced into California in 1891. Green Sunfish were first introduced into Washington through stocking of a batch of a mixed sunfish species. Presently, Green Sunfish are found in a small number of eastern Washington lakes. This species occurs in the Snake River drainage below Shoshone Falls in Idaho[6] and in portions of the Bear River drainage.

Size and Longevity

Growth of most Green Sunfish is slow and the ultimate size small, depending on specific ecosystems' features. Stunting occurs in many populations. Lengths at the end of each year for Michigan Green Sunfish ages one to eight were, in mm (in), approximately 25.4 (1.0), 50.8 (2.0), 76.2 (3.0), 104 (4.1), 125 (4.9), 152 (6.0), 160 (6.3), and 185 (7.3). In a Utah spring-fed pond, lengths of Green Sunfish at the end of each year for fish ages one to six were, in mm (in), 50.8 (2.0), 76.2 (3.0), 102 (4.0), 122 (4.8), 137 (5.4), and 163 (6.4). Green Sunfish may rarely reach the age of 10 or 11 years. The largest Green Sunfish caught in Utah weighed 0.3 kg (10.6 oz)2. The angling record for Green Sunfish in Washington is a 254 mm (10.0 in), 0.36 kg (12.7 oz) individual taken from Bailey Lake in 1994[6]. The international all-tackle hook-and-line record for Green Sunfish is 0.96 kg (2 lb, 2 oz), taken from Stockton Lake, Missouri, in 1971[9]. An alternative record of 1.4 kg (3 lb, 1 oz) was taken from the York River, Ontario, Canada, in 1974[6].

Food and Feeding

Green Sunfish feed on a variety of prey, depending upon their availability in a particular ecosystem. Green Sunfish fry feed on zooplankton but very soon convert to larger food organisms as they grow. As they become larger, Green Sunfish feed on forage fish when they are available. However, in waters where forage fish are absent or of low abundance, terrestrial insects may form a large part of the diet. Dietary items identified for this species include insects, such as dragonflies; damselfly nymphs, midges, and caddisfly larvae; crustaceans; and occasionally mollusks, higher plants, and fishes. Due to their large mouths, Green Sunfish are able to eat fishes almost half their size[6].

Green Sunfish in northern Utah commonly feed on insects in early winter, crustaceans and fishes in mid-winter, and freshwater shrimp in late winter. Generally, this species feeds on insects, mollusks, and small fishes. Green Sunfish are able to eat large hard-body organisms and fairly large fish. They prey readily on large active invertebrates, and as they increase in size they depend more on crayfish, fishes, and large aquatic and terrestrial insects[2].

Limiting Factors

Because of their small adult size, Green Sunfish probably fall prey to a wide variety of warm-water fishes, particularly the basses, in a given ecosystem. Parasites identified in Green Sunfish include protozoans, trematodes, cestodes, nematodes, acanthocephalans, leeches, mollusks, and crustaceans. In nature Green Sunfish may hybridize with other members of the genus *Lepomis*, including Longear Sunfish, Bluegill, Orangespotted Sunfish, Pumpkinseed, and Redbreast Sunfish, in ecosystems where the species are sympatric[3].

Breeding Habits

Green Sunfish may mature as early as age one but typically mature at age two, and some may not mature until age three in certain ecosystems. Males have been reported to mature when they were only 46 mm (1.8 in) long and females when they were only 66 mm (2.6 in) long. Generally, spawning periods are from May to August, with a peak in June when water temperatures are 15.6 to 16.7°C (60 to 62°F). Oval nests constructed by the male measure approximately 15 by 31 cm (6 by 12 in). Nests are typically depressions formed in shallow water less than 38 cm (15 in) deep, where the bottom is smooth and clean, generally with a gentle slope. Green Sunfish eggs hatch in two days at 24°C (75°F). Swim-up fry are about 4.5 mm (0.18 in) long. Not only do males build the nest, but they also protect the site until the eggs are deposited and guard the nest until the young leave. Nests often occur in colonies at high densities, up to 25 nests per 4.5 m2 (50 ft2). Hybridization is common between Green Sunfish and other species of *Lepomis*[2,6].

Habits and Habitat

Green Sunfish typically inhabit slow-moving streams and weed beds of warm-water lake and reservoir ecosystems. Green Sunfish tolerate a wide range of water temperatures up to 32°C (90°F) and can survive in water temperatures of 33 to 34.4°C (92 to 94°F) for days. Green Sunfish are also successful in surviving ecosystem conditions such as drought, where they are able to survive in residual pools of intermittent streams. Green Sunfish are solitary in habit due to their aggressive behavior in defending territories. Individuals generally occupy small forage areas along the edges of the shore, typically under or in association with cover such as vegetation, rocks, logs, or exposed tree roots. Green Sunfish form a social hierarchy in which the dominant individual is light in color and subordinate fishes are darker. In some instances, dominant and intruding individuals meet head-to-head and lock jaws, twisting and rolling as they settle to the bottom. Fins of submissive fishes, particularly caudal fins, become tattered in appearance from being bitten by dominant fishes. During spawning season, Green Sunfish remain aggressive but congregate at suitable sites where smaller territories of only about 0.19 m^2 (2 ft^2) are maintained[6].

Management and Conservation Status

Green Sunfish are considered common, wide spread, and abundant rangewide (G5). A state conservation ranking is not applicable (SNA) since Green Sunfish are an introduced species. They are classified as a game fish in Idaho[21]. Similar to other Centrarchidae, suitable habitat, with weedy areas for spawning and nursery areas, is necessary for the survival of Green Sunfish. Green Sunfish are preyed upon by other warm-water fishes. Because they typically do not qualify as a valuable sportfish, their welfare is not generally a fisheries concern, unless they are in an area fished heavily by youngsters. Green Sunfish may adversely affect populations of native species if stocked over the native species. No specific management protocols have been developed for Green Sunfish in Idaho. Status of Green Sunfish in Idaho is undocumented, but their range appears to be expanding.

"Here is your country. Cherish these natural wonders, cherish the natural resources, cherish the history and romance as a sacred heritage, for your children and your children's children. Do not let selfish men or greedy interests skin your country of its beauty, its riches or its romance."

Theodore Roosevelt

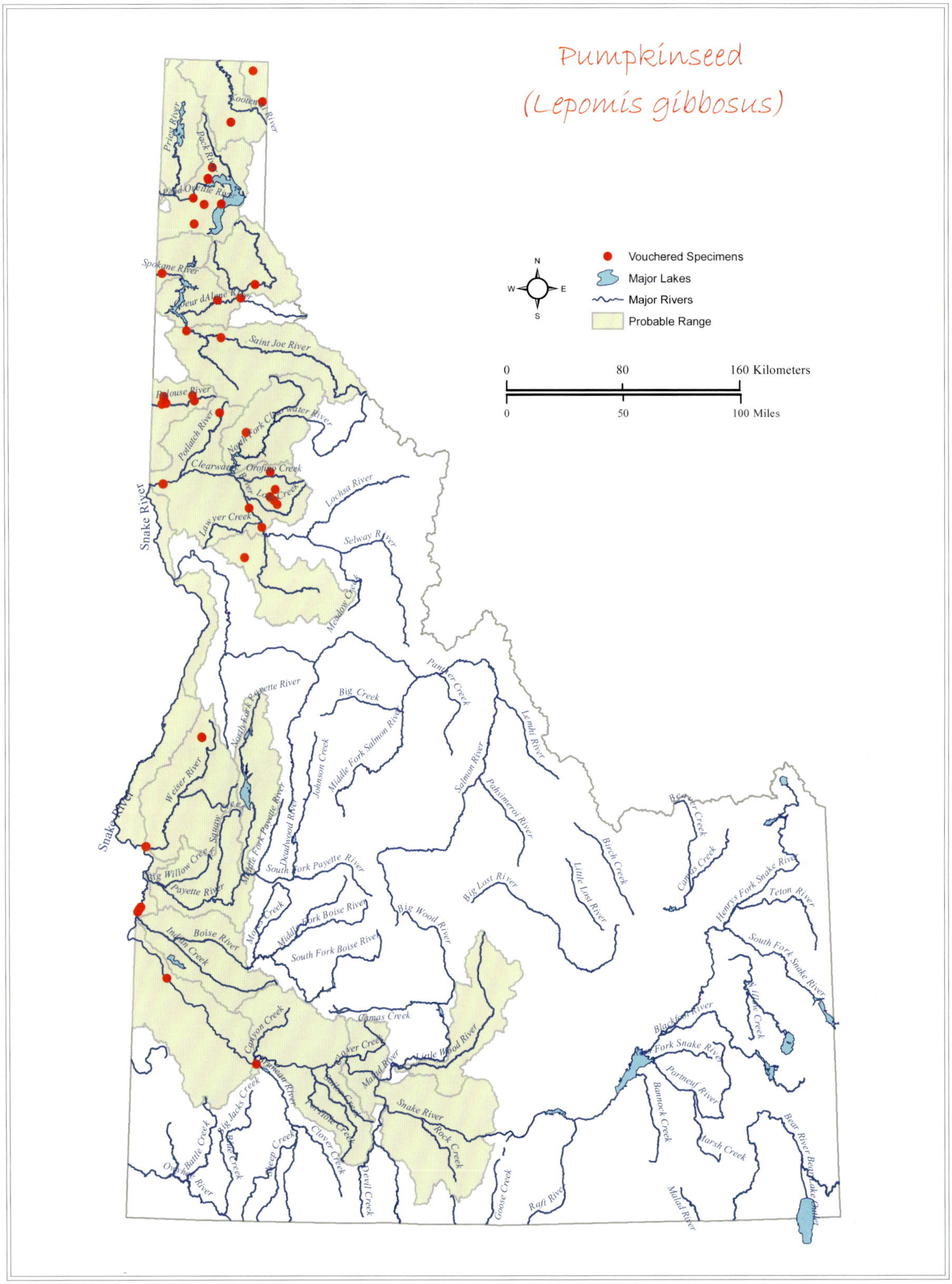
Pumpkinseed
(Lepomis gibbosus)
Vouchered Specimens
Major Lakes
Major Rivers
Probable Range
N
S
E
W
0
80
160 Kilometers
0
50
100 Miles
Priest River
Pack River
Spokane River
Saint Joe River
Potlatch River
Lochsa River
Snake River
Lemhi River
Big Creek
Middle Fork Salmon River
Johnson Creek
Salmon River
Pahsimeroi River
Birch Creek
Little Lost River
Big Lost River
Big Wood River
Weiser River
Boise River
Canyon Creek
South Fork Boise River
Teton River
South Fork Snake River
Portneuf River
Bannock Creek
Marsh Creek
Rock Creek
Goose Creek
Raft River
Malad River
Devil Creek
Clover Creek

Pumpkinseed

Lepomis gibbosus
(Linnaeus, 1758)

Etymology

Lepomis = scaled operculum; *gibbosus* = gibbous, or like a full moon, referring to the body shape.

Importance

Because Pumpkinseed grow slowly in the relatively cool waters of the Pacific Northwest, they do not generally reach an acceptable size for anglers in many ecosystems. They may provide some forage for Largemouth Bass and Smallmouth Bass, Walleye, Yellow Perch, and Northern Pike[6].

In some ecosystems, however, particularly in Canada, Pumpkinseed occur in large numbers in shallow, sheltered habitats close to shore. Because they attack even small pieces of live bait viciously and fight strongly, they are a delight to younger anglers. They are, however, generally ignored by adult anglers. Flesh of Pumpkinseed is white, flaky, sweet, and delicious. They may provide considerable sport, being readily available when and where larger fishes are not[3].

Description

The dorsal surfaces of the head, body, and upper sides of Pumpkinseed are golden brown to olive; the lower sides are golden, with irregular, wavy interconnecting blue-green lines; the ventral surface is bronze to red-orange; the sides of the body and the head are flecked with spots of olive, orange, or red and with blue, emerald, or green reflections. The sides of adult Pumpkinseed have several vague vertical bars, and the sides of the head have prominent wavy blue-green stripes. Female Pumpkinseed are paler; the vertical bars are more prominent. All colors are intensified in breeding males. The young have much less color pattern and only green to olive background color[3].

Pumpkinseed are very deep-bodied, laterally compressed, and almost disk-like fish. The mouth is terminal, only slightly oblique, and rather small. There are many short needlelike teeth in brush-like patches on the jaws and a single row on the vomer; there are no palatine teeth. The lower pharyngeal teeth are on pads. Gill rakers are short and stubby, numbering about eight on the lower limb and four on the upper limb.

Branchiostegal rays are six or seven. The two dorsal fins are broadly joined and appear as one. The first dorsal fin is long, spiny, and of moderate height, with 10 or 11 spines but usually 10; the second dorsal is soft rayed, higher, but with the base only two-thirds that of the first dorsal; rays number 10 to 12 but usually 11, and the edge is shallowly rounded. The caudal fin is only moderately broad and shallowly forked, with rounded tips. The anal fin has three sharp spines preceding eight to 11 but typically nine soft rays; the base of the spiny portion is about one-third the length of the base of the soft portion. Pelvic fins are thoracic, and the origin is under the first or second dorsal spine; one spine and five soft rays are present; the fin is moderately long, with a pointed tip and a square edge. Pectoral fins are moderately high on the body and not overly broad but long and pointed, with 12 to 14 but typically 13 rays[3].

Scales are typically ctenoid, not large, and crowded dorsally front and rear; the lateral line is complete and high, with a long, shallow arch over the pectoral fins; there are 35 to 47 lateral line scales. The peritoneum is silvery, and the intestine is well differentiated; there are seven or eight moderately long pyloric caeca. Vertebrae almost always number 29 but rarely 28. No nuptial tubercles are present, but colors are intensified during spawning. Palatine teeth are typically absent. The sides are densely covered with dusky spots, either scattered or in rows, over a background of brassy yellow or olive green. Orange spots may also be present on the sides, and young specimens often have dark vertical bars[3,4].

Pumpkinseed exhibit sexual dimorphism with respect to color. Adult Pumpkinseed of both sexes are highly colored, but the color pattern in breeding females is noticeably more subdued than in males. When viewed through water, the light-blue trailing edge of the soft dorsal and caudal fins is conspicuous in breeding males and inconspicuous in females, while the dark vertical bands are more prominent in breeding females than in breeding males[5].

Range and Distributional Status

This species is not native to Idaho. Native distribution of Pumpkinseed is from eastern North Dakota and southern Saskatchewan, Canada, east to Maine and southern Québec, south to Georgia, and northwest to eastern Nebraska, eastern South Dakota, and eastern North Dakota[6].

Pumpkinseed have been widely introduced into many other parts of the United States. Presumably, Pumpkinseed were included among the sunfish fry released in the Pacific Northwest by the US Fish Commission into the Willamette River near Salem, Oregon, in 1893. They are the common sunfishes found in many western Washington lakes[6]. In Idaho Pumpkinseed have been documented in the Kootenai, Pend Oreille, Spokane, Clearwater, and Snake River drainages below Shoshone Falls.

Size and Longevity

Pumpkinseed are reported to live to age 12 in captivity and to age 10 in a Midwest holding facility. Individuals from most populations do not live beyond age six. In Washington Pumpkinseed rarely weigh 0.23 kg (8 oz). In Washington Pumpkinseed generally live to age five and six in most ecosystems, but in some ecosystems they live to age seven. Pumpkinseed longer than 152 mm (6.0 in) would be considered large, but some individuals reach lengths of 305 mm (12.0 in). In Ohio the largest Pumpkinseed reported was 224 mm (8.8 in) long and weighed 0.3 kg (10.6 oz). The angling record for Washington Pumpkinseed was a 229 mm (9.0 in), 0.34 kg (12.0 oz) fish from Hicks Lake, caught in 1977[6].

The all-tackle hook-and-line record for Pumpkinseed is a 0.63 kg (1 lb, 6 oz) from Mexico, New York[9]. An alternative record fish was reported as a 1.0 kg (2 lb, 3 oz) individual caught in the North Saluda River, South Carolina, in 1997[6].

Food and Feeding

Although Pumpkinseed are morphologically specialized for foraging on hard-bodied prey, their diet varies with age, prey availability, habitat, season, and the presence (or absence) of other fish species. In some lake ecosystems lacking Bluegill, two behavioral and morphological forms of Pumpkinseed occur—an open-water (limnetic) form that feeds on plankton and an inshore (littoral) form that forages on snails and the nymphs of aquatic insects. In the absence of other sunfishes, aquatic insects often dominate the diet of Pumpkinseed. Juvenile Pumpkinseed forage on softer-bodied prey than adults, and the larvae feed on plankton. Pumpkinseed foraging stops when water temperatures drop below 6.5°C (44°F)[5].

Pumpkinseed eat aquatic insects, small mollusks, and crustaceans. California populations of Pumpkinseed are reported to feed primarily on mayfly nymphs and cladocerans as well as small numbers of dragonfly larvae, amphipods, clams, and snails. Pumpkinseed are site feeders that forage primarily during daylight, but some individuals also feed at night. Foraging activity by Pumpkinseed declined substantially at water temperatures below 20.6°C (69°F). Foraging activity also declines for up to two weeks in late spring in males that are actively guarding nests[6].

Within their native range, Pumpkinseed are reported to feed most heavily on micro crustaceans, such as cladocerans and copepods, as well as chironomid dipteran larvae. Pumpkinseed are also reported to feed heavily on snails as adults. To some extent, these findings indicate a primarily benthic-oriented feeding behavior[4].

Limiting Factors

Small Pumpkinseed form part of the prey of many predatory fishes. Pumpkinseed are eaten by various basses, Walleye, Yellow Perch, Northern Pike, and Muskellunge. Pumpkinseed are parasitized by a variety of organisms, such as protozoans, trematodes, cestodes, nematodes, acanthocephalans, leeches, mollusks, crustaceans, and linguatulans. Pumpkinseed are known to hybridize in nature with many other sunfishes, including Warmouth, Redbreast Sunfish, Green Sunfish, Orangespotted Sunfish, Bluegill, and Longear Sunfish[3,6].

Breeding Habits

Pumpkinseed are usually sexually mature at age two or three. Pumpkinseed generally spawn in late spring and early summer when water temperatures reach about 15.6°C (60°F). In some ecosystems, faster-growing males may mature at only age one. As with other sunfishes, male Pumpkinseed construct the nest as a shallow depression about 31 cm (1 ft) in diameter in gravel, sand, or mud bottoms. Nests are typically constructed in water that is 18 to 69 cm (7 to 27 in) deep, at sites with rooted aquatic vegetation. Pumpkinseed are not considered to be colonial nesters by some investigators, but nearly 200 males were observed on nests at depths of 0.9 m (3 ft) or less in a small Ontario, Canada, pond in mid-June. Several female Pumpkinseed may spawn with a single male. Egg production is dependent on the size of the female: two-year-old females with an average length of 73 mm (2.9 in) produce an average of 1,030 eggs; five-year-old females with an average length of 94 mm (3.7 in) produce an average of 2,436 eggs. Pumpkinseed eggs hatch in about three days at 28°C (82°F); incubation periods are longer at cooler water temperatures. Fry are about 3 mm (0.1 in) long at hatching. Pumpkinseed nests are defended by the male until the young leave. Small non-nesting males may engage in "cuckoldry," racing into the nests of larger males to fertilize eggs. Males guard the eggs until fry have hatched (three to 10 days) and dispersed[4,6].

Natural hybrids are known with Warmouth, Bluegill, Green Sunfish, Orangespotted Sunfish, Redbreast Sunfish, and Longear Sunfish[4].

Habits and Habitat

Pumpkinseed have a preference for clear, quiet water with dense aquatic vegetation and are generally found in weedy ponds and lake ecosystems as well as in sloughs and backwaters of slow-moving river ecosystems. Pumpkinseed are typically found in a denser aquatic vegetation than are Bluegill and are not as abundant in open deeper waters. Pumpkinseed do not do well in waters where summer temperatures do not reach at least 21°C (70°F). Pumpkinseed tend to occupy home ranges[6]. A case of what appears to be mutualism has been reported and occurs with the association of Chain Pickerel and Pumpkinseed. Chain Pickerel remained undisturbed over the Pumpkinseed nest and preyed on Golden Shiner that moved in to eat the Pumpkinseed eggs in the absence of the male Pumpkinseed guard[3].

Management and Conservation Status

Pumpkinseed are considered common, wide spread, and abundant rangewide (G5). A state conservation ranking is not applicable (SNA) since Pumpkinseed are an introduced species. They are classified as a game fish in Idaho[21]. No specific management protocols have been developed for Pumpkinseed in Idaho. Status of Pumpkinseed in Idaho is unknown.

Side Falls, Cauldron Linn, Snake River, @ > 18,000 cfs, spring, 2017-. Photo Courtesy John Sigler.

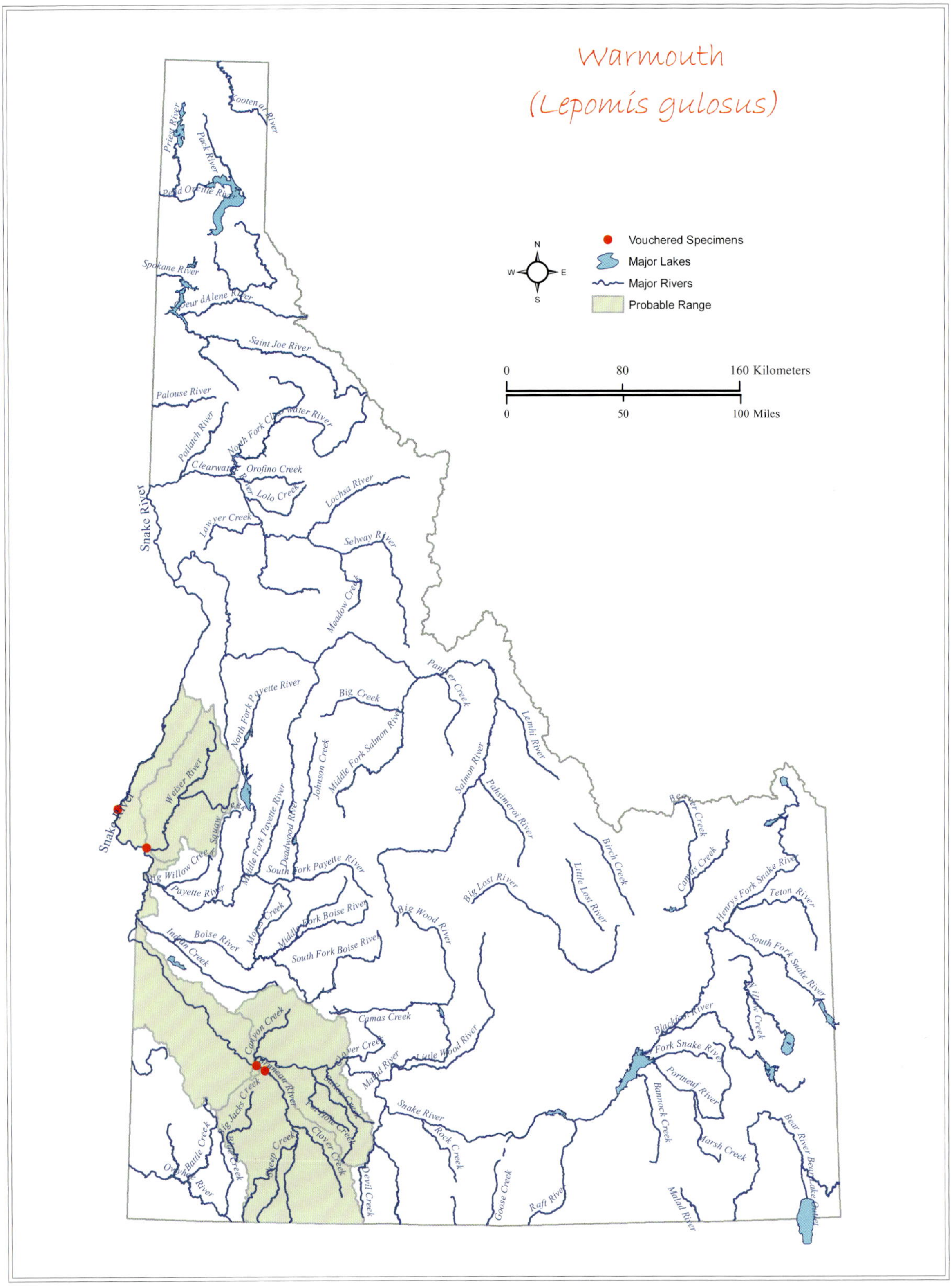
Warmouth
(Lepomis gulosus)
N
W
E
S
Vouchered Specimens
Major Lakes
Major Rivers
Probable Range
0
80
160 Kilometers
0
50
100 Miles
Pack River
Spokane River
Saint Joe River
Palouse River
Potlatch River
Orofino Creek
Lolo Creek
Lochsa River
Snake River
Selway River
Big Creek
Johnson Creek
Middle Fork Salmon River
Salmon River
Lemhi River
Pahsimeroi River
Birch Creek
Little Lost River
Big Lost River
Big Wood River
Weiser River
South Fork Payette River
Payette River
Boise River
South Fork Boise River
Camas Creek
Canyon Creek
Little Wood River
Teton River
South Fork Snake River
Henrys Fork Snake River
Portneuf River
Bannock Creek
Marsh Creek
Bear River
Rock Creek
Goose Creek
Raft River
Malad River
Clover Creek
Devil Creek
Salmon Falls Creek

Warmouth

Lepomis gulosus
(Cuvier, 1829)

Etymology

Lepomis = scaled operculum; *gulosus* = large mouthed.

Importance

Small Warmouth are probably eaten by other warm-water fishes such as Largemouth Bass. Their initial strike at bait is generally vicious, but when hooked they fight for only a short time, making them less desirable than some other species. Although they do take bait, flies, or lures readily, Warmouth are not easily caught because of the difficulty in angling in the dense vegetation cover they typically occupy[6].

Description

Warmouth have a background color that is dark olivaceous to brassy yellow, with profuse dark mottling on the body and fins; the lower sides often have horizontal streaks or rows of small spots. Cheeks and opercles have three to four distinctive dark bars radiating posteriorly from the eyes; the eyes may be reddish, especially in breeding males. Young-of-year have vertical dark bars with pale centers. Lateral line scales number 36 to 44. The dorsal fin has 10 (range of nine to 11) spines and typically 10 soft rays. Anal-fin soft rays number nine to 10. Pectoral-fin rays number 12 to 13. Gill rakers are nine to 13; the length of the longest rakers is about four times (adults) to six times (young-of-year) their basal width. Palatine teeth are present[4].

Warmouth have three spines in the anal fin, a patch of tiny teeth near the tip of the tongue, and fewer than 45 scales in the lateral line series. They differ from Rock Bass that have five or more spines in the fin. Green Sunfish are similar, with three anal spines, but lack the teeth on the tip of the tongue and have more than 45 scales in the lateral line[6].

Range and Distributional Status

This species is not native to Idaho. Warmouth are found from Minnesota, southwest through most of Texas and southeastern New Mexico, east to Florida, and north to New Jersey, but they are absent from the

Appalachian Mountains. Warmouth were introduced into California in 1891, and they were subsequently introduced into Loon Lake, Washington, but apparently did not establish there. Warmouth were stocked in the Boise River, Idaho, and the Willamette River, Oregon, in the Columbia River drainage in the early 1890s[6].

In Idaho Warmouth have been documented in the Weiser River and portions of the Snake River below the Weiser River confluence. They are present in the lower reaches of the Bruneau River, near its confluence with the Snake River (CJ Strike Reservoir).

Size and Longevity

Warmouth are not a large sunfish. In Washington the angling record for Warmouth is a 330 mm (13.0 in), 0.24 kg (8.5 oz) fish taken from Silver Lake in 1996. Warmouth rarely exceed 254 mm (10.0 in) in length, and most are less than 0.23 kg (8.1 oz) in weight. In some ecosystems, Warmouth may become stunted and grow only to about 102 mm (4.0 in). As with many of the other sunfishes, Warmouth may show particularly good growth when first introduced into new waters. Warmouth are relatively long-lived sunfishes and have been known to live for at least seven years. Nine-year-old fishes have been collected from Silver Lake, Washington[6].

In Illinois young Warmouth, ages one through eight, averaged TLs, in mm (in), of 40 (1.6), 84 (3.3), 125 (4.9), 162 (6.4), 188 (7.4), 203 (8.0), 213 (8.4), and 216 (8.5), with males growing faster than females. Similar patterns were found in southern Georgia Warmouth populations, except that growth was more accelerated in the initial years[4].

The Idaho record Warmouth weighed 0.27 kg (9.5 oz) and was 223 mm (8.8 in), taken from the Snake River in 1988[10]. The all-tackle world record for Warmouth was 1.1 kg (2 lb, 7 oz), taken in 1985 from Guess Lake, Yellow River, Holt, Florida[9].

Food and Feeding

Young Warmouth feed primarily on micro crustaceans and aquatic insect larvae. Larger specimens feed heavily on crayfish and other larger crustaceans, such as freshwater shrimp (*Palaemonetes*), and a few small fish. These prey items indicate a bottom-oriented feeding behavior. There are no reports of Warmouth feeding on terrestrial insects or other evidence of surface feeding[4].

In laboratory tests, post-larval Warmouth fed on protozoa and bacteria. At larger sizes, they ate larger organisms such as zooplankton. Warmouth as small as 19 mm (0.75 in) have been reported to feed on fish. Larger Warmouth feed on larvae of aquatic insects, such as mayflies, damselflies, dragonflies, and caddisflies; crayfish; and small fishes. Larger Warmouth are more piscivorous than other sunfishes, as documented in studies that found more fish in their stomachs[6].

Limiting Factors

Small Warmouth are presumed to be eaten by a variety of other warm-water fishes such as Largemouth Bass. Warmouth are presumed to be infested with a range of parasites similar to other sunfishes.

Breeding Habits

Warmouth reach sexual maturity at a length of 76 to 102 mm (3.0 to 4.0 in). Populations of Warmouth in Illinois spawn from mid-May to August, with a peak in June at water temperatures that are near 21°C (70°F). Males choose the nest sites, which have rocks, logs, or vegetation and are usually in shallow water, less than 1.5 m (5 ft) deep, where they dig a rounded depression in a soft bottom or fan away debris in a rubble area. Males remain on the nest to court females and may spawn with several females. Generally, all eggs in a nest are of the same stage of development, indicating that males spawn during only a short period of time. After spawning is completed, males remain on the nest to fan and aerate the developing eggs and guard their territory. Females are apparently intermittent spawners, that is, they may spawn several times during the breeding season. A Park Lake, Illinois, population had 89 mm (3.5 in) females that produced 4,500 eggs, and a 180 mm (7.1 in) female produced 37,500 eggs. Warmouth eggs hatch in about two to two and a half days at 24°C (75°F) and in about one and a third days at 25 to 25.6°C (77 to 78°F)[6].

Habits and Habitat

Warmouth are inhabitants of sluggish streams and lakes and are usually associated with areas of dense cover, such as debris or weed beds. They also occur in brackish coastal systems[4].

Warmouth are generally abundant in clear lakes that have thick growth of submergent plants. Occasionally, Warmouth occur in turbid waters without vegetation. Generally, they are found over substrates, such as soft mud, allowing the dense growth of rooted aquatic vegetation preferred by this species. In California ecosystems where they are sometimes abundant, Warmouth appear to tolerate turbid water better than other sunfishes. In Silver Lake, Washington, Warmouth are sparsely scattered in shallow waters among willow roots and artificial brush shelters[6].

Management and Conservation Status

Warmouth are considered common, wide spread, and abundant rangewide (G5). A state conservation ranking is not applicable (SNA) since Warmouth are an introduced species. They are classified as a game fish in Idaho[21]. No specific management protocols have been developed for Warmouth in Idaho. Status of Warmouth in Idaho is unknown.

Kackley Springs on Rocky Mountain Power lands before it enters the Sagebrush Steppe Land Trust Kackley Springs Preserve, near Grace, ID. --Photo Courtesy John Sigler.

Orangespotted Sunfish

Lepomis humilis
(Girard, 1858)

Etymology

Lepomis = scaled operculum; *humilis* = humble, perhaps in reference to the small size of this species.

Importance

Over most of its native and transplanted range, Orangespotted Sunfish are likely too small to be of much importance. It is presumed that in some ecosystems, they may provide forage for other warm-water fishes.

Description

Orangespotted Sunfish have lateral line scale counts of 32 to 42. The dorsal fin has 10 (range of nine to 11) spines and 10 to 11 soft rays. Anal fin soft rays number nine. Pectoral fin rays are 15 (14 to 15). Gill rakers are 10 to 15, with the length of the longest rakers about five times their basal width (in young and adults). Vertebrae number 29 to 30. Palatine teeth are present. Unique among *Lepomis*, Orangespotted Sunfish have a pair of sensory pores between the eyes, situated in depressions or pits. Nuptial males may have a bright-yellow or orange breast and belly. The opercular lobe is conspicuously outlined, with a broad white or pale-green border. Most specimens, particularly juveniles, will appear more subdued, having only scattered dark and orange spots. Juveniles have dusky bars on the sides[4].

Range and Distributional Status

This species is not native to Idaho. Orangespotted Sunfish are native to the central United States, from the upper Mississippi River Basin and southern Great Lakes south, and in Gulf Coastal drainages from Mobile Bay through much of Texas. Orangespotted Sunfish are locally common in turbid overflow swamps and backwaters of sluggish streams in western Tennessee and also abundant in natural lakes of the Mississippi floodplain. They have apparently found a niche in reservoirs, as they occur in mainstream impoundments of the Tennessee and Cumberland Rivers[4]. Orangespotted Sunfish were reported to have been collected in the Bruneau River at the Bruneau Hot Springs in 1986[11]. No collections of this species are known to have occurred since, therefore we consider them to be extirpated.

Size and Longevity

Growth data for Louisiana and Iowa specimens indicate young grow to about 30 mm (1.2 in) TL at age one and grow only about 10 to 12 mm (0.4 to 0.5 in) per year thereafter through age four. Populations from western Tennessee may approximate this growth rate; specimens over age one (based on scale-annulus formation) approaching the end of the second growing season (August) had attained an average of about 50 mm (2.0 in) TL, and two-, three-, and four-year-old fish had reached 60, 75, and 85 mm (2.4, 3.0, and 3.3 in), respectively. Young of year were not present in the collection. Gonads had matured in 60 mm (2.4 in) two-year-old specimens[4].

Food and Feeding

Previous studies indicated a diet composed chiefly of micro crustaceans and aquatic insect larvae, principally chironomid dipterans. Stomachs of Tennessee specimens contained both chironomid larvae and terrestrial insects, possibly indicating a varied feeding behavior between surface and bottom[4].

Limiting Factors

Orangespotted Sunfish are likely preyed upon by other warm-water fishes and give their small size are presumably infested with the same general range of parasites as other sunfishes.

Breeding Habits

Spawning occurs from May to August. Females collected from western Tennessee in August still contained many ripe ova. Small nests are constructed by the male in shallow, quiet water areas. Sound production may be important during mating. After spawning the male remains with and maintains the nest until hatching has occurred. Natural hybrids are known with Bluegill, Pumpkinseed, Longear Sunfish, and Green Sunfish[4].

Habits and Habitat

Orangespotted Sunfish inhabit turbid, sluggish waters. They are locally common in turbid overflow swamps and backwaters of sluggish streams in western Tennessee and are also abundant in natural lakes of the Mississippi floodplain. They also have apparently found a niche in reservoirs, as they occur in mainstream impoundments of the Tennessee and Cumberland Rivers[4].

Management and Conservation Status

No specific management protocols have been developed for Orangespotted Sunfish in Idaho. Status of Orangespotted Sunfish in Idaho is unknown, but they are believed to be extirpated.

"It is that range of biodiversity that we must care for – the whole thing – rather than just one or two stars."

David Attenborough

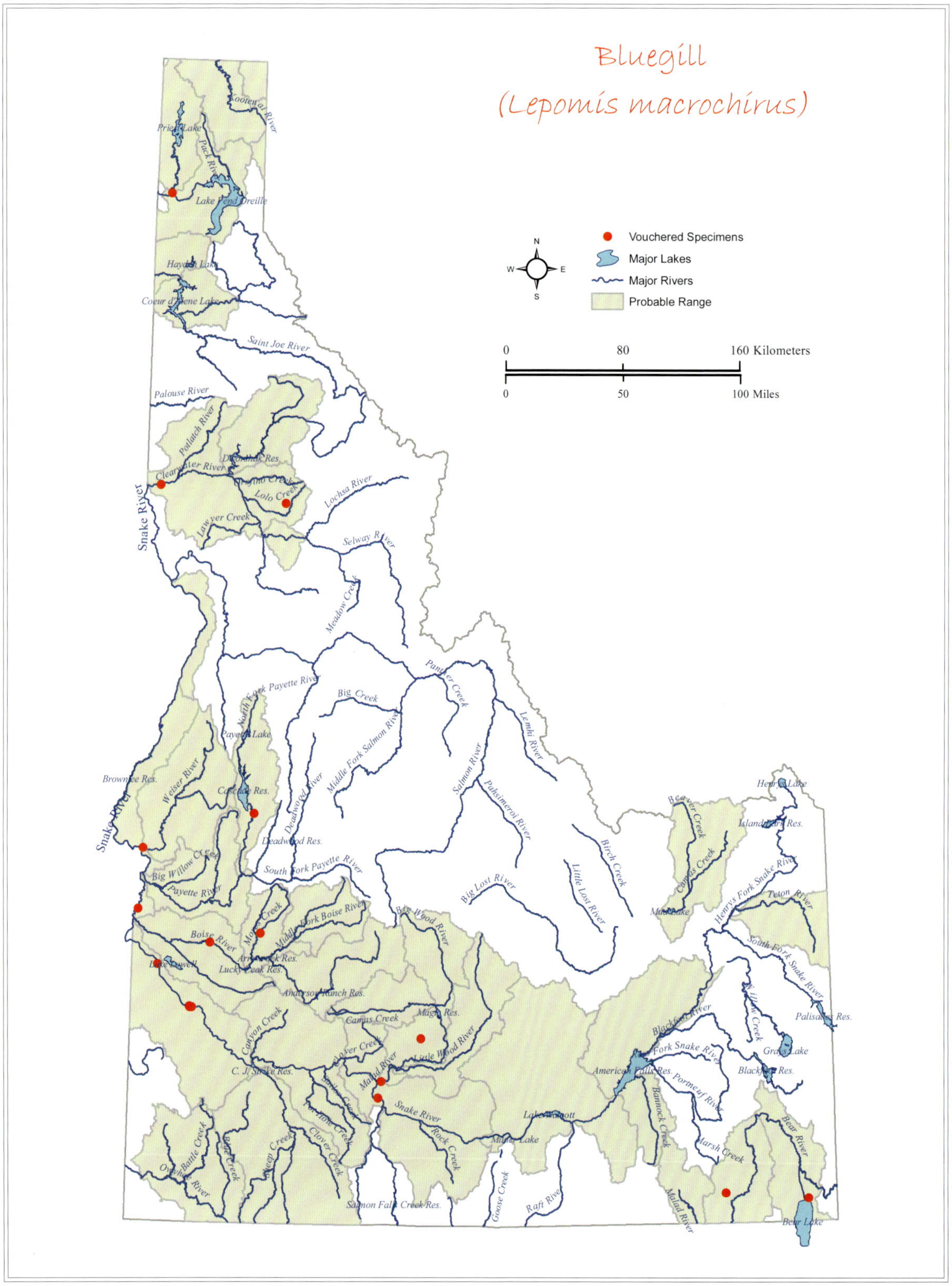
Bluegill
(Lepomis macrochirus)
N
W
E
S
Vouchered Specimens
Major Lakes
Major Rivers
Probable Range
0
80
160 Kilometers
0
50
100 Miles
Priest Lake
Pack River
Kootenai River
Lake Pend Oreille
Coeur d'Alene Lake
Saint Joe River
Palouse River
Potlatch River
Clearwater River
Lolo Creek
Lawyer Creek
Lochsa River
Selway River
Snake River
Meadow Creek
Big Creek
Middle Fork Salmon River
Payette Lake
Cascade Res.
Brownlee Res.
Weiser River
Deadwood Res.
South Fork Payette River
Big Willow Creek
Payette River
Boise River
Middle Fork Boise River
Lucky Peak Res.
Anderson Ranch Res.
Canyon Creek
Camas Creek
Magic Res.
Little Wood River
Big Wood River
C. J. Strike Res.
Malad River
Snake River
Rock Creek
Salmon Falls Creek Res.
Goose Creek
Raft River
Salmon River
Lemhi River
Pahsimeroi River
Big Lost River
Little Lost River
Birch Creek
Beaver Creek
Camas Creek
Mud Lake
Henrys Lake
Island Park Res.
Henrys Fork Snake River
Teton River
South Fork Snake River
Palisades Res.
Willow Creek
Grays Lake
Blackfoot River
Blackfoot Res.
American Falls Res.
Portneuf River
Bannock Creek
Lake Walcott
Marsh Creek
Malad River
Bear River
Bear Lake

Bluegill

Lepomis macrochirus

Rafinesque, 1819

Etymology

Lepomis = scaled operculum; *macrochirus* = large hand, probably in reference to the body shape.

Importance

Bluegill, both young and smaller adults, are common in the diet of a wide variety of predaceous fishes, particularly Largemouth Bass and Northern Pike. In lake ecosystems that contain large populations of Bluegill, this species can be a serious competitor for bottom organisms with other fishes in the ecosystem. Calculations indicate that Bluegill often consume up to six times their weight in food and may seriously deplete the bottom fauna. Bluegill may be considered to be enjoyable simply as an active, colorful, and obvious part of life in many freshwaters of North America. This species is inquisitive and fearless, coming readily to the surface or near shore to be fed with a variety of bread crumbs or other foods. In some locations they have commercial value. For sportfishing, the combination of a very light fly rod, wet flies, dry flies, small poppers, or bugs to pursue Bluegill is an angling delight. Bluegill attack bait in groups and bite and fight hard. Bluegill can provide considerable enjoyment for younger anglers using small hooks and small pieces of live bait[3].

Anglers enjoy Bluegill because of their ease of capture, scrappy fight, and abundance. Bluegill, and to a lesser extent Pumpkinseed, are the main prey for Largemouth Bass. Bluegill provide popular warm-water fisheries for harvest-oriented anglers in many waters. For either of these species to grow to an acceptable size to anglers, there must be considerable predation on their young. By managing Largemouth Bass with a quality or trophy regulation, the increased density of basses reduces young Bluegill densities and allows for improved growth.

Description

Bluegill have a dorsal surface that is green, olive to almost brown, with several vague vertical bands extending down the sides. The upper sides are brown to green, shading into brown, orange, or pink; the lower sides and abdomen behind the pelvic fins are silvery to white; there is a yellow breast in females and males most of the year. The opercular flap is black and often iridescent blue at the anterior edge. All fins except the pectorals usually have some black on the lower parts the membranes. Bluegill have a very deep and compressed body. There are brush-like teeth on the jaws and the vomer, and the lower pharyngeal teeth are numerous, conical, and bluntly pointed. Gill rakers are moderately long, slender, and typically number 12 on the lower limb and four on the upper limb. Branchiostegal rays usually number six, rarely six and five. There are two dorsal fins, but they are broadly joined, appearing as one usually, with 10 but rarely with 11 spines; the second dorsal fin is soft rayed and a very little bit higher than the first dorsal; rays usually number 11 but frequently 10 to 12. There is a distinct dark blotch at the posterior end of the second dorsal fin. The caudal fin is moderately broad and shallowly forked, and the tips are rounded points. The anal fin has a moderately long base with three graduated sharp spines and usually 11 rays, rarely 8 to 10 rays. Pelvic fins are thoracic, with their origin under the second dorsal spine and with one small spine and five rays. Pectorals are moderately high on the body and rather long, with square to slightly rounded and pointed tips; there are 13 or rarely 14 rays. Scales are mostly ctenoid, rather small, and less crowded on the breast than in some other sunfishes. The lateral line is complete and high except on the peduncle, with a long, shallow arch over the pectoral fin. There are 40 to 44 lateral line scales. The peritoneum is silvery, and the intestine is well differentiated; there are typically six prominent caeca. Vertebrae usually number 28 or 29. No nuptial tubercles are present, but color intensifies in breeding males[3].

Bluegill demonstrate sexual dimorphism with respect to color. The breast and ventral parts of the body in front of the pectoral and pelvic fins develop a copper-orange color in breeding males. This area is yellowish in breeding females and nonbreeding males. Outside the breeding season, the shape of the urogenital opening differs in males and females[5].

Range and Distributional Status

This species is not native to Idaho. Native distribution of the Bluegill is from Minnesota and southern Ontario eastward to the Atlantic states, south to Georgia, west to eastern Texas and northeastern Mexico, and north through the Midwestern states to Minnesota. Bluegill have been widely introduced into various parts of the United States. Bluegill may have been introduced into the lower Columbia River during 1880 and 1881. Bluegill have been captured in reservoirs on both the Columbia River and the Snake River and in the various waters of Puget Sound[6].

In Idaho Bluegill have been reported in the Priest River, Kootenai River, Clark Fork River, Spokane River, and Coeur d'Alene River of northern Idaho as well as the Clearwater River, portions of the Boise River, the Little Wood River near its confluence with the Big Wood River, portions of the Snake River in south-central Idaho, and portions of the Bear River drainage.

Size and Longevity

Bluegill have a rather long life span and grow well in most parts of the United States. However, Bluegill often become stunted in some lake ecosystems, particularly in waters that are infertile or have dense vegetation. In Washington Bluegill live to 11 years in some lake ecosystems; average length was 196 mm (7.7 in)[6].

Newly hatched Bluegill are approximately 6.3 mm (0.25 in) long. By the end of the first year, individuals have reached lengths of about 38 to 50.8 mm (1.5 to 2.0 in) and may grow about 25 mm (1 in) each year thereafter. They may live to nine or more years. In lower Snake River reservoirs, reported mean total lengths of Bluegill ages one to seven were, in mm (in), 35.6 (1.4), 71 (2.8), 122 (4.8), 150 (5.9), 175 (6.9), 193 (7.6), and 211 (8.3)[2].

The all-tackle world record for Bluegill weighed 2.15 kg (4 lb, 12 oz) and was taken from Ketona Lake, Alabama, in 1950[9]. The Idaho record, taken from CJ Strike Reservoir in 1966, weighed 1.6 kg (3 lb, 8 oz)[10].

Food and Feeding

Bluegill fry eat zooplankton, principally crustaceans such as copepods and cladocerans. As individuals grow in size, they consume increasing proportions of various aquatic insects. Other items identified as Bluegill food include mollusks, small crayfish, amphipods, fish eggs, and larval or small juvenile fishes. At times, terrestrial insects such as grasshoppers and crickets may be important in the diet. During the summer, Bluegill may eat plants such as algae as well as some rooted aquatic vegetation. Bluegill continue to feed in winter, but the consumption of food decreases dramatically when water temperatures drop below 12.8°C (55°F). This habit makes them easy to take by angling through the ice in some ecosystems. The diet of juvenile Bluegill in one Washington lake contained 42 percent aquatic Diptera (primarily chironomids), 23 percent cladocerans, and 11 percent amphipods. Adult Bluegill fed on 30 percent aquatic Diptera, 22 percent amphipods, and 9 percent water mites[6].

Bluegill begin feeding at about 5 to 6 mm (0.2 to 0.24 in), and their first food consists of rotifers and copepodites; however, as Bluegill grow, cladocerans become the dominant food item. Juveniles and adults may feed primarily on the larvae, pupae, and adults of insects (mainly chironomids), macroinvertebrates (e.g., amphipods), and zooplankton (particularly *Daphnia*)[5].

Limiting Factors

Low oxygen levels, below 0.60 mg/l (ppm), in winter are lethal to Bluegill. Temperatures in the 30 to 31°C (high 80s°F) are reported as near lethal as well. However, Bluegill have been reported to tolerate temperatures up to 35°C (95°F). Bluegill are host to a number of parasites, including flukes, tapeworms, roundworms, spiny-headed worms, fish lice, and snail larvae. Natural mortality from all causes is generally higher than angling mortality, though stunting may reduce the numbers of desirable-size fish but not total numbers[2].

Breeding Habits

Bluegill generally mature when age two or three. Maturity, however, appears to be related to growth, with the fastest-growing fishes maturing sooner than fishes that are growing more slowly. Bluegill spawn in the spring when water temperature is above 19.4°C (67°F), and spawning may occur throughout their entire growing season. In some waters, Bluegill spawn when water temperatures are about 22 to 26°C (72 to 79°F). Male Bluegill abandoned their nests when water temperatures dropped in a Utah population but nested again when water temperatures increased[6].

Bluegill males generally form hollows or saucer-shaped depressions for nests on a sandy bottom in shallow water. Number of eggs produced by females varies from 2,360 to 81,104 depending on female size; larger fishes produce more eggs. After spawning, males vigorously protect the nest and keep eggs clean and aerated by fanning them with their fins. Males also protect the nest and fry for several days until they disperse. Newly hatched Bluegill are about 2.5 mm (0.1 in) long[6].

When male Bluegill construct nests, it is generally on a sand or gravel bar in water that is 31 to 102 cm (12 to 40 in) deep, and the depressions that are created are 20 to 31 cm (8 to 12 in) or more in diameter and 5 to 15 cm (2 to 6 in) deep. Males build nests in close proximity to one another and defend their territories, but male Bluegill are more tolerant of each other than most sunfishes. Prespawning females may stay in schools in the general area of the nest. When ready to spawn, females approach a nest area and are in turn approached by a male, generally the largest in the vicinity. Males attract females to the nest and spawn. Courtship movements are accompanied by distinctive grunting sounds. Bluegill eggs are adhesive and attach to debris on the bottom of the nest, hatching in two to three days at temperatures of 20°C (68°F) water. When fry move from the nest to aquatic vegetation or other cover, they may remain pelagic for a period of one or more months[2].

Habits and Habitat

Bluegill generally inhabit warm, shallow lakes with rooted vegetation. All sizes and ages of Bluegill exhibit strong orientation to habitat with cover or structure. Bluegill travel in small, loose schools while feeding, particularly in early morning or in the evening. Because Bluegill are sight feeders, they feed primarily during daylight. Bluegill grow fastest at water temperatures of 15.6 to 27°C (60 to 80°F). Aggressive behavior is exhibited by male Bluegill during nest construction, courtship, and spawning and increases in crowded nesting situations, behavior typical for the family Centrarchidae[6].

In Nebraska male Bluegill selected emergent vegetation during April, June, and July, whereas females showed no preference for emergent vegetation and used all available habitat types. No evidence of diel habitat shifts in females was found. In contrast, however, adult male Bluegill spent significantly less time in macrophytes at night than during the day[5].

Management and Conservation Status

Bluegill are considered common, wide spread, and abundant rangewide (G5). A state conservation ranking is not applicable (SNA) since Bluegill are an introduced species. They are classified as a game fish in Idaho. In some lowland lakes with high angling effort, anglers have been able to crop off the larger Bluegill, preventing them from exceeding 203 mm (8 in). Through dispersal or unlawful introductions, Bluegill have become established in several north Idaho lakes. In most cases, they support popular fisheries and have not generally been detrimental. They appear to have successfully outcompeted Pumpkinseed Sunfish in nearly all north Idaho waters where they have been introduced. Hybrids between the two species exist in some lakes. Bluegill have, however, replaced perch in some north Idaho lakes.

"Most of the world is covered by water. A fisherman's job is simple: Pick out the best parts."

Charles Waterman

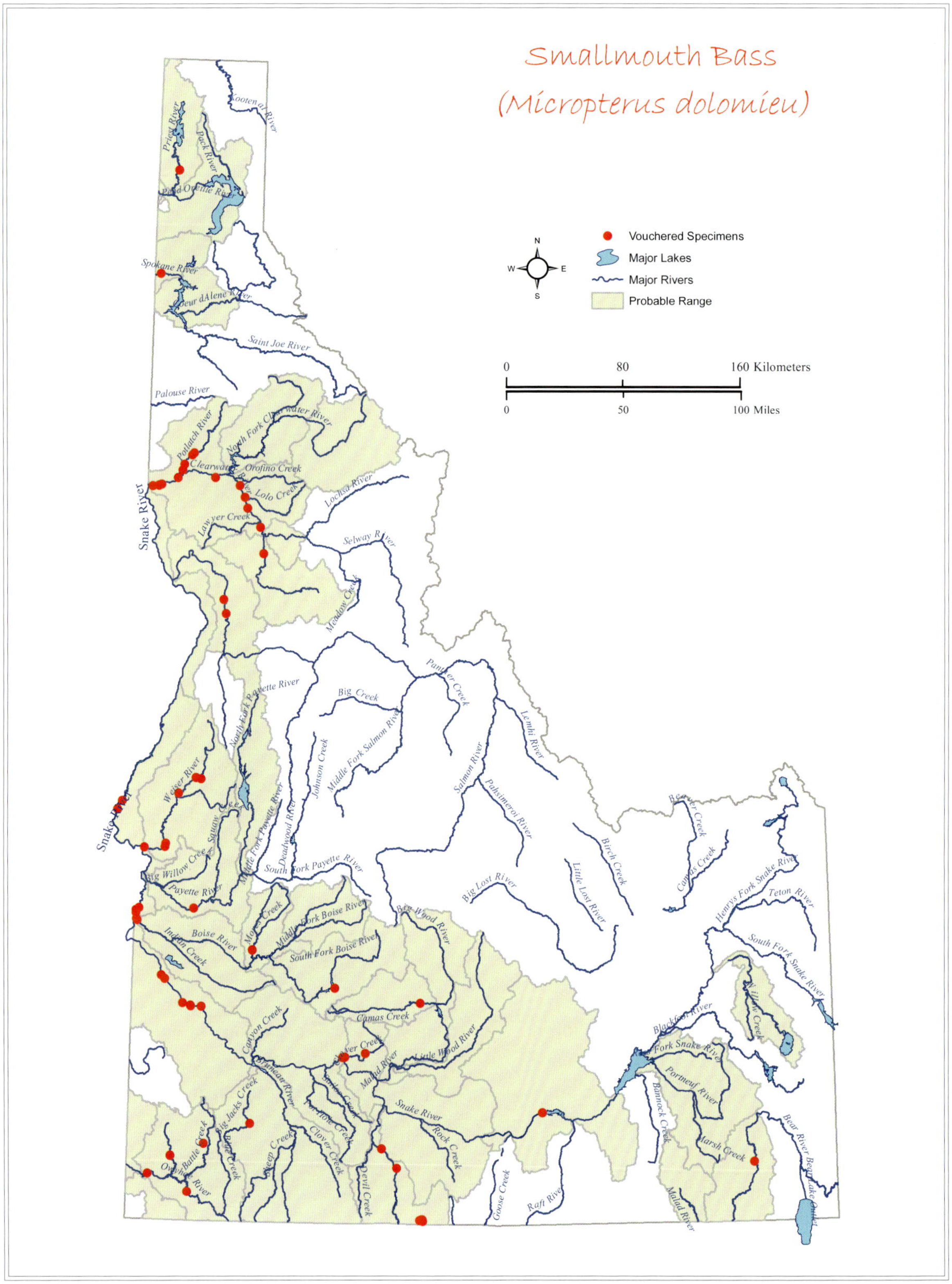
Smallmouth Bass
(Micropterus dolomieu)
Vouchered Specimens
Major Lakes
Major Rivers
Probable Range
N
W
E
S
0
80
160 Kilometers
0
50
100 Miles
Priest River
Pack River
Spokane River
Saint Joe River
Palouse River
Potlatch River
Orofino Creek
Lolo Creek
Lochsa River
Snake River
Selway River
Meadow Creek
Big Creek
Johnson Creek
Middle Fork Salmon River
Salmon River
Lemhi River
Pahsimeroi River
Birch Creek
Little Lost River
Big Lost River
Camas Creek
Teton River
South Fork Snake River
Weiser River
Deadwood River
Middle Fork Payette River
Payette River
Boise River
Indian Creek
South Fork Boise River
Camas Creek
Canyon Creek
Little Wood River
Malad River
Snake River
Rock Creek
Clover Creek
Devil Creek
Deep Creek
Big Jacks Creek
Battle Creek
Goose Creek
Raft River
Bannock Creek
Portneuf River
Marsh Creek
Malad River
Bear River

Smallmouth Bass

Micropterus dolomieu
Lacepède, 1802

Etymology

Micropterus = small fin (apparently the second dorsal fin for the type specimen for this genus was deformed, leading Lacepède to think there was a short fin at the rear of the dorsal); *dolomieu* after M. Dolomieu. See Sigler and Sigler[2] for the colorful original description and its English translation.

Importance

Both the magnitude and the antiquity of interest in Smallmouth Bass are obvious when one is presented with the facts: that angling results were mentioned in Fothergills' 1836–37 *Account Of The Natural History Of Eastern Canada,* that introductions in North America began in the 1850s, and that fish culture practices in Canada began in 1884[3]. The attraction of Smallmouth Bass for anglers is legendary. Techniques for capturing this species, along with many other "stories," are the subject of an untold number of popular books and articles. Smallmouth Bass are one of the most sought-after game fishes in North America[2]. Smallmouth Bass are an excellent game fish and a favorite among anglers, in part because they stage a spectacular aerial show when hooked, providing a much better fight than Largemouth Bass in general. As early as 1881, Smallmouth Bass were described "inch for inch and pound for pound [as] the gamiest fish that swims"[6]. Millions of dollars are spent annually in the United States on equipment, boats, travel, and other expenses by anglers in pursuit of this species, although a large portion of these expenses is spent by those in search of Largemouth Bass. Smallmouth Bass are renowned for their fighting abilities and relatively good pan qualities; they may be taken by a number of methods[4].

Description

Smallmouth Bass vary in color based on size, condition, and the particular ecosystem in which they exist. In clear, vegetated water or stained water, Smallmouth Bass are darker, with pronounced, contrasting markings. In turbid and lighter waters, they are lighter, with vague markings. The dorsal surface of the back and head is brown to golden brown through olive green, whereas the sides are lighter and more golden, with gold flecks on most scales. The ventral surface is cream to milky white, and the sides of adults have eight to 15 pronounced to vague thin vertical bars that are sometimes broken. The eye is usually red or orange, and fins are dusky to amber. However, the pectorals are clear, while the other fins are opaque, with some black on rays, spines, or membranes. Young Smallmouth Bass appear much like the adults; however, the vertical bars or rosy spots are very prominent, and the caudal fin is unmistakably marked with orange at the base, followed by a black band and white to yellow tips[3].

Smallmouth Bass are a moderately large and robust fish; only the Largemouth Bass is larger in this family. The body of Smallmouth Bass is laterally compressed, and the head is large and long; the dorsal surface is rounded or has a shallow depression over the eyes. The mouth is terminal and slightly oblique, with the lower jaw slightly longer than the upper jaw. There are brush-like teeth on both jaws, palatines, and the vomer. Lower pharyngeal teeth are on long, narrow pads and are numerous, fine, and uniform in size. Gill rakers typically number eight on the lower limb and three on the upper limb. Branchiostegal rays are 6, 6 and 7, 7. There are two dorsal fins, but they are joined and appear more like one: the first dorsal fin is rather low, with 10 stout spines of approximately the same length; the second dorsal is higher, with 12 to 15 soft rays and rounded edges. The caudal fin is not markedly long but is broad and moderately forked, with tips that are blunt to round. The anal fin has three graduated spines and 10 to 12 (typically 11) soft rays and a rounded edge. Pelvic fins are thoracic, and the origin is ahead of the origin of the dorsal fin; rays are joined by membranes. The pelvics are not long and have one spine and five rays; the tip is rounded. Pectoral fins are not long but are broad and rounded, with 13 to 15 rays. Scales are ctenoid and slightly smaller than Largemouth Bass, typically numbering 11 to 13 rows from the lateral line to the dorsal-fin origin and 19 to 23 rows from the lateral line to the anal-fin origin. The lateral line is complete, high, and only moderately arched, with 67 to 78 scales in the lateral line series. The peritoneum is silvery, and the intestine is well differentiated. Vertebrae number 31 or 32. There are no nuptial tubercles, but colors darken on spawning males and the color pattern of females intensifies3. Juvenile Smallmouth Bass have thin, vertical bars along the midline, a conspicuous orange mark at the base of the tail, and radiating cheek stripes[5].

Range and Distributional Status

This species is not native to Idaho. Smallmouth Bass's native range is from the middle of Minnesota east to southern Quebec, southeast to northern Georgia, west to Oklahoma, and north to Minnesota. In general, the original Smallmouth Bass range was restricted to the freshwaters of eastern central North America[3,6]. Smallmouth Bass were first introduced into California in 1874, and the first introduction of Smallmouth Bass in Washington was in 1924. In the Columbia River system, Smallmouth Bass density is greatest in the Snake River reservoirs[6]. In Idaho Smallmouth Bass are known from the Pend Oreille, Spokane, Snake (above and below Shoshone Falls), and Bear River drainages.

Size and Longevity

Smallmouth Bass are the second-largest member of the sunfish family. In the Clearwater River, Smallmouth Bass ages one to nine had lengths, in mm (in), of 84 (3.3), 142 (5.6), 201 (7.9), 241 (9.5), 269 (10.6), 295 (11.6), 315 (12.4), 325 (12.8), and 338 (13.3). At hatching young are approximately 6 mm (0.2 in) long, and after the yolk sac is absorbed they are 8.4 mm (0.3 in) long. Few Smallmouth Bass live to be older than age 15[2]. The Idaho record Smallmouth Bass was taken from Dworshak Reservoir in 2006 and weighed 4.4 kg (9 lb, 11 oz); this individual was 603 mm (23.7 in) long[10]. The all-tackle world record for Smallmouth Bass is 5.41 kg (11 lb, 15 oz), taken from Dale Hollow Lake in Tennessee in 1955[9].

Food and Feeding

Smallmouth Bass begin to feed extensively when water temperature rises above 10°C (50°F), but they feed most actively at their preferred temperature of 20°C (68°F). Smallmouth Bass eat crustaceans such as copepods and cladocerans. While still in the 25.4 to 50 mm (1.0 to 2.0 in) size range, Smallmouth Bass switch to a diet of insects and begin to eat larval fishes. Adult Smallmouth Bass feed on insects, crayfish, and other fishes. In waters with abundant crayfish, they frequently constitute up to two-thirds of the diet for this species[6].

Crayfish, fish, and insects, in that order, are the dominant foods of adult Smallmouth Bass in Little Goose Reservoir on the lower Snake River. Smallmouth Bass take a variety of food from the surface to the bottom; dead food is generally but not always rejected. In some ecosystems, crayfish may make up to 60 to 90 percent of the diet, fish 10 to 30 percent, and aquatic and terrestrial insects up to 10 percent[2].

Smallmouth Bass are noted for their voracious appetites, and diet is strongly influenced by the range of available prey. In summer months, adult Smallmouth Bass show activity peaks in the morning and evening, being less active during the day than Largemouth Bass. Young Smallmouth Bass start life foraging on zooplankton and gradually add insects to their diet. Smallmouth Bass may start taking small fishes at sizes of about 50 mm (2.0 in), and the proportion of fish in the diet increases with size[5].

Limiting Factors

Parasites of Smallmouth Bass include protozoans, trematodes, cestodes, nematodes, acanthocephalans, leeches, mollusks, and crustaceans. They may also be infested with the bass tapeworm, which can limit reproduction in this species. Smallmouth Bass may hybridize with Spotted Bass *(Micropterus punctulatus)* in ecosystems where they are sympatric[3].

Smallmouth Bass cannot prosper in warm turbid stream ecosystems with very low flows or in turbid shallow lakes or reservoirs. Fluctuating water levels or high winds just after spawning may destroy or cause desertion of the nest. Smallmouth Bass are sensitive to domestic, industrial, and thermal pollution. In the northern part of their range, Smallmouth Bass may be limited by the availability of young-of-the-year overwintering habitat[2].

Breeding Habits

Smallmouth Bass mature at age three or four. Spawning generally occurs in spring when water temperatures are 12.8 to 18°C (55 to 65°F); spawning may cease when water temperatures reach 18.9°C (66°F). In the Columbia River, Smallmouth Bass have been observed spawning in sloughs where water temperatures were in the 12.8 to 15.6°C (55 to 60°F) range, warmer than the main river channel, which was 7 to 10°C (45 to 50°F)[6].

Male Smallmouth Bass excavate shallow depressions for a nest and guard eggs and young. Nests are generally about 61 cm (2 ft) in diameter and 5 to 10 cm (2 to 4 in) deep. In some instances, Smallmouth Bass males guarded the fry for nearly one month after they left the nest. Predation on the eggs and young may be severe at this time. Male Smallmouth Bass may desert their nest if there is a substantial, rapid decline in water temperature or water level or if the eggs become covered with fungus. Female Smallmouth Bass produce between 2,000 and 20,825 eggs. A female 249 mm (9.8 in) long contained 5,148 eggs, and a female 432 mm (17.0 in) long contained 20,825 eggs[6].

Typical of all centrarchids, male Smallmouth Bass provide sole parental care for developing offspring. In early spring, when water temperatures reach approximately 15°C (59°F), male Smallmouth Bass excavate shallow saucer-shaped nests in the gravel substrate of the littoral zone. Once nest construction is complete, females are courted, and spawning occurs shortly thereafter. After egg deposition, females leave the area, whereas the male remains to fan the eggs and protect developing offspring from brood predators. Care provided by nesting males can persist for as long as five weeks, at which point the young are free swimming and independent (i.e., they can forage and avoid predators on their own). If a male Smallmouth Bass leaves his brood unattended, even for a short period, before the offspring develop to the independent free-swimming stage, the offspring are likely to be consumed by predators. Angling for Smallmouth Bass during the nesting stage can remove a brood-guarding male from his nest, resulting in some cases in nest abandonment. After catch-and-release angling, however, some parental male Smallmouth Bass resume brood defense, guarding offspring until they have reached independence. As a result, many anglers target basses during the nesting period under the assumption that catch-and-release angling has no negative effect—an assumption, however, that is incorrect. Mechanisms controlling parental care decisions associated with abandonment of broods among male Smallmouth Bass after catch-and-release angling remain unidentified. As the extent of brood loss increases, the value of that brood to the male decreases. At some amount of brood loss, a male should cease investment in that reproductive bout and conserve resources for future reproductive efforts[12].

During the breeding season, courtship activities of Smallmouth Bass involve an initial phase away from the nest site and a final phase at the nest site. Spawning occurs after this final courtship phase, typically at either dusk or dawn. Eggs are deposited in a series of spawning bouts. Males will aggressively attack intruders into their territory[5].

Habits and Habitat

Smallmouth Bass inhabit cool, clear stream or lake ecosystems. Generally, Smallmouth Bass are found in lake ecosystems in the northern part of the range and more in stream ecosystems in the southern part of their range. Smallmouth Bass are most abundant in stream ecosystems that are more than 11 m (36 ft) wide with a gradient of 21 to 132 m/km (4 to 25 ft/mi) over rocks, gravel, or boulder substrate. In lake ecosystems, Smallmouth Bass are found around rocky reefs, littoral zone drop-offs, and gravel bars. In stream ecosystems, Smallmouth Bass have definite home ranges and do not travel long distances. Smallmouth Bass prefer water temperatures of 21 to 27°C (70 to 80°F)[6].

Although differences in habitat use of adult Largemouth Bass and Smallmouth Bass have been well documented, less is known about habitat uses of smaller juveniles, particularly age-0 fish. Because both species feed predominantly on invertebrates and are highly vulnerable to predation when small, they may share preferences for habitats with available prey and suitable refuge. Observations of age-0 Largemouth Bass and Smallmouth Bass indicated a high degree of habitat partitioning in natural systems in preferred habitat types. On average, more than 99 percent of Largemouth Bass in three study lakes were observed in vegetated habitats, and none was observed in areas of cobble substrate. Among Smallmouth Bass, 94 percent were observed in cobble habitats, and only 3 percent were observed in vegetated habitats. Bass showed clear differences in habitat use. These patterns were observed in late June and correspond closely with preferred spawning habitats of both species. Largemouth Bass and Smallmouth Bass appear to partition habitats at all life stages. Results of laboratory experiments further suggest that habitat partitioning among juveniles reflects behavioral differences between the two species[13].

With respect to both Largemouth Bass and Smallmouth Bass, there are some concerns that when these species are released following tournament capture, behavioral disruptions occur. Only 14 percent of tournament-caught Largemouth Bass and 32 percent of Smallmouth Bass returned to their sites of original capture. Fifty-one percent of Largemouth Bass and 26 percent of Smallmouth Bass dispersed less than 1.6 km (1 mi) from their release sites. On average, Smallmouth Bass dispersed distances of 7.3 km (4.5 mi) from their release sites, while Largemouth Bass dispersed shorter average distances of 3.5 km (2.2 mi). There was no difference in dispersal distances for fish captured and released in rivers versus those released in lakes and reservoirs. Twenty-two percent of Largemouth Bass and 15 percent of Smallmouth Bass caught and released in fishing tournaments were subsequently recaptured by anglers. Previous studies recommended modifications to tournament release practices, but little has changed[14].

Management and Conservation Status

Smallmouth Bass are considered common, wide spread, and abundant rangewide (G5). A state conservation ranking is not applicable (SNA) since Smallmouth Bass are an introduced species. They are classified as a game fish in Idaho[21]. Smallmouth Bass populations are rapidly expanding in some waters where they are negatively impacting native species as well as other popular warm-water fisheries. Smallmouth Bass were intentionally introduced in Hayden Lake in 1983, but were illegally moved to Coeur d'Alene Lake in about 1990. They have now spread upstream to most of the lateral lakes; the lower Coeur d'Alene, Saint Joe, and Saint Maries Rivers; as well as downstream to the Spokane River. Montana Fish Wildlife and Parks introduced Smallmouth Bass into Noxon Reservoir on the Clark Fork River in 1983 and 1986, and they are now well established in Lake Pend Oreille and the Pend Oreille River and have moved upstream into the Priest Lake system. Smallmouth Bass prey on juvenile Westslope Cutthroat Trout and Bull Trout, and they have significantly reduced many of the native minnow species in these waters. In some waters, Smallmouth Bass are also impacting popular Largemouth Bass, crappie, and perch fisheries. In most northern Idaho waters, Smallmouth Bass are managed with liberal (six-fish) bag limits and no size restrictions. In the Salmon River and Clearwater River, and in the Snake River below Hells Canyon Dam, the IDFG also permits the harvest of any size bass in an attempt to reduce predation on salmon and Steelhead juveniles.

By creating regulations that protect bass from any form of angling (either intentional or accidental) during the highly vulnerable nest-guarding stage, abandonment rates should be reduced. This protection could be accomplished through regulations that restrict angling through seasonal closures or through the use of no-fishing sanctuaries[12].

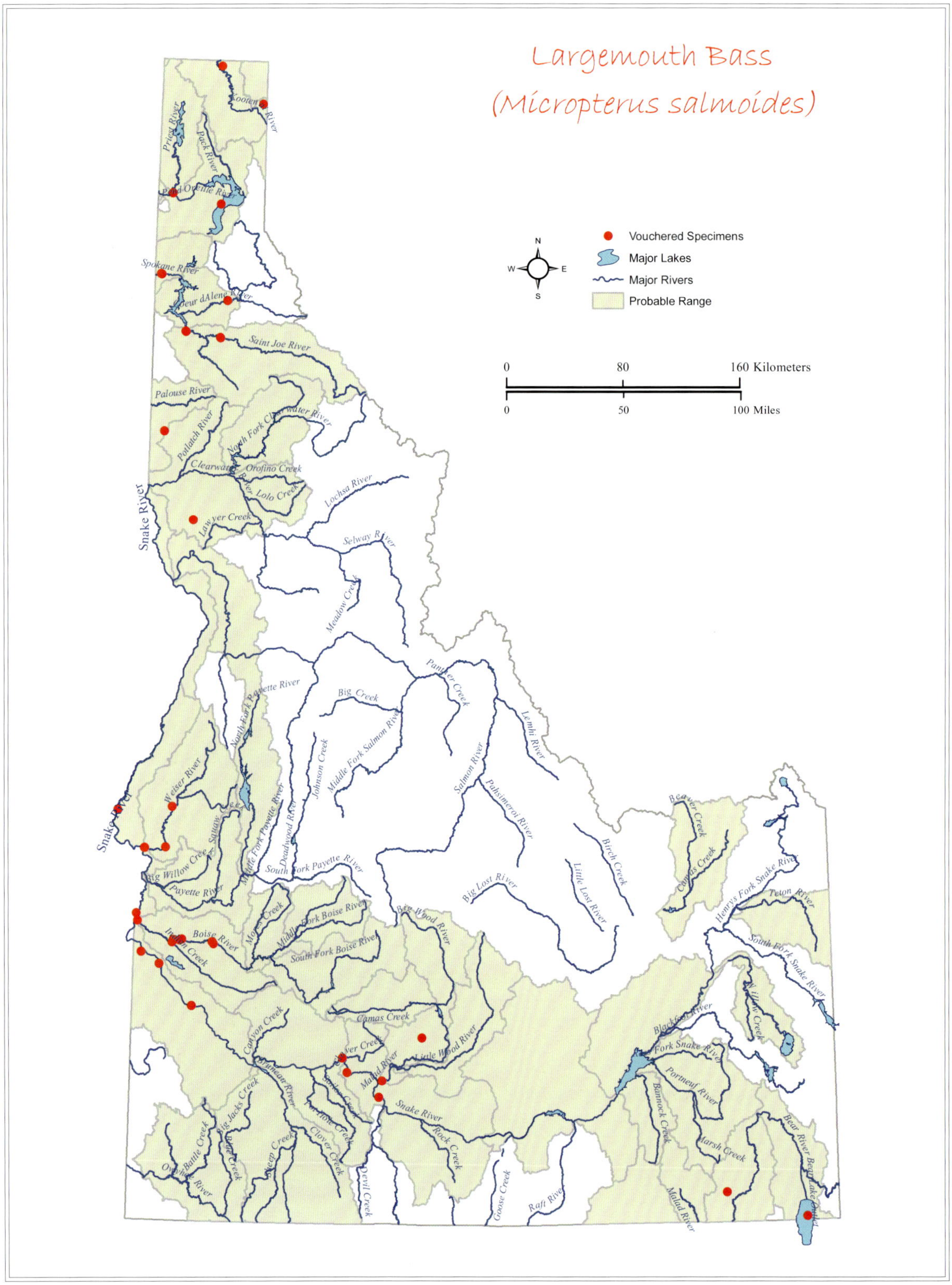
Largemouth Bass
(Micropterus salmoides)
Vouchered Specimens
Major Lakes
Major Rivers
Probable Range
N
S
E
W
0
80
160 Kilometers
0
50
100 Miles
Snake River
Saint Joe River
Palouse River
Potlatch River
Lochsa River
Selway River
Meadow Creek
Big Creek
Johnson Creek
Middle Fork Salmon River
Salmon River
Pahsimeroi River
Lemhi River
Birch Creek
Little Lost River
Big Lost River
Big Wood River
Little Wood River
Camas Creek
Weiser River
Payette River
Boise River
South Fork Boise River
Indian Creek
Canyon Creek
Snake River
Rock Creek
Goose Creek
Raft River
Malad River
Marsh Creek
Portneuf River
Bannock Creek
Bear River
Teton River
Henrys Fork Snake River
South Fork Snake River
Beaver Creek
Camas Creek
Willow Creek
Blackfoot River
Devil Creek
Salmon Falls Creek
Big Jacks Creek
Owyhee River
Lolo Creek
Orofino Creek
Lawyer Creek
South Fork Payette River
Deadwood River
Middle Fork Payette River
North Fork Payette River
Big Willow Creek
Spokane River

Largemouth Bass

Micropterus salmoides

(Lacepède, 1802)

Etymology

Micropterus = small fin (apparently the second dorsal fin for the type specimen for this genus was deformed, leading Lacepède to think there was a short fin at the rear of the dorsal); *salmoides* = from Salmo, meaning trout-like.

Importance

In a fashion similar to anglers' fascination with Smallmouth Bass, Largemouth Bass are one of the major sportfishes in North America. Largemouth Bass, with their tolerance for high temperatures and slight turbidity, their fast growth rate, and their fish diet, are a favorite for stocking warm-water farm ponds as well as lakes and reservoirs both for food and for sport[3].

Over all of their range, Largemouth Bass are highly prized by anglers. Partly because Largemouth Bass prosper over a wide variety of habitats from large, cool reservoirs to warm farm ponds, they do well in many different ecosystem types. They are arguably the most important warm-water (14.4 to 28.8°C [58 to 82°F]) game fish in the United States[2].

Largemouth Bass are one of the most popular spiny-rayed fishes in Washington. Although trout can be overharvested from a lake, it is much more difficult to fish out bass. Angling pressure on Largemouth Bass has increased in different parts of the United States in concert with increased numbers of bass fishing tournaments[6].

Description

Juvenile Largemouth Bass lack the orange mark at the base of the tail and the radiating cheek stripes of Smallmouth Bass. Juvenile Largemouth Bass have a conspicuous dark mid-lateral stripe that extends from the snout through the eye and backward to the base of the caudal fin[5].

Largemouth Bass are bright green to olive on the dorsal surface of the body and head; the sides are almost as dark in the largest individuals or golden green. The ventral surface is milk white to yellow. Dorsal and caudal fins are opaque and green to olive, while the anal and pelvic fins are green to olive with some white; pectoral fins are amber and clear. Populations in clear, weedy aquatic ecosystems are darker, and the black pigment is more obvious, while those in the darker or turbid water are paler green overall[3].

Largemouth Bass are a moderately large, robust fish with a laterally compressed body that is deeper than that of a Smallmouth Bass; a cross-section is a broad oval. The head is large and long. Largemouth Bass have a terminal mouth that is slightly oblique and is large and wide; the lower jaw is slightly longer than the upper jaw. There are fine brush-like teeth on both jaws, palatines, and the vomer. Lower pharyngeal teeth are on long, narrow pads. Gill rakers are usually stiff, numbering 68 on the lower limb and with only two rudiments on the upper limb. Branchiostegal rays are six, six and seven, seven. The two dorsal fins are joined, but the separation is more obvious than in Smallmouth Bass; the first dorsal is rather low, with 10 stout spines; the second dorsal fin is higher, with 12 or 14 (typically 12) rays, and the fin edge is rounded. The caudal fin is not long but broad and shallowly forked with rounded tips; the rounded anal fin has three spines and 10 to 12 (typically 11) soft rays. Pelvic fins are thoracic, with the origin under that of the first dorsal fin. Pelvic fins are short and rounded and have one spine and five soft rays. Pectoral fins are rather short and broad, with rounded tips and 13 to 15 rays. Scales are ctenoid and are larger than those on Smallmouth Bass, usually with about eight rows from the lateral line to the dorsal origin and 14 to 18 rows from the lateral line to the anal origin. The lateral line is complete, high, and a little arched, with 60 to 68 scales in the lateral line series. The peritoneum is silvery, and the intestine is well differentiated, with 24 to 28 thinner pyloric caeca, at least six to 10 of which are branched. Vertebrae number 30 to 32. There are no nuptial tubercles, but colors darken on spawning males[3].

Range and Distributional Status

This species is not native to Idaho. Native range of the Largemouth Bass extends from southern Quebec south to and including Florida, west to central Texas and northeastern Mexico, and north to Minnesota and southern Manitoba. Largemouth Bass were introduced into California in 1874, and they were widely introduced into Oregon and Washington in the period from 1890 to 1895 by the US Bureau of Fisheries[6].

Largemouth Bass now extend over the 48 contiguous states and to some extent into Canada and south through much of Central America. Introductions have been made in England, Scotland, Germany, France, South Africa, Hong Kong, the Philippines, and Brazil[2]. Largemouth Bass occur in all major Idaho river drainages.

Size and Longevity

Largemouth Bass are both fast growing and long lived. They are the largest member of this family. In the southern edge of their range, Largemouth Bass may grow year-round, while individuals in colder climates grow slower and live longer. Largemouth Bass ages one to seven in lower Snake River reservoirs had mean total lengths, in mm (in), of 84 (3.3), 140 (5.5), 231 (9.1), 292 (11.5), 358 (14.1), 419 (16.5), and 442 (17.4)[2].

The Idaho record Largemouth Bass is a 4.96 kg (10 lb, 15 oz) individual from Anderson Lake[10]. The all-tackle world record for Largemouth Bass is a tie between two 10.12 kg (22 lb, 5 oz) fishes; one was taken in 1932 from Montgomery Lake, Georgia, and one from Lake Biwa, Japan, in 2009[9].

Food and Feeding

The diet of Largemouth Bass fry is primarily small crustaceans and insects. They may eat primarily copepods, cladocerans, and midge larvae but also feed on other crustaceans, such as amphipods, ostracods, and isopods, as well as insects, such as nymphs of mayflies, dragonflies, and damselflies. When Largemouth Bass reach lengths of approximately 76 to 103 mm (3.0 to 4.0 in), they begin to eat fishes. Largemouth Bass fry as small as 10 to 23 mm (0.4 to 0.9 in) fed on Fathead Minnow fry in Florida, and fish longer than 103 mm (4.1 in) at other locations fed primarily on various fish species[6].

Adult Largemouth Bass are largely fish-eating predators, but food type changes with size from plankton, to insects, to fish, crayfish, and frogs. Food is taken at the surface both morning and evening and in the water mass during the day as well as from the bottom. Largemouth Bass are sight feeders and often feed in schools near shore, usually close to vegetation. Feeding may be restricted when water temperatures are below 10°C (50°F) and also decreases in winter and during spawning. Calculations have been made noting that it requires 1.8 kg (4 lb) of food to produce 0.45 kg (1 lb) of fish. Fishes consumed by Largemouth Bass include Gizzard Shad, Common Carp, Bluntnose Minnow, Silvery Minnow, Golden Shiner, Yellow Perch, Pumpkinseed, Bluegill, Largemouth Bass, and Silversides. Cannibalism is reported to be higher in this species than in Smallmouth Bass[3,5].

As with Smallmouth Bass, Largemouth Bass are noted for their voracious appetites. In summer months, Largemouth Bass show peak feeding periods in the morning and evening, but they are more active during the day than are Smallmouth Bass[5].

Limiting Factors

Largemouth Bass embryos and fry are preyed upon by other fishes, a variety of birds, crayfish, dragonfly nymphs, and predaceous diving beetles. In some lake ecosystems, Bluegill have been reported to be the primary decimating factor of young Largemouth Bass; in alternate years, Bluegill were removed and Largemouth Bass spawning was successful. Drops in water temperatures during spawning, heavy wave action, nest desertion, fluctuating water levels, parasite-caused sterility, and food availability are strong decimating factors for young[2,3].

Availability of invertebrate food of appropriate size, and at the right time and water temperature, governs the size of young during early August and determines their size and overwinter survival rate. This availability creates strong and weak year classes that are eventually reflected in the harvest, that is, year class strength is established by young-of-the-year-class strength, which is impacted by overwintering capacity. Overfishing generally is not a serious mortality factor except in the northern edges of the range where global growth rates are slow. Largemouth Bass do not do well in areas of high turbidity, sparse cover, low oxygen concentrations, or high total alkalinity (900 mg/l or ppm)[2].

North American lakeshores have been and continue to be highly developed and altered by human activity. Coinciding with the increase in development has been a change in the types of alterations to shorelines. Small often seasonal cabins are being replaced by larger permanent homes. These properties commonly have large highly managed and manicured yards and beach areas. The ultimate consequence of shoreline development is a decline in diversity of riparian and littoral habitat. Previous research found that the amount of coarse woody debris along developed shorelines in Wisconsin and Michigan was substantially less than along undeveloped shorelines. Dramatic declines in native vegetation, particularly of riparian brush and emergent aquatic vegetation, have also been linked to housing development in some ecosystems[15].

Breeding Habits

When Largemouth Bass spawn, males select a spawning site and may return to the same general area in successive years. Nests are saucer-shaped depressions in the bottom, typically at water depths of 1 m (3.3 ft) or less. Typically, nests are located near cover, often including stumps or logs, and males actively guard the area. Nests of adjacent males are generally more than 2 m (6.6 ft) apart. Most spawning occurs at dawn or dusk[5].

Age at sexual maturity for Largemouth Bass is dependent upon the rate of growth; they may mature when only one year old in fast-growing populations in southern waters such as Florida, Alabama, Kentucky, and California. They may mature at age two in New York and generally at age three or four in Wisconsin. The number of eggs produced by Largemouth Bass ranges between 2,000 and 109,314 depending upon age and size of the female. Older and larger females produce more eggs than younger, smaller fish. Depending on water temperature, young hatch in three to seven days. Fry are 3 mm (0.12 in) long at hatching and about 7 mm (0.28 in) long at swim up in several north-central US waters. Fry remain in tight schools along the shore until they are about 33 mm (1.3 in) long, when they disperse from the nest site. Larger Largemouth Bass spawn earlier than smaller fish, so their larvae grow faster during a longer growing season and likely contribute more to recruitment[6].

Largemouth Bass generally spawn in water 0.3 to 1.2 m (1 to 4 ft) deep or often as deep as 2.1 to 2.4 m (7 to 8 ft) over sand, gravel, or rubble bottoms. Often spawning is initiated first in shallow bays, where the water warms earlier than in the main lake ecosystem. In Duck Lake, British Columbia, Largemouth Bass spawn from May 20 through June 24 in water averaging 0.7 m (28.9 in) deep, with depth ranges of 0.5 to 1.0 m (20 to 41 in) and when water temperatures were between 20 and 23°C (68 and 73°F). Males dig the nest, which consists of a depression that may be 0.61 m (2 ft) in diameter and 0.15 m (0.5) ft deep. Males defend the nest against intruders until fry disperse. Males are especially vulnerable to angling during this period because they tend to strike at anything that comes near the nest[6].

Largemouth Bass in Lake Powell, Utah-Arizona, spawn from late April to mid-June. Spawning activity commences when water temperatures reach 14.4 to 15°C (58 to 59°F). Nest construction may begin earlier at 11.7 to 13.9°C (53 to 57°F). Nests are often built near cover. Eggs are adhesive and attach to the substrate; eggs hatch within five to six days at water temperatures of 16.7 to 18.3°C (62 to 65°F)[2].

In some ecosystems, emergent and floating-leaf vegetation in littoral areas adjacent to developed shorelines was less abundant than along undeveloped shorelines (66 percent reduction in vegetative cover when development was present). Other human-induced changes in aquatic ecosystems include declines in the composition and density of macrophytes and declines in the size and uniformity of substrates. Freshwater fish depend on littoral zone habitat for at least a portion of their life history. Developed shoreline was the only attribute that had a strongly significant effect, with Largemouth Bass avoiding developed shorelines for nesting[15].

Habits and Habitat

Largemouth Bass inhabit the upper levels of warm water of small and shallow lake ecosystems, shallow bays of larger lakes, and, more rarely, larger slow rivers. Largemouth Bass are rarely caught at depths over 6.1 m (20 ft). They are almost universally found in association with soft bottoms, stumps, and extensive growth of a variety of emergent and submerged vegetation, particularly water lilies, cattails, and various species of pond weed. They are very rarely found in rocky situations that are characteristic of Smallmouth Bass habitat. Habitat of Largemouth Bass often contains a variety of other sunfishes, Muskellunge, Northern Pike, Yellow Perch, Brown Bullhead, and a variety of minnows. Habitat usage by Smallmouth Bass and Largemouth Bass seldom overlaps, even when the two species occur in the same aquatic ecosystem. Movement by Largemouth Bass is not extensive, typically amounting to less than 8 km (5 mi), and summer territories are small[3].

Largemouth Bass can tolerate higher water temperatures than Smallmouth Bass. Temperature preference is in the range of 26.6 to 27.7°C (79.8 to 81.9°F), but experimental determinations indicate temperature preferences are from 30 to 32°C (86 to 89.6°F), with some dependency on acclimation temperature. Higher temperatures later in the summer result in Largemouth Bass inactivity while individuals rest in the shade of aquatic or shore vegetation. Largemouth Bass have a low tolerance for reduced oxygen concentrations and tend to avoid areas with oxygen levels less than 1.5 mg/l (ppm). Largemouth Bass move to the bottom of their ecosystem to winter and are more active than Smallmouth Bass; therefore, they can be taken through the ice in several northern US states[3].

Although differences in habitat use of adult Largemouth Bass and Smallmouth Bass have been well documented, less is known about habitat uses of smaller juveniles, particularly age-0 fish. Because both species feed predominantly on invertebrates and are highly vulnerable to predation when small, they may share preferences for habitats with available prey and suitable refuge. Observations of age-0 Largemouth Bass and Smallmouth Bass indicated a high degree of habitat partitioning in natural systems. On average, more than 99 percent of Largemouth Bass in three study lakes were observed in vegetated habitats, and none was observed in areas of cobble substrate. Among Smallmouth Bass, 94 percent were observed in cobble habitats, and only 3 percent were observed in vegetated habitats. Bass showed clear differences in habitat use. These patterns were observed in late June and correspond closely with preferred spawning habitats of both species. Largemouth Bass and Smallmouth Bass appear to partition habitats at all life stages. Results of laboratory experiments further suggest that habitat partitioning among juveniles reflects behavioral differences between the two species[13].

With respect to both Largemouth Bass and Smallmouth Bass, there are some concerns that when these species are released following tournament capture, behavioral disruptions occur. Only 14 percent of tournament-caught Largemouth Bass and 32 percent of Smallmouth Bass returned to their sites of original capture. Fifty-one percent of Largemouth Bass and 26 percent of Smallmouth Bass dispersed less than 1.6 km (1 mi) from their release sites. On average, Smallmouth Bass dispersed distances of 7.3 km (4.5 mi) from their release sites, while Largemouth Bass dispersed shorter average distances of 3.5 km (2.2 mi). There was no difference in dispersal distances for fish captured and released in rivers versus those released in lakes and reservoirs. Twenty-two percent of Largemouth Bass and 15 percent of Smallmouth Bass caught and released in fishing tournaments were subsequently recaptured by anglers. Previous studies recommended modifications to tournament release practices, but little has changed[14].

Management and Conservation Status

Largemouth Bass are considered common, wide spread, and abundant rangewide (G5). A state conservation ranking is not applicable (SNA) since Largemouth Bass are an introduced species. They are classified as a game fish in Idaho[21]. Largemouth Bass and Smallmouth Bass were some of the very first nonnative species introduced into Idaho, and they now support many popular fisheries. Bass are prolific enough to produce adequate numbers of young fish without stocking. However, the growing season for basses in Idaho is generally short due to the high altitude and northern latitude. Research studies indicate that bass growth is regulated primarily by water temperature and not food availability, so bass fisheries should focus on regulations that allow basses to live longer. Largemouth Bass are generally most successful in smaller ponds and lakes that get warm, where vegetation is present, and where there is an abundant forage base of fish. However, 12,748 ha (31,500 ac) Coeur d'Alene Lake and the eight connecting "lateral lakes" support a very popular fishery for Largemouth Bass. Growth rate of Largemouth Bass in Idaho is limited primarily by water temperature and is generally much slower than areas of the country where basses are native. Due to their slow growth, Largemouth Bass are susceptible to overharvest. Despite slow growth rates and low-productivity water in many areas of the state, Idaho anglers enjoy good bass fishing from a combination of restrictive regulations and voluntary support for catch-and-release fishing.

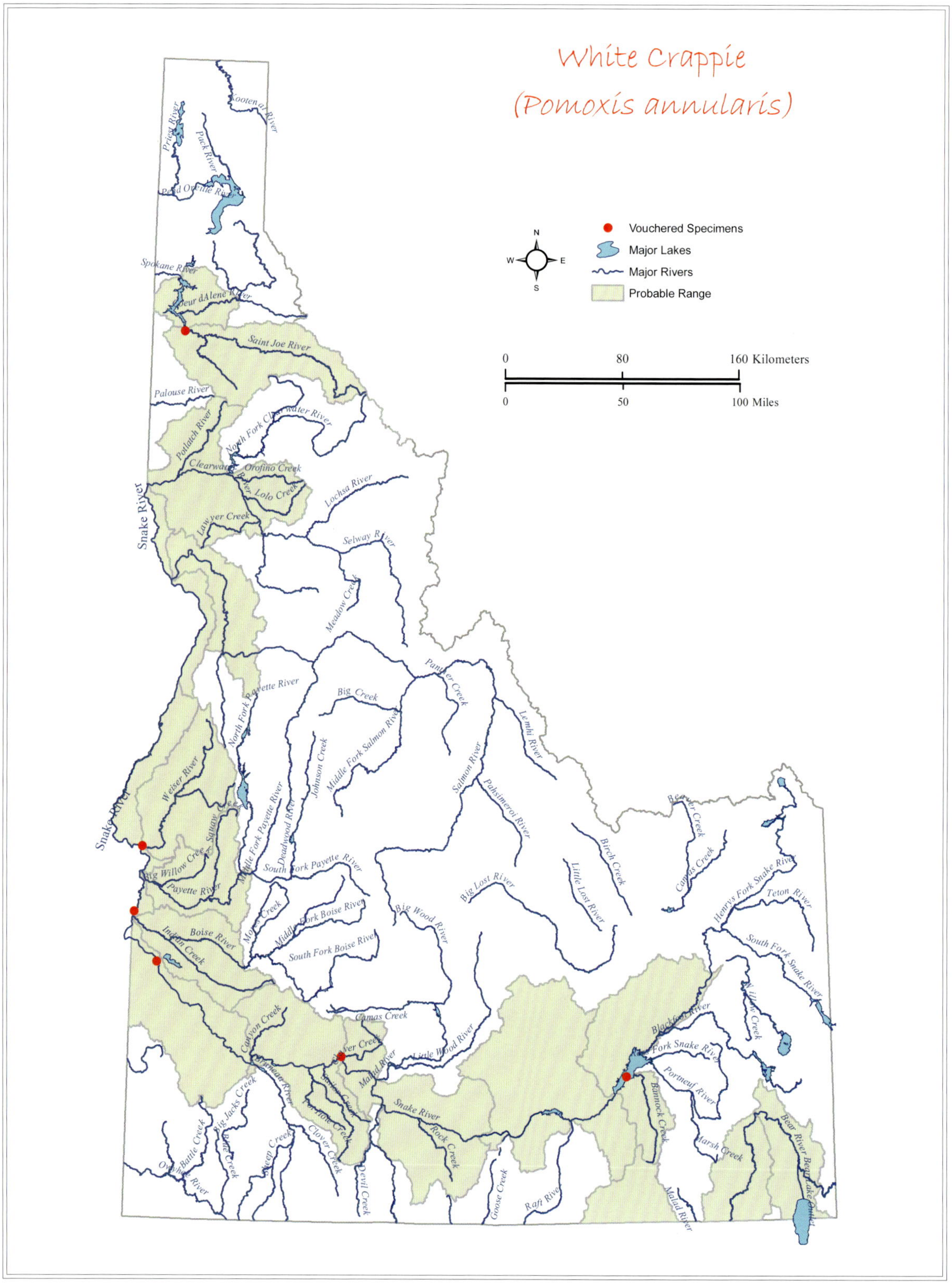
White Crappie
(Pomoxis annularis)
N
W
E
S
Vouchered Specimens
Major Lakes
Major Rivers
Probable Range
0
80
160 Kilometers
0
50
100 Miles
Pack River
Spokane River
Saint Joe River
Palouse River
Potlatch River
Orofino Creek
Lolo Creek
Lochsa River
Snake River
Selway River
Meadow Creek
Big Creek
Johnson Creek
Middle Fork Salmon River
Salmon River
Lemhi River
Pahsimeroi River
Birch Creek
Little Lost River
Big Lost River
Big Wood River
Teton River
South Fork Snake River
Weiser River
Payette River
Boise River
South Fork Boise River
Canyon Creek
Camas Creek
Snake River
Rock Creek
Clover Creek
Devil Creek
Goose Creek
Portneuf River
Bannock Creek
Marsh Creek
Malad River
Bear River

White Crappie

Pomoxis annularis

Rafinesque, 1818

Etymology

Pomoxis = sharp opercle, alluding to the fact that the opercle ends in a point rather than a flap; *annularis* = having rings, in reference to the vague vertical bars on the body.

Importance

White Crappie are important as game fish over most of their native range. They are caught in greater numbers than Largemouth Bass and contribute greater weight to the creel than do the smaller sunfishes. Their flesh is white, flaky, and quite palatable[2].

Black Crappie and White Crappie support important recreational fisheries in the United States, with more than 8 million crappie anglers encompassing 28 percent of total freshwater anglers (excluding the Great Lakes). Popular fisheries occur in a variety of water bodies, including flood-control or water-storage reservoirs, hydroelectric reservoirs with short retention times, and natural lakes[16].

Description

White Crappie have a body with a dorsal surface that is dark green, olive to brownish, with a blue-green or silver overcast; upper sides are lighter and more iridescent green, shading to silver. There are five to 10 vague vertical bands of black, sometimes extending down to the anal fin, but more obvious on the lower back and upper sides. The ventral surface is silvery to milky white; opercular bones may be iridescent pink or green. The eye is yellow to green. Caudal and anal fins have black vermiculations or rows of distinct black spots; pelvic fins are opaque white, and pectoral fins are transparent amber to yellow; there is no black on the paired fins[3].

White Crappie are a moderately large, deep-bodied sunfish that are more elongate than Black Crappie and have more extreme lateral compression. The mouth is terminal and somewhat oblique, and the rather large lower jaw is longer than the upper jaw. There are fine teeth on both jaws and palatines; the lower pharyngeal pads are very long and narrow with fine brush-like teeth. Gill rakers are very numerous, long, and then usually 22 to 24 on the lower limb and six to eight on the upper limb. Branchiostegal rays are seven. There are two dorsal fins, but they are so joined and the rays so graduated in length as to appear as a single dorsal fin. The base of the first dorsal fin is shorter than the base of the second dorsal fin and has six or rarely seven spines of graduated length. The second dorsal fin is longer, higher, and soft rayed, with 13 to 15, but typically 14 or 15, rays and a rounded edge. The caudal fin is moderately long and shallowly forked with rounded tips; the anal fin has a long base with six or seven graduated spines, followed by 16 to 18 soft rays but usually six spines and 17 rays with a square to rounded edge. Pelvic fins are thoracic, and the origin is under the pectoral fins, far in advance of the origin of the dorsal fin; the edge is square to round; there is one spine and five rays. Pectoral fins are high and not long or overly broad but rounded, typically with 13 rays. Scales are mostly ctenoid and rather large, although markedly smaller and crowded on the anterior dorsal surface. The lateral line is complete, high, and shallowly arched over the pectoral fin. There are 34 to 44 lateral line scales. The peritoneum is silvery, and the intestine is well differentiated, usually with 10 thin but well-marked caeca. Vertebrae number 30 to 32. No nuptial tubercles are present, but dark pigment and iridescent colors intensify on the males during breeding[3].

Range and Distributional Status

This species is not native to Idaho. Native range of the White Crappie is from central Minnesota east to eastern New York, south to Alabama, west to central Texas and northeastern Mexico, and north to eastern South Dakota. In the early 1890s, various warm-water fish species, potentially including White Crappie, were stocked into the Willamette River near Salem, Oregon, by the US Fish Commission. This species occurs in reservoirs on the Columbia and Snake Rivers. The most abundant populations occur in the reservoirs and backwater sloughs of the mid–Columbia River, the Lower Snake River, and other portions of the Columbia Basin[6] near American Falls Reservoir. In Idaho White Crappie have been documented in the Saint Joe River, the Snake River near the confluence of the Boise and Weiser Rivers, and in the upper Snake River.

Size and Longevity

White Crappie do not attain old age; most do not live to be more than six years of age, although some have been reported to live to 10 years. A 76 mm (3.0 in) White Crappie will average about (0.005 kg) (0.18 oz), a 280 mm (11.0 in) fish will weigh nearly 0.5 kg (1 lb, 2 oz), and a 330 mm (13.0 in) fish will average about 0.7 kg (1 lb, 9 oz). Lengths of 330 mm (13.0 in) may be obtained by White Crappie in some ecosystems by the third year of life, but in other waters individuals may not reach this size by the 10th year[6].

In northern waters, White Crappie reach lengths each year for ages one to nine as follows, in mm (in): 66 (2.6), 137 (5.4), 193 (7.6), 239 (9.4), 269 (10.6), 287 (11.3), 312 (12.3), 318 (12.5), and 363 (14.3). White Crappie may grow at about the same rate in western ecosystems but live to only seven years of age[2].

White Crappie in lower Snake River reservoir ecosystems have been reported with the following total length for fishes ages one to six, in mm (in): 68.6 (2.7), 150 (5.9), 218 (8.6), 244 (9.6), 259 (10.2), and 269 (10.6). Sexes grow at about the same rate[2].

The Idaho record White Crappie is a 1.7 kg (3.8 lbs), 445 mm (17.5 in) catch from Crane Creek Reservoir in 2012[10]. The all-tackle world record for White Crappie is a 2.35 kg (5 lb, 2.9 oz) catch from Enid Dam, Mississippi, in 1957[9].

Food and Feeding

Feeding activity of White Crappie is greatest from May through October. Young White Crappie feed primarily during daylight, mostly during the early morning through early afternoon. Subadult and adult White Crappie feed most readily at night during the summer but at dawn and dusk (in twilight) in other seasons. Zooplankton such as *Daphnia* and *Cyclops* are the dominant food item in the diet of small White Crappie. As individuals grow larger, they convert to larger food organisms such as insects. Both subadult and adult White Crappie are primarily piscivorous. Fish species consumed by White Crappie include various species of shiners, darters, catfish, and other sunfishes. White Crappie populations in a particular ecosystem may become stunted if forage fishes are not available[6].

White Crappie are omnivorous and opportunistic in their feeding behavior. White Crappie feed where forage is most abundant, in shallow water where insects are emerging, or in deeper waters where forage fish are available. White Crappie in lower Snake River reservoirs eat cladocerans, fishes, and aquatic insects. Cladocerans make up of 43.2 percent of the total and are eaten by 83 percent of White Crappie. Fishes represent 46.2 percent of total food volume, and 13 percent of identifiable species are suckers. Young White Crappie feed on rotifers, copepods, and other zooplankton. Feeding mechanisms of White Crappie are unusual in that they have fine, long gill rakers that are suitable for straining small zooplankton from the water, but this feature is combined with a large protrusible mouth capable of ingesting large prey, including various species of fishes[2].

Limiting Factors

White Crappie are parasitized by protozoans, trematodes, cestodes, and leeches over much of their range. White Crappie are also occasionally attacked by the virus lymphocystis. They may hybridize with Black Crappie[2]. White Crappie may be preyed upon by a variety of avian species and other fishes in some ecosystems.

Breeding Habits

White Crappie typically reach sexual maturity at age two to three and at lengths of 178 to 203 mm (7.0 to 8.0 in). Upon reaching maturity, males become darker in color on the sides of the head, lower jaw, and breast. White Crappie spawning occurs in the spring, at water temperatures around 18 to 20°C (64 to 68°F). White Crappie spawn near objects such as brush piles, stumps, and rocky outcrops; they may deposit their eggs on or near dead or living vegetation. White Crappie prepare nests that are typically 13 to 18 cm (5 to 7 in) in diameter; males guard the nest until fry leave the nest site. White Crappie males in South Dakota ponds defended territories of about 1 m^2 (11 ft^2). White Crappie females contain from 970 eggs for a 147 mm (5.8 in) fish to 213,000 eggs for a 330 mm (13.0 in) fish. White Crappie eggs hatch in about two days (roughly 42 hours) at temperatures of 23°C (73°F). Larvae are about 1.2 to 2.0 mm (0.05 to 0.08 in) long at hatching, and swim-up fry are about 4.5 mm (0.18 in)[6] long.

White Crappie spawning takes place for approximately one month in late spring to early summer at water temperatures of approximately 13.9 to 22.8°C (57 to 73°F), but this species is most active reproductively at temperatures of 16.1 to 20°C (61 to 68°F). Males, which are somewhat territorial, arrive at the spawning grounds first and clean out nests. Nests are not well defined but are approximately 13 to 31 cm (5 to 12 in) in diameter and have only a slight depression; they are constructed in 20 to 152 cm (8 in to 5 ft) of water. There is some numerical disparity between the sexes on the spawning sites, with females being more abundant. White Crappie generally nest in colonies of 35 to 50, with nests 0.61 to 1.21 m (2 to 4 ft) apart. Spawning generally occurs in the mornings, with little of the active, circular swimming characteristic of some other sunfishes. Spawning acts last from two to five seconds, with the same pair spawning as many as 50 times, typically with one to 50 minutes between spawning acts. Only a small portion of eggs present in a female are released at one time, and the same females may spawn in the nests of several males. Egg numbers per female are highly variable: a 152 mm (6.0 in) female had 1,908 eggs, while one 330 mm (13.0 in) female had 325,677 eggs. White Crappie eggs are 0.7 mm (0.03 in) in diameter and are colorless, adhesive, and demersal, adhering to the substrate and plants. Males guard the nest and fan the eggs. Eggs hatch in 17.5 to 24 hours at temperatures of 21.1 to 23.3°C (70 to 74°F). Young, which are transparent, remain on the nest only a short time and may leave as early as four days after hatching. Breeding colors of males may be lost as early as June[2].

Habits and Habitat

White Crappie are generally most abundant in warm, somewhat turbid lakes, reservoirs, and stream backwaters. They may also occur in ecosystems of rather silted streams, lakes, and ponds and in muddy, slow-moving rivers. White Crappie have a greater tolerance than Black Crappie for high turbidity, alkaline waters, current, high temperatures, and lack of aquatic vegetation or other cover. White Crappie tend to school loosely and are often only locally distributed. Individuals in some ecosystems may move considerable distances, but most are sedentary. Optimum temperatures for White Crappie appear to be 26.7 to 28.9°C (80 to 84°F). During daylight hours, White Crappie tend to congregate near cover in quiet water 1.9 to 4.0 m (6 to 13 ft) deep. White Crappie may move out into open water to feed during evening and early morning, being more nocturnal than diurnal[2].

Management and Conservation Status

White Crappie are considered common, wide spread, and abundant rangewide (G5). A state conservation ranking is not applicable (SNA) since White Crappie are an introduced species. They are classified as a game fish in Idaho[21]. Crappie are one of the most popular panfish species in Idaho due to excellent table qualities and high-yield populations. Unfortunately, crappie are probably the most difficult warm-water species for IDFG to manage successfully for anglers. In smaller water bodies in southern Idaho, crappie tend to stunt and do not reach a size preferred by anglers. Better population structure is generally found in larger, more productive lakes and reservoirs where crappie can provide tremendous harvest opportunity. Populations can fluctuate greatly from year to year, depending on the survival rate of young crappie. In northern Idaho, a lack of large fish may be caused by the short growing season and inconsistent recruitment. Crappie are usually most vulnerable to anglers when concentrated near shoreline structures during the spring spawning season. Despite this vulnerability, angler exploitation does not generally exceed 30 percent of the adult population, even in the most intense fisheries. During other times of the year, they suspend in open water areas, making them more difficult to catch.

Hayden Lake in north Idaho has consistently low crappie recruitment and slow growth and is the only quality crappie fishery in the state being managed with a 254 mm (10-in) minimum-size limit and a six-fish limit. In most other crappie fisheries, harvest does not play a large role in structuring populations, so there is no statewide bag limit for crappie. At Brownlee Reservoir, for example, with no bag limit, anglers typically harvest just 25–30 percent of crappie larger than 203 mm (8 in) in a given year. This is a very sustainable harvest rate, and restricting harvest would not result in noticeably better size or catch rates in this fishery.

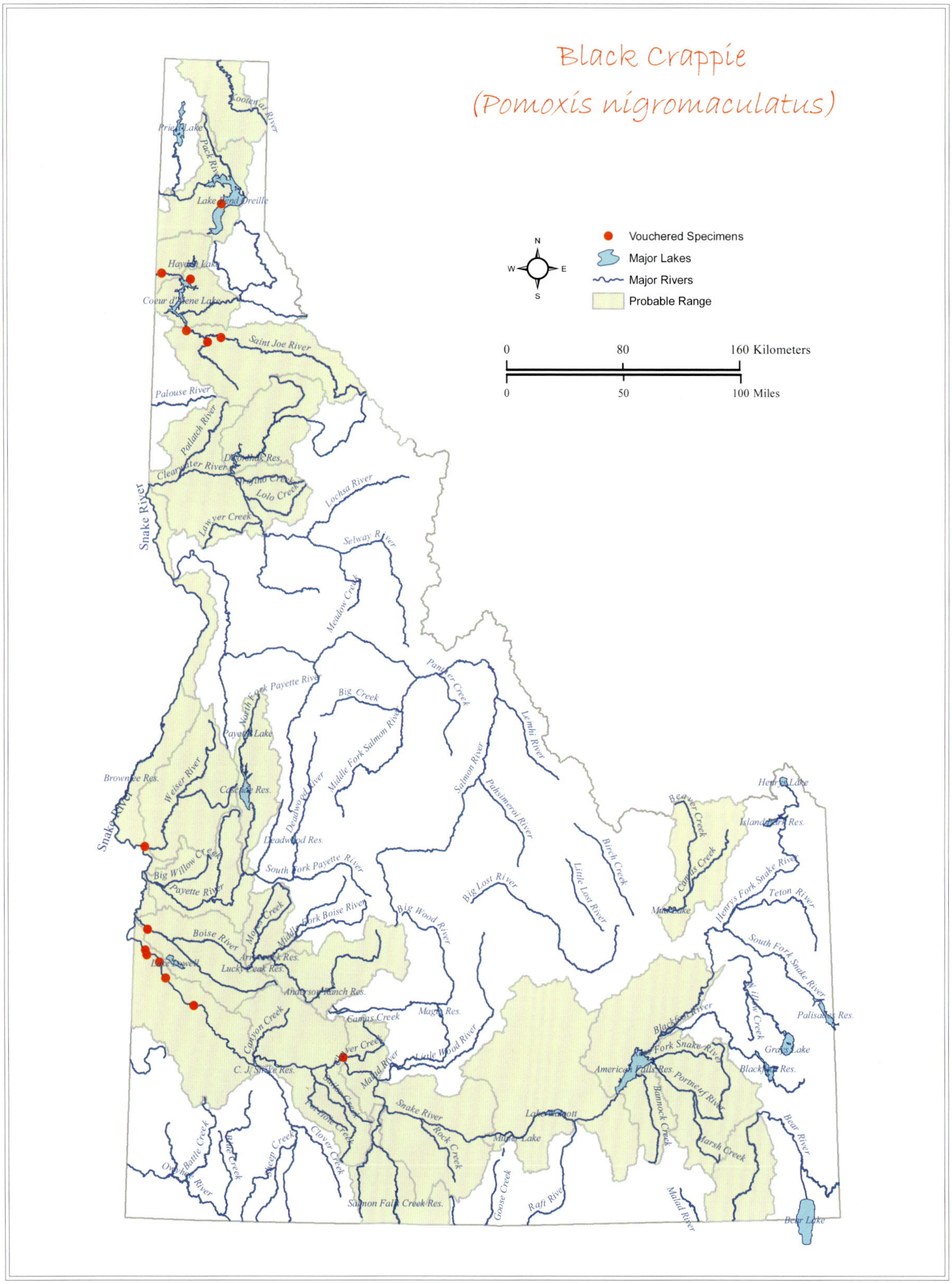
Black Crappie
(Pomoxis nigromaculatus)
Vouchered Specimens
Major Lakes
Major Rivers
Probable Range
N
S
E
W
0
80
160 Kilometers
0
50
100 Miles
Saint Joe River
Palouse River
Potlatch River
Lochsa River
Selway River
Lawyer Creek
Lolo Creek
Snake River
Big Creek
Lemhi River
Salmon River
Pahsimeroi River
Birch Creek
Little Lost River
Big Lost River
Big Wood River
Weiser River
Payette River
Boise River
Lucky Peak Res.
Anderson Ranch Res.
Canyon Creek
Camas Creek
Magic Res.
Teton River
South Fork Snake River
Palisades Res.
Grays Lake
Bannock Creek
Marsh Creek
Bear River
Raft River
Goose Creek
Malad River
Rock Creek
Clover Creek
Salmon Falls Creek Res.
C. J. Strike Res.

Black Crappie

Pomoxis nigromaculatus

(Lesueur, 1829)

Etymology

Pomoxis = sharp opercle—the opercle ends in a point rather than a flap; *nigromaculatus* = black spotted.

Importance

Black Crappie are popular with anglers because they are a medium-size, easy-to-catch panfish. Catch rates for Black Crappie may be as high as several fish per hour. Flesh is white, flaky, and very palatable[2].

Description

Black Crappie exhibit sexual dimorphism during breeding when the dark coloration of the males becomes a deep velvety black and color patterns are more vivid than in females. Color patterns on adults on the dorsal, anal, and caudal fins are striking—irregular pale (yellow to light-green) spots against a black background[5].

Black Crappie have a dorsal surface on the back and head that is olive, metallic green to gold, and brown, with an overcast of silver or blue; the sides are lighter, iridescent green to silvery, and the head and sides have an irregular mosaic of black blotches. There are no vertical rows of color, but there are vague horizontal groupings of body pigment that vary in intensity based on ecosystem type and fish size. The dorsal, caudal, and anal fins have striking black vermiculations, which form round or oblong yellow to pale-green spots in the centers. Pelvic fins are opaque, with some black on the tips of the membranes, and the pectoral fins are dusky and transparent. Black Crappie from clear, vegetated ecosystems have a darker contrasting pattern, while fish from turbid ecosystems have a bleached look. In breeding males the black coloration, particularly on the head, becomes darker and velvety[3].

Black Crappie are a large, deep-body sunfish, somewhat less elongate than the White Crappie, and they exhibit extreme lateral compression. Black Crappie have fine teeth on both jaws and the palatines, and the lower pharyngeal pads are very long, very narrow, and with fine brush-like teeth. Gill rakers are very numerous, long and slender, usually numbering 22 or 23 on the lower limb and five or six on the upper. Branchiostegal rays are seven. There are two dorsal fins, but they are so joined and the rays so graduated in length as to appear as a single fin. The first dorsal fin has a shorter base than the second and has seven or eight spines of graduated length, the last almost as high as the second dorsal. The second dorsal fin is longer, higher, and soft rayed, with 14 to 16 rays but often with eight spines and 14 rays. The caudal fin is long, broad, and shallowly forked, with rounded tips. The anal fin has six or seven graduated spines, followed by 16 to 18 soft rays but usually six spines and 17 rays; the edge is rounded to squarish. Pelvic fins are thoracic, with their origin under the pectoral fins, and they have a square edge with a rounded tip and one fine spine and five rays. The pectoral fins are moderately high on the body and moderately long, broad, and rounded, usually with 13 to 15 rays. Scales on Black Crappie are mostly ctenoid, rather large, but smaller and crowded on the anterior dorsal surface and the breast. The lateral line is complete, high, and shallowly arched over the pectoral fin, and there are 36 to 41 lateral line scales. The peritoneum is silvery, and the intestine is well differentiated; there are usually eight caeca, and vertebrae number 31 to 33. There are no nuptial tubercles, but dark pigment and iridescent colors intensify on males during breeding season[3].

Black Crappie are deep bodied and less elongate than White Crappie. Black Crappie have a long head, almost one-third their body length, with a narrow depression over the eyes. The back is rounded, and the snout is upturned, producing an S-shaped profile. Black Crappie have a short snout that is deeper than that of White Crappie, and the terminal mouth is large and oblique, the lower jaw extending beyond the upper jaw. The maxillary reaches to the posterior edge of the pupil[2].

Range and Distributional Status

This species is not native to Idaho. Native distribution of the Black Crappie is from eastern North Dakota and southeastern Manitoba, Canada; east to southern Quebec and Vermont; southeast to North Carolina, although absent from the Atlantic coastal areas; south to Florida; west to eastern Texas; and north to eastern South Dakota. Black Crappie were successfully introduced to areas in the West throughout the late 1800s and into California in 1908. In Washington Black Crappie are found in the reservoirs on the Columbia and Snake Rivers[6].

In Idaho Black Crappie are found along the western border of the state and particularly in the Clark Fork, Spokane, Saint Joe, and Coeur d'Alene Rivers as well as farther south in the Boise and Snake Rivers. They have been reported in the upper Snake River in Idaho and in the Bear River drainage.

Size and Longevity

Both sexes of Black Crappie grow at about the same rate. Individuals up to nine years of age have been reported in Lake Washington, Washington, but fishes in most ecosystems do not usually live beyond age five or six[6].

In northern waters, Black Crappie, ages one to nine, had total lengths, in mm (in), as follows: 64 (2.5), 140 (5.5), 188 (7.4), 216 (8.5), 249 (9.8), 264 (10.4), 295 (11.6), 302 (11.9), and 310 (12.2). In lower Snake River reservoir ecosystems, Black Crappie, ages one to four, had mean total lengths, in mm (in), as follows: 74 (2.9), 145 (5.7), 196 (7.7), and 218 (8.6). Growth rates may vary enormously in a given ecosystem based on population status and ecosystem productivity[2].

The all-tackle world record for Black Crappie was a 2.26 kg (5 lb, 0 oz) individual caught in a private lake in Missouri in 2006[9]. The Idaho record was a 1.7 kg (3 lb, 12 oz), 445 mm (17.5 in) individual caught from Brownlee Reservoir in 2003[10].

Food and Feeding

Young Black Crappie feed principally on zooplankton, but as they grow individuals tend to feed more on small larval aquatic insects; large fishes are generally piscivorous. Large fishes are an indicator that forage fishes are available in ecosystems where Black Crappie are prevalent. Prey fishes noted for Black Crappie include shad, Chinook Salmon, Goldfish, and Bluegill[6].

Black Crappie have the same feeding mechanisms as do White Crappie, being primarily mid-water feeders and eating zooplankton and small dipteran larvae when young. Larger Black Crappie eat fishes and aquatic insects, and they need at least a partially piscivorous diet to reach their maximum growth potential. Black Crappie may feed at temperatures as low as 6.7°C (44°F) and in some ecosystems may therefore feed year-round. Black Crappie will feed anytime, day or night, but feeding generally peaks at noon, midnight, and early morning. Cladocerans are reported to be the single most important food item for Black Crappie in the lower Snake River reservoirs, where 65.8 percent of the total food eaten was cladocerans and aquatic insects constituted 16 percent of total food volume[2].

Limiting Factors

Black Crappie may grow slower and develop stunted populations in high-turbidity ecosystems or where there is a lack of forage fishes. Growth rates are higher in ecosystems where there is an abundance of insects rather than copepods or other small invertebrates. Lacking adequate breeding sites, Black Crappie populations may experience reduced reproduction. Black Crappie may be parasitized by protozoans, trematodes, cestodes, and nematodes over most of their range. Black Crappie and White Crappie may hybridize[2].

North American lakeshores have been and continue to be highly developed and altered by human activity. Coinciding with the increase in development has been a change in the types of alterations to shorelines. Small often seasonal cabins are being replaced with larger permanent homes. These properties commonly have large highly managed and manicured yards and beach areas. The ultimate consequence of shoreline development is a decline in diversity of riparian and littoral habitat. Previous research found that the amount of coarse woody debris along developed shorelines in Wisconsin and Michigan was substantially less than along undeveloped shorelines. Dramatic declines in native vegetation, particularly of riparian brush and emergent aquatic vegetation, have also been linked to housing development in some ecosystems.

In some ecosystems, emergent and floating-leaf vegetation in littoral areas adjacent to developed shorelines was less abundant than along undeveloped shorelines (66 percent reduction in vegetative cover when development was present). Other human-induced changes in aquatic ecosystems include declines in the composition and density of macrophytes and declines in the size and uniformity of substrates. Freshwater fish depend on littoral-zone habitat for at least a portion of their life history.

Breeding Habits

Black Crappie generally reach sexual maturity at age two to three and lengths of 178 to 203 mm (7.0 to 8.0 in). In some ecosystems, faster-growing individuals may mature at an earlier age. Black Crappie may become stunted due to overcrowding and spawn at much smaller sizes. Black Crappie spawn in the spring during May and early June in most ecosystems at water temperatures of about 14 to 18°C (58 to 64°F). Males dig shallow depressions about 25 to 61 cm (10 in to 2 ft) in diameter in soft mud bottoms at depths usually less than 2.4 m (8 ft). Depending on size, females produce between 11,000 and 188,000 eggs, with egg number being dependent on female size. Black Crappie eggs are whitish and slightly less than 1 mm (0.04 in) in diameter and are demersal and adhesive. At about 18°C (65°F), eggs hatch in two to three days. Males very aggressively guard the nest until the fry leave the site[6].

Male Black Crappie build nests, as shallow depressions, on bottoms ranging from mud to gravel to debris. Generally, nests are built in water that is 0.15 to 6.1 m (0.5 to 20 ft) deep near or in beds of aquatic vegetation. Males may construct new nests close to neighboring ones. Females spawn with several males, releasing only a small portion of their eggs at any one time, laying a total of 10,000 to 200,000 eggs, with egg number being roughly related to size[2].

Black Crappie use littoral habitat to construct nests for spawning. Black Crappie use areas of firm substrate usually associated with some type of macrophyte.

In research of habitat associated with 478 Black Crappie nests, nest depths ranged from 0.25 to 1.5 m (0.8 to 5.0 ft). Black Crappie nest site selection indicated a high degree of habitat specificity. Results indicated that Black Crappie showed a preference to nest near shorelines with canopy, emergent vegetation, and understory, while avoiding submergent vegetation and developed shorelines. Examination of proportions of nests in each lake associated with this habitat factor indicated that avoidance of submerged vegetation was consistent across the lake ecosystems evaluated.

Habits and Habitat

Black Crappie generally inhabit clear waters of large streams, reservoirs, and some medium-size lake ecosystems. They prefer dense aquatic vegetation over bottoms of sand, mud, or organic debris. Black Crappie in general inhabit cooler waters than White Crappie. Black Crappie feed most actively in spring when they occur in weedy areas with water usually less than 3.1 m (10 ft) deep. During the summer, Black Crappie move into deeper water and are less available to anglers in some ecosystems[6]. Generally, Black Crappie are found in shallow water during evening and at night in the spring and during the day and evening in the summer in some ecosystems.

Movement patterns range from 0 to 584 m/h (0 to 1,916 ft/h) and differ seasonally and within diel periods[5].

Black Crappie and Largemouth Bass use littoral habitat to construct nests for spawning. Black Crappie use areas of firm substrate usually associated with some type of macrophyte. Largemouth Bass, while more cosmopolitan in their nest-site selection, also have specific habitat needs. Shallow, protected sites in bays, among emergent vegetation or rocks, logs, and stumps, are the preferred Largemouth Bass spawning habitat. Any changes or alterations to those habitats could affect spawning success and recruitment.

Better population structure is generally found in larger, more productive lakes and reservoirs where crappie can provide tremendous harvest opportunity. Populations can fluctuate greatly from year to year, depending on the survival rate of young crappie. In northern Idaho, a lack of large fishes may be caused by the short growing season and inconsistent recruitment. These species eat primarily zooplankton when small, then become more opportunistic when they reach a large size. Crappie are usually most vulnerable to anglers when concentrated near shoreline structures during the spring spawning season. Despite this vulnerability, angler exploitation does not generally exceed 30 percent of the adult population, even in the most intense fisheries. During other times of the year, they hold in open water areas, making them more difficult to catch.

Management and Conservation Status

Black Crappie are considered common, wide spread, and abundant rangewide (G5). A state conservation ranking is not applicable (SNA) since Black Crappie are an introduced species. They are classified as a game fish in Idaho[21]. Crappie are one of the most popular panfish species in Idaho due to excellent table qualities and high-yield populations. Unfortunately, crappie are probably the most difficult warm-water species for IDFG to manage successfully for anglers. In smaller water bodies in southern Idaho, crappie tend to stunt and do not reach a size preferred by anglers.

Hayden Lake in north Idaho has consistently low crappie recruitment and slow growth and is the only quality crappie fishery in the state being managed with a 254 mm (10.0 in) minimum-size limit and a six-fish limit. In most other crappie fisheries, harvest does not play a large role in structuring populations, so there is no statewide bag limit for crappie. At Brownlee Reservoir, for example, with no bag limit, anglers typically harvest just 25–30 percent of crappie larger than 203 mm (8 in) in a given year. This is a very sustainable harvest rate, and restricting harvest would not result in noticeably better size or catch rates in this fishery.

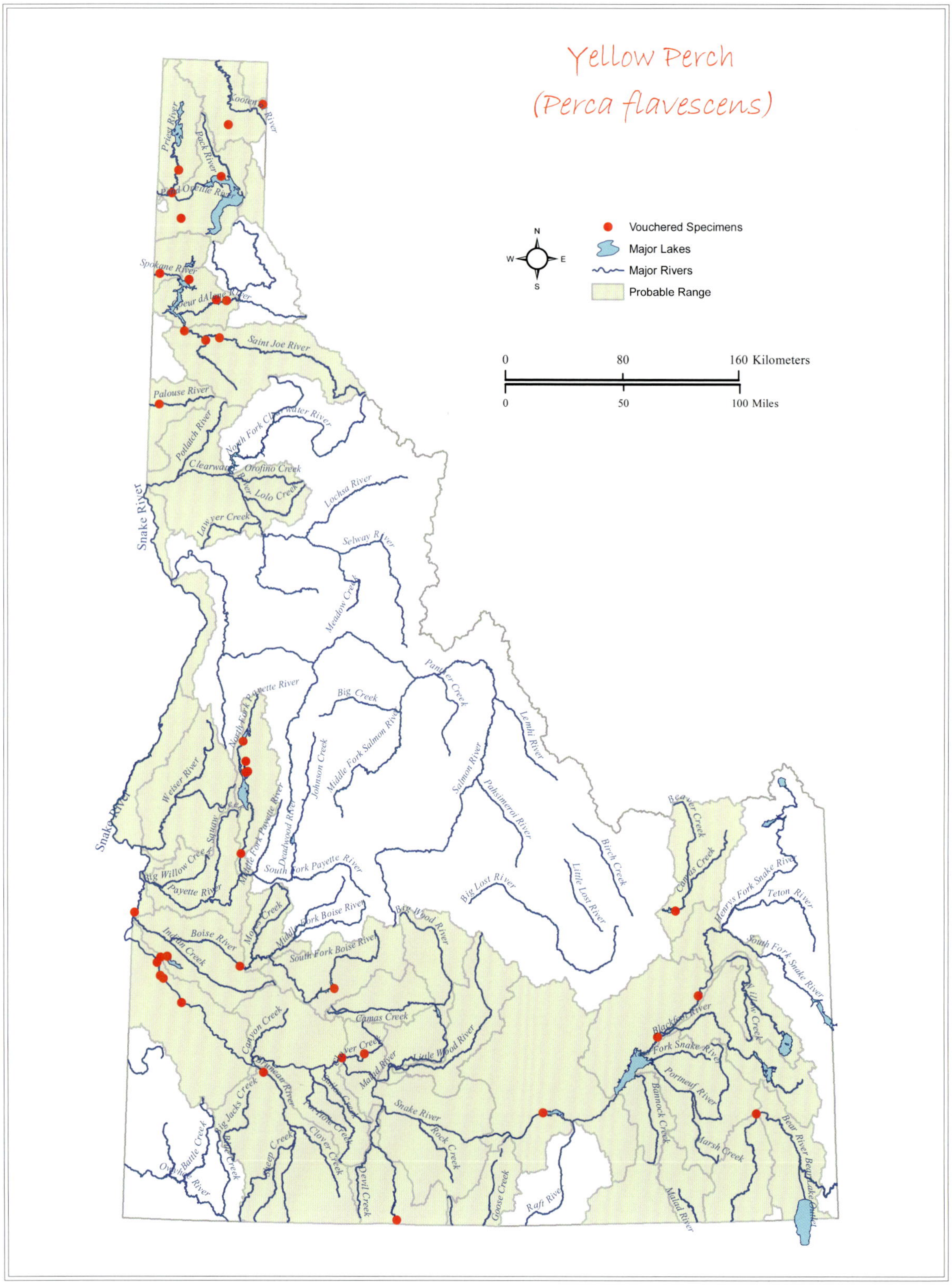
Yellow Perch
(Perca flavescens)
N
S
E
W
Vouchered Specimens
Major Lakes
Major Rivers
Probable Range
0
80
160 Kilometers
0
50
100 Miles
Priest River
Pack River
Spokane River
Saint Joe River
Palouse River
Potlatch River
Orofino Creek
Lolo Creek
Lochsa River
Lawyer Creek
Snake River
Selway River
Meadow Creek
Big Creek
Middle Fork Salmon River
Johnson Creek
Salmon River
Lemhi River
Pahsimeroi River
Birch Creek
Little Lost River
Big Lost River
Weiser River
Big Willow Creek
Payette River
Middle Fork Payette River
Deadwood River
South Fork Payette River
Boise River
Indian Creek
Mores Creek
South Fork Boise River
Big Wood River
Camas Creek
Little Wood River
Canyon Creek
Bruneau River
Big Jacks Creek
Clover Creek
Deep Creek
Devil Creek
Snake River
Rock Creek
Goose Creek
Raft River
Teton River
Henrys Fork Snake River
South Fork Snake River
Beaver Creek
Camas Creek
Blackfoot River
Portneuf River
Bannock Creek
Marsh Creek
Malad River
Bear River

Family Percidae

Perches occur in freshwater in the Northern Hemisphere. Members of this family have two dorsal fins that are separate or narrowly joined and two anal spines. Pelvic fins are thoracic, with one spine and five soft rays; branchiostegal rays are five to eight. Ten genera contain 201 species (187 in North America)[1].

Earliest fossils tentatively assigned to perches today are those of the Eocene (30 mya), known from the extensive Green River formations of the Utah-Wyoming region[4].

In Idaho there are three species in two genera:

Perca flavescens	(Mitchill, 1814)	Yellow Perch
Sander canadensis	(Griffith & Smith, 1834)	Sauger
Sander vitreus	(Mitchill, 1818)	Walleye

Yellow Perch

Perca flavescens

(Mitchill, 1814)

Etymology

Perca = dusky, and possibly the ancient common name of the Eurasian perch; *flavescens* = yellow.

Importance

Yellow Perch have long been of great importance to humans because of their vast distribution, their wide variety of habitat use, and the fact that they school and congregate near shore in the spring. These factors combined make them readily available to both commercial fishing and recreational anglers[3]. Yellow Perch feed actively and can be caught year-round. They take most lures rapidly and readily, providing a good sport fishery. Yellow Perch rate high in palatability[2].

Description

Adult female Yellow Perch demonstrate some aspects of sexual dimorphism and are, on average, larger than adult males. In the breeding season, females' pectoral and pelvic fins are less colorful than those of males[5].

Yellow Perch vary in color with both size and ecosystem type. The dorsal surfaces of the back and head are generally bright green to olive to golden brown. The sides to below the pectoral fins are yellow-green to yellow, and the color of the back extends down the sides in about seven tapering bars. Dorsal and caudal fins are yellow to green, and the edge of the first dorsal fin is often black; there is black on the membranes between the first and second spines and the last four or five membranes in adults. Pectoral fins are amber and transparent, while the anal and pelvic fins are opaque and yellow to silvery white. Colors in spawning males are more intense, shading to bronze-green or bright yellow with darker bars. Females are less highly colored, and the young are transparent and then silvery or dull pale green[3].

Yellow Perch have an elongate body that is oval rather than tubular or subcylindrical and is laterally compressed. The snout is moderately long and blunt, not extending beyond the lower jaw. The premaxillaries are protractile, and the mouth is moderately large, terminal, and slightly oblique. Teeth occur in brush-like bands on the jaws, palatines, and vomer. Gill rakers are fine and comb-like and usually number 12 to 16 on the lower limb and four to eight on the upper limb. Branchiostegal rays are 6, 7; 7, 7; 7, 8; or8, 8; but typically 7, 7. There are two dorsal fins that are obviously separated. The first dorsal fin is spiny, high, and rounded, with strong spines, usually numbering 13 to 15. The second dorsal fin is smaller but of about the same height, with one or two spines and 12 to 15 rays. The caudal peduncle is long and narrow, and the caudal fin is shallowly forked with rounded tips. The anal fin has two spines and six to eight rays. Pelvic fins are thoracic and of moderate length, a little longer than the pectoral fins and square, with one spine and five rays. Pectoral fins are broad and rounded, with 13 to 15 rays. Scales are ctenoid, and the cheeks are scaled; the opercles are mostly naked, and the breast and belly are scaled. The lateral line is complete and high, and scales number 51 to 61 but usually 53 to 57. The peritoneum is silvery, and the intestine is well differentiated; there are three short, thick pyloric caeca. Vertebrae number 38 to 41, and there are no nuptial tubercles[3].

Range and Distributional Status

This species is not native to Idaho. Native range of Yellow Perch is from north-central Saskatchewan east to Newfoundland in Canada, south to South Carolina, and northwesterly to western North Dakota. They were absent from the Appalachian Mountains from Pennsylvania south to South Carolina. Yellow Perch now occur in all reservoirs in the mid- and lower Columbia River and lower Snake River. As recently as 1999, Yellow Perch were considered one of the most abundant fishes in Lake Washington, Washington[6].

In Idaho Yellow Perch are fairly well distributed. Yellow Perch occur in the Priest River, Coeur d'Alene River, Spokane River, and Saint Joe River in the north. They occur in the Boise River, the lower Snake River, and the Payette River on the western side of the state. They occur in the upper Snake River and the Bear River in the southeastern portion of the state.

Size and Longevity

Growth rates in Yellow Perch are variable with ecosystem type and habitat, size of the individual fish, and competition for food. Northern populations grow slower and live longer than do populations in southern ecosystems. Yearly size for Klamath River, California, Yellow Perch ages one to five, in mm (in), is as follows: 89 (3.5), 150 (5.9), 201 (7.9), 231 (9.1), and 269 (10.6). In Green Bay, Wisconsin, yearly size for Yellow Perch ages one to seven is, in mm (in), as follows: 71 (2.8), 114 (4.5), 155 (6.1), 188 (7.4), 221 (8.7), 246

(9.7), and 274 (10.8). Yellow Perch rarely live more than nine years, and females grow faster than males and achieve larger sizes[2].

The all-tackle world record Yellow Perch is a 1.91 kg (4 lb, 3 oz) fish taken from Bordertown, New Jersey, in 1865[9]. Lake Cascade produced the Idaho record, 1.23 kg (2 lb, 11.4 oz), 394 mm (15.5 in), in 2014[10].

Food and Feeding

Young Yellow Perch feed in shallow areas on zooplankton, particularly copepods and cladocerans. As size increases, individuals begin to feed on immature insects, and large Yellow Perch feed on forage fishes when they are available. Switching to piscivory results in rapid growth in Yellow Perch. The diet of Yellow Perch in Box Canyon Reservoir on the Pend Oreille River contained 22 percent cladocerans (primarily *Daphnia*), 20 percent chironomids, 11 percent caddisflies, 10 percent mayflies, 7 percent dragonflies and damselflies, 7 percent amphipods, 6 percent snails and limpets, 5 percent copepods, 4 percent water mites, 4 percent seed shrimp, 2 percent flies, excluding chironomids, 1 percent aquatic earthworms, and 1 percent fish[6].

Yellow Perch use their comblike gill rakers to effectively strain small live animals out of the water. They feed in open water both near the surface and off the bottom. Adults may browse among patches of aquatic vegetation, eating a variety of crayfish, dragonfly larvae, and snails. Most adult Yellow Perch feeding occurs during the day, with peak activity in the morning and at dusk, when they move closer inshore and travel parallel to shore. Larger Yellow Perch prefer piscivory but readily shift to invertebrates in some ecosystems[2].

Limiting Factors

Yellow Perch are afflicted by a wide variety of parasites, including protozoans, trematodes, cestodes, nematodes, acanthocephalans, leeches, mollusks, and crustaceans. Yellow Perch suffer as well from a number of fish diseases and such pathological conditions as tumors. Yellow Perch are preyed upon by almost all other warm- to cool-water predatory fishes, such as basses, sunfishes, crappies, Walleye, Sauger, other Yellow Perch, Northern Pike, Muskellunge, and to some extent, where ecosystems have overlapping populations, Lake Trout. A variety of waterbirds, such as gulls, mergansers, loons, and kingfishers, eat Yellow Perch young and adults. Yellow Perch may compete for food with Brook Trout, a variety of ciscoes, Lake Whitefish, basses, crappies, and Bluegill. Their high reproductive potential, voracious appetite, and effectiveness in feeding can, in some ecosystems, lead to serious competition with more valued game species. Additionally, the same characteristics are probably the cause of many populations of stunted Yellow Perch in a variety of ecosystems[3].

Yellow Perch spawning and therefore reproduction may be reduced or eliminated if there is a sharp and protracted temperature drop just before spawning is initiated. Yellow Perch also decline in areas where aquatic vegetation has been greatly reduced or is gone[2].

Breeding Habits

Yellow Perch generally spawn in April or May at water temperatures of 7 to 11°C (45 to 52°F). Males and females school separately during this time, and males precede females to spawning areas; females remain in the spawning area only for a short time when ready to spawn. Yellow Perch spawning occurs on vegetation or submerged brush and other objects over a variety of bottom types, including sand, gravel, and rubble. Yellow Perch deposit eggs in a ribbonlike gelatinous mass that is about 5 to 8 cm (2 to 3 in) wide and 61 to 215 cm (2 to 7 ft) long. The number of eggs produced is directly related to the size and age of the female. Female Yellow Perch from Lake Washington contained between 18,354 and 139,565 eggs for fish ranging from 168 to 375 mm (6.6 and 14.8 in) TL. Males reach sexual maturity at age one or two, and females reach sexual maturity between ages two and three[6].

When ready to spawn, Yellow Perch females make a series of rapid turns or other quick movements to attract males. Females then swim rapidly, releasing a long string of eggs encased in a gelatinous sheath. While the total number of eggs deposited may be as high as 76,000 or more, it is generally between 4,000 and 40,000 and related to the size and age of the female. Male Yellow Perch follow females and release sperm, which envelop the eggs. The sticky bands of eggs are draped across vegetation or over the bottom and hatch in 10 to 20 days at temperatures of 10 to 16.7°C (50 to 62°F). The larvae, which are 0.5 mm (0.2 in) long, begin feeding on zooplankton in shallow water soon after hatching, although some reserve food is available in the yolk sac[2].

Spawning in Yellow Perch is initiated at water temperatures above 7°C (45°F). Spawning is preceded by the movement of adults into littoral areas, and gamete release is usually associated with vegetation or submerged woody debris. However, Yellow Perch can spawn over almost any substrate. Spawning occurs at night or at dawn[5].

Habits and Habitat

Yellow Perch are highly adaptable, able to utilize a wide variety of warm to cooler ecosystem habitats, ranging from large lakes to ponds to quiet rivers. Yellow Perch are generally most abundant in the open water of lakes with moderate vegetation, clear water, and bottoms of mud to sand and gravel. Increases in turbidity will result in decreases in numbers of Yellow Perch in a given ecosystem. Yellow Perch are apparently more tolerant of low oxygen levels than are most sunfishes. Yellow Perch are usually found at temperatures of 19 to 21°C (66.2 to 69.8°F). Seasonal vertical movements suggest Yellow Perch move to follow the 20°C (68°F) isotherm, and upper lethal temperature has been experimentally determined at 26.5°C (79.7°F). Yellow Perch are usually considered shallow-water fishes, and anglers usually take them above 9.2 m (30 ft). However, Yellow Perch have been taken as deep as 45.7 m (150 ft) in some ecosystems. Both adults and young are gregarious, often moving about in loose aggregations of 50 to 200 individuals that are segregated by size, with young in shallower water and nearer shore than the adults. They are often found in mixed schools that include minnows. Yellow Perch tend to be inactive at night and rest on the bottom. Migratory movements in the spring are common, as are movements inshore and out or up and down over a period of 24 hours as well as seasonally into and out of deeper water in response to temperature changes and likely due to food distribution. Yellow Perch are generally active all winter under the ice in shallow water[3], making them a primary target for ice anglers.

Management and Conservation Status

Yellow Perch are considered common, wide spread, and abundant rangewide (G5). A state conservation ranking is not applicable (SNA) since Yellow Perch are an introduced species. They are classified as a game fish in Idaho[21]. Yellow Perch populations are more likely to stunt in cold, infertile ecosystems than in warm, eutrophic ecosystems. Large lakes are most likely to produce predators that prey on, and therefore produce, through age class reductions, desirable-size Yellow Perch. Yellow Perch are intensely competitive, and populations in some ecosystems tend to stunt, making them of questionable value as either a sportfish or a forage fish in some ecosystems. Stable habitats, especially at spawning time, are an important requirement for this species.

Yellow Perch can produce important sport fisheries in Idaho's larger reservoirs and lakes (e.g., Cascade Reservoir). Many ice fisheries in northern Idaho's lowland lakes are supported by Yellow Perch. Yellow Perch are notoriously difficult for IDFG to manage for consistent size structure and abundance. The species tends to have cyclic year-class strength, where the formation of strong year classes can dominate and suppress subsequent year classes.

Stable Yellow Perch populations and fisheries are associated with productive waters generally larger than 4,050 ha (10,000 ac) that have complex fish communities. Complex fish communities are thought to be necessary to maintain adequate levels of predation to prevent stunting and, at the same time, provide alternative food items for other predators. Yellow Perch are extremely fecund, producing up to 40,000 eggs per female, and can easily stunt because of overpopulation or sometimes because of suppressed food supplies caused by poor water quality. In a mixed fishery, young Yellow Perch are an important food source for other predators. When Yellow Perch become established in trout fisheries, trout growth can be severely impacted.

IDFG has renovated fisheries because of illegal introductions of Yellow Perch into ecosystems managed for trout. Once Yellow Perch are introduced, it is extremely difficult and expensive to eliminate them to allow a more desirable fish species to become established.

Lake Coeur d' Alene, near Harrison, ID. Photo Courtesy John Sigler.

Sauger

Sander canadensis
(Griffith & Smith, 1834)

Etymology

Sander = obscure, but perhaps related to one or more of the common names when this fish was referred to as *Stizostedion canadense; canadensis* = of Canada, for the type location of the species.

Prior to the release of the sixth edition of Common and Scientific Names of Fishes from the United States, Canada and Mexico in 2004[17], the American Fisheries Society names committee published the rationale for changing the generic name of this species from *Stizostedion* to *Sander*[18]. In short, the Committee found that *Sander* had priority over *Stizostedion* due to prior use. For the specific name, *canadense* was replaced with *canadensis* for gender agreement with the generic name.

Importance

Sauger are important sportfishes. In various ecosystems they may also be important commercially; average Sauger taken in the commercial catches weigh about 0.45 kg (1 lb, 0 oz). Sauger have firm, white flesh that is preferred by some to other species such as Walleye. Sauger are typically caught still fishing or drift fishing with dead or live minnows[3].

Description

Sauger coloration is variable with habitat and, to a lesser extent, with size. Background color of Sauger is most often sandy or dull brown or occasionally dull gray, with darker brown markings. Markings may be absent from Sauger that occur in ecosystems with clear water. The dorsal surface is brown, and the sides are paler; the back has three or four dark-brown saddles, some of which extend horizontally on the sides. The ventral surface is milky white, and the head has dark bars. The first dorsal fin has dark pigment on the edge of the membranes in rows of distinct black spots. The second dorsal has narrow dark bands of spots on the membranes, and the caudal fin has rows of dark-brown spots. Young Sauger may have a sandy-colored background and an inconspicuous dark-brown, almost continuous, saddle under the dorsal fin[3].

The body form of Sauger is elongate, almost cylindrical; there is little or no lateral compression. The caudal peduncle is of moderate length, shallow, and round. The preopercle is strongly serrate at least at the bottom, and there is a short spine at the upper edge of the tip of the opercle. The jaws are equal. There are teeth on the jaws, the head of the vomer, and the palatines. There are sharp, large canines on the vomer, palatines, and jaw. The pharyngeal teeth are large, sharp, and recurved. Gill rakers are moderately long, typically six to eight on the lower limb and three to five on the upper limb. Branchiostegal rays are six, eight; six, seven; seven, seven; or seven, eight. There are two dorsal fins that are obviously separated. The first dorsal fin is spiny, high, and long, with strong spines typically numbering 13 to 15. The second dorsal is as high or higher as the first, but the base is somewhat shorter, typically with one, or rarely two, fine spines and 16 to 21 (typically 18) rays. The second dorsal is square to slightly emarginated. Caudal fin is of moderate size and long but not overly broad, well forked, with tips that are rather pointed. The anal fin has two spines and 11 to 14 rays. Pelvic fins are thoracic and under the pectoral fins, with one spine and five rays. Pectoral fins are only moderately broad, with rounded tips and 12 to 14 rays. Sauger scales are strongly ctenoid, and the cheeks are mostly scaled; the opercle is mostly naked, and the breast and belly are typically scaled. The lateral line is complete, high, and slightly curved, with rather small scales. There are 82 to 100 (usually 88 to 95), including three or four pitted scales in the area where the lateral line extends onto the scaled base of the caudal fin. The peritoneum is white, the intestine is well differentiated, and pyloric caeca may number three to nine but typically four to six. Vertebrae number 43 to 45, and there are no nuptial tubercles[3,4].

Range and Distributional Status

This species was not native to Idaho. Sauger are native to freshwaters of North America. Sauger occur from the St. Lawrence–Champlain system in the South, west of the Appalachian Mountains to the Tennessee River in Alabama, southwest to northern Louisiana, northwest through eastern Oklahoma to central Montana and central Alberta, and east below James Bay to Québec[3,4].

Recent surveys suggest that Sauger populations are declining throughout much of their native range. Many Sauger populations in the Great Plains have declined in association with construction of reservoirs that have inundated long reaches of rivers and affected downstream habitat[19]. Relatively little is known about the habitat associations of Sauger in small high-elevation rivers in the upper Missouri River watershed[19].

In Idaho Sauger were reported from the Oneida Narrows Reservoir to Black Canyon in the lower Bear River drainage. Sauger are now believed to be extirpated from Idaho.

Size and Longevity

Sauger in Tennessee reservoirs are 84 to 106 mm (3.3 to 4.2 in) at age one, 126 to 156 mm (5.0 to 6.1 in) at age two, 150 to 174 mm (5.9 to 6.6 in) at age three, 170 to 180 mm (6.7 to 7.1 in) at age four, and 190 to 200 mm (7.5 to 7.9 in) at age five[4].

The all-tackle world record Sauger weighed 3.96 kg (8 lb, 12 oz) and was taken from Lake Sakakawea, North Dakota, in 1971[9].

Food and Feeding

Larvae are 5 mm (0.2 in) TL and soon begin feeding on cladocerans, copepods, and midge larvae: however, Sauger juveniles soon switch to a diet that is almost exclusively fishes. In some ecosystems, small Gizzard Shad and Threadfin Shad are primary prey, with Emerald Shiners and some mayfly nymphs important food in ecosystems where they are abundant[4].

Limiting Factors

In ecosystems where other species such as Northern Pike, Walleye, and Lake Whitefish occur, they may be the chief food competitors of Sauger. Effects of wind, waves, and temperature exert greater influence on hatching success and early growth and subsequent population control than competition and predation. Sauger of various sizes are preyed upon by other Sauger, Walleye, Northern Pike, and probably Yellow Perch. Parasites of Sauger include cestodes, nematodes, acanthocephalans, mollusks, crustaceans, and leeches. Sauger and Walleye may hybridize in nature[3].

Breeding Habits

In late winter, Sauger began congregating near their spawning areas. Spawning activity actually begins in late April, as water temperatures approach 6°C (43°F) in the northern portion of Sauger range. Sauger may cover considerable distances in spawning migrations, as much as 380 km (235 mi). In the southern part of their range, Sauger begin to spawn earlier (late March) at higher water temperatures of 11°C (52°F). Nest construction and parental care do not occur. In a typical event, several males attend each spawning female. Successful spawning requires firm substrates. Sauger spawning occurs at night and extends over a two-week period. Large females may produce more than 100,000 eggs annually, however, most produce 20,000 to 60,000 eggs. Mature eggs are 1.3 mm (0.05 in) in diameter. Sauger eggs are adhesive immediately after spawning, but shortly after spawning they lose their adhesiveness and may be widely dispersed by currents. Hatching occurs in about 21 days at temperatures of 8.3°C (47°F)[4].

Habits and Habitat

Sauger typically inhabit large, often turbid rivers but have been successful in many reservoirs. Sauger are more tolerant of turbidity than are Walleye, and in ecosystems where both species occur, one is typically far more abundant. Sauger adults disperse to deeper waters after spawning. Young-of-year Sauger are pelagic during the early part of their first summer[4].

Sauger occur naturally in large rivers and the lower portions of their tributaries, and adult Sauger have been described as preferring turbid river segments that have deep low-velocity pools and runs with sand or silt substrates and cover features that provide refuge from currents. Summer thermal preferences of Sauger are 20 to 28°C (68 to 82°F), and it is likely that these relatively warm temperature needs govern the northern and western boundaries of the species' range[19].

Previous research in Wyoming identified factors affecting occurrence of adult Sauger throughout the Wind River basin upstream from Boysen Reservoir, on the periphery of the species' natural distribution in Wyoming. There was a contiguous distribution of Sauger over 179 km (111 mi) of streams among four rivers in the watershed, and it was determined that upstream boundaries consisted of low summer water temperatures, high channel slopes, and water diversion dams that created barriers to upstream movement[19].

In Wyoming studies of Sauger distribution within the Wind River Range, maximum depth was the instream habitat feature that exerted the greatest influence on Sauger biomass. However, biomass also tended to be greater in pools than in runs. As the size of the dominant substrate declined, biomass of adult Sauger tended to increase, but small substrates were probably a secondary characteristic of low-velocity pools. Overall, findings reinforced previous research suggesting that adult Sauger prefer warm, deep pools and runs with low current velocities[19].

Management and Conservation Status

The IDFG no longer manages Oneida Narrows Reservoir or the Bear River for Sauger.

"Everybody is a genius. But if you judge a fish by its ability to climb a tree, it will live its whole life believing that it is stupid."

Albert Einstein

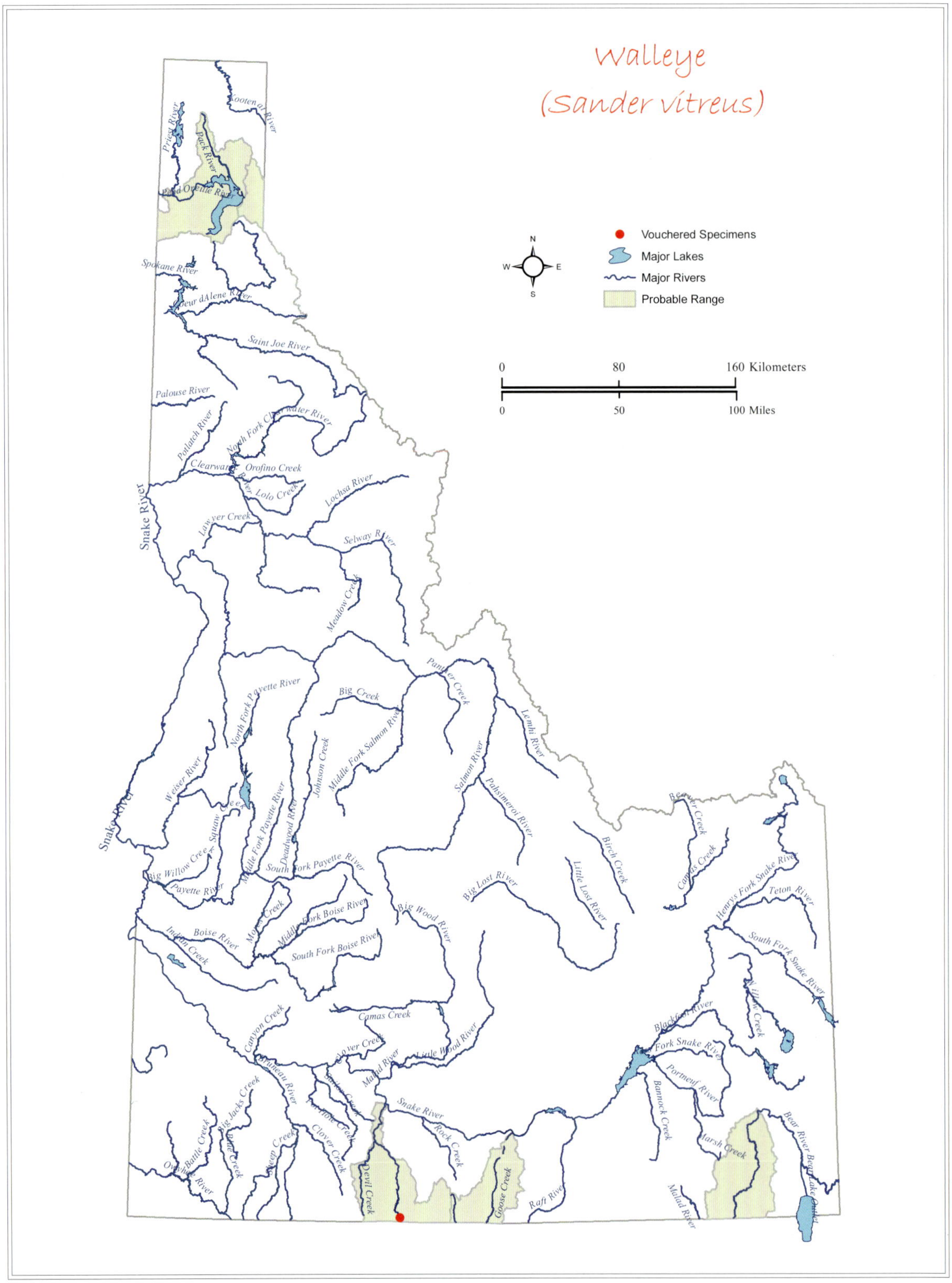
Walleye
(Sander vitreus)
N
W
E
S
Vouchered Specimens
Major Lakes
Major Rivers
Probable Range
0
80
160 Kilometers
0
50
100 Miles
Saint Joe River
Palouse River
Potlatch River
Orofino Creek
Lolo Creek
Lochsa River
Snake River
Selway River
Meadow Creek
Big Creek
Johnson Creek
Middle Fork Salmon River
Salmon River
Lemhi River
Pahsimeroi River
Birch Creek
Little Lost River
Big Lost River
Big Wood River
Weiser River
Big Willow Creek
Payette River
Middle Fork Payette River
Deadwood River
South Fork Payette River
Boise River
Indian Creek
South Fork Boise River
Camas Creek
Canyon Creek
Bruneau River
Clover Creek
Snake River
Rock Creek
Goose Creek
Raft River
Devil Creek
Teton River
South Fork Snake River
Willow Creek
Portneuf River
Bannock Creek
Marsh Creek
Malad River
Bear River
Bear Lake

Walleye

Sander vitreus

(Mitchill, 1818)

Etymology

Sander = obscure, but perhaps related to one or more of the common names when this fish was in the genus *Stizostedion* with the Sauger; previously *vitreum* = glassy, referring to the large, glassy eyes; *vitreus* = modification of glassy eyes; Walleye = this species' eyes point outward.

Prior to the release of the sixth edition of Common and Scientific Names of Fishes from the United States, Canada and Mexico in 2004[17], the Committee published the rationale for changing the generic name of this species from *Stizostedion* to *Sander*[18]. In short, the Committee found that *Sander* had priority over *Stizostedion* due to prior use. For the specific name, *vitreum* was replaced with *vitreus* for gender agreement with the generic name.

Importance

In Canada's inland waters, Walleye are probably the most economically valuable species. This species is important in Canada as both a commercial fish and a sportfish. In many locations they are an important species in the summer sport fishery and for ice anglers in the winter. Walleye are not spectacular fighters, but they do provide a good battle, always seeking to go to the bottom[3]. Because of their fine table quality and desirable size, Walleye are sought by anglers wherever the species occurs. Walleye are somewhat difficult to catch until anglers learn how to fish for them in a particular ecosystem. They are typically found in deep water, and anglers must work to get the lure down to them, making trolling somewhat effective. Walleye may effectively control overabundant sunfishes in small shallow ponds or lakes[6].

Description

Overall coloration of Walleye is highly variable with ecosystem type and to a lesser extent with size. Turbid waters result in paler individuals that are less marked with the obvious black pattern than do clear waters, where individuals are more vividly marked. The background of Walleye is usually olive-brown to golden brown to yellow, while the dorsal surface of the head and back is darker and the sides are paler, often with golden flecks on the scales. The ventral surface is milky white or yellowish white. Smaller individuals that are 102 to 356 mm (4.0 to 14.0 in) in length have vague to obvious dark vertical bands across the back and down the sides, a characteristic typically absent in adults. The dorsal fin is dusky, clear, or vaguely speckled and has no definite rows of spots. There is a noticeable black blotch at the base of the last few membranes on the dorsal, and the second dorsal and caudal fins have speckles or tiny spots in rather regular rows. The lower lobe of the caudal and the tip of the anal fin are milky white. The pelvic fins are yellow or orange-yellow, and the pectoral fins are dark or pale olive, with a dark blotch at the base. Eyes are silvery from light reflected by a light-sensitive layer[3].

Walleye have an elongate body that is subcylindrical in young and slightly compressed, the compression being more obvious in large adults. The caudal peduncle is moderately long and deep and somewhat compressed. Walleye have a long snout that is bluntly pointed and does not extend beyond the upper jaw. Maxillaries are very long, extending to the posterior edge of the eye. There are strong teeth on the premaxillaries, jaws, head of the vomer, and palatines. There are large, often recurved, canine teeth on the head of the vomer, the palatines, jaws, and premaxillaries. Pharyngeal teeth are large, sharp, and recurved. Gill rakers are moderately long and tooth bearing on the outer edge of the gill arch. There are six to eight rakers on the lower arch and four or five on the upper arch. Branchiostegal rays are seven, seven or seven, eight. There are two dorsal fins that are obviously separated. The first dorsal fin is spiny, high, long, and rounded, with 12 to 16 strong spines, while the second dorsal fin is as high as or higher than the first, with one fine spine and 18 to 22 (typically 19 to 21) rays, and it is square to slightly emarginated. The caudal fin is long and not overly broad but well forked, and the tips are rounded points. The anal fin has two spines, with 11 to 14 (typically 12 or 13) soft rays. The pelvic fins are thoracic and widely spaced, and the length is about equal to the pectoral fins; they have rounded tips, with one spine and five rays. Pectoral fins are only moderately broad, with rounded tips and 13 to 16 (typically 14) rays. Scales are strongly ctenoid, but cheeks are scaleless or slightly scaled; the opercles are mostly naked, and the breast and belly are normally scaled. The lateral line is complete, high, and a little curved; lateral line scales are rather small, numbering 83 to 104 (typically 86 to 92), including three or four pitted scales of the lateral line. The peritoneum is white, and the short intestine is well differentiated, while the pyloric caeca number three, each about the same length as the stomach. Vertebrae number 44 to 48, and there are no nuptial tubercles, making differentiation between sexes difficult[3].

Range and Distributional Status

This species is not native to Idaho. Native distribution of Walleye is in the central and northern Mississippi River system, east to North Carolina, northward to the Hudson Bay region and Labrador, and northwest to the Mackenzie River and Saskatchewan River systems in Canada. In the 1990s, Walleye were captured throughout the lower Columbia River downstream to at least Vancouver, Washington, and Portland, Oregon. They were captured from the mouth of the Snake River upstream to Ice Harbor Dam. Later in the 1990s, Walleye were captured at Little Goose Dam on the Snake River, 70 miles upstream from its confluence with the Columbia River[6]. Walleye have been reported in Salmon Falls Creek and the Bear River drainage of southern Idaho, as well as the Clark Fork and Hayden Lake of northern Idaho.

Size and Longevity

Walleye reach lengths of 1,041 mm (41.0 in) and may weigh as much as 11.3 kg (24 lb, 15 oz). Walleye in British Columbia may reach lengths of 760 mm (30 in) and weights of 5 kg (11 lb, 0 oz). Individual

Walleye weighing more than 4.5 kg (10 lb, 0 oz) have been taken from Banks Lake in Washington. Washington's angling record for Walleye was caught in the Columbia River in 2002 and weighed 8.6 kg (19 lb, 0 oz). Upon reaching sexual maturity, females are larger than males of corresponding ages[6].

The all-tackle world record Walleye was taken from Old Hickory Lake, Tennessee, in 1960 and weighed 11.34 kg (25 lb, 0 oz)[9]. The Idaho record Walleye was taken from Lower Goose Creek Reservoir in 2011 and weighed 8.1 kg (17 lb, 14 oz) and was 876 mm (34.5 in) long[10].

Food and Feeding

Initially, Walleye fingerlings begin to feed on planktonic crustaceans and gradually change to a diet of insects, such as midge larvae, as they increase in size. Upon reaching sizes of about 76 mm (3.0 in) long, Walleye young gradually convert to larval and juvenile fishes and are feeding primarily on fishes by the time they reach lengths of 152 mm (6.0 in). Fish that have been reported in Walleye diets include Freshwater Drum, Gizzard Shad, whitefishes, ciscoes, Goldeneye, sticklebacks, sculpins, suckers, minnows, and juvenile salmon and Steelhead[6].

Walleye are somewhat opportunistic feeders, and fish prey are vulnerable. As Walleye are strong swimmers, they can catch most prey species. In ecosystems where there is a large diversity of forage species, they appear to select Yellow Perch and Freshwater Drum. Walleye also take crayfish, snails, frogs, and occasionally small mammals[2].

Limiting Factors

Northern Pike are probably the most important predator of Walleye over much of their overlapping range. Muskellunge also prey on Walleye in some portions of their overlapping range. Northern Pike may also be important competitors, as they are the only other major shallow-water predator in some ecosystems. A variety of other warm-water species prey on young Walleye, as do other predatory fishes. A range of fish-eating birds and mammals, as well as lampreys, prey on Walleye, but they are probably not an important predatory factor. Factors important in controlling Walleye populations include water temperature, stream flow, and winds at spawning time. A major controlling factor in Walleye is fry mortality, which may be up to 99 percent. Walleye are host to a wide variety of parasites, including protozoans, trematodes, cestodes, nematodes, acanthocephalans, leeches, mollusks, crustaceans, and in some ecosystems in Canada tapeworms. In some ecosystems, Walleye may hybridize with Sauger[3].

Breeding Habits

During spawning season, Walleye males arrive at spawning sites before females, usually outnumber the females on the spawning sites, and remain on-site for a few days after females depart. Genetic evidence implies some degree of natal homing. In lakes Walleye prefer to spawn in shallow water, generally over rocky, wave-washed shoal water, while in contrast their preferred substrate in rivers appears to be gravel and cobble. Spawning occurs primarily at dusk, but observations are rare. There is no site preparation, but some form of courtship precedes spawning and may be initiated by either sex. Courtship consists of an approach (from behind, from the side, or by drifting backward) and contact that consists of pushing with another fish. The first dorsal fin is alternately raised and lowered during these approaches. As courtship activity increases in frequency and intensity, individual fishes begin to dart forward and upward to the surface, and eventually one or more males and one or more females form a compact group and rush upward to the surface, swimming vigorously around and then stopping and releasing gametes near the surface[5].

Sexual maturation of Walleye varies in different ecosystems. In Lake Erie, females become mature between ages three and five at lengths of at least 432 mm (17.0 in), while males usually mature about one year earlier. Walleye spawn early in the spring with water temperatures at about 3.3 to 6.8°C (38 to 44°F). In general, Walleye spawn on gravel and stream riffles or on rocky shoals of lakes. However, Walleye have also been reported to spawn over sand or silt bottoms. They may travel long distances to reach the spawning grounds. In Michigan populations, females 406 to 432 mm (16.0 to 17.0 in) long contain an average of 44,000 eggs. Walleye broadcast eggs over the spawning area; eggs are adhesive when first broadcast but later become firm and free-floating after water hardening. In some Columbia River populations, Walleye spawn at night in water that is 0.8 to 0.9 m (2 to 3 ft) deep over rock or gravel substrates. Egg development varies with water temperature, requiring about 26 days at 4.4°C (40°F), 21 days at 10 to 13°C (50 to 55°F), and seven days at about 14°C (57°F). At emergence Walleye fry are about 13 mm (0.5 in) long. High mortality of young occurs, however, as they are eaten by all fish-eating species, including larger Walleye[6].

Habits and Habitat

Walleye are generally found in large lake and stream ecosystems. Typically, Walleye travel in loose schools and in open water near or on the bottom. For spawning Walleye typically choose bottoms consisting of rubble, gravel, and occasionally bedrock. In summer Walleye feed in shallow water at night and move to deeper water during daylight. They are typically found in water temperatures of 0 to 32°C (32 to 90°F), but they seem to prefer summer temperature maximums of 25°C (77°F). The Walleye habit of feeding throughout the year makes them catchable both through the ice in winter and on their spawning grounds[6].

Walleye occur in river ecosystems as well as lake and reservoir ecosystems but appear to prefer slow water. Despite this preference for slow water, some Walleye populations undertake lengthy migrations. Walleye in the Canadian portion of the Columbia River system migrate more than 250 km (155 mi) to spawn in or near Roosevelt Reservoir in Washington. One tagged fish in northern Alberta moved 600 km (375 mi). In lake ecosystems, adult Walleye avoid well-lit areas and are found in moderately shallow water of 1 to 15 m (3 to 50 ft), close to cover; however, Walleye may participate in diurnal movements into shallow water as day fades and into deeper water at sunrise[5].

Walleye populations thrive in moderately fertile ecosystems. Walleye tolerate a wide range of environmental conditions, from acid bog lake ecosystems to oligotrophic lakes to eutrophic hard-water lake ecosystems. The most favorable conditions appear to exist in large, fertile, shallow, turbid lakes or large stream or river ecosystems, provided they are deep or turbid enough to afford cover[2].

Management and Conservation Status

Walleye are considered common, wide spread, and abundant rangewide (G5). A state conservation ranking is not applicable (SNA) since Walleye are an introduced species. They are classified as a game fish in Idaho[21]. Walleye are extremely voracious predators. This characteristic may allow them to be used as effective predators for panfish control in some ecosystems. There is also some evidence that angling does not harm populations, though there are some indications that it does. Existing ecological conditions at spawning and rearing time usually dictate the strength of a Walleye year class. A variety of size limits may be used to help manage take by anglers, because if they are fished year-round by experienced Walleye anglers, populations may be decimated.

Walleye are one of the most controversial introduced species in the western United States. In general, waters of the western United States do not have the diverse and abundant forage base needed to support these prolific keystone predators. As a result, Walleye introduced in the western United States typically exhibit poor growth and often create conflict with management of other game species. The Idaho Fish and Game Commission approved a policy in the 1980s to introduce Walleye only in closed systems where they have no chance of escaping to other waters. Walleye were first introduced into Salmon Falls Creek Reservoir in south-central Idaho in the mid-1970s. IDFG currently manages Walleye in Salmon Falls Creek Reservoir, Lower Goose Creek Reservoir, and Oneida Reservoir.

A reproducing population and limited target fishery now exist in Lake Pend Oreille and the Pend Oreille River. These populations originated from Walleye that were introduced illegally or drifted in from elsewhere in the drainages. IDFG will monitor these populations and, if possible, eliminate them or control their expansion. In 2006 IDFG removed bag limits on Walleye in all waters statewide where they are not being specifically managed.

Walleye can significantly impact native fish populations and existing sport fisheries. IDFG will not restrict harvest or permit catch-and-release fishing contests or tournaments on Walleye in waters where unauthorized introductions have occurred. IDFG will not introduce Walleye into new waters during this planning period.

Hybrid Perches

Sander canadensis × *Sander vitreus* Sauger × Walleye

Family Cichlidae

Family Cichlidae—Cichlids occur in freshwater and occasionally brackish water throughout South and Central Americas. These species generally have 20 to 50 scales in the lateral line series. There are seven to 25 spines and five to 30 soft rays in the dorsal fin and three to 15 spines and four to 15 soft rays in the anal fin. There are about 112 genera and 1,350 species1.

Cichlids form an important group of relatively large and often colorful aquarium fishes. Species of this family have highly organized breeding activities and often provide care for their young. Two types of parental care are recognized: mouth brooders, which are usually polygamous and usually only the female carries fertilized eggs and newly hatched eggs in her mouth, or substrate brooders, which are usually monogamous and both sexes may care for the eggs1.

Members of this family are problematic in Idaho and exist in only a very few locations. In most cases, these species are part of the large aquarium fish industry and have been released by owners who no longer care to keep them; in some cases, there have been releases in an attempt to help control aquatic weeds, often with no reflection on long-term ecological consequences. Because of their nonnative status and limited distribution in Idaho, we have elected to provide only cursory information on these species. Five species in this family are discussed in brief below.

Amatitlania nigrofasciata	(Günther, 1867)	Convict Cichlid
Astronotus ocellatus	(Agassiz, 1831)	Oscar
Oreochromis aureus	(Steindachner, 1864)	Blue Tilapia
Oreochromis mossambicus	(Peters, 1852)	Mozambique Tilapia
Tilapia zillii	(Gervais, 1848)	Redbelly Tilapia

In general, cichlids (Cichlidae) are superficially similar to sunfishes and basses (*Lepomis* and *Micropterus* of the Centrarchidae family). Cichlids can be distinguished from centrarchids by a single nostril opening on each side of the head (vs. two openings in centrarchids) and the presence of a discontinuous or two-part lateral line (vs. a continuous lateral line in centrarchids). Individuals interested in additional life history information, or on potential aquarium species, could make a visit to a local library or pet store or perform an online search for individual species of this family. Much of the information presented below on these species can be found in the Nonindigenous Aquatic Species fact sheets which can be found at https://nas.er.usgs.gov/queries/factsheet.aspx.

Convict Cichlid

Amatitlania nigrofasciata

(Günther, 1867)

Native range of Convict Cichlid includes Guatemala, El Salvador, Costa Rica, Panama, and other Central American locations. Convict Cichlid are an important aquarium fish in North America. However, in the wild in Idaho ecosystems, they are a completely negative factor due to their competitiveness with native species. Convict Cichlid get their common name from the visibly distinctive color pattern: eight or nine vertical stripes over dark-brown to bluish back and mouse-gray sides. Convict Cichlids reach a maximum size of about 150 mm (6 in) and probably live to age four or five[20]. Any additional transfers of Convict Cichlids into Idaho ecosystems for any reason are strongly discouraged.

Oscar

Astronotus ocellatus

(Agassiz, 1831)

Oscar are native to South America including Orinoco and Amazon basins; also to French Guiana, and to the northern part of Paraguay drainage, Parana basin. Oscar are a popular aquarium species across the United States. Oscar may also be known as Velvet Cichlid, Red Oscar, and Marble Cichlid. However, in the wild in Idaho, they, like all the species in this group, are a completely negative ecological factor due to their competitiveness with native species. They may grow to 400 mm (15.8 in). Presence of Oscar, sometimes only one specimen, has been noted from Alaska to Florida and some southern US states have introduced this species for a variety of reasons. Oscar are established in Florida, Hawaii, and Puerto Rico but introductions have failed elsewhere. Any additional transfers of Oscar into Idaho ecosystems for any reason are strongly discouraged.

Blue Tilapia

Oreochromis aureus

(Steindachner, 1864)

Blue Tilapia are native to tropical and subtropical Africa, and Middle East. Their native range includes Senegal, Niger, and many smaller drainages and lakes in Africa and the Middle East. Blue Tilapia are a popular aquarium species across the United States. However, in the wild in Idaho, they, like all the species in this group, are an undesirable species. They may grow to 510 mm (20.0 in). This species has been cultured in Idaho in the Hagerman Valley, Twin Falls County, and may have become established following escape into the Snake River near natural thermal outflows. Presence of this species has been noted across the United States. Blue Tilapia are established or possibly established in ten states: Arizona, California, Florida, Nevada, North Carolina, Texas, Colorado, Idaho, Oklahoma, and Pennsylvania. Reported from Alabama, Georgia, and Kansas. For more than a decade it has been considered the most widespread foreign fish in Florida. Any additional transfers of Blue Tilapia into Idaho ecosystems for any reason are strongly discouraged.

Mozambique Tilapia

Oreochromis mossambicus

(Peters, 1852)

The native range of the Mozambique Tilapia is tropical and subtropical Africa and southern Africa from lower Zambezi to Brak River, and Limpopo system; they occur in both fresh and brackish waters. They are established or locally established in seven states including Arizona, California, Colorado, Florida, Hawaii, Idaho, and Texas. As with other species in this group, this species is suspected to be a threat to native species in some ecosystems. Mozambique Tilapia may impact native fishes through competition for food and/or space, or through secondary effects. A reproducing population was found in Idaho in Barney Hot Springs and at the upper end of Barney Creek in Custer County, Little Lost River Valley, in September 1985. The species was listed by Idaho Fish and Game as being introduced into and confined to geothermal waters of the Snake River below Shoshone Falls. Mozambique Tilapia have been reported in several states as individuals or small groups. They may reach sizes up to 400 mm (15.8 in). Any additional transfers of Mozambique Tilapia into Idaho ecosystems for any reason are strongly discouraged.

Redbelly Tilapia

Tilapia zillii

(Gervais, 1848)

Redbelly Tilapia have been introduced into various ecosystems in the United States to control aquatic vegetation. Where they have been stocked in the wild, they generally affect native species adversely by destroying aquatic vegetation. Redbelly Tilapia are native to rivers and lakes in northern and central Africa to Jordan. They were at one time successfully established in irrigation canals in the Imperial Valley, California, and in man-made ponds in Nevada. Redbelly Tilapia live to age six or more and reach sizes of 140 to 254 mm (5.5 to 10.0 in) by the end of their third year of life. Redbelly Tilapia are omnivorous, which is part of the reason introductions into Idaho ecosystems are strongly discouraged[20].

Chapter Citations

1. Nelson, 2006. 2. Sigler and Sigler, 1996. 3. Scott and Crossman, 1973. 4. Etnier and Starnes, 1993. 5. McPhail, 2007. 6. Wydoski and Whitney, 2003. 7. Brinsmead and Fox, 2002. 8. U.S. Department of Commerce and Labor, 1907. 9. International Game Fish Association, 2015. 10. Idaho Department of Fish and Game, 2015. 11. Courtenay, et al., 1987. 12. Suski, et al., 2003. 13. Olson, Young, and Blinkoff, 2003. 14. Wilde, 2003. 15. Reed and Pereira, 2009. 16. Allen, Hoyer, and Canfield, 1998. 17. Nelson, et al., 2004. 18. Nelson, et al., 2003. 19. Amadio, et al., 2006. 20. Sigler and Sigler, 1987.

Additional readings for this chapter are listed in the Supplemental References section.

Appendix A. Checklist of fish taxa reported in Idaho, excluding hybrids. Superscript 1 indicates native taxa. Superscript 2 indicates taxa believed to be currently extirpated from Idaho.

Common Name	Scientific Name	Author
	Petromyzontidae - Lampreys	
Pacific Lamprey	[1]*Entosphenus tridentatus*	(Gairdner in Richardson, 1836)
	Acipenseridae - Sturgeons	
White Sturgeon	[1]*Acipenser transmontanus*	Richardson, 1836
	Clupeidae - Herrings	
American Shad	*Alosa sapidissima*	(Wilson, 1811)
	Cyprinidae - Carps and Minnows	
Chiselmouth	[1]*Acrocheilus alutaceus*	Agassiz & Pickering in Agassiz, 1855
Goldfish	*Carassius auratus*	(Linnaeus, 1758)
Lake Chub	[1]*Couesius plumbeus*	(Agassiz, 1850)
Grass Carp	*Ctenopharyngodon idella*	(Valenciennes in Cuvier and Valenciennes, 1844)
Common Carp	*Cyprinus carpio*	Linnaeus, 1758
Utah Chub	[1]*Gila atraria*	(Girard, 1856)
Northern Leatherside Chub	[1]*Lepidomeda copei*	(Jordan & Gilbert, 1881)
Peamouth	[1]*Mylocheilus caurinus*	(Richardson, 1836)
Golden Shiner	*Notemigonus crysoleucas*	(Mitchill, 1814)
Spottail Shiner	*Notropis hudsonius*	(Clinton, 1824)
Fathead Minnow	*Pimephales promelas*	Rafinesque, 1820
Northern Pikeminnow	[1]*Ptychocheilus oregonensis*	(Richardson, 1836)
Longnose Dace	[1]*Rhinichthys cataractae*	(Valenciennes in Cuvier and Valenciennes, 1842)
Leopard Dace	[1]*Rhinichthys falcatus*	(Eigenmann & Eigenmann, 1893)
Speckled Dace	[1]*Rhinichthys osculus*	(Girard, 1856)
Umatilla Dace	[1]*Rhinichthys umatilla*	(Gilbert & Evermann, 1894)
Redside Shiner	[1]*Richardsonius balteatus*	(Richardson, 1836)
Tui Chub	*Siphateles bicolor*	(Girard, 1856)
Tench	*Tinca tinca*	(Linnaeus, 1758)
	Catostomidae - Suckers	
Utah Sucker	[1]*Catostomus ardens*	Jordan & Gilbert, 1881
Longnose Sucker	[1]*Catostomus catostomus*	(Forster, 1773)
Largescale Sucker	[1]*Catostomus macrocheilus*	Girard, 1856
Snake River Sucker	[1]*Chasmistes muriei*	Miller & Smith, 1981
Cascadian Sucker	[1]*Pantosteus bondi*	Smith et al. 2013
Bridgelip Sucker	[1]*Pantosteus columbianus*	(Eigenmann & Eigenmann, 1893)
Mountain Sucker	[1]*Pantosteus platyrhynchus*	(Cope, 1874)
Green Sucker	[1]*Pantosteus virescens*	Cope, 1875
	Cobitidae - Loaches	
Oriental Weatherfish	*Misgurnus anguillicaudatus*	(Cantor, 1842)

Common Name	Scientific Name	Author
Ictaluridae - North American Catfishes		
Black Bullhead	*Ameiurus melas*	(Rafinesque, 1820)
Yellow Bullhead	*Ameiurus natalis*	(Lesueur, 1819)
Brown Bullhead	*Ameiurus nebulosus*	(Lesueur, 1819)
Blue Catfish	*Ictalurus furcatus*	(Valenciennes in Cuvier and Valenciennes, 1840)
Channel Catfish	*Ictalurus punctatus*	(Rafinesque, 1818)
Tadpole Madtom	*Noturus gyrinus*	(Mitchill, 1817)
Flathead Catfish	*Pylodictis olivaris*	(Rafinesque, 1818)
Osmeridae - Smelts		
[2]Rainbow Smelt	*Osmerus mordax*	(Mitchill, 1814)
Salmonidae - Whitefishes, Chars, Trouts and Salmons		
Lake Whitefish	*Coregonus clupeaformis*	(Mitchill, 1818)
Bear Lake Whitefish	[1]*Prosopium abyssicola*	(Snyder, 1919)
Pygmy Whitefish	[1]*Prosopium coulterii*	(Eigenmann & Eigenmann, 1892)
Bonneville Cisco	[1]*Prosopium gemmifer*	(Snyder, 1919)
Bonneville Whitefish	[1]*Prosopium spilonotus*	(Snyder, 1919)
Mountain Whitefish	[1]*Prosopium williamsoni*	(Girard, 1856)
Arctic Grayling	*Thymallus arcticus*	(Pallas, 1776)
Sunapee Trout	*Salvelinus alpinus oquassa*	(Girard, 1854)
Bull Trout	[1]*Salvelinus confluentus*	(Suckley, 1859)
Brook Trout	*Salvelinus fontinalis*	(Mitchill, 1814)
Lake Trout	*Salvelinus namaycush*	(Walbaum in Artedi, 1792)
[2]Atlantic Salmon	*Salmo salar*	Linnaeus, 1758
Brown Trout	*Salmo trutta*	Linnaeus, 1758
Golden Trout	*Oncorhynchus aquabonita*	(Jordan, 1892)
Cutthroat Trout	[1]*Oncorhynchus clarkii*	(Richardson, 1836)
Yellowstone Cutthroat Trout	[1]*Oncorhynchus clarkii bouvieri*	(Bendire, 1882)
Snake River Fine-Spotted Cutthroat Trout	[1]*Oncorhynchus clarkii behnkei*	Montgomery, 1995
Westslope Cutthroat Trout	[1]*Oncorhynchus clarkii lewisi*	(Girard, 1856)
Bonneville Cutthroat Trout	[1]*Oncorhynchus clarkii utah*	(Suckley, 1874)
[2]Chum Salmon	*Oncorhynchus keta*	(Walbaum in Artedi, 1792)
Coho Salmon	[1]*Oncorhynchus kisutch*	(Walbaum, 1792)
Redband and Steelhead	[1]*Oncorhynchus mykiss*	(Walbaum, 1792)
Sockeye Salmon/Kokanee	[1]*Oncorhynchus nerka*	(Walbaum *in* Artedi, 1792)
Chinook Salmon	[1]*Oncorhynchus tshawytscha*	(Walbaum *in* Artedi, 1792)
Esocidae - Pikes		
Northern Pike	*Esox lucius*	Linnaeus, 1758
Percopsidae - Trout-perches		
Sand Roller	[1]*Percopsis transmontana*	(Eigenmann and Eigenmann in Eigenmann, 1892)

Common Name	Scientific Name	Author
	Gadidae Cods	
Burbot	[1]*Lota lota*	(Linnaeus, 1758)
	Fundulidae - Topminnows	
Banded Killifish	*Fundulus diaphanus*	(Lesueur, 1817)
	Poeciliidae - Livebearers	
Western Mosquitofish	*Gambusia affinis*	(Baird & Girard, 1853)
Shortfin Molly	*Poecilia mexicana*	Steindachner, 1863
Guppy	*Poecilia reticulata*	Peters, 1859
Green Swordtail	*Xiphophorus hellerii*	Heckel, 1848
Platy	*Xiphophorus* spp.	Heckel, 1848
	Cottidae - Sculpins	
Mottled Sculpin complex	[1]*Cottus bairdii* complex	Girard, 1850
Paiute Sculpin	[1]*Cottus beldingii*	Eigenmann & Eigenmann, 1891
Slimy Sculpin	[1]*Cottus cognatus*	Richardson, 1836
Shorthead Sculpin	[1]*Cottus confusus*	Bailey & Bond, 1963
Bear Lake Sculpin	[1]*Cottus extensus*	Bailey & Bond, 1963
Shoshone Sculpin	[1]*Cottus greenei*	(Gilbert and Culver in Jordan and Evermann, 1898)
Columbia Sculpin	[1]*Cottus hubbsi*	Bailey & Dimick, 1949
Wood River Sculpin	[1]*Cottus leiopomus*	Gilbert & Evermann, 1894
Torrent Sculpin	[1]*Cottus rhotheus*	(Smith, 1882)
Cedar Sculpin	[1]*Cottus schitsuumsh*	LeMoine et al. 2014
	Centrarchidae - Sunfishes	
[2]Rock Bass	*Ambloplites rupestris*	(Rafinesque, 1817)
Green Sunfish	*Lepomis cyanellus*	Rafinesque, 1819
Pumpkinseed	*Lepomis gibbosus*	(Linnaeus, 1758)
Warmouth	*Lepomis gulosus*	(Cuvier in Cuvier and Valenciennes, 1829)
[2]Orangespotted Sunfish	*Lepomis humilis*	(Girard, 1858)
Bluegill	*Lepomis macrochirus*	Rafinesque, 1819
Smallmouth Bass	*Micropterus dolomieu*	Lacepède, 1802
Largemouth Bass	*Micropterus salmoides*	(Lacepède, 1802)
White Crappie	*Pomoxis annularis*	Rafinesque, 1818
Black Crappie	*Pomoxis nigromaculatus*	(Lesueur in Cuvier and Valenciennes, 1829)
	Percidae - Perches	
Yellow Perch	*Perca flavescens*	(Mitchill, 1814)
[2]Sauger	*Sander canadensis*	(Griffith & Smith, 1834)
Walleye	*Sander vitreus*	(Mitchill, 1818)
	Cichlidae - Cichlids	
Convict Cichlid	*Amatitlania nigrofasciata*	(Günther, 1867)
Oscar	*Astronotus ocellatus*	(Agassiz *in* Spix and Agassiz, 1831)
Blue Tilapia	*Oreochromis aureus*	(Steindachner, 1864)

Common Name	Scientific Name	Author
Mozambique Tilapia	*Oreochromis mossambicus*	(Peters, 1852)
Redbelly Tilapia	*Tilapia zillii*	(Gervais, 1848)

GLOSSARY

abdomen The belly. The area between the throat and the anus.

abdominal On the belly. When fins are located on the belly, behind the pectoral fins, they are termed abdominal.

acre-foot A volume of water equal to an acre (43,560 square feet) covered with one foot of water—43,560 cubic feet of water (ac-ft) or 325,821 gallons.

adfluvial Refers to fish life histories for species that spawn in tributary streams where the young rear from 1 to 4 years before migrating to a lake system, where they grow to maturity and live as adults.

adipose fin A small, fleshy fin, typically without rays that occurs between the dorsal fin and caudal fin of some families of fish (salmonids, percopsids, and catfishes).

air bladder A gas-filled sac beneath the vertebral column; also known as the swim bladder or gas bladder.

alevin Yolk sac fish which have emerged from spawning gravel.

allopatry or allopatric Geographically separated, as different populations of fish species that cannot interact.

alluvial A deposit of sand, mud, or other material left when flowing water recedes.

ammocoetes Filter feeding, worm-like larval stage of lampreys.

amphidromous Fishes that migrate between fresh and salt water at some stages of their life.

anadromous Life history characteristic of some fishes (for example, salmonids) whereby individuals migrate from fresh to salt water, grow for a period of time, and return to fresh water to spawn.

anal fin Fin located behind the anus on the ventral median line.

anguilliform Eel-or snake-like; long, slender, and flexible.

anterior Preceding or in front of, opposite of posterior.

autochthonous Produced within the system under discussion.

axillary process Fleshy tab in the axillae of the pelvic fins.

axillae Area immediately above the paired fins.

barbels Fleshy, thin projections, usually on the head, on some fishes (for example, catfishes and some minnows).

basibranchials Three median bones on the floor of the gill chamber.

benthic On or near the bottom, as benthic feeders.

binomial nomenclature Taxonomic names of all species are composed of two names, one for genus and one for species.

bony fishes Species of fishes that have hard, calcified skeletons as opposed to cartilaginous ones.

branchiostegal rays Blade-shaped bones in series below the subopercle. On each side they form a curved fan that reaches from the ceratohyals under the throat to close to the back and bottom of the branchial chamber while water is being sucked in through the mouth for feeding and respiration. The number is invariably 3 on a side in minnows and suckers, but more in, for example, Yellow Perch, which usually have 7 on a side, the number may vary by 1 or 2 on a side. They are easily counted without dissection and are often useful in identification.

breast An area of indefinite extent between the pectoral fins and the isthmus, sometimes called chest.

breeding tubercles Hardened protuberances developing on the fins and body of adults, generally more on males. Also nuptial tubercles.

buccal Of the mouth.

buccal cavity Space behind the jaws and in front of the pharynx, the oral cavity.

canine teeth Long, sharp, slender teeth.

cardiform teeth Arranged in two dimensional patches.

carinated With a ridge

carnivorous Eaters of meat, feeding or preying on live animals.

cartilaginous Noncalcified, referring to skeleton.

caudal fin Tail fin.

caudal peduncle Slender, tapering portion of the body of a fish behind the last ray of the anal fin and forward of the flare of the tail or caudal fin.

char or charr Common names for fishes belonging to the genus *Salvelinus*.

cheek The area of a fish's face between the eye and the preopercular bone.

chironomid Midge (member of a family [Chironomidae] of true flies [Diptera]).

cloaca A chamber in the lower part of the gut into which the ducts from the kidney and reproductive organs empty, with one external opening instead of separate vent and urogenital openings.

cladistics A method of biological classification in which organisms are grouped together based on shared unique characteristics that come from their last common ancestor and are not present in more distant ancestors.

compressed Flattened from side to side, producing a form higher than wide.

convoluted Coiled or twisted.

copepods Small organisms in the Class Crustacea.

cranium Portion of the skull that encloses the brain.

ctenoid scales Scales possessing comb like posterior margins of tiny spines or ctenii on posterior surface (for example, in sunfishes). Rough to the touch.

cycloid scales Scales possessing a smooth, evenly curved posterior margin (for example, in suckers). Smooth to the touch.

decimate Destroy, literally reduce by 10 percent.

decurved Downward curving.

decurved lateral line A lateral line that dips deeply toward the belly, as in the Golden Shiner.

dentary One of a pair of large bones forming the front of the mandible or lower jaw.

degree-day Measure of time/temperature with reference to egg incubation or seasonality. One degree-day is a 24-hour period in which the temperature is 1 degree above freezing (33 F, 0.56 C).

demersal Characteristic of eggs and plankton material that is heavier than water and therefore sinks. On the bottom.

depressed Flattened from top to bottom, producing a form that is wider than high, e.g. Sculpins.

desiccation Drying out, as in large lakes.

desmids Charophyta, a division of green algae that forms a sister group to the land plants.

detritus Organic debris deposited on the bottom of an aquatic habitat or suspended in the water column.

diploid Two similar complements of genetic material.

diadromous Fishes that are migratory between salt and fresh water.

diatoms A major group of phytoplankton (algae) with distinctive, transparent cell walls made of hydrated silica.

dorsal Of the back or upper side of the body.

dorsal fin Fin, or fins (often separate fins) attached on the midline of the back (upper) surface of a fish. May consist of soft rays only or of both soft and spiny rays.

effluent Outflow.

ecosystem All of the organisms and their physical environment that interact as an ecological unit; including the entire biological and physical contents of the community and their interactions.

elasmobranch Taxonomic group of cartilaginous fishes that includes sharks and rays.

electrophoresis Technique for separating proteins based on molecular weight and electrical charge.

electrical conductivity Conductivity of water is based on the amount of dissolved solids present and significantly effects electrofishing efficiency.

electrical current The movement of electric charge from one location to another.

electrotetany A muscular rigidity induced in fish subjected to an electric field.

elongate Long and narrow.

emarginate More weakly forked, crescent-shaped, as in a fin outline. Characterized by a shallow notch at the margins as in a weakly forked caudal fin.

endemic Occurring only in a given location or region. For example, Bonneville Cisco (*Prosopium gemmifer*) is a Bear Lake endemic species.

epilimnion Uppermost layer of circulating water in a lake, separated from the hypolimnion by the thermocline.

eutrophic Nutrient rich or polluted, opposite of oligotrophic. (Eutrophication—the process of becoming nutrient rich or polluted).

exoskeleton Hard outer shell.

exotic Foreign, not native to a drainage or ecosystem, region or political unit. Fishes that are introduced or nonnative.

extant Remaining either in nature or being maintained under artificial conditions. Opposite of extinct.

extinct Entirely extirpated or exterminated. No longer present on Earth.

extirpate Loss of a population of a species as opposed to loss of the entire species.

extirpation Local or regional elimination of a species

eyed eggs Stage of fish egg development in which eye spots are visible.

FL Fork length is a measurement of fish length from the tip of the snout to the fork in the tail.

F_{ST} A measure of population differentiation due to genetic structure.

falcate Deeply emarginate dorsal or anal fin, as in *Rhinichthys falcatus.*

faunal Animal.

fecundity Measure of potential reproductive capacity using production of mature eggs.

filamentous Thrcadlikc.

fin ray Flexible and usually branched strut of small rectangular bones supporting the membrane of a fin. The number of rays in a fin is usually counted from the first fully elongated ray to the last one, which is branched to its base and counted as one.

fin spine Unbranched, sharp, bony struts supporting fins; usually sharp and inflexible, except, for example, in sculpins, which have soft, unbranched spines in the dorsal and anal fins.

fish or fishes Cold-blooded vertebrate animals adapted for aquatic existence by means of gills that remove oxygen from the water and pass waste materials out of the system. When used in the plural form (fishes) it indicates two or more taxa.

fluvial Rivers or streams and the animals that live in them.

forage To seek food.

FL Length of a fish as measured from the anterior most margin of the head to the top of the middle ray (or fork) of the caudal fin. Contrasted with (TL).

form A division of subspecies with no formal taxonomic status.

frenum A fleshy bridge connecting the upper lip to the tip of the snout. (for example in Longnose Dace).

fry Fish that are newly hatched but beyond alevin stage.

galvanotaxis An artificial swimming motion induced in fish subjected to electrical stimulus.

ganoid scales Plate like; thick, hardened, interlocking diamond-shaped scale (for example, in gars).

gape Extent of the opening of the mouth; measured between the posterior clefts of the lips.

gastropod Taxonomic group of mollusks to which all snails belong.

genotype Genetic constitution of an organism.

gill Structures for gas exchange and food capture.

gill arch Curved supports for gill filaments and gill rakers, extending from the neurocranium to the throat, behind the eye and jaws.

gill cover Paired plates made up of opercle, preopercle, interopercle, subopercle, and branchiostegal rays, covering the gills.

gill filaments The vascularized, posterior projections from gill arches, for exchange of oxygen, carbon dioxide, and ions.

gill rakers The anterior, fingerlike projections from gill arches, usually toothed, for catching and sorting food items. 7 – 9 + 11 – 13 would be interpreted as 7 to 9 gill rakers on the upper limb of the first gill arch and 11 to 13 gill rakers on the lower limb.

gonopodium An organ adapted for sperm transfer (for example, in male Western Mosquitofish).

gradient Change from one level or type to another over time, space, or other dimension.

gravid Female fish with mature eggs or embryos.

herbivorous Feeding on living plant (vegetative) materials.

heterocercal Descriptor for tail or caudal fin that has an upper lobe larger than the lower lobe. The vertebral column extends into the upper lobe.

Holarctic Palaearctic (Old World) and Nearctic (New World) biogeographical regions of the North Temperate Zone.

holotyde A single type specimen that is the basis of the description and name of a new species.

homocercal Descriptor for tail or caudal fin that has more or less equal upper and lower lobes.

hybrid Product of sexual union between two genera or two species.

hyoid teeth Teeth found on the tongue, as in trout and salmon.

hypocercal Descriptor for tail or caudal fin that has dissimilar lobes; the lower lobe is longer.

hypolimnion The layer of water below the thermocline in lakes; may stagnate during summer heat.

indigenous See native.

inferior Refers to the lower jaw of a fish when it is shorter than the snout.

introgression (genetic) Movement of genes between species through hybridization.

insertion (of fin) The posterior point (trailing edge) of attachment of a fin to the body. See origin.

interorbital width Narrowest width across the head between the bony edges of the orbits; or distance across the head between the fleshy orbital margins.

interspecific Between two species.

lacustrine In reference to lakes.

landlocked Describes the condition of fish that do not migrate to the ocean in any part of their life cycle because of barriers to migration (for example Kokanee).

lateral On the side, as in lateral line.

lateral line The series of sense organ pores located along each side of a fish.

lateral line scales The row(s) of scales, from the shoulder to the base of the tail, each with a tube housing sensory structures.

lateral series Diagonal rows of scales along the fish's sides, from the top of the gill cover of scales to the caudal base. The number of rows, usually about the same as the number of scales in the lateral line (if present), is counted to help identify many fish.

limnetic Of, or relating to or living in open water of oceans or lakes;

length Standard length (SL) is a measurement of fish length from the tip of the snout to the crease at the base of the caudal fin behind the last vertebra; fork length (FL) is a measurement of fish length from the tip of the snout to the fork in the tail; total length (TL) is a measurement of fish length from the tip of the snout to the end of the tail.

lentic Water that is standing, as in a lake or pond; opposite of lotic or flowing.

littoral zone Inshore area of lakes and reservoirs. Well lit and defined by presence of macrophytes.

live bearer Fishes in which development through the larval stage occurs within the mother, following internal fertilization.

lotic Running water (rivers, streams, creeks) opposite of lentic.

macroinvertebrate Aquatic invertebrate fauna retained on a 500 micrometer net or sieve.

macropthalmia Juvenile stage of lamprey.

mandible Lower jaw.

margin Edge or border (of a fin) that is farthest from the points of attachment.

maxillae Paired bones forming the back part of the upper jaw edge. It is toothed in some families of soft-rayed fishes.

membranes, fin Skin webbing between fin rays or spines.

median fins Unpaired fins; adipose, anal, caudal, dorsal fins

meristic Relative to a count of serial parts of fishes; fin rays, lateral line scales, vertebrae.

mesotrophy A nutrient state of water enrichment between eutrophy and oligotrophy.

metalimnion Layer of very steep thermal gradient (high rate of change or temperature) (see thermocline).

monophyletic group In cladistic usage, a monophyletic group is a taxon (group of organisms) which forms a clade, meaning that it consists of an ancestral species and all its descendants. Monophyletic groups are typically characterized by shared derived characteristics.

molar teeth Broad, flattened teeth, as on the pharyngeal bones of Common Carp and Peamouth, for example. Flat teeth typically used for grinding or crushing.

mya Standard geological abbreviation for duration in millions of years, e.g., the Cutthroat Trout lineage has been in the West for 10 m.y.

naked Without scales or other structures on the skin surface, as in North American catfishes.

nape Area along the back between the posterior end of the head and the dorsal origin.

narcosis A state of immobility in fish resulting from slackened muscles in an electrical field.

nasal Of or relating to nostrils and the area surrounding them.

native Naturally occurring in an ecosystem, ecosystem or region; also indigenous.

nonnative Not naturally occurring (in an ecosystem), neither endemic nor indigenous to a place or ecosystem. Exotic, introduced.

nuchal The anterior most part of the back, just behind the head. Enlarged with massive muscles in some Colorado River *Gila* or a bony keel in *Xyrauchen* of the Colorado River.

nuptial tubercles Breeding tubercles.

oligotrophic Said of waters, generally lakes, that are nutrient poor, contrasted with eutrophic, nutrient rich.

opercle The bone that is the main gill cover. Surrounded by the preopercle (in front) and the interopercle and subopercle (below).

opercular flap Extension of the gill cover, usually bearing a spot used in social displays, as in sunfishes.

otoliths Calcareous "earstones" used for hearing and balance in the middle ear of bony fishes. Daily and annual growth rings are counted to age the fish or analyzed chemically to determine diet and temperature during life.

oviparous Said of fish that lay eggs that are externally fertilized.

ovoviviparous Said of fish that give birth to live young hatched from eggs that are fertilized internally and develop inside the body cavity.

origin of fin The anterior point of attachment of the leading edge of a fin to a fish's body.

palatine bones Paired bones, often bearing teeth, on the sides of the roof of the mouth internal to (mesial to) the maxillae and lateral to the vomer.

parapatric Occupying contiguous but non-overlapping distributions in an ecosystem, so that species interbreeding is prevented. See allopatric, sympatric.

papillae Small, raised clusters of taste buds in fish mouths and on sucker lips.

parr A juvenile and freshwater stage of salmonids, usually in reference to individuals in their second, or later growing season.

parr marks Vertical dark marks on the sides of juveniles of many species of salmonids.

pectoral fins Paired fins attached to shoulder bones, directly behind the head, close behind or below the gill openings.

peduncle, caudal Body section between the anal and caudal fins. Relative depth can be diagnostic, as in Torrent Sculpin, Chiselmouth, Tui Chub, or Tench.

pelagic Of or relating to deep water, open water away from shore.

pelvic fins Paired ventral fins behind or below the pectoral fins (in spiny-rayed fishes, for example) or below the dorsal fin or midbody (in soft-rayed fishes, for example).

peritoneum Membrane lining the abdominal cavity. Color is often diagnostic.

pharyngeal bones Paired or fused bones lying in the throat behind the gills and ahead of the esophagus; the fifth pharyngeal arch.

pharyngeal teeth Teeth on the posterior gill arches in the pharynx, below or beside the esophagus and between the shoulder bones. These are readily dissected with a scalpel or dental tool, for examination. Each species has unique pharyngeal teeth.

phenotype Physical expression of the interaction of individuals genotype and their environment.

physoclistic Fish having a swim bladder that is isolated from the esophagus (species cannot swallow air).

phytoplankton Small plants that float or drift without capability for locomotion.

physostomous Fish having a swim bladder connected to the esophagus by a duct (these species can swallow air).

piscivorous An organism that feeds on fish.

pluvial Wet periods in non-glaciated regions that correspond to a glacial maximum in ice-covered regions.

polymorphic Occurring in more than one shape; a species with more than one morphological type.

polyphyletic Of or relating to a taxonomic group derived from more than one ancestor.

polytypic In a taxonomic group, having many subdivisions. (A species having many subspecies and geographical races, such as Cutthroat Trout).

posterior Toward the tail.

premaxillae Paired bones forming the front of the upper jaw. Premaxilla or premaxillary.

preopercle Half-moon-shaped cheekbone below and behind the eye and in front of the opercle; the site of insertion of jaw muscles. The posterior edge may be serrated or spiny in cottids and some percoids.

prickles Small bony spines in the skin of sculpins, especially in Bear Lake Sculpin and Torrent Sculpin, for example.

principal rays The fin rays counted from the upper leading edge of the fin to the last, excluding the small rays anterior to the long one reaching the outer edge of the fin.

pseudobranch False gill arch with gill filaments located on the inside surface of the gill cover.

pyloric caecae Finger-like pouches arising from the intestinal tract immediately behind the stomach.

putative Generally considered to reputed to be (as extant or present).

ray One of the supporting structures within the fin membranes; either flexible or segmented (soft rayed) or stiff and unsegmented (spines), (fin ray).

relict Surviving organism or population, generally of somewhat or highly restricted distribution where the species was formerly more widespread.

riffle A generally shallow, turbulent part of a stream with high to moderate slope.

redd Nest prepared by most females of the genera *Oncorhynchus, Salmo*, and *Salvelinus*, as a repository for fertilized eggs.

resident Refers to fish that do not migrate but remain in a defined home range for their entire life.

riverine Pertaining to rivers.

rosette of lower lip papillae Half-moon-shaped pattern of anterior sensory papillae on the lower lip of *Pantosteus platyrhynchus*.

run A relatively deep, moderate gradient area of stream or river habitat with little surface turbulence. May also refer to groups of fish as they migrate—generally anadromous species such as Steelhead (A Run).

SL Standard length. Body length measured from the anterior tip of the head to the base of the caudal fin, excluding caudal rays.

salmonids Members of the Salmonidae family of fishes.

sawtooth belly scales Hard, sharp-edged, and pointed scales on the belly of herring and shad.

scales Overlapping bony plates within the skin of fishes. Usually cycloid, without teeth, in soft-rayed fishes, and ctenoid, with an exposed tooth patch, in spiny-rayed fishes.

scutes A bony plate (for instance on sturgeon).

Secchi disk A mechanical optical device, generally a round plate, which is lowered into the water column to estimate turbidity.

serrated spine The pectoral spines of catfishes and dorsal and anal spines of carp and goldfish, bearing bony hooks or posterior sawteeth, respectively.

sinus A cavity within a bone structure.

slot limits Definition of fish lengths [for example, 305 to 356 mm (12 to 14 in)] for which all fish in that size range are legally defined as "keep" or "return."

species concepts Biological species are reproductively isolated from others. Phylogenetic species are diagnosed as unique, monophyletic lineages. Evolutionary species have independent histories.

spine An unsegmented, unbranched, usually rigid body part supporting fin membranes, or a sharp, bony projection; generally on the head.

subterminal mouth Mouth position slightly below the tip of the head, so that the lower jaw is well below the snout and extends slightly less anteriorly than the upper jaw. A close look at the size and position of the mouth is crucial for the identification of fish, especially minnows.

sympatry Living in the same place. Ability of two species to coexist and retain genetic identity. See allopatric, parapatric.

swim-up fry Also referred to as "yolk-sac fry." An indeterminate stage of life for fishes with egg yolk sacs. Fry are "swim-up" when they emerge from the gravel (redds) and enter the water column and begin to feed just prior to complete absorption of the yolk sac.

TL Total length, total length is a measurement of fish length from the tip of the snout to the end of the tail.

terete-bodied Cylindrical, streamlined, and tapered body shape.

terminal mouth Mouth position at the end of the head, so that upper and lower jaws both reach the leading point of the fish.

thermocline Zone of rapidly changing gradient in water temperature separating the epilimnion (uppermost) from the hypolimnion (lower) layers of water in a lake water circulation event.

truncate Characterized as an edge that is vertical or straight, cut off squarely, as in caudal fins of bullheads.

tubercle Conical projection (generally one of many) with a keratinized (cornified) tip occurring on the fins, head, and or body (generally breeding individuals).

type In taxonomy, specifying the first-named individual of a species or genus (type specimen). Type locality, geographic location of the individual(s) used to make the initial description for a species. Typically refers to the holotype system.

vent Synonymous with anus.

ventral Away from the neurocranium or axial skeleton toward the throat or belly position. The belly of a fish.

ventral fins The pelvic fins in fish.

ventral mouth Mouth position well below snout, on ventral profile of the head and body, as in most suckers.

vertebrae counts The number of vertebrae between the head and the caudal fin, useful in some identifications; often expressed as thoracic (with ribs) and caudal vertebrae.

vermiculate Marked with irregular lines or blotches, as in Brook Trout.

viviparous Bearing live young.

vomer Anterior bone on the roof of the mouth.

vomerine teeth Teeth on the vomer bone lying in the front part of the roof the buccal cavity behind the upper jaw.

water column A section of water from the surface to the bottom or some other definable depth.

watershed The entire land surface (drainage) draining into a stream course or basin.

water year The period from 1 October to 30 September during which precipitation is measured and calculated in many western states.

Weberian apparatus Modifications of the first four or five vertebrae and their parts into paired chains of bones that connect the air bladder to the inner ear in minnows, suckers, catfishes, and relatives.

zooplankton Minute aquatic animals (invertebrates) that swim, float, or drift in water currents.

Glossary Source Materials

Sigler and Sigler, 1996. Scott and Crossman, 1973. Miller, Minckley, and Norris, 2005. Wydoski and Whitney, 2003. Hubbs and Lagler, 2004.

LITERATURE CITED

Adams, S. B. and Schmetterling, D. A. Freshwater sculpins: phylogenetics to ecology. *Transactions of the American Fisheries Society* **136,** 1736–1741 (2007). DOI:10.1577/T07-023.1.

Adams, W. J., Behmer, D. J. and Weingarten, W. O. Recovery of shocked common shiner, Notropis cornutus, related to electric energy. *Transactions of the American Fisheries Society* **101,** 552–555. DOI:10.1577/1548-8659 (1972)101<552:ROSCSN>2.0.CO;2.

Agassiz, L. Synopsis of the ichthyological fauna of the Pacific slope of North America, chiefly from the collections made by the U. S. Expl. Exped. under the command of Capt. C. Wilkes, with recent additions and comparisons with eastern types. *American Journal of Science and Arts* **19,** 71–99, 215–231 (1855). Online at http://www.biodiversitylibrary.org/item/92734#page/1/mode/1up.

Agassiz, L. and Cabot, E. *Lake Superior: Its Physical Character, Vegetation, and Animals, Compared with those of Other and Similar Regions* (Gould, Kendall, and Lincoln, Boston, Massachusetts, 1850). Online at http://dx.doi.org/10.5962/bhl.title.29144.

Albrecht, B., Robinson, B. J., Luecke, C. and Kennedy, B. *Bioenergetics Simulation Assessing the Consumption Demand of Different Stocking Regimes of Sterile Lake Trout in Bear Lake, Utah/Idaho: Final Report.* (Idaho Department of Fish and Game, Boise, Idaho, 2004).

Allen, M. S., Hoyer, M. V. and Canfield, D. E. Factors related to black crappie occurrence, density, and growth in Florida lakes. *North American Journal of Fisheries Management* **18,** 864–871 (1998). DOI: 10.1577/1548-8675(1998)018<0864:FRTBCO>2.0.CO;2.

Amadio, C. J., Hubert, W. A., Johnson, K., Oberlie, D. and Dufek, D. Abundance of adult saugers across the Wind River watershed, Wyoming. *North American Journal of Fisheries Management* **26,** 156–162 (2006). DOI:10.1577/M05-092.1.

Ambrose, S. E. and Abell, S. *Lewis and Clark: Voyage of Discovery* (National Geographic Society, Washington D. C., 2002).

Anders, P. J., Richards, D. L. and Powell, M. S. The first endangered white sturgeon population: repercussions in an altered large river-floodplain ecosystem. *Pages 67–82 in Biology, Management, and Protection of North American Sturgeon: Proceedings of the Symposium Biology, Management, and Protection of Sturgeon* (American Fisheries Society, Bethesda, Maryland, 2002).

Andreasen, J. K. and Barnes, J. R. Reproductive life history of *Catostomus ardens* and *C. discobolus* in the Weber River, Utah. Copeia **1975,** 643–648 (1975). Online at http://www.jstor.org/stable/1443317.

Apperson, K. A. and Anders, P. J. *Kootenai River White Sturgeon Investigations and Experimental Culture, Annual Progress Report FY 1989* (U. S. Bonneville Power Administration, Division of Fish and Wildlife, Portland, Oregon, 1990). Online at http://www.fishsciences.net/projects/columbia_sturgeon/_pdfs/03-Koot_wh-sturg_inv_AnProgFY1989.pdf.

Bailey, J. E. Life history and ecology of the sculpin *Cottus bairdi punctulatus* in southwestern Montana. Copeia **1952,** 243–255 (1952). Online at http://www.jstor.org/stable/1439271.

Bailey, R. M. and Bond, C. E. *Four New Species of Freshwater Sculpins, Genus Cottus, from Western North America* (University of Michigan Museum of Zoology, Ann Arbor, Michigan, 1963). Online at https://deepblue.lib.umich.edu/bitstream/handle/2027.42/57070/OP634.pdf.

Bailey, R. M. and Dimick, M. F. *Cottus hubbsi, a New Cottid Fish from the Columbia River System in Washington and Idaho* (Occasional Papers of the Museum of Zoology number 513, University of Michigan, Ann Arbor, Michigan, 1949). Online at http://hdl.handle.net/2027.42/56951.

Bailey, R. M., Fitch, J. E., Herald, E. S., Lachner, E. A., Lindsey, C. C., Robins, C. R. and Scott, W. B. A *List of Common and Scientific Names of Fishes from the United States and Canada, Third edition* (Special Publication No. 6, American Fisheries Society, Washington, D. C., 1970).

Bajkow, A. D. Migration of white sturgeon (*Acipenser transmontanus*) in the Columbia River. *Oregon Fish Commission Research Briefs* **3,** 8–21 (1951). Online at http://www.fishsciences.net/projects/columbia_sturgeon/_pdfs/04-Migration_w-sturgeon_Columbia.pdf.

Barton, B. A, and Dwyer, W. P. Physiological stress effects of continuous- and pulsed- DC electroshock on juvenile bull trout. *Journal of Fish Biology* **51,** 998–1008 (1997). DOI:10.1111/j.1095-8649.1997.tb01538.x.

Bates, P. Juvenile indexing and juvenile abundance estimates of white sturgeon in the Snake River between Bliss and C. J. Strike dams in *Snake River White Sturgeon Conservation Plan – Implementation of Measures for the White Sturgeon in the Shoshone Falls to C. J. Strike and Hells Canyon Reaches of the Snake River.* (Idaho Power Company, Boise, Idaho, 2013).

Baxter, G. T. and Stone, M. D. *Fishes of Wyoming.* (Wyoming Game and Fish Department, Cheyenne, Wyoming, 1995)

Beamesderfer, R. and Anders, P., eds. *Columbia Basin White Sturgeon Planning Framework.* (Northwest Power and Conservation Council, Portland, Oregon, 2013). Online at https://www.nwcouncil.org/media/6948703/ColumbiaBasinWhiteSturgeonPlanningFramework2013Dec.pdf.

Beamesderfer, R. C. P., Rien, T. A. and Nigro, A. A. Differences in the dynamics and potential production of impounded and unimpounded white sturgeon populations in the lower Columbia River. *Transactions of the American Fisheries Society* **124,** 857–872 (1995). DOI:10.1577/1548-8659(1995)124<0857:DITDAP>2.3.CO;2.

Bean, T. H. Description of new species of fishes (*Uranidea marginata, Potamocottus bendirei*) and of *Myctophum crenulare* J. and G. *Proceedings of the United States National Museum* **4,** 27–28 (1881). Online at http://www.biodiversitylibrary.org/page/15663163.

Becker, G. C. *Fishes of Wisconsin* (University of Wisconsin Press, Madison, Wisconsin, 1983). Online at http://digicoll.library.wisc.edu/cgi-bin/EcoNatRes/EcoNatRes-idx?id=EcoNatRes.FishesWI.

Behnke, R. J. *About Trout: The Best of Robert J. Behnke from Trout Magazine* (Lyons Press, Guilford, Connecticut, 2007).

Behnke, R. J. Brown Trout. *Trout* **27,** 42–47 (1986).

Behnke, R. J. *Monograph of the Native Trouts of the Genus Salmo of Western North America* (U. S. Department of Agriculture, Forest Service, Rocky Mountain Region, Lakewood, Colorado, 1979). Online at http://www.nativefishlab.net/library/textpdf/18137.pdf.

Behnke, R. J. Personal communication to Russ Thurow. Wood River Redband Trout (Professor of Fishery and Wildlife Biology, Colorado State University, 1988).

Behnke, R. J. *Native Trout of Western North America* (American Fisheries Society, Bethesda, Maryland, 1992).

Behnke, R. J. *Trout and Salmon of North America* (Free Press, New York, New York, 2002).

Bell, R. Personal communication to Russ Thurow. Big Wood River Redband Trout (1988).

Benson, L. V., Lund, S. P., Smoot, J. P., Rhode, D. E., Spencer, R. J., Verosub, K. L., Louderback, L. A., Johnson, C. A., Rye, R. O. and Negrini, R. M. The rise and fall of Lake Bonneville between 45 and 10.5 ka. *Quaternary International* **235,** 57–69 (2011). DOI:10.1016/j.quaint.2010.12.014.

Bentz, B. Personal communication John W. Sigler. White Sturgeon. (Biologist, Idaho Power, Boise, Idaho, 2014).

Bentz, B. and Lepla, K. Report C. Population survey of white sturgeon between C. J. Strike and Swan Falls dams. Pages *69–92 in Snake River White Sturgeon Conservation Plan-Implementation of Measures for White Sturgeon in the Shoshone Falls to C. J. Strike and Hells Canyon Reaches of the Snake River* (Idaho Power Company, Boise, Idaho, 2013). Online at http://www.worldcat.org/title/trout-and-salmon-of-north-america/oclc/49626102andreferer=brief_results.

Beranek, L. P., Link, P. K. and Fanning, C. M. Miocene to Holocene landscape evolution of the western Snake River Plain region, Idaho: using the SHRIMP detrital zircon provenance record to track eastward migration of the Yellowstone hotspot. *Geological Society of America Bulletin* **118,** 1027–1050 (2006). DOI:10.1130/B25896.1.

Bestland, E. A., Link, P. K., Lanphere, M. A. and Champion, D. E. Paleoenvironments of sedimentary interbeds in the Pliocene and Quaternary Big Lost Trough, eastern Snake River Plain, Idaho. *Geological Society of America Special Papers* **353,** 27–44 (2002). DOI:10.1130/0-8137-2353-1.27.

Bevan D., Harville P., and Bergman, P. Summary Report, National Marine Fisheries Service, Snake River salmon recovery team: Final recommendations to the Summary Report, Portland, OR. 25 pages. (1994).

Billman, E. J., Wagner, E. J. and Arndt, R. E. Reproductive ecology and spawning substrate preference of the northern leatherside chub. *North American Journal of Aquaculture* **70,** 273–280 (2008). Online at http://www.tandfonline.com.dist.lib.usu.edu/doi/full/10.1577/A07-044.1.

Blakney, J. R., Loxterman, J. L. and Keeley, E. R. Range-wide comparisons of northern leatherside chub populations reveal historical and contemporary patterns of genetic variation. *Conservation Genetics* **15,** 757–770 (2014). DOI:10.1007/s10592-014-0576-1.

Bonar, S. A., Hubert, W. A. and Willis, D. W. *Standard Methods for Sampling North American Freshwater Fishes* (American Fisheries Society, Bethesda, Maryland, 2009).

Bond, C. E. *Biology of Fishes* (Saunders College Publishing, Philadelphia, Pennsylvania, 1979).

Bond, C. E. *Distribution and Ecology of Freshwater Sculpins, Genus Cottus, in Oregon* (University of Michigan, Ann Arbor, Michigan, 1963).

Botsford, L. W., Castilla, J. C. and Peterson, C. H. The management of fisheries and marine ecosystems. *Science* **277,** 509–515 (1997). DOI:10.1126/science.277.5325.509.

Bouchard, D. P. *Quaternary Bear River Paleohydrogeography Reconstructed from the* $^{87}Sr/^{86}Sr$ *Composition of Lacustrine Fossils* (Utah State University, Logan, Utah, 1997). Online at http://digitalcommons.usu.edu/etd/4424/.

Bouchard, D. P., Kaufman, D. S., Hochberg, A. and Quade, J. Quaternary history of the Thatcher Basin, Idaho, reconstructed from the $^{87}Sr/^{86}Sr$ and amino acid composition of lacustrine fossils: implications for the diversion of the Bear River into the Bonneville Basin. *Palaeogeography, Palaeoclimatology, Palaeoecology* **141,** 95–1 14 (1998).

Bouck, G. R. and Ball, R. C. Influence of capture methods on blood characteristics and mortality in the rainbow trout (*Salmo gairdneri*). *Transactions of the American Fisheries Society* **95,** 170–176 (1966). DOI: 10.1577/1548-8659(1966)95[170:IOCMOB]2.0.CO;2.

Bouwes, N. and Luecke, C. The fate of Bonneville cisco eggs in Bear Lake: evaluating mechanisms of egg loss. *Transactions of the American Fisheries Society* **126,** 240–247 (1997). DOI:10.1577/1548-8659(1997)126<0240:TFOBCE>2.3.CO;2.

Brannon, E., Brewer, S., Stetter, A., Miller, M. Utter, F. and Hersberger, W. *Columbia River White Sturgeon (Acipenser transmontanus) Population Genetics and Early Life History Study-Final Report* (U. S. Bonneville Power Administration, Division of Fish and Wildlife, Portland, Oregon, 1986). Online at http://www.fishsciences.net/projects/columbia_stur¬geon/_pdfs/15-Col_R_w-st_pop_gen_1_1_86-12_31_86.pdf.

Brannon, E. L. and Setter, A. L. *Movements of White Sturgeon in Lake Roosevelt: Final Report 1988-1991* (Bonneville Power Administration, Portland, Oregon, 1992). Online at http://www.osti.gov/scitech/biblio/10109586-BHhzQI/webviewable/.

Bretz, J. H. Glacial drainage on the Columbia Plateau. GSA *Bulletin* **34,** 573–608 (1923a). DOI:10.1130/GSAB-34-573.

Bretz, J. H. The channeled scabland of the Columbia Plateau. *Journal of Geology* **31,** 617–649. (1923b). Online at http://wpg.forestry.oregonstate.edu/sites/wpg/files/seminars/1923_Bretz.pdf.

Bright, R. C. *Pleistocene Lakes Thatcher and Bonneville, Southeastern Idaho* (University of Minnesota, Minneapolis, Minnesota, 1963).

Bright, R. C. and Ore, H. T. Evidence for the spillover of Lake Bonneville, southeastern Idaho. *Pages 143–146 in Centennial Field Guide 2 -Rocky Mountain Section* (Geological Society of America, Boulder, Colorado, 1987).

Brinkman, S. and Woodling, J. Zinc toxicity to the mottled sculpin (*Cottus bairdi*) in high-hardness water. *Environmental Toxicology and Chemistry* **24,** 1515-1517 (2005). DOI:10.1897/04-235R.1.

Brinsmead, J. and Fox, M. G. Morphological variation between lake- and stream-dwelling rock bass and pumpkinseed populations. *Journal of Fish Biology* **61,** 1619–1638 (2002). DOI:10.1111/j.1095-8649.2002.tb02502.x.

Broughton, J. M., Madsen, D. B. and Quade, J. Fish remains from Homestead Cave and lake levels of the past 13,000 years in the Bonneville Basin. *Quaternary Research* **83,** 392–401 (2000). DOI:10.1006/qres.2000.2133.

Brunner, P. C., Douglas, M. R., Osinov, A., Wilson, C. C. and Bernatchez, L. Holarctic phylogeography of arctic charr (*Salvelinus alpinus L.*) inferred from mitochondrial DNA sequences. *Evolution* **55,** 573–586 (2001). DOI:10.1111/j.0014-3820.2001.tb00790.x.

Bryne, A. B. *Idaho Steelhead Status Update-Session One, 2014 Pacific Coast Steelhead Management Meeting* (Pacific States Marine Fisheries Commission, Portland, Oregon, 2014). Online at http://www.psmfc.org/steelhead/past-2014.html.

Buddington, R. K. and Christofferson, J. P. Digestive and feeding characteristics of the chondrosteans. *Environmental Biology of Fishes* **14,** 31–41 (1985). DOI:10.1007/BF00001574.

Budy, P., Thiede, G. P. and McHugh, P. Quantification of the vital rates, abundance, and status of a critical, endemic population of Bonneville Cutthroat Trout. *North American Journal of Fisheries Management* **27,** 593–604 (2007). DOI:10.1577/M06-085.1.

Burger, C. V., Parkin, J. W., O'Farrell, M. and Murphy, A. Barrier technology helps deter fish at hydro facilities. *Hydro Review* **34,** 50–57 (2015). Online at http://www.hydroworld.com/articles/hr/print/volume-34/issue-5/articles/barrier-technology-helps-deter-fish-at-hydro-facilities.html

Burgner, R. L. Life history of sockeye salmon (*Oncorhynchus nerka*). *Pages 3–117 in Pacific Salmon Life Histories* (University of British Columbia Press, Vancouver, British Columbia, 1991). Online at http://swrcb2.swrcb.ca.gov/waterrights/water_issues/programs/bay_delta/deltaflow/docs/exhibits/nmfs/spprt_docs/nmfs_exh4_healeyi991.pdf.

Burner, C. J. Characteristics of spawning nests of Columbia River salmon. *Pages 97–110 in An Annotated Bibliography on the Biology of Pacific Tuna* (U. S. Department of the Interior, Fish and Wildlife Service, Washington D. C., 1951). Online at http://www.biodiversitylibrary.org/item/22630#page/111/mode/1up.

Burns, T. A. and Lantz, K. Physiological effects of electrofishing on largemouth bass. *The Progressive Fish-Culturist* **40,** 148–150 (1978). DOI:10.1577/1548-8659(1978)40[148:PEOEOL]2.0.CO;2.

Butts, A. E., Koenig, M., Kozfkay, J. R. and Porter, N. *Fisheries Management Annual Report, Southwest Region, 2014* (Idaho Department of Fish and Game, Southwest Region, Nampa, Idaho, 2016). Online at http://fishandgame.idaho.gov/public/docs/fishReportsNewsletters/SWReport15.pdf.

Campbell, M. R. *Genetic Diversity and Differentiation of Shoshone Sculpin Cottus greenei* (Idaho Department of Fish and Game, Boise, 2011). Online at http://idahodocs.cdmhost.com/cdm/singleitem/collection/p16293coll7/id/241797/rec/19.

Campbell, M. R., Kozfkay, C. C., Meyer, K. A., Powell, M. S. and Williams, R. N. Historical influences of volcanism and glaciation in shaping mitochondrial DNA variation and distribution in Yellowstone cutthroat trout across its native range. *Transactions of the American Fisheries Society* **140,** 91–107 (2011). DOI:10.1080/00028487.2011.557001.

Cannamela, D. A. *Autumn and Spring Food Habits of the Shorthead Sculpin, Cottus confusus, in the Big Lost River, Idaho* (Masters Thesis, Murray State University, Murray, Kentucky, 1979).

Carrey, J. and Conley, C. T*he Middle Fork and the Sheepeater War* (Backeddy Books, Cambridge, Idaho, 1977).

Cavender, T. M. Taxonomy and distribution of the bull trout, Salvelinus confluentus (Suckey) from the American northwest. *California Fish and Game* **64,** 139–174 (1978).

Chapman, D. and Chandler, J. A. Historical abundance of anadromous fish upstream of the Hells Canyon complex. *Pages 1–26 in Feasibility of Reintroduction of Anadromous Fish Above or Within the Hells Canyon Complex, Technical Report Appendix E.3. 1-2.* (Idaho Power Company, Boise, Idaho, 2003). Online at https://www.idahopower.com/pdfs/Relicensing/hellscanyon/hellspdfs/techappendices/Aquatic/e31_02_ch06.pdf.

Chase, B. C. The spawning habitat of anadromous rainbow smelt: trouble at the tidal interface. *Pages 859–862 in Challenges for Diadromous Fishes in a Dynamic Global Environment* (American Fisheries Society, Bethesda, Maryland, 2009). Online at http://www.mass.gov/eea/docs/dfg/dmf/publications/chase-859-862.pdf.

Christiansen, R. L. *The Quaternary and Pliocene Yellowstone Plateau Volcanic Field of Wyoming, Idaho, and Montana* (U. S. Geological Survey Professional Paper 729-G, Denver, Colorado, 2001). Online at http://pubs.usgs.gov/pp/pp729g/.

Clemens, B. J. *The Physiology Ecology and Run Diversity of Adult Pacific Lamprey, Entosphenus tridentatus, During the Fresh ter Spawning Migration* (Oregon State University, Corvallis, Oregon, 2011). Online at http://ir.library.oregonstate.edu/xmlui/handle/1957/23870.

Clinton, D. Description of a new species of fish from the Hudson River. *Annals of the Lyceum of Natural History of New York* **1,** 49–50 (1824). Online at http://www.biodiversitylibrary.org/item/54048#page/7/mode/1up.

Close, D. A., Currens, K. P., Jackson, A., Wildbill, A. J., Hansen, J., Bronson, P. and Aronsuu, K. Lessons from the reintroduction of a noncharismatic, migratory fish: Pacific Lamprey in the upper Umatilla River, Oregon. *Pages 233–253 in Biology and Conservation of Lampreys in North America* (American Fisheries Society, Bethesda, Maryland, 2009). Online at http://citeseerx.ist.psu.edu/viewdoc/download?doi=10.1.1.476.9355andrep=rep1andtype=pdf.

Coburn, M. M. and Cavender, T. M. Interrelationships of North American cyprinid fishes. *Pages 328–373 in Systematics, Historical Ecology, and North American Freshwater Fishes* (Stanford University Press, Stanford, California 1992).

Cole, D. D. *Ecomorphological and Genetic Investigations into the Utah Lake, UT Sucker Complex with Comparisons to the Jackson Lake, WY Sucker Complex* (Utah State University, Logan, Utah, 2014). Online at http://digitalcommons.usu.edu/etd/2122/.

Cole, D. D., Mock, K. E., Cardall, B. L. and Crowl, T. A. Morphological and genetic structuring in the Utah Lake sucker complex. *Molecular Ecology* **17,** 5189–5204 (2008). DOI:10.1111/j.1365-294X.2008.03990.x.

Colman, S. M., Kaufman, D. S., Bright, J., Heil, C., King, J. W., Dean, W. E., Rosenbaum, J. G., Forester, R. M., Bischoff, J. L., Perkins, M. and McGeehin, J. P. Age model for a continuous, ca 250-ka Quaternary lacustrine record from Bear Lake, Utah-Idaho. *Quaternary Science Reviews* **25,** 2271–2282 (2006). DOI:10.1016/j.quascirev.2005.10.015.

Columbia River Basin Lamprey Technical Workgroup. *Proceedings for the 1st Current Status of Lamprey Research in the Pacific Northwest Workshop.* (Northwest Power and Conservation Council, Vancouver Water Resources Center, Vancouver, Washington, March 8, 2004).

Committee on the Status of Endangered Wildlife in Canada. COSEWIC *Assessment and Status Report on the Columbia Sculpin, Cottus hubbsi, in Canada* (Committee on the Status of Endangered Wildlife in Canada, Ottawa, Ontario, 2010). Online at https://www.registrelep-sararegistry.gc.ca/virtual_sara/files/cosewic/sr_columbia_sculpin_0911_eng.pdf.

Connolly, P. J. *Life History of Shoshone Sculpin, Cottus greenei, in Southcentral Idaho* (University of Idaho, Moscow, Idaho, 1983).

Cope, E. D. Report on the recent reptiles and fishes of the survey, collected by Campbell Carrington and C. M. Dawes. *Page 467–476 in Preliminary Report of the United States Geological Survey of Montana and Portions of Adjacent Territories, Fifth Annual Report of Progress* (U. S. Department of the Interior, Washington, D. C. 1 872a). Online at https://archive.org/details/preliminaryrepor00geol.

Cope, E. D. and Yarrow, H. C. Report upon the collections of fishes made in portions of Nevada, Utah, California, Colorado, New Mexico, and Arizona, during the years of 1871, 1872, 1873, and 1874. *Pages 635–703 in Report upon United States Geographical Surveys West One Hundredth Meridian, Volume 5.* (U.S. Geographical Survey, Washington D.C., 1875). Online at https://babel.hathitrust.org/cgi/pt?id=njp.32101076035672.

Corley, D. R. *Snorkel Trend Counts of Fish in the Middle Fork–1971* (Idaho Department of Fish and Game, Division of Fisheries, Boise, Idaho, 1972). Online at http://idahodocs.cdmhost.com/cdm/compoundobject/collection/p16293coll7/id/231923/rec/2#img_view_container.

Courtenay, W. R. Jr., Robins, C. R., Bailey, R. M. and Deacon, J. E. Records of exotic fishes from Idaho and Wyoming. *The Great Basin Naturalist* **47,** 523–526 (1987). Online at https://ojs.lib.byu.edu/spc/index.php/wnan/article/view/28914.

Craig, J. F. A short review of pike ecology. *Hydrobiologia* **601,** 5–16 (2008). DOI:10.1007/s10750-007-9262-3.

Crozier, L. G., Hendry, A. P., Lawson, P. W., Quinn, T. P., Mantua, N. J., Battin, J., Shaw, R. G. and Huey, R. B. Potential responses to climate change in organisms with complex life histories: evolution and plasticity in Pacific salmon. *Evolutionary Applications* **1,** 252–270 (2008). DOI:10.1111/j.1752-4571.2008.00033.x.

Cucherousset, J., Paillisson, J.M., Carpentier, A., Eybert, M.C. and Olden, J. D. Habitat use of an artificial wetland by the invasive catfish *Ameiurus melas. Ecology of Freshwater Fish* **15,** 589–596 (2006). DOI:10.1111/j.1600-0633.2006.00199.x.

Curet, T. The Unwritten Laws of Fisheries with Revisions and Additions (Unpublished document of the Idaho Chapter of the American Fisheries Society, Boise, Idaho, 2012).

Curet, T. Personal communication to John Sigler. Big Wood River Redband Trout (Fish biologist, Idaho Department of Fish and Game, 2014).

Dalbey, S. R., McMahon, T. E. and Fredenberg, W. Effect of electrofishing pulse shape and electrofishing-induced spinal injury on long-term growth and survival of wild rainbow trout. *North American Journal of Fisheries Management* **16,** 560–569 (1996). DOI:10.1577/1548-8675(1996)016<0560:EOEPSA>2.3.CO;2.

Deacon, J. E., Williams, J. D., Kobetich, G. and Contreras, S. Fishes of North America endangered, threatened, or of special concern: 1979. *Fisheries* **4,** 29–44 (1979). DOI:10.1577/1548-8446-4-2.

DeMarais, B. D. *Genetic Relationships Among Fishes Allied to the Genus Gila (Teleostei: Cyprinidae) from the American Southwest.* (Arizona State University, Tempe, Arizona, 1992).

Dillon, J. and Grunder, S. A. *Management Plan for the Conservation of Snake River White Sturgeon in Idaho* (Idaho Department of Fish and Game, Boise, Idaho, 2008). Online at http://fishandgame.idaho.gov/public/fish/planSnakeWhiteSturgeon.pdf.

Dillon, J. C. and Myers, D. *American Shad Introduction to Brownlee Reservoir, Evaluation of Proposal* (Idaho Department of Fish and Game, Boise, Idaho, 1990). Online at http://idahodocs.cdmhost.com/cdm/singleitem/collection/p16293coll7/id/235399/rec/1#img_view_container.

Dowling, T. E., Tibbets, C. A., Minckley, W. L. and Smith, G. R. Evolutionary relationships of the plagopterins (Teleostei: Cyprinidae) from cytochrome b sequences. *Copeia* **2002,** 665–678 (2002). Online at http://www.jstor.org/stable/1448148.

Duke, S., Anders, P., Ennis, G., Hallock, R., Hammond, J., Ireland, S., Laufle, J., Lauzier, R., Lockhard, L., Marotz, B., Paragamian, V. L. and Westerhof, R. Recovery plan for Kootenai River white sturgeon (*Acipenser transmontanus*). *Journal of Applied Ichthyology* **15,** 157–163 (1999). DOI:10.1111/j.1439-0426.1999.tb00226.x.

Dunham, J. B., Adams, S. B., Schroeter, R. E. and Novinger, D. C. Alien invasions in aquatic ecosystems: toward an understanding of brook trout invasions and potential impacts on inland cutthroat trout in western North America. *Reviews in Fish Biology and Fisheries* **12,** 373–39 1 (2002). DOI:10.1023/A:1025338203702.

Dunham, J., Baxter, C., Fausch, K., Fredenberg, W., Kitano, S., Koizumi, I., Morita, K., Nakamura, T., Rieman, B., Savvaitova, K., Stanford, J., Taylor, E. and Yamamoto, S. Evolution, ecology, and conservation of dolly varden, white spotted char, and bull trout. *Fisheries* **33,** 537–550 (2008). DOI:10.1577/1548-8446-33.11.537.

Eades, R. T., Neal, J. W., Lang, T. J., Hunt, K. M., and Pajak, P. *Urban and Community Fisheries Programs: Development, Management, and Evaluation* (American Fisheries Society, Bethesda, Maryland, 2008).

Eaton, J. G. and Scheller, R. M. Effects of climate warming on fish thermal habitat in streams of the United *Limnology and Oceanography* **41,** 1109–1115 (1996). DOI:10.4319/lo.1996.41.5.1109.

Eddy, S. *How to Know the Freshwater Fishes* (Wm. C. Brown Company, Dubuque, Iowa, 1957).

Edson, M. The sturgeon story. *Idaho Wildlife Review* **9,** 4–9 (1956).

Egan, T. *The Big Burn: Teddy Roosevelt and the Fire that Saved America* (Mariner Books, Boston, Massachusetts, 2010).

Eigenmann, C. H. The percopsidae on the Pacific slope. *Science* **20,** 233–234 (1892). DOI:10.1126/science.ns-20.507.233.

Eigenmann, C. H. and Eigenmann, R. S. *Cottus beldingii*, sp. nov. *The American Naturalist* **25,** 1132 (1891). DOI:10.1086/275458.

Eigenmann, C. H. and Eigenmann, R. S. Preliminary descriptions of new fishes from the northwest. *The American Naturalist* **27,** 151–154 (1893). Online at http://www.journals.uchicago.edu/doi/pdfplus/10.1086/275676.

Eiting, T. P. and Smith, G. R. Miocene salmon (*Oncorhynchus*) from western North America: gill raker evolution correlated with plankton productivity in the Eastern Pacific. *Palaeogeography, Palaeoclimatology, Palaeoecology* **249,** 412–424 (2007). DOI:10.1016/j.palaeo.2007.02.011.

Elz, A. E. A *Hierarchical Analysis of Historical Processes and Phylogeographic Patterns in Salvelinus (Pisces: Salmonidae)* (University of British Columbia, Vancouver, British Columbia, 2003). Online at https://circle.ubc.ca/handle/2429/14632.

Etnier, D. A. and Starnes, W. C. *The Fishes of Tennessee* (University of Tennessee Press, Knoxville, Tennessee, 1993).

Everhart, W. H. and Waters, C. A. Life history of the blueback trout (Arctic char, Salvelinus (Linnaeus)) in Maine. T*ransactions of the American Fisheries Society* **94,** 393–397 (1965). DOI:10.1577/1548-8659(1965)94[393:LHOTBT]2.0.CO;2.

Federal Register. *Endangered and Threatened Wildlife and Plants; 12-Month Finding for a Petition to List the Yellowstone Cutthroat Trout as Threatened.* (U. S. D. I. Fish and Wildlife Service, FR Document No. 06-1539, 2006).

Flinders, J., Hansen, J., White, M., Beller, B. and Curet, T. *Fishery Management Investigations, Salmon Region 2012* (Idaho Department of Fish and Game, Boise, Idaho, 2013). Online at https://collaboration.idfg.idaho.gov/FisheriesTechnicalReports/Mgt.13-123Flinders2012%20Fishery%20Management%20Annual%20Report%20Salmon%20Region%202012.pdf.

Forster, J. R. An account of some curious fishes, sent from Hudson's Bay. *Philosophical Transaction of the Royal Society* **62,** 149–160 (1773). Online at http://dx.doi.org/10.5962/bhl.title.38012.

Franzin, W. G. and Clayton, J. W. A biochemical genetic study of zoogeography of Lake Whitefish (*Coregonus clupeaformis*) in western Canada. *Journal of the Fisheries Research Board of Canada* **34,** 617–625 (1977). DOI:10.1139/f77-097.

Fuller, P., Cannister, M. and Neilson, M. *Salvelinus alpinus oquassa* (U. S. Geological Survey, Nonindigenous Aquatic Species Database, Gainesville, Florida). Online at http://nas.er.usgs.gov/queries/FactSheet.aspx?speciesID=936.(2012).

Fuller, P., Cannister, M. and Neilson, M. *Thymallus arcticus* (U. S. Geological Survey, Nonindigenous Aquatic Species Database, Gainesville, Florida). Online at http://nas.er.usgs.gov/queries/factsheet.aspx?SpeciesID=943. (2014).

Fulton, L. A. *Spawning Areas and Abundance of Chinook Salmon (Oncorhynchus tshawytscha) in the Columbia River Basin - Past and Present.* (Special Scientific Report – Fisheries No. 571, United States Fish and Wildlife Service, Washington, D.C., 1968). Online at https://www.nwfsc.noaa.gov/assets/26/6638_08042010j45107_Fulton.1968-rev.pdf.

Gadomski, D. M. and Parsley, M. J. Effects of turbidity, light level, and cover on predation of white sturgeon larvae by prickly sculpins. *Transactions of the American Fisheries Society* **134,** 369–374 (2005). DOI:10.1577/T03-213.1.

Gamett, B. L. and Bartell, J. A. *The Status of Fishes on the Middle Fork Ranger District, Salmon-Challis National Forest (2002-2008)* (Trout Unlimited, Idaho Water Project, Boise, Idaho, 2011). Online at http://www.fs.fed.us/rm/boise/AWAE/projects/ClimateShield/downloads/Bibliography/13_Gamett-MiddleForkFishPopulationReport.pdf.

Gamett, B. L. Personal communication to John Sigler. Bull Trout in the Lost River drainages. (Fish Biologist, U. S. Forest Service, Salmon-Challis National Forest, 2015).

Gasser, K. W., Cannamela, D. A. and Johnson., D. W. Contributions to the life history of the shorthead sculpin, *Cottus confusus,* in the Big Lost River, Idaho: age, growth, and fecundity. *Northwest Science* **55,** 174–181. (1981).

Gebhards, S. V. *Columbia River Fisheries Development Program: Salmon River Preliminary Planning Report* (Idaho Department of Fish and Game, Boise, Idaho, 1959). Online at http://idahodocs.contentdm.oclc.org/cdm/compoundobject/collection/p16293coll7/id/13489/rec/15.

Geslin, J. K., Link, P. K. and Fanning, C. M. High-precision provenance determination using detrital-zircon ages and petrography of Quaternary sands on the eastern Snake River Plain, Idaho. *Geology* **27,** 295–298 (1999). DOI:10.1130/0091-7613(1999)027<0295:HPPDUD>2.3.CO;2.

Giddings, E. M and Stephens, D. *Selected Aquatic Biological Investigations in the Great Salt Lake Basins, 1875-1998, National Water-Quality Assessment Program.* (Water-Resource Investigations Report 99-4132, U. S. Geological Survey, West Valley City, Utah, 1999). Online at http://pubs.usgs.gov/wri/wri994132/pdf/WRI994132.pdf.

Gilbert, C. H. and Culver., G. B. *Uranidea greenei. Page 1965 in The Fishes of North and Middle America: A Descriptive Catalogue of the Species of Fish-Like Vertebrates Found in the Waters of North America, North of the Isthmus of Panama* (Bulletin 47, Part II, Smithsonian Institution, United States National Museum, Washington, D. C., 1898). Online at http://dx.doi.org/10.5962/bhl.title.46755.

Gilbert, C. H. and Evermann, B. W. A report upon investigations in the Columbia River basin, with descriptions of four new species of fishes. *Pages 169-207 in Investigations in the Columbia River Basin* (Bulletin of the United States Fish Commission, Washington D. C., 1895) Online at http://fishbull.noaa.gov/14-1/gilbert1.pdf.

Gilbert, G. K. *Lake Bonneville* (United States Geological Survey, Washington, D. C., 1890). Online at https://archive.org/details/lakebonneville00gilb.

Gill, T. N. Observations on the genus *Cottus,* and descriptions of two new species (abridged from the forthcoming report of Capt. J. H. Simpson). *Proceedings of the Boston Society of Natural History* **8,** 40–42 (1862).

Girard, C. F. A monograph of the fresh water *Cottus* of North America. *Pages 409–411 in Proceedings of the American Association for the Advancement of Science: Second Meeting* (Henry Flanders and Company, Boston, Massachusetts, 1850). Online at http://books.google.com/books?id=5RsfAAAAMAAJ&printsec=titlepage.

Girard, C. F. Researches upon the cyprinoid fishes inhabiting the fresh waters of the United States west of the Mississippi valley, from specimens in the museum of the Smithsonian Institution. *Proceedings of the Academy of Natural Sciences of Philadelphia* **8,** 165–213 (1856). Online at http://biodiversitylibrary.org/page/1935220.

Godsey, H. S., Currey, D. R. and Chan, M. A. New evidence for an extended occupation of the Provo shoreline and implications for regional climate change, Pleistocene Lake Bonneville, Utah, USA. *Quaternary Research* **63,** 212–223 (2005). DOI:10.1016/j.yqres.2005.01.002.

Golder Associates Ltd. *White Sturgeon Spawning at Waneta, 2009 Investigations* (Golder Associates, Ltd, Castlegar, British Columbia, 2010).

Goto, A., Yokoyama, R. and Sideleva., V. G. Evolutionary diversification in freshwater sculpins (Cottoidea): a review of two major adaptive radiations. *Environmental Biology of Fishes* **98,** 307–335 (2015). DOI:10.1007/s10641-014-0262-7.

Grunder, S. A. Personal communication to John Sigler. Northern Leatherside Chub. (Fish Biologist, Idaho Department of Fish and Game, 2014a).

Grunder, S. A. Personal communication to John Sigler. Westslope Cutthroat Trout. (Fish Biologist, Idaho Department of Fish and Game, 2014b).

Grunder, S. A., McArthur, T. J., Clark, S. and Moore, V. K. *Idaho Department of Fish and Game 2003 Economic Survey Report* (Idaho Department of Fish and Game, Boise, Idaho, 2008). Online at https://collaboration.idfg.idaho.gov/FisheriesTechnicalReports/Mgt08-129Grunder2003%20Economic%20Survey%20Report.pdf.

Gunnell, K., Tada, M. K., Hawthorne, F. A., Keeley, E. R. and Ptacek, M. B. Geographic patterns of introgressive hybridization between native Yellowstone cutthroat trout (*Oncorhynchus clarkii bouvieri*) and introduced rainbow trout (*O. mykiss)* in the South Fork of the Snake River watershed, Idaho. *Conservation Genetics* **9,** 49–64 (2008). DOI:10.1007/s10592-007-9302-6.

Guy, C. S. and Brown, M. L. *Analysis and Interpretation of Freshwater Fisheries Data* (American Fisheries Society, Bethesda, Maryland, 2007).

Halverson, A. An *Entirely Synthetic Fish: How Rainbow Trout Beguiled America and Overran the World* (Yale University Press, New Haven, Connecticut, 2010). Online at http://www.jstor.org/stable/j.ctt1nq8bk.

Halverson, M. A. Stocking trends: a quantitative review of governmental fish stocking in the United States, 1931 to 2004. *Fisheries* **33,** 69–75 (2008). DOI:10.1577/1548-8446-33.2.69.

Hasselman, D. J., Ricard, D. and Bentzen, P. Genetic diversity and differentiation in a wide ranging anadromous fish, American shad *(Alosa sapidissima)*, is correlated with latitude. *Molecular Ecology 22,* 1558–1573 (2013). DOI:10.1111/mec.12197.

Hatten, J. R. and Parsley, M. J. A spatial model of white sturgeon rearing habitat in the lower Columbia River, USA. *Ecological Modelling* **220,** 3638–3646 (2009). DOI:10.1016/j.ecolmodel.2009.03.006.

Hauck, F. *A Survey of Fish Losses in Irrigation Diversions of the Big Wood River in Determining the Placement of an Experimental Fish Screen* (Idaho Department of Fish and Game, Boise, Idaho, 1949).

Haynes, J. M., Gray, R. H. and Montgomery, J. C. Seasonal movements of white sturgeon *(Acipenser trans-montanus)* in the mid-Columbia River. *Transactions of the American Fisheries Society* **107,** 275–280 (1978). DOI:10.1577/1548-8659(1978)107<275:SMOWSA>2.0.CO;2.

Healey, M. C. Life History of chinook salmon *(Oncorhynchus tshawytscha). Pages 311–393 in Pacific Salmon Life Histories* (UBC Press, University of British Columbia, Vancouver, British Columbia, 1991). Online at http://swrcb2.swrcb.ca.gov/waterrights/water_issues/programs/bay_delta/delta-flow/docs/exhibits/nmfs/spprt_docs/nmfs_exh4_healeyi991.pdf.

Hearst, J. M. and Smith, G. R. Fishes of the Late Pleistocene Duck Point Local Fauna, Power County, Idaho: equable climate indicator or reworked assemblage? *Pages 131–140 in And Whereas... Papers on the Vertebrate Paleontology of Idaho Honoring John A. White.* (Occasional Paper 37, Idaho Museum of Natural History, Pocatello, Idaho, 2002).

Hebdon, L. Personal communication to John Sigler. Chinook Salmon. (Fish biologist, Idaho Department of Fish and Game, 2014).

Hebdon, L. Personal communication to John Sigler. Sand Roller. (Fish biologist, Idaho Department of Fish and Game, 2015a).

Hebdon, L. Personal communication to John Sigler. Sand Roller abundance. (Fish biologist, Idaho Department of Fish and Game, 2015b).

Hesse, J. Personal communication to John Sigler. Fish hatchery management. (Fish biologist, Nez Perce Tribe, 2015)

Helfman, G., Collette, B. B., Facey, D. E. and Bowen, B. W. *The Diversity of Fishes: Biology, Evolution, and Ecology.* (John Wiley & Sons, Hoboken, New Jersey, 2009).

Hermann, L. Eine wirkung galvanischer strome auf organismen. *Pflügers Archiv - European Journal of Physiology* **37,** *457–460* (1885). DOI:10.1007/BF01752435.

Hermann, L. and Matthias, F. Der galvanotropismus der von Rana temporia und der fische. *Pflügers Archiv - European Journal of Physiology* **57,** 391–405 (1894). DOI:10.1007/BF01662194.

Higgins, S. N. and Vander Zanden, M. J. What a difference a species makes: a meta-analysis of dreissenid mussel impacts on freshwater ecosystems. *Ecological Monographs* **80,** 179–196 (2010). Online at http://dx.doi.org/10.1890/09-1249.1.

Hillyard, R. W. and Keeley, E. R. Temperature-related changes in habitat quality and use by Bonneville cutthroat trout in regulated and unregulated river segments. *Transactions of the American Fisheries Society* **141,** 1649–1663 (2012). DOI:10.1080/00028487.2012.717517.

Hodges, M. K. V., Link, P. K. and Fanning, C. M. Drainage disruption by basaltic lava fields along the Snake River Plain hotspot track: detrital zircon evidence from drillcore constrains the course of the Pliocene Big Lost River. *Journal of Volcanology and Geothermal Research* **188,** 237–249 (2009). DOI:10.1016/j.jvolgeores.2009.08.019.

Holliman, F. M., Reynolds, J. B. and Holmes, J. A. S. Effects of healing time and pulsed-DC waveform on injury detection and incidence in electroshocked juvenile Chinook salmon. *North American Journal of Fisheries Management* **30,** 1413–1419 (2010). DOI:10.1577/M10-049.1.

Hopken, M. W., Douglas, M. R. and Douglas, M. E. Stream hierarchy defines riverscape genetics of a North American desert fish. *Molecular Ecology* **22,** 956–971 (2013). DOI:10.1111/mec.12156.

Houston, D. D., Shiozawa, D. K., Smith, B. T. and Riddle, B. R. Investigating the effects of Pleistocene events on genetic divergence within *Richardsonius balteatus*, a widely distributed western North American minnow. BMC *Evolutionary Biology* **14,** 1–18 (2014). DOI:10.1186/1471-2148-14-111.

Howell, M. D. and McLellan, J. G. *Lake Roosevelt White Sturgeon Recovery Project.* (Bonneville Power Administration, Portland, Oregon, 2013).

Hrabik, T. R., Carey, M. P. and Webster, M. S. Interactions between young-of-the-year exotic rainbow smelt and native yellow perch in a northern temperate lake. *Transactions of the American Fisheries Society* **130,** 568–582 (2001). DOI:10.1577/1548-8659(2001)130<0568:IBYOTY>2.0.CO;2.

Hubbs, C. L. and Lagler, K. F. *Fishes of the Great Lakes Region, Revised edition* (University of Michigan Press, Ann Arbor, Michigan, 2004).

Hubbs, C. L. and Miller, R. R. Correlation between fish distribution and hydrographic history of the desert basins of the western United States. *Pages 17–166 in The Great Basin, With Emphasis on Glacial and Postglacial Times* (University of Utah, Salt Lake City, Utah, 1948). Online at http://content.lib.utah.edu/cdm/ref/collection/uspace/id/7188.

Hubbs, C. L. and Schultz, L. P. *Cottus tubulatus, a New Sculpin from Idaho* (University of Michigan Museum of Zoology, Ann Arbor, Michigan, 1932). Online at https://deepblue.lib.umich.edu/bitstream/handle/2027.42/56681/OP242.pdf.

Hubert, W. A., Pope, K. L. and Dettmers, J. M. Passive capture techniques. *Pages 223–265 in Fisheries Techniques, Third edition* (American Fisheries Society, Bethesda, Maryland, 2012). Online http://digitalcommons.unl.edu/ncfwrustaff/111/.

Idaho Department of Fish and Game. Go Fish Idaho! An Angler's Guide to the Gem State. (Undated).

Idaho Department of Fish and Game. *Anadromous Fish Management Plan 1992–1996* (Idaho Department of Fish and Game, Boise, Idaho, 1992). Online at http://idahodocs.cdmhost.com/cdm/ref/collection/p16293coll7/id/235465.

Idaho Department of Fish and Game. *Fisheries Management Plan 1996-2000.* (Fisheries Bureau, Boise, Idaho, 1996).

Idaho Department of Fish and Game. *Idaho Comprehensive Wildlife Conservation Strategy* (Idaho Conservation Data Center, Idaho Department of Fish and Game, Boise, Idaho, 2005) Online at http://fishandgame.idaho.gov/cms/tech/CDC/cwcs.cfm.

Idaho Department of Fish and Game. *Idaho Fisheries Management Plan: 1991–1995* (Idaho Department of Fish and Game, Boise, Idaho, 1990). Online at http://www.chelanpud.org/relicense/study/refer/4354_1.pdf.

Idaho Department of Fish and Game. *Idaho Fisheries Management Plan 2013–2018: A Comprehensive Guide to Managing Idaho's Fisheries Resources* (Idaho Department of Fish and Game, Boise, Idaho, 2013). Online at https://fishandgame.idaho.gov/public/fish/planFisheries.pdf.

Idaho Department of Fish and Game. *Idaho Record Fish* (Idaho Department of Fish and Game, Boise, Idaho). Online at http://fishandgame.idaho.gov/public/fish/?getPage=82. (2015).

Idaho Department of Fish and Game. *Mountain Lakes of Idaho* (Idaho Fish and Game, Boise, Idaho, 1973).

Idaho Department of Fish and Game. *Mountain Whitefish Conservation and Management Plan for the Big Lost River Drainage, Idaho* (Idaho Department of Fish and Game, Boise, Idaho, 2007). Online at http://fishandgame.idaho.gov/public/fish/planBigLostRiverWhitefish.pdf.

Idaho Department of Fish and Game. *Reintroduction of Salmon and Steelhead into Portions of the Clearwater River Drainage July 1, 1967 to June 30, 1968: Annual Project Closing Report* (Idaho Department of Fish and Game, Boise, Idaho, 1969). Online at http://idahodocs.cdmhost.com/cdm/compoundobject/collection/p16293coll7/id/230603/rec/12.

Idaho Department of Fish and Game. *Status of Pacific Lamprey Entosphenus tridentatus in Idaho* (Idaho Department of Fish and Game, Boise, Idaho, 2011). Online at http://idahodocs.cdmhost.com/cdm/singleitem/collection/p16293coll7/id/245182/rec/11.

Idaho Power Company. *Snake River White Sturgeon Conservation Plan* (Idaho Power Company, Boise, Idaho, 2005). Online at https://www.idahopower.com/pdfs/ourEnvironment/WSCP_Final.pdf.

Idaho State Historical Society. *Alexander Ross's 1824 Route into Stanley Basin* (Idaho State Historical Society, Boise, Idaho, 1964). Online at http://history.idaho.gov/sites/default/files/uploads/reference-series/0281.pdf.

International Game and Fish Association. Online World Record Fish. (2014). Online at http://wrec.igfa.org/.

International Game and Fish Association. Online World Record Fish. (2015). Online at http://wrec.igfa.org/.

Irizarry, R. A. *The Effects of Stream Alterations in Idaho* (Idaho Department of Fish and Game, Boise, Idaho, 1969).

Irvine, R. L., Schmidt, D. C. and Hildebrand, L. R. Population status of white sturgeon in the lower Columbia River within Canada. *Transactions of the American Fisheries Society* **136,** 1472–1479 (2007). DOI:10.1577/T06-1901.

Israel, J., Drauch, A. and Gingras, M. *Life History Conceptual Model for White Sturgeon (Acipenser transmontanus)* (Bay Delta Ecosystem Restoration and Improvement Program, Sacramento, California, 2009). Online at https://nrm.dfg.ca.gov/FileHandler.ashx?DocumentID=28423.

Jager, H. I., Chandler, J. A., Lepla, K. B. and Van Winkle, W. Theoretical study of river fragmentation by dams and its effects on white sturgeon populations. *Environmental Biology of Fishes* **60,** 347–361 (2001). DOI:10.1023/A:1011036127663.

Janecke, S. U. and Oaks, R. Q. Jr. New insights into the outlet conditions of Late Pleistocene Lake Bonneville, southeastern Idaho, USA. *Geosphere* **7,** 1369–1391 (2011). DOI:10.1130/GES00587.1.

Jeppson, P. and Ball, K. *Regional Fisheries Management Investigations: Region 6 Streams Investigations. Job Performance Report. Project F-71-R-3, Job VI-C* (Idaho Department of Fish and Game, Boise, Idaho 1979).

Johnson, B. M. and Goettl, J. P. Food web changes over fourteen years following introduction of rainbow smelt into a Colorado reservoir. *North American Journal of Fisheries Management* **19,** 629–642 (1999). DOI:10.1577/1548-8675(1999)019<0629:FWCOFY>2.0.CO;2.

Johnson, D. W., Cannamela, D. A. and Gasser., K. W. Food habits of the shorthead sculpin (*Cottus confusus*) in the Big Lost River, Idaho. *Northwest Science* **57,** 229–239 (1983).

Johnson, J. B. Evolution after the flood: phylogeography of the desert fish Utah Chub. *Evolution* **56**, 948–960 (2002). DOI:10.1111/j.0014-3820.2002.tb01407.x.

Johnson, J. B. and Belk, M. C. Effects of predation on life-history evolution in Utah chub *(Gila atraria). Copeia* **1999,** 948–957 (1999). Online at http://www.jstor.org/stable/1447970.

Johnson, J. B., Dowling, T. E. and Belk, M. C. Neglected taxonomy of rare desert fishes: congruent evidence for two species of leatherside chub. *Systematic Biology* **53,** 841–855 (2004). DOI:10.1080/10635150490522557.

Jordan, D. S. *A Catalogue of the Fishes of the Fresh Waters of North America* (Extracted from the Bulletin of the Survey, volume 4, number 2., U. S. Geological and Geographical Survey, Washington, D. C., 1878). Online at http://hdl.handle.net/2027/hvd.hwfk58.

Jordan, D. S. Description of a new species of salmon, *Oncorhynchus kamloops*, from the lakes of British Columbia. *Forest and Stream* **39**, 405–406 (1892). Online at http://biodiversitylibrary.org/page/46152580.

Jordan, D. S. Description of the golden trout of Kern County, California (*Salmo mykiss aqua-bonita*). *Pages 481–483 in Volume 15 of the Proceedings of the U. S. National Museum* (Smithsonian Institution, Washington D. C., 1893). Online at http://www.biodiversitylibrary.org/item/53607#page/653/mode/1up.

Jordan, D. S. and Evermann, B. M. *American Food and Game Fishes: A Popular Account of All Species Found in America North of the Equator* (Doubleday, Garden City, New York, 1904). Online at http://dx.doi.org/10.5962/bhl.title.1191.

Jordan, D. S., Evermann, B. M. and Clark, H. W. *Check List of the Fishes and Fishlike Vertebrates of North and Middle America, North of the Northern Boundary of Venezuela and Columbia* (A Reprint of Appendix X to the Report of the U. S. Commissioner of Fisheries for the fiscal year 1928, Washington D. C. 1930). Online at http://hdl.handle.net/2027/uc1.31822000621789.

Jordan, D. S. and Gilbert, C. H. Notes on a collection of fishes from Utah Lake. *Proceedings of the United States National Museum* **3**, 459–464 (1880). Online at http://biodiversitylibrary.org/page/14917559.

Keeley, E. Personal communication to John Sigler. Characteristics of Westslope Cutthroat Trout meristics. (Professor of Biological Sciences, Idaho State University, Pocatello, Idaho, 2015).

Keil, W. *Annual Report of the Fish Commissioner for Idaho* (Idaho Fish Commissioner, Idaho Fish and Game, Boise, Idaho, 1929).

Kendall, W. C. *The Fishes of New England, The Salmon Family, Part I - The Trout or Charrs* (Memoirs of the Boston Society of Natural History, Boston, Massachusetts, 1914).

Kennedy, B. M., Thompson, B. W. and Luecke, C. Ecological differences between two closely related morphologically similar benthic whitefish (*Prosopium spilonotus* and *Prosopium abyssicola*) in an endemic whitefish complex. *Canadian Journal of Fisheries and Aquatic Sciences* **63**, 1700–1709 (2006). DOI:10.1139/f06-065.

Kennedy P. and Meyer, K. A. 2015. Trends in abundance and the influence of bioclimatic factors on westslope cutthroat trout in Idaho. *Journal of Fish and Wildlife Management* **6**, 305–317 (2015). DOI:10.3996/012015-JFWM-001.

Kennedy, P. A. *The Effect of Irrigation Diversions on the Mountain Whitefish Population (Prosopium williamsoni) in the Big Lost River* (Utah State University, Logan, Utah, 2009). Online at http://digitalcommons.usu.edu/etd/512/.

Kimmel, P. G. Fishes of the Miocene-Pliocene Deer Butte Formation, Southeast Oregon. *Pages 69-87 in Claude W. Hibbard Memorial Volume 5* (Papers on Paleontology 14. University of Michigan, Museum of Paleontology, Ann Arbor, Michigan, 1975). Online at http://www.nativefishlab.net/library/textpdf/12524.pdf.

Kinziger, A. P., Wood, R. W. and Neely, D. A. Molecular systematics of the genus *Cottus* (Scopaeniformes: Cottidae). *Copeia* **2005**, 303–311 (2005). DOI: http://dx.doi.org/10.1643/CI-03-290R1.

Kircheis, F. W., Kornfield, I. and Seyoum, S. Genetic identity of transplanted arctic char: implications for restocking programs. *North American Journal of Fisheries Management* **15**, 54–59 (1995). DOI:10.1577/1548-8675(1995)015<0054:GIOTAC>2.3.CO;2.

Klemetsen, A., Amundsen, P. A., Dempson, J. B., Johsson, B., Jonsson, N., O'Connell, M. F. and Mortensen, E. Atlantic salmon *Salmo salar* L., brown trout *Salmo trutta* L. and arctic charr *Salvelinus alpinus* (L.): a review of aspects of their life histories. *Ecology of Freshwater Fish* **12,** 1–59 (2003). DOI:10.1034/j.1600-0633.2003.00010.x.

Kline, P. A. and Flagg, T. A. Putting the red back in Redfish Lake, 20 years of progress toward saving the Pacific northwest's most endangered salmon population. *Fisheries* **39**, 488–500 (2014). DOI:10.1080/03632415.2014.966087.

Knudsen, R. D., Stapp, D., Hackenberger, D., Lipe, W. D. and Rossillon, M. *A cultural resource reconnaissance in the Middle Fork Salmon River basin, Idaho, 1978* (U. S. Department of Agriculture, Forest Service, Intermountain Region, Ogden, Utah, 1982).

Kohler, C. C. and Hubert, W. A. *Inland Fisheries Management in North America: Second edition* (American Fisheries Society, Bethesda, Maryland, 1999).

Kootenai Tribe of Idaho and Montana Fish Wildlife and Parks. *Kootenai Subbasin Plan, Part I: Kootenai River Subbasin Assessment* (Northwest Power and Conservation Council, Portland, Oregon, 2004). Online at https://www.nwcouncil.org/media/119099/Kootenai_Assessment.pdf.

Kozfkay, C. C., Campbell, M. R., Meyer, K. A. and Schill, D. J. Influences of habitat and hybridization on the genetic structure of redband trout in the upper Snake River Basin, Idaho. *Transactions of the American Fisheries Society* **140**, 282–295 (2011). DOI:10.1080/00028487.2011.567837.

Kozfkay, C. C., Campbell, M. R., Yundt, S. P., Peterson, M. P. and Powell., M. S. Incidence of hybridization between naturally sympatric westslope cutthroat trout and rainbow trout in the Middle Fork Salmon River Drainage, Idaho. *Transactions of the American Fisheries Society* **136**, 624–638 (2007). DOI:10.1577/T06-085.1.

Kuda, D. B. and Griffith, J. S. Establishment of Shoshone sculpin (*Cottus greenei*) in a spring inhabited by mottled sculpin (*C. bairdi*). *Great Basin Naturalist* **53**, 190–193 (1993). Online at http://www.jstor.org/stable/41712773.

Lamarque, P. Electrophysiology of fish in electric fields. *Pages 4-33 in Fishing with Electricity: Applications in Freshwater Management.* (Fishing News Books, Oxford, U.K., 1990).

La Rivers, I. L. *Fishes and Fisheries of Nevada* (Biological Society of Nevada Memoirs, Nevada State Fish and Game Commission, Carson City, Nevada, 1962).

Lawson, M. *Fly-Fishing Guide to the Henrys Fork* (Stackpole Books, Mechanicsburg, Pennsylvania, 2012).

Lebreton, G. T. O. and Beamish, F. W. H. Growth, bioenergets and age. *Pages 195–216 in Sturgeons and paddlefish of North America* (Dordrecht, the Netherlands, 2004). DOI10.1007/1-4020-2833-4_10.

Leeseberg, C. A. and Keeley, E. R. Prey size, prey abundance, and temperature as correlates of growth in stream populations of cutthroat trout. *Environmental Biology of Fishes* **97**, 599–614 (2014). DOI:10.1007/s10641-014-0219-x.

LeMoine, M., Young, M. K., McKelvey, K. S., Eby, L., Pilgrim, K. L., and Schwartz, M. K. *Cottus schitsuumsh*, a new species of sculpin (Scorpaeniformes: Cottidae) in the Columbia River basin, Idaho-Montana, USA. *Zootaxa* **3755**, 241–258 (2014). Online at http://dx.doi.org/10.11646/zootaxa.3755.3.3.

Leuzinger, D. *Summary of Recreational Float Use-Middle Fork Salmon River. Salmon-Challis National Forest* (Salmon Challis National Forest, Salmon, Idaho, 2015).

Liljeblad, S. S. *Indian Peoples in Idaho* (Idaho State College, Pocatello, Idaho, 1957).

Linder, A. D. Idaho's alien fishes. *Tebiwa: The Journal of the Idaho State University Museum* **6**, 12–15 (1963).

Lindland, R. L. and Bowler, B. *Clearwater River Development of Spring Chinook and Steelhead Stocks. Columbia River Fisheries Development Program Annual Project Closing Report Period Covered: October 1, 1986 to September. 30, 1987* (Idaho Department of Fish and Game, Boise, Idaho, 1988). Online at http://idahodocs.cdmhost.com/cdm/singleitem/collection/p16293coll7/id/235163/rec/17.

Link, P. K., Fanning, C. M. and Beranek, L. P. Reliability and longitudinal change of detrital-zircon age spectra in the Snake River system, Idaho and Wyoming: an example of reproducing the bumpy barcode. *Sedimentary Geology* **182,** 101–142 (2005). DOI:10.1016/j.sedgeo.2005.07.012.

Link, P. K., Kaufman, D. S. and Thackray, G. D. Field guide to Pleistocene Lakes Thatcher and Bonneville and the Bonneville Flood, southeastern Idaho. *Pages 251-266 in Hughes, S. S. and Thackray, G. D., eds. Guidebook to the Geology of Eastern Idaho* (Idaho Museum of Natural History, Pocatello, Idaho, 1999)

Linnaeus, C., Langen, J. J. and Halle, C. J. J. *Systema Naturae, per Regnatria Naturae Secundum Classes, Ordines, Genera, Species, cum Characteribus, Differentiis, Synonmyis, Locis, Tenth edition* (Salvius, Stockholm, 1758). Online at http://dx.doi.org/10.5962/bhl.title.542.

Loomis, J. Use of survey data to estimate economic value and regional economic effects of fishery improvements. *North American Journal of Fisheries Management* **26**, 301–307 (2006). DOI:10.1577/M05-116.1.

Loxterman, J. L. and Keeley, E. R. Watershed boundaries and geographic isolation: patterns of diversification in cutthroat trout from western North America. *BMC Evolutionary Biology* **12**, 1–16 (2012). DOI:10.1186/1471-2148-12-38.

Loxterman, J. L., Keeley, E. R. and Njoroge, Z. M. Evaluating the influence of stocking history and barriers to movement on the spatial extent of hybridization between westslope cutthroat trout and rainbow trout. *Canadian Journal of Fisheries and Aquatic Sciences* **71**, 1050–1058 (2014). DOI:10.1139/cjfas-2013-0424.

Luzier, C. W., Schaller, H. A., Brostrom, J. K., Cook-Tabor, C., Goodman, D. H., Nelle, R. D., Ostrand, K. and Streif, B. *Pacific Lamprey (Entosphenus tridentatus) Assessment and Template for Conservation Measures* (U. S. Fish and Wildlife Service, Portland, Oregon, 2011). Online at http://www.fws.gov/columbiariver/publications/USFWS_Pacific_Lamprey_Assessment_and%20_Template_for_Conservation_Measures_2011.pdf.

Lynas, M. *God Species: How the Planet Can Survive the Age of Humans* (National Geographic, Washington, D. C., 2011).

MacCrimmon, H. R. and Campbell, J. S. World distribution of brook trout, *Salvelinus fontinalis. Journal of the Fisheries Research Board of Canada* **26,** 1699-1725 (1969). Online at https://doi.org/10.1139/f69-159.

Mach, E. *Grundlinien der Lehre von den Bewegungsempfindungen.* (Engelmann, Leipzig, Germany, 1875).

Mack, R. N., Simberloff, D., Mark, L. W., Evans, H. Clout, M. and Bazzaz, F. A. Biotic invasions: causes, epidemiology, global consequences, and control. *Ecological Applications* **10**, 689–710 (2000). http://dx.doi.org/10.1890/1051-0761(2000)010[0689:BICEGC]2.0.CO;2.

MacPhee, C. Influence of differential angling mortality and gradient on fish abundance in a trout-sculpin biotope. *Transactions of the American Fisheries Society* **95**, 381–387 (1966). DOI:10.1577/1548-8659(1966)95[381:IODAMA]2.0.CO;2.

Malde, H. E. *The Catastrophic Late Pleistocene Bonneville Flood in the Snake River Plain, Idaho* (Professional Paper 596, U. S. Geological Survey, Washington D. C., 1968). Online at http://pubs.usgs.gov/pp/0596/report.pdf.

Mallet, J. *Middle Fork Salmon River Trout Fisheries Investigations: Annual Progress Report* (Idaho Department of Fish and Game, Boise, Idaho, 1960).

Mallet, J. *Savings Idaho's Westslope Cutthroat Trout Fisheries* (Idaho Department of Fish and Game, Boise, Idaho, 2013). Online at https://collaboration.idfg.idaho.gov/FisheriesTechnicalReports/Res13-14Mallet%20Saving%20the%20Westslope%20Cutthroat%20Trout.pdf.

Maret, T. R. and MacCoy, D. E. Fish assemblages and environmental variables associated with hard-rock mining in the Coeur d'Alene River basin, Idaho. *Transactions of the American Fisheries Society* **131**, 865–884 (2002). DOI:10.1577/1548-8659(2002)131<0865:FAAEVA>2.0.CO;2.

Markle, D. F. *A Guide to Freshwater Fishes of Oregon.* (Oregon State University Press, Corvallis, Oregon, 2016).

Markle, D. F. and Hill, D. L. Taxonomy and distribution of the Malheur mottled sculpin, *Cottus bendirei. Northwest Science* **74**, 202–211 (2000). Online at http://hdl.handle.net/2376/1056.

Matthews, G. M. and Waples. R. S. *Status Review for Snake River Spring and Summer Chinook Salmon* (National Marine Fisheries Service, Seattle, Washington, 1991). Online at http://www.westcoast.fisheries.noaa.gov/publications/status_reviews/salmon_steelhead/chinook/sr1991-chinook1.pdf.

Matthews, M. *Hidden History: How the Malad River Got its Name* (Twin Falls Times-News January 31). Online at http://magicvalley.com/news/local/hidden-history-how-the-malad-river-got-its-name/article_ec73e625-b28b-5d97-98b3-8ef79c271db9.html. (2013).

Maughan, O. E. A survey of fishes of the Clearwater River. *Northwest Science* **50**, 76–86. (1976).

Maughan, O. E. *Distribution and Geographic Variation of Sculpins in the Clearwater Basin* (Washington State University, Pullman, Washington, 1972).

May, B. E. *Westslope Cutthroat Trout Status Update Summary 2009* (Wild Trout Enterprises, Bozeman, Montana, 2009). Online at http://www.streamnet.org/files/459/2009statusupdatereport.pdf.

May, B. E. and Albeke, S. E. *Range-Wide Status Assessment for Yellowstone Cutthroat Trout (Oncorhynchus clarkii bouvieri): 2006* (Yellowstone Cutthroat Trout Interagency Coordination Group, 2007). Online at http://www.fws.gov/mountain-prairie/species/fish/yct/YCTStatusReviewReport.pdf.

McAdam, S. O. Effects of substrate condition on habitat use and survival by white sturgeon (*Acipenser transmontanus*) larvae and potential implications for recruitment. *Canadian Journal of Fisheries and Aquatic Sciences* **68**, 812–822 (2011). DOI:10.1139/f2011-021.

McCabe, G. T. and Tracy, C. A. Spawning and early life history of white sturgeon, *Acipenser transmontanus*, in the lower Columbia River. *Fishery Bulletin* **92**, 760–772 (1994). Online at http://fishbull.noaa.gov/924/mccabe.pdf.

McClane, A. J. *McClane's New Standard Fishing Encyclopedia and International Angling Guide* (Gramercy Books, New York, New York, 1998).

McConnell, W. J., Clark, W. J. and Sigler, W. F. *Bear Lake: Its Fish and Fishing* (Utah State Department of Fish and Game, Salt Lake City, Utah, 1957).

McDaniel, J. *Fly Fishing the Harriman Ranch of the Henry's Fork of the Snake River: Lessons Learned and Friends Made Sight Fishing to Selective Trout* (Whitefish Press, Cincinnati, Ohio, 2012).

McDevitt, T. J. *Idaho's Malad Valley: A History* (Little Red Hen, Pocatello, Idaho, 2001).

McDonald, M. The salmon fisheries of the Columbia River basin. *Bulletin of the United States Fish Commission* **14**,153–168 (1895). Online at http://fishbull.noaa.gov/14-1/mcdonald.pdf.

McDonald, R., Nelson, J., Paragamian, V. and Barton, G. Modeling the effect of flow and sediment transport on white sturgeon spawning habitat in the Kootenai River, Idaho. *Journal of Hydraulic Engineering* **136**, 1077–1092 (2010). Online at http://dx.doi.org/10.1061/(ASCE)HY.1943-7900.0000283.

McMichael, G. A. Examination of electrofishing injury and short-term mortality in hatchery rainbow trout. *North American Journal of Fisheries Management* **13**, 229–233. DOI:10.1577/1548-8675 (1993) 013<0229:EOEIAS>2.3.CO;2.

McPhail, J. D. and Paragamian, V. L. Burbot biology and life history. *Pages 11–23 in Burbot Biology, Ecology, and Management* (Fisheries Management Section of the American Fisheries Society, Spokane, Washington, 2000). Online at https://collaboration.idfg.idaho.gov/FisheriesTechnicalReports/Burbot%20Biology%20and%20Life%20History.pdf.

McPhail, J. D. and Paragamian, V. L. Burbot biology and life history. *Pages 11-23 in Burbot Biology, Ecology, and Management* (Fisheries Management Section of the American Fisheries Society, Spokane, Washington, 2000) Online at https://collaboration.idfg.idaho.gov/FisheriesTechnicalReports/Burbot%20Biology%20and%20Life%20History.pdf.

McPhail, J. D. *The Freshwater Fishes of British Columbia* (University of Alberta Press, Edmonton, Alberta, 2007).

McQuarrie, N. and Rodgers, D. W. Subsidence of a volcanic basin by flexure and lower crustal flow: the eastern Snake River Plain, Idaho. *Tectonics* **17**, 203–220 (1998). DOI:10.1029/97TC03762.

Mervin, J. *The Ultimate Guide to Fly Fishing for Carp* (Field and Stream) Online at http://www.fieldandstream.com/articles/fishing/2006/05/ultimate-guide-fly-fishing-carp (2015).

Meyer, K. A., Cassinelli, J. D. and Elle, F. S. Life history characteristics of the Wood River sculpin, *Cottus leiopomus* (Cottidae), in Idaho. *Copeia* **2008**, 648–655 (2008). Online at http://www.jstor.org/stable/25140827.

Meyer, K. A., Elle, F. S. and Lamansky, J. A. Environmental factors related to the distribution, abundance, and life history characteristics of mountain whitefish in Idaho. *North American Journal of Fisheries Management* **29**, 753–767 (2009). DOI:10.1577/M08-015.1.

Meyer, K. A., Garton, E. O. and Schill, D. J. Bull trout trends in abundance and probabilities of persistence in Idaho. *North American Journal of Fisheries Management* **34**, 202–214 (2014). DOI:10.1080/02755947.2013.869280.

Meyer, K. A., Lamansky, J. A., Schill, D. J. and Zaroban, D. W. *Non-game Fish Species Distribution and Habitat Associations in the Upper Snake River Basin of Idaho* (Idaho Department of Fish and Game, Boise, Idaho, 2012). Online at http://idahodocs.cdmhost.com/cdm/ref/collection/p16293coll7/id/243345.

Miller, R. G. *The Natural History of Lake Tahoe Fishes* (Stanford University, Stanford, California, 1951).

Miller, R. R. *Snyderichthys*, a new generic name for the leatherside chub of the Bonneville and upper Snake drainages in western United States. *Journal of the Washington Academy of Sciences* **35**, 28 (1945). Online at http://biodiversitylibrary.org/page/39700658.

Miller, R. R. and Hubbs, C. L. *The Spiny-rayed Cyprinid Fishes (Plagopterini of the Colorado River System.* (Miscellaneous Publications, 115 of the University of Michigan Museum of Zoology, Ann Arbor, Michigan, 1960). Online at http://hdl.handle.net/2027.42/56359.

Miller, R. R. and Smith, G. R. *Distribution and Evolution of Chasmistes (Pisces: Catostomidae) in Western North America* (Occasional Papers of the Museum of Zoology, University of Michigan, Ann Arbor, Michigan, 1981). http://hdl.handle.net/2027.42/57132.

Miller, R. R. and Smith, G. R. *New Fossil Fishes from Plio-Pleistocene Lake Idaho* (Occasional Papers of the Museum of Zoology, University of Michigan, Ann Arbor, Michigan, 1967). Online at http://hdl.handle.net/2027.42/57090.

Miller, R. R., Minckley, W. L. and Norris, S. M. *Freshwater Fishes of Mexico* (University of Chicago Press, Chicago, Illinois, 2005).

Miller, R. R., Williams, J. D. and Williams, J. E. Extinctions of North American fishes during the past century. *Fisheries* **14**, 22–38 (1989). DOI:10.1577/1548-8446(1989)014<0022:EONAFD>2.0.CO;2.

Mock, K. E., Evans, R. P., Crawford, M., Cardall, B. L., Janecke, S. U. and Miller, M. P. Rangewide molecular structuring in the Utah sucker *(Catostomus ardens)*. *Molecular Ecology* **15,** 2223–2238 (2006). DOI:10.1111/j.1365-294X.2006.02932.x.

Montana Natural Heritage Program and Montana Fish, Wildlife and Parks. *Rocky Mountain Sculpin – Cottus bondi* (Montana Field Guide, Montana.Gov). Online at http://fieldguide.mt.gov/speciesDetail.aspx?elcode=AFC4E02380. (2015a).

Montana Natural Heritage Program and Montana Fish, Wildlife and Parks. *Westslope Cutthroat Trout-Oncorhynchus clarkii lewisi* (Montana Field Guide, Montana.Gov). Online at http://FieldGuide.mt.gov/speciesDetail.aspx?elcode=AFCHA02088. (2015b).

Montgomery, M. R. *Many Rivers to Cross: Of Good Running Water, Native Trout, and the Remains of Wilderness* (Simon and Schuster, New York, New York, 1996).

Moodie, G. E. E. *Some Factors Affecting the Distribution and Abundance of the Chiselmouth (Acrocheilus alutaceus)* (University of British Columbia, Vancouver, British Columbia, 1966). Online at http://hdl.handle.net/2429/37204.

Mooney, H. A. and Cleland, E. E. The evolutionary impact of invasive species. *Proceedings of the National Academy of Sciences* **98**, 5446–5451 (2001). DOI:10.1073/pnas.091093398.

Moser, M. L. and Close, D. A. Assessing Pacific lamprey status in the Columbia River Basin. *Northwest Science* **77**, 116–125 (2003).

Moyle, P. B. *Fish: An Enthusiast's Guide* (University of California Press, Berkeley, California, 1995). Online at http://www.jstor.org/stable/10.1525/j.ctt1pngpn.

Moyle, P. B. *Inland Fishes of California, Revised and Expanded* (University of California Press, Berkeley, California, 2002).

Moyle, P. B. and Cech, J. J. *Fishes: An Introduction to Ichthyology, Fifth edition* (Pearson Prentice Hall, Upper Saddle River, New Jersey, 2004).

Muhlfeld, C. C., Bennett, D. H., Steinhorst, R. K., Marotz, B. and Boyer, M. Using bioenergetics modeling to estimate consumption of native juvenile salmonids by nonnative northern pike in the upper Flathead River system, Montana. *North American Journal of Fisheries Management* **28**, 636–648 (2008). DOI:10.1577/M07-004.1.

Muhlfeld, C. C., McMahon, T. E., Belcer, D. and Kershner, J. L. Spatial and temporal spawning dynamics of native westslope cutthroat trout, *Oncorhynchus clarkii lewisi*, introduced rainbow trout, *Oncorhynchus mykiss*, and their hybrids. *Canadian Journal of Fisheries and Aquatic Sciences* **66**, 1153–1168 (2009). DOI:10.1139/F09-073.

National Marine Fisheries Service. *Draft Biological Opinion of the Federal Columbia River Power System, Including the Juvenile Fish Transportation Program and the Bureau of Reclamation's 31 Dams, Including The Entire Columbia Basin Project* (National Marine Fisheries Service, Seattle, Washington, 2000).

Nechako White Sturgeon Recovery Initiative. *Recovery plan for Nechako White Sturgeon* (Golder Associates Ltd., Edmonton, Alberta, 2004). Online at http://nechakowhitesturgeon.org/uploads/files/Recovery%20Plan%20July07%202004.pdf.

Neely, D. A. *A Systematic and Taxonomic Reassessment of the Mottled Sculpin Species Complex, Cottus Bairdii Girard (Teleostei: Cottidae)* (University of Alabama, Tuscaloosa, Alabama, 2003).

Nelson, J. S., Crossman, E. J., Espinosa-Pérez, H., Findley, L. T., Gilbert, C. R., Lea, R. N. and Williams, J. D. *Common and Scientific Names of Fishes from the United States, Canada, and Mexico, Sixth edition* (Special Publication No. 29, American Fisheries Society, Bethesda, Maryland, 2004).

Nelson, J. S. *Fishes of the World, Fourth edition* (John Wiley, Hoboken, New Jersey, 2006).

Nelson, J. S., Crossman, E. J., Espinosa-Pérez, H., Findley, L. T., Gilbert, C. R., Lea, R. N. and Williams, J. D. The 'Names of Fishes' list, including recommended changes in fish names: Chinook salmon for chinook salmon, and *Sander* to replace *Stizostedion* for the sauger and walleye. *Fisheries* **28,** 38–39 (2003). DOI:10.1577/1548-8446-28-7.

Neuhold, J. M. Age and growth of the Utah chub, *Gila atraria* (Girard) in Panguitch Lake and Navajo Lake, Utah, from scales and opercular bones. *Transactions of the American Fisheries Society* **85**, 217–233 (1957). DOI:10.1577/1548-8659(1955)85[217:AAGOTU]2.0.CO;2.

Nevermann, D. and Wurtsbaugh, W. A. The thermoregulatory function of diel vertical migration for a juvenile fish, *Cottus extensus*. *Oecologia* **98**, 247–256 (1994). DOI:10.1007/BF00324211.

Nickum, M. J., Mazik, P. M., Nickum, J. G. and MacKinlay, D. D., editors. P*ropagated Fish in Resource Management* (American Fisheries Society, Bethesda, Maryland, 2005).

Northcote, T. G. Comparative biology and management of Arctic and European grayling (*Salmonidae, Thymallus*). *Reviews in Fish Biology and Fisheries* **5**, 141–194 (1995). DOI:10.1007/BF00179755.

O'Connor, J. E. Hydrology, hydraulics, and geomorphology of the Bonneville flood: *GSA Special Paper* **274**, 1-84, 1993). DOI:10.1130/SPE274-p1.

Olson, M. H., Young, B. P. and Blinkoff, K. D. Mechanisms underlying habitat use of juvenile largemouth bass and smallmouth bass. *Transactions of the American Fisheries Society* **132**, 398–405 (2003). DOI: 10.1577/1548-8659(2003)132<0398:MUHUOJ>2.0.CO;2.

Ore, H. T., Reid, R. V. and Link, P. K. Pre-Bonneville-level, catastrophic overflow of Plio-Pleistocene Lake Bonneville, south of Rockland, Idaho. *Pages 1–15 (1996). Geology of the Crook and Snake River Plain, Twin Falls and Vicinity, Idaho. Proceedings of the 21st Annual TRGS Field Conference.* (College of Southern Idaho, Twin Falls, Idaho, 1996).

Orma J. Smith Museum of Natural History. *Fish Collection: Catalog Numbers CIDA -81,646, CIDA -81,647, CIDA-101,186, CIDA -101,187.* (The College of Idaho, Caldwell, Idaho, 2015).

Othberg, K. L., Kauffman, J. D., Gillerman, V. S. and Garwood, D. L. *Geologic Map of the Twin Falls 30 X 60 Minute Quadrangle, Idaho: Scale 1:100,000* (Idaho Geological Survey Map 49, University of Idaho, Moscow, Idaho, 2012). Online at http://www.idahogeology.org/PDF/Maps_(M)/Geologic_Maps_(GM)/PDF/Twin_Falls__GM-49-B.pdf.

Oviatt, C. G. Lake Bonneville fluctuations and global climate change. *Geology* **25**, 155–158. (1997). DOI:10.1130/0091-7613(1997)025<0155:LBFAGC>2.3.CO;2.

Oviatt, C. G., Currey, D. R. and Sack, D. Radiocarbon chronology of Lake Bonneville, Eastern Great Basin, USA. *Palaeogeography, Palaeoclimatology, Palaeoecology* **99**, 225–241 (1992). DOI:10.1016/0031-0182(92)90017-Y.

Oviatt, C. G. and Miller, D. New explorations along the northern shores of Lake Bonneville. *Brigham Young University Geology Studies* **42**, 345–371. (1997). Online at http://geology.byu.edu/Home/sites/default/files/volume-42-part-2-oviatt-miller-sack-kaufman.pdf.

PacifiCorp, U. S. Fish and Wildlife Service, U. S. Bureau of Land Management, U. S. Park Service, U. S. Forest Service, Shoshone-Bannock Tribes, Idaho Department of Environmental Quality, Idaho Department of Fish and Game, Idaho Department of Parks and Recreation, Idaho Council of Trout Unlimited, Idaho Rivers United, Greater Yellowstone Coalition and American Whitewater. Settlement Agreement Resolving the Relicensing of the Bear River Hydroelectric Projects. (Pacific Power, Rocky Mountain Power, Portland, Oregon, 2002). Online at http://www.pacificorp.com/content/dam/pacificorp/doc/Energy_Sources/Hydro/Hydro_Licensing/Bear_River/Bear_River_Final_Settlement_Agreement.pdf.

Page, L. M. and Burr, B. M. *Peterson Field Guides: Freshwater Fishes.* (Houghton Mifflin Company, Boston, Massachusetts, 1991)

Page, L. M., Espinosa-Pérez, H., Findley, L. T., Gilbert, C. R., Lea, R. N., Mandrak, N. E., Mayden, R. L. and Nelson, J. S.. *Common and Scientific Names of Fishes from the United States, Canada, and Mexico, Seventh edition* (Special Publication No. 34, American Fisheries Society, Bethesda, Maryland, 2013).

Palacios, P., Luecke, C. and Robinson, J. Fish of Bear Lake, Utah. *Natural Resources and Environmental Issues* **14**, 67–76 (2007). Online at http://digitalcommons.usu.edu/nrei/vol14/iss1/15.

Paquet, P. J., Flagg, T., Appleby, A., Barr, L., Blankenship, L., Campton, D., Delarm M., Evelyn, T., Fast, D., Gislason, J., Kline, P., Maynard, D., Mobrand, L., Nandor, G., Seidel, P., and Smith, S., Hatcheries, conservation, and sustainable fisheries: achieving multiple goals: results of the Hatchery Scientific Review Group's Columbia River basin review. *Fisheries* **36**, 547–561 (2011). DOI:10.1080/03632415.2011.626661.

Paragamian, V. L., Beamesderfer, R. C. P. and Ireland, S. C. Status, population dynamics, and future prospects of the endangered Kootenai River white sturgeon population with and without hatchery intervention. *Transactions of the American Fisheries Society* **134**, 518–532 (2005). DOI:10.1577/T03-011.1.

Paragamian, V. L. and Kruse, G. Kootenai River white sturgeon spawning migration behavior and a predictive model. *North American Journal of Fisheries Management* **21**, 10–21 (2001). DOI:10.1577/1548-8675(2001)021<0010:KRWSSM>2.0.CO;2.

Paragamian, V. L., McDonald, R., Nelson, G. J. and Barton, G. Kootenai River velocities, depth, and white sturgeon spawning site selection-a mystery unraveled? *Journal of Applied Ichthyology* **25**, 640–646 (2009). DOI:10.1111/j.1439-0426.2009.01364.x.

Paragamian, V. L., Pyper, B. J., Daigneault, M. J., Beamesderfer, R. C. P. and Ireland, S. C. Population dynamics and extinction risk of burbot in the Kootenai River, Idaho, USA and British Columbia, Canada. *Pages 213–234 in Burbot: Ecology, Management, and Culture* (American Fisheries Society, Bethesda, Maryland, 2008).

Paragamian, V. L., Wakkinen, V. D. and Kruse, G. Spawning locations and movement of Kootenai River white sturgeon. *Journal of Applied Ichthyology* **18,** 608-616 (2002). DOI:10.1046/j.1439--0426.2002.00397 .x.

Parkhurst, Z. E., Bryant, F. G. and Nielson, R. S. *Survey of the Columbia River and its Tributaries, Part 3* (U. S. Department of the Interior, Fish and Wildlife Service, Washington, D. C., 1950). Online at http://hdl.handle.net/2027/uiug.30112106601625.

Parkos, J. J., Santucci, V. J., and Wahl, D. H. Effects of adult common carp (*Cyprinus carpio)* on multiple trophic levels in shallow mesocosms. *Canadian Journal of Fisheries and Aquatic Sciences* **60**, 182–192 (2003). DOI:10.1139/f03-011.

Parsley, M. J. and Popoff, N. D. *Site Fidelity, Habitat Associations, and Behavior During Dredging Operations of White Sturgeon at Three Tree Point in the Lower Columbia River: Final Report* (Environmental Resource Branch, U. S. Army Corps of Engineers, Portland District, Portland, Oregon, 2004).

Parsley, M. J., Popoff, N. D., Wright, C. D. and van der Leeuw, B. K. Seasonal and diel movements of white sturgeon in the lower Columbia River. *Transactions of the American Fisheries Society* **137**, 1007–1017 (2008). DOI:10.1577/T07-027.1.

Parsley, M. J., Beckman, L. G. and McCabe, G. T. Spawning and rearing habitat use by white sturgeons in the Columbia River downstream from McNary Dam. *Transactions of the American Fisheries Society* **122**, 217–227 (1993). DOI:10.1577/1548-8659(1993)122<0217:SARHUB>2.3.CO;2.

Paukert, C. P., Klammer, J. A., Pierce, R. B. and Simonson, T. D. An Overview of northern pike regulations in North America. *Fisheries* **26**, 6–13 (2001). DOI:10.1577/1548-8446(2001)026<0006:AOONPR>2.0.CO;2.

Paukert, C. P. and Willis, D. W. Population characteristics and ecological role of northern pike in shallow natural lakes in Nebraska. *North American Journal of Fisheries Management* **23**, 313–322 (2003). DOI:10.1577/1548-8675(2003)023<0313:PCAERO>2.0.CO;2.

Pearsons, T. N. and Temple, G. M. Changes to rainbow trout abundance and salmonid biomass in a Washington watershed as related to hatchery salmon supplementation. *Transactions of the American Fisheries Society* **139**, 502–520 (2010). DOI:10.1577/T08-094.1.

Pederson, J. L., Janecke, S. U., Reheis, M. C., Kaufman, D. S. and Oaks, R. Q. The Bear River's history and diversion—constraints, unsolved problems, and implications for the Lake Bonneville record. *Pages 28–58 in Lake Bonneville: A Scientific Update, 1st edition* (Elsevier, Inc., Atlanta, Georgia, 2016).

Perrin, C. J., Rempel, L. L. and Rosenau, M. L. White sturgeon spawning habitat in an unregulated river: Fraser River, Canada. *Transactions of the American Fisheries Society* **132**, 154–165 (2003). DOI:10.1577/1548-8659(2003)132<0154:WSSHIA>2.0.CO;2.

Peters, E. J. The *Effects of Highway Construction on the Fish Populations in the Weber River near Henefer, Summit County, Utah.* (Brigham Young University, Provo, Utah, 1974).

Peterson, J. T., Thurow, R. F. and Guzevich, J. W. An evaluation of multipass electrofishing for estimating the abundance of stream-dwelling salmonids. *Transactions of the American Fisheries Society* **133,** 462–475 (2004). Online at http://www.fs.fed.us/rm/pubs_other/rmrs_2004_peterson_j001.pdf.

Petty, J. T. and Grossman, G. D. Size-dependent territoriality of mottled sculpin in a southern Appalachian stream. *Transactions of the American Fisheries Society* **136**, 1750–1761 (2007). DOI:10.1577/T06-034.1.

Pierce, K. L. and Morgan, L. A. Is the track of the Yellowstone hotspot driven by a deep mantle plume? Review of volcanism, faulting and uplift in light of new data. *Journal of Volcanology and Geothermal Research* **188**, 1–25 (2009). DOI:10.1016/j.jvolgeores.2009.07.009.

Pierce, K. L. and Morgan, L. A. The track of the Yellowstone hot spot: volcanism, faulting, and uplift. *Pages 1–53 in Regional Geology of Eastern Idaho and Western Wyoming* (Geological Society of America Memoir 179, U. S. Geological Survey, Denver, Colorado, 1992). Online at http://nrmsc.usgs.gov/files/norock/products/PierceMorgan1992txt.pdf.

Potter, K. E. Fluvial sands at 6000 ft in the Kimama Drill Core (Project Hotspot, central Snake River Plain, ID): Heise Field volcanogenic zircons and the late Miocene paleo-Wood River. *Pages 37–38 in Late Cenozoic to Recent Geologic and Biotic History of the Snake River, Workshop March 24 to 26, 2014* (Idaho State University, Pocatello, Idaho, 2014). Online at http://geology.isu.edu/Papers/SRPProceedings.pdf.

Probst, W. N. and Eckmann, R. The influence of light on the diel vertical migration of young-of-the-year burbot *Lota lota* in Lake Constance. *Journal of Fish Biology* **74**, 150–166 (2009). DOI:10.1111/j.1095-8649.2008.02120.x.

Quadri, S. U. Taxonomic status of the *Salvelinus alpinus* complex. *Journal of the Fisheries Research Board of Canada* **31**, 1355–1361 (1974). DOI:10.1139/f74-160.

Quintela, J. G. *Habitat Variables Affecting the Distribution and Density of Torrent Sculpin (Cottus rhotheus) in Northern Idaho* (University of Idaho, Moscow, Idaho, 2004).

Rafinesque, C. S. Description of the silures or catfishes of the River Ohio. *Quarterly Journal of Science, Literature, and the Arts* **9**, 48–52 (1820). Online at http://hdl.handle.net/2027/hvd.32044106374739?urlappend=%3Bseq=72.

Rafinesque, C. S., Hunt, W. G. and Brevoort, J. C. *Ichthyologia ohiensis, or Natural History of the Fishes Inhabiting the River Ohio and its Tributary Streams, preceded by a Physical Description of the Ohio and its Branches* (W.G. Hunt, Lexington, Kentucky, 1820). Online at http://dx.doi.org/10.5962/bhl.title.6892.

Raymond, H. L. Effects of dams and impoundments on migrations of juvenile chinook salmon and steelhead from the Snake River, 1966 to 1975. *Transactions of the American Fisheries Society* **108**, 505–529 (1979). DOI:10.1577/1548-8659(1979)108<505:EODAIO>2.0.CO;2.

Reed, J. R. and Pereira, D. L. Relationships between shoreline development and nest site selection by black crappie and largemouth bass. *North American Journal of Fisheries Management* **29**, 943–948 (2009). DOI:10.1577/M06-234.1.

Reheis, M., Laabs, B. J. C. and Kaufman, D. S. Geology and geomorphology of Bear Lake Valley and upper Bear River, Utah and Idaho. *Pages 15–48 Paleoenvironements of Bear Lake, Utah and Idaho, and its Catchment* (Geological Society of America, Boulder, Colorado, 2009). DOI:10.1130/2009.2450(02).

Reiser, D. W. and Wesche, T. A. *Determination of Physical and Hydraulic Preferences of Brown and Brook Trout in the Selection of Spawning Locations* (University of Wyoming, Laramie, Wyoming, 1977).

Reynolds, J. B. and Harlan, L. *Quick Start Guide to Electrofishing.* (Smith-Root, Inc., Vancouver, Washington, 2011).

Reynolds, J. B. and Kolz, A. L. Electrofishing. *Pages 305-361 in Fisheries Techniques, Third Edition.* (American Fisheries Society, Bethesda, Maryland, 2012).

Richardson, J. *Fauna Boreali-Americana: or The Zoology of the Northern Parts of British America: Part Third: The Fish* (Richard Bentley, London, 1836). Online at http://biodiversitylibrary.org/page/27206778.

Richter, A. and Kolmes, S. A. Maximum temperature limits for chinook, coho, and chum salmon, and steelhead trout in the Pacific Northwest. *Reviews in Fisheries Science* **13**, 23–49 (2005). DOI:10.1080/10641260590885861.

Rieman, B. E. and Allendorf, F. W. Effective population size and genetic conservation criteria for bull trout. *North American Journal of Fisheries Management* **21**, 756–764 (2001). DOI:10.1577/1548-8675(2001)021<0756:EPSAGC>2.0.CO;2.

Rieman, B. E. and Apperson, K. A. *Status and Analysis of Salmonid Fisheries: Westslope Cutthroat Trout Synopsis and Analysis of Fishery Information* (Idaho Department of Fish and Game, Boise, Idaho, 1989). Online at https://collaboration.idfg.idaho.gov/FisheriesTechnicalReports/Res-Rieman1989%20Status%20and%20Analysis%20of%20Salmonid%20Fisheries--Westslope%20Cutthroat%20Trout%20Synopsis%20and%20Analysis.pdf.

Rieman, B. E., Peterson, J. T. and Myers, D. L. Have brook trout (*Salvelinus fontinalis*) displaced bull trout (*Salvelinus confluentus*) along longitudinal gradients in central Idaho streams? *Canadian Journal of Fisheries and Aquatic Sciences* **63**, 63–78 (2006). DOI:10.1139/f05-206.

Robb, D. H. F., O'Callaghan, M., Lines, J. A. and Kestin, S. C. Electrical stunning of rainbow trout (*Oncorhynchus mykiss*): factors that affect stun duration. *Aquaculture* **205**, 359–371 (2002). DOI:10.1016/S0044-8486(01)00677-9.

Robbins, C. R., Bailey, R. M., Bond, C. E., Brooker, J. R., Lachner, E. A., Lea, R. N. and Scott, W. B. *Common and Scientific Names of Fishes from the United States and Canada. Fifth edition* (Special Publication No. 20, American Fisheries Society, Bethesda, Maryland, 1991).

Robbins, C. R., Bailey, R. M., Bond, C. E., Brooker, J. R., Lachner, E. A., Lea, R. N. and Scott, W. B. *A List of Common and Scientific Names of Fishes from the United States and Canada, Fourth edition* (Special Publication No. 12, American Fisheries Society, Bethesda, Maryland, 1980).

Rodgers, D. W., Hackett, W. R. and Ore, H. T. Extension of the Yellowstone Plateau, eastern Snake River Plain, and Owyhee Plateau. *Geology* **18**, 1138–1141. (1990). DOI:10.1130/0091-7613(1990)018<1138:EOTYPE>2.3.CO;2.

Rodgers, D. W., Ore, H. T., Bobo, R. T., McQuarrie, N. and Zentner, N. Extension and subsidence of the eastern Snake River Plain, Idaho. *Pages 121–156 in Tectonic and Magmatic Evolution of the Snake River Plain Volcanic Province* (Idaho Geological Survey Bulletin 30, University of Idaho, Moscow, Idaho, 2002). Online at http://www.idahogeology.org/PDF/Bulletins_(B)/Snake_River_Plain_B-30.pdf.

Rogers, S. M., Gagnon, V. and Bernatchez, L. Genetically based phenotype-environment association for swimming behavior in Lake Whitefish ecotypes (*Coregonus clupeaformis* Mitchill). *Evolution* **56**, 2322–2329 (2002). DOI:10.1111/j.0014-3820.2002.tb00155.x.

Rosenfeld, J., Porter, M., Pearson, M., Wicks, B., van Dishoeck, P., Patton, T., Parkinson, E., Haas, G. and McPhail, D. The influence of temperature and habitat on the distribution of chiselmouth, *Acrocheilus alutaceus*, in British Columbia. *Environmental Biology of Fishes* **62,** 401–413 (2001). DOI:10.1023/A:1012222415961.

Rosenbaum, J. G. and Kaufman, D. S. *Paleoenvironments of Bear Lake, Utah and Idaho, and its Catchment* (Geological Society of America, Boulder, Colorado, 2009). Online at http://specialpapers.gsapubs.org/content/450.

Rust, P. and Wakkinen, V. D. *Kootenai River White Sturgeon Spawning and Recruitment Evaluation: Annual Progress Report April 1, 2005-March 31, 2006* (Idaho Department of Fish and Game, Boise, Idaho, 2007). Online at https://collaboration.idfg.idaho.gov/FisheriesTechnicalReports/Res07-23Rust2006%20Kootenai%20River%20White%20Sturgeon%20Spawning%20and%20Recruitment%20Evaluation.pdf.

Ruzycki, J. R. and Wurtsbaugh, W. A. *Examination of the Factors Controlling the Abundance of Bear Lake Forage Fishes: Piscivore Populations and Consumption Estimates: Final Report* (Idaho Department of Fish and Game, Boise, Idaho, 1995).

Ruzycki, J. R., Wurtsbaugh, W. A. and Lay, C. Reproductive ecology and early life history of a lacustric sculpin, *Cottus extensus*. *Environmental Biology of Fishes* **53**, 117–127 (1998). DOI:10.1023/A:1007436502285.

Ruzycki, J. R., Wurtsbaugh, W. A. and Luecke, C. Salmonine consumption and competition for endemic prey fishes in Bear Lake, Utah-Idaho. *Transactions of the American Fisheries Society* **130**, 1175–1189 (2001). DOI:10.1577/1548-8659(2001)130<1175:SCACFE>2.0.CO;2.

Ryan, R., Gutknecht, E. and Megargle, D. *Fishery Management Annual Report, Magic Valley Region 2006* (Idaho Department of Fish and Game, Boise, Idaho, 2008). Online at https://collaboration.idfg.idaho.gov/FisheriesTechnicalReports/Mgt08-136Ryan2006%20Fishery%20Management%20Annual%20Report%20Magic%20Valley%20Region.pdf.

Saint John, S. Personal communication to John Sigler. Smallmouth Bass. (U. S. Geological Survey, Western Fisheries Research Center, 2014).

Sass, G. G. and Allen, M. S. *Foundations of Fisheries Science* (American Fisheries Society, Bethesda, Maryland, 2014).

Scarola, J. F. *Freshwater Fishes of New Hampshire* (New Hampshire Fish and Game Department, Division of Inland and Marine Fisheries, Concord, New Hampshire, 1973).

Scarpella, R. and S. Dudley. Torrent sculpin (*Cottus rhotheus*). *Pages 24–26 in Rare Fish Species Status Reviews: Proceedings From a Workshop held March 15, 1995* (Idaho Chapter of the American Fisheries Society, Boise, Idaho, 1995).

Schade, C. B. and Bonar, S. A. Distribution and abundance of nonnative fishes in streams of the western United States. *North American Journal of Fisheries Management* **25**, 1386–1394 (2005). DOI:10.1577/M05-037.1.

Scheerer, P. D. and Jacobs, S. E. Status and distribution of the sand roller (*Percopsis transmontana*) in the Willamette Basin, Oregon. *Northwestern Naturalist* **87**, 233–239 (2006). Online at http://www.jstor.org/stable/4501967.

Scheerer, P. D., Scheu, M., Bangs, B. L. and Jacobs, S. E. *Sand Roller Investigations in the Willamette River Drainage. Annual Progress Report* (Oregon Department of Fish and Wildlife, Salem, Oregon, 2005). Online at http://oregonstate.edu/dept/ODFW/NativeFish/pdf_files/SandRoller2005.pdf.

Schoby, G. P. and Keeley, E. R. Home range size and foraging ecology of bull trout and westslope cutthroat trout in the upper Salmon River Basin, Idaho. *Transactions of the American Fisheries Society* **140**, 636–645 (2011). DOI:10.1080/00028487.2011.585578.

Schrank, A. J. and Rahel, F. J. Factors influencing summer movement patterns of Bonneville cutthroat trout (*Oncorhynchus clarkii utah*). *Canadian Journal of Fisheries and Aquatic Sciences* **63**, 660–669 (2006). DOI:10.1139/f05-250.

Schrank, A. J. and Rahel, F. J. Movement patterns in inland cutthroat trout (*Oncorhynchus clarki utah*): management and conservation implications. *Canadian Journal of Fisheries and Aquatic Sciences* **61**, 1528–1537 (2004). DOI:10.1139/f04-097.

Scott, O. A. *Pioneer Days on the Shadowy St. Joe* (Caxton Printers, Coeur d'Alene, Idaho, 1968).

Scott, W. B. and Crossman, E. J. *Freshwater Fishes of Canada* (Fisheries Research Board of Canada, Ottawa, Ontario, 1973).

Scott, W. E., McCoy, W. D., Shroba, R. R. and Rubin, M. Reinterpretation of the exposed record of the last two cycles of Lake Bonneville, western U. S. *Quaternary Research* **20**, 261–285. (1983). DOI:10.1016/0033-5894(83)90013-3.

Semakula, S. N. *The Age and Growth of the White Sturgeon (Acipenser transmontanus Richardson) of the Fraser River, British Columbia, Canada* (University of British Columbia, Vancouver, British Columbia, 1963).

Sexauer, H. Personal communication to John Sigler. Lake Alice Bonneville Cutthroat Trout. (Fish biologist, Wyoming Department of Game and Fish, 2015)

Sharber, N. G., Carothers, S. W., Sharber, J. P., De Vos, J. C. and House, D. A. Reducing electrofishing-induced injury of rainbow trout. *North American Journal of Fisheries Management* **14**, 340–346 (1994). DOI:10.1577/1548-8675(1994)014<0340:REIIOR>2.3.CO;2.

Shepard, B. B., May, B. E. and Urie, W. *Status of Westslope Cutthroat Trout (Oncorhynchus clarki lewisi) in the United States: 2002* (Montana Fish, Wildlife, and Parks, Helena, Montana, 2003). Online at fwp.mt.gov/fwpDoc.html?id=7538.

Shepard, B. B., Spoon, R. and Nelson, L. A native westslope cutthroat trout population responds positively after brook trout removal and habitat restoration. *Intermountain Journal of Sciences* **8**, 193–214 (2003).

Shroder, J. F. and Lowndes, T. Threshold modeling, slope failure, and the Bonneville Flood. *Geological Society of America Abstracts with Program* **21**, 60 (1989).

Sigler, J. W. *Investigation of the Algal Productivity of Selected and Limited Sites Along the Western Shore of Bear Lake, Utah-Idaho* (Utah State University, Logan, Utah, 1972). Online at http://digitalcommons.usu.edu/etd/333/.

Sigler, J. W. and Sigler, W. F. History of fish hatchery development in the Great Basin states of Utah and Nevada. *The Great Basin Naturalist* **46**, 583–594 (1986). Online at http://www.biodiversitylibrary.org/item/33077#page/609/mode/1up.

Sigler, W. F. Age and growth of the brown trout, *Salmo trutta fario* Linnaeus, in Logan River, Utah. *Transactions of the American Fisheries Society* **81**, 171–178 (1952). DOI:10.1577/1548-8659(1951)81[171:AAGOTB]2.0.CO;2.

Sigler, W. F. *Bear Lake and Its Future: Twenty-sixth Honor Lecture* (Utah State University, Logan, Utah, 1962). Online at http://digitalcommons.usu.edu/honor_lectures/22/.

Sigler, W. F. *The Ecology and Use of Carp in Utah* (Agricultural Experiment Station, Utah State University, Logan, Utah, 1958a).

Sigler, W. F. Fish life history series: the whitefishes of Bear Lake. *Utah Fish and Game Magazine* **14**, 20–21 (1958b).

Sigler, W. F. *The Life History and Management of the Mountain Whitefish Prosopium williamsoni (Girard) in Logan River, Logan, Utah* (Agricultural Experiment Station, Utah State University, Logan, Utah, 1951). Online at http://digitalcommons.usu.edu/uaes_bulletins/308/.

Sigler, W. F. *The Rainbow Trout in Relation to Other Fish in Fish Lake, Utah* (Agricultural Experiment Station, Utah State University, Logan, Utah, 1953).

Sigler, W. F., Helm, W. T., Kucera, P. A., Vigg, S. and Workman, G. W. Life history of the lahontan cutthroat trout, *Salmo clarki henshawi,* in Pyramid Lake, Nevada. *Great Basin Naturalist* **43**, 1–29 (1983). Online at http://www.biodiversitylibrary.org/item/35769#page/15/mode/1up.

Sigler, W. F. and Kennedy, J. F. *Pyramid Lake Ecological Study* (W.F. Sigler and Associates Inc., Logan, Utah, 1978).

Sigler, W. F. and Sigler, J. W. *Fishes of the Great Basin: A Natural History* (University of Nevada Press, Reno, Nevada, 1987).

Sigler, W. F and Sigler, J. W. *Fishes of the Great Basin: A Natural History, Revised edition* (University of Nevada Press, Reno, Nevada, 2014).

Sigler, W. F. and Sigler, J. W. *Fishes of Utah: A Natural History* (University of Utah Press, Salt Lake City, Utah, 1996).

Sigler, W. F. and Sigler, J. W. *Recreational Fisheries: Management, Theory, and Application* (University of Nevada Press, Reno, Nevada, 1990).

Sigler, W. F., Vigg, S. and Bres, M. Life history of the cui-ui, *Chasmistes cujus,* Cope, in Pyramid Lake Nevada: a review. *The Great Basin Naturalist* **45**, 571–603 (1985). http://www.biodiversitylibrary.org/item/35771#page/601/mode/1up.

Sigler, W. F. and Workman, G. W. *The Bonneville Cisco of Bear Lake, Utah-Idaho* (Agricultural Experiment Station, Utah State University, Logan, Utah, 1978).

Simons, A. M. and Mayden, R. L. The phylogenetic relationships of the creek chubs and spinefins: an enigmatic group of North American cyprinid fishes (*Actinopterygi* : *Cyprinidae*). *Cladistics* **13**, 187–205 (1997). DOI:10.1111/j.1096-0031.1997.tb00315.x.

Simpson, J. C. and Wallace, R. L. *Fishes of Idaho* (University Press of Idaho, Moscow, Idaho, 1978).

Sirois, P. and Dodson, J. J. Influence of turbidity, food density and parasites on the ingestion and growth of larval rainbow smelt *Osmerus mordax* in an estuarine turbidity maximum. *Marine Ecology Progress Series* **193**, 167–197 (2000). DOI:10.3354/meps193167.

Smith, E. *A History of the Salmon National Forest* (U. S. Forest Service, Washington D. C., 1973). Online at http://www.foresthistory.org/ASPNET/Publications/region/4/salmon/index.htm.

Smith, G. R. *Distribution and Evolution of the North American Catostomid Fishes of the Subgenus Pantosteus, Genus Catostomus* (Museum of Zoology, University of Michigan, Ann Arbor, Michigan, 1966). Online at http://deepblue.lib.umich.edu/bitstream/handle/2027.42/56373/MP129.pdf?sequence=1andisAllowed=y.

Smith, G. R. Fishes of the Pliocene Glenns Ferry Formation, Southwest Idaho. *Pages 1–68 in Claude W. Hibbard Memorial Volume 5* (Papers on Paleontology 14, University of Michigan, Museum of Paleontology, Ann Arbor, Michigan, 1975). Online at https://deepblue.lib.umich.edu/handle/2027.42/48614.

Smith, G. R. Personal communication to John Sigler. Unique meristic characteristics of Bridgelip Sucker in the Wood River and Palouse River basins. (Curator Emeritus of Fishes, Museum of Zoology; Curator Emeritus of Lower Vertebrates, Museum of Paleontology, University of Michigan, Ann Arbor, 2016).

Smith, G. R. Phylogeny and biogeography of the Catostomidae freshwater fishes of North America and Asia, *Pages 778–826 in Systematics, Historical Ecology and North American Freshwater Fishes* (Stanford University Press, 1992).

Smith, G. R., Chow, J., Unmack, P. J., Markle, D. F. and Dowling, T. E. Fishes of the Mio-Pliocene western Snake River plain and vicinity. II. Evolution of the *Rhinichthys osculus* complex (Teleosei: Cyprinidae) in western North America. (Miscellaneous Publications, Museum of Zoology, University of Michigan, 204, no. 2, 2017).

Smith, G. R., Stewart, J. D. and Carpenter, N. E. *Fossil and Recent Mountain Suckers, Pantosteus, and Significance of Introgression in Catostomin Fishes of Western United States* (Museum of Zoology, University of Michigan, Ann Arbor, Michigan, 2013).

Smith, G. R., Stokes, W. L. and Horn, K. F. Some late pleistocene fishes of Lake Bonneville. *Copeia* **1968**, 807–816 (1968). DOI:10.2307/1441848.

Smith, G. R., Zaroban, D. W., High, B., Sigler, J. W., Krabbenhoft, T. and Dowling, T. E. Hybrid tetraploid lake suckers (Teleostei, Catostomidae) in western North America. (Miscellaneous Publications, Museum of Zoology, University of Michigan, 204, no. 3, in preparation).

Smith, R. Description of a new species of Uranidea (*Uranidea rhothea*) from Spokane River, Washington territory. *Proceedings of the United States National Museum* **5**, 347-348 (1882). http://dx.doi.org/10.5479/si.00963801.286.347.

Snyder, J. O. Notes on certain catostomids of the Bonneville system, including the type of *Pantosteus virescens*. *Proceedings of the U.S. National Museum* **64:18**, 1–4 (1924). DOI:10.5479/si.00963801.2508

Snyder, J. O. Three new whitefishes from Bear Lake, Idaho and Utah. *Pages 1–10 in Bulletin of the United States Bureau of Fisheries Volume 36* (U. S. Department of Commerce, Washington, D. C., 1919). http://biodiversitylibrary.org/page/27335546.

Sparks, R. E., Barkley, T. L., Creque, S. M., Dettmers, J. M. and Stainbrook, K. M. Evaluation of an electric fish dispersal barrier in the Chicago Sanitary and Ship Canal. *Pages 139–161 in Invasive Asian Carps in North American* (Symposium 74, American Fisheries Society, Bethesda, Maryland, 2011).

Spencer, J. E., Smith, G. R. and Dowling, T. E. Middle to late Cenozoic geology, hydrography, and fish evolution in the American Southwest. *Geological Society of America* Special Paper **439**, 279–299 (2008). DOI:10.1130/2008.2439(12).

Stamford, M. D. and Taylor, E. B. Phylogeographical lineages of arctic grayling (*Thymallus arcticus*) in North America: divergence, origins and affinities with Eurasian *Thymallus*. *Molecular Ecology* **13**, 1533–1549 (2004). DOI:10.1111/j.1365-294X.2004.02174.x.

Stanton, S. and Megargle, D. *Fishery Management Annual Report, Magic Valley 2009* (Idaho Department of Fish and Game, Boise, Idaho, 2014). Online at https://collaboration.idfg.idaho.gov/FisheriesTechnicalReports/Mgt.13-124Stanton2009%20Fishery%20Management%20Annual%20Report%20Magic%20Valley%202009.pdf.

Stroud, R. H. *Fish Culture in Fisheries Management* (American Fisheries Society, Bethesda, Maryland, 1986).

Suckley, G. Description of several new species of salmonidae, from the north-west coast of America. *Pages 1–10 in Annals of the Lyceum of Natural History, Volume 7* (John Wiley, New York, New York, 1862). Online at http://biodiversitylibrary.org/page/16227474.

Suckley, G. On the North American species of salmon and trout. *Pages 91–160 in Report of the Commissioner for 1872 and 1973* (U. S. Fish Commissioner, Washington, D. C., 1874).

Suski, C. D., Svec, J. H., Ludden, J. B., Phelan, F. J. S. and Philipp, D. P. The effect of catch-and-release angling on the parental care behavior of male smallmouth bass. *Transactions of the American Fisheries Society* **132**, 210–218 (2003). DOI:10.1577/1548-8659(2003)132<0210:TEOCAR>2.0.CO;2.

Swink, W. D. Effectiveness of an electrical barrier in blocking a sea lamprey spawning migration on the Jordan River, Michigan. *North American Journal of Fisheries Management* **19**, 397-405 (1999). DOI: 10.1577/1548-8675(1999)019<0397:EOAEBI>2.0.CO;2.

Tabor, R. A., Fresh, K. L., Paige, D. K., Warner, E. J. & Peters, R. J. Distribution and habitat use of cottids in the Lake Washington basin. *Pages 135–150 in Status, Distribution, and Conservation of Native Freshwater Fishes of Western North America.* (Symposium 53, American Fisheries Society, Bethesda, Maryland, 2007).

Tabor, R. A., Wagner, E. J. and Hager, S. An oriental weatherfish (*Misgurnus anguillicaudatus*) population established in Washington State. *Northwest Science* **75**, 72–76 (2001).

Taylor, E. B., Darveau, C.-A. and Schulte, P. M. Setting conservation priorities in a widespread species: phylogeographic and physiological variation in the lake chub, *Couesius plumbeus* (Pisces: *Cyprinidae*). *Diversity* **5**, 149–165 (2013). DOI:10.3390/d5020149.

Taylor, E. B., Foote, C. J. and Wood, C. C. Molecular genetic evidence for parallel life-history evolution within a Pacific salmon (sockeye salmon and kokanee *Oncorhynchus nerka*). *Evolution* **50**, 401–416 (1996). Online at http://www.jstor.org/stable/2410810.

Taylor, E. B., Gow, J. L., Witt, J. and Zemlak, R. Connectivity among populations of pygmy whitefish (*Prosopium coulterii*) in northwestern North America inferred from microsatellite DNA analyses. *Canadian Journal of Zoology* **89**, 255–266 (2011). DOI:10.1139/z10-114.

Teuscher, D. Personal communication to John Sigler. Bonneville Cutthroat Trout and Bear Lake. (Fish biologist, Idaho Department of Fish and Game, 2015).

Teuscher, D. and Capurso, J. *Management Plan for Conservation of Bonneville Cutthroat Trout in Idaho* (Idaho Department of Fish and Game, Boise, Idaho, 2007). Online at https://fishandgame.idaho.gov/public/fish/planBonCutthroat.pdf.

Thackray, G. D., Rodgers, D. W. and Drabick, A. Neogene drainage development of Marsh and Portneuf valleys, eastern Idaho. *Pages 89–101 in Geologic Field Trips to the Basin and Range, Rocky Mountains, Snake River Plain, and Terranes of the U. S. Cordillera* (Geological Society of America, Boulder, Colorado, 2011). DOI:10.1130/2011.0021(04).

Thurow, R. F. *Middle Fork Salmon River Fisheries Investigations* (Idaho Department of Fish and Game, Boise, Idaho, 1982).

Thurow, R. F. *Middle Fork Salmon River Fisheries Investigations* (Idaho Department of Fish and Game, Boise, Idaho, 1985).

Thurow, R. F. *Salmon and wilderness: sharing the responsibility, sharing the benefits. Newsletter of the Chief's Wilderness Advisory Group* **Winter**, 6,11 (2015). Online at http://www.wilderness.net/NWPS/documents/FS/WAG_Newsletter_Winter2015.pdf.

Thurow, R. F. *Wood River Fisheries Investigations* (Idaho Department of Fish and Game, Boise, Idaho, 1987).

Thurow, R. F. *Wood River Fisheries Investigations* (Idaho Department of Fish and Game, Boise, Idaho, 1988a).

Thurow, R. F. Effects of stream alterations on rainbow trout in the Big Wood River, Idaho. *Pages 175-188 in Proceedings of the Western Association of Fish and Wildlife Agencies.* (Alburquerque, New Mexico, 1988b)

Thurow, R. F. *Wood River Fisheries Investigations* (Idaho Department of Fish and Game, Boise, Idaho, 1990).

Thurow, R. F., Lee, D. C., and Rieman, B. E. Status and distribution of chinook salmon and steelhead in the interior Columbia River basin and portions of the Klamath River basin. *Pages 133–160 in Sustainable Fisheries Management: Pacific Salmon* (CRC Press, Taylor and Francis Group, 2000). DOI:10.1201/9781439822678.ch12.

Thurow, R. F., Rieman, B. E., Lee, D. C., Howell, P. J. and Perkinson, R. D. *Distribution and Status of Redband Trout in the Interior Columbia River Basin and Portions of the Klamath River and Great Basins* (Oregon Chapter, American Fisheries Society, Portland, Oregon, 2007). Online at http://www.treesearch.fs.fed.us/pubs/29180.

Tolentino, S. and Teuscher, D. M. *Bear Lake Fisheries Management Plan* (Idaho Department of Fish and Game, Boise, Idaho, 2010).

Tolonen, A., Kjellman, J. and Lapalainen, J. Diet overlap between burbot (*Lota lota* (L.)) and whitefish (*Coregonus lavaretus* (L.)) in a subarctic lake. *Annales Zoologici Fennici* **36**, 205–214 (1999). Online at http://www.sekj.org/PDF/anzf36/anzf36-205p.pdf.

Trotter, P. *Cutthroat: Native Trout of the West, Second edition* (University of California Press, Berkley, California, 2008).

Trushenski, J. T., Blankenship, H. L., Bowker, J. D., Flagg, T. A., Hesse, J. A., Leber, K. M., MacKinlay, D. D., Maynard, D. J., Moffitt, C. M., Mudrak, V. A., Scribner, K. T., Stuewe, S. F., Sweka, J. A., Whelan, G. E. and Young-Dubovsky, C. Hatcheries and Management of Aquatic Resources (HaMAR): considerations for use of hatcheries and hatchery-origin fish. *North American Journal of Aquaculture* **77,** 327–342 (2015). DOI:10.1080/15222055.2015.1017130.

Trushenski, J. T., Flagg, T. and Kohler, C. C. Use of hatchery fish for conservation, restoration, and enhancement of fisheries. *Pages 261–293 in Inland Fisheries Management in North America, Third edition* (American Fisheries Society, Bethesda, Maryland, 2010).

U . S. vs Oregon Technical Advisory Committee. *Biological assessment of incidental impacts on salmon species listed under the Endangered Species Act in the 2008-2017 non-Indian and treaty Indian fisheries in the Columbia River basin.* Final 2008 BA 04-21-08.doc. (2008).

U. S. Department of Commerce. *Local Population Estimates.* (Bureau of the Census Series P-26, No. 84-W-SC, 1984).

U. S. Department of Commerce and Labor. *The Distribution of Food Fishes During the Fiscal Year 1906* (Bureau of Fisheries Document No. 613, Department of Commerce and Labor, Washington, D. C., 1907). Online at http://hdl.handle.net/2027/uiug.30112101028121.

U. S. Fish and Wildlife Service. *Recovery Plan for the White Sturgeon (Acipenser transmontanus): Kootenai River Population* (U. S. Fish and Wildlife Service, Portland, Oregon, 1999). Online at http://www.fws.gov/montanafieldoffice/Endangered_Species/Recovery_and_Mgmt_Plans/White_Sturgeon_Recovery_Plan.pdf.

U. S. Forest Service. *Frank Church-River of No Return Wilderness Draft Programmatic Management Plan, Bitterroot, Boise, Nez Perce, Payette and Salmon-Challis National Forests* (U. S. Forest Service, Northern Region, Missoula, Montana, 1998).

Unmack, P. J., Dowling, T. E., Laitinen, N. J., Secor, C. L., Mayden, R. L., Shiozawa, D. K. and Smith, G. R. Influence of introgression and geological processes on phylogenetic relationships of western North American mountain suckers (*Pantosteus*, Catostomidae). PLoS ONE **9**, 17 (2014). DOI:10.1371/journal.pone.0090061.

Urquhart, A. N. and Koetsier, P. Pectoral fin morphology as a reliable field sexing characteristic in populations of the invasive oriental weatherfish (*Misgurnus anguillicaudatus*). *Copeia* **201**, 296–300 (2011). DOI:10.1643/CI-10-027.

Utah Division of Wildlife Resources and Idaho Department of Fish and Game. *Bear Lake Fisheries Management Plan, Boise and Salt Lake City* (Utah Division of Wildlife Resources, Salt Lake City, Utah, 2010).

Uyeno, T. *Osteology and Phylogeny of American Cyprinid Fishes Allied to the Genus Gila.* (University of Michigan, Ann Arbor, Michigan, 1960).

Van Houdt, J. K., Hellemans, B. and Volckaert, F. A. M. Phylogenetic relationships among Palearctic and Nearctic burbot (*Lota lota*): Pleistocene extinctions and recolonization. *Molecular Phylogenetics and Evolution* **29**, 599–612 (2003). DOI:10.1016/S1055-7903(03)00133-7.

Van Kirk, R. Personal communication to John Sigler. Overview of Yellowstone Cutthroat Trout status. (Senior Scientist, Henrys Fork Foundation, Ashton, Idaho, 2014).

Van Kirk, R. and Gamblin, M. History of fisheries management in the upper Henry's Fork watershed. *Intermountain Journal of Sciences* **6**, 263–284 (2000).

Van Kirk, R. and Griffin, C. B. Building a collaborative process for restoration: Henrys Fork of Idaho and Wyoming. *Pages 256–273 in Watershed Restoration: Principles and Practices* (American Fisheries Society, Bethesda, Maryland, 1997).

Varley, J. D. and Gresswell, R. E. Ecology, status, and management of the Yellowstone cutthroat trout. *Pages 13–24 in Status and Management of Interior Stocks of Cutthroat Trout* (American Fisheries Society, Bethesda, Maryland, 1988).

Venturelli, P. A. and Tonn, W. M. Diet and growth of northern pike in the absence of prey fishes: initial consequences for persisting in disturbance-prone lakes. *Transactions of the American Fisheries Society* **135**, 1512–1522 (2006). DOI:10.1577/T05-228.1.

Vibert, R. Neurophysiology of electric fishing. *Transactions of the American Fisheries Society* **92**, 265–275 (1963). DOI:10.1577/1548-8659(1963)92[265:NOEF]2.0.CO;2.

Waitt, R. B. Jr. and Thorson, R. M. The Cordilleran ice sheet in Washington, Idaho, and Montana. *Pages 53-70 in Late-Quaternary Environments of the United States* (University of Minnesota Press, Minneapolis/St. Paul, Minnesota, 1983).

Waitt, R. B. Jr. Case for periodic, colossal jökulhlaups from Pleistocene glacial Lake Missoula. *Geological Society of America Bulletin* **96**, 1271–1286. (1985). DOI:10.1130/0016-7606(1985)96<1271:CFPCJF>2.0.CO;2

Wallace, R. L., Griffith, J. S., Connolly, P. J., Daley, D. M. and Beckham, G. B. *Distribution, Relative Abundance, Life History and Habitat Preferences of Shoshone Sculpin: Final Report* (U. S. Fish and Wildlife Service, Boise, Idaho, 1982).

Wallace, R. L., Griffith, J. S., Daley, D. M. Connolly, P. J. Beckham, G. B. Distribution of the Shoshone sculpin (*Cottus greenei*: Cottidae) in the Hagerman Valley of south central Idaho. *The Great Basin Naturalist* **44**, 324–326 (1984). Online at http://www.jstor.org/stable/41712079.

Wallace, R. L. and Zaroban, D. W. *Native Fishes of Idaho* (American Fisheries Society, Bethesda, Maryland, 2013).

Walrath, J. D., Quist, M. C. and Firehammer, J. A. Trophic ecology of nonnative Northern Pike and the effect on conservation of native westslope Cutthroat Trout. *North American Journal of Fisheries Management* **35**, 158-177 (2015). DOI:10.1080/02755947.2014.970678.

Walsh, M. G., O'Gorman, R., Strang, T., Edwards, W. H. and Rudstam, L. G. Fall diets of alewife, rainbow smelt, and slimy sculpin in the profundal zone of southern Lake Ontario during 1994-2005 with an emphasis on occurrence of *Mysis relicta*. *Aquatic Ecosystem Health and Management* **11**, 368–376 (2008). DOI:10.1080/14634980802516128.

Wang, Y. L., Binkowski, F. P. and Doroshow, S. I. Effect of temperature on early development of white and lake sturgeon, *Acipenser transmontanus* and *A. fulverscens*. *Pages 43–50 in North American Sturgeons: Biology and Aquaculture Potential* (Dordrecht, The Netherlands, 1985). DOI:10.1007/BF00001575.

Waples, R. S., Gustafson, R. G., Weitkamp, L. A., Myers, J. M., Johnson, O. W., Busby, P. J., Hard, J. J., Bryant, G. J., Waknitz, F. W., Neely, K., Teel, D., Grant, W. S., Winans, G. A., Phelps, S., Marshall, A. and Baker, B. M. Characterizing diversity in salmon from the Pacific Northwest. *Journal of Fish Biology* **59 Supplement sA**, 1–41 (2001). DOI:10.1111/j.1095-8649.2001.tb01376.x.

Ward, A. *Morphometric Evaluation of the Whitefish Complex in Bear Lake, Utah/Idaho* (Utah State University, Logan, Utah, 2001). Online at http://digitalcommons.usu.edu/etd/4428/.

Webber, P. A., Thompson, P. D. and Budy, P. Status and structure of two populations of the bluehead Sucker (*Catostomus discobolus*) in the Weber River, Utah. *The Southwestern Naturalist* **57**, 267–276 (2012). DOI:http://dx.doi.org/10.1894/0038-4909-57.3.267.

Weber, M. J. and Brown, M. L. Relationships among invasive common carp, native fishes and physicochemical characteristics in upper midwest (USA) lakes. *Ecology of Freshwater Fish* **20**, 270–278 (2011). DOI:10.1111/j.1600-0633.2011.00493.x.

Weigel, D. E., Peterson, J. T. and Spruell, P. Introgressive hybridization between native cutthroat trout and introduced rainbow trout. *Ecological Applications* **13**, 38–50 (2003). Online at http://dx.doi.org/10.1890/1051-0761(2003)013[0038:IHBNCT]2.0.CO;2.

Whitehead, R. L. *Geohydrologic Framework of the Snake River Plain Aquifer System, Idaho and Eastern Oregon*. (Professional Paper 1408-B, U.S. Geological Survey, Washington D.C., 1992). Online at http://pubs.usgs.gov/pp/1408b/report.pdf.

Wilde, G. R. Dispersal of tournament-caught black bass. *Fisheries* **28**, 10–17 (2003). DOI:10.1577/1548 8446(2003)28[10:DOTBB]2.0.CO;2.

Williams, K., Horne, J. K. and Punt, A. E. Examining influences of environmental, trawl gear, and fish population factors on midwater trawl performance using acoustic methods. *Fisheries Research* **164**, 94–101 (2015). DOI:10.1016/j.fishres.2014.11.001.

Williams, R. N. *Return to the River: Restoring Salmon to the Columbia River* (Elsevier Academic Press, Boston, Massachusetts, 2006). Online at http://www.sciencedirect.com/science/book/9780120884148.

Williams, R. N. and Shiozawa, D. K. *Genetic Analysis of Rainbow Trout from the Big Wood River and Trail Creek, Blaine County, Idaho* (Idaho Department of Fish and Game, Boise, Idaho, 1993). Online at http://idahodocs.contentdm.oclc.org/cdm/singleitem/collection/p16293coll7/id/237840/rec/1.

Willson, M. F. and Halupka, K. C. Anadromous fish as keystone species in vertebrate communities. *Conservation Biology* **9**, 489–497 (1995). DOI:10.1046/j.1523-1739.1995.09030489.x.

Wilson, K. Personal communication to John Sigler. Northern Leatherside Chub status.(Fish biologist, Utah Division of Wildlife Resources, Salt Lake City, Utah,2013).

Wilson, K. and Belk, M. C. Current distribution and habitat use of leatherside chub (*Gila copei*) in the Sevier and Beaver River drainages in south central Utah. (*Final Report*, Utah Division of Wildlife Resources, Salt Lake City, Utah, 1996).

Wilson, K. W. and Belk, M. C. Habitat characteristics of leatherside chub (*Gila copei*) at two spatial scales. *Western North American Naturalist* **61,** 36–42 (2001). https://ojs.lib.byu.edu/spc/index.php/wnan/article/view/28100.

Wolnicki, J., Myszkowski, L. and Kamiński, R. The influence of water temperature on the growth, survival and biological quality of juvenile burbot, (*Lota lota* (L.)). *Archives of Polish Fisheries* **9**, 79–86 (2001). Online at http://www.infish.com.pl/wydawnictwo/Archives/Fasc/work_pdf/Vol9Fasc1/Abstract07.html.

Wood, J. and Budy, P. The role of environmental factors in determining early survival and invasion success of exotic brown trout. *Transactions of the American Fisheries Society* **138**, 756–767 (2009). DOI:10.1577/T08-123.1.

Wurtsbaugh, W. A. and Hawkins, C. *Trophic Interactions Between Fish and Invertebrates in Bear Lake, Utah-Idaho* (Utah State University, Department of Fisheries and Wildlife, Logan, Utah, 1990). Online at http://utah.ptfs.com/awweb/awarchive?type=fileanditem=31127.

Wurtsbaugh, W. A. and Luecke, C. Limnological relationships and population dynamics of fishes in Bear Lake, Utah/Idaho. (*Final Report*, Utah Division of Wildlife Resources, Salt Lake City, Utah, 1998).

Wydoski, R. S. and Whitney, R. R. *Inland Fishes of Washington, Second edition* (American Fisheries Society in association with University of Washington Press, Bethesda, Maryland, 2003).

Young, M. K. Personal communication to Don Zaroban. Pacific Northwest sculpin genetics. (Research Fisheries Biologist, Rocky Mountain Research Station, U. S. Forest Service, Missoula, Montana, 2016).

Zale, A. V., Parrish, D. L. and Sutton, T. M. *Fisheries Techniques, Third edition* (American Fisheries Society Bethesda, Maryland, 2012).

Zambrano, L., Martínez-Meyer, E., Menezes, N. and Peterson, A. T. Invasive potential of common carp (*Cyprinus carpio*) and Nile tilapia (*Oreochromis niloticus*) in American freshwater systems. *Canadian Journal of Fisheries and Aquatic Sciences* **63**, 1903–1910 (2006). DOI:10.1139/f06-088.

Zaroban, D. W. Considerations for wood river sculpin conservation: historical occurrence and sampling efficiency. *Western North American Naturalist* **70**, 446–456 (2011a). Online at https://ojs.lib.byu.edu/spc/index.php/wnan/article/view/27371.

Zaroban, D. W., Mulvey, M. P., Maret, T. R., Hughes, R. M. and Merritt, G. D. Classification of species attributes for Pacific Northwest freshwater fishes. *Northwest Science* **73**, 81–93 (1999).

Zurstadt, C. F. and Stephans, K. Seasonal migration of westslope cutthroat trout in the Middle Fork Salmon River drainage, Idaho. *Northwest Science* **78**, 278–285 (2004).

SUPPLEMENTAL REFERENCES

Chapter 2

Artedi, P., Walbaum, J. J. and von Linne, C. *Bibliotheca et Philosophia Ichthyologica, pars 3: Genera Piscium in Quibus Systema Totum Ichthyologiæ Proponitur Cum Classibus Ordinibus, Generum Characteribus, Specierum Differentiis, Observationibus Plurimis Ichthyologiæ* (Impensis A. F. Roese, Grypeswaldiae, 1792)

Batzer, D. P., Pusateri, C. R. and Vetter, R. Impacts of fish predation on marsh invertebrates: direct and indirect effects. *Wetlands* **20,** 307–312 (2000). DOI:10.1672/0277-5212(2000)020[0307:IOFPOM]2.0.CO;2.

Baxter, G. T. and Simon, J. R. *Wyoming Fishes* (Bulletin 4, Wyoming Game and Fish Commission, Cheyenne, Wyoming, 1970).

Billman, E. J., Kreitzer, J. D., Creighton, J. C., Habit, E., McMillan, B. and Belk, M. C. Habitat enhancement and native fish conservation: can enhancement of channel complexity promote the coexistence of native and introduced fishes? *Environmental Biology of Fishes* **96,** 555–566 (2013). DOI:10.1007/s10641-012-0041-2.

Cope, E. D. Recent reptiles and fishes: report on the reptiles and fishes obtained by the naturalists of the expedition. *Pages 432–442 in Preliminary Report of the United States Geological Survey of Wyoming and Portions of Contiguous Territories, Second Annual Report of Progress* (U. S. Geological Survey, Washington D. C., 1872). Online at http://biodiversitylibrary.org/page/40322515.

Cuvier, G. and Valenciennes, *M. Histoire Naturelle des Poissons t. 17* (Libraire de la Societe Geologique de France, Paris, France, 1844). Online at http://biodiversitylibrary.org/page/4443970#page/7/mode/1up.

Dunham, J. B., Rosenberger, A. E., Luce, C. H. and Rieman, B. E. Influences of wildfire and channel reorganization on spatial and temporal variation in stream temperature and the distribution of fish and amphibians. *Ecosystems* **10,** 335–346 (2007). DOI:10.1007/s10021-007-9029-8.

Fischer, J. R., Krogman, R. M. and Quist, M. C. Influences of native and non-native benthivorous fishes on aquatic ecosystem degradation. *Hydrobiologia* **711,** 187–199 (2013). DOI:10.1007/s10750-013-1483-z.

Flinders, C. A., Ragsdale, R. L. and Hall, T. J. Patterns of fish community structure in a long-term watershed-scale study to address the aquatic ecosystem effects of pulp and paper mill discharges in four US receiving streams. *Integrated Environmental Assessment and Management* **5,** 219–233 (2009). DOI:10.1897/IEAM_2008-057.1.

Helmus, M. R. and Sass, G. G. The rapid effects of a whole-lake reduction of coarse woody debris on fish and benthic macroinvertebrates. *Freshwater Biology* **53,** 1423–1433 (2008). DOI:10.1111/j.1365-2427.2008.01974.x.

Jelks, H. L., Walsh, S. J., Burkhead, N. M., Contreras-Balderas, S., Diaz-Pardo, E., Hendrickson, D. A., Lyons, J., Mandrak, N. E., McCormick, F., Nelson, J. S., Platania, S. P., Porter, B. A., Renaud, C. B., Schmitter-Soto, J. J., Taylor, E. B. and Warren, M. L. Jr. Conservation status of imperiled North American freshwater and diadromous fishes. *Fisheries* **33,** 372–407 (2008). DOI:10.1577/1548-8446-33.8.372.

Jordan, D. S. and Henshaw, H. W. Report upon the fishes collected during the years 1875, 1876, and 1877, in California and Nevada. *Pages 187–200 in the 1878 Annual Report Geological Survey West of 100th Meridian, in California, Nevada, Nebraska, Utah, Arizona, Colorado, New Mexico, Wyoming, and Montana* (United States Geological Survey, Washington, D. C., 1878).

Knapp, R. A. Non-native trout in natural lakes of the Sierra Nevada: an analysis of their distribution and impacts on native aquatic biota. *Pages 363–407 in Sierra Nevada Ecosystem Project: Final Report to Congress. Volume III: Assessments, Commissioned Reports, and Background Information* (University of California, Centers for Water and Wildland Resources, Davis, California, 1996). Online at http://knapplab.msi.ucsb.edu/pdfs/Knapp_SNEP_VIII_C08.pdf.

Koel, T. M., Bigelow, P. E., Doepke, P. D., Ertel, B. D. and Mahony, D. L. Nonnative lake trout result in Yellowstone cutthroat trout decline and impacts to bears and anglers. *Fisheries* **30,** 10–19 (2005). DOI:10.1577/1548-8446(2005)30[10:NLTRIY]2.0.CO;2.

Latta, W. C., Breck, J. E. and Marshall Duchon, E. R. Species-area and latitudinal patterns for Michigan fishes. *American Midland Naturalist* **159,** 349–363 (2008). Online at http://www.jstor.org/stable/20491339.

Lee, D. S., Gilbert, C. R., Hocutt, C. H., Jenkins, R. E., McAllister, D. E. and Stauffer, J. R. Jr. *Atlas of North American Freshwater Fishes* (North Carolina Museum of Natural History, Raleigh, North Carolina, 1980).

Lewis, M. and Clark, W. *The Definitive Journals of Lewis and Clark* (University of Nebraska Press, Lincoln, Nebraska, 2002).

Magoulick, D. D. and Lewis, L. C. Predation on exotic zebra mussels by native fishes: effects on predator and prey. *Freshwater Biology* **47,** 1908–1918 (2002). DOI:10.1046/j.1365-2427.2002.00940.x.

Meador, M. R., Brown, L. R. and Short, T. Relations between introduced fish and environmental conditions at large geographic scales. *Ecological Indicators* **3,** 81–92 (2003). DOI:10.1016/S1470-160X(03)00013-X.

Minckley, W. L. and Marsh, P. C. *Inland Fishes of the Greater Southwest: Chronicle of a Vanishing Biota* (University of Arizona Press, Tucson, Arizona, 2009).

Mitchill, S. L. and Gill, T. Report in Part of Samuel L. Mitchill, M.D., Professor of Natural History, etc., on the Fishes of New York (Printed for Theodore Gill, Columbian University, Washington, 1898). Online at http://hdl.handle.net/2027/chi.72742111.

Mitchill, S. L. The fishes of New-York, described and arranged. *Pages 355–492 in the Transactions of the Literary and Philosophical Society of New York* (New York, 1815). Online at http://biodiversitylibrary.org/page/41884397.

Naish, K. A. and Hard, J. J. Bridging the gap between the genotype and the phenotype: linking genetic variation, selection and adaptation in fishes. *Fish and Fisheries* **9,** 396–422 (2008). DOI:10.1111/j.1467-2979.2008.00302.x.

Olden, J. D. and Poff, N. L. Long-term trends of native and non-native fish faunas in the American Southwest. *Animal Biodiversity and Conservation* **28,** 75–89 (2005). Online at http://abc.museucien-ciesjournals.cat/files/ABC-28-1-pp-75-89.pdf.

Propst, D. L. and Gido, K. B. Responses of native and nonnative fishes to natural flow regime mimicry in the San Juan River. *Transactions of the American Fisheries Society* **133,** 922–931 (2004). DOI:10.1577/T03-057.1.

Quist, M. C., Hubert, W. A. and Isaak, D. J. Fish assemblage structure and relations with environmental conditions in a Rocky Mountain watershed. *Canadian Journal of Zoology* **82,** 1554–1565 (2004). DOI:10.1139/z04-134.

Quist, M. C., Hubert, W. A. and Rahel, F. J. Elevation and stream-size thresholds affect distributions of native and exotic warmwater fishes in Wyoming. *Journal of Freshwater Ecology* **19,** 227–236 (2004). DOI:10.1080/02705060.2004.9664536.

Saunders, D. L., Meeuwig, J. J. and Vincent, A. C. J. Freshwater protected areas: strategies for conservation. *Conservation Biology* **16,** 30–41 (2002). DOI:10.1046/j.1523-1739.2002.99562.x.

Schoenebeck, C. W. and Hansen, M. J. Electrofishing catchability of walleyes, largemouth bass, smallmouth bass, northern pike, and muskellunge in Wisconsin Lakes. *North American Journal of Fisheries Management* **25,** 134 1–1352 (2005). DOI:10.1577/M04-125.1.

Sigler, W. F. and Miller, R. R. *Fishes of Utah* (Utah State Department of Fish and Game, Salt Lake City, Utah, 1963).

Smith, G. R. Introgression in fishes: significance for paleontology, cladistics, and evolutionary rates. *Systematic Biology* **41,** 41–57 (1991). DOI:10.1093/sysbio/41.1.41.

Søndergaard, M. and Jeppesen, E. Anthropogenic impacts on lake and stream ecosystems, and approaches to restoration. *Journal of Applied Ecology* **44,** 1089–1094 (2007). DOI:10.1111/j.1365-2664.2007.01426.x.

Suckley, G. and Cooper, J. G. *The natural history of Washington Territory and Oregon: with much relating to Minnesota, Nebraska, Kansas, Utah, and California between the thirty-sixth and forty-ninth parallels of latitude: being those parts of the final reports on the survey of the Northern Pacific Railroad route, relating to the natural history of the regions explored, with full catalogues and descriptions of the plants and animals collected from 1853 to 1860* (Bailliere Brothers, New York, New York, 1860). Online at http://www.sos.wa.gov/legacy/publications_detail.aspx?p=120.

Tyus, H. M. and Saunders, J. F. Nonnative fish control and endangered fish recovery: lessons from the Colorado River. *Fisheries* **25,** 17–24 (2000). DOI:10.1577/1548-8446(2000)025<0017:NFCAEF>2.0.CO;2.

University of California Museum of Paleontology. *Carl Linnaeus (1 707-1 778)* (University of California Museum of Paleontology, Berkeley, California). Online at http://www.ucmp.berkeley.edu/history/linnaeus.html. (2000).

Ward, D. L., Schultz, A. A. and Matson, P. G. Differences in swimming ability and behavior in response to high water velocities among native and nonnative fishes. *Environmental Biology of Fishes* **68,** 87–92 (2003). DOI:10.1023/A:1026031128486.

Ward, N. E. and Ward, D. L. Resident fish in the Columbia River Basin. *Fisheries* **29,** 10–18 (2004). DOI :10.1577/1548-8446(2004)29[10:RFITCR]2.0.CO;2.

Webb, J. F., Montgomery, J. C. and Mogdans, J. Bioacoustics and the lateral line system of fishes. *Pages 145–182 in Fish Bioacoustics* (Springer, New York, New York, 2008). DOI:10.1007/978-0-387-73029-5_5.

Weber, E. D. and Fausch, K. D. Interactions between hatchery and wild salmonids in streams: differences in biology and evidence for competition. *Canadian Journal of Fisheries and Aquatic Sciences* **60,** 1018–1036 (2003). DOI:10.1139/F03-087.

Wei, A., Chow-Fraser, P. and Albert, D. Influence of shoreline features on fish distribution in the Laurentian Great Lakes. *Canadian Journal of Fisheries and Aquatic Sciences* **61,** 1113–1123 (2004). DOI:10.1139/f04-061.

Wiley, R. W. The 1962 rotenone treatment of the Green River, Wyoming and Utah, revisited: lessons learned. *Fisheries* **33,** 611–617 (2008). DOI:10.1577/1548-8446-33.12.611.

Williams, J. E. *Habitat Relationships of Native and Non-Native Fishes of the Willamette River, Oregon* (Oregon State University, Corvallis, Oregon, 2014). Online at http://ir.library.oregonstate.edu/xmlui/handle/1957/49883.

Winemiller, K. O., Tarim, S., Shormann, D. and Cotner, J. B. Fish assemblage structure in relation to environmental variation among Brazos River oxbow lakes. *Transactions of the American Fisheries Society* **129,** 451–468 (2000). DOI:10.1577/1548-8659(2000)129<0451:FASIRT>2.0.CO;2.

Wurtsbaugh, W. A., Luecke, C., Ward, A. E., Ruzycki, J. and Lay, C. *Limnological Relationships and Population Dynamics of Fishes in Bear Lake Utah/Idaho* (Utah State University, Department of Fisheries and Wildlife, Logan, Utah, 1990). Online at http://utah.ptfs.com/awweb/awarchive?item=31125.

Zeug, S. C. and Winemiller, K. O. Evidence supporting the importance of terrestrial carbon in a large-river food web. *Ecology* **89,** 1733–1743 (2008). Online at http://dx.doi.org/10.1890/07-1064.1.

Chapter 3

Allison, R. R. *Climatic, Volcanic, and Tectonic Influences on Late Pleistocene Sedimentation Along the Snake River and in Market Lake: Bonneville, Jefferson, and Madison Counties, Idaho* (Idaho State University, Pocatello, Idaho, 2001). Online at http://geology.isu.edu/thesis/Allison.Richard.2001.pdf.

Bohacs, K. M., Lamb-Wozniak, K., Demko, T. M., Eleson, J., McLaughlin, O., Lash, C., Cleveland, D. M. and Kaczmarek, S. Vertical and lateral distribution of lacustrine carbonate lithofacies at the parasequence scale in the Miocene Hot Spring limestone, Idaho: an analog addressing reservoir presence and quality. *American Association of Petroleum Geologists Bulletin* **97,** 1967–1995 (2013). DOI:10.1306/07081312192.

Bonnichsen, B. and Godchaux, M. M. Late Miocene, Pliocene and Pleistocene geology of southwestern Idaho with emphasis on basalts in the Bruneau-Jarbidge, Twin Falls, and western Snake River Plain regions. *Pages 233–312 in Tectonic and Magmatic Evolution of the Snake River Plain Volcanic Province* (Idaho Geological Survey Bulletin 30, University of Idaho, Moscow, Idaho, 2002). Online at http://www.idahogeology.org/PDF/Bulletins_(B)/Snake_River_Plain_B-30.pdf.

Bright, J. Ostracode endemism in Bear Lake, Utah and Idaho. *Pages 197–216 in Paleoenvironments of Bear Lake, Utah and Idaho, and its Catchment* (Geological Society of America, Boulder, Colorado, 2009). DOI:10.1130/2009.2450(08).

Broughton, J. M. Terminal Pleistocene fish remains from Homestead Cave, Utah, and implications for fish biogeography in the Bonneville Basin. *Copeia* **2000,** 645–65 6 (2000).

Byers, D. A. and Broughton, J. M. Holocene environmental change, artiodactyl abundances, and human hunting strategies in the Great Basin. *American Antiquity* **69,** 235–25 5 (2004). Online at http://www.jstor.org/stable/4128418.

Carr, W. J. and Trimble, D. E. *Geology of the American Falls Quadrangle Idaho* (Geological Survey Bulletin 1 121-G, U. S. Geological Survey, Washington, D. C., 1963). Online at http://pubs.usgs.gov/bul/1121g/report.pdf.

Chetel, L. M., Janecke, S. U., Carroll, A. R., Beard, B. L., Johnson, C. M. and Singer, B. S. Paleogeographic reconstruction of the Eocene Idaho River, North American Cordillera. *Geological Society of America Bulletin* **123,** 71–88 (2011). DOI:10.1130/B30213.1.

Cope, E. D. On the geology and paleontology of the Cretaceous stata of Kansas. *Annual Report of the U. S. Geological Survey of the Territories* **5,** 318-349 (1871).

Eversole, E. K. *Soldier Bar Landslide and Subsequent Late Pleistocene Lakes, Big Creek, Valley County, Idaho* (Idaho State University, Pocatello, Idaho, 2008). Online at http://geology.isu.edu/thesis/Eversole_Elijah_2008.pdf.

Fritz, W. J. and Sears, J. W. Tectonics of the Yellowstone hotspot wake in southwestern Montana. *Geology* **21,** 427–430 (1993). http://dx.doi.org/10.1130/0091-7613(1993)021%3C0427:TOTYHW%3E2.3.CO;2.

Gianniny, G. L., Thackray, G. D., Kaufman, D. S. Forman, S. L., Sherbondy, M. J. and Findeisen, D. Late quaternary highstands in the Mud Lake and Big Lost Trough subbasins of Lake Terreton, Idaho. *GSA Special Papers* **353,** 77–90 (2002). DOI:10.1130/0-8137-2353-1.77.

Gillerman, V. S. *Geologic Report on the 1993 Bliss Landslide, Gooding County, Idaho* (Idaho Geological Survey, University of Idaho, Moscow, Idaho, 2001). Online at http://idahodocs.cdmhost.com/cdm/ref/collection/p 1 6293coll6/id/226652.

Hughes, S. S., Smith, R. P., Hackett, W. R. and Anderson, S. R. Mafic volcanism and environmental geology of the eastern Snake River plain, Idaho. *Pages 143–168 in Guidebook to the Geology of Eastern Idaho* (Idaho Museum of Natural History, Pocatello, Idaho, 1999). Online at http://geology.isu.edu/Digital_Geology_Idaho/papers/gsac3p9.pdf.

Janecke, S. U. and Oaks, R. Q. Jr. Diversion(s) of the lower Bear River in Gem Valley, southeast Idaho, by a tectono-volcanic valve. *Pages 27–28 in Abstracts of the workshop on Late Cenozoic to Recent Geologic and Biotic History of the Snake River, March 24–26, 2014* (Idaho State University, Pocatello, Idaho, 2014). Online at http://geology.isu.edu/SnakeRiver/Proceedings.pdf.

Janecke, S. U., VanDenburg, C. J., Blankenau, J. J. and M'Gonigle, J. W. Long-distance longitudinal transport of gravel across the Cordilleran thrust belt of Montana and Idaho. *Geology* **28,** 439–442 (2000). DOI:10.1130/0091-7613(2000)28<439:LLTOGA>2.0.CO;2.

Kenworthy, M. K., Rittenour, T. M., Pierce, J. L., Sutfin, N. A. and Sharp, W. D. Luminescence dating without sand lenses: An application of OSL to coarse-grained alluvial fan deposits of the Lost River Range, Idaho, USA. *Quaternary Geochronology* **23,** 9–25 (2014). DOI:10.1016/j.quageo.2014.03.004.

Link, P. K., Crosby, B. T., Lifton, Z. M., Eversole, E. A. and Rittenour, R. M. The late Pleistocene (17 ka) Soldier Bar landslide and Big Creek Lake, Frank Church-River of No Return Wilderness, central Idaho. *Rocky Mountain Geology* **49,** 17–31. (2014). DOI:10.2113/gsrocky.49.1.17.

Link, P. K., McDonald., H. G., Fanning, C. M. and Godfrey, A. E. Detrital zircon evidence for Pleistocene drainage reversal at Hagerman Fossil Beds National Monument, central Snake River Plain, Idaho. *Pages 105–119 in Tectonic and Magmatic Evolution of the Snake River Plain Volcanic Province* (Idaho Geological Survey Bulletin 30, University of Idaho, Moscow, Idaho, 2002). Online at http://www.idahogeology.org/PDF/Bulletins_(B)/Snake_River_Plain_B-30.pdf.

Long, S. P., Link, P. K., Janecke, S. U., Perkins, M. E. and Fanning, C. M. Multiple phases of Tertiary extension and synextensional deposition of the Miocene-Pliocene Salt Lake Formation in an evolving supradetachment basin, Malad Range, southeast Idaho, U. S.A. *Rocky Mountain Geology* **41,** 1–27 (2006). DOI:10.2113/gsrocky.41.1.1.

Malde, H. E. Quaternary geology and structural history of the Snake River Plain, Idaho and Oregon. *Pages 251–280 Quaternary Nonglacial Geology, Conterminous U. S.* (The Geology of North America, K-2, Geological Society of America, Boulder, Colorado, 1991). DOI:10.1130/DNAG-GNA-K2.251.

Malde, H. E. Stratigraphy of the Glenns Ferry formation from Hammett to Hagerman, Idaho. U. S. *Geological Survey Bulletin* **1331,** D1–D19 (1982). Online at http://pubs.usgs.gov/bul/1331d/report.pdf.

McDonald, H. G., Link, P. K. and Lee, D. E. An overview of the geology and paleontology of the Pliocene Glenns Ferry Formation, Hagerman Fossil Beds National Monument. *Pages 16–45 in Geology of the Crook and Snake River Plain, Twin Falls and Vicinity, Idaho, Proceedings of the 21st Annual TRGS Field Conference* (College of Southern Idaho, Twin Falls, Idaho, 1996).

Meyer, G. A. and Leidecker, M. E. Fluvial terraces along the Middle Fork Salmon River, Idaho, and their relation to glaciation, landslide dams, and incision rates: A preliminary analysis and river-mile guide. *Pages 219–236 in Guidebook to the Geology of Eastern Idaho* (Idaho Museum of Natural History, Pocatello, Idaho, 1999). Online at http://www.imnh.isu.edu/digitalatlas/geo/gsa/papers/gsac4p13.pdf.

Miller, S., Glanzman, D., Doran, S., Parkinson, S., Buffington, J. and Milligan, J. *Geomorphology of the Hells Canyon Reach of the Snake River* (Idaho Power Company, Boise, Idaho, 2003). Online at http://www.idahopower.com/pdfs/Relicensing/hellscanyon/hellspdfs/techappendices/Sediment/e01_02.pdf.

Palmer, T. *The Snake River: Window to the West* (Island Press, Washington, D. C., 1991).

Pierce, K. L. and Scott, W. E. Pleistocene episodes of alluvial-gravel deposition, southern Idaho. *Pages 685-702 in Cenozoic Geology of Idaho* (Bulletin 26, Idaho Bureau of Mines and Geology, Moscow, Idaho, 1982). Online at http://geology.isu.edu/Digital_Geology_Idaho/papers/B-26ch13-2.pdf.

Repenning, C. A., Weasma, T. R. and Scott, G. R. *The Early Pleistocene (Latest Blancan-earliest Irvingtonian) Froman Ferry Fauna and History of the Glenns Ferry Formation, Southwestern Idaho* (U. S. Geological Survey, Denver, Colorado, 1995). Online at http://pubs.usgs.gov/bul/2105/report.pdf.

Scott, W. E., Pierce, K. L., Bradbury, J. P. and Forester, R. M. Revised quaternary stratigraphy and chronology in the American Falls Area, Southeastern Idaho. *Pages 581–595 in Cenozoic Geology of Idaho* (Bulletin 26, Idaho Bureau of Mines and Geology, Moscow, Idaho, 1982). Online at http://geology.isu.edu/Digital_Geology_Idaho/papers/B-26ch10-1.pdf.

Shafer, A. B. A., Cullingham, C. I., Côté, S. D. and Coltman, D. W. Of glaciers and refugia: a decade of study sheds new light on the phylogeography of northwestern North America. *Molecular Ecology* **19,** 4589–4621 (2010). DOI:10.1111/j.1365-294X.2010.04828.x.

Shapley, M., Finney, B. and Barnes, B. Varved sediments in landslide-formed lakes: slope failures and natural eutrophication promote formation of high-resolution sediment-delivery records in central Idaho, USA. *Page 201 in Sixth International Limnogeology Congress, Abstract volume, Reno, Nevada, June 15-19, 2015* (Open-File Report 2015-1092, U. S. Geological Survey, Reston, Virginia, 2015). DOI:10.3133/ofr20151092.

Sharma, S., Legendre, P., De Cáceres, M. and Boisclair, D. The role of environmental and spatial processes in structuring native and non-native fish communities across thousands of lakes. *Ecography* **34,** 762–771 (2011). DOI:10.1111/j.1600-0587.2010.06811.x.

Smith, C. T., Nelson, R. J., Wood, C. C. and Koop, B. F. Glacial biogeography of North American coho salmon *(Oncorhynchus kisutch). Molecular Ecology* **10,** 2775–2785 (2001). DOI:10.1046/j.1365-294X.2001.t01-1-01405.x.

Smith, G. R. Biogeography of Intermountain fishes. *Great Basin Naturalist Memoirs* **2,** 17–42 (1978). Online at https://journals.lib.byu.edu/spc/index.php/gbnmem/article/view/30422/28904.

Smith, G. R., Dowling, T. E., Gobalet, K. W., Lugaski, T., Shiozawa, D. K. and Evans, R. P. Biogeography and timing of evolutionary events among Great Basin fishes. *Pages 1 75–234 in Great Basin Aquatic Systems History* (Smithsonian Contributions to Earth Sciences 33, Smithsonian Institution Press, Washington D. C., 2002). Online at http://lsa.umich.edu/ummz/fishes/publications/pdf/biogeography-timing-evol-events-greatbasin-fish.pdf.

Smith, G. R., Morgan, N. and Gustafson, E. *Fishes of the Mio-Pliocene Ringold Formation, Washington: Pliocene Capture of the Snake River by the Columbia River* (University of Michigan Museum of Paleontology, Ann Arbor, Michigan, 2000). Online at http://deepblue.lib.umich.edu/handle/2027.42/48664.

Smith, G. R., Swirydczuk, K., Kimmel, P. G. and Wilkinson, B. H. Fish biostratigraphy of the late Miocene to Pleistocene sediments of the western Snake River Plain, Idaho. *Pages 519–541 in Cenozoic Geology of Idaho* (Idaho Bureau of Mines and Geology, Moscow, Idaho, 1982). Online at http://geology.isu.edu/Digital_Geology_Idaho/papers/B-26ch9-1.pdf.

Stearns, H. T., Crandall, L. and Stewart, W. G. *Geology and Ground-Water Resources of the Snake River Plain in Southeastern Idaho* (Water-Supply Paper 774, U. S. Geological Survey, 1938). Online at http://pubs.usgs.gov/wsp/0774/report.pdf.

Stroup, C. N., Link, P. K. and Fanning, C. M. Provenance of late Miocene fluvial strata of the Sixmile Creek Formation, southwest Montana: evidence from detrital zircon. *Northwest Geology* **37,** 69–84 (2008).

Vogl, J. J., Min, K., Carmenate, A., Foster, D. A., and Marsellos, A. Miocene regional hotspot-related uplift, exhumation, and extension north of the Snake River Plain: evidence from apatite (U-Th)/He thermochronology. *Lithosphere* **6,** 108–123 (2014). DOI:10.1130/L308.1.

Waters, J. M., Craw, D., Youngson, J. H. and Wallis, G. P. Genes meet geology: fish phylogeographic pattern reflects ancient, rather than modern, drainage connections. *Evolution* **55,** 1844–1851 (2001). DOI:10.1111/j.0014-3820.2001.tb00833.x.

Wood, S. H. and Clemens, D. M. Geologic and tectonic history of the western Snake River Plain, Idaho and Oregon. *Pages 69–104 in Tectonic and Magmatic Evolution of the Snake River Plain Volcanic Province* (Idaho Geological Survey Bulletin 30, University of Idaho, Moscow, Idaho, 2002). Online at http://www.idahogeology.org/PDF/Bulletins_(B)/Snake_River_Plain_B-30.pdf.

Wood, S. H. Seismic expression and geological significance of a lacustrine delta in Neogene deposits of the western Snake River Plain, Idaho. *American Association of Petroleum Geologists Bulletin* **78,** 102–121 (1994). Online at https://earth.boisestate.edu/swood/files/2010/08/Wood-1994-AAPG.pdf.

Chapter 6

Clemens, B. J., van der Wetering, S., Sower, S. A. and Schreck, C. B. Maturation characteristics and life-history strategies of the Pacific lamprey, *Entosphenus tridentatus. Canadian Journal of Zoology* **91,** 775–788 (2013). DOI:10.1139/cjz-2013-0114.

Close D. A., Fitzpatrick, M. S. and Li, H. W. The ecological and cultural importance of a species at risk of extinction, pacific lamprey. *Fisheries* **July,** 19–25 (2002). DOI:10.1577/1548-8446(2002)027<0019:TEACIO>2.0.CO;2.

Conservation Agreement of Pacific Lamprey Team. *Conservation Agreement of Pacific Lamprey (Entosphenos tridentatus) in the States of Alaska, Washington, Oregon, Idaho and California* (U. S. Fish and Wildlife Service, Pacific Region, Portland, Oregon, 2012). Online at http://www.fws.gov/Pacific/fisheries/sphabcon/lamprey/pdf/Pacific_Lamprey_CI.pdf.

Lower Columbia Fish Recovery Board. Pacific lamprey. *Pages 3:1–3:17 in Lower Columbia Salmon and Steelhead Recovery and Subbasin Plan: Technical Foundation, Volume III: Other Species, Draft* (Northwest Power and Conservation Council, Portland, Oregon, 2004). Online at http://www.nwcouncil.org/media/21157/Vol._III_Ch._3__Lamprey.pdf.

McIlraith, B. J., Caudill, C. C., Kennedy, B. P., Peery, C. A. and Keefer, M. L. Seasonal migration behaviors and distribution of adult Pacific lampreys in unimpounded reaches of the Snake River basin. *North American Journal of Fisheries Management* **35,** 123–134 (2015). DOI:10.1080/02755947.2014.986344.

Swink, W. D. Effectiveness of an electrical barrier in blocking a sea lamprey spawning migration on the Jordan River, Michigan. *North American Journal of Fisheries Management* **19,** 397–405 (2011). DOI:10.1577/1548-8675(1999)019<0397:EOAEBI>2.0.CO;2.

Chapter 7

Hughes, J. B. and Bates, P. Juvenile indexing and juvenile abundance estimate of white sturgeon in the Snake River between Bliss and C.J. Strike dams. *Pages 7–32 in Snake River White Sturgeon Conservation Plan–2013 Implementation and Reporting* (Unpublished Annual Report to Idaho Power Company, Boise, Idaho, 2013). Online at https://www.idahopower.com/pdfs/ourEnvironment/SturgeonReports/2013AnnualReport.pdf.

Pacific States Marine Fisheries Commission. *White Sturgeon Management Framework Plan* (Pacific States Marine Fisheries Commission, Portland, Oregon, 1992). Online at http://www.calwater.ca.gov/Admin_Record/C-051912.pdf.

Paragamian, V. L., Kruse, G. and Wakkinen, V. Spawning habitat of Kootenai River white sturgeon, post-Libby Dam. *North American Journal of Fisheries Management* **21,** 22–33 (2001). DOI:10.1577/1548-8675(2001)021<0022:SHOKRW>2.0.CO;2.

Parsley, M. J., Anders, P. J., Miller, A. I., Beckman, L. G. and McCabe, G. T. Recovery of white sturgeon populations through natural production: understanding the influence of abiotic and biotic factors on spawning and subsequent recruitment. *Pages 55–66 in Biology, Management, and Protection of North American Sturgeon* (American Fisheries Society, Bethesda, Maryland, 2002).

Schreier, A. D., Mahardja, B. and May, B. Patterns of population structure vary across the range of the white sturgeon. *Transactions of the American Fisheries Society* **142,** 1273–1286 (2013). DOI:10.1080/00028487.2013.788554.

Chapter 9

Billman, E. J., Lee, J. B., Young, D. O., McKell, M. D., Evans, R. P. and Shiozawa, D. K. Phylogenetic divergence in a desert fish: differentiation of speckled dace within the Bonneville, Lahontan, and Upper Snake River Basins. *Western North American Naturalist* **70,** 39–47 (2010). Online at https://ojs.lib.byu.edu/spc/index.php/wnan/article/view/27432.

Brown, J. H. *The Life History and Ecology of the Northern Lake Chub (Couesius plumbeus) in the La Ronge Region of Saskatchewan* (University of Saskatchewan, Saskatoon, Saskatchewan, 1969). Online at http://ecommons.usask.ca/handle/10388/etd-10132010-084453.

Clark, W. H. A third locality for the tui chub, *Gila bicolor* (Cyprinidae), in Idaho with habitat description. *Journal of the Idaho Academy of Science* **19,** 1-6 (1983).

Findlay, C. S., Bert, D. G. and Zheng, L. Effect of introduced piscivores on native minnow communities in Adirondack lakes. *Canadian Journal of Fisheries and Aquatic Sciences* **57,** 570–580 (2000). DOI:10.1139/f99-276.

Fresh, K. L., Schroder, S. L. and Carr, M. I. Predation by northern pikeminnow on hatchery and wild coho salmon smolts in the Chehalis River, Washington. *North American Journal of Fisheries Management* **23,** 1257–1264 (2003). DOI:10.1577/M02-030.

Gadomski, D. M., Barfoot, C. A., Bayer, J. M. and Poe, T. P. Early life history of the northern pikeminnow in the lower Columbia River Basin. *Transactions of the American Fisheries Society* **130,** 250–262 (2001). DOI:10.1577/1548-8659(2001)130<0250:ELHOTN>2.0.CO;2.

Gilbert, C. H. and Burgess, G. H. *Notropis hudsonius* (Clinton) spottail shiner. *Page 275 in Atlas of North American Freshwater Fishes* (North Carolina Museum of Natural History, Raleigh, North Carolina, 1980).

Hänfling, B., Bolton, P., Harley, M. and Carvalho, G. R. A molecular approach to detect hybridisation between crucian carp *(Carassius carassius)* and non-indigenous carp species *(Carassius spp. and Cyprinus carpio). Freshwater Biology* **50,** 403–417 (2005). DOI:10.1111/j.1365-2427.2004.01330.x.

Harris, P. M. *Systematic Studies of the Genus Siphateles (Ostariophysi: Cyprinidae) From Western North America* (Oregon State University, Corvallis, Oregon, 2000). Online at http://ir.library.oregonstate.edu/xmlui/handle/1957/9829.

Houston, D. D., Shiozawa, D. K. and Riddle, B. R. Phylogenetic relationships of the western North American cyprinid genus *Richardsonius,* with an overview of phylogeographic structure. *Molecular Phylogenetics and Evolution* **55,** 259–273 (2010). DOI:10.1016/j.ympev.2009.10.017.

Jackson, Z. J., Quist, M. C., Downing, J. A. and Larscheid, J. G. Common carp (*Cyprinus carpio*), sport fishes, and water quality: ecological thresholds in agriculturally eutrophic lakes. *Lake and Reservoir Management* **26,** 14–22 (2010). DOI:10.1080/07438140903500586.

Johnson, J. B., Belk, M. C. and Shiozawa, D. K. Age, growth, and reproduction of leatherside chub (*Gila copei*). *Great Basin Naturalist* **55,** 183–187 (1995).

Johnson, J. B. and Jordan, S. Phylogenetic divergence in leatherside chub (*Gila copei*) inferred from mitochondrial cytochrome b sequences. *Molecular Ecology* **9,** 1029–1035 (2000). DOI:10.1046/j.1365-294x.2000.00960.x.

Jordan, D. S. Report of Exploration in Colorado and Utah during the summer of 1889, with an account of the fishes found in each of the river basins examined. *Bulletin of the United States Fish Commission* **9,** 1-40 (1889).

Kim, D. and Conway, K. W. Phylogeography of *Rhinichthys cataractae (Teleostei: Cyprinidae)*: pre-glacial colonization across the Continental Divide and Pleistocene diversification within the Rio Grande drainage. *Biological Journal of the Linnean Society* **111,** 317–333 (2014). DOI:10.1111/bij.12209.

Kloskowski, J. Impact of common carp *Cyprinus carpio* on aquatic communities: direct trophic effects versus habitat deterioration. *Fundamental and Applied Limnology / Archiv für Hydrobiologie* **178,** 245–255 (2011). Online at http://dx.doi.org/10.1127/1863-9135/2011/0178-0245.

Koç, H. T., Erdogan, Z., Tinkci, M. and Treer, T. Age, growth and reproductive characteristics of chub, *Leuciscus cephalus* (L., 1758) in the İkizcetepeler dam lake (Balikesir), Turkey. *Journal of Applied Ichthyology* **23,** 19–24 (2007). DOI:10.1111/j.1439-0426.2006.00787.x.

Koehn, J. D. Carp (*Cyprinus carpio*) as a powerful invader in Australian waterways. *Freshwater Biology* **49,** 882–894 (2004). DOI:10.1111/j.1365-2427.2004.01232.x.

Lee, D. S. *Notemigonus crysoleucas* (Mitchill) golden shiner. *Page 219 in Atlas of North American Freshwater Fishes* (North Carolina Museum of Natural History, Raleigh, North Carolina, 1980).

Oakey, D. D., Douglas, M. E. and Douglas, M. R. Small fish in a large landscape: diversification of *Rhinichthys osculus* (Cyprinidae) in western North America. *Copeia* **2004,** 207–221 (2004). Online at http://www.jstor.org/stable/1448559.

Reese, C. D. and Harvey, B. C. Temperature-dependent interactions between juvenile steelhead and Sacramento pikeminnow in laboratory streams. *Transactions of the American Fisheries Society* **131,** 599–606 (2002). DOI:10.1577/1548-8659(2002)131<0599:TDIBJS>2.0.CO;2.

Remple, S. L. *Taxonomy and Systematic Relationships of Tui Chubs (Siphateles: Cyprinidae) from Oregon's Great Basin* (Oregon State University, Corvallis, Oregon, 2013). Online at http://hdl.handle.net/1957/38120.

Scheerer, P. D. Implications of floodplain isolation and connectivity on the conservation of an endangered minnow, Oregon chub, in the Willamette River, Oregon. *Transactions of the American Fisheries Society* **131,** 1070–1080 (2002). DOI:10.1577/1548-8659(2002)131<1070:IOFIAC>2.0.CO;2.

Wallace, R. L. *Gila copei* (Jordan and Gilbert) leatherside chub. *Page 163 in Atlas of North American freshwater fishes* (North Carolina Museum of Natural History, Raleigh, North Carolina, 1980).

Wallace, R. L. *Richardsonius balteatus* (Richardson) redside shiner. P*age 358 in Atlas of North American freshwater fishes* (North Carolina Museum of Natural History, Raleigh, North Carolina, 1980).

Walser, C. A., Belk, M. C. and Shiozawa, D. K. Habitat use of leatherside chub (*Gila copei*) in the presence of predatory brown trout *(Salmo trutta). Western North American Naturalist* **59,** 272–277 (1999). Online at https://ojs.lib.byu.edu/spc/index.php/wnan/article/view/28197.

Wydoski, R. G. and Wydoski, R. S. Age, growth, and reproduction of mountain suckers in Lost Creek Reservoir, Utah. *Transactions of the American Fisheries Society* **131,** 320–328 (2002). DOI:10.1577/1548-8659(2002)131<0320:AGAROM>2.0.CO;2.

Chapter 10

Braig, E. C. and Johnson, D. L. Impact of black bullhead (*Ameiurus melas*) on turbidity in a diked wetland. *Hydrobiologia* **490,** 11–21 (2003). DOI:10.1023/A:1023405823216.

Breckels, R. D. and Neff, B. D. Pollution-induced behavioural effects in the brown bullhead *(Ameiurus nebulosus). Ecotoxicology* **19,** 1337–1346 (2010). DOI:10.1007/s10646-010-0520-1.

Cucherousset, J., Paillisson, J.-M. and Carpentier, A. Is mass removal an efficient measure to regulate the North American catfish *Ameiurus melas* outside of its native range? *Journal of Freshwater Ecology* **21,** 699–704 (2006). DOI:10.1111/j.1600-0633.2006.00199.x.

Daugherty, D. J. and Sutton, T. M. Seasonal movement patterns, habitat use, and home range of flathead catfish in the lower St. Joseph River, Michigan. *North American Journal of Fisheries Management* **25,** 256–269 (2005). DOI:10.1577/M03-252.2.

Eggleton, M. A. and Schramm Jr, H. L. Feeding ecology and energetic relationships with habitat of blue catfish, *Ictalurus furcatus*, and flathead catfish, *Pylodictis olivaris,* in the lower Mississippi River, U. S. A. *Environmental Biology of Fishes* **70,** 107–121 (2004). DOI:10.1023/B:EBFI.0000029341.45030.94.

Grabowski, T. B., Isely, J. J. and Weller, R. R. Age and growth of flathead catfish, *Pylodictus olivaris* Rafinesque, in the Altamaha River system, Georgia. *Journal of Freshwater Ecology* **19,** 411–417 (2004). DOI:10.1080/02705060.2004.9664914.

Murie, D. J., Parkyn, D. C., Loftus, W. F. and Nico, L. G. Variable growth and longevity of yellow bullhead *(Ameiurus natalis)* in the Everglades of south Florida, USA. *Journal of Applied Ichthyology* **25,** 740–745 (2009). DOI:10.1111/j.1439-0426.2009.01300.x.

Sakaris, P. C., Irwin, E. R., Jolley, J. C. and Harrison, D. Comparison of native and introduced flathead catfish populations in Alabama and Georgia: growth, mortality, and management. *North American Journal of Fisheries Management* **26,** 867–874 (2006). DOI:10.1577/M05-135.1.

Sakaris, P. C., Jesien, R. V. and Pinkney, A. E. Brown bullhead as an indicator species: seasonal movement patterns and home ranges within the Anacostia River, Washington, D. C. *Transactions of the American Fisheries Society* **134,** 1262–1270 (2005). DOI:10.1577/T04-086.1.

Weller, R. R. and Winter, J. D. Seasonal variation in home range size and habitat use of flathead catfish in Buffalo Springs Lake, Texas. *North American Journal of Fisheries Management* **21,** 792–800 (2001). DOI:10.1577/1548-8675(2001)021<0792:SVIHRS>2.0.CO;2.

Wright, J. J. Adaptive significance of venom glands in the tadpole madtom *Noturus gyrinus* (Siluriformes: Ictaluridae). *Journal of Experimental Biology* **215,** 1816–1823 (2012). DOI:10.1242/jeb.068361.

Chapter 12

Adams, S. B., Frissell, C. A. and Rieman, B. E. Changes in distribution of nonnative brook trout in an Idaho drainage over two decades. *Transactions of the American Fisheries Society* **131,** 561–568 (2002). DOI:10.1577/1548-8659(2002)131<0561:CIDONB>2.0.CO;2.

Alaska Department of Fish and Game. *Dolly Varden (Salvelinus malma) Species Profile.* (Alaska Department of Fish and Game, Juneau, Alaska, 2008). Online at http://www.adfg.alaska.gov/index.cfm?adfg=dollyvarden.main.

Allendorf, F. W. and Leary, R. F. Conservation and distribution of genetic variation in a polytypic species, the cutthroat trout. *Conservation Biology* **2,** 170–184 (1988). DOI:10.1111/j.1523-1739.1988.tb00168.x.

Anderson, J. H. and Quinn, T. P. Movements of adult coho salmon (*Oncorhynchus kisutch*) during colonization of newly accessible habitat. *Canadian Journal of Fisheries and Aquatic Sciences* **64,** 1143–1154 (2007). DOI:10.1139/F07-087.

Beauchamp, D. A. and Van Tassell, J. J. Modeling seasonal trophic interactions of adfluvial bull trout in Lake Billy Chinook, Oregon. *Transactions of the American Fisheries Society* **130,** 204–216 (2001). DOI:10.1577/1548-8659(2001)130<0204:MSTIOA>2.0.CO;2.

Beecher, H. A., Caldwell, B. A. and DeMond, S. B. Evaluation of depth and velocity preferences of juvenile coho salmon in Washington streams. *North American Journal of Fisheries Management* **22,** 785–795 (2002). DOI:10.1577/1548-8675(2002)022<0785:EODAVP>2.0.CO;2.

Behnke, R. J. The systematics of salmonid fishes of recently glaciated lakes. *Journal of the Fisheries Research Board of Canada* **29,** 639–671 (1972). Online at http://www.nrcresearchpress.com.dist.lib.usu.edu/doi/abs/10.1139/f72-112#.VhBvtBNViko.

Bernatchez, L., Colombani, F. and Dodson, J. J. Phylogenetic relationships among the subfamily *Coregoninae* as revealed by mitochondrial DNA restriction analysis. *Journal of Fish Biology* **39 Supplement sA,** 283–290 (1991). DOI:10.1111/j.1095-8649.1991.tb05091.x.

Cheever, B. M. and Simon, K. S. Seasonal influence of brook trout and mottled sculpin on lower trophic levels in an Appalachian stream. *Freshwater Biology* **54,** 524–535 (2009). DOI:10.1111/j.1365-2427.2008.02128.x.

Connor, W. P., Burge, H. L. and Bennett, D. H. Detection of PIT-tagged subyearling chinook salmon at a Snake River dam: implications for summer flow augmentation. *North American Journal of Fisheries Management* **18,** 530–536 (1998). DOI:10.1577/1548-8675(1998)018<0530:DOPTSC>2.0.CO;2.

Crespi, B. J. and Teo, R. Comparative phylogenetic analysis of the evolution of semelparity and life history in salmonid fishes. *Evolution* **56,** 1008–1020 (2002). DOI:10.1111/j.0014-3820.2002.tb01412.x.

Danzmann, R. G., Cairney, M., Davidson, W. S., Ferguson, M. M., Gharbi, K., Guyomard, R., Holm, L. E., Leder, E., Okamoto, N., Ozaki, A., Rexroad, C. E. III, Sakamoto, T., Taggart, J. B. and Woram, R. A. A comparative analysis of the rainbow trout genome with 2 other species of fish (Arctic charr and Atlantic salmon) within the tetraploid derivative Salmonidae family (subfamily: Salmoninae). *Genome* **48,** 1037–1051 (2005). DOI:10.1139/g05-067.

Downs, C. C., Horan, D., Morgan-Harris, E. and Jakubowski, R. Spawning demographics and juvenile dispersal of an adfluvial bull trout population in Trestle Creek, Idaho. N*orth American Journal of Fisheries Management* **26,** 190–200 (2006). DOI:10.1577/M04-180.1.

Drinan, D. P., Zale, A. V., Webb, M. A. H., Taper, M. L., Shepard, B. B. and Kalinowski, S. T. Evidence of local adaptation in westslope cutthroat trout. *Transactions of the American Fisheries Society* **141,** 872–880 (2012). DOI:10.1080/00028487.2012.675907.

Drinan, D. P., Kalinowski, S. T., Vu, N. V., Shepard, B. B., Muhlfeld, C. C. and Campbell, M. R. Genetic variation in westslope cutthroat trout *Oncorhynchus clarkii lewisi:* implications for conservation. *Conservation Genetics* **12,** 1513–1523 (2011). DOI:10.1007/s10592-011-0249-2.

Dunham, J., Rieman, B. and Chandler, G. Influences of temperature and environmental variables on the distribution of bull trout within streams at the southern margin of its range. *North American Journal of Fisheries Management* **23,** 894–904 (2003). DOI:10.1577/M02-028.

DuPont, J. M., Brown, R. S. and Geist, D. R. Unique allacustrine migration patterns of a bull trout population in the Pend Oreille River drainage, Idaho. *North American Journal of Fisheries Management* **27,** 1268–1275 (2007). DOI:10.1577/M06-137.1.

Fayram, A. H. and Sibley, T. H. Impact of predation by smallmouth bass on sockeye salmon in Lake Washington, Washington. *North American Journal of Fisheries Management* **20,** 81–89 (2000). DOI:10.1577/1548-8675(2000)020<0081:IOPBSB>2.0.CO;2.

Fritts, A. L. and Pearsons, T. N. Effects of predation by nonnative smallmouth bass on native salmonid prey: the role of predator and prey size. *Transactions of the American Fisheries Society* **135,** 853–860 (2006). DOI:10.1577/T05-014.1.

Fritts, A. L. and Pearsons, T. N. Smallmouth bass predation on hatchery and wild salmonids in the Yakima River, Washington. *Transactions of the American Fisheries Society* **133,** 880–895 (2004). DOI:10.1577/T03-003.1.

Froufe, E., Knizhin, I. and Weiss, S. Phylogenetic analysis of the genus *Thymallus* (Grayling) based on mtDNA control region and ATPase 6 genes, with inferences on control region constraints and broad-scale Eurasian phylogeography. *Molecular Phylogenetics and Evolution* **34,** 106–117 (2005). DOI:10.1016/j.ympev.2004.09.009.

Fuller, P. and Neilson, M. *Salvelinus namaycush* (U. S. Geological Survey, Nonindigenous Aquatic Species Database, Gainesville, Florida). Online at http://nas.er.usgs.gov/queries/FactSheet.aspx?SpeciesID=942. (2014).

Gall, B. G. and Mathis, A. Innate predator recognition and the problem of introduced trout. *Ethology* **116,** 47–58 (2010). DOI:10.1111/j.1439-0310.2009.01718.x.

Gallagher, S. P. and Gallagher, C. M. Discrimination of chinook salmon, coho salmon, and steelhead redds and evaluation of the use of redd data for estimating escapement in several unregulated streams in northern California. *North American Journal of Fisheries Management* **25,** 284–300 (2005). DOI:10.1577/M04-016.1.

Garrett, J. W., Bennett, D. H., Frost, F. O. and Thurow, R. F. Enhanced incubation success for kokanee spawning in groundwater upwelling sites in a small Idaho Stream. *North American Journal of Fisheries Management* **18,** 925–930 (1998). DOI:10.1577/1548-8675(1998)018<0925:EISFKS>2.0.CO;2.

Giannico, G. R. and Hinch, S. G. The effect of wood and temperature on juvenile coho salmon winter movement, growth, density and survival in side-channels. *River Research and Applications* **19,** 219–231 (2003). DOI:10.1002/rra.723.

Goniea, T. M., Keefer, M. L., Bjornn, T. C., Peery, C. A., Bennett, D. H. and Stuehrenberg, L. C. Behavioral thermoregulation and slowed migration by adult fall chinook salmon in response to high Columbia River water temperatures. *Transactions of the American Fisheries Society* **135,** 408–419 (2006). DOI:10.1577/T04-113.1.

Gresh, T., Lichatowich, J. and Schoonmaker, P. An estimation of historic and current levels of salmon production in the northeast Pacific ecosystem: evidence of a nutrient deficit in the freshwater systems of the Pacific northwest. *Fisheries* **25,** 15–21 (2000). DOI:10.1577/1548-8446(2000)025<0015:AEOHAC>2.0.CO;2.

Gresswell, R. E. Biology, status, and management of the Yellowstone cutthroat trout. *North American Journal of Fisheries Management* **31,** 782–812 (2011). DOI:10.1080/02755947.2011.608980.

Gresswell, R. E. *Status and Management of Interior Stocks of Cutthroat Trout* (American Fisheries Society, Bethesda, Maryland, 1988).

Gunckel, S. L., Hemmingsen, A. R. and Li, J. L. Effect of bull trout and brook trout interactions on foraging habitat, feeding behavior, and growth. *Transactions of the American Fisheries Society* **131,** 1119–1130 (2002). DOI:10.1577/1548-8659(2002)131<1119:EOBTAB>2.0.CO;2.

Haas, G. R. and McPhail, J. D. Systematics and distributions of dolly varden (*Salvelinus malma*) and bull trout (*Salvelinus confluentus*) in North America. *Canadian Journal of Fisheries and Aquatic Sciences* **48,** 2191–2211 (1991). DOI:10.1139/f91-259.

Haas, G. R. and McPhail, J. D. The post-Wisconsinan glacial biogeography of bull trout (*Salvelinus confluentus*): a multivariate morphometric approach for conservation biology and management. *Canadian Journal of Fisheries and Aquatic Sciences* **58,** 2189–2203 (2001). DOI:10.1139/f01-139.

Harper, D. D. and Farag, A. M. Winter habitat use by cutthroat trout in the Snake River near Jackson, Wyoming. *Transactions of the American Fisheries Society* **133,** 15–25 (2004). DOI:10.1577/T02-072.

Helm, W. T. *Endemic Whitefishes of Bear Lake, Utah-Idaho: A Problem in Systematics* (Utah State University, Logan, Utah, 1974). Online at http://digitalcommons.usu.edu/etd/4341/.

High, B., Meyer, K. A., Schill, D. J. and Mamer, E. R. J. Distribution, abundance, and population trends of bull trout in Idaho. *North American Journal of Fisheries Management* **28,** 1687–1701 (2008). DOI:10.1577/M06-164.1.

High, B., Peery, C. A. and Bennett, D. H. Temporary staging of Columbia River summer steelhead in coolwater areas and its effect on migration rates. *Transactions of the American Fisheries Society* **135,** 519–528 (2006). DOI:10.1577/T04-224.1.

Hitt, N. P., Frissell, C. A., Muhlfeld, C. C. and Allendorf, F. W. Spread of hybridization between native westslope cutthroat trout, (*Oncorhynchus clarki lewisi*) and nonnative rainbow trout, (*Oncorhynchus mykiss*). *Canadian Journal of Fisheries and Aquatic Sciences* **60,** 1440–1451 (2003). DOI:10.1139/f03-125.

Isaak, D. J., Thurow, R. F., Rieman, B. E. and Dunham, J. B. Chinook salmon use of spawning patches: relative roles of habitat quality, size, and connectivity. *Ecological Applications* **17,** 352–364 (2007). Online at http://dx.doi.org/10.1890/05-1949.

Johnson, E. L., Clabough, T. S., Bennett, D. H., Bjornn, T. C., Peery, C. A., Caudill, C. C. and Stuehrenberg, L. C. Migration depths of adult spring and summer chinook salmon in the lower Columbia and Snake Rivers in relation to dissolved gas supersaturation. *Transactions of the American Fisheries Society* **134,** 1213–1227 (2005). DOI:10.1577/T04-116.1.

Jonsson, B. and Jonsson, N. Polymorphism and speciation in Arctic charr. *Journal of Fish Biology* **58,** 605–638 (2001). DOI:10.1111/j.1095-8649.2001.tb00518.x.

Jordan, D. S. *Salmon and Trout of the Pacific Coast* (California Fish and Game Commission, Sacramento, California, 1892).

Jordan, D. S. and Evermann, B. W. The fishes of North and Middle America: a descriptive catalog of the species of fish-like vertebrates found in the waters of North America, north of the Isthmus of Panama. *Bulletin of the United States National Museum* **47,** Part I (1896).

Luecke, C. and MacKinnon, P. Landscape effects on growth of age-0 arctic grayling in tundra streams. *Transactions of the American Fisheries Society* **137,** 236–243 (2008). DOI:10.1577/T05-039.1.

Martinez,P. J., Bigelow, P. E., Deleray, M. A., Fredenberg, W. A., Hansen, B. S., Horner, N. J., Lehr, S. K., Schneidervin, R. W., Tolentino, S. A. and Viola, A. E. Western lake trout woes. *Fisheries* **34,** 424–442 (2009). DOI:10.1577/1548-8446-34.9.424.

McDermid, J. L., Ihssen, P. E., Sloan, W. N. and Shuter, B. J. Genetic and environmental influences on life history traits in lake trout. *Transactions of the American Fisheries Society* **136,** 1018–1029 (2007). DOI:10.1577/T06-189.1.

McDermid, J. L., Shuter, B. J. and Lester, N. P. Life history differences parallel environmental differences among North American lake trout *(Salvelinus namaycush)* populations. *Canadian Journal of Fisheries and Aquatic Sciences* **67,** 314–325 (2010). DOI:10.1139/F09-183.

Mirza, R. S. and Chiers, P. Attraction of slimy sculpins to chemical cues of brook charr eggs. *Journal of Fish Biology* **61,** 532–539 (2002). DOI:10.1111/j.1095-8649.2002.tb00894.x.

Muhlfeld, C. C., Bennett, D. H. and Marotz, B. Fall and winter habitat use and movement by Columbia River redband trout in a small stream in Montana. *North American Journal of Fisheries Management* **21,** 170–177 (2001). DOI:10.1577/1548-8675(2001)021<0170:FAWHUA>2.0.CO;2.

Muhlfeld, C. C., Bennett, D. H. and Marotz, B. Summer habitat use by Columbia River redband trout in the Kootenai River drainage, Montana. *North American Journal of Fisheries Management* **21,** 223–235 (2001). DOI:10.1577/1548-8675(2001)021<0223:SHUBCR>2.0.CO;2.

Muhlfeld, C. C., Albeke, S. E., Gunckel, S. L., Writer, B. J., Shepard, B. B. and May, B. E. Status and conservation of interior redband trout in the western United States. *North American Journal of Fisheries Management* **35,** 31–53 (2015). DOI:10.1080/02755947.2014.951807.

Naughton, G. P., Bennett, D. H. and Newman, K. B. Predation on juvenile salmonids by smallmouth bass in the lower Granite Reservoir system, Snake River. *North American Journal of Fisheries Management* **24,** 534–544 (2004). DOI:10.1577/M02-177.1.

Neraas, L. P. and Spruell, P. Fragmentation of riverine systems: the genetic effects of dams on bull trout *(Salvelinus confluentus)* in the Clark Fork River system. *Molecular Ecology* **10,** 1153–1164 (2001). DOI:10.1046/j.1365-294X.2001.01269.x.

O'Gorman, R., Elrod, J. H., Owens, R. W., Schneider, C. P., Eckert, T. H. and Lantry, B. F. Shifts in depth distributions of alewives, rainbow smelt, and age-2 lake trout in southern Lake Ontario following establishment of dreissenids. *Transactions of the American Fisheries Society* **129,** 1096–1106 (2000). DOI:10.1577/1548-8659(2000)129<1096:SIDDOA>2.0.CO;2.

Peterson, D. P., Fausch, K. D. and White, G. C. Population ecology of an invasion: effects of brook trout on native cutthroat trout. *Ecological Applications* **14,** 754–772 (2004). Online at http://dx.doi.org/10.1890/02-5395.

Pientka, B. and Parrish, D. L. Habitat selection of predator and prey: Atlantic salmon and rainbow smelt overlap, based on temperature and dissolved oxygen. *Transactions of the American Fisheries Society* **131,** 1180–1193 (2002). DOI:10.1577/1548-8659(2002)131<1180:HSOPAP>2.0.CO;2.

Reese, C. D. and Harvey, B. C. Temperature-dependent interactions between juvenile steelhead and Sacramento pikeminnow in laboratory streams. *Transactions of the American Fisheries Society* **131,** 599–606 (2002). DOI:10.1577/1548-8659(2002)131<0599:TDIBJS>2.0.CO;2.

Rosenfeld, J., Porter, M. and Parkinson, E. Habitat factors affecting the abundance and distribution of juvenile cutthroat trout *(Oncorhynchus clarki)* and coho salmon *(Oncorhynchus kisutch). Canadian Journal of Fisheries and Aquatic Sciences* **57,** 766–774 (2000). DOI:10.1139/f00-010.

Ruetz, C. R., Hurford, A. L. and Vondracek, B. Interspecific interactions between brown trout and slimy sculpin in stream enclosures. *Transactions of the American Fisheries Society* **132,** 611–618 (2003). DOI:10.1577/1548-8659(2003)132<0611:IIBBTA>2.0.CO;2.

Shepard, B. B. Factors that may be influencing nonnative brook trout invasion and their displacement of native westslope cutthroat trout in three adjacent southwestern Montana streams. *North American Journal of Fisheries Management* **24,** 1088–1100 (2004). DOI:10.1577/M03-105.1.

Shepard, B. B., May, B. E. and Urie, W. Status and conservation of westslope cutthroat trout within the western United States. *North American Journal of Fisheries Management* **25,** 1426–1440 (2005). DOI:10.1577/M05-004.1.

Sigler, J. W., Bjornn, T. C. and Everest, F. H. Effects of chronic turbidity on density and growth of steelheads and coho salmon. *Transactions of the American Fisheries Society* **113,** 142–150 (1984). DOI:10.1577/1548-8659(1984)113<142:EOCTOD>2.0.CO;2.

Smith, C. T., Nelson, R. J., Wood, C. C. and Koop, B. F. Glacial biogeography of North American coho salmon *(Oncorhynchus kisutch). Molecular Ecology* **10,** 2775–2785 (2001). DOI:10.1046/j.1365-294X.2001.t01-1-01405.x.

Smith, G. R. and Stearley, R. F. The classification and scientific names of rainbow trout and cutthroat trouts. *Fisheries* **14,** 4–10 (1989). DOI:10.1577/1548-8446(1989)014<0004:TCASNO>2.0.CO;2.

Snake River Recovery Team. *Final Recommendations to the National Marine Fisheries Service* (National Marine Fisheries Service, Portland, Oregon, 1994).

Spangler, R. E. and Scarnecchia, D. L. Summer and fall microhabitat utilization of juvenile bull trout and cutthroat trout in a wilderness stream, Idaho. *Hydrobiologia* **452,** 145–154 (2001). DOI:10.1023/A:1011988313707.

Stakėnas, S., Copp, G. H. and Scott, D. M. Tagging effects on three nonnative fish species in England *(Lepomis gibbosus, Pseudorasbora parva, Sander lucioperca)* and of native *Salmo trutta. Ecology of Freshwater Fish* ***18,*** 167–176 (2009). DOI:10.1111/j.1600-0633.2008.00339.x.

Tabor, R. A., Footen, B. A., Fresh, K. L., Celedonia, M. T., Mejia, F., Low, D. L. and Park, L. Smallmouth bass and largemouth bass predation on juvenile chinook salmon and other salmonids in the Lake Washington basin. *North American Journal of Fisheries Management* **27,** 1174–1188 (2007). DOI:10.1577/M05-221.1.

Taylor, E. B., Stamford, M. D. and Baxter, J. S. Population subdivision in westslope cutthroat trout (*Oncorhynchus clarki lewisi)* at the northern periphery of its range: evolutionary inferences and conservation implications. *Molecular Ecology* **12,** 2609–2622 (2003). DOI:10.1046/j.1365-294X.2003.01937.x.

Thurow, R. F. Effects of stream alterations on rainbow trout in the Big Wood River, Idaho. *Pages 175–188 in Western Proceedings: 68th Annual Conference of the Western Association of Fish and Wildlife Agencies* (Western Association of Fish and Wildlife Agencies, Boise, Idaho, 1988). Online at http://db.lib.uidaho.edu/FG/Vol/079/Article_12.pdf.

U. S. vs Oregon Technical Advisory Committee. *Biological Assessment of Incidental Impacts on Salmon Species Listed Under the Endangered Species Act in the 2008-2017 Non-Indian and Treaty Indian Fisheries in the Columbia River Basin. Final April 21, 2008* (Technical Advisory Committee, Portland, Oregon, 2008).

Van Kirk, R. W. and Benjamin, L. Status and conservation of salmonids in relation to hydrologic integrity in the Greater Yellowstone Ecosystem. *Western North American Naturalist* **61,** 359–374 (2001). Online at https://ojs.lib.byu.edu/spc/index.php/wnan/article/view/28077.

Vehanen, T., Hyvärinen, P. and Huusko, A. Food consumption and prey orientation of piscivorous brown trout (*Salmo trutta*) and pikeperch (*Stizostedion lucioperca*) in a large regulated lake. *Journal of Applied Ichthyology* **14,** 15–22 (1998). DOI:10.1111/j.1439-0426.1998.tb00608.x.

Vidergar, D. T. *Population Estimates, Food Habits and Estimates of Consumption of Selected Predatory Fishes in Lake Pend Oreille, Idaho* (Masters Thesis, University of Idaho, Moscow, Idaho, 2000). Online at http://collaboration.idfg.idaho.gov/FisheriesTechnicalReports/ResVidergar 1999%20Population%20 Estimates,%20Food%20Habits,%20Estimates%20of%20Consumption%20of%20Selected%20 Predatory%20Fishes%20in%20LPO.pdf.

Wells, B. K., Rieman, B. E., Clayton, J. L., Horan, D. L. and Jones, C. M. Relationships between water, otolith, and scale chemistries of westslope cutthroat trout from the Coeur d'Alene River, Idaho: the potential application of hard-part chemistry to describe movements in freshwater. *Transactions of the American Fisheries Society* **132,** 409–424 (2003). DOI:10.1577/1548-8659(2003)132<0409:RBWOAS>2.0.CO;2.

Witt, J. D. S., Zemlak, R. J. and Taylor, E. B. Phylogeography and the origins of range disjunctions in a north temperate fish, the pygmy whitefish *(Prosopium coulterii)*, inferred from mitochondrial and nuclear DNA sequence analysis. *Journal of Biogeography* **38,** 1557–1569 (2011). DOI:10.1111/j.1365-2699.2011.02497.x.

Chapter 13

Klefoth, T., Kobler, A. and Arlinghaus, R. The impact of catch-and-release angling on short-term behaviour and habitat choice of northern pike (*Esox lucius L.*). *Hydrobiologia* **601,** 99–110 (2008). DOI:10.1007/s10750-007-9257-0.

McMahon, T. E. and Bennett, D. H. Walleye and northern pike. *Fisheries* **21,** 6–13 (1996). DOI:10.1577/1548-8446(1996)021<0006:WANP>2.0.CO;2.

Paukert, C. P., Stancill, W., DeBates, T. J. and Willis, D. W. Predatory effects of northern pike and largemouth bass: bioenergetic modeling and ten years of fish community sampling. *Journal of Freshwater Ecology* **18,** 13–24 (2003). DOI:10.1080/02705060.2003.9663947.

Pierce, R. B. and Tomcko, C. M. Interrelationships among production, density, growth, and mortality of northern pike in seven north-central Minnesota lakes. *Transactions of the American Fisheries Society* **132,** 143–153 (2003). DOI:10.1577/1548-8659(2003)132<0143:IAPDGA>2.0.CO;2.

Walrath, J. D., Quist, M. C. and Firehammer, J. A. Trophic ecology of nonnative Northern Pike and the effect on conservation of native westslope Cutthroat Trout. *North American Journal of Fisheries Management* **35,** 158-177 (2015). DOI:10.1080/02755947.2014.970678.

Chapter 14

Gray, R. H. and Dauble, D. D. Biology of the sand roller in the central Columbia River. *Transactions of the American Fisheries Society* **108,** 646–649 (1979). DOI:10.1577/1548-8659(1979)108<646:BOTSRI>2.0.CO;2.

Chapter 15

Hardewig, I., Pörtner, H. O. and van Dijk, P. How does the cold stenothermal gadoid Lota lota survive high water temperatures during summer? *Journal of Comparative Physiology B* **174,** 149–156 (2004). DOI:10.1007/s00360-003-0399-8.

Stapanian, M. A., Paragamian, V. L., Madenjian, C. P., Jackson, J. R., Lappalainen, J. Evenson, M. J. and Neufeld, M. D. Worldwide status of burbot and conservation measures. *Fish and Fisheries* **11,** 34–56 (2010). DOI:10.1111/j.1467-2979.2009.00340.x.

Chapter 17

Breen, M. J., Ruetz III, C. R., Thompson, K. J. and Kohler, S. L. Movements of mottled sculpins (*Cottus bairdii*) in a Michigan stream: how restricted are they? *Canadian Journal of Fisheries and Aquatic Sciences* **66,** 31–41 (2009). DOI:10.1139/F08-189.

Cheever, B. M. and Simon, K. S. Seasonal influence of brook trout and mottled sculpin on lower trophic levels in an Appalachian stream. *Freshwater Biology* **54,** 524–535 (2009). DOI:10.1111/j.1365-2427.2008.02128.x.

Edwards, P. A. and Cunjak, R. A. Influence of water temperature and streambed stability on the abundance and distribution of slimy sculpin (*Cottus cognatus*). *Environmental Biology of Fishes* **80,** 9–22 (2007). DOI:10.1007/s10641-006-9102-8.

Gray, M. A., Curry, R. A. and Munkittrick, K. R. Impacts of nonpoint inputs from potato farming on populations of slimy sculpin (*Cottus cognatus*). *Environmental Toxicology and Chemistry* **24,** 2291–2298 (2005). DOI:10.1897/04-615R.1.

Grossman, G. D., Ratajczak, R. E. Jr., Petty, T., Hunter, M. D., Peterson, J. T. and Grenouillet, G. Population dynamics of mottled sculpin (*Pisces*) in a variable environment: information theoretic approaches. *Ecological Monographs* **76,** 217–234 (2006). Online at http://dx.doi.org/10.1890/04-1623.

Jenkins, T. *The Population Genetic Structure of Paiute Sculpin (Cottus beldingi) in the Marys River Basin: A Step Towards Ecosystem Management* (University of Nevada, Reno, Nevada, 2006).

Kanter, M. J. and Coombs, S. Rheotaxis and prey detection in uniform currents by Lake Michigan mottled sculpin *(Cottus bairdi)* . *Journal of Experimental Biology* **206,** 59–70 (2003). DOI:10.1242/jeb.00056.

Meyer, K. A., Lamansky J. A., Schill, D. J. and Zaroban, D. W. Nongame fish species distribution and habitat associations in the Snake River Basin of southern Idaho. *Western North American Naturalist* **73,** 20–34 (2013). Online at https://ojs.lib.byu.edu/spc/index.php/wnan/article/view/27210.

Meyer, K. A., Schill, D. J., Campbell, M. R. and Kozfkay, C. C. Distribution, abundance, and genetic population structure of Wood River sculpin, *Cottus leiopomus. Western North American Naturalist* **68,** 508–520 (2008). Online at https://ojs.lib.byu.edu/spc/index.php/wnan/article/view/27539.

Montana Natural Heritage Program and Montana Fish, Wildlife and Parks. *Rocky Mountain Sculpin – Cottus bondi* (Montana Field Guide, Montana.Gov). Online at http://fieldguide.mt.gov/species-De-tail.aspx?elcode=AFC4E02380. (2015).

Quist, M. C., Hubert, W. A. and Isaak, D. J. Factors affecting allopatric and sympatric occurrence of two sculpin species across a Rocky Mountain watershed. *Copeia* **2004,** 617–623 (2004). Online at http://www.jstor.org/stable/1448481.

Richardson, J. *Cottus cognatus* (Richardson) Bear Lake bull-head. *Pages 40-43 in Fauna Boreali-Americana; or the Zoology of the Northern Parts of British America: Part Three: The Fish* (R. Bentley, London, England, 1831) Online at http://www.biodiversitylibrary.org/bibliography/39293#/summary.

Ruetz, C. R., Hurford, A. L. and Vondracek, B. Interspecific interactions between brown trout and slimy sculpin in stream enclosures. *Transactions of the American Fisheries Society* **132,** 611–618 (2003). DOI:10.1577/1548-8659(2003)132<0611:IIBBTA>2.0.CO;2.

Smith, C. N., Crowley, J. M., Shiozawa, D. K. and Evans, R. P. *Phylogenetic relationships of Cottus beldingi in the basin and range and Colorado Plateau of western North America. American Geophysical Union, Spring Meeting, Abstract* **NB52C-02** (2005). Online at http://adsabs.harvard.edu/abs/2005AGUSMNB52C.02S.

Zaroban, D. W. *Distribution, Habitat Associations and Conservation Opportunities for the Wood River Sculpin* (Doctoral Dissertation, University of Idaho, Moscow, Idaho, 2011b).

Zimmerman, J. K. H. and Vondracek, B. Interactions between slimy sculpin and trout: slimy sculpin growth and diet in relation to native and nonnative trout. *Transactions of the American Fisheries Society* **136,** 1791–1800 (2007). DOI:10.1577/T06-020.1.

Chapter 18

Andraso, G. M. Summer food habits of pumpkinseeds *(Lepomis gibbosus)* and bluegills *(Lepomis macrochirus)* in Presque Isle Bay, Lake Erie. *Journal of Great Lakes Research* **31,** 397–404 (2005). DOI:10.1016/S0380-1330(05)70271-9.

Beard, T. D. and Kampa, J. M. Changes in bluegill, black crappie, and yellow perch populations in Wisconsin during 1967–1991. *North American Journal of Fisheries Management* **19,** 1037–1043 (1999). DOI:10.1577/1548-8675(1999)019<1037:CIBBCA>2.0.CO;2.

Bennett, D. H. and McArthur, T. J. Predicting success of walleye stocking programs in the United States and Canada. *Fisheries* **15,** 19–23(1990). DOI:10.1577/1548-8446(1990)015<0019:PSOWSP>2.0.CO;2.

Boschung, H. T. and Mayden, R. L. Redbelly tilapia: *Tilapia zillii* (Gervais). Page 622 in Fishes of Alabama. (Smithsonian Books, Washington D.C., 2004).

Bunnell, D. B., Thomas, S. E. and Stein, R. A. Prey resources before spawning influence gonadal investment of female, but not male, white crappie. *Journal of Fish Biology* **70,** 1838–1854 (2007). DOI:10.1111/j.1095-8649.2007.01459.x.

Bunt, C. M., Cooke, S. J., Schreer, J. F. and Philipp, D. P. Effects of incremental increases in silt load on the cardiovascular performance of riverine and lacustrine rock bass, *Ambloplites rupestris. Environmental Pollution* **128,** 437–444 (2004). DOI:10.1016/j.envpol.2003.09.015.

Burness, G., Casselman, S. J., Schulte-Hostedde, A. I., Moyes, C. D. and Montgomerie, R. Sperm swimming speed and energetics vary with sperm competition risk in bluegill *(Lepomis macrochirus). Behavioral Ecology and Sociobiology* **56,** 65–70 (2004). DOI:10.1007/s00265-003-0752-7.

Burness, G., Moyes, C. D. and Montgomerie, R. Motility, ATP levels and metabolic enzyme activity of sperm from bluegill *(Lepomis macrochirus). Comparative Biochemistry and Physiology Part A: Molecular and Integrative Physiology* **140,** 11–17 (2005). DOI:10.1016/j.cbpb.2004.09.021.

Curry, R. A., Currie, S. L., Arndt, S. K. and Bielak, A. T. Winter survival of age-0 smallmouth bass, *Micropterus dolomieu,* in northeastern lakes. *Environmental Biology of Fishes* **72,** 111–122 (2005). DOI:10.1007/s10641-004-6582-2.

Dockendorf, K. J. and Allen, M. S. Age-0 black crappie abundance and size in relation to zooplankton density, stock abundance, and water clarity in three Florida lakes. *Transactions of the American Fisheries Society* **134,** 172–183 (2005). DOI:10.1577/FT03-135.1.

Fayram, A. H. and Sibley, T. H. Impact of predation by smallmouth bass on sockeye salmon in Lake Washington, Washington. *North American Journal of Fisheries Management* **20,** 81–89 (2000). DOI:10.1577/1548-8675(2000)020<0081:IOPBSB>2.0.CO;2.

Fox, M. G. and Crivelli, A. J. Life history traits of pumpkinseed *(Lepomis gibbosus)* populations introduced into warm thermal environments. *Archiv für Hydrobiologie* **150,** 561–580 (2001).

Frey, A. P., Bozek, M. A., Edwards, C. J. and Newman, S. P. Diet overlap and predation between smallmouth bass and walleye in a north temperate lake. *Journal of Freshwater Ecology* **18,** 43–54 (2003). DOI:10.1080/02705060.2003.9663950.

Fritts, A. L. and Pearsons, T. N. Effects of predation by nonnative smallmouth bass on native salmonid prey: the role of predator and prey size. *Transactions of the American Fisheries Society* **135,** 853–860 (2006). DOI:10.1577/T05-014.1.

Fritts, A. L. and Pearsons, T. N. Smallmouth bass predation on hatchery and wild salmonids in the Yakima River, Washington. *Transactions of the American Fisheries Society* **133,** 880–895 (2004). DOI:10.1577/T03-003.1.

Galinat, G. F., Willis, D. W., Blackwell, B. G. and Hubers, M. J. Influence of a saugeye (sauger × walleye) introduction program on the black crappie population in Richmond Lake, South Dakota. *North American Journal of Fisheries Management* **22,** 1416–1424 (2002). DOI:10.1577/1548-8675(2002)022<1416:IOASSW>2.0.CO;2.

Goldstein, R. J., Harper, R. W. and Edwards, R. *American Aquarium Fishes* (W. L. Moody Jr. Natural History Series, Texas A and M University Press, College Station, Texas, 2000).

Graeb, B. D. S., Dettmers, J. M., Wahl, D. H. and Cáceres, C. E. Fish size and prey availability affect growth, survival, prey selection, and foraging behavior of larval yellow perch. *Transactions of the American Fisheries Society* **133,** 504–514 (2004). DOI:10.1577/T03-050.1.

Haponski, A. E. and Stepien, C. A. Phylogenetic and biogeographical relationships of the Sander pikeperches (Percidae: Perciformes): patterns across North America and Eurasia. *Biological Journal of the Linnean Society* **110,** 156–179 (2013). DOI:10.1111/bij.12114.

Hayward, R. S. and Arnold, E. Temperature dependence of maximum daily consumption in white crappie: implications for fisheries management. *Transactions of the American Fisheries Society* **125,** 132–138 (1996). DOI:10.1577/1548-8659(1996)125<0132:TDOMDC>2.3.CO;2.

Henderson, B. A., Collins, N., Morgan, G. E. and Vaillancourt, A. Sexual size dimorphism of walleye *(Stizostedion vitreum vitreum). Canadian Journal of Fisheries and Aquatic Sciences* **60,** 1345–1352 (2003). DOI:10.1139/f03-115.

Henderson, B. A., Trivedi, T. and Collins, N. Annual cycle of energy allocation to growth and reproduction of yellow perch. *Journal of Fish Biology* **57,** 122–133 (2000). DOI:10.1111/j.1095-8649.2000.tb00780.x.

Januszkiewicz, A. J. and Robinson, B. W. Divergent walleye (*Sander vitreus*)-mediated inducible defenses in the centrarchid pumpkinseed sunfish (*Lepomis gibbosus*). *Biological Journal of the Linnean Society* **90,** 25–36 (2007). DOI:10.1111/j.1095-8312.2007.00708.x.

Johnston, T. A., Lysack, W. and Leggett, W. C. Abundance, growth, and life history characteristics of sympatric walleye (*Sander vitreus*) and sauger (*Sander canadensis*) in Lake Winnipeg, Manitoba. *Journal of Great Lakes Research* **38, Supplement 3,** 35–46 (2012). DOI:10.1016/j.jglr.2010.06.009.

Karchesky, C. M. and Bennett, D. H. Winter habitat use by adult largemouth bass in the Pend Oreille River, Idaho. *North American Journal of Fisheries Management* **24,** 577–585 (2004). DOI:10.1577/M02-175.1.

Liao, H., Pierce, C. L. and Larscheid, J. G. Diet dynamics of the adult piscivorous fish community in Spirit Lake, Iowa, USA 1995–1997. *Ecology of Freshwater Fish* **11,** 178–189 (2002). DOI:10.1034/j.1600-0633.2002.00015.x.

McCairns, R. J. S. and Fox, M. G. Habitat and home range fidelity in a trophically dimorphic pumpkinseed sunfish (*Lepomis gibbosus*) population. *Oecologia* **140,** 27 1–279 (2004). DOI:10.1007/s00442-004-1580-9.

McMahon, T. E. and Bennett, D. H. Walleye and northern pike. *Fisheries* **21,** 6–13 (1996). DOI:10.1577/1548-8446(1996)021<0006:WANP>2.0.CO;2.

Miranda, L. E. and Allen, M. S. Use of length limits to reduce variability in crappie fisheries. *North American Journal of Fisheries Management* **20,** 752–758 (2000). DOI:10.1577/1548-8675(2000)020<0752:UOLLTR>2.3.CO;2.

Naughton, G. P., Bennett, D. H. and Newman, K. B. Predation on juvenile salmonids by smallmouth bass in the lower Granite Reservoir system, Snake River. *North American Journal of Fisheries Management* **24,** 534–544 (2004). DOI:10.1577/M02-177.1.

Nico, L. G., Fuller, P. and Neilson, M. *Astronotus ocellatus. U. S. Geological Survey Nonindigenous Aquatic Species Database* (U. S. Geological Survey, Gainesville, Florida). Online at http://nas.er.usgs.gov/queries/FactSheet.aspx?speciesID=436(2013).

Paukert, C. P., Stancill, W., DeBates, T. J. and Willis, D. W. Predatory effects of northern pike and largemouth bass: bioenergetic modeling and ten years of fish community sampling. *Journal of Freshwater Ecology* **18,** 13–24 (2003). DOI:10.1080/02705060.2003.9663947.

Peake, S., McKinley, R. S. and Scruton, D. A. Swimming performance of walleye *(Stizostedion vitreum). Canadian Journal of Zoology* **78,** 1686–1690 (2000). DOI:10.1139/z00-097.

Pope, K. L. and Willis, D. W. Early life history and recruitment of black crappie *(Pomoxis nigromaculatus)* in two South Dakota waters. *Ecology of Freshwater Fish* 7, 56–68 (1998). DOI:10.1111/j.1600-0633.1998.tb00172.x.

Pope, K. L. and Willis, D. W. Environmental characteristics of black crappie *(Pomoxis nigromaculatus)* nesting sites in two South Dakota waters. *Ecology of Freshwater Fish* **6,** 183–189 (1997). DOI:10.1111/j.1600-0633.1997.tb00161.x.

Rasmus, R. A., Phelps, Q. E., Duehr, J. P. and Berry, C. R. Population characteristics of lotic orangespotted sunfish. *Journal of Freshwater Ecology* **23,** 459–461 (2008). DOI:10.1080/02705060.2008.9664224.

Sammons, S. M., Bettoli, P. W. and Greear, V. A. Early life history characteristics of age-0 white crappies in response to hydrology and zooplankton densities in Normandy Reservoir, Tennessee. *Transactions of the American Fisheries Society* **130,** 442–449 (2001). DOI:10.1577/1548-8659(2001)130<0442:ELHCOA>2.0.CO;2.

Sanderson, B. L., Hrabik, T. R., Magnuson, J. J. and Post, D. M. Cyclic dynamics of a yellow perch *(Perca flavescens)* population in an oligotrophic lake: evidence for the role of intraspecific interactions. *Canadian Journal of Fisheries and Aquatic Sciences* **56,** 1534–1542 (1999). DOI:10.1139/f99-077.

Sharma, S. and Jackson, D. A. Predicting smallmouth bass *(Micropterus dolomieu)* occurrence across North America under climate change: a comparison of statistical approaches. *Canadian Journal of Fisheries and Aquatic Sciences* **65,** 471–481 (2008). DOI:10.1139/f07-178.

Stakėnas, S., Copp, G. H. and Scott, D. M. Tagging effects on three nonnative fish species in England (*Lepomis gibbosus, Pseudorasbora parva, Sander lucioperca)* and of native Salmo trutta. *Ecology of Freshwater Fish* **18,** 167–176 (2009). DOI:10.1111/j.1600-0633.2008.00339.x.

Standen, E. M. and Lauder, G. V. Dorsal and anal fin function in bluegill sunfish *Lepomis macrochirus:* three-dimensional kinematics during propulsion and maneuvering. *Journal of Experimental Biology* **208,** 2753–2763 (2005). DOI:10.1242/jeb.01706.

Sweka, J. A. and Hartman, K. J. Reduction of reactive distance and foraging success in smallmouth bass, *Micropterus dolomieu,* exposed to elevated turbidity levels. *Environmental Biology of Fishes* **67,** 341–347 (2003). DOI:10.1023/A:1025835031366.

Tabor, R. A., Footen, B. A., Fresh, K. L., Celedonia, M. T., Mejia, F., Low, D. L. and Park, L. Smallmouth bass and largemouth bass predation on juvenile chinook salmon and other salmonids in the Lake Washington basin. *North American Journal of Fisheries Management* **27,** 1174–1188 (2007). DOI:10.1577/M05-221.1.

Travnichek, V. H., Maceina, M. J. and Dunham, R. A. Hatching time and early growth of age-0 black crappies, white crappies, and their naturally produced F1 hybrids in Weiss Lake, Alabama. *Transactions of the American Fisheries Society* **125,** 334–337 (1996). DOI:10.1577/1548-8659(1996)125<0334:HTAEGO>2.3.CO;2.

Travnichek, V. H., Maceina, M. J., Smith, S. M. and Dunham, R. A. Natural hybridization between black and white crappies (*Pomoxis*) in 10 Alabama reservoirs. *American Midland Naturalist* **135,** 310–316 (1996). DOI:10.2307/2426714.

Travnichek, V. H., Maceina, M. J., Wooten, M. C. and Dunham, R. A. Symmetrical hybridization between black crappie and white crappie in an Alabama reservoir based on analysis of the cytochrome-b gene. *Transactions of the American Fisheries Society* **126,** 127–132 (1997). DOI:10.1577/1548-8659(1997)126<0127:SHBBCA>2.3.CO;2.

Tyson, J. T. and Knight, R. L. Response of yellow perch to changes in the benthic invertebrate community of western Lake Erie. *Transactions of the American Fisheries Society* **130,** 766–782 (2001). DOI:10.1577/1548-8659(2001)130<0766:ROYPTC>2.0.CO;2.

Vehanen, T., Hyvärinen, P. and Huusko, A. Food consumption and prey orientation of piscivorous brown trout (*Salmo trutta*) and pikeperch (*Stizostedion lucioperca*) in a large regulated lake. *Journal of Applied Ichthyology* **14,** 15–22 (1998). DOI:10.1111/j.1439-0426.1998.tb00608.x.

Zanden, M. J. V., Olden, J. D., Thorne, J. H. and Mandrak, N. E. Predicting occurrences and impacts of smallmouth bass introductions in north temperate lakes. *Ecological Applications* **14,** 132–148 (2004). Online at http://dx.doi.org/10.1890/02-5036.

INDEX

italics* = taxonomic key page, **Bold = first page of taxon account, or aquatic gem account

G

H

I

J

K

L

M

N

O

T

U

V

W

X

Y

Z